IN CASE OF ACCIDENT[1]

In case of accident notify the laboratory instructor ***immediately.***

FIRE

Burning Clothing. Prevent the person from running and fanning the flames. Rolling the person on the floor will help extinguish the flames and prevent inhalation of the flames. If a safety shower is nearby hold the person under the shower until flames are extinguished and chemicals washed away. Do not use a fire blanket if a shower is nearby. The blanket does not cool and smouldering continues. Remove contaminated clothing. Wrap the person in a blanket to avoid shock. Get prompt medical attention.

Do not, under any circumstances, use a carbon tetrachloride (toxic) fire extinguisher and be very careful using a CO_2 extinguisher (the person may smother).

Burning Reagents. Extinguish all nearby burners and remove combustible material and solvents. Small fires in flasks and beakers can be extinguished by covering the container with an asbestos-wire gauze square, a big beaker, or a watch glass. Use a dry chemical or carbon dioxide fire extinguisher directed at the base of the flames. **Do not use water.**

Burns, either Thermal or Chemical. Flush the burned area with cold water for at least 15 min. Resume if pain returns. Wash off chemicals with a mild detergent and water. Current practice recommends that no neutralizing chemicals, unguents, creams, lotions, or salves be applied. If chemicals are spilled on a person over a large area quickly remove the contaminated clothing while under the safety shower. Seconds count and time should not be wasted because of modesty. Get prompt medical attention.

CHEMICALS IN THE EYE: Flush the eye with copious amounts of water for 15 min using an eye-wash fountain or bottle, or by placing the injured person face up on the floor and pouring water in the open eye. Hold the eye open to wash behind the eyelids. After 15 min of washing obtain prompt medical attention, regardless of the severity of the injury.

CUTS: Minor Cuts. This type of cut is most common in the organic laboratory and usually arises from broken glass. Wash the cut, remove any pieces of glass, and apply pressure to stop the bleeding. Get medical attention.

Major Cuts. If blood is spurting place a pad directly on the wound, apply firm pressure, wrap the injured to avoid shock, and get **immediate** medical attention. Never use a tourniquet.

POISONS: Call 800 information (1-800-555-1212) for the telephone number of the nearest Poison Control Center, which is usually also an 800 number.

[1] Adapted from *Safety in Academic Chemistry Laboratories*, prepared by the American Chemical Society Committee on Chemical Safety, March 1974.

Microscale Organic Experiments

Microscale Organic Experiments

Kenneth L. Williamson
Mount Holyoke College

D. C. HEATH AND COMPANY
Lexington, Massachusetts Toronto

Cover photograph: Ken O'Donoghue

Published simultaneously in Canada.

Printed in the United States of America.

International Standard Book Number: 0-669-14922-5

Library of Congress Catalog Card Number: 86-82832

Preface

Microscale Organic Experiments is the result of five and one-half years of careful development. It is based on Fieser and Williamson's *Organic Experiments,* which has been continuously refined through numerous editions. The objective of *Microscale Organic Experiments* is to introduce undergraduates to the basic techniques of the organic laboratory in preparation for carrying out a wide range of meaningful experiments that exemplify the principles of organic chemistry. The emphasis throughout is on observation and deductive reasoning.

A prime consideration in this guide is laboratory safety above and beyond the safety that is automatically conferred by carrying out experiments on one-tenth to one-thousandth the scale previously employed. These experiments have been extensively tested by undergraduates at Mount Holyoke College and most recently at Dartmouth College. During this time I have also coauthored the National Research Council's *Prudent Practices for the Disposal of Chemicals from Laboratories* and acted as consultant for laboratory safety on NASA's Space Station.

Safety

Microscale means carrying out experiments on one-tenth to one-thousandth the scale now commonly used in the undergraduate organic chemistry laboratory. Students receive one-tenth to one-thousandth the exposure to substances that are:

- Flammable — They burn.
- Irritants — They irritate the eyes, lungs, and skin.
- Toxic — They are poisonous in either the short or long term.
- Carcinogenic — They cause cancer.
- Teratogenic — They cause defects in the unborn fetus.
- Mutagenic — They cause genetic mutations.
- Explosive — They explode, usually on being mixed with air.
- Corrosive — They burn the eyes, lungs, and skin.

Avoiding these problems is not only desirable, it is becoming mandatory under the Occupational Safety and Health Act (OSHA).

Savings

Working on one-tenth to one-thousandth the scale presently used means cutting down on chemical costs by the same factors. Reagents, solvents, and even the solvents used to clean glassware are reduced by factors of ten to one thousand.

One can no longer dump chemicals down the drain or put them in the trash. Disposing of hazardous laboratory waste to a sanitary landfill or authorized incinerator is expensive. Microscale experimentation slashes these costs.

The cost of upgrading ventilation to meet new federal requirements can also be enormous. In the climate of New York City it costs $1200 to operate one hood for one year (primarily due to heat losses). Microscale experimentation reduces air pollution and levels of toxic substances in the air to the point where such upgrading may be unnecessary.

Other Advantages

Students are exposed to a wide variety of experiments. The experiments are fast, so students can do more experiments per term. They can use more exotic and more expensive reagents. They often rely on instrumental methods for analysis—gas chromatography, and infrared, ultraviolet, and to some extent nuclear magnetic resonance spectroscopy.

Microscale Organic Experiments features sections taken from Fieser and Williamson's *Organic Experiments*. These are entitled "On a Larger Scale" and show how the same operation—for example distillation, extraction, sublimation, or crystallization—is carried out on a scale ten to one thousand times larger. In this way students will become familiar with the older, more common apparatus and techniques, which they may encounter in research or in a job. These sections can be used to help integrate macroscale with microscale experimentation.

Apparatus

A system of apparatus has been designed and is sold by Kontes Scientific Glassware/Instruments. This apparatus embodies a number of innovations, in particular the use of a 10 × 100 mm reaction tube in which to carry out experiments.

The glass parts of the Williamson apparatus have a tooled end that fits into a chemical- and heat-resistant elastomeric connector (Santoprene) to connect the pieces of apparatus. This has a number of advantages over standard-taper apparatus: the connection is inexpensive, flexible, self-supporting, and has no male or female parts so the apparatus can be joined together in combinations impossible with standard-taper joints. The interior diameter of the apparatus is cylindrical and unobstructed; there is no place for flooding to occur during distillation or for crystals to hang up when being removed from the apparatus. There is no possibility for contamination by grease from a joint.

A support rod is molded into one of the connectors. The piece functions as both connector and clamp, thus obviating the necessity for a separate three-prong micro clamp.

Cost of Glassware

The glassware to be used in *Microscale Organic Experiments* costs less than half that of other microscale glassware presently available. It has no ground glass joints, so it is inexpensive; does not stick together; does not break easily because the joints have some flexibility; requires no lubricants, which might contaminate reactions; and is self-supporting, with its integral clamp.

The Williamson/Kontes glassware is so inexpensive that students can carry out duplicate or triplicate experiments, such as the Grignard reaction, because they are issued several reaction tubes. Since the glassware has no male or female joints one piece can serve several purposes, for example when mounted "upside down." And with a reinforced lip and low mass the glassware does not break easily when dropped.

Cost of Supporting Equipment

A number of items including centrifuges, magnetic stirrers, automatic pipettes, and refractometers are *not required* in the Williamson approach to the microscale organic laboratory. This saves several thousand dollars compared to other approaches.

Since the Craig tube is not used for filtration, centrifuges are not required. *Microscale Organic Experiments* employs a unique Pasteur pipette method or the conventional Hirsch funnel for filtrations.

This approach does not *require* a stirrer, although it is suggested for two heterogeneous reactions (catalytic hydrogenation using Pd on C and Corey oxidation using pyridinium chlorochromate) but even these will work satisfactorily without mechanical stirring. The use of automatic micropipettes and refractometers is not a part of this laboratory program.

The Pasteur Pipette Method for Filtration

Crystals are in the bottom of the long, narrow reaction tube. A Pasteur pipette is forced through the crystals, and the filtrate is withdrawn. Filtration occurs between the bottom of the tube and the pipette tip. This method is very fast, material can be kept on ice while filtering, washings can be made while cooling the product on ice, and the product can be dried in the reaction tube under vacuum. There is no waiting for centrifuges, and no expensive or delicate Craig tube to handle. Because the reaction tube in which recrystallization is carried out is long and narrow (10 × 100 mm) it can be held in the hand while heating. Furthermore, it will not tip over easily in an ice bath.

In those cases where the crystals are too fine to filter this way they are filtered on a micro Büchner funnel, a 6-mm diameter funnel that uses a removable polyethylene frit or a Hirsch funnel equipped with a removable polyethylene frit and molded adapter that fits the 25-mL filter flask (both are items unique to the Williamson/Kontes kit of microscale equipment).

Distillation

Simple and fractional distillations are carried out in conventional apparatus. A 5-mL round-bottomed flask, a packed distilling column, and a distillation head bearing a thermometer are used. Simple distillation can be carried out by eliminating the packed column. Meaningful fractional and simple distillations are carried out on 3 mL of liquid. Less material can be distilled (as little as 0.5 mL) using a high-boiling chaser. Steam distillation of natural products and substances such as aniline can easily be carried out using an addition port to add water to the distilling flask, generating steam *in situ*. The apparatus is readily converted to a unique microscale Dean-Stark apparatus, allowing easy microscale synthesis of esters using Dowex-50 as the acid catalyst.

Advantages of the Reaction Tube

The reaction tube is a key piece of apparatus for microscale experimentation. It is *not* a test tube. The diameter-to-length ratio is 1 to 10; thus it is longer and narrower than any test tube on the market. The upper part of the tube can be grasped between the fingers because it remains cool. Vapors reflux on the cool upper part of the tube, a result of the long and narrow design. The reaction tube has a large surface area compared to the area where heat is applied. It is, in fact, so difficult to distil from the tube that 5-mL round-bottomed flasks are provided for this purpose.

Cost. The reaction tube is a flask and condenser combined. It has no ground-glass joint to buy, break, leak, or contaminate products with grease. It has no small-diameter neck to restrict the passage of vapors, liquids, or solids. With no joint between flask and condenser there is no joint to stick together or come loose and no need to match it in size to other apparatus. Since the reaction tube is inexpensive, students have several in their kit and can carry out several reactions simultaneously.

Crystallization in same tube. When products crystallize out the solvent is removed and crystallization is conducted in the reaction tube. It can be held in the hand while boiling 1–2 mL of crystallization solvent. Repeated crystallizations can be carried out in the tube without the necessity for transferring crystals. It can be cooled in ice while removing mother liquor and washings from crystals using the Pasteur pipette. The long, narrow design facilitates removal of solvent with the Pasteur pipette. Crystals can then be dried under vacuum in the reaction tube.

Low heat capacity—fast heating and cooling. The Williamson reaction tube has low mass and low heat capacity, unlike heavy-walled vials. One milliliter of nitrobenzene (bp 210°C) can be brought to a boil in 10 sec and 1 mL of benzene (mp 4°C) can be frozen in 10 sec.

Extraction. The long, narrow design of Williamson's reaction tube facilitates the removal of either the upper or lower layer in an extraction using a Pasteur pipette. The tube has a capacity of 4.5 mL. If a product does not crystallize on completion of a reaction the necessary extraction can be carried out in the reaction tube.

Inert atmosphere reactions. These reactions are easily carried out in the tube. When capped with a rubber septum the tube is ideal for preparing Grignard reagents and for making ferrocene or trityl free radical. Using the two-needle technique the tube is easily flushed of oxygen. The rubber septa are inexpensive and reusable. These septa are much cheaper and more effective (they can be punctured many times before leaking) than Teflon-lined septa used on screw cap vials.

Instant microscale distillation. The Williamson/Kontes reaction tube is ideally suited, because of its diameter-to-length ratio, to this unique technique: the air is expelled from a Pasteur pipette, it is thrust into the hot vapor refluxing halfway up a reaction tube, and the hot vapor is pulled into the cold pipette, where it condenses. The liquid is expelled into another reaction tube. In this way enough pure material can be distilled to determine a boiling point, run a spectrum, make a derivative, or carry out a reaction.

The microscale chromatography column. The microscale column with funnel and stopcock can easily be cut to any convenient length. The bottom of this column doubles as a microscale Büchner funnel just 6 mm in diameter. This funnel has a removable polypropylene frit, which can easily be replaced when it becomes contaminated. Similarly the Hirsch funnel is fitted with a polypropylene frit that renders it much easier to use than the conventional porcelain funnel. Gas transfer is carried out in polyethylene tubing rather than in expensive and fragile bent glass tubes.

Syringes are used extensively throughout *Microscale Organic Experiments* where an addition funnel would be used in macroscale experiments. These syringes of polypropylene have the necessary inertness to reagents and are unbreakable.

Range of Experiments

Microscale Organic Experiments includes a wide range of experiments and experimental procedures:

A unique procedure has been devised for the cracking of dicyclopentadiene on a micro scale (Chapter 28). One-half gram of the dimer gives the monomer in 84% yield. The experiment can be conducted outside of the hood, since the odor of the dienes cannot be detected.

Pelletized Norit is introduced (Chapter 3). It is easily handled on a micro scale; the filtrate can be removed with a Pasteur pipette.

Gases, such as the butenes, are generated and collected in apparatus that costs just a few cents (Chapter 12). Gas evolution is monitored easily to determine when a reaction is complete.

Catalytic hydrogenation with H_2 gas generated from zinc and acid and collected over water can be carried out quantitatively (Chapter 62).

Oxidative coupling of acetylenes is carried out on a micro scale (Chapter 50).

Corrosive gases such as SO_2 from thionyl chloride and HCl from the Friedel-Crafts reaction are trapped by absorption of the gas in moist cotton (Chapters 37, 38, 53).

Enzymes are used to carry out chiral reduction of a ketone (Chapter 64).

Several polymers are synthesized on a micro scale (Chapter 67). The reaction vessel, a reaction tube, is inexpensive and thus expendable.

Diffusion of moisture into the reaction mixture is hindered by use of a syringe needle and polyethylene tubing (not a drying tube).

Several products are purified in sublimation apparatus that requires no running water and is fast and easy to set up (Chapters 7, 29).

Photolysis in a thin-walled, small ampoule is rapid because UV light is not absorbed by thick glass and thick layers of solvent (Chapter 44).

A unique synthesis of ferrocene on a micro scale is presented (Chapter 29).

Acknowledgments

I would like to acknowledge the help of many classes of Chemistry 302 students at Mount Holyoke in developing and refining the experiments in this text. Thanks go to Diane Miller, Maria Dulay, Joanne Dalpe, and Pamela Schaefer for their aid in special development tasks, but especially to Heidi Hulse, R.N. and Mount Holyoke English major, who made manifold experimental contributions to every aspect of this text.

For the privilege of teaching a select and able group of Chemistry 57 and 58 students at Dartmouth College during the 1986–87 academic year and for their support and ideas, I wish to thank my Hanover colleagues Tom Spencer, Dave Lemal, Gordon Gribble, and Steve Teeter. The generosity and inspiration of Dana Mayo is gratefully acknowledged. More than any other single person he has awakened the organic chemistry community to the possibilities of microscale experimentation.

For her typing skills, preparation of the index, and organization of the *Instructor's Guide,* and for her marvelous tolerance of the writing process, I thank my wife Louise. I am appreciative of the efforts of the entire staff of D. C. Heath, but especially those of Mary Le Quesne, who recognized the possibilities for this book and got it off to a fast start.

Contents

Introduction 1

PRELAB EXERCISE: *Study the glassware diagrams and be prepared to identify the reaction tube, fractionating column, distilling head, addition port, Hirsch funnel, and micro Büchner funnel.*

Organic Reactions

Welcome to the organic chemistry laboratory! This laboratory manual presents a new method for carrying out organic experiments on a scale one-tenth to one-hundredth of that previously used. Microscale experiments have been chosen primarily for reasons of safety. You will be exposed to much smaller amounts of toxic, flammable, explosive, carcinogenic, and teratogenic material than your predecessors. A pleasant benefit is that the experiments can be conducted in much less time than large-scale efforts; you will have the opportunity to explore a wide range of organic chemical experiments and should something go awry you will have time to repeat the procedure.

Synthesis and structure determination are two major concerns of the organic chemist, and both are dealt with in this book. The rational synthesis of an organic compound, whether it involves the transformation of one functional group into another or a carbon-carbon bond forming reaction, starts with a *reaction*.

Organic reactions usually take place in the liquid phase and are *homogeneous*, in that the reactants are all in one phase. The reactants can be solids and/or liquids dissolved in an appropriate solvent to mediate the reaction. Some reactions are *heterogeneous*—one of the reactants is in the solid phase —and thus require stirring or shaking to bring the reactants in contact with one another. A few heterogeneous reactions involve the reaction of a gas, such as oxygen, carbon dioxide, or hydrogen, with material in solution. Examples of all of these will be found among the experiments in this book.

Effect of temperature

In an *exothermic* organic reaction, simply mixing the reactants will produce the products because the reaction evolves heat. If it is highly exothermic, one reactant is added slowly to the other and heat is removed by external cooling. Most organic reactions are, however, mildly *endothermic*, which means the reaction mixture must be heated to increase the rate of the reaction. A very useful rule of thumb is that *the rate of an organic reaction doubles with a 10°C rise of temperature*. The late Louis Fieser, a chemist and author at Harvard University, introduced the idea of changing the traditional solvents of many reactions to high-boiling solvents in order to reduce reaction times. Throughout this book we will use solvents such as triethylene glycol, with a boiling point (bp) of 290°C, to replace ethanol (bp 78°C) and triethylene glycol dimethyl ether (bp 222°C) to replace dimethoxyethane (bp 85°C). These high-boiling solvents greatly increase the rates of many reactions.

Running an organic reaction is usually the easiest part of a synthesis. The challenge comes in isolating and purifying the product from the reaction because organic reactions seldom give quantitative yields of one pure substance.

"Working up the reaction"

In some cases the solvent and concentrations of reactants are chosen so that, after the reaction mixture has been cooled, the product will *crystallize*. It is then collected by *filtration* and the crystals are washed with an appropriate solvent. If sufficiently pure at that point, the product is dried and collected; otherwise it is purified by the process of recrystallization or, less commonly, by *sublimation*.

If the product of reaction does not crystallize from the reaction mixture, it is often isolated by the process of *extraction*. This involves adding a solvent to the reaction mixture that will dissolve the product and will be immiscible with the solvent used in the reaction. Shaking the mixture will cause the product to dissolve in the extracting solvent, after which the two layers of liquid are separated and the product isolated from the extraction solvent.

If the product is a liquid, it is isolated by *distillation*, usually after extraction. Occasionally the product can be isolated by the process of *steam distillation* from the reaction mixture.

Apparatus

The apparatus used for these operations is relatively simple. Reactions are carried out in a *reaction tube* (Fig. 1.1) in which the reactants are dissolved in an appropriate solvent and, depending on the reaction, heated or cooled as the reaction proceeds. The mass of the reaction tube is small enough that a milliliter of nitrobenzene (bp 210°C) will boil in 10 s and a milliliter of benzene (mp 5°C) will crystallize in the same period of time. Cooling is effected by simply shaking the tube in a small beaker of ice water, and heating by immersing the reaction tube to the appropriate depth in an electrically heated sand bath.

The sand bath, an electrically heated 100-mL flask heater filled with sand, is a versatile heat source. The relatively poor heat conduction of sand results in a very large temperature difference between the top of the sand and the bottom. Thus, depending on the immersion depth in the sand, a similarly wide temperature range will be found in the reaction tube. Because the area of the tube exposed to heat is fairly small, it is difficult to transfer enough heat to the contents of the tube to cause solvents to boil away. The reaction tube is 100 mm long so that the upper part of the tube can function as an air condenser. Solvents, such as water and ethanol, are boiled and as the hot vapor ascends to the upper part of the tube it condenses and runs back down the tube. This process is called *refluxing* and is the most common method for conducting a reaction at a constant temperature, the boiling point of the solvent. On a larger scale the reaction tube is replaced by a flask connected via a ground glass joint to a water-cooled condenser (Fig 1.2).

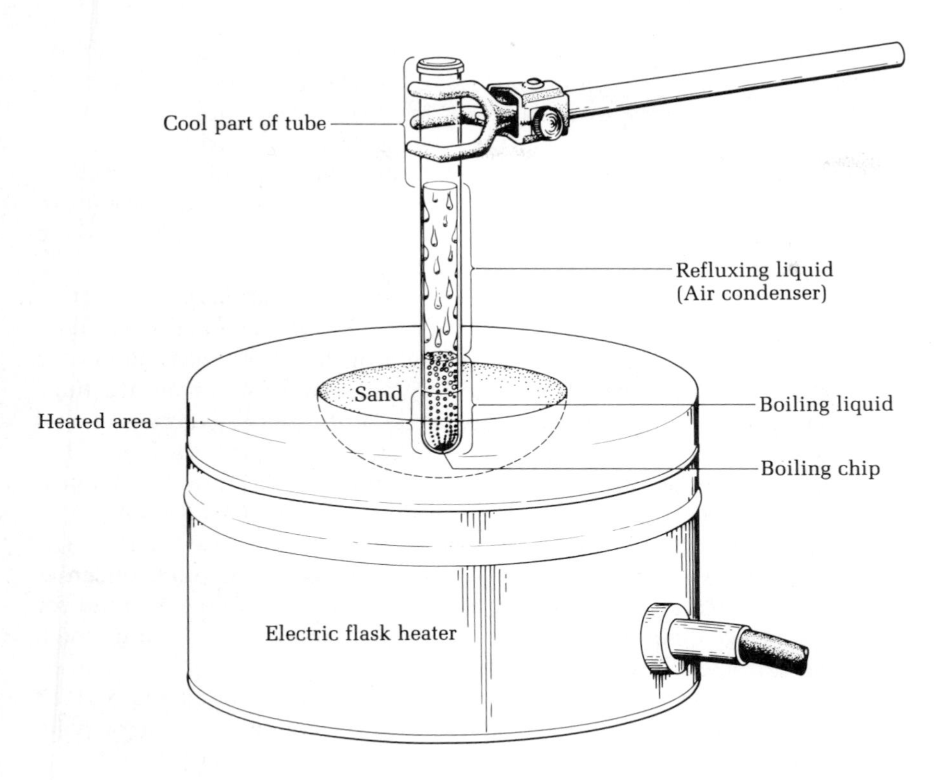

FIG. 1.1 Reaction tube being heated on hot sand bath in a flask heater. The area of the tube exposed to heat is small. The liquid boils and condenses on the cool upper portion of the tube, which functions as an air condenser.

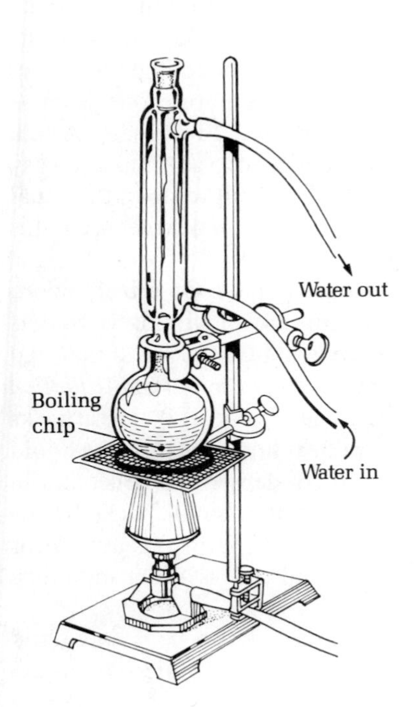

FIG. 1.2 Reflux apparatus for larger reactions. Liquid boils in flask and condenses on cold inner surface of water-cooled condenser.

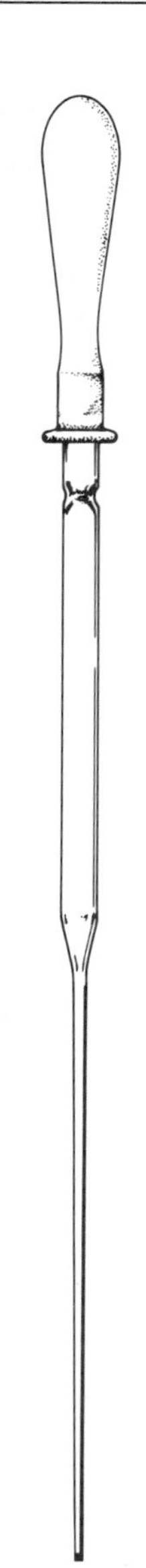

FIG. 1.3 Filtration using the Pasteur pipette and reaction tube.

If the product of a reaction crystallizes from the reaction mixture on cooling, it is isolated by *filtration*. Using the reaction tube this is accomplished by inserting a *Pasteur pipette* (Fig. 1.3) to the bottom of the tube, expelling the air, and withdrawing the solvent. Very effective filtration occurs between the square tip of the pipette and the bottom of the tube. This method of filtration has several advantages. The mixture of crystals and solvent can be kept on ice during the entire process. This minimizes the solubility of the crystals in the solvent. There are no transfer losses of material because an external filtration device is not used. Several recrystallizations can be carried out in the same tube, with final drying of the product under vacuum. Knowing the tare weight of the tube (weight of the empty tube) allows the weight of the product to be determined without removing it from the tube. In this way a compound can be synthesized, purified by crystallization, and dried without ever being taken from the reaction tube. After the removal of material for analysis, the compound in the tube can be used for the next reaction. This technique is used for several experiments in this book.

Occasionally a solid can be purified by the process of *sublimation*. The solid is heated, usually under vacuum, and the vapor of the solid condenses on a cold surface to form crystals. Sublimation is an effective method for isolating micro quantities of material because transfer losses are small. Caffeine extracted from a single tea bag will be purified in this manner. Mixtures of solids and occasionally of liquids can be separated and purified by *column chromatography*. The chromatography column is made of polypropylene and thus can be cut to any length desired (Fig. 1.4).

Sometimes the product of a reaction will not crystallize out. It may be a liquid, it may be a mixture of compounds, or it may be too soluble in the solvent being used. In this case an immiscible solvent is added, the two layers are shaken to effect *extraction*, and after the layers separate one layer is removed with a Pasteur pipette and the process repeated if necessary. A tall, thin column of liquid such as that produced in the reaction tube makes it easy to remove one layer selectively. This is much more difficult to do in the usual test tube because the height/diameter ratio is too small. On a larger scale this process is carried out in a *separatory funnel*.

Some of the compounds to be synthesized are liquids. On a truly micro scale the best way to separate and purify a mixture of liquids is by gas chromatography, but this technique is limited to less than 100 mg of material on the usual gas chromatograph. For larger quantities of material *distillation* is used. For this purpose *small distilling flasks* are used (Fig. 1.5). These flasks have a large surface area, which allows sufficient heat input to cause the liquid to vaporize rapidly. It can then be distilled and condensed for collection in a receiver. *Fractional distillation* is carried out using a small packed *fractionating column* in the apparatus depicted in Fig. 1.6. With this apparatus 2–4 mL of a liquid can be fractionally distilled, and one or more milliliters can be purified by *simple distillation*.

Some liquids with a relatively high vapor pressure can be isolated and purified by *steam distillation*, a process in which the organic compound codistills with water at a temperature below the boiling point of water.

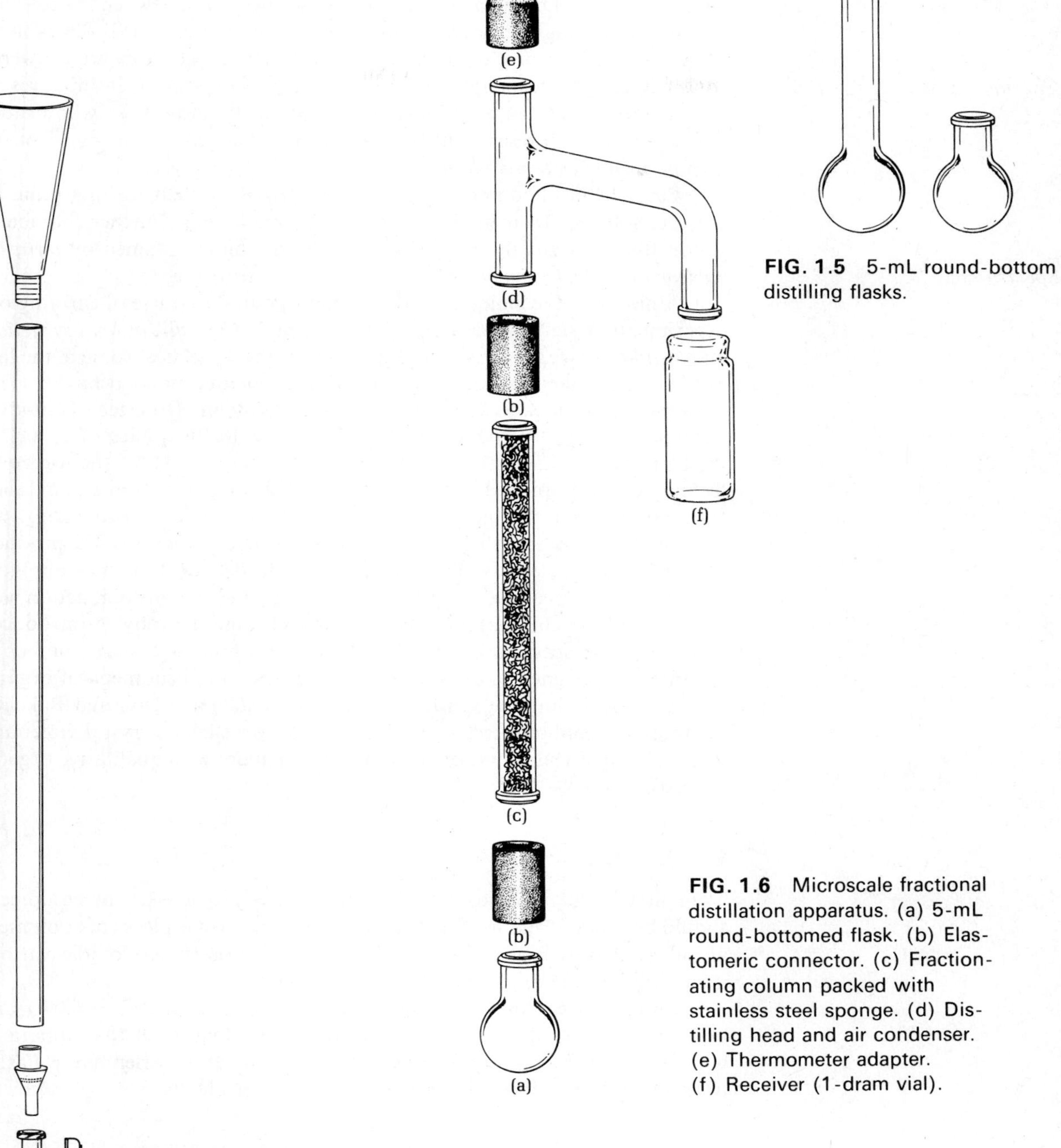

FIG. 1.5 5-mL round-bottom distilling flasks.

FIG. 1.6 Microscale fractional distillation apparatus. (a) 5-mL round-bottomed flask. (b) Elastomeric connector. (c) Fractionating column packed with stainless steel sponge. (d) Distilling head and air condenser. (e) Thermometer adapter. (f) Receiver (1-dram vial).

FIG. 1.4 Chromatography column consisting of funnel, tube, base fitted with polyethylene frit, and Leur valve.

Analysis and structure determination are other major concerns of the organic chemist. During a synthesis it is desirable to know when the reaction is over and during the isolation and purification processes it is desirable to know whether the product is being purified. These questions are answered routinely by the technique of *thin-layer chromatography*. If mixtures of products are known to result, it is often desirable to know exactly how much of each product is present, an analysis performed by *gas chromatography* and *high-performance liquid chromatography*.

Chromatography

Pure chemical compounds are characterized by their melting points if they are solids or their boiling points if they are liquids. Further characterization that aids greatly in structure determination is obtained by *infrared, nuclear magnetic resonance*, and *ultraviolet spectroscopy*.

Spectroscopy

A number of techniques will be introduced in the course of carrying out experiments designed to exemplify the method. *Crystallization, extraction*, and *distillation* are the most important and are the subjects of some of the first experiments, coupled with an examination of melting point behavior. This information is put to use in the extraction of caffeine from tea. Thin-layer chromatography is introduced with a variety of colorful applications, such as the extraction of the red coloring matter of tomatoes. All of the foregoing techniques are employed in the separation of the ingredients in a pain killer. Infrared, nuclear magnetic resonance and ultraviolet spectroscopy are presented next because they will be used to characterize many of the products of reaction. Because radical chlorination and the dehydration of alcohols are often two of the first reactions studied, they are presented in conjunction with analysis by gas chromatography. Column chromatography is introduced along with the acetylation of cholesterol. There follows a large number of experiments designed to exemplify the reactions and phenomena of organic chemistry, including the synthesis of polymers such as nylon and Bakelite, luminol (a chemiluminescent substance) and dyes such as crystal violet and methyl orange. The course of experiments concludes with qualitative organic analysis.

Check In

Your first duty will be to check in to your assigned desk. The equipment should be checked against the inventory list. The various pieces of equipment are illustrated in Figs. 1.1–1.6 and 1.7–1.8. The usefulness of the various items will become apparent in later experiments.

Check to see that your thermometer is correct (20°C = 68°F) and replace any flasks that have star-shaped cracks. Remember that apparatus with graduations is expensive; small-scale apparatus, Erlenmeyer flasks, beakers, and test tubes are, by comparison, fairly cheap.

Washing and Drying Laboratory Equipment

Clean apparatus immediately

Considerable time can be saved by cleaning each piece of equipment soon after use, for you will know at that point what contaminant is present and

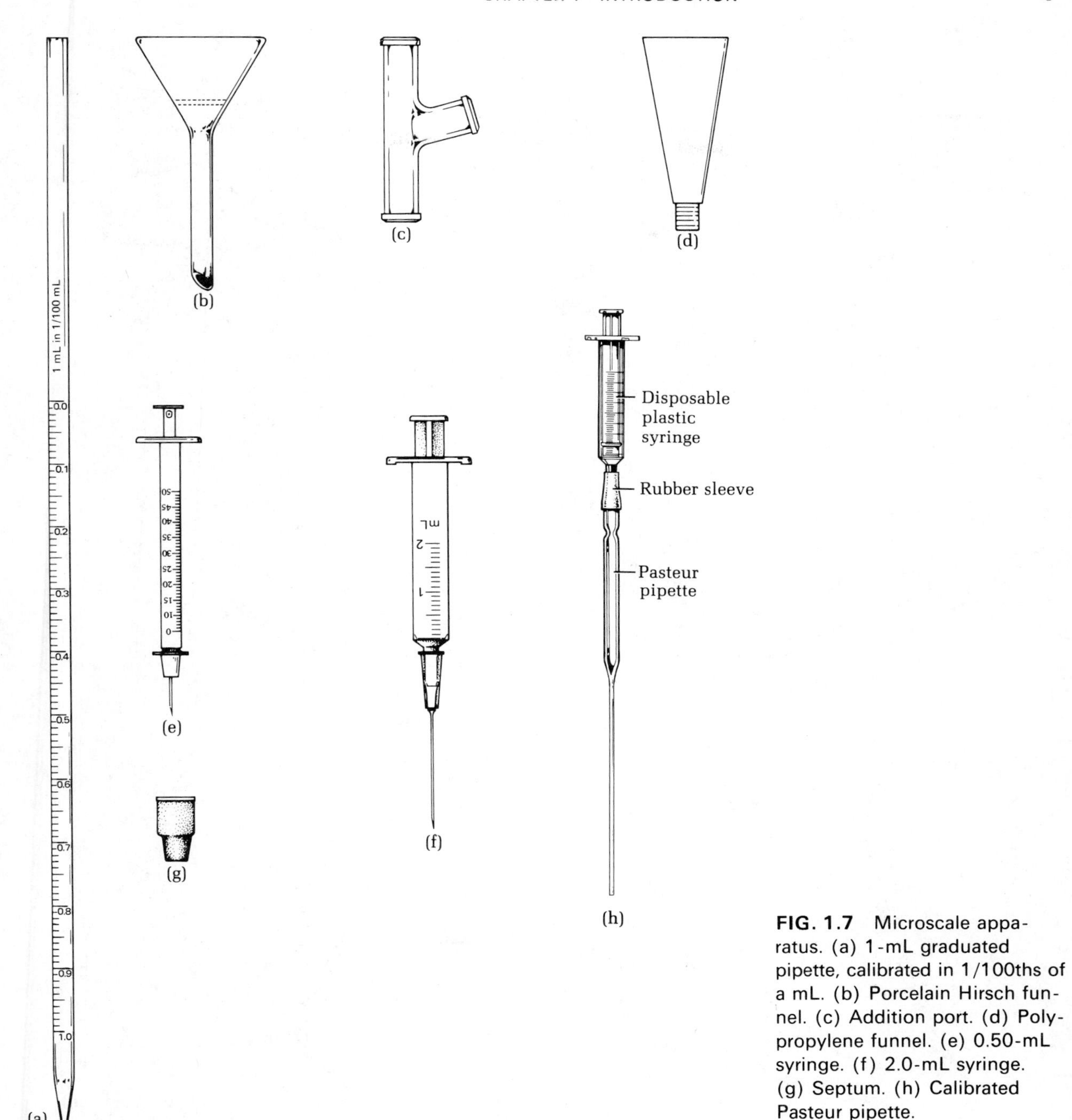

FIG. 1.7 Microscale apparatus. (a) 1-mL graduated pipette, calibrated in 1/100ths of a mL. (b) Porcelain Hirsch funnel. (c) Addition port. (d) Polypropylene funnel. (e) 0.50-mL syringe. (f) 2.0-mL syringe. (g) Septum. (h) Calibrated Pasteur pipette.

be able to select the proper method for removal. A residue is easier to remove before it has dried and hardened. A small amount of organic residue usually can be dissolved with a few milliliters of an appropriate organic solvent. Acetone (bp 56.1°C) has great solvent power and is often effective. Because

FIG. 1.8 Miscellaneous apparatus. (a) 50-mL filter flask fitted with Neoprene adapter and plastic Hirsch funnel. (b) Polyethylene wash bottle. (c) Single-pan electronic balance with automatic zeroing and digital readout. 100 g ± 0.001 g capacity. (d) Glass scorer. (e) Erlenmeyer flask with approximate graduations. (f) Electric flask heater. (g) Solid state control for electric flask heater.

it is miscible with water and vaporizes readily, it is easy to remove from the vessel. Cleaning after an operation often can be carried out while another experiment is in process.

A polyethylene bottle (shown in Fig. 1.8) is a convenient wash bottle for acetone. The name, symbol, or formula of a solvent can be written on a bottle with a Magic Marker or wax pencil. For crystallizations, extractions, and quick cleaning of apparatus, it is convenient to have a bottle for each frequently used solvent—95% ethanol, ligroin, dichloromethane, and diethyl ether. A pinhole opposite the spout, which is covered with the finger in use, will prevent the spout dribbling the solvent.

Pasteur pipettes (see Fig. 1.3) are very useful for transferring small quantities of liquid, adding reagents dropwise, and carrying out crystallizations.

Sometimes a reaction tube or flask will not be clean after a washing with detergent and acetone. At that point try an abrasive household cleaner and as a last resort a powerful oxidizing agent. In this process, rinse the flask with water, let it drain, and **carefully** add about 1 mL of concentrated sulfuric acid and 0.2 mL of concentrated nitric acid. Let the mixture remain in the flask for a time if there is a vigorous reaction, then heat on the steam bath. After the reaction is over, you will find that, on decantation of the acids and washing with water, the flask is clean.

Caution!

To dry a piece of apparatus rapidly, rinse with a few millimeters of acetone and invert over a beaker to drain. **Do not use compressed air**, which contains droplets of oil, water, and particles of rust. Instead draw a slow stream of air through the apparatus using the suction of your water aspirator.

Insertion of a glass tube into a rubber connector or adaptor or hose is easy if the glass is lubricated with a very small drop of glycerol. Grasp the tube very close to the end to be inserted; if it is grasped at a distance, especially at the bend, the pressure applied for insertion may break the tube and result in a serious cut.

If a glass tube or thermometer should become stuck to a rubber connector, it can be removed by painting on glycerol and forcing the pointed tip of an 18-cm spatula between the rubber and glass. Another method is to select a cork borer that fits snugly over the glass tube, moisten it with glycerol, and slowly work it through the connector. When the stuck object is valuable, such as a thermometer, the best policy is to cut the rubber with a sharp knife.

Heat Source

A 10°C rise in temperature will approximately double the rate of an organic reaction. The processes of distillation, sublimation, and crystallization all require heat, which is most conveniently and safely applied from an electrically heated sand bath. A flask heater (see Fig. 1.8) is filled with sand, the temperature of which depends on the setting of the controller. This heater is small in diameter, giving good access to small apparatus, and the air above the heater is not hot. It is possible to hold a reaction tube containing refluxing ethanol in the fingers of the hand. Because sand is a fairly poor conductor of heat, there is a very large variation in temperature in the sand bath depending on its depth. The heater is easily capable of producing temperatures in excess of 300°C; therefore do not leave the controller at its maximum setting. Because the flask heater can provide high temperatures, your equipment need not include a Bunsen burner. Similarly there will be no need for a steam bath, the traditional flameless method for applying heat up to 90°C.

Transfer of a Solid

A convenient funnel

A plastic funnel that fits the top of the reaction tube is most convenient for transfer of solids to the reaction tube or to small Erlenmeyer flasks. It is also the top of the chromatography column (see Fig. 1.4).

Weighing and Measuring

The single-pan electronic balance (see Fig. 1.8) capable of weighing to ± 0.001 g and having a capacity of 100 g is the single most important instrument making small-scale organic experiments possible. Most of the quantitative measurements made in this laboratory will use the balance. Weighing is a pleasure with these balances.

A container such as a reaction tube in a beaker or flask is placed on the pan. At the touch of a bar the digital readout registers zero and the desired quantity of reagent (solid or liquid) can be added to the reaction tube as the weight is measured periodically to the nearest milligram.

It is often convenient to weigh reagents on glossy weighing paper and then transfer the chemical to the reaction container. The success of an experiment often depends on using certain amounts of starting materials and reagents. Inexperienced workers might think that if one-tenth of a millimeter of a reagent will do the job, then two-tenths of a milliliter will do the job twice as well. Such assumptions are usually erroneous.

Liquids can be measured by either volume or weight according to the relationship

$$\text{Volume (mL)} = \frac{\text{weight (g)}}{\text{density (g/mL)}}$$

Modern Erlenmeyer flasks and beakers have approximate volume calibrations fused into the glass, but these are *very* approximate. Somewhat more accurate volumetric measurements are made in the 10-mL graduated cylinders. For volumes less than about 4 mL, use a graduated pipette. **Never** apply suction to a pipette by mouth. The pipette can be fitted with a small rubber bulb. A Pasteur pipette can be converted into a calibrated pipette with the addition of a plastic syringe body (see Fig. 1.8). You will find among your

Never pipette by mouth

equipment a 1-mL pipette, calibrated in hundredths of a milliliter. Determine whether it is designed to *deliver* 1 mL or to *contain* 1 mL between the top and bottom calibration marks. For our purposes the latter is the better pipette.

Because the viscosity, surface tension, and wetting characteristics of organic liquids are different from those of water, the so-called automatic pipette (designed for aqueous solutions) gives poor accuracy in measuring organic liquids. *Disposable polyethylene syringes* (see Fig 1.7), on the other hand, are quite useful and frequent use will be made of them. The disposable insulin syringe is marked in "units" that correspond to hundredths of a milliliter. Several reactions that require especially dry or oxygen-free atmosphere will be run in sealed systems. Reagents can be added to the system via syringe through a rubber *septum*.

Tare Weights

The tare weight of a container is its weight when empty. Throughout this laboratory course it will be necessary to know the tare weights of containers so that the weights of the compounds within can be calculated. If identifying marks can be placed on the containers (e.g., with a diamond stylus) you may want to record tare weights for frequently used containers in your laboratory notebook.

Tare wt. = wt. of empty container

The Laboratory Notebook

A complete, accurate record is an essential part of laboratory work. Failure to keep such a record means laboratory labor lost. An adequate record includes the procedure (what was done), observations (what happened), and conclusions (what the results mean).

Use a lined, paperbound, $8\frac{1}{2} \times 11$ in. notebook and record all data in ink. Allow space at the front for a table of contents, number the pages throughout, and date each page as you use it. Reserve the left-hand page for calculations and numerical data, and use the right-hand page for notes. Never record **anything** on scraps of paper to be recorded later in the notebook. Do not erase, remove, or obliterate notes; simply draw a single line through incorrect entries.

Never record anything on scraps of paper

The notebook should contain a statement or title for each experiment followed by *balanced equations* for all principal and side reactions, and, where relevant, mechanisms of the reactions. Consult your textbook for supplementary information on the class of compounds or type of reaction involved. Give a *reference to the procedure* used; do not copy verbatim the procedure in the laboratory manual.

Before coming to the lab to do preparative experiments, prepare a *table* (in your notebook) *of reagents* to be used and the *products* expected, with their *physical properties*. (An illustrative table appears with the first preparative equipment, the preparation of 1-bromobutane.) From your table, use the molar ratios of reactants and determine the *limiting reagent* and calculate the *theoretical yield* (in grams) of the desired product (see Chapter 15). Enter all data in your notebook (left-hand page).

Include an outline of the procedure and method of purification of the product in a *flow sheet*, which lists all possible products, by-products, unused reagents, solvents, etc., that appear in the crude reaction mixture. On the flow sheet diagram indicate how each of these is removed, for example by extraction, various washing procedures, distillation, or crystallization. With this information entered in the notebook before coming to the laboratory, you will be ready to carry out the experiments with the utmost efficiency. Plan your time before the laboratory period. Often two and three experiments can be run simultaneously.

The laboratory notebook
What you did.
How you did it.
What you observed.
Your conclusions.

When working in the laboratory, record everything you do and everything you observe **as it happens**. The recorded observations constitute the most important part of the laboratory record, as they form the basis for the conclusions you will draw at the end of each experiment. Record the physical properties of the product, the yield in grams, and the percentage yield. Analyze your results. When things do not turn out as expected, explain why. When your record of an experiment is complete, another chemist should be able to understand your account and determine what you did, how you did it, and what conclusions you reached. In other words, from the information in your notebook a chemist should be able to repeat your work.

2 Laboratory Safety

PRELAB EXERCISE: *Locate the emergency eye-wash station, safety shower, and fire extinguisher in your laboratory. Check your safety glasses or goggles for size and transparency. Learn which reactions must be carried out in the hood. Learn to use your laboratory fire extinguisher; learn how to summon help and how to put out a clothing fire. Learn first aid procedures for acid and alkali spills on the skin. Learn how to tell if your laboratory hood is working properly. Learn which operations under reduced pressure require special precautions. Check to see that compressed gas cylinders in your lab are firmly fastened to benches or walls. Learn the procedures for properly disposing of solid and liquid waste in your laboratory.*

Small-scale organic experiments are much safer to conduct than their counterparts run on a scale 10 to 100 times larger. However, the organic chemistry laboratory is an excellent place to learn and practice safety. Commonsense procedures practiced here also apply to the shop, kitchen, and studio.

Know the safety rules of your particular laboratory. Know the locations of emergency eye washes and safety showers. Never eat, drink, or smoke in the laboratory. Don't work alone. Perform no unauthorized experiments and don't distract your fellow workers; horseplay has no place in the laboratory.

Eye protection
Don't wear contact lenses

Eye protection is extremely important. Safety glasses of some type must be worn at all times. Contact lenses should not be worn because reagents can get under a lens and cause damage to the eye before the lens can be removed. It is very difficult to remove a contact lens from the eye after a chemical splash has occurred.

Ordinary prescription eyeglasses don't offer adequate protection. Laboratory safety glasses should be of plastic or tempered glass. If you do not have such glasses, wear goggles that afford protection from splashes and objects coming from the side as well as the front. If plastic safety glasses are used, they should have side shields (see Fig. 2.1).

Dress sensibly

Dress sensibly in the laboratory. Wear shoes, not sandals or cloth-top sneakers. Confine long hair and loose clothes. Don't use mouth suction to fill a pipette, and wash your hands before leaving the laboratory. Don't use a solvent to remove chemicals from skin. This will only hasten the absorption of the chemical through the skin.

Working with Flammable Substances

Relative flammability of organic solvents

Flammable substances are the most common hazard of the organic laboratory, but two factors make this laboratory much safer than its predecessor:

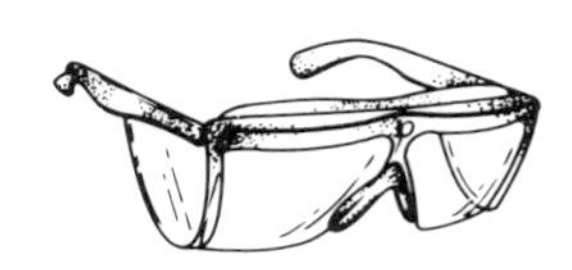

FIG. 2.1 Safety goggles and safety glasses.

the scale of the experiments is small and burners are not used. Even so, carbon disulfide (bp 46°C) can be ignited by the heat from a warm hot plate; it has an ignition temperature of 80°C. Diethyl ether (bp 35°C), the most flammable substance you will usually work with in this course, has an ignition temperature of 160°C, which means that a hot plate at that temperature will cause it to burn. For comparison, *n*-hexane (bp 69°C), a constituent of gasoline, has an ignition temperature of 225°C. The flash points of these three organic liquids—that is, the temperatures at which they will catch fire if exposed to a flame or spark—range from −20 to −45°C. These are three very flammable liquids; however, if you are careful, they are not difficult to work with. Except for water, almost all of the liquids you will use in the laboratory will be flammable.

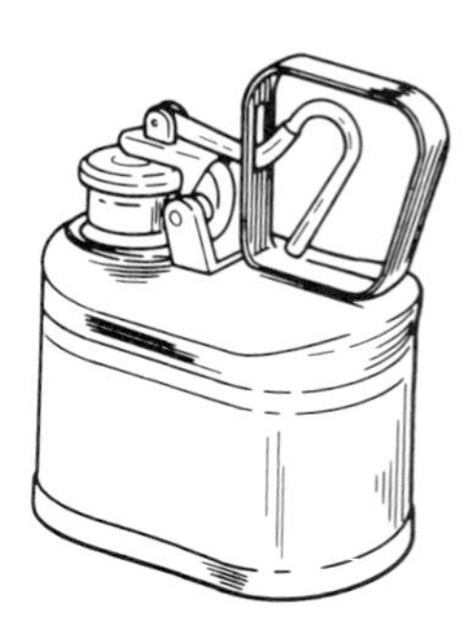

FIG. 2.2 Solvent safety can.

Bulk solvents should be stored in and dispensed from *safety cans* (see Fig. 2.2). These and other liquids will burn in the presence of the proper amount of their flammable vapors, oxygen, and a source of ignition (most commonly a flame or spark). It is usually difficult to remove oxygen, although it is possible to put out a fire in a beaker or a flask by simply covering the vessel with a flat object, thus cutting off the supply of air. Your lab notebook might do in an emergency. The best solution is to pay close attention to sources of ignition—open flame, sparks, and hot surfaces. Remember the vapors of flammable liquids are **always** heavier than air and thus will travel along bench tops and down drain troughs and will remain in sinks. For this reason all flames within the vicinity of a flammable liquid must be extinguished. Adequate ventilation is one of the best ways to prevent flammable vapors from accumulating. Work in an exhaust hood when manipulating large quantities (> 10 mL) of flammable liquids.

Flammable vapors travel along bench tops

Should a person's clothing catch fire, first knock the person down and roll him or her over to extinguish the flames. It is extremely important to prevent the victim from running or standing because the greatest harm comes from breathing the hot vapors that rise past the mouth. The safety shower might then be used to extinguish glowing cloth that is no longer aflame. A so-called fire blanket should not be used—it tends to funnel flames past the victim's mouth, and clothing continues to char beneath it. However, it is useful for retaining warmth to ward off shock after the flames are out.

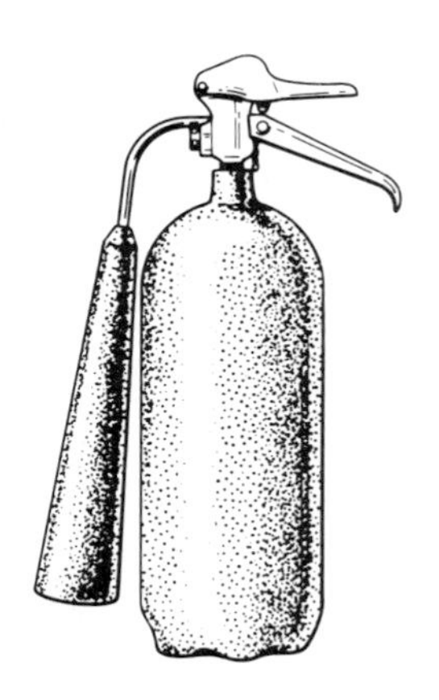

FIG. 2.3 Carbon dioxide fire extinguisher.

An organic chemistry laboratory should be equipped with a carbon dioxide or dry chemical (monoammonium phosphate) *fire extinguisher* (see Fig. 2.3). To use this type of extinguisher, lift it from its support, pull the ring to break the seal, raise the horn, aim it at the base of the fire, and squeeze the handle. Do not hold onto the horn because it will become extremely cold. Do not replace the extinguisher; report the incident so it can be refilled.

When disposing of certain chemicals, be alert for the possibility of *spontaneous combustion.* This may occur in oily rags; organic materials exposed to strong oxidizing agents such as nitric acid, permanganate ion, and peroxides; alkali metals such as sodium; or very finely divided metals such as zinc dust and platinum catalysts.

Working with Hazardous Chemicals

If you do not know the properties of a chemical you will be working with, it is wise to regard the chemical as hazardous. The *flammability* of organic substances poses the most serious hazard in the organic laboratory, although the danger is greatly reduced by small-scale work. There is the possibility, however, that storage containers in the laboratory may contribute to a fire. Large quantities of organic solvents should not be stored in glass bottles. Use safety cans.

A flammable liquid can often be vaporized to form, with air, a mixture that is *explosive* in a confined space. The beginning chemist is sometimes surprised to learn that diethyl ether is more likely to cause a laboratory fire or explosion than a worker's accidental anesthesia. The chances of being confined in a laboratory with a high enough concentration of ether to cause loss of consciousness are extremely small. A spark in such a room would probably eradicate the building.

Flammable vapors plus air in a confined space are explosive

The probability of forming an explosive mixture of volatile organic liquids with air is much greater than that of producing an explosive solid or liquid. The chief functional groups that render compounds explosive are the *peroxide*, *acetylide*, *azide*, *diazonium*, *nitroso*, *nitro*, and *ozonide* groups (see Fig. 2.4). Not all members of these groups are equally sensitive to shock or heat. You would find it difficult to detonate trinitrotoluene (TNT) in the laboratory, but nitroglycerine is treacherously explosive. Peroxides present special problems that are dealt with below.

You will need to contend with the corrosiveness of many of the reagents you will handle. The danger here is principally to the eyes. Proper eye protection is *mandatory* and small-scale experiments can be hazardous to the eyes. It takes only a single drop of a corrosive reagent to do lasting damage.

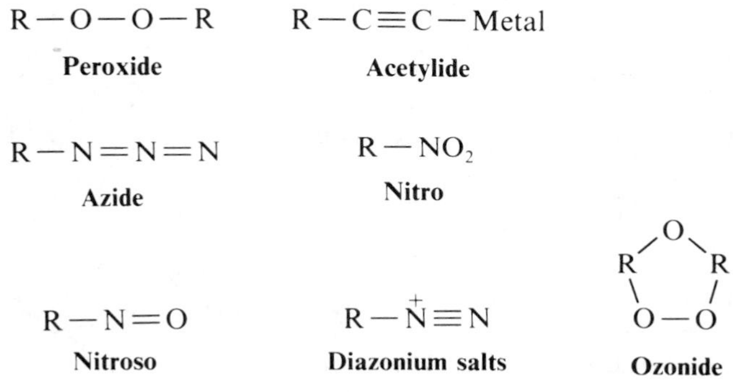

Safety glasses must *be worn at all times*

$R{-}O{-}O{-}R$ **Peroxide**

$R{-}C{\equiv}C{-}Metal$ **Acetylide**

$R{-}N{=}N{=}N$ **Azide**

$R{-}NO_2$ **Nitro**

$R{-}N{=}O$ **Nitroso**

$R{-}\overset{+}{N}{\equiv}N$ **Diazonium salts**

Ozonide: R–O–R bridged ring with O–O (five-membered ring: R, O, R, O, O)

Ozonide

FIG. 2.4 Functional groups that can be explosive in some compounds.

Handling concentrated acids and alkalis, dehydrating agents, and oxidizing agents calls for commonsense care to avoid spills and splashes and to avoid breathing the often corrosive vapors.

Certain organic chemicals present problems with acute toxicity from short-duration exposure and chronic toxicity from long-term or repeated exposure. Exposure can come about through ingestion, contact with the skin, or, most commonly, inhalation. Currently great attention is being focused on chemicals that are teratogens (chemicals that often have no effect on a pregnant woman but cause abnormalities in a fetus), mutagens (chemicals causing changes in the structure of the DNA, which can lead to mutations in offspring), and carcinogens (cancer-causing chemicals). Small-scale experiments reduce these hazards greatly but do not eliminate them.

Peroxides

Ethers form explosive peroxides

Certain functional groups can make an organic molecule become sensitive to heat and shock, such that it will explode. Chemists work with these functional groups only when there are no good alternatives. One of these functional groups, the *peroxide* group, is particularly insidious because it can form spontaneously when oxygen and light are present (see Fig. 2.5). Ethers, especially *cyclic ethers* and those made from primary or secondary alcohols (such as tetrahydrofuran, diethyl ether, and diisopropyl ether), form peroxides. Other compounds that form peroxides are *aldehydes*, alkenes that have allylic hydrogen atoms (such as cyclohexene), compounds having benzylic hydrogens on a tertiary carbon atom (such as isopropyl benzene), and vinyl compounds (such as vinyl acetate). Peroxides are low-power explosives but are extremely sensitive to shock, sparks, light, heat, friction, and impact. The biggest danger from peroxide impurities comes when the peroxide-forming compound is distilled. The peroxide has a higher boiling point than the parent compound and remains in the distilling flask as a residue that can become overheated and explode. This is one reason why it is very poor practice to distill anything to dryness.

Don't distill to dryness

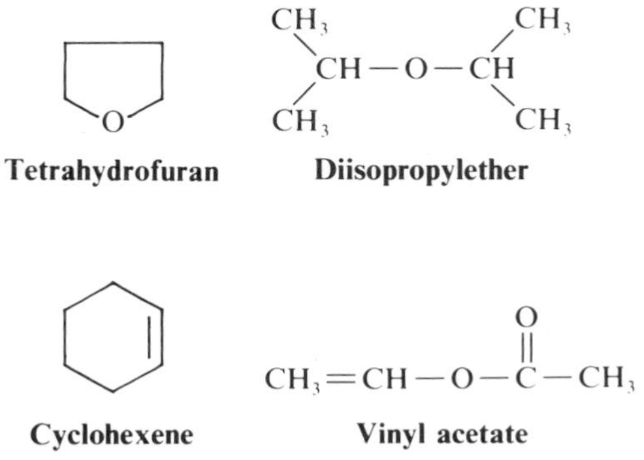

Tetrahydrofuran **Diisopropylether**

Cyclohexene **Vinyl acetate**

$CH_3{=}CH{-}O{-}C(=O){-}CH_3$

FIG. 2.5 Some compounds that form peroxides.

Detection of peroxides	Removal of Peroxides
To a solution of 0.01 g sodium iodide in 0.1 mL of glacial acid, add 0.1 mL of the liquid suspected of containing a peroxide. If the mixture turns brown, a high concentration of peroxide is present; if it turns yellow, a low concentration of peroxide is present.	Pouring the solvent through a column of activated alumina will simultaneously remove peroxides and dry the solvent. Do not allow the column to dry out while in use. When the alumina column is no longer effective, wash the column with 5% aqueous ferrous sulfate and discard it.

Problems with peroxide formation are especially critical for ethers. Ethers form peroxides readily and, because they are frequently used as solvents, they are often used in quantity and then removed to leave reaction products. Cans of diethyl ether should be dated when opened and if not used within one month should be treated for peroxides.

In this course you may have occasion to use *30% hydrogen peroxide*. This material causes severe burns if it contacts the skin, and it decomposes violently if contaminated with metals or their salts. Be particularly careful not to contaminate the reagent bottle.

Working with Corrosive Substances

Handle strong acids, alkalis, dehydrating agents, and oxidizing agents carefully so as to avoid contact with the skin and eyes and to avoid breathing the corrosive vapors that attack the respiratory tract. All strong concentrated acids attack the skin and eyes. *Concentrated sulfuric acid* is both a dehydrating agent and a strong acid and will cause very severe burns. *Nitric acid* and *chromic acid* (used in cleaning solutions) also cause bad burns. *Hydrofluoric acid* is especially harmful, causing deep, painful, and slow-healing wounds. It should be used only after thorough instruction.

Sodium, potassium, and ammonium hydroxides are common bases you will encounter. The first two are extremely damaging to the eye, and ammonium hydroxide is a severe bronchial irritant. Like sulfuric acid, sodium hydroxide, phosphorous pentoxide, and calcium oxide are powerful dehydrating agents. Their great affinity for water will cause burns to the skin. Because they release a great deal of heat when they react with water, they should always be added to water rather than water being added to them.

Add H_2SO_4, P_2O_5, CaO, and $NaOH$ to water, not the reverse

You will receive special instruction when it comes time to handle metallic *sodium* and *lithium aluminum hydride*, two substances that can react explosively with water.

Among the strong oxidizing agents *perchloric acid* is probably the most hazardous. It can form heavy metal and organic *perchlorates* that are *explosive*, and it can react explosively when organic compounds are added to it.

Should one of these substances get on the skin or in the eyes, wash the affected area with very large quantities of water, using the safety shower and/or eye-wash fountain (Fig. 2.6). Do not attempt to neutralize the reagent chemically. Remove contaminated clothing so that thorough washing can take place. Take care to wash the reagent from under the fingernails.

Wipe up spilled hydroxide pellets rapidly

When you are using very small quantities of these reagents, no particular safety equipment is needed except *safety glasses*. Take care not to let the reagents, such as sulfuric acid, run down the outside of a bottle or flask and come in contact with the fingers. Wipe up spills immediately with a very damp sponge, especially in the area around the balances. Pellets of sodium and potassium hydroxide are very hygroscopic and will dissolve in the water they pick up from the air; therefore they should be wiped up very quickly. When working with larger quantities of these corrosive chemicals, wear protective gloves; with still larger quantities, use a face mask, gloves, and a Neoprene apron. The corrosive vapors can be avoided by carrying out work in a good exhaust hood.

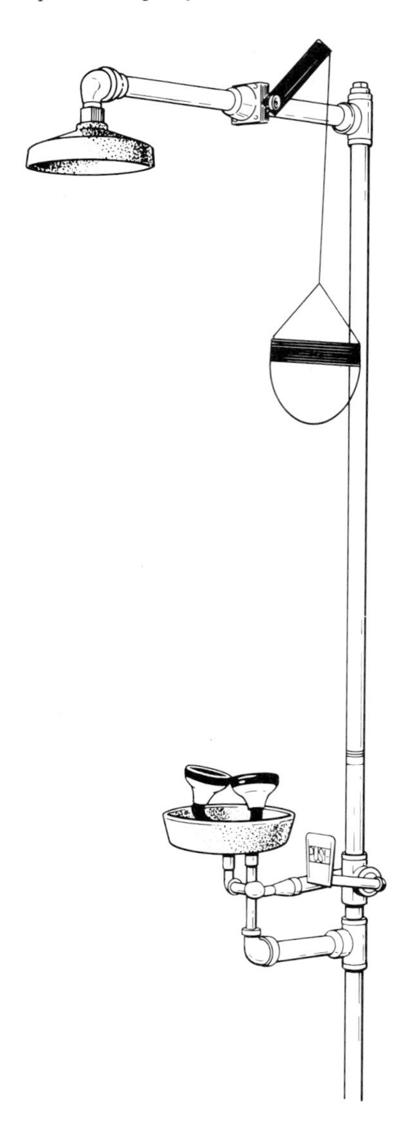

FIG. 2.6 Emergency shower and eye-wash station.

Working with Toxic Substances

Many chemicals have very specific toxic effects. They interfere with the body's metabolism in a known way. For example, the cyanide ion combines irreversibly with hemoglobin to form cyanomethemoglobin, which can no longer carry oxygen. Aniline acts in the same way. Carbon tetrachloride and some other halogenated compounds cause liver and kidney failure. Carcinogenic and mutagenic substances deserve special attention because of their long-term insidious effects. The ability of certain carcinogens to cause cancer is very great; for example, special precautions are needed in handling aflatoxin B_1. In other cases, such as with dioxane, the hazard is so low that no special precautions are needed beyond reasonable normal care in the laboratory.

Women of child-bearing age should be careful when handling any substance of unknown properties. Certain substances are highly suspect teratogens and will cause abnormalities in an embryo or fetus. Among these are benzene, toluene, xylene, aniline, nitrobenzene, phenol, formaldehyde, dimethylformamide (DMF), dimethyl sulfoxide (DMSO), polychlorinated biphenyls (PCBs), estradiol, hydrogen sulfide, carbon disulfide, carbon monoxide, nitrites, nitrous oxide, organolead and mercury compounds, and the notorious sedative thalidomide. Some of these substances will be used in subsequent experiments. Use care. Of course, the leading known cause of embryotoxic effects is ethyl alcohol in the form of maternal alcoholism. The amount of ethanol vapor inhaled in the laboratory or absorbed through the skin is so small it is unlikely to have these morbid effects.

It is impossible to avoid handling every known or suspected toxic substance, so it is wise to know what measures should be taken. Because the eating of food or the consumption of beverages in the laboratory is strictly forbidden and because one should never taste material in the laboratory, the possibility of poisoning by mouth is remote. Be more careful than your predecessors—the hallucinogenic properties of LSD and **all** artificial sweeteners were discovered by accident. The two most important measures to be taken then are avoiding skin contact by wearing protective gloves and avoiding inhalation by working in a good exhaust hood. A very thorough treatment of ventilation in the organic laboratory is found in *Microscale Organic Laboratory*, by D. W. Mayo, R. M. Pike, and S. S. Butcher.

Because you have not had previous experience working with organic chemicals, most of the experiments you will carry out in this course will not involve the use of known carcinogens, although you will work routinely with flammable, corrosive, and toxic substances. A few experiments involve the use of substances that are suspected of being carcinogenic, such as hydrazine. If you pay proper attention to the rules of safety, you should find working with these substances no more hazardous than working with ammonia or nitric acid. The single, short-duration exposure you might receive from a suspected carcinogen, should an accident occur, would probably have no long-term consequences. The reason for taking the precautions noted in each experiment is to learn, from the beginning, good safety habits.

Using the Laboratory Hood

Modern practice dictates that in laboratories where workers spend most of their time working with chemicals, there should be one exhaust hood for every two people. This precaution is often not possible in the beginning organic chemistry laboratory, however. In this course you will find that for some experiments the hood must be used and for others it is advisable; in these instances it may be necessary to schedule experimental work around access to the hoods. However, many experiments formerly carried out in the hood can now be carried out at the desk because the concentration of vapors will be minimal when working at a small scale.

The hood offers a number of advantages for work with toxic and flammable substances. Not only does it draw off the toxic and flammable fumes, it also affords an excellent physical barrier on all four sides of a reacting system when the sash is pulled down. And should a chemical spill occur, it is nicely contained within the hood.

Keep the hood sash closed

It is your responsibility each time you use a hood to see that it is working properly. You should find some type of indicating device that will give you this information on the hood itself. A simple propeller on a cork works well (Fig. 2.7). The hood is a back-up device. Don't use it alone to dispose of chemicals by evaporation; use an aspirator tube. Toxic and flammable fumes should be trapped or condensed in some way and disposed of in the prescribed manner. Except when you are actually carrying out manipulations on the experimental apparatus, the sash should be pulled down. The water, gas, and electrical controls should be on the outside of the hood so it is not necessary to open the hood to adjust them. The ability of the hood to remove vapors is greatly enhanced if the apparatus is kept as close to the back of the

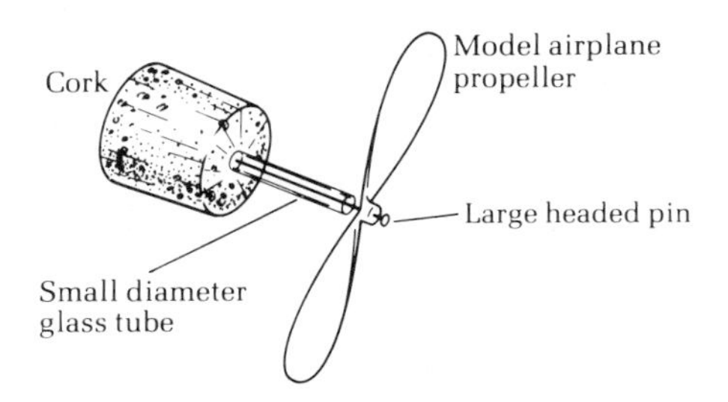

FIG. 2.7 Air flow indicator for hoods. The indicator should be permanently mounted in the hood and should be spinning whenever the hood is in operation.

FIG. 2.9 Dewar flask with safety net in place.

hood as possible. Everything should be at least 10 cm back from the hood sash. Chemicals should not be stored permanently in the hood but should be removed to ventilated storage areas. If the hood is cluttered with chemicals, you will not have good, smooth air flow and adequate room for experiments.

Working at Reduced Pressure

Implosion

Whenever a vessel or system is evacuated, an implosion could result from atmospheric pressure on the empty vessel. It makes little difference whether the vacuum is perfect or just 10 mm Hg; the pressure difference is almost the same (760 mm Hg versus 750 mm Hg). An implosion may occur if there is a star crack in a flask, or if the flask is scratched or etched. Only with heavy walled flasks specifically designed for vacuum filtration is the use of a safety shield (Fig. 2.8) ordinarily unnecessary. The chances of implosion of the apparatus used for small-scale experiments are very small.

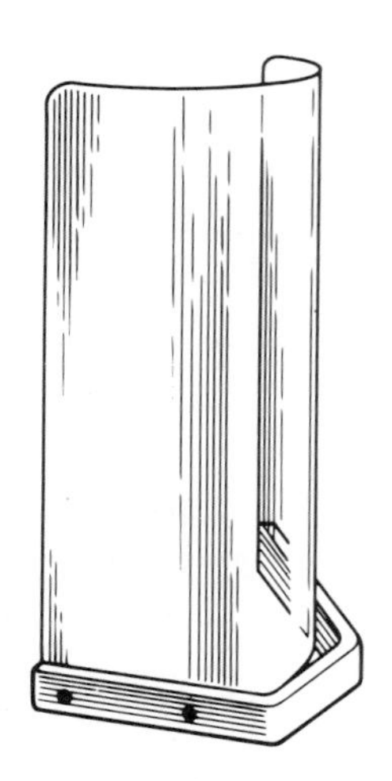

FIG. 2.8 Safety shield.

Dewar flasks (thermos bottles) are often found in the laboratory without shielding. They should be wrapped with friction tape or covered with plastic net to prevent the glass from flying about in case of an implosion (Fig. 2.9). Similarly, vacuum desiccators should be wrapped with tape before being evacuated.

Working with Compressed Gas Cylinders

Many reactions are carried out under an inert atmosphere so that the reactants and/or products will not react with oxygen or moisture in the air. Nitrogen and argon are the inert gases most frequently used. Oxygen is widely used both as a reactant and to provide a hot flame for glassblowing and welding. It is used in the oxidative coupling of alkynes (Chapter 50). Helium is the carrier gas used in gas chromatography. Some other gases commonly used in the laboratory are ammonia, often used as a solvent; chlorine, used for chlorination reactions; acetylene, used in combination with oxygen for welding; and hydrogen used for high- and low-pressure hydrogenation reactions.

Always *clamp gas cylinders*

Certain rules apply to all compressed gases: Compressed gas cylinders should be firmly secured at all times. For temporary use, a clamp that attaches to the laboratory bench top and has a belt for the cylinder will suffice

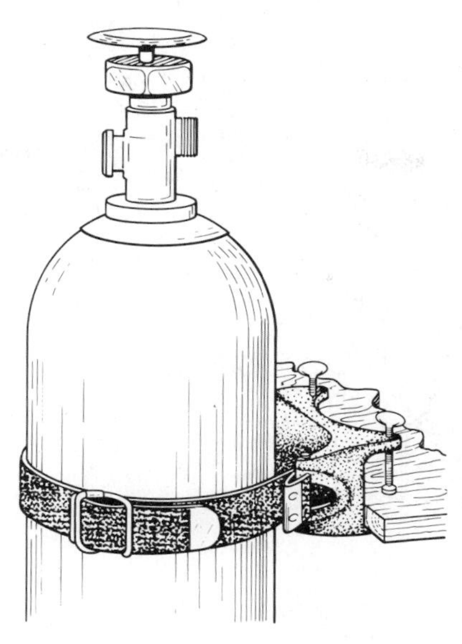

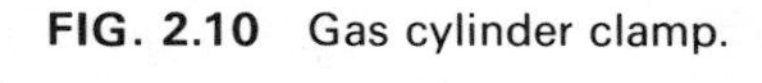

FIG. 2.10 Gas cylinder clamp.

(Fig. 2.10). Eyebolts and chains should be used to secure cylinders in permanent installations.

A variety of outlet threads are used on gas cylinders to prevent incompatible gases from becoming mixed because of an interchange of connections. Both right- and left-handed external and internal threads are used. Left-handed nuts are notched to differentiate them from right-handed nuts. Right-handed threads are used on nonfuel and oxidizing gases, and left-handed threads are used on fuel gases, such as hydrogen.

Cylinders come equipped with caps that should be left in place during storage and transportation. These caps can be removed by hand. Under these caps is a hand wheel valve. It can be opened by turning the wheel counterclockwise; however, because most compressed gases in full cylinders are under very high pressure (commonly up to 3000 lb in.^{2}), a pressure regulator must be attached to the cylinder. This pressure regulator is almost always of the diaphragm type and has two gauges, one indicating the pressure in the cylinder, the other the outlet pressure (Fig. 2.11). On the outlet, low-pressure side of the regulator is located a small needle valve and then the outlet connector. After connecting the regulator to the cylinder, unscrew the diaphragm valve (turn it counterclockwise) before opening the hand wheel valve on the top of the cylinder. This valve should be opened only as far as necessary. For most gas flow rates in the laboratory, this will be a very small amount. The gas flow and/or pressure is increased by turning the two-flanged diaphragm valve **clockwise**. When the apparatus is not being used, turn off the hand wheel valve (clockwise) on the top of the cylinder. Before removing the regulator from the cylinder, reduce the flow or pressure to zero. Cylinders should never be emptied to zero pressure and left with the valve open because the residual contents will become contaminated with air. Empty cylinders

Clockwise *movement of diaphragm valve handle* increases *pressure*

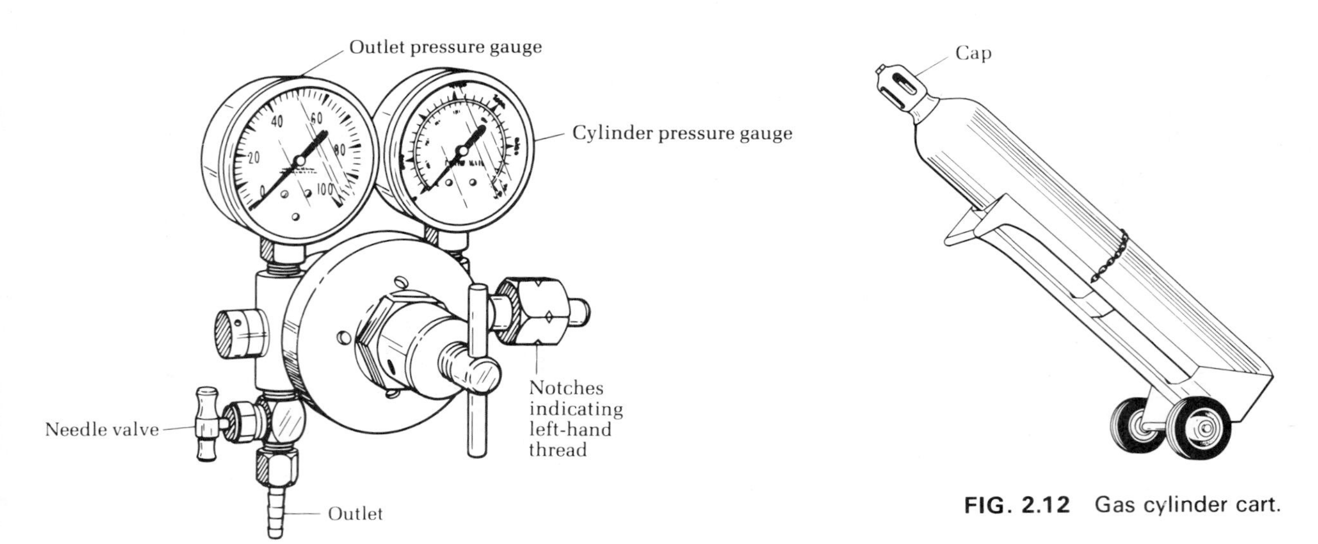

FIG. 2.11 Gas pressure regulator. Turn *clockwise* to increase outlet pressure.

FIG. 2.12 Gas cylinder cart.

should be labeled "empty," capped, and returned to the storage area, separated from full cylinders. Gas cylinders should never be dragged or rolled from place to place but should be fastened into and moved in a car designed for the purpose (Fig. 2.12).

Disposing of Chemicals in the Laboratory

Today there is an ever-increasing awareness of the effects of various waste substances on our fellow citizens and on the environment. The organic chemistry laboratory is an ideal place to learn how to dispose of some of these substances, but the principles apply equally well to the disposal of such substances as paint thinners, acid etching baths, photo lab chemicals, degreasers, lubricating oils, cleaning fluids, and garden chemicals generated in the home, the art studio, the machine shop, the farm, and the garage.

Water-Soluble Liquid Wastes

Dilution is the solution, for some waste disposal

In the past the sanitary sewer system was used to dispose of much laboratory waste; however, this is poor practice and should be limited to water-soluble substances that are not toxic and are so dilute as not to be flammable. Strong acids and bases should be diluted to the pH 3–11 range before being poured down the sink. Lachrymators (tear-producing) and malodorous substances should not be disposed of in the sewer system because their vapors may come up in some other laboratory. Similarly, very toxic water-soluble substances such as cyanide ion should not be disposed of down the drain. The incompatibility of chemicals poured down the drain should be taken into account. For example, if someone pours sulfide ion into a sink and then acid is added, hydrogen sulfide gas will be generated. Cyanide ion and acid generate deadly hydrogen cyanide gas. The number of water-soluble organic chemicals that may legally be disposed of through the sanitary sewer system is becoming restricted as more is learned about their effects on the environment. Two notable examples of substances that should not be disposed of in this way are acetone and 1-butanol.

Sulfide ion + acid → Hydrogen sulfide

Cyanide ion + acid → Hydrogen cyanide

Both very poisonous

Water-Insoluble Liquid Wastes

In general, water-insoluble liquid wastes will be various flammable solvents. There should be at least two labeled containers for their disposal, one for nonchlorinated and one for chlorinated solvents (Fig. 2.13). The chlorinated solvents call for special disposal measures because they generate hydrogen chloride on combustion. The solvents for disposal should be free of solids and reactive substances.

Chlorinated and nonchlorinated organic solvents go in separate waste containers

Solid Wastes

Solid wastes should be placed in appropriately labeled containers for disposal. If these containers are glass bottles, they should be placed in buckets. It will be the primary responsibility of the instructor to label the containers, but you too should be aware of the problems of mixing incompatible chemicals.

FIG. 2.13 Waste disposal can.

Cleaning up Chemical Spills

Spilled solids should simply be swept up and placed in the appropriate solid waste container. This should be done promptly because many solids are hygroscopic and become difficult if not impossible to sweep up in a short time. This is particularly true of sodium hydroxide and potassium hydroxide.

Spilled acids should be neutralized. Use sodium carbonate or, for larger spills, cement or limestone. For bases use sodium bisulfate. If the spilled material is very volatile, clear the area and let it evaporate, provided there is no chance of igniting flammable vapors. Other liquids can be taken up into such absorbents as vermiculite, diatomaceous earth, dry sand, or paper towels. Be particularly careful in wiping up spills with paper towels. If a strong oxidizer is present, the towels can later ignite. Bits of sodium metal will also cause paper towels to ignite. Sodium metal is best destroyed with *n*-butyl alcohol. If the spilled liquid is toxic, then wear gloves when using paper towels or a sponge to remove the liquid.

Clean up spills rapidly

Mercury requires special measures—see instructor

Questions

1. Write a balanced equation for the reaction between iodide ion, a peroxide, and hydrogen ion. What causes the orange or brown color?
2. Why does the horn of the carbon dioxide fire extinguisher become cold when the extinguisher is used?
3. Why is water not used to put out most fires in the organic laboratory?

3 Crystallization

PRELAB EXERCISE: *Write an expanded outline for the seven-step process of crystallization*

Crystallization: the most important purification method for small-scale experiments

Crystallization is perhaps the most important method for the purification of organic compounds. A crystalline organic substance is made up of a three-dimensional array of molecules held together primarily by van der Waals forces. These intramolecular attractions are fairly weak; most organic solids melt in the range of room temperature to 250°C.

Crystals can be grown from the molten state just as water is frozen into ice, but it is not easy to remove impurities from crystals made in this way. Thus most purifications in the laboratory involve dissolving the material to be purified in the appropriate hot solvent. As the solvent cools, the solution becomes saturated with respect to the substance, which crystallizes. As the perfectly regular array of a crystal is formed, foreign molecules are excluded and thus the crystal is one pure substance. Soluble impurities stay in solution because they are not concentrated enough to saturate the solution. The crystals are collected by *filtration*, the surface of the crystals is washed with cold solvent to remove the impurities, and then the crystals are dried. This process is carried out on an enormous scale in the commercial purification of sugar.

In small-scale organic experiments, crystallization is the most rapid and convenient method for purifying the products of a reaction. Initially you will be told which solvent to use to crystallize a given substance and how much of it to use; later on you will judge how much solvent is needed, and finally the choice of both the solvent and its volume will be left to you. It takes both experience and knowledge to pick the correct solvent for a given purification.

The process of crystallization can be broken into seven discrete steps: choosing the solvent, dissolving the *solute*, *decolorizing* the solution, filtering suspended solids, crystallizing the solute, collecting and washing the crystals and drying the product. The process involves dissolving the impure substance in an appropriate hot solvent, removing some impurities by decolorizing and/or filtering the hot solution, allowing the substance to crystallize as the temperature of the solution falls, removing the crystallization solvent, and drying the resulting purified crystals.

How crystallization starts

Crystallization is initiated at some nuclear center—a seed crystal, a speck of dust or a scratch on the wall of the test tube if the solution is supersaturated with respect to the substance being crystallized (the solute). Supersaturation will occur as a hot, saturated solution cools. Large crystals, which are easy to isolate, are formed by slow cooling of the hot solvent.

1. Choosing the Solvent and Solvent Pairs

Similia similibus solvuntur

In choosing the solvent the chemist is guided by the dictum "like dissolves like." Even the nonchemist knows that oil and water do not mix and that sugar and salt dissolve in water but not in oil. Hydrocarbon solvents such as hexane will dissolve hydrocarbons and other nonpolar compounds, and hydroxylic solvents such as water and ethanol will dissolve polar compounds. Often it is difficult to decide, simply by looking at the structure of a molecule, just how polar or nonpolar it is and therefore which solvent would be best. Therefore the solvent is often chosen by experimentation.

The ideal solvent

The best crystallization solvent (and none is ideal) will dissolve the solute when the solution is hot but not when the solution is cold, it will either not dissolve the impurities at all or it will dissolve them very well (so they won't crystallize out along with the solute), it will not react with the solute, and it will be nonflammable, nontoxic, inexpensive, and very volatile (so it can be removed from the crystals).

Some common solvents and their properties are presented in Table 3.1 in order of decreasing polarity of the solvent. Solvents adjacent to each other

TABLE 3.1 Crystallization Solvents

Solvent	Boiling Point (°C)	Remarks
Water (H_2O)	100	The solvent of choice because it is cheap, nonflammable, nontoxic, and will dissolve a large variety of polar organic molecules. Its high boiling point makes it somewhat difficult to remove from the crystals.
Acetic acid (CH_3COOH)	118	Will react with alcohols and amines. Difficult to remove. Not a common solvent for recrystallizations, although used as a reaction solvent when carrying out oxidations.
Dimethyl sulfoxide (DMSO) Methyl sulfoxide (CH_3SOCH_3)	189	Also not a commonly used solvent for crystallization, but used for reactions.
Methanol (CH_3OH)	64	A very good solvent, used often for crystallization. Will dissolve molecules of higher polarity than will the other alcohols.
95% Ethanol (CH_3CH_2OH)	78	One of the most commonly used crystallization solvents. Its high boiling point makes it a better solvent for the less polar molecules than methanol. Evaporates readily from the crystals. Esters may undergo interchange of alcohol groups on recrystallization.

Note: The solvents in this table are listed in decreasing order of polarity. Adjacent solvents in the list will in general be miscible with each other.

TABLE 3.1 *Cont.*

Solvent	Boiling Point (°C)	Remarks
Acetone (CH_3COCH_3)	56	An excellent solvent, but its low boiling point means there is not much difference in solubility of a compound at its boiling point and at room temperature.
2-Butanone, Methyl ethyl ketone, MEK ($CH_3COCH_2CH_3$)	80	An excellent solvent with many of the most desirable properties of a good crystallization solvent.
Ethyl acetate ($CH_3COOC_2H_5$)	78	Another excellent solvent that has about the right combination of moderately high boiling point and yet the volatility needed to remove it from crystals.
Dichloromethane, methylene chloride (CH_2Cl_2)	40	Although a common extraction solvent, dichloromethane boils too low to make it a good crystallization solvent. It is useful in a solvent pair with ligroin.
Diethyl ether, ether ($CH_3CH_2OCH_2CH_3$)	35	Its boiling point is too low for crystallization, although it is an extremely good solvent and fairly inert. Used in a solvent pair with ligroin.
Dioxane, $C_4H_8O_2$	101	A very good solvent, not too difficult to remove from crystals, but a mild carcinogen.
Toluene ($C_6H_5CH_3$)	111	An excellent solvent that has replaced the widely used benzene (a weak carcinogen) for crystallization of aryl compounds. Because of its boiling point it is not easily removed from crystals.
Pentane (C_5H_{12})	36	A widely used solvent for nonpolar substances. Not often used alone for crystallization, but good in combination with a number of other solvents as part of a solvent pair.
Hexane (C_6H_{14})	69	Frequently used to crystallize nonpolar substances. It is inert and has the correct balance between boiling point and volatility. Often used as part of a solvent pair.
Cyclohexane (C_6H_{12})	81	Similar in all respects to hexane.
Petroleum ether	30–60	A mixture of hydrocarbons of which pentane is a chief component. Used interchangeably with pentane because it is cheap. Unlike diethyl ether, it is not an ether in the modern chemical sense.
Ligroin	60–90	A mixture of hydrocarbons with the properties of hexane and cyclohexane. A very commonly used crystallization solvent.

Miscible: capable of being mixed

in the list will dissolve in each other, i.e., they are miscible with each other, and each solvent will, in general, dissolve substances that are similar to it in chemical structure. These solvents are used both for crystallization and as solvents in which reactions are carried out.

Picking a solvent

To pick a solvent for crystallization, put a few crystals of the impure solute in a reaction tube and add a very small drop of the solvent. Allow it to flow down the side of the tube and onto the crystals. If the crystals dissolve instantly at room temperature, that solvent cannot be used for crystallization because too much of the solute will remain in solution at low temperatures. If the crystals do not dissolve at room temperature, warm the tube on the hot sand bath and observe the crystals. If they do not go into solution, add a drop more solvent. If the crystals go into solution at the boiling point of the solvent and then crystallize when the tube is cooled, you have found a good crystallization solvent. If not, remove the solvent by evaporation and try another solvent. In this trial-and-error process it is easiest to try low-boiling solvents first, because they can be removed most easily. Occasionally no single satisfactory solvent can be found so mixed solvents, or *solvent pairs*, are used.

Solvent pairs

To use a mixed solvent dissolve the crystals in the better solvent and add the poorer solvent to the hot solution until it becomes cloudy and the solution is saturated with the solute. The two solvents must, of course, be miscible with each other. Some useful solvent pairs are given in Table 3.2

TABLE 3.2

	SOLVENT PAIRS
acetic acid–water	ethyl acetate–cyclohexane
ethanol–water	acetone–ligroin
acetone–water	ethyl acetate–ligroin
dioxane–water	diethyl ether–ligroin
acetone–ethanol	dichloromethane–ligroin
ethanol–diethyl ether	toluene–ligroin

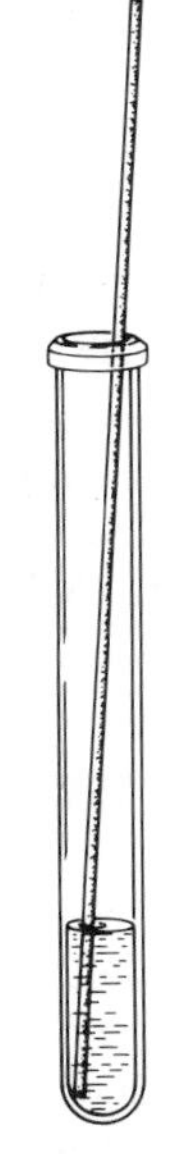

FIG. 3.1 Wood applicator stick in reaction tube to promote even boiling.

2. Dissolving the Solute

Once a crystallization solvent has been found, the impure crystals are placed in a reaction tube, solvent is added, the crystals are stirred with a micro spatula or a small glass rod, and the tube warmed on a hot plate or sand bath until the crystals dissolve. A solution of this type can become *superheated*, i.e., heated above its boiling point without actually boiling. When boiling does suddenly occur it can happen with almost explosive violence. To prevent this from happening a *wood applicator stick* can be added to the solution (Fig. 3.1). Air trapped in the wood comes out of the stick and forms the nuclei on which even boiling can occur. Porous porcelain *boiling chips* work in the same way. Never add a boiling chip or boiling stick to a hot solution, as it may boil over.

Prevention of bumping

Inspect the hot solution. If it contains no undissolved impurities, you can simply let it cool and collect the crystals. On the other hand, if it contains solid impurities it must be filtered before crystallization takes place, and if colored it must be treated with activated (decolorizing) charcoal and then filtered before crystallization.

To remove solid impurities (insoluble by-products of reaction, lint, dust, etc.), it is necessary to dilute the solution with excess solvent (the better solvent if a solvent mixture is being used), carry out the filtration near room temperature, and then evaporate the solvent to a point at which the hot solution is once more saturated with respect to the solute so that crystallization can take place in the usual way.

3. Decolorizing the Solution

Activated charcoal = decolorizing carbon = Norit

The vast majority of pure organic chemicals are colorless or a light shade of yellow. Occasionally a chemical reaction will produce high-molecular-weight by-products that are highly colored. The impurities can be adsorbed onto the surface of activated charcoal by simply boiling the solution with charcoal. Activated charcoal has an extremely large surface area per gram (several hundred square meters) and can bind a large number of molecules to this surface. On a commercial scale the impurities in brown sugar are adsorbed onto charcoal in the process of refining sugar.

Pelletized Norit

In the past laboratory manuals have advocated the use of finely powdered activated charcoal for removal of colored impurities. This has two drawbacks. Because the charcoal is so finely divided it can only be separated from the solution by filtration through paper, and even then some of the finer particles pass through the filter paper. And the presence of the charcoal completely obscures the color of the solution, so that adding the correct amount of charcoal is mostly a matter of luck. If too little charcoal is added, the solution will still be colored after filtration, making repetition necessary; if too much is added, it will absorb some of the product in addition to the impurities. We have found that charcoal extruded as short cylindrical pieces measuring about 0.8×3 mm made by the Norit Company solves both of these problems. It works just as well as the finely divided powder and it does not obscure the color of the solution. It can be added in small portions until the solution is decolorized and the size of the pieces makes it easy to remove from the solution.[1]

To use decolorizing charcoal, simply add a small amount (0.1% of the solute weight is sufficient) to the colored solution and then boil the solution for a few minutes. Be careful not to add the charcoal pieces to a superheated solution; the charcoal functions like hundreds of boiling chips and will cause the solution to boil over. Dilute the solution with solvent so the product will not crystallize when the solvent cools during the process of filtration.

1. Available from Aldrich Chemical Co., 940 West St. Paul Ave., Milwaukee, WI 53233 or American Norit Co., Inc., 6301 Glidden Way, Jacksonville, FL 32208 as Norit RO 0.8. This form of Norit is an extrudate 0.8 mm dia. It has a surface area of 1000 m^2/g, a total pore volume of 1.1 mL/g, and it will adsorb 0.22 g of methylene blue per gram and 1.1 g of iodine per gram.

Perusal of the current research literature reveals that decolorizing charcoal is rarely used these days. Colored impurities are instead removed by adsorption on silica gel or alumina during the process of column chromatography (Chapter 10) because more of the product can be recovered.

4. Filtering Suspended Solids

The filtration of a small volume of a dilute solution to remove solid impurities or charcoal can be done in a number of ways. With a volume less than 10 mL, the funnel and fluted filter paper technique conventionally used with larger volumes is inappropriate because a large proportion of the sample would be lost on the surface of the apparatus and into the filter paper.

Filter pipette

One way to filter 0.1–3 mL of a solution is to force a bit of cotton into a Pasteur pipette, put the solution to be filtered into this filter pipette using another Pasteur pipette, and then force the liquid through the filter using air pressure from a pipette bulb (Fig. 3.2). Fresh solvent can then be added to rinse the pipette and cotton.

If the solid impurities are large in size, they can be removed by the same process described in detail below: Filter the liquid through the small space between the square end of a Pasteur pipette and the bottom of a reaction tube. Powdered charcoal is too fine for this technique, but granular charcoal is easily removed in this way.

Micro Büchner

A general technique applicable to volumes of 0.1–10 mL requires a micro Büchner funnel, which is made of polyethylene and fitted with a porous *polyethylene frit* 6 mm in diameter. An extension tube is added to the funnel to hold the solution to be filtered. It is usually necessary to apply pressure, using a pipette bulb, to force the solution through the filter (Fig. 3.3). The micro Büchner funnel can also be used to collect very small quantities ($<$ 50 mg) of crystalline products too fine to be isolated by the Pasteur pipette method. The polyethylene or porcelain *Hirsch funnel* is the next largest funnel and can be used to collect 50–500 mg of material.

Filtration Using the Pasteur Pipette

The best and fastest micro filtration technique

The most important technique to be used in small-scale organic experiments is filtration using the Pasteur pipette. About 90% of the crystalline products from the experiments can be isolated in this way. The others will be isolated by filtration on the Hirsch funnel.

After crystallization has taken place, the usual practice is to cool the solution in ice to decrease the solubility of the product and cause more material to crystallize. The cold crystalline matrix is stirred with the Pasteur pipette and, while air is being expelled, the pipette tip is forced to the bottom of the reaction tube. The bulb is released and the solvent is drawn into the pipette through the very small space between the square tip of the pipette and the curved bottom of the reaction tube (Fig. 3.4). When

(*box continued on page 32*)

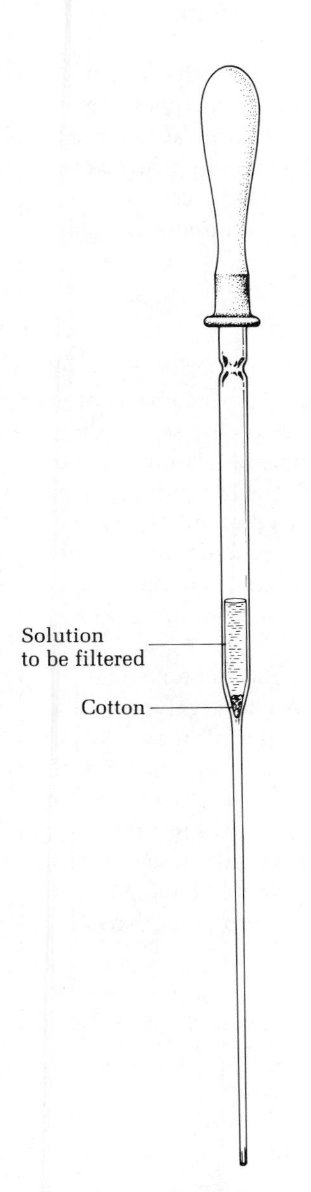

FIG. 3.2 Filtration of a solution in a Pasteur pipette.

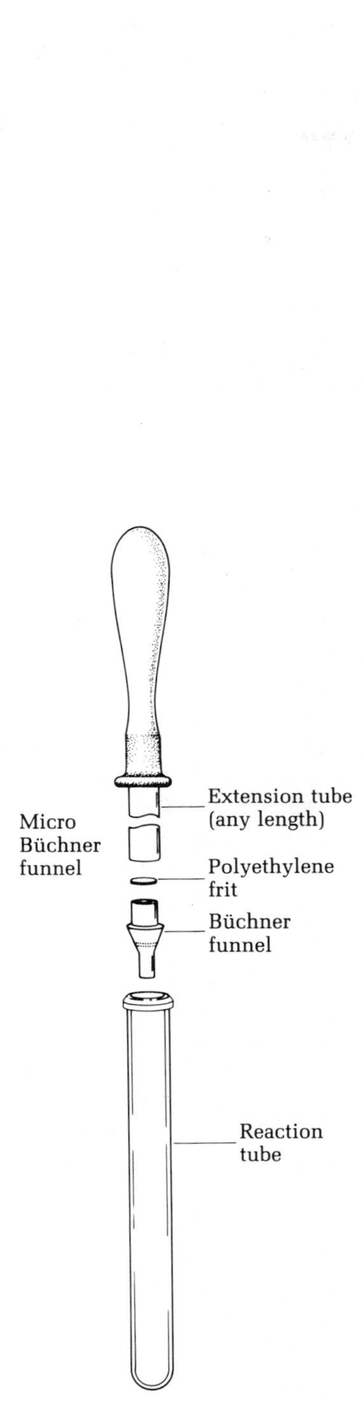

FIG. 3.3 Micro Büchner funnel.

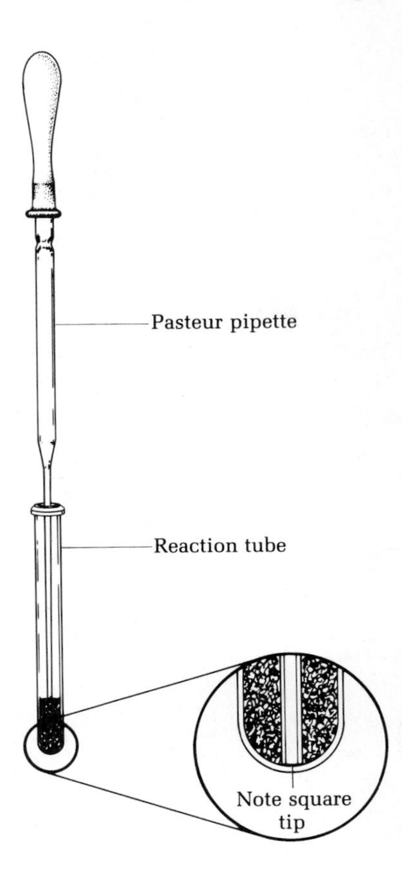

FIG. 3.4 Filtration using the Pasteur pipette and reaction tube.

all of the solvent has been withdrawn, it is expelled into another reaction tube held next to the tube containing the crystals. It is sometimes useful to rap the tube containing the wet crystals against a hard surface to pack them down so that more solvent can be removed. The tube is returned to the ice bath, a few drops of cold solvent are added to the crystals, the mixture stirred to wash the crystals, and the solvent again removed. This process can be repeated as many times as necessary.

Filtration on the Hirsch Funnel

Used for larger amounts of material

When the quantity of material is larger than can be conveniently filtered using the Pasteur pipette technique, or the crystals are so fine they pass into the pipette, you can collect them by vacuum filtration on the Hirsch funnel. This funnel, made of either polypropylene or porcelain, has a perforated plate at the bottom into which is placed a piece of 1-cm filter paper (Fig. 3.5). The funnel is fitted through a *Neoprene adapter* to a 25-mL *filter flask* (clamped so it won't tip over), which is connected via heavy-wall rubber tubing to a trap, which in turn is connected to a *water aspirator* (Fig. 3.6).

Use a trap

Water flowing through the aspirator will produce a vacuum equal to its vapor pressure (17 torr at 20°C, 5 torr at 4°C). A check valve is built into the aspirator, but even so when the water is turned off it may back into the evacuated system. For this reason a *trap* is always installed in the line. (Fig. 3.7). *The water passing through the aspirator should always be turned on full force.* The system can be opened to the atmosphere by removing the hose from the small filter flask or by opening the screw clamp on the trap. Open the system, then turn off the water to avoid having water suck back into the filter trap. Thin rubber tubing on the top of the trap will

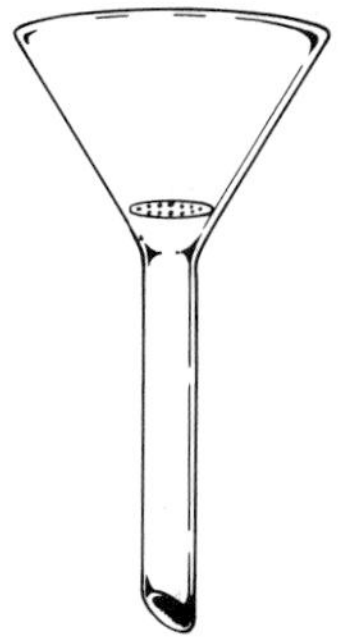

FIG. 3.5 Porcelain Hirsch funnel.

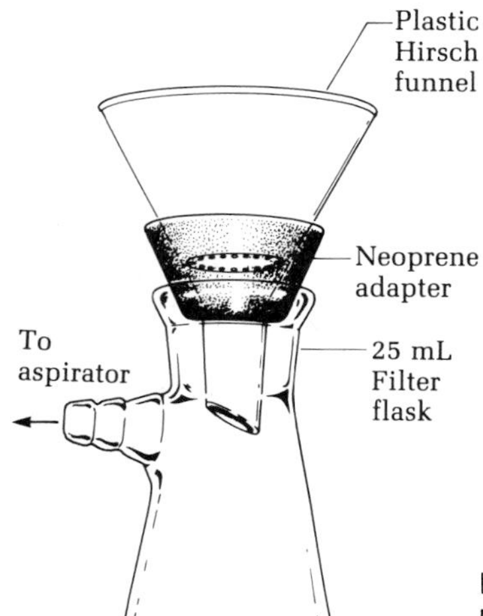

FIG. 3.6 Plastic Hirsch funnel, Neoprene adapter, and 25-mL filter flask.

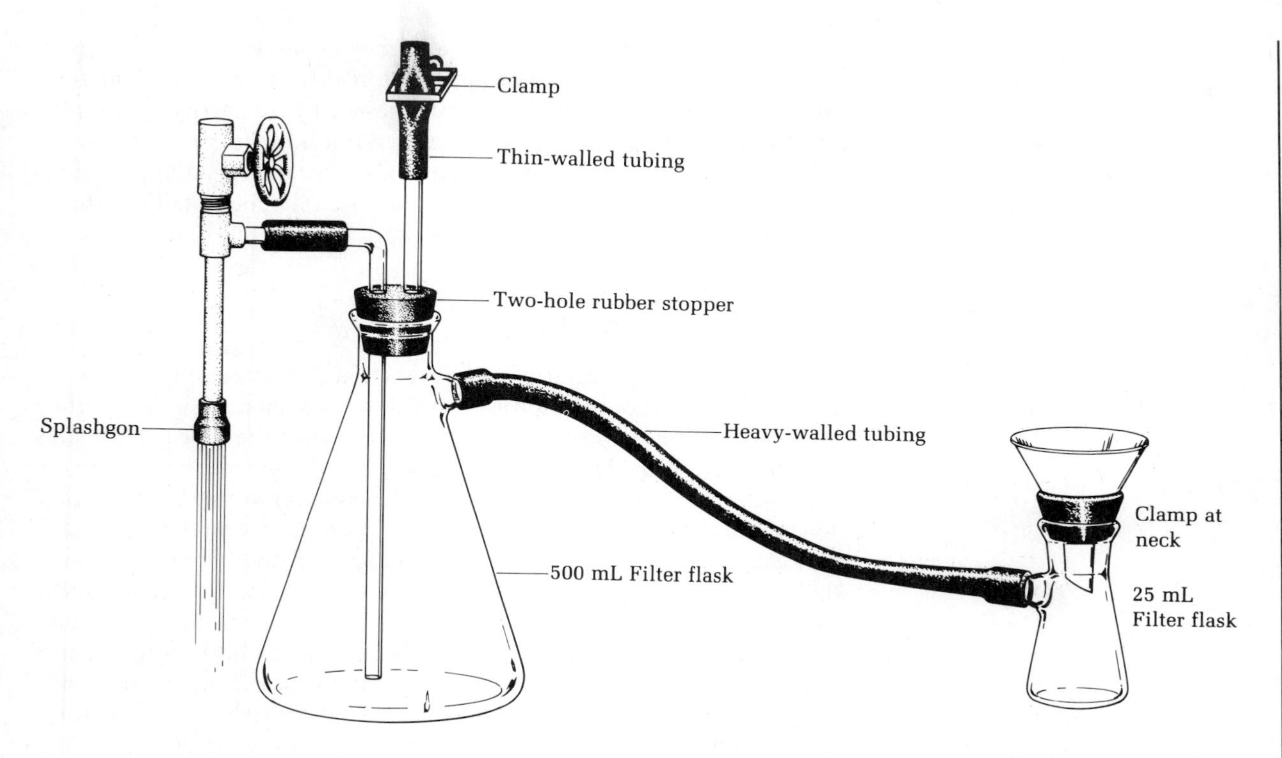

FIG. 3.7 Filter traps functioning as pressure gauges, (left) under vacuum, and (right) at atmospheric pressure.

collapse and bend over when a good vacuum is established. You will, in time, learn to hear the difference in the sound of an aspirator when it is pulling a vacuum and when it is working on an open system.

Filtration technique using a Hirsch funnel

To carry out the filtration turn on the aspirator full force, place a piece of filter paper on the funnel, wet it with some solvent and press it down, so that crystals can't escape around the edge and under the filter paper. Transfer the crystals to the funnel and just as soon as the filtrate has passed through, break the vacuum either at the valve on the trap or by disconnecting the hose from the filter flask. In this way impurities won't dry on the crystals. Use the filtrate to complete the transfer of material, breaking the vacuum each time. To wash the crystals, place cold wash solvent on the crystals with the vacuum off, then apply the vacuum just long enough to pull the solvent through. Once the washing has been completed, allow air to flow through the crystals in order to dry them. Finally turn the crystals out onto a piece of filter paper if they are still damp and squeeze out the excess solvent between sheets of filter paper, then scrape them onto a tared *watch glass* to dry completely.

The crystals left in the reaction tube after removal of the solvent can be dried under vacuum in the reaction tube and, if the tare weight of the tube is known, the weight of the crystals can be obtained by weighing the tube plus crystals. Another crystallization can be carried out if melting point determination shows the material is not of sufficient purity, or the crystalline material can be used to carry out another reaction, because the tube is designed to be used for reactions as well as crystallizations.

5. Crystallizing the Solute

Dissolving the crystals

Crystallization or recrystallization in the reaction tube is exceptionally easy because the upper part of the tube is designed to function as an air condenser while a liquid boils in the lower part of the tube. When the tube is heated on a sand bath, the top of the tube is cool enough to be grasped in the fingers while the solvent refluxes in the process of dissolving crystals. It is good practice to add a boiling stick to the tube to promote even boiling.

Any extra solvent added so that the solution can be filtered for impurities will have to be removed. This can be done by simply allowing the excess solvent to evaporate as the solution boils, but the process is greatly speeded up by directing a stream of air or, better, nitrogen at the hot solution. Once the hot solution is saturated with the compound just below the boiling point of the solvent, you can let it cool slowly to room temperature. Crystallization should begin immediately. If it does not, add a seed crystal or scratch the inside of the tube with a glass rod at the liquid-air interface. On the small scale being used for these experiments, it is best to allow the tube to cool in a beaker filled with cotton or paper towels, which act as insulation so cooling takes place slowly. Even with this protection the small reaction tube will cool to room temperature within a few minutes. Slow cooling guarantees the formation of large crystals, which are easily separated by filtration and easily washed free of adhering impure solvent. On a small scale it is difficult to obtain crystals that are too large and occlude impurities. Once the tube has cooled to room temperature without disturbance, it can be cooled in ice to maximize the amount of product that comes out of solution. The crystals are then separated from the *mother liquor* (the *filtrate*) by filtration.

Insulation of reaction tube allows slow cooling

Slow cooling = large crystals

6. Collecting and Washing the Crystals

The alternative to collecting crystals by the Pasteur pipette method is to filter them off on a small funnel using vacuum filtration (see Fig. 3.6). On a small scale the relative loss of material during the transferal from the crystallization container to the funnel is considerable. Furthermore, it is very difficult to keep the solution being filtered ice-cold. On the other hand the crystals can be isolated easily from the filter paper.

Another method for isolating small quantities of crystals is to centrifuge them in a Craig tube (Fig. 3.8). Again it is impossible to keep the solution cold during the initial filtration and subsequent washing, and the process is

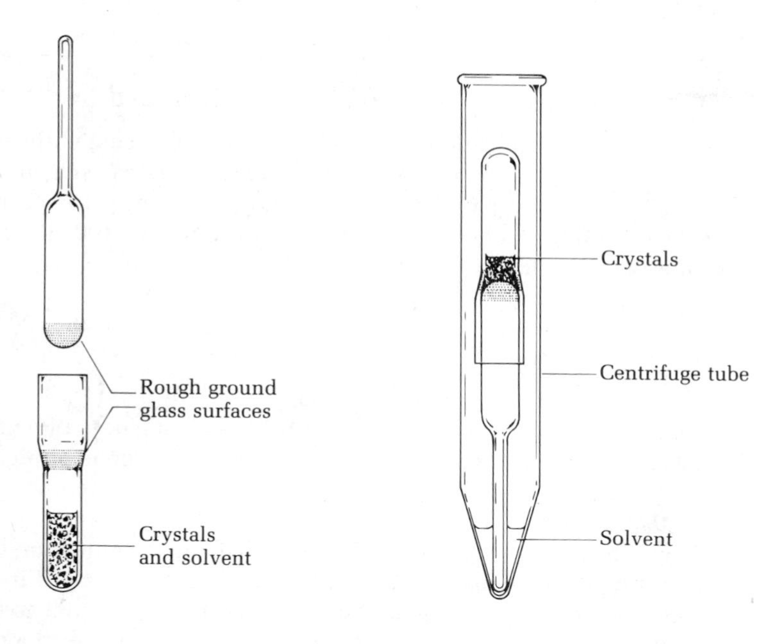

FIG. 3.8 Craig tube filtration apparatus. Filtration occurs between the rough ground glass surfaces when the apparatus is centrifuged.

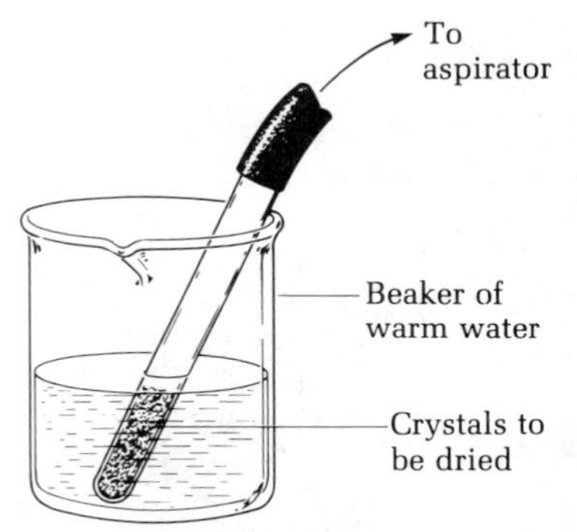

FIG. 3.9 Drying crystals under reduced pressure in a reaction tube.

slowed by the time it takes for the centrifuge to come to speed and then to stop. The advantage of the Craig tube is that the crystals are easily recovered from the apparatus and are drier than those isolated using the Pasteur pipette method of recovery.

Collecting a Second Crop of Crystals

Regardless of the method used to collect the crystals, the filtrate and washings can be combined and evaporated to the point of saturation to obtain a second crop of crystals. This second crop will increase the overall yield, but the crystals will not be as pure as the first crop.

7. Drying the Product

Vacuum drying

If possible, dry the product in the reaction tube after removal of the solvent using a Pasteur pipette. This can be done by simply connecting the tube to the water aspirator. If the tube is clamped in a beaker of hot water, the solvent will evaporate more rapidly under vacuum, but take care not to melt the product (Fig. 3.9). Scrape the product out onto a watch glass and allow it to dry to constant weight, which will indicate if all of the solvent is gone. If the product is collected on the Hirsch funnel, the last bit of solvent can be removed by squeezing the crystals between sheets of filter paper before drying them on the watch glass.

Experiments

1. Crystallization of Pure Phthalic Acid

The process of crystallization can be observed readily using phthalic acid. In the reference book *The Handbook of Chemistry and Physics*, in the table "Physical Constants of Organic Compounds," the entry for phthalic acid gives the following solubility data (in grams of solute per 100 mL of solvent). The superscripts refer to temperature in °C:

Water	Alcohol	Ether, etc.
0.54^{14} 18^{99}	11.71^{18}	0.69^{15} eth., i. chl.

The large difference in solubility in water as a function of temperature suggests this as the solvent of choice. The solubility in alcohol is high at room temperature. Ether is difficult to use because it is so volatile; the compound is insoluble in chloroform (i. chl.).

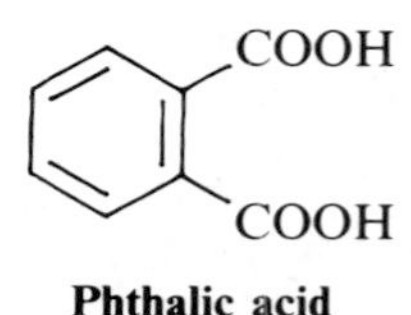

Phthalic acid

Crystallize 20 mg (0.020 g) of phthalic acid from the minimum volume of water, using the above data to calculate the required volume. First turn on the electrically heated sand bath. Add the solid to a 10 × 100 mm reaction tube and then, using a Pasteur pipette, add water dropwise. Use the calibration marks found in the end papers of this book to measure the volume of water in the pipette and the reaction tube. Add a boiling stick (a wood applicator stick) to facilitate even boiling and prevent bumping. After a portion of the water has been added, gently heat the solution to boiling on a hot *sand bath* in the electric heater. The deeper the tube is placed in the sand, the hotter it will be. As soon as boiling begins, continue to add water dropwise until all the solid just dissolves. Cork the tube and clamp it as it cools and observe the phenomenon of crystallization.

After the tube reaches room temperature, cool it in ice, stir the crystals with a Pasteur pipette, and expel the air from the pipette as the tip is pushed to the bottom of the tube. When the tip is firmly and squarely seated in the bottom of the tube, release the bulb and withdraw the water. Rap the tube sharply on a wood surface to compress the crystals and remove as much of the water as possible with the pipette. Using the stainless steel spatula, scrape the crystals onto a piece of filter paper, fold the paper over the crystals, and squeeze out excess water before allowing the crystals to dry to constant weight. Weigh the dry crystals and calculate the percent recovery of product.

2. Decolorizing a Solution with Decolorizing Charcoal

Decolorizing using pellitized Norit

Into a reaction tube place 1.0 mL of a solution of methylene blue dye that has been made up at a concentration of 100 mg per 100 mL of water. Add to the tube four pieces of decolorizing charcoal, shake and observe the color over a period of a minute or two. Heat the contents of the tube to boiling (reflux) and observe the color by holding the tube in front of a piece of white paper from time to time. How rapidly is the color removed?

3. Crystallization of Benzoic Acid from Water and a Solvent Pair

Crystallize 50 mg of benzoic acid from water in the same way phthalic acid was crystallized. Then in a dry reaction tube dissolve another 50-mg sample of benzoic acid in the minimum volume of hot toluene and add cyclohexane to the hot solution dropwise. When the hot solution becomes cloudy and crystallization has started, allow the tube to cool slowly to room temperature and collect the crystals. Compare crystallization in water to that in the solvent pair.

COOH

Benzoic acid

4. Purification of an Unknown

The seven-step crystallization procedure

Bearing in mind the seven-step crystallization procedure:

1. Choose the solvent
2. Dissolve the solute
3. Decolorize the solution (if necessary)
4. Filter suspended solids (if necessary)
5. Crystallize the solute
6. Collect and wash the crystals
7. Dry the product

You are to purify 100 mg of a crude unknown, provided by the instructor. Conduct tests for solubility and crystallizability in several organic solvents, solvent pairs, and water. If only a drop or two of solvent is used, the solvent can be evaporated by heating the tube on the sand bath and the residue can be used for another test. Submit as much pure product as possible after determining the melting point (Chapter 4) with evidence for the purity of the compound. From the posted list identify the unknown.

FIG. 3.10 Swirling a liquid in an Erlenmeyer flask.

On a Larger Scale. The crystallization of material on a 1–50 g scale follows the same seven steps as that on a smaller scale; however, the filtration method chosen to remove decolorizing charcoal and insoluble impurities is different. The material to be crystallized is dissolved in an Erlenmeyer flask (Fig. 3.10), decolorizing charcoal added if necessary, and then it and the insoluble impurities are removed by filtration through a fluted filter paper supported in a stemless funnel over another Erlenmeyer flask (Fig. 3.11). Once crystals have formed, they are collected on a Büchner funnel in exactly the same way as for collection of small amounts of product on the Hirsch funnel.[2]

FIG. 3.11 Gravity filtration of hot solution through fluted filter paper.

2. For complete details, see Chapter 5 of *Organic Experiments*, 6th ed., D. C. Heath and Co., Lexington, MA, 1987.

Crystallization Problems and Their Solutions

Induction of Crystallization

Seeding

Scratching

Occasionally a sample will not crystallize from solution on cooling, even though the solution is saturated with the solute at elevated temperature. The easiest method for inducing crystallization is to add to the supersaturated solution a seed crystal that has been saved from the crude material (if it was crystalline before recrystallization was attempted). In a probably apocryphal tale, the great sugar chemist Emil Fischer merely had to wave his beard over a recalcitrant solution and the appropriate seed crystals would drop out, causing crystallization to occur. In the absence of seed crystals, crystallization can often be induced by scratching the inside of the flask with a stirring rod at the air–liquid interface. One theory holds that part of the freshly scratched glass surface has angles and planes corresponding to the crystal structure, and crystals start growing on these spots. Often crystallization is very slow to begin and placing the sample in a refrigerator overnight will bring success. Other expedients are to change the solvent (usually to a poorer one) and to place the sample in an open container where slow evaporation and dust from the air may help induce crystallization.

Oils and "Oiling Out"

Crystallize at a lower temperature

Some saturated solutions, especially those containing water, when they cool deposit not crystals but small droplets referred to as oils. Should these droplets subsequently crystallize and be collected they will be found to be rather impure. Should the temperature of the saturated solution be above the melting point of the solute when it starts to come out of solution the solute will, of necessity, be deposited as an oil. Similarly, the melting point of the desired compound may be depressed to a point such that a low-melting eutectic mixture of the solute and the solvent comes out of solution. The simplest remedy for this latter problem is to lower the temperature at which the solution becomes saturated with the solute by simply adding more solvent. In extreme cases it may be necessary to lower this temperature well below room temperature by cooling the solution with dry ice.

Crystallization Summary

1. **Choosing the solvent.** "Like dissolves like." Some common solvents are water, methanol, ethanol, ligroin, and toluene. When you use a solvent pair, dissolve the solute in the better solvent and add the poorer solvent to the hot solution until saturation occurs. Some common solvent pairs are ethanol–water, diethyl ether–ligroin, and toluene–ligroin.
2. **Dissolving the solute.** To the crushed or ground solute in an Erlenmeyer flask add solvent; heat the mixture to boiling and swirl. Add more solvent as necessary to obtain a hot, saturated solution.

3. **Decolorizing the solution.** If it is necessary to remove colored impurities, cool the solution to near room temperature and add more solvent to prevent crystallization from occurring. Add decolorizing charcoal to the cooled solution, and then heat it to boiling for a few minutes, taking care to swirl the solution to prevent bumping.
4. **Filtering suspended solids.** If it is necessary to remove suspended solids or decolorizing charcoal, dilute the hot solution slightly to prevent crystallization from occurring during filtration. Filter the hot solution through the micro Büchner funnel. Add solvent if crystallization begins in the funnel. Concentrate the filtrate to obtain a saturated solution.
5. **Crystallizing the solute.** Let the hot saturated solution cool spontaneously to room temperature. Do not disturb the solution. Then cool the tube in ice. If crystallization does not occur, scratch the inside of the tube or add seed crystals.
6. **Collecting and washing the crystals.** Collect the crystals using the Pasteur pipette method or by vacuum filtration on a Hirsch funnel. If the latter technique is employed, wet the filter paper with solvent, apply vacuum, break vacuum, add crystals and liquid, apply vacuum until solvent just disappears, break vacuum, add cold wash solvent, apply vacuum, and repeat until crystals are clean and filtrate comes through clear.
7. **Drying the product.** Press the product on the filter to remove solvent. Then remove it from the filter, squeeze it between sheets of filter paper to remove more solvent, and spread it on a watch glass to dry.

Questions

1. A sample of naphthalene, which should be pure white, was found to have a greyish color after the usual purification procedure. The melting point was correct and the melting point range small. Explain the grey color.

2. How many milliliters of boiling water are required to dissolve 25 g of phthalic acid? If the solution were cooled to 14°C, how many grams of phthalic acid would crystallize out?

3. What is the reason for using activated carbon during a crystallization?

4. If a little activated carbon does a good job removing impurities in a crystallization, why not use a lot?

5. Under what circumstances is it wise to use a mixture of solvents to carry out a crystallization?

4 Melting Points and Boiling Points

PRELAB EXERCISE: *Predict what the melting points of the three urea–cinnamic acid mixtures will be.*

Part 1. Melting Points

Melting points—a micro technique

The melting point of a pure solid organic compound is one of its characteristic physical properties, along with molecular weight, boiling point, refractive index, and density. A pure solid will melt reproducibly over a narrow range of temperatures, typically less than 1°C. The process of determining this melting "point" is done on a truly micro scale using less than 1 mg of material; the apparatus is very simple, consisting of a thermometer, a capillary tube to hold the sample, and a heating bath.

Characterization

Melting points are determined for three reasons. If the compound is a known one the melting point will help to characterize the sample in hand. If the compound is new then the melting point is recorded in order to allow future characterization by others. And finally the range of the melting point is indicative of the purity of the compound; an impure compound will melt over a wide range of temperatures. Recrystallization of the compound will purify it and the melting point range will decrease. In addition, the entire range will be displaced upward. For example, an impure sample might melt from 120–124°C and after recrystallization melt at 125–125.5°C. A solid is considered pure if the melting point does not rise after recrystallization.

An indication of purity

A crystal is an orderly arrangement of molecules in a solid. As heat is added to the solid, the molecules will vibrate and perhaps rotate but still remain a solid. At a characteristic temperature it will suddenly acquire the necessary energy to overcome the forces that attract one molecule to another and it will undergo translational motion—in other words, it will become a liquid.

The forces by which one molecule is attracted to another include ionic attraction, van der Waals forces, hydrogen bonds, and dipole–dipole attraction. Most, but by no means all, organic molecules are covalent in nature and melt at temperatures below 300°C. Typical inorganic compounds are ionic and have much higher melting points, e.g., sodium chloride melts at 800°C.

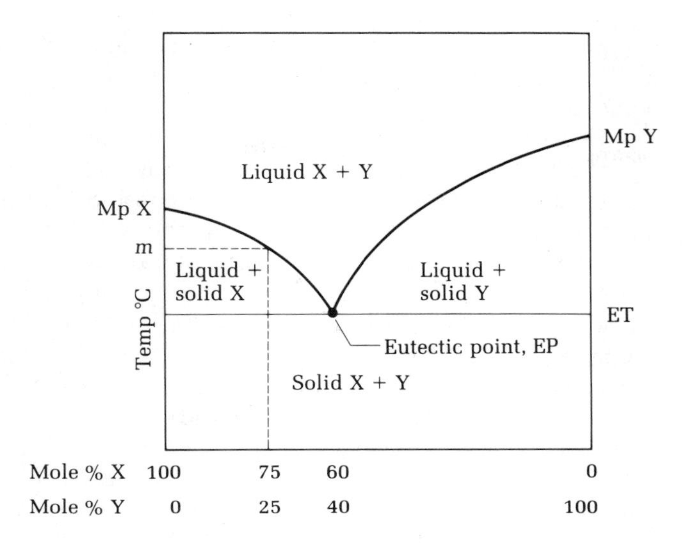

FIG. 4.1 Melting point–composition diagram for mixtures of the solids X and Y.

Ionic organic molecules often decompose before melting, as do compounds having strong hydrogen bonds such as sucrose.

Other factors being equal, larger molecules melt at higher temperatures than smaller ones. Among structural isomers the more symmetrical will have the higher melting point. Among optical isomers the R and S enantiomers will have the same melting points; but the racemate, the mixture of equal parts of R and S, will usually have a different melting point. Molecules that can form hydrogen bonds will usually have higher melting points than their counterparts of similar molecular weight.

A phase diagram

The melting point behavior of impure compounds is best understood by consideration of a simple binary mixture of compounds X and Y (Fig. 4.1). This melting point–composition diagram shows the melting point behavior as a function of composition. The melting point of a pure compound is the temperature at which the vapor pressures of the solid and liquid are equal. But in dealing with a mixture the situation is different. Consider the case of a mixture of 75% X and 25% Y. At a temperature below ET, the eutectic temperature, the mixture is solid Y and solid X. At the eutectic temperature the solid begins to melt. The melt is a solution of Y dissolved in liquid X. The vapor pressure of the solution of X and Y together is less than that of pure X at the melting point; therefore the temperature at which X will melt is lower when mixed with Y. This is an application of Raoult's Law (Chapter 5). As the temperature is raised, more and more of solid X melts until it is all gone at point M (temperature m). The melting point range is thus from ET to m. In practice it is very difficult to detect point ET when a melting point is determined in a capillary because it represents the point at which an infinitesimal amount of the liquid solution has started to melt.

Melting point depression

In this hypothetical example the liquid solution becomes saturated with Y at point EP. This is the point at which X and Y and their liquid solutions

The eutectic point

are in equilibrium. A mixture of X and Y containing 60% X will appear to have a sharp melting point at temperature ET. This point, EP, is the eutectic point.

In general the melting point range of a mixture of compounds is broad and the breadth of the range is an indication of purity. The chances of accidentally coming upon the eutectic composition are small. Recrystallization will enrich the predominant compound while excluding the impurity and therefore the melting point range will decrease.

It should be apparent that the impurity must be soluble in the compound, so an insoluble impurity such as sand or charcoal will not depress the melting point. The impurity does not need to be a solid. It can be a liquid such as water (if it is soluble) or an organic solvent, such as the one used to recrystallize the compound; hence the necessity for drying the compound before determining the melting point.

Mixed melting points

Advantage is taken of the depression of melting points of mixtures to prove whether two compounds having the same melting points are identical. If X and Y are identical, then a mixture of the two will have the same melting point; but if X and Y are not identical, then a small amount of X in Y or of Y in X will cause the melting point to be lowered.

Apparatus

The apparatus needed for determining an accurate melting point need not be elaborate; the same results are obtained on the simplest as on the most complex devices.

Thomas-Hoover Uni-Melt

The Thomas-Hoover Uni-Melt apparatus (Fig. 4.2) will accommodate seven capillaries in a small magnified, lighted beaker of high-boiling silicone oil that is stirred and heated electrically. The heating rate is controlled with a variable transformer that is part of the apparatus. The rising mercury column of the thermometer can be observed with an optional traveling periscope device so the eye need not move away from the capillary. For industrial analytical and control work there is even an apparatus (Mettler) that automatically determines the melting point and displays the result in digital form.

Mel-Temp

The Mel-Temp apparatus (Fig. 4.3) consists of an electrically heated aluminum block that accommodates three capillaries. The sample is illuminated through the lower port and observed with a 6-power lens through the upper port. The heating rate can be controlled, and with a special thermometer the apparatus can be used up to 500°C, far above the useful limit of silicone oil (about 350°C).

Test tube

Using a 4-in test tube half filled with silicone oil and a stirrer made from heavy wire, melting points can be determined on an electric heater filled with sand (Fig. 4.4). Once the sand is warmed up it has a very large temperature gradient from top to bottom, so the temperature and heating rate can be adjusted according to the depth of the test tube in the sand.

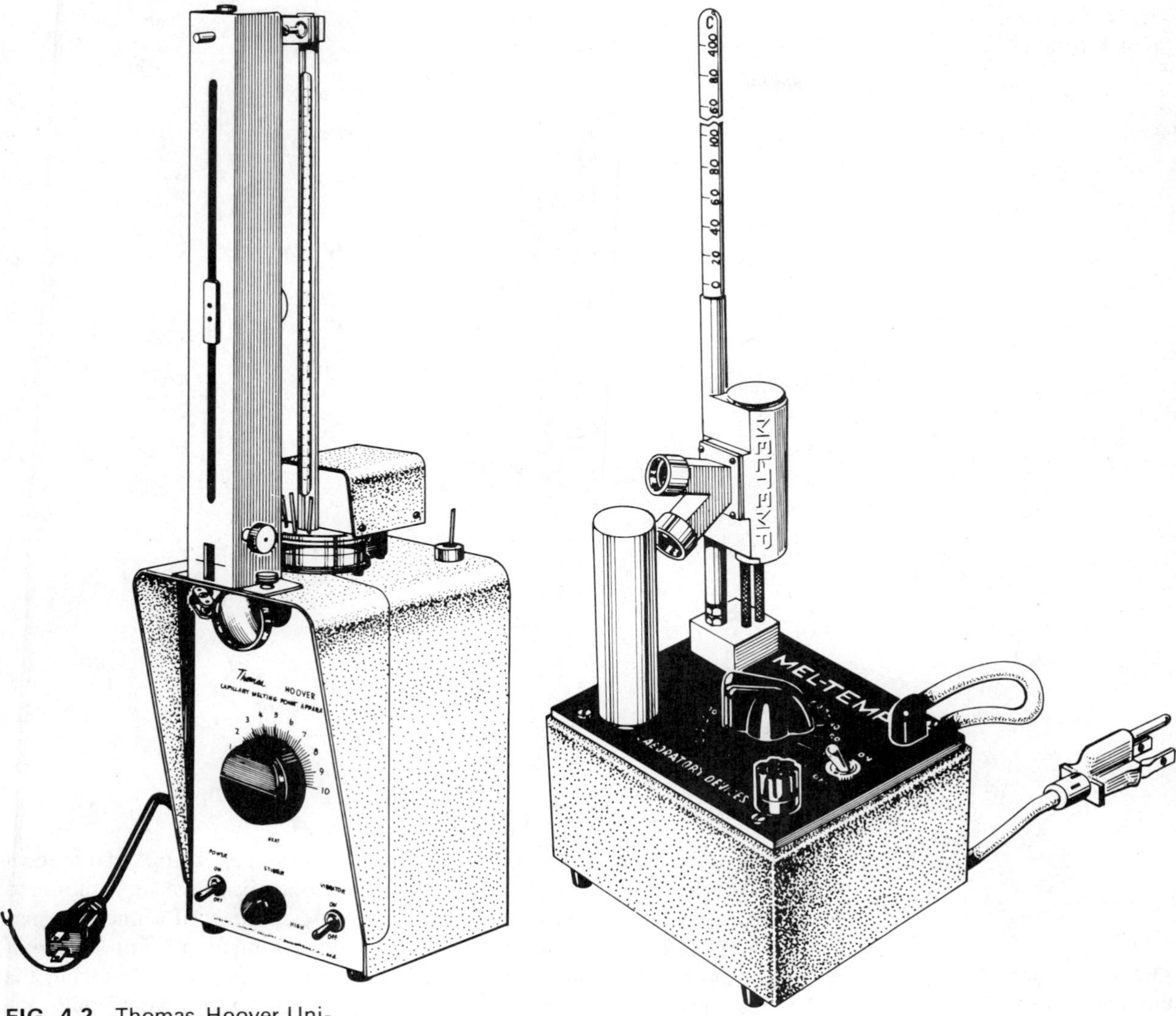

FIG. 4.2 Thomas-Hoover Uni-Melt melting point apparatus.

FIG. 4.3 Mel-Temp melting point apparatus.

A simple device is illustrated in Fig. 4.5. The thermometer is fitted through a cork, a section of which is cut away for pressure release and so that the scale is visible. A single-edge razor blade is convenient for cutting, and the cut can be smoothed or deepened with a triangular file. The curvature of the walls of the flask causes convection currents in the heating liquid to rise evenly along the walls and then descend and converge at the center; hence, the thermometer must be centered in the flask. The long neck prevents

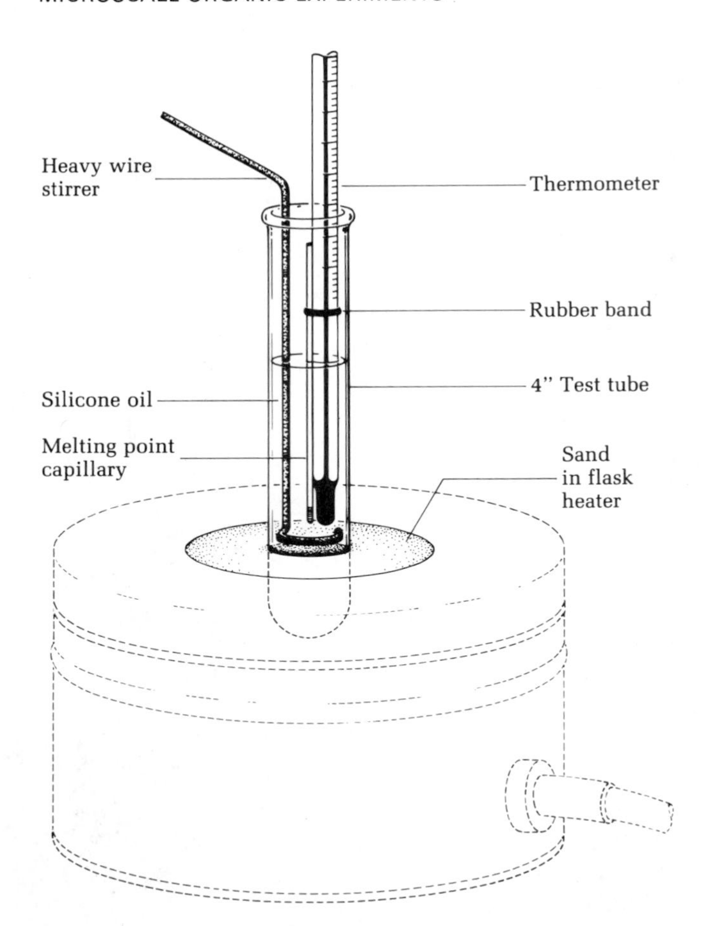

FIG. 4.4 Test tube melting point apparatus.

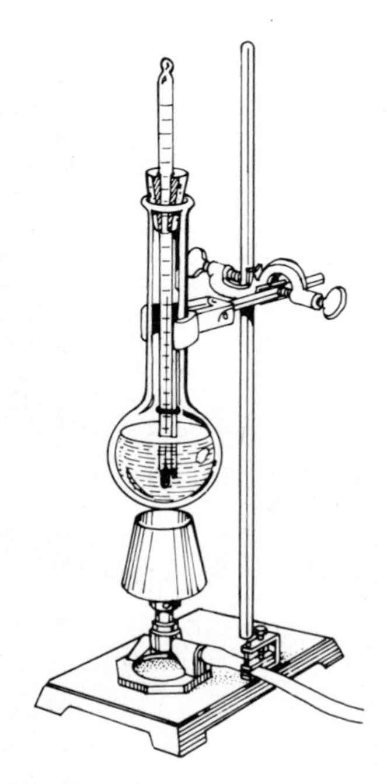

FIG. 4.5 Apparatus for melting point determination.

spilling and fuming and minimizes error due to stem exposure. The bulb of the flask (dry!) is three-quarters filled with silicone oil. The flask is mounted as shown in Fig. 4.5, with the bulb close to the chimney of a microburner. Careful control of heat input required in taking a melting point is accomplished both by regulating the gas supply and by raising or lowering the heating bath. This same apparatus can be heated more safely with the electrically heated sand bath.

Thiele tube

The Thiele apparatus (Fig. 4.6) achieves stirring and uniform heat distribution by convection. It is filled to the base of the neck with silicone oil (the oil expands on heating) and outfitted with the same type of slotted cork described above. The tube is heated at the base of the bend. The bulb of the thermometer should be halfway down the tube, as shown in Fig. 4.6, to assure uniform heating.

Beaker

Melting points are also easily determined in a beaker as seen in Fig. 4.7. The glass rod used for stirring has a circular base and a handle as seen in the figure. Alternatively the beaker can be heated on a hot plate and stirred magnetically using a Teflon-covered stirring bar.

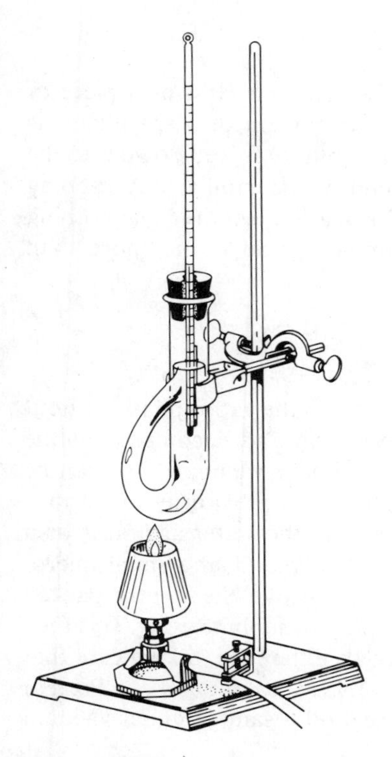

FIG. 4.6 Thiele apparatus.

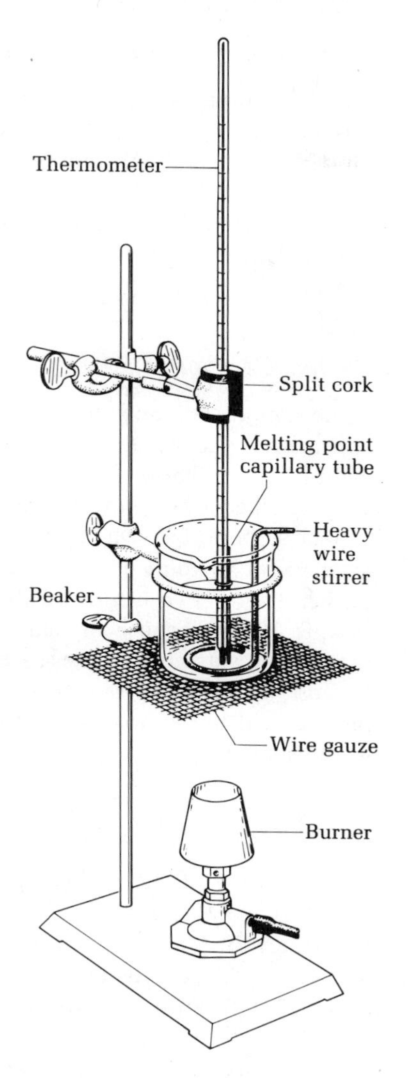

FIG. 4.7 Beaker melting point apparatus.

Do not discard the oil used in the apparatus because it will be necessary to determine a number of melting points in future experiments.

Capillaries can be obtained commercially or can be made by drawing out 12-mm soft-glass tubing. The tubing is rotated in the hottest part of the Bunsen burner flame until it is very soft and begins to sag. It should not be drawn out during heating, but is removed from the flame and after a moment's hesitation drawn steadily and not too rapidly to arm's length. With some practice it is possible to produce 10–15 good tubes in a single drawing. The long capillary tube can be cut into 100-mm lengths with a glass scorer. Each tube is sealed by rotating the end in the edge of a small flame, as seen in Fig. 4.8.

FIG. 4.8 Sealing a melting point capillary tube.

Filling Melting Point Capillaries

The dry sample is ground to a fine powder on a watch glass or a piece of glassine paper on a hard surface using the flat portion of a spatula. It is formed into a small pile and the melting point capillary forced down into the pile. The sample is shaken into the closed end of the capillary by rapping sharply on a hard surface or by dropping it down a 2-ft length of glass tubing onto a hard surface. The height of the sample should be no more than 2–3 mm.

Sealed Capillaries

Samples that sublime

Some samples sublime (go from the solid directly to the vapor phase without appearing to melt) or undergo rapid air oxidation and decompose at the melting point. These samples should be sealed under vacuum. This can be accomplished by forcing a capillary through a hole previously made in a rubber septum and evacuating the capillary using the water aspirator or a mechanical vacuum pump (Fig. 4.9). Using the flame from a small micro-burner the tube is gently heated about 15 mm above the tightly packed sample. This will cause any material in this region to sublime away. It is then heated more strongly in the same place to collapse the tube, taking care that the tube is straight when it cools. It is also possible to seal the end of a Pasteur pipette, add the sample, pack it down, and seal off a sample under vacuum in the same way.

Determining the Melting Point

The accuracy of the melting point depends on the accuracy of the thermometer, so the first exercise in this experiment will be to calibrate the thermometer. Melting points of pure, known compounds will be determined and deviations recorded so that a correction can be applied to future melting points. Be forewarned however that the thermometers are usually fairly accurate.

Rate of heating at the melting point: 1°C/min

The most critical factor in determining an accurate melting point is the rate of heating. At the melting point the temperature rise should not be greater than 1°C per minute. This may seem extraordinarily slow, but it is necessary in order that heat from the bath be transferred equally to the sample and to the glass and mercury of the thermometer.

From experience you know the rate at which ice melts. Consider doing a melting point experiment on an ice cube. Because water melts at 0°C, you would need to have a melting point bath a few degrees below zero. To observe the true melting point of the ice cube you would need to raise the temperature extraordinarily slowly. The ice cube would appear to begin to melt at 0°C and, if you waited for temperature equilibrium to be established, it would all be melted at 0.5°C. If you were impatient and raised the temperature too rapidly, the ice might appear to melt over the range 0 to 20°C. Similarly melting points determined in capillaries will not be accurate if the rate of heating is too fast.

The rate of heating is the most important factor in obtaining accurate melting points.

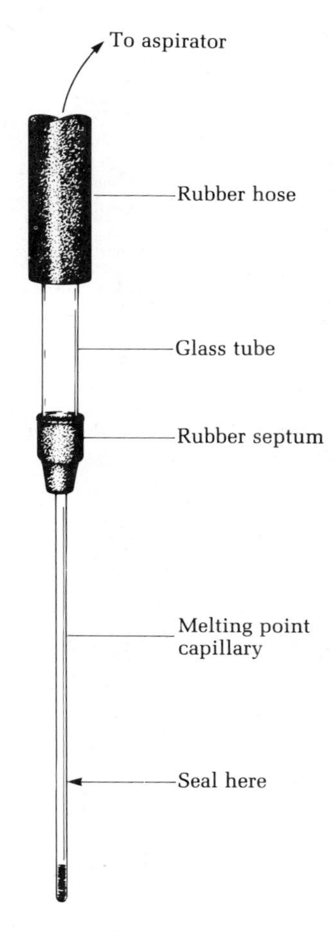

FIG. 4.9 Evacuation of a melting point capillary prior to sealing.

Experiments

1. Calibration of the Thermometer

Determine the melting point of standard substances (Table 4.1) over the temperature range of interest. The difference between the values found and those expected constitutes the correction that must be applied to future

TABLE 4.1 Melting Point Standards

Compound	Structure	Melting Point (°C)
Naphthalene		80–82
Urea	O ‖ H_2NCNH_2	132.5–133
Sulfanilamide	NH_2 … SO_2NH_2	164–165
4-Toluic Acid	COOH … CH_3	180–182
Anthracene		214–217
Caffeine (evacuated capillary)	O, CH_3, H_3C, N, N, N, O, N, CH_3	234–236.5

temperature readings. If the thermometer has been calibrated previously, then determine one or more melting points of known substances to familiarize yourself with the technique. If the determinations do not agree within 1°C, then repeat the process.

2. Melting Points of Pure Urea and Cinnamic Acid

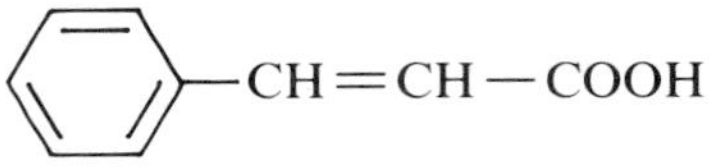

C_6H_5—CH=CH—COOH

Cinnamic acid

Using a metal spatula crush the sample to a fine powder on a hard surface such as a watch glass. Push a melting point capillary into the powder and force the powder down in the capillary by tapping the capillary or by dropping it through a long glass tube held vertically and resting on a hard surface. The column of solid should be no more than 2–3 mm in height and it should be tightly packed.

Except for the Thomas-Hoover and Mel-Temp apparatus, the capillary is held to the thermometer with a rubber band made by cutting a slice off the end of a piece of $\frac{3}{16}$-in rubber tubing. This rubber band must be above the level of the oil bath; otherwise it will break in the hot oil. Insertion of a fresh tube under the rubber band is facilitated by leaving the used tube in place. The sample should be close to and on a level with the center of the thermometer bulb.

Heat rapidly to within 20°C of the melting point

If the approximate melting temperature is known, the bath can be heated rapidly until the temperature is about 20°C below this point, but the heating during the last 15–20°C should slow down considerably so that the rate of heating at the melting point is no more than 1°C per minute while the sample is melting. As the melting point is approached the sample may shrink because of crystal structure changes. However, the melting process begins when the first drops of liquid are seen in the capillary and it ends when the last trace of solid disappears. For a pure compound this whole process may occur over a range of only 0.5°C; hence the necessity of having the temperature rise slowly during the determination.

Two or three melting points at once

If determinations are to be done on two or three samples that differ in melting point by as much as 10°C, two or three capillaries can be secured to the thermometer together and the melting points observed in succession without removal of the thermometer from the bath. As a precaution against interchange of tubes while they are being attached, use some system of identification, such as one, two, and three dots made with a marking pencil.

Determine the melting point of either urea (mp 132.5–133°C) or cinnamic acid (mp 132.5–133°C). Repeat the determination and if the two determinations do not check within 1°C, do a third one.

3. Melting Points of Urea–Cinnamic Acid Mixtures

Make mixtures of urea and cinnamic acid in the approximate proportions 1 : 4, 1 : 1, and 4 : 1 by putting side by side the correct number of equal-sized small piles of the two substances and then mixing them. Grind the mixture thoroughly for at least a minute on a watch glass using a metal spatula. Note

TABLE 4.2 Melting Point Unknowns

Compound	Melting Point (°C)
Benzophenone	49–51
Maleic anhydride	54–56
Naphthalene	80–82
Acetanilide	113.5–114
Benzoic acid	121.5–122
Urea	132.5–133
4-Nitrotoluene	148–149
Salicylic acid	158.5–159
Sulfanilamide	165–166
Succinic acid	184.5–185
3,5-Dinitrobenzoic acid	205–207
p-Terphenyl	210–211

the ranges of melting of the three mixtures and use the temperatures of complete liquefaction to construct a rough diagram of mp versus composition.

4. Unknowns

Determine the melting point of one or more of the following unknowns to be selected by the instructor (Table 4.2) and on the basis of the melting point identify the substance. Prepare two capillaries of each unknown. Run a very fast determination on the first sample to ascertain the approximate melting point and then cool the melting point bath to just below the melting point and make a slow, careful determination.

Part 2. Boiling Points

The boiling point of a pure organic liquid is one of its characteristic physical properties, just like the density, molecular weight, and refractive index and the melting point of a solid. The boiling point is used to characterize a new organic liquid and knowledge of the boiling point helps to compare one organic liquid with another, as in the process of identifying an unknown organic substance.

Comparison of boiling points with melting points is instructive. The process of determining the boiling point is more complex than that for the melting point: it requires more material and because it is affected less by impurities, it is not as good an indication of purity. Boiling points can be determined on a few microliters of a liquid, but on a small scale it is difficult to determine the boiling point *range*. This requires enough material to distill, about 1 to 2 mL. Like the melting point, the boiling point of a liquid is affected by the forces that attract one molecule to another—ionic attraction, van der Waals forces, dipole–dipole interactions, and hydrogen bonding.

Structure and Boiling Point

In a homologous series of molecules the boiling point increases in a perfectly regular manner. The normal saturated hydrocarbons have boiling points ranging from −162°C for methane to 330°C for *n*-$C_{19}H_{40}$. It is convenient to remember that *n*-heptane with a molecular weight of 100 has a boiling point near 100°C (98.4°C). A spherical molecule such as 2,2-dimethylpropane has a lower boiling point than *n*-pentane because it cannot have as many points of attraction to adjacent molecules. For molecules of the same molecular weight, those with dipoles, such as carbonyl groups, will have higher boiling points than those without and molecules that can form hydrogen bonds will boil even higher. The boiling point of such molecules depends on the number of hydrogen bonds that can be formed, so that an alcohol with one hydroxyl group will boil lower than one with two if they both have the same molecular weight. A number of other generalizations can be made about boiling point behavior as a function of structure; you will learn of these throughout your study of organic chemistry.

Boiling Point as a Function of Pressure

Since the boiling point of a pure liquid is defined as the temperature at which the vapor pressure of the liquid exactly equals the pressure exerted upon it, the boiling point will be a function of atmospheric pressure. At an altitude of 14,000 ft the boiling point of water is 81°C. At pressures near that of the atmosphere at sea level (760 mm), the boiling point of most liquids decreases about 0.5°C for each 10-mm decrease in atmospheric pressure. This generalization does not hold for greatly reduced pressures because the boiling point decreases as a nonlinear function of pressure (see Fig. 5.1). Under these conditions a nomograph relating observed boiling point, boiling point at 760 mm, and pressure in mm should be consulted (see Fig. 7.11). This nomograph is not highly accurate; the change in boiling point as a function of pressure also depends on the type of compound (polar, nonpolar, hydrogen bonding, etc.). Consult the *Handbook of Chemistry and Physics* for the correction of boiling points to standard pressure.

The Laboratory Thermometer

Most laboratory thermometers have a mark around the stem that is three inches (76 mm) from the bottom of the bulb. This is the immersion line; the thermometer will record accurate temperatures if immersed to this line. If the thermometer happens to be of the total immersion type then the readings will need to be corrected when taking melting and boiling points because only a few inches of the thermometer are being heated. Again the procedure for making this "emergent stem correction for liquid-in-glass thermometers" can be found in the *Handbook of Chemistry and Physics*. Should you break a mercury thermometer immediately inform your instructor, who will use special apparatus to clean up the mercury. Mercury vapor is very toxic.

Boiling Sticks and Boiling Stones

A very clean liquid in a very clean vessel will superheat and not boil when subjected to a temperature above its boiling point. This means that a thermometer placed in the liquid will register a temperature higher than the boiling point of the liquid. If boiling does occur under these conditions, it occurs with explosive violence. To avoid this problem boiling stones are always added to liquids before heating them to boiling—whether to determine a boiling point or to carry out a reaction. The stones provide the nuclei on which the bubble of vapor indicative of a boiling liquid can form. Some boiling stones, also called boiling chips, are porous unglazed porcelain. This material is filled with air in numerous fine capillaries. On heating this air expands to form the fine bubbles on which even boiling can take place. Once the liquid cools it will fill these capillaries and the boiling chip becomes ineffective, so another must be added each time the liquid is heated to boiling. Sticks of wood, so-called applicator sticks about 1.5 mm in dia., also promote even boiling and unlike stones are easy to remove from the solution.

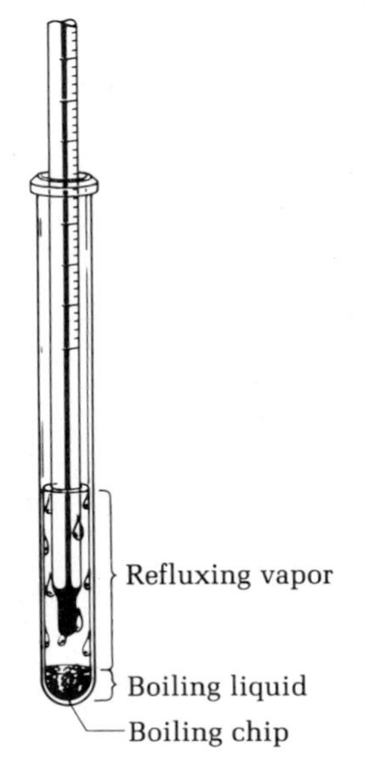

FIG. 4.10 Small-scale boiling point. Clamp thermometer and reaction tube so that they do not touch.

Apparatus and Technique

When enough material is available the best method for determining the boiling point of a liquid is to distill it (Chap. 5). Distillation allows the boiling range to be determined and thus gives an indication of purity.

For smaller quantities of material the apparatus is very similar to that used for determining melting points. A very simple method is to place 0.2 mL of the liquid along with a boiling stone in a 10×100 mm reaction tube, clamp a thermometer so the bulb is just above the level of the liquid, and then heat the liquid with a sand bath (Fig. 4.10). It is very important that no part of the thermometer touch the reaction tube. Heating is regulated so that the boiling liquid refluxes (condenses and drips down) about one or two inches up the thermometer, but does not boil out the top of the tube. Droplets of liquid must drip from the thermometer bulb in order to heat the mercury thoroughly. The boiling point is the highest temperature recorded by the thermometer and maintained over about a 1-min time interval.

For smaller quantities the tube is attached to the side of the thermometer (Fig. 4.11) and heated with a liquid bath. The tube, which can be from tubing 3–5 mm in diameter, contains a small inverted capillary. This is made by cutting a 6-mm piece from the sealed end of a melting point capillary,

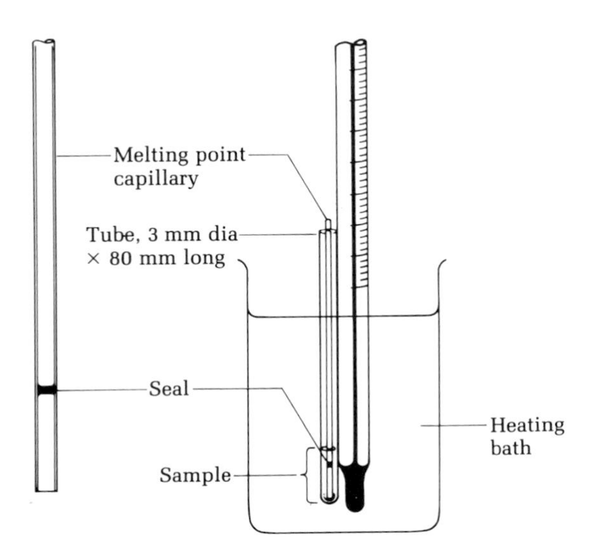

FIG. 4.11 Smaller-scale boiling point.

inverting it, and sealing it again to the capillary. A cm/mm ruler is printed on the inside cover of this book.

When the sample is heated in this device the air in the inverted capillary will expand and an occasional bubble will escape. At the boiling point a continuous and rapid stream of bubbles will emerge from the inverted capillary. At this point the heating is stopped and the bath allowed to cool. A time will come when bubbling ceases and the liquid just begins to rise in the inverted capillary. The temperature at which this happens is recorded. The liquid is allowed to partially fill the small capillary and the heat is applied carefully until the first bubble comes from the capillary, and that temperature is recorded. The two temperatures approximate the boiling point range for the liquid. The explanation: As the liquid was being heated the air expanded in the inverted capillary and was replaced by vapor of the liquid. The liquid was actually slightly superheated when rapid bubbles emerged from the capillary, but on cooling the point was reached at which the pressure on the inside of the capillary matched the outside (atmospheric) pressure. This is, by definition, the boiling point.

It is possible, with care, to make this same boiling point device on an even smaller scale. A Pasteur pipette is heated in a microburner flame until the glass at one point is very soft. The pipette is lifted from the flame and after a moment's hesitation drawn out steadily and not too rapidly for a foot or so to produce a very fine capillary. One end of this capillary is sealed; a piece about 6 mm long is broken off and the sealed end attached to a 3-in piece of capillary. The sample is added to an ordinary melting point capillary, shaken or centrifuged to the bottom, and the fine capillary just made is added to the melting point capillary. The boiling point is determined as described above, using a Thomas-Hoover or similar liquid bath melting point apparatus.

Questions

1. What effect would poor circulation of the melting point bath liquid have on the observed melting point?
2. What is the effect of an insoluble impurity, such as sodium sulfate, on the observed melting point of a compound?
3. Three test tubes, labeled A, B, and C, contain substances with approximately the same melting points. How could you prove the test tubes contain three different chemical compounds?
4. One of the most common causes of inaccurate melting points is too rapid heating of the melting point bath. Under these circumstances how will the observed melting point compare with the true melting point?
5. Strictly speaking, why is it incorrect to speak of a melting *point*?
6. What effect would the incomplete drying of a sample (e.g., the incomplete removal of a recrystallization solvent) have on the melting point?
7. Which would be expected to have the higher boiling point, *t*-butyl alcohol (2-methyl-2-propanol) or *n*-butyl alcohol (1-butanol)?

Reference

A. Weissberger and B. W. Rossiter (eds). *Physical Methods of Chemistry*, Vol. 1, Part V, Wiley-Interscience, New York, 1971.

5 Distillation

PRELAB EXERCISE: *Predict what a plot of temperature vs. volume of distillate will look like for the simple distillation and the fractional distillation of (a) a cyclohexane–toluene mixture, (b) an ethanol–water mixture.*

The origins of distillation are lost in antiquity as man in his thirst for more potent beverages found that dilute solutions of alcohol from fermentation could be separated into alcohol-rich and water-rich portions by heating the solution to boiling and condensing the vapors above the boiling liquid—the process of distillation. Since ethyl alcohol, ethanol, boils at 78°C and water boils at 100°C, one might naively assume that heating a 50 : 50 mixture of ethanol and water to 78°C would cause the ethanol molecules to leave the solution as a vapor that could be condensed to give pure ethanol. Such is not the case. A mixture of 50 : 50 ethanol : water boils near 87°C and the vapor above it is not 100% ethanol.

Consider a simpler mixture, cyclohexane and toluene. The vapor pressures as a function of temperature are plotted in Fig. 5.1. When the vapor pressure of the liquid equals the applied pressure the liquid boils, so this diagram shows that at 760 mm pressure, standard atmospheric pressure, these pure liquids boil at 78 and 111°C respectively. If one of these pure liquids were to be distilled, it would be found that the boiling point of the liquid would equal the temperature of the vapor and that the temperature of the vapor would remain constant throughout the distillation.

A mixture of two liquids will usually have a bp that is between the bps of the pure liquids

Raoult's Law

Figure 5.2 is a boiling point–composition diagram for the cyclohexane-toluene system. If a mixture of 75 mole percent toluene and 25 mole percent cyclohexane is heated, we find from Fig. 5.2 it boils at 100°C, or point A. Above a binary mixture of cyclohexane and toluene the vapor pressure has contributions from each component. Raoult's law states that the vapor pressure of the cyclohexane is equal to the product of the vapor pressure of pure cyclohexane and the mole fraction of cyclohexane in the liquid mixture:

$$P_c = P_c^\circ N_c$$

where P_c is the partial pressure of cyclohexane, P_c° is the vapor pressure of pure cyclohexane at the given temperature, and N_c is the mole fraction of cyclohexane in the mixture. Similarly for toluene:

$$P_t = P_t^\circ N_t$$

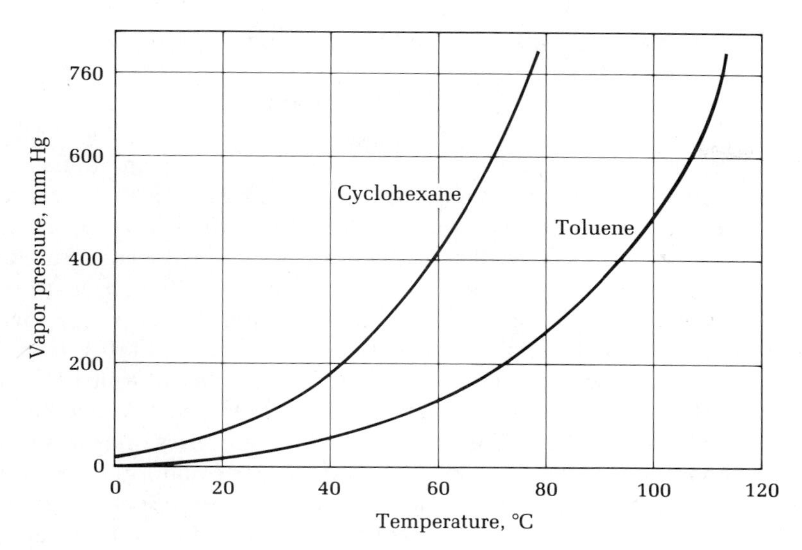

FIG. 5.1 Vapor pressure vs. temperature for cyclohexane and toluene.

and the total vapor pressure above the solution, P_{Tot}, is given by the sum of the partial pressures due to cyclohexane and toluene:

$$P_{\text{Tot}} = P_c + P_t$$

Dalton's law states that the mole fraction of cyclohexane in the vapor at a given temperature is equal to the partial pressure of the cyclohexane at

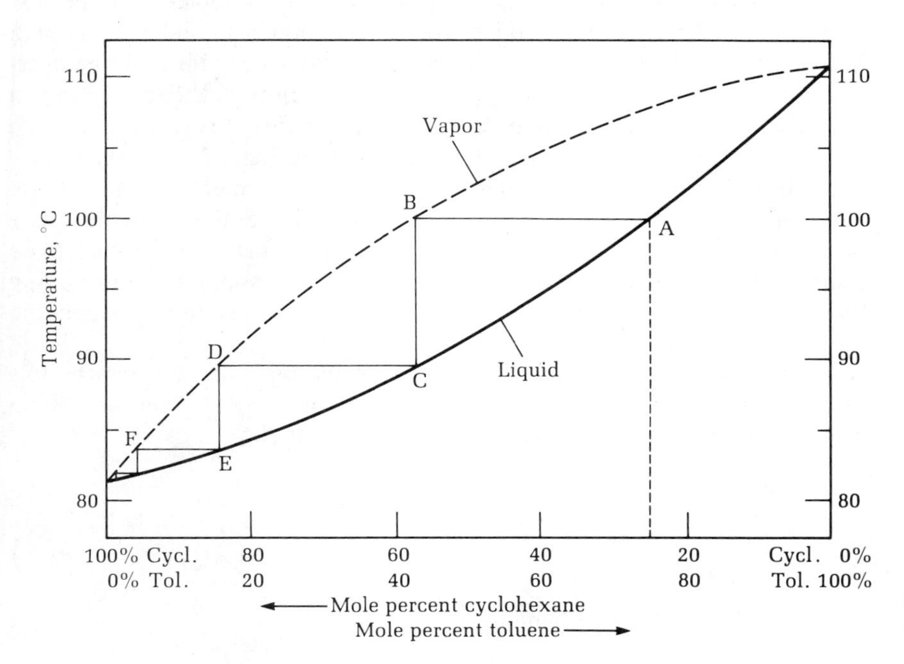

FIG. 5.2 Boiling point–composition curves for cyclohexane–toluene mixtures.

that temperature divided by the total pressure:

$$X_c = P_c/\text{total vapor pressure}$$

At 100°C cyclohexane has a partial pressure of 433 mm and toluene a partial pressure of 327 mm; the sum of the partial pressures is 760 mm and so the liquid boils. If some of the liquid in equilibrium with this boiling mixture were condensed and analyzed, it would be found to be 433/760 or 57 mole percent cyclohexane (point B, Fig. 5.2). This is the best separation that can be achieved on simple distillation of this mixture. As the simple distillation proceeds, the boiling point of the mixture moves toward 110°C along the line from A, and the vapor composition becomes richer in toluene as it moves from B to 110°C. In order to obtain pure cyclohexane, it would be necessary to condense the liquid at B and redistil it. When this is done it is found that the liquid boils at 90°C (point C) and the vapor equilibrium with this liquid is about 85 mole percent cyclohexane (point D). So to separate a mixture of cyclohexane and toluene, a series of fractions would be collected and each of these partially redistilled. If this fractional distillation were done enough times the two components could be separated.

Fractional distillation

This series of redistillations can be done "automatically" in a fractionating column. Perhaps the easiest to understand is the bubble cap column used to fractionally distill crude oil. These columns dominate the skyline of oil refineries, some being 150 ft high and capable of distilling 200,000 barrels of crude oil per day. The crude oil enters the column as a hot vapor (Fig. 5.3). Some of this vapor with high boiling components condenses on one of the plates. The more volatile substances travel through the bubble cap to the next higher plate where some of the less volatile components condense. As high boiling liquid material accumulates on a plate it descends through the overflow pipe to the next lower plate and vapor rises through the bubble cap to the next higher plate. The temperature of the vapor that is rising through a cap is above the boiling point of the liquid on that plate. As bubbling takes place heat is exchanged and the less volatile components on that plate vaporize and go on to the next plate. The composition of the liquid on a plate is the same as that of the vapor coming from the plate below. So on each plate a simple distillation takes place. At equilibrium vapor containing low boiling material is ascending and high boiling liquid is descending through the column.

Figure 5.2 shows that the condensations and redistillations in a bubble cap column consisting of three plates corresponds to moving on the boiling point–composition diagram from point A to point E.

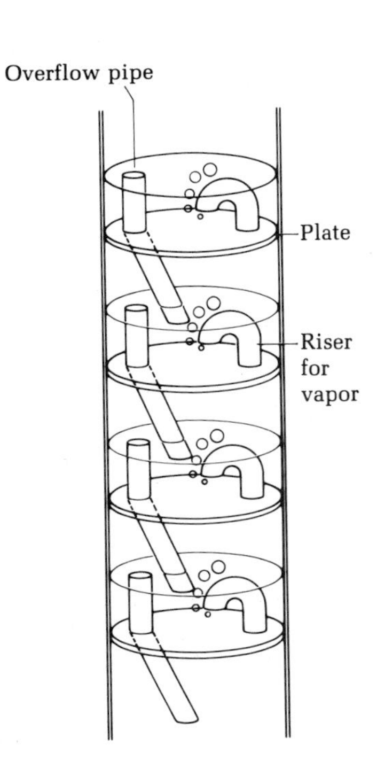

FIG. 5.3 Bubble plate distilling column.

In the laboratory the successive condensations and distillations that occur in the bubble cap column take place in a distilling column. The column is packed with some material on which heat exchange between ascending vapor and descending liquid can take place. A large surface area for this packing is desirable, but the packing cannot be so dense that pressure changes take place within the column causing nonequilibrium conditions. Also if the column packing has a very large surface area, it will absorb (hold up) much of the material being distilled. A number of different packings for distilling columns have been tried—glass beads, glass helices, carborundum chips, etc. We find one of the best packings is a stainless steel sponge (Chore Boy). It is easy to put into the column, does not come out of the column as beads do, has a large surface area, good heat transfer characteristics, and low holdup. It can be used in both the microscale and macroscale apparatus.

Heat exchange between ascending vapor and descending liquid

Column packing

The ability of different column packings to separate two materials of differing boiling point is evaluated by calculating the number of theoretical plates, each theoretical plate corresponding to one distillation and condensation as discussed above. Other things being equal, the number of theoretical plates is proportional to the height of the column, so various packings are evaluated according to the height equivalent to a theoretical plate (HETP); the smaller the HETP, the more plates the column will have and the more efficient it will be. The calculation is made by analyzing the proportion of lower to higher boiling material at the top of the column and in the distillation pot.[1]

Height equivalent to a theoretical plate (HETP)

Although not obvious, the most important variable contributing to a good fractional distillation is the rate at which the distillation is carried out. A series of simple distillations take place within a fractionating column and it is important that complete equilibrium be attained between the ascending vapors and the descending liquid. This process in not instantaneous. It should be an adiabatic process, that is, heat should be transferred from the ascending vapor to the descending liquid with no net loss or gain of heat. In larger, more complex distilling columns (Fig. 5.4), a means is provided for adjusting the ratio between the amount of material that boils up and condenses (refluxes) and is returned to the column (thus allowing equilibrium to take place) and the amount that is removed as distillate. A reflux ratio of 30 : 1 or 50 : 1 would not be uncommon for a 40-plate column; distillation would take several hours.

Equilibration is slow

Good fractional distillation takes a long time

Carrying out a fractional distillation on the truly micro scale (< 1 mg) is impossible, and even impossible on a small scale (10–400 mg). In the present experiment the distillation will be carried out on a 4-mL scale. As seen in later chapters various types of chromatography are employed for the separation of micro and semimicro quantities of material while distillation is the best method for separating more than a few grams of material.

1. See *Technique of Organic Chemistry*, Vol. IV, "Distillation," A. Weissberger (ed.), Wiley-Interscience, New York, 1951.

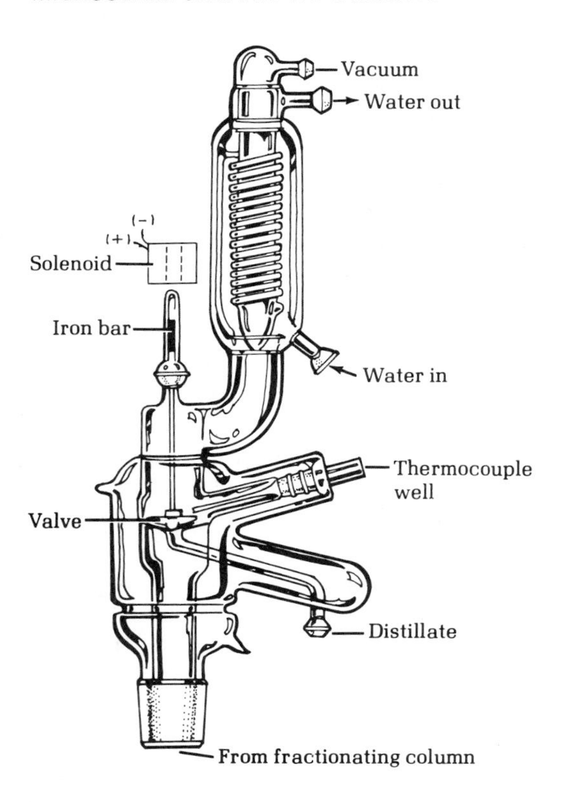

FIG. 5.4 Total reflux fractionating head, vacuum jacketed.

Azeotropes

The ethanol/water azeotrope

Not all liquids form ideal solutions and conform to Raoult's law. Ethanol and water are such liquids. Because of molecular interaction, a mixture of 95.5% (by weight) of ethanol and 4.5% of water boils *below* (78.15°C) the boiling point of pure ethanol (78.3°C). Thus, no matter how efficient the distilling apparatus, 100% ethanol cannot be obtained by distillation of a mixture of, say, 75% water and 25% ethanol. A mixture of liquids of a certain definite composition that distils at a constant temperature without change in composition is called an azeotrope; 95% ethanol is such an azeotrope. The boiling point–composition curve for the ethanol-water mixture is seen in Fig. 5.5. To prepare 100% ethanol the water can be removed chemically (reaction with calcium oxide) or by removal of the water as an azeotrope (with still another liquid). An azeotropic mixture of 32.4% ethanol and 67.6% benzene (bp 80.1°C) boils at 68.2°C. A ternary azeotrope (bp 64.9°C) contains 74.1% benzene, 18.5% ethanol, and 7.4% water. Absolute alcohol (100% ethanol) is made by addition of benzene to 95% alcohol and removal of the water in the volatile benzene-water-alcohol azeotrope.

The ethanol and water form a minimum boiling azeotrope. Other substances, such as formic acid (bp 100.7°C) and water (bp 100°C), form

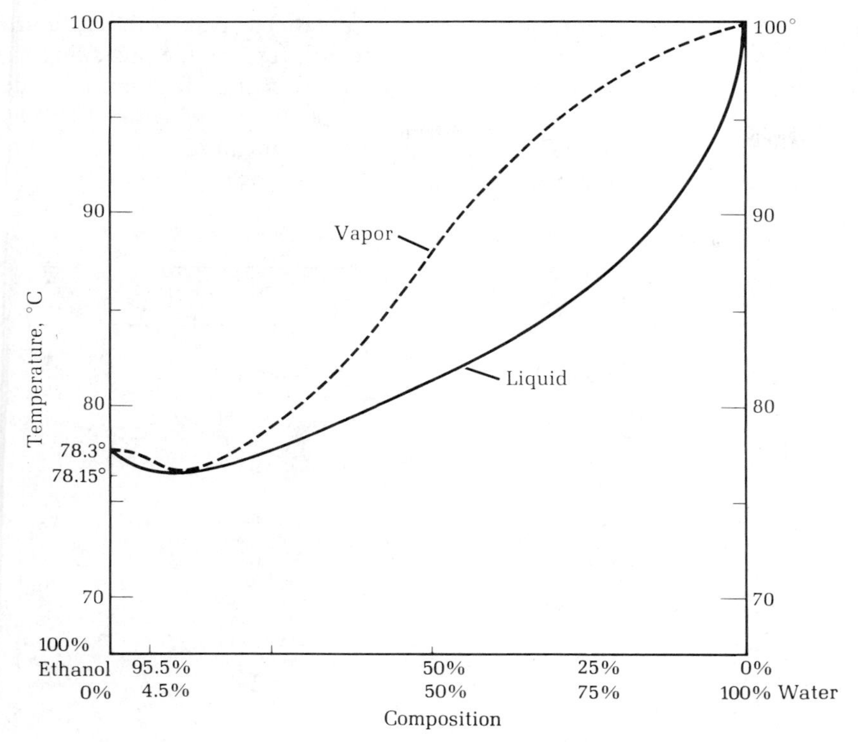

FIG. 5.5 Boiling point–composition curves for ethanol–water mixtures.

maximum boiling azeotropes. For these two compounds the azeotrope boils at 107.3°C.

A pure liquid has a constant boiling point. A change in boiling point during distillation is an indication of impurity. The converse proposition, however, is not always true, and constancy of a boiling point does not necessarily mean that the liquid consists of only one compound. For instance, two miscible liquids of similar chemical structure that boil at the same temperature individually will have nearly the same boiling point as a mixture. And, as noted previously, azeotropes have constant boiling points that can be either above or below the boiling points of the individual components.

A constant bp on distillation does not guarantee *that the distillate is one pure compound*

When a solution of sugar in water is distilled, the boiling point recorded on a thermometer located in the vapor phase is 100°C (at 760 torr) throughout the distillation, whereas the temperature of the boiling sugar solution itself is initially somewhat above 100°C and continues to rise as the concentration of sugar in the remaining solution increases. The vapor pressure of the solution is dependent upon the number of water molecules present in a given volume; and hence with increasing concentration of nonvolatile sugar molecules and decreasing concentration of water, the vapor pressure at a given temperature decreases and a higher temperature is required for boiling. However, sugar molecules do not leave the solution, and the drop clinging to the thermometer is pure water in equilibrium with pure water vapor.

Bp changes with pressure

When a distillation is carried out in a system open to the air and the boiling point is thus dependent on existing air pressure, the prevailing barometric pressure should be noted and allowance made for appreciable deviations from the accepted boiling point temperature (see Table 5.1). Distillation can also be done at the lower pressures that can be achieved by an oil pump or an aspirator with substantial reduction of boiling point.

TABLE 5.1 Variation in Boiling Point with Pressure

Pressure (mm)	Water (°C)	Benzene (°C)
780	100.7	81.2
770	100.4	80.8
760	100.0	80.1
750	99.6	79.9
740	99.2	79.5
584*	92.8	71.2

* Instituto de Quimica, Mexico City, altitude 7700 ft (2310 m).

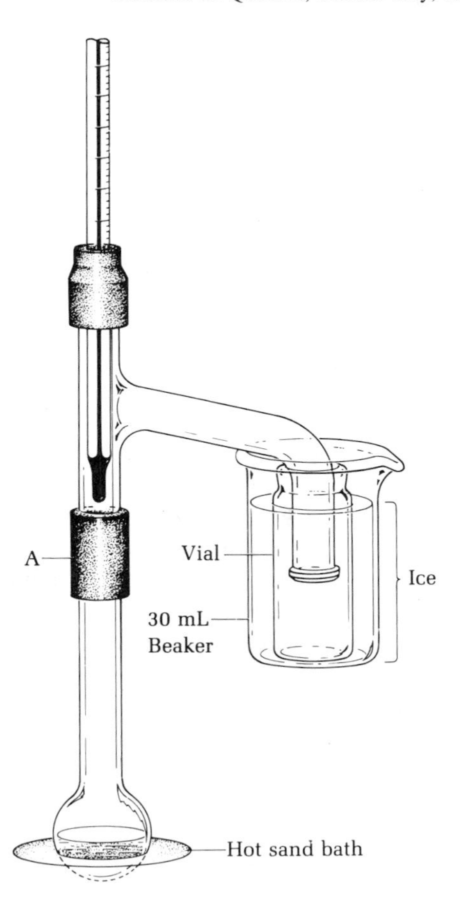

FIG. 5.6 Small-scale simple distillation apparatus. Apply miniature three-prong clamp at point A. This apparatus can be adapted for fractional distillation by packing the long neck with a stainless steel sponge.

Apparatus

In this experiment the two liquids to be separated are placed in a 5-mL round-bottomed long-necked flask that is fitted to a distilling head (Fig. 5.6). The flask has a larger surface area exposed to heat than the reaction tube and so the necessary thermal energy can be put into the system to cause the materials to distill. The hot vapor rises and completely envelops the bulb of the thermometer before passing over and down toward the receiver. The downward-sloping portion of the distilling head functions as an air condenser; to ensure complete condensation the receiver, a vial, is cooled in ice. The apparatus is conveniently clamped at point A, Fig. 5.6, with a miniature three-prong clamp. The beaker is supported with a large three-prong clamp.

Apparatus for simple distillation

By carrying out a careful distillation of cyclohexane and toluene in this apparatus and then packing the long neck of the flask with a stainless steel sponge and again carrying out the distillation, the effect of a column packing on separation efficiency can be evaluated. Alternatively the fractional distillation can be carried out using the 10-cm packed column (Fig. 5.7), which is even more efficient than the 5-cm packed column. A plot of volume of distillate versus boiling point allows this evaluation to be made.

Although their separation efficiency is not great, a number of devices have been devised to distil very small quantities of liquids. One of these, which might make an elementary glassblowing project (see Chapter 69 on glassblowing), is shown in Fig. 5.8.

Micro distillation apparatus

On a Larger Scale. Apparatus that can be used for the fractional distillation of 25–200 mL or more is shown in Fig. 5.9. The flask is more safely heated electrically. A variety of column packings can be used; glass helices have proved to be most efficient.

One problem with packed columns is the amount of liquid they retain that does not distil, termed the holdup. A very efficient column (low HETP) with low holdup is the spinning band column. A motor-driven metal or Teflon spiral band forces vapors back to the pot, thus allowing complete equilibration to take place and minimizing the holdup.

Instant Microscale Distillation

Frequently a very small quantity of freshly distilled material is needed in an experiment. For example, two compounds that need to be freshly distilled are aniline, which turns dark black because of the formation of oxidation products, and benzaldehyde, a liquid that easily oxidizes to solid benzoic acid. The impurities in both of these compounds have much higher boiling points than the parent substance, so a very simple distillation suffices to separate them. This is accomplished as follows.

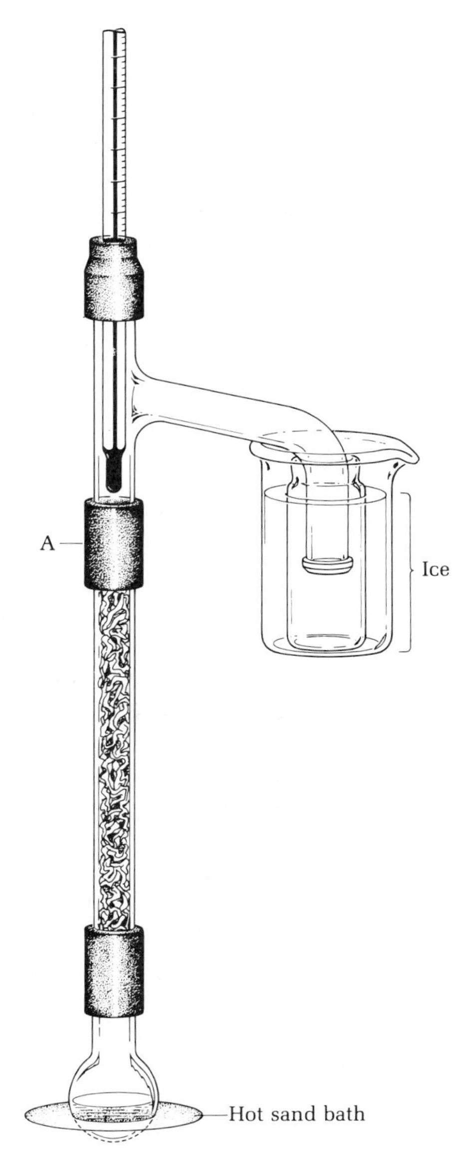

FIG. 5.7 Small-scale fractional distillation apparatus. The 10-cm column is packed with a stainless steel sponge. Clamp at point A.

Place a few drops of the impure liquid in a reaction tube along with a boiling chip. Clamp the tube in a hot sand bath and adjust the heat so that the liquid refluxes about halfway up the tube. Expel the air from a Pasteur pipette, thrust it down in the hot vapor, and then pull the hot vapor into the cold upper portion of the pipette. The vapor will immediately condense and it can then be expelled into another reaction tube held adjacent to the hot one (Fig. 5.10). In this way enough pure material can be distilled to determine a boiling point, run a spectrum, make a derivative, or carry out a reaction.

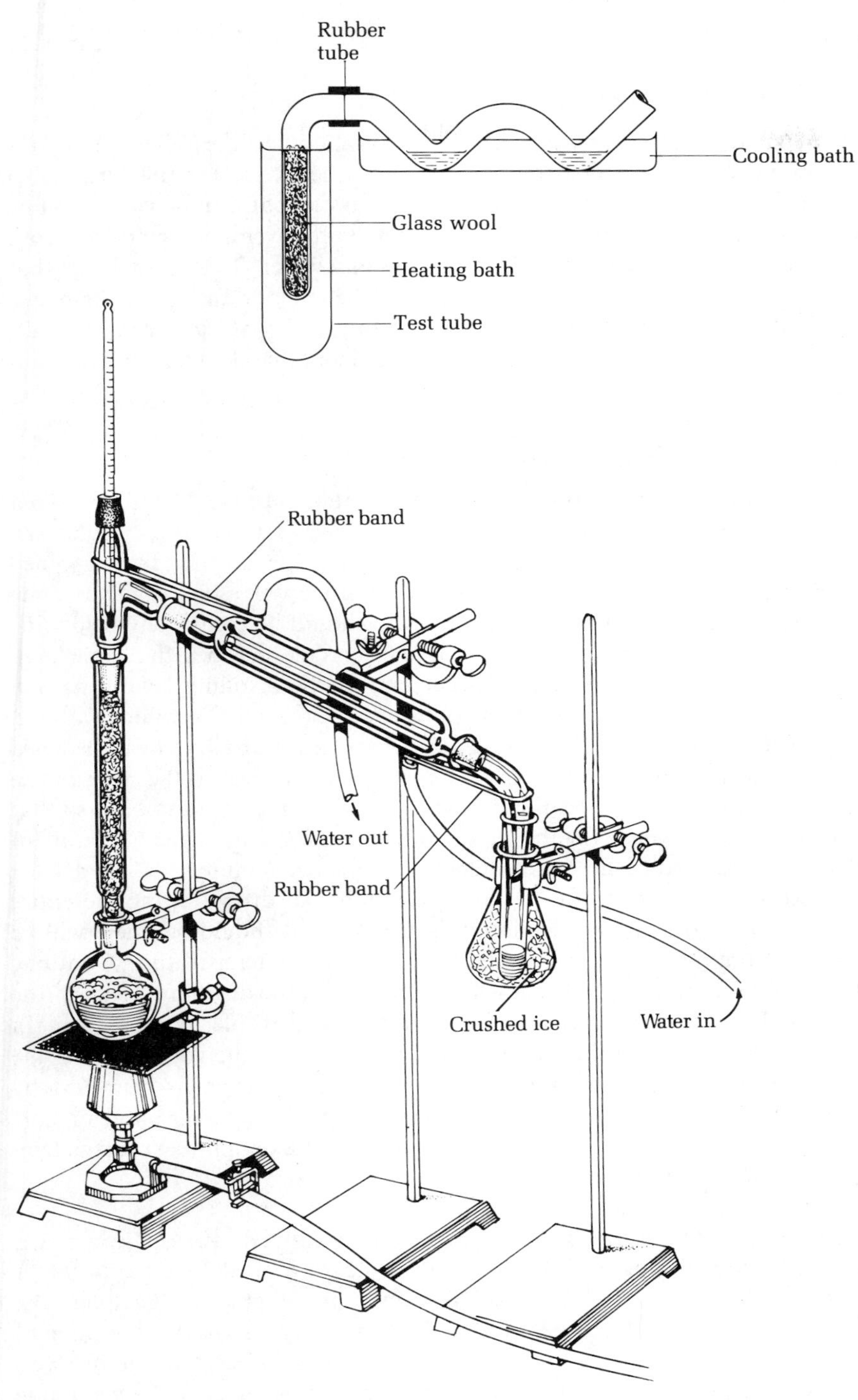

FIG. 5.8 Microscale distillation apparatus.

FIG. 5.9 Apparatus for fractional distillation of a volatile substance.

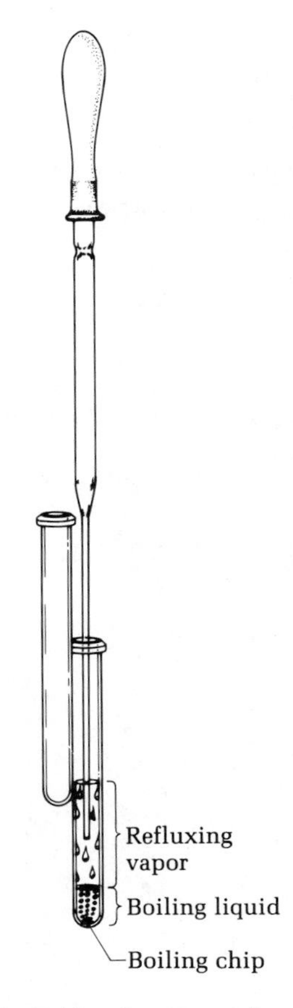

FIG. 5.10 Apparatus for instant microscale distillation.

Experiments

1. Calibration of Thermometer

If you have not previously carried out a calibration, test the 0°C point of your thermometer with a well-stirred mixture of crushed ice and distilled water. To check the 100°C point, put 2 mL of water in a reaction tube with a boiling chip to prevent bumping and boil the water gently over a hot sand bath with the thermometer in the vapor from the boiling water. Take care to see that the thermometer does not touch the side of the reaction tube. Then immerse the bulb of the thermometer in the liquid and see if you can observe superheating. Check the atmospheric pressure to determine the true boiling point of the water.

2. Simple Distillation

The thermometer bulb must be completely below *the side arm*

Slow *distillation, 2 drops/min*

(A) Simple Distillation of a Cyclohexane–Toluene Mixture. To a 5-mL long-necked round-bottomed flask is added 2.0 mL of dry cyclohexane and 2.0 mL of dry toluene and a boiling chip (see Fig. 5.6). This flask is joined by means of a connector to a distilling head fitted with a thermometer using a rubber connector. The thermometer bulb should be completely below the side arm of the Claisen head so that the mercury reaches the same temperature as the vapor that distils. The end of the distilling head dips well down into a vial, which rests on the bottom of a 30-mL beaker filled with ice. The distillation is started by raising a fairly warm sand bath to heat the flask. As soon as boiling starts the vapors can be seen to rise up the neck of the flask. Adjust the rate of heating by raising or lowering the sand bath so that it takes *several minutes* for the vapor to rise to the thermometer. The rate of distillation should be no faster than two drops per minute.

Record the temperature versus the number of drops during the entire distillation. If the rate of distillation is as slow as it should be, there will be sufficient time between drops to read and record the temperature. Continue the distillation until only about 0.4 mL remains in the distilling flask. At the end of the distillation measure as accurately as possible, perhaps using a syringe, the volume of the distillate and, after it cools, the volume left in the pot; the difference is the holdup of the column if none has been lost by evaporation. Note the barometric pressure, make any thermometer corrections necessary, and make a plot of milliliters (drop number) versus temperature for the distillation.

(B) Simple Distillation of an Ethanol–Water Mixture. In a 5-mL round-bottomed long-necked flask place 4 mL of a 10–20% ethanol–water mixture. It could come from the fermentation of glucose (Chapter 51). Assemble the apparatus as described above and carry out the distillation until you believe a representative sample of ethanol has collected in the receiver. In the hood place three drops of this sample on a watch glass and try to ignite it with the blue cone of a microburner flame. Does it burn? Is any unburned residue observed? There was a time when alcohol-water mixtures were mixed with gunpowder and ignited to give proof that the alcohol had not been diluted. One hundred proof alcohol is 50% ethanol by volume.

3. Fractional Distillation

Apparatus

Assemble the apparatus shown in Figs. 5.6 or 5.7. In the first, the neck of the long-necked flask is packed tightly with a piece of stainless steel sponge; in the second, the 10-cm column is packed with a sponge and connected to the 5-mL short-necked flask. The column should be perfectly vertical and care should be taken to ensure that the bulb of the thermometer does not touch the side of the distilling head. The column, but not the distilling head, will be insulated with glass wool at the appropriate time to ensure that the process is adiabatic.

(A) Fractional Distillation of a Cyclohexane–Toluene Mixture. This mixture can be fractionally distilled in one of two ways. It can be be distilled through a short packed column in which the results can be compared directly with those obtained in the simple distillation, which uses the same apparatus with no packing. Or the mixture can be distilled through the packed column twice as long, which should give a better separation of the components.

To the long-necked flask is added 2.0 mL of cyclohexane and 2.0 mL of toluene and a boiling chip. The column is packed tightly with the stainless steel sponge. Alternatively the short-necked 5-mL round-bottomed flask is charged with the same mixture and then connected to the 10-cm distilling column, which is packed tightly with a stainless steel sponge. In either case the mixture is brought to a boil over a hot sand bath. Observe the ring of condensate that should rise slowly through the column; if you cannot at first see this ring, locate it by touching the column with the fingers. Reduce or remove the heat and wrap the column, but not the distilling head, with glass wool.

Insulate the distilling column, but not the head

The distilling head and the thermometer function as a small reflux condenser. Again apply the heat and as soon as the vapor reaches the thermometer bulb reduce the heat by lowering the sand bath. Distill the mixture at a rate no faster than two drops per minute and record the temperature as a function of the number of drops. Stop the distillation when only about 0.4 mL remains in the flask and measure the volume of distillate and the pot residue as before. Make a plot of boiling point versus milliliters of distillate (drops) and compare it to the simple distillation carried out in the same apparatus. Compare your results with those of Fig. 5.11.

(B) Fractional Distillation of an Ethanol–Water Mixture. Distil 4 mL of the same ethanol–water mixture used in the simple distillation experiment following the procedure used for cyclohexane–toluene with either the short or the long distilling column. Remove what you regard to be the ethanol fraction and repeat ignition test. Is any difference noted?

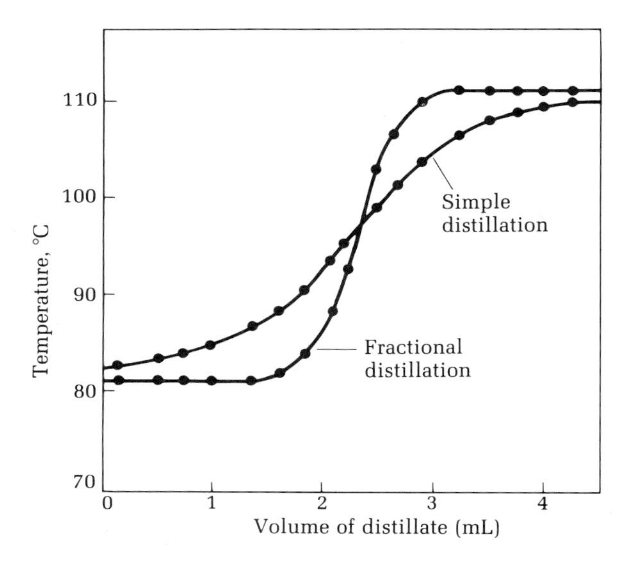

FIG. 5.11 Simple and fractional distillation curves for cyclohexane and toluene.

4. Unknowns

You will be supplied with an unknown, prepared by the instructor, that is a mixture of two solvents from those listed in Table 5.2. The solvents in the mixture will be mutually soluble and differ in boiling point by more than 20°C. Fractionate the unknown and identify the components from the boiling points. Prepare a distillation curve.

TABLE 5.2 Some Properties of Common Solvents

Solvent	Boiling Point
Acetone	56.5
Methanol	64.7
Hexane	68.8
1-Butanol	117.2
2-Methyl-2-propanol	82.2
Water	100.0
Toluene*	110.6

* Methanol and toluene form an azeotrope, bp 63.8°C (69% methanol).

Questions

1. In the simple distillation experiment (2A or 2B), can you account for the boiling point of your product in terms of the known boiling points of the pure components of your mixture?
2. From the plot of boiling point versus volume of distillate in the simple distillation experiment, what can you conclude about the purity of your product?
3. From the boiling point versus volume of distillate plot in the fractional distillation of the cyclohexane–toluene mixture (3A), what conclusion can you draw about the homogeneity of the distillate?
4. From the boiling point versus volume of distillate in the fractional distillation of the ethanol–water mixture (3B), what conclusion can you draw about the homogeneity of the distillate? Does it have a constant boiling point? Is it a pure substance because it has a constant boiling point?
5. Using the data of another student, compare the efficiencies of the 5- and 10-cm packed columns. What are the advantages and disadvantages of each?
6. What is the effect on the boiling point of a solution (e.g., water) produced by a soluble nonvolatile substance (e.g., sodium chloride)? What is the effect of an insoluble substance such as sand or charcoal? What is the temperature of the vapor above these two boiling solutions?
7. Calculate the weight of toluene vapor required to fill the distillation flask and the long distillation column (assume a volume of 5 mL for this and a temperature of 110°C, the boiling point of toluene).
8. In the distillation of a pure substance (e.g., water), why does not all the water vaporize at once when the boiling point is reached?
9. In fractional distillation, liquid can be seen running from the bottom of the distillation column back into the distilling flask. What effect does this returning condensate have on the fractional distillation?
10. Why is it dangerous to attempt to carry out a distillation in a completely closed apparatus, one with no vent to the atmosphere?
11. Why is better separation of two liquids achieved by slow rather than fast distillation?
12. Explain why a packed fractionating column is more efficient than an unpacked one.
13. In the distillation of the cyclohexane–toluene mixture, the first few drops of distillate may be cloudy. Explain.

6 Steam Distillation

PRELAB EXERCISE: *If a mixture of toluene and water is distilled at 97°, what weight of water would be necessary to carry over 1 g of toluene?*

Each liquid exerts its own vapor pressure

The bp of immiscible liquids is below *the bp of either pure liquid*

When a mixture of cyclohexane and toluene are distilled (Experiments 5.2 and 5.3) the boiling point of these two miscible liquids is *between* the boiling points of each of the pure components. By contrast, if a mixture of benzene and water (immiscible liquids) is distilled, the boiling point of the mixture will be found *below* the boiling point of each pure component. Since the two liquids are essentially insoluble in each other, the benzene molecules in a droplet of benzene are not diluted by water molecules from nearby water droplets, and hence the vapor pressure exerted by the benzene is the same as that of benzene alone at the existing temperature. The same is true of the water present. Because they are immiscible, the two liquids independently exert pressures against the common external pressure, and when the sum of the two partial pressures equals the external pressure boiling occurs. Benzene has a vapor pressure of 760 torr at 80.1°, and if it is mixed with water the combined vapor pressure must equal 760 torr at some temperature below 80.1°. This temperature, the boiling point of the mixture, can be calculated from known values of the vapor pressures of the separate liquids at that temperature. Vapor pressures found for water and benzene in the range 50–80° are plotted in Fig. 6.1. The dotted line cuts the two curves at points where the sum of the vapor pressures is 760 torr; hence this temperature is the boiling point of the mixture (69.3°).

Practical use can sometimes be made of the fact that many water-insoluble liquids and solids behave as benzene does when mixed with water, volatilizing at temperatures below their boiling points. Thus, naphthalene, a solid, boils at 218° but distils with water at a temperature below 100°. Since naphthalene is not very volatile, considerable water is required to entrain it and the conventional way of conducting the distillation is to pass steam into a boiling flask containing naphthalene and water. The process is called *steam distillation*. With more volatile compounds, or with a small amount of material, the substance can be heated with water in a simple distillation flask and the steam generate *in situ*.

Some high boiling substances decompose before the boiling point is reached and, if impure, cannot be purified by ordinary distillation. However, they can be freed from contaminating substances by steam distillation at a lower temperature at which they are stable. Steam distillation also offers

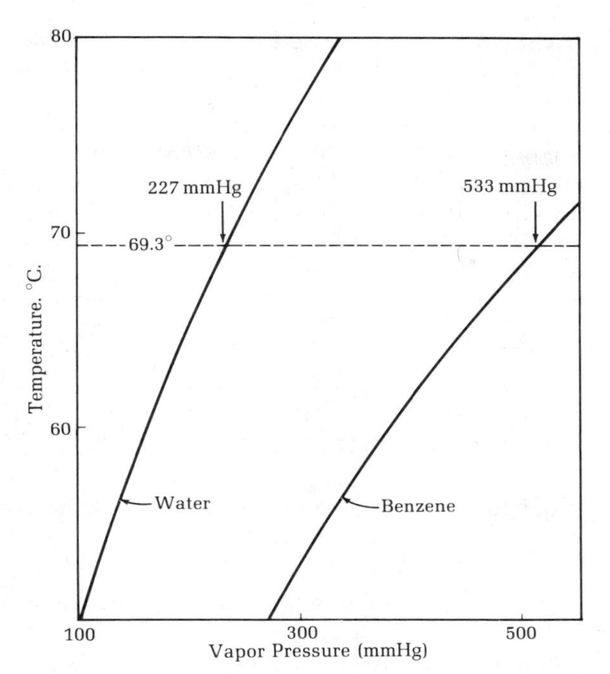

FIG. 6.1 Vapor pressure vs. temperature curves for water and benzene.

the advantage of selectivity, since some water-insoluble substances are volatile with steam and others are not, and some volatilize so very slowly that sharp separation is possible. The technique is useful in processing natural oils and resins, which can be separated into steam-volatile and nonsteam-volatile fractions. It is useful for recovery of a nonsteam-volatile solid from its solution in a high boiling solvent such as nitrobenzene, bp 210°; all traces of the solvent can be eliminated and the temperature can be kept low.

Used for isolation of perfume and flavor oils

The boiling point remains constant during a steam distillation so long as adequate amounts of both water and the organic component are present to saturate the vapor space. Determination of the boiling point and correction for any deviation from normal atmospheric pressure permits calculation of the amount of water required for distillation of a given amount of organic substance. According to Dalton's law the molecular proportion of the two components in the distillate is equal to the ratio of their vapor pressures (p) in the boiling mixture; the more volatile component contributes the greater number of molecules to the vapor phase Thus,

Dalton's law

$$\frac{\text{Moles of water}}{\text{Moles of substance}} = \frac{p_{\text{water}}}{p_{\text{substance}}}$$

The vapor pressure of water (p_{water}) at the boiling temperature in question can be found by interpolation of the data of Table 6.1 and that of the organic substance is, of course, equal to $760 - p_{\text{water}}$. Hence, the weight of water

TABLE 6.1 Vapor Pressure of Water in mm of Mercury

t°	p	t°	p	t°	p	t°	p
60	149.3	70	233.7	80	355.1	90	525.8
61	156.4	71	243.9	81	369.7	91	546.0
62	163.8	72	254.6	82	384.9	92	567.0
63	171.4	73	265.7	83	400.6	93	588.6
64	179.3	74	277.2	84	416.8	94	610.9
65	187.5	75	289.1	85	433.6	95	633.9
66	196.1	76	301.4	86	450.9	96	657.6
67	205.0	77	314.1	87	468.7	97	682.1
68	214.2	78	327.3	88	487.1	98	707.3
69	223.7	79	341.0	89	506.1	99	733.2

required per gram of substance is given by the expression

$$\begin{array}{c}\text{Wt. of water per} \\ \text{g of substance}\end{array} = \frac{18 \times p_{\text{water}}}{\text{MW of substance} \times (760 - p_{\text{water}})}$$

From the data given in Fig. 6.1 for benzene–water, and the molecular weight 78.11 for benzene, the water required for steam distillation of 1 g of benzene is only $227 \times \frac{18}{533} \times \frac{1}{78} = 0.10$ g. Nitrobenzene (bp 210°, MW 123.11) steam distils at 99° and requires 4.0 g of water per gram. The low molecular weight of water makes water a favorable liquid for two-phase distillation of organic compounds.

Steam distillation will be employed in a number of experiments in this text. On a small scale steam is generated in the flask that contains the substance to be steam distilled by simply boiling a mixture of water and the immiscible substance.

Apparatus

The apparatus (see Fig. 5.11) consists of a 5-mL short-necked boiling flask, addition port, and distilling head leading to an ice-cooled receiver. As the steam distillation proceeds it may be necessary to add more water to the flask. This is done via syringe through the septum on the addition port.

Isolation of Citral

Terpenes

Citral is an example of a very large group of *natural products* called *terpenes*. They are responsible for the characteristic odors of plants such as eucalyptus, pine, mint, peppermint, and lemon. The odors of camphor, menthol, lavender, rose and hundreds of other fragrances are due to terpenes, which have ten carbon atoms with double bonds, and aldehyde, ketone, or alcohol functional groups. See Fig. 6.2.

FIG. 6.2 Structures of some terpenes.

Citral

Isopentenyl pyrophosphate

Isoprene (2-Methyl-1,3-butadiene)

Neral

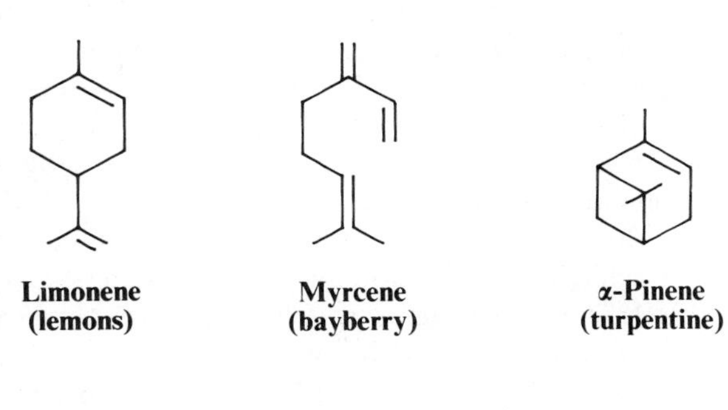

Limonene (lemons)

Myrcene (bayberry)

α-Pinene (turpentine)

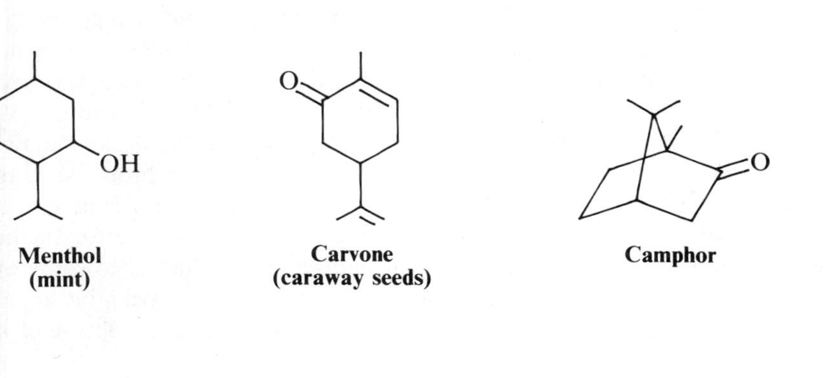

Menthol (mint)

Carvone (caraway seeds)

Camphor

1,8-Cineole (eucalyptus)

Pulegone (pennyroyal oil)

In nature these terpenes all arise from a common precursor, isopentenyl pyrophosphate. At one time they were thought to come from the simple diene, isoprene (2-methyl-1,3-butadiene), because the skeletons of terpenes can be dissected into isoprene units, having five carbon atoms arranged as in 2-methylbutane. These isoprene units are almost always arranged in a "head-to-tail fashion."

The isoprene rule

In the present experiment citral is isolated by steam distillation of lemongrass oil, which is used to make lemongrass tea, a popular drink in Mexico. The distillate contains 90% citral and 10% neral, the isomer about the 2,3-bond. Citral is used in a commercial synthesis of vitamin A.

Lemongrass oil contains a number of substances; simple or fractional distillation would not be a practical method for obtaining pure citral. And because it boils at 229°C, it has a tendency to polymerize, oxidize, and decompose during distillation. For example, heating with potassium bisulfate, an acidic compound, converts citral to 1-methyl-4-isopropylbenzene (*p*-cymene).

Steam distillation: the isolation of heat-sensitive compounds

Citral and neral are geometric isomers

Steam distillation is thus a very gentle method for isolating citral. The distillation takes place below the boiling point of water. The distillate consists of a mixture of citral (and some neral) and water. It is isolated by shaking the mixture with ether. The citral dissolves in the ether, which is immiscible with water, and the two layers are separated. The ether is dried (it dissolves some water) and evaporated to leave citral.

Store the citral in the smallest possible container in the dark for later characterization. The homogeneity of the substance can be investigated using thin-layer chromatography (Chapter 9), gas chromatography (Chapter 12) or high-performance liquid chromatography (Chapter 16). Chemically the molecule can be characterized by reaction with bromine and also permanganate, which shows it contains double bonds, and by the Tollens test, which confirms the presence of an aldehyde (Chapter 30). An infrared spectrum (Chapter 19) would confirm the presence of the aldehyde; an ultraviolet spectrum (Chapter 21) would indicate it is a conjugated aldehyde; and an nmr spectrum (Chapter 20) would clearly show the aldehyde proton, the three methyl groups, and the two olefinic protons. Finally the molecule can be reacted with another substance to form a crystalline derivative for further identification (Chapter 70). You will probably not be required to carry out all of these analyses; however, these are the major tools available to the organic chemist for ascertaining purity and determining the structure of an organic molecule.

Experiment

Into a 5-mL short-necked round-bottomed flask place a boiling chip, 0.5 mL of lemongrass oil (*not* lemon oil), and 3 mL of water. Assemble the apparatus as depicted in Fig. 6.3 and distill as rapidly as possible, taking care that all of the distillate condenses. Using the 3-mL syringe inject water dropwise

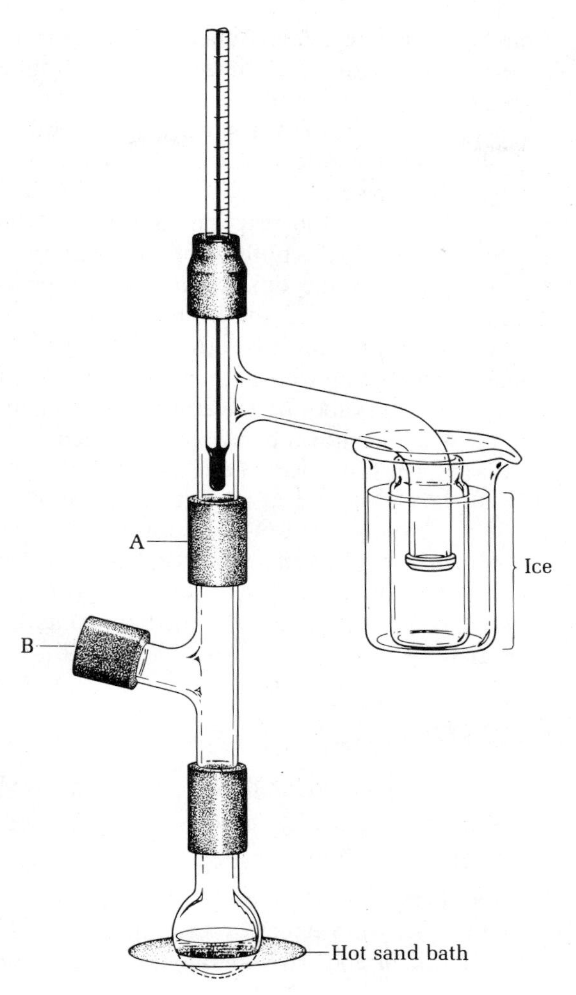

FIG. 6.3 Small-scale steam distillation apparatus. Clamp at A with three-prong micro clamp. Add water via syringe at B.

through the septum to keep the volume in the flask constant. Distill into an ice-cooled vial, as shown, or into a test tube until no more oily drops can be detected, about 10–12 mL. Transfer the product as necessary from the receiver using a Pasteur pipette.

Extraction of Citral

See the next chapter for information on the theory and practice of extraction.

In a reaction tube place 2.5 mL of ether and then almost fill the tube with some of the cold distillate. Shake, let the layers separate, and then draw off most of the lower aqueous layer with a Pasteur pipette leaving the ether in the reaction tube. Continue this extraction process by again adding the cold distillate to the reaction tube, shaking, and separating to remove the citral from the steam distillate. When the last portion of the distillate has been added to the reaction tube, rinse the 6-in test tube with about 0.5 mL of ether to recover adhering citral.

Dry the ether

Carefully remove the last traces of water from the reaction tube containing ether and then add about 2 g of anhydrous sodium sulfate to the ether. Stopper the tube and shake it over a period of 5–10 min. This will remove water adhering to the reaction tube and dissolved in the ether. Force a Pasteur pipette to the bottom of the reaction tube, expel the air from the pipette, draw up the ether and expel it into another tared (previously weighed) reaction tube. Add fresh ether to the drying agent, which will serve to wash it off, and add that ether to the tared reaction tube. Add a boiling stick or boiling chip to the ether in the reaction tube, place it in a beaker of hot water, and evaporate the ether by boiling and drawing the ether vapors into the water aspirator or by blowing a gentle stream of air into the tube (in the hood). The last traces of ether can be removed by connecting the reaction tube directly to the water aspirator. The residue should be a clear, fragrant oil. If it is cloudy it is wet. Determine the weight of the citral and calculate the percentage of citral recovered from the lemongrass oil, assuming the density of the lemongrass oil is the same as that of citral (0.89). Transfer the product to the smallest possible air-tight container and store it in the dark for later analyses.

Questions

1. A mixture of ethyl iodide (C_2H_5I, bp 72.3°C) and water boils at 63.7°C. What weight of ethyl iodide would be carried over by 1 g of steam during steam distillation?

2. Iodobenzene (C_6H_5I, bp 188°C) steam distils at a temperature of 98.2°C. How many molecules of water are required to carry over one molecule of iodobenzene? How many grams per gram of iodobenzene?

3. The condensate from a steam distillation contains 8 g of an unknown compound and 18 g of water. The mixture steam distilled at 98°C. What is the molecular weight of the unknown?

On a Larger Scale. Steam is often available in an organic chemistry laboratory. It is passed into a flask containing water and the substance to be steam distilled. This flask is heated to prevent the steam from condensing in the flask and filling it.

In the assembly shown in Fig. 6.4 steam is passed into a 250-mL, round-bottomed flask through a section of 6-mm glass tubing fitted into a stillhead with a piece of 5-mm rubber tubing connected to a trap, which in turn is connected to the steam line. The trap serves two purposes: it allows water, which is in the steam line, to be removed before it reaches the round-bottomed flask, and adjustment of the clamp on the hose at the bottom of the trap allows precise control of the steam flow. The stopper in the trap should be wired on, as shown, as a precaution. A bent adapter attached to a *long* condenser delivers the condensate into a 250-mL Erlenmeyer flask.

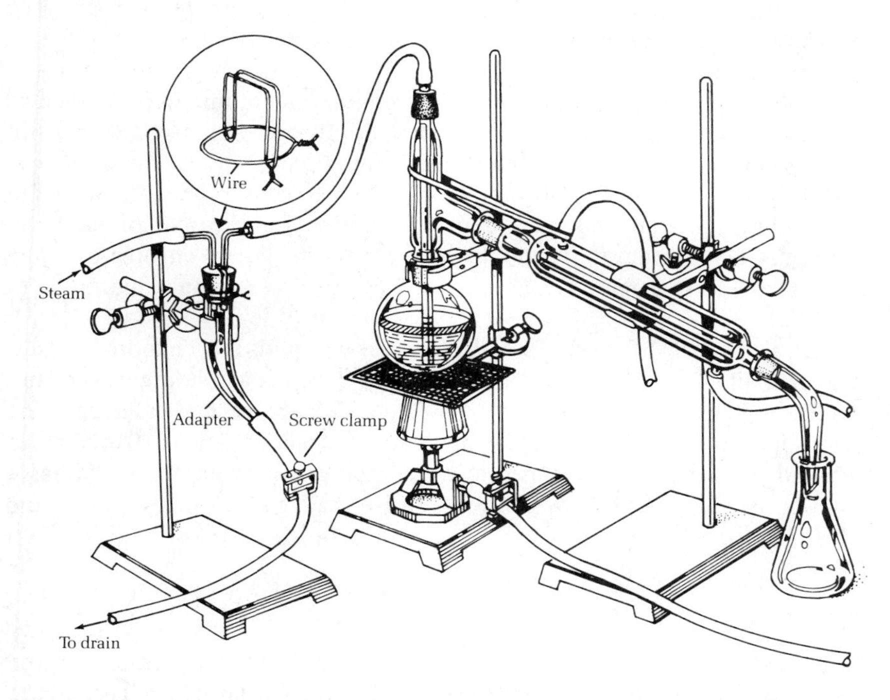

FIG. 6.4 Steam distillation apparatus.

7 Vacuum Distillation and Sublimation

Part 1. Vacuum Distillation

Vacuum distillation is used to distill substances that will decompose or change in some way if distilled at atmospheric pressure. Substances that oxidize, rearrange, dehydrate, or undergo some other pyrolytic process are candidates for distillation under reduced pressure. These include most compounds that boil much above 200°C at atmospheric pressure. Reducing the pressure reduces the boiling point and thus decreases the possibility for decomposition.

On a truly micro scale (< 10 mg) simple distillation is not practical because of mechanical losses. On a small scale (10–500 mg), distillation is still not a good method for isolating material, again because of mechanical losses; but very crude bulb-to-bulb distillations do serve to rid a sample of all low-boiling material and nonvolatile impurities. In research, preparative-scale gas chromatography and high-performance liquid chromatography (HPLC) have supplanted vacuum distillation of very small quantities of material.

Fractional distillation of small quantities of liquids is even more difficult. The necessity for maintaining an equilibrium between ascending vapors and descending liquid means there will inevitably be large mechanical losses. The best apparatus for the fractional distillation of about 0.5 to 20 mL under reduced pressure is the so-called micro spinning band column, which has a hold-up of about 0.1 mL. Again preparative-scale gas chromatography and HPLC are the best means for separating mixtures of liquids.

Apparatus

Fig. 7.1 illustrates a very simple piece of homemade apparatus for the vacuum distillation of a small quantity of a liquid. The liquid is placed in the tube, which can be about 4 to 6 mm in diameter. Then the tube is packed loosely with glass wool, which serves to prevent the material from bumping and splashing over into the receiver. The receiver is simply a tube bent into a shallow W shape. By monitoring the temperature of the heating bath and the pressure, you can determine an approximate boiling point as the material distils.

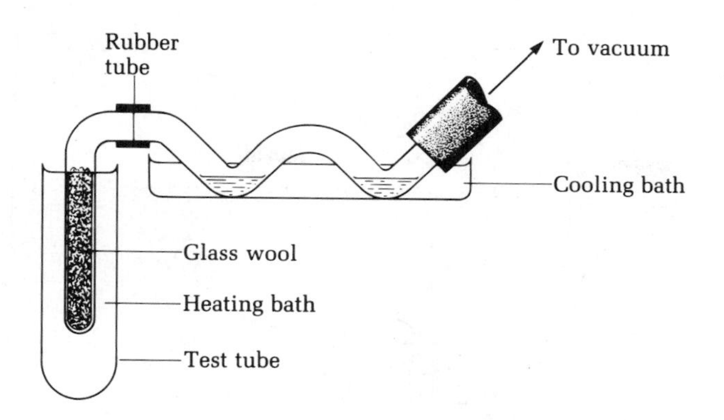

FIG. 7.1 Microscale vacuum distillation apparatus.

On a Larger Scale. For the vacuum distillation of 10 to 200 mL, a typical apparatus is illustrated in Fig. 7.2. It is constructed of a round-bottomed flask (often called the "pot") containing the material to be distilled, a Claisen distilling head fitted with a hair-fine capillary mounted through a rubber tubing sleeve, and a thermometer with the bulb extending below the side arm

Carry out vacuum distillation behind a safety shield. Check apparatus for scratches and cracks

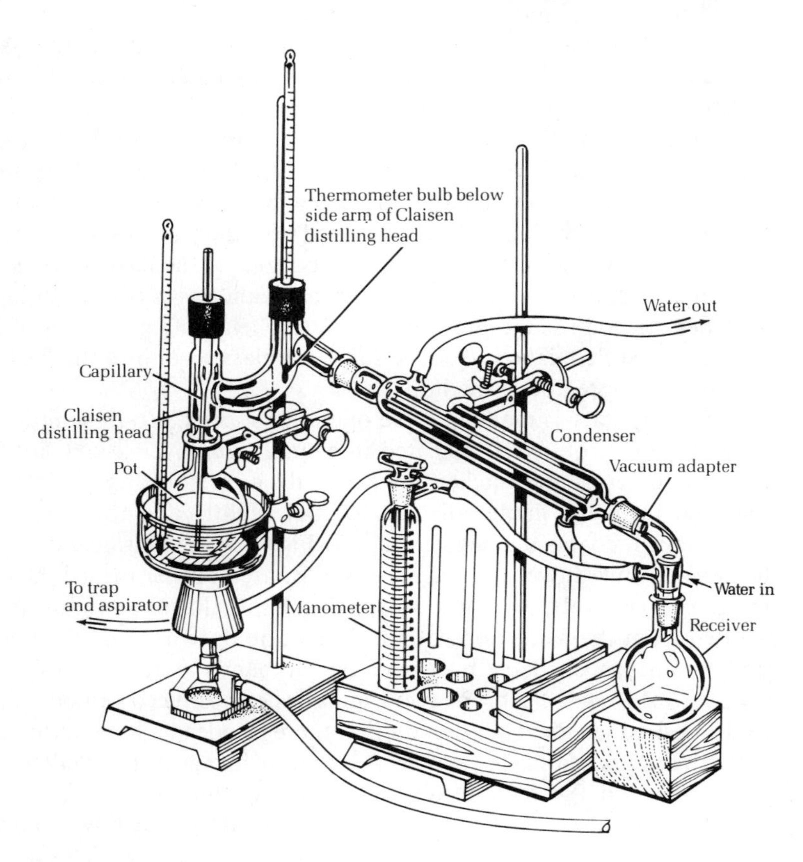

FIG. 7.2 Vacuum distillation apparatus.

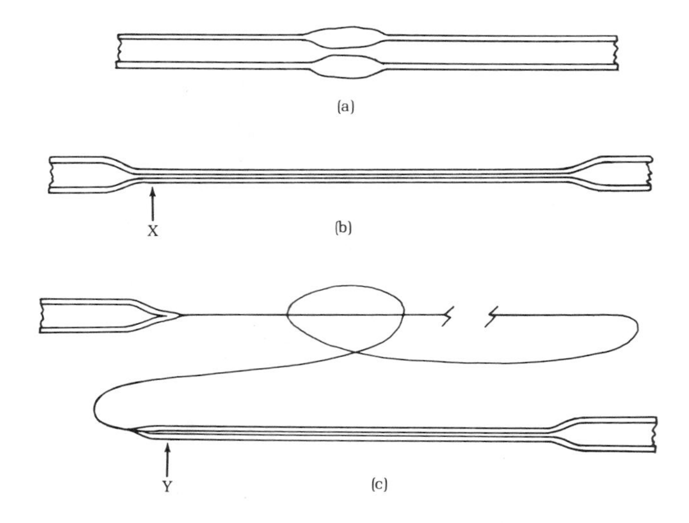

FIG. 7.3 Capillary for vacuum distillation.

Prevention of bumping

opening. The condenser fits into a vacuum adapter that is connected to the receiver and, via heavy-walled rubber tubing, to a mercury manometer and thence to the trap and water aspirator.

Liquids usually bump vigorously when boiled at reduced pressure and most boiling stones lose their activity in an evacuated system; it is therefore essential to make special provision for controlling the bumping. This is done by allowing a fine stream of air bubbles to be drawn into the boiling liquid through a glass tube drawn to a hair-fine capillary. The capillary should be so fine that even under vacuum only a few bubbles of air are drawn in each second; smooth boiling will be promoted and the pressure will remain low. The capillary should extend to the very bottom of the flask and it should be slender and flexible so that it will whip back and forth in the boiling liquid. Another method used to prevent bumping, when small quantities of material are being distilled, is to introduce sufficient glass wool into the flask to fill a part of the space above the liquid.

Making a hair-fine capillary

The capillary is made in three operations. First, a 6-in. length of 6-mm glass tubing is rotated and heated in a small area over a *very hot* flame to collapse the glass and thicken the side walls as seen in Fig. 7.3. The tube is removed from the flame, allowed to cool slightly and then drawn into a thick-walled, coarse diameter capillary (Fig. 7.3(b)). This coarse capillary is heated at point X over the wing top of a Bunsen burner turned 90° C. When the glass is very soft, but not so soft as to collapse the tube entirely, the tubing is lifted from the flame and without hesitation drawn smoothly and rapidly into a hair-fine capillary by stretching the hands as far as they will reach (about 2 m) (Fig. 7.3(c)). The two capillaries so produced can be snapped off to the desired length. To ascertain that there is indeed a hole in the capillary, place the end beneath a low-viscosity liquid such as acetone or ether and blow in the large end. A stream of very small bubbles should be seen. If the stream of bubbles is not extremely fine make a new capillary. The flow of air through the capillary *cannot* be controlled by attaching a rubber tube and clamp to the top of the capillary tube. Should the right-hand capillary of Fig. 7.3(c) break when in use, it can be fused to a scrap of glass (for use as a handle) and heated again at point Y (Fig. 7.3(c)). In this way the capillary can be redrawn many times.

Heating baths

The pot is heated with a heating bath or, better, an electric flask heater rather than a free flame to promote even boiling and make possible accurate determination of the boiling point. The bath is filled with a suitable heat transfer liquid (water, cottonseed oil, silicone oil, or molten metal) and heated to a temperature about 20°C higher than that at which the substance in the flask distils. The bath temperature is kept constant throughout the distillation. The surface of the liquid in the flask should be below that of the heating medium, for this condition lessens the tendency to bump. Heating of the flask is begun only after the system has been evacuated to the desired pressure; otherwise the liquid might boil too suddenly on reduction of the pressure.

Changing fractions

To change fractions the following must be done in sequence: Remove the source of heat, release the vacuum, change the receiver, restore the vacuum to the former pressure, resume heating.

A better vacuum distillation apparatus is shown in Fig. 7.4. The distillation neck of the Claisen adapter is longer than other adapters and has a series of indentations made from four directions, so that the points nearly meet in the center. These indentations increase the surface area over which

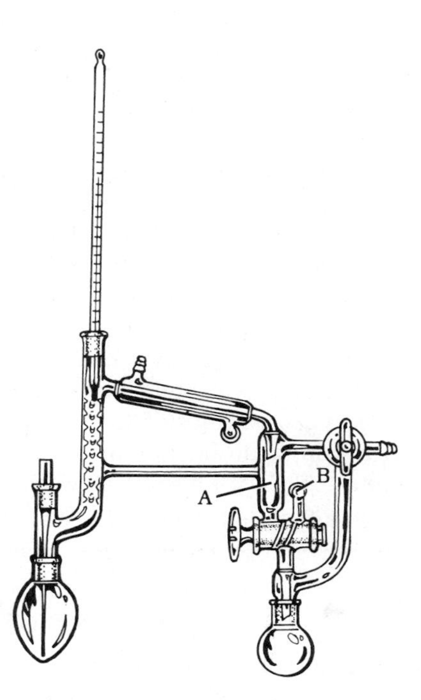

FIG. 7.4 Vacuum distillation apparatus with Vigreaux column and fraction collector.

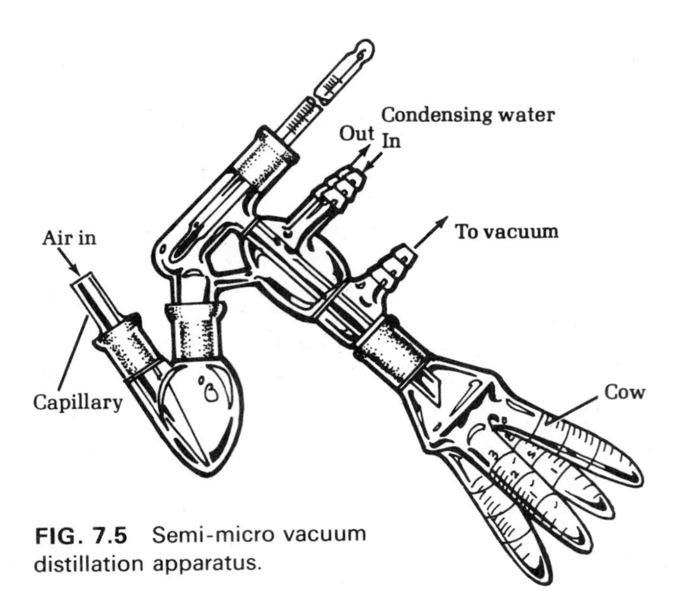

FIG. 7.5 Semi-micro vacuum distillation apparatus.

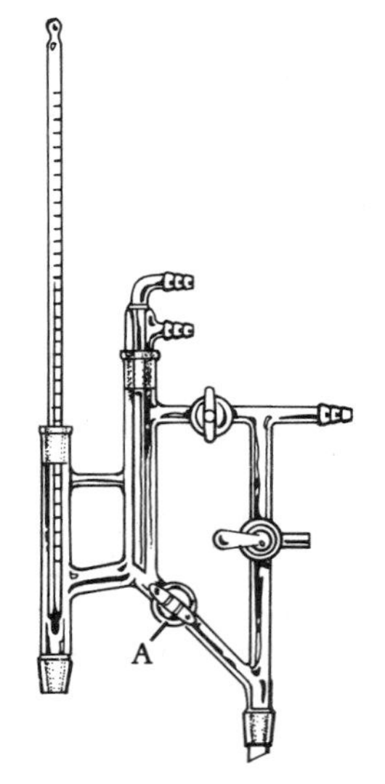

FIG. 7.6 Vacuum distillation head.

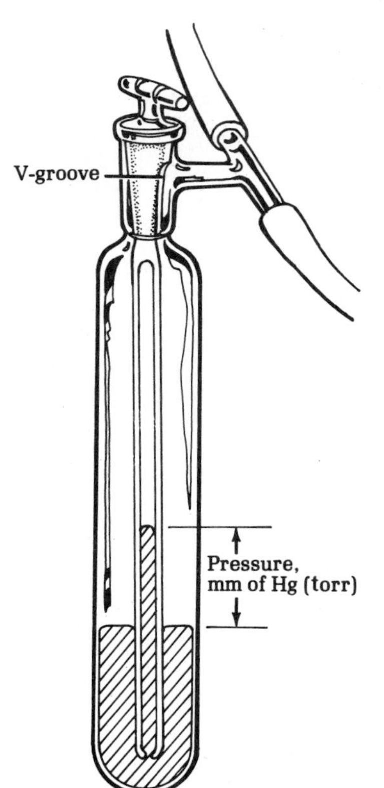

FIG. 7.7 Closed-end mercury manometer.

rising vapor can come to equilibrium with descending liquid and it then serves as a fractionating column (a Vigreaux column). A column packed with a metal sponge has a great tendency to become filled with liquid (flood) at reduced pressure. The apparatus illustrated in Fig. 7.4 also has a fraction collector, which allows the removal of a fraction without disturbing the vacuum in the system. While the receiver is being changed the distillate collects in the small reservoir A. The clean receiver is evacuated by another aspirator at tube B before being connected again to the system.

If only a few milliliters of a liquid are to be distilled, the apparatus shown in Fig. 7.5 has the advantage of low hold-up, that is, not much liquid is lost wetting the surface area of the apparatus. The fraction collector illustrated is known as a "cow." Rotation of the cow about the standard taper joint will allow four fractions to be collected without interrupting the vacuum.

A distillation head of the type shown in Fig. 7.6 allows fractions to be removed without disturbing the vacuum, and it also allows control of the reflux ratio (Chapter 5) by manipulation of the condenser and stopcock A. These can be adjusted to remove all material that condenses or only a small fraction, with the bulk of the liquid being returned to the distilling column to establish equilibrium between descending liquid and ascending vapor. In this way liquids with small boiling-point differences can be separated.

1. Pressure Measurement—the Manometer

The pressure of the system is measured with a closed-end mercury manometer. The manometer (Fig. 7.7) is connected to the system by turning the stopcock until the V-groove in the stopcock is aligned with the side arm. To avoid contamination of the manometer it should be connected to the system only when a reading is being made. The pressure, in mm Hg, is given by the height, in mm, of the central mercury column above the reservoir of mercury and represents the difference in pressure between the nearly perfect vacuum in the center tube (closed at the top, open at the bottom) and the large volume of the manometer, which is at the pressure of the system.

Mercury manometer

2. The Water Aspirator in Vacuum Distillation

A water aspirator in good order gives a vacuum nearly corresponding to the vapor pressure of the water flowing through it. Polypropylene aspirators give good service and are not subject to corrosion as are the brass ones. If a manometer is not available, and the assembly is free of leaks and the trap and lines clean and dry, an approximate estimate of the pressure can be made by measuring the water temperature and reading the pressure from Table 7.1.

TABLE 7.1 Vapor Pressure of Water at Different Temperatures

t (°C)	p (mm Hg)	t (°C)	p (mm Hg)	t (°C)	p (mm Hg)	t (°C)	p (mm Hg)
0	4.58	20	17.41	24	22.18	28	28.10
5	6.53	21	18.50	25	23.54	29	29.78
10	9.18	22	19.66	26	24.99	30	31.55
15	12.73	23	20.88	27	26.50	35	41.85

3. The Rotary Oil Pump

To obtain pressures below 10 mm Hg a mechanical vacuum pump of the type illustrated in Fig. 7.8(a) is used. A pump of this type in good condition can give pressures as low as 0.1 mm Hg. These low pressures are measured with a tilting type McLeod gauge (Fig. 7.8(b)). When a reading is being made the gauge is tilted to the vertical position shown and the pressure is read as the difference between the heights of the two columns of mercury. Between readings the gauge is rotated clockwise 90°.

Dewar should be wrapped with tape or otherwise protected from implosion

Never use a mechanical vacuum pump before placing a mixture of dry ice and isopropyl alcohol in a Dewar flask (Fig. 7.8(c)) around the trap and never pump corrosive vapors (e.g., HCl gas) into the pump. Should this happen change the pump oil immediately. With care, it will give many years of good service. The dry ice trap condenses organic vapors and water vapor, both of which would otherwise contaminate the vacuum pump oil and exert enough vapor pressure to destroy a good vacuum.

For an exceedingly high vacuum (5×10^{-8} mm Hg) a high-speed three-stage mercury diffusion pump is used (Fig. 7.9).

FIG. 7.8 (a) Rotary oil pump, (b) Tilting McLeod gauge, (c) Vacuum system.

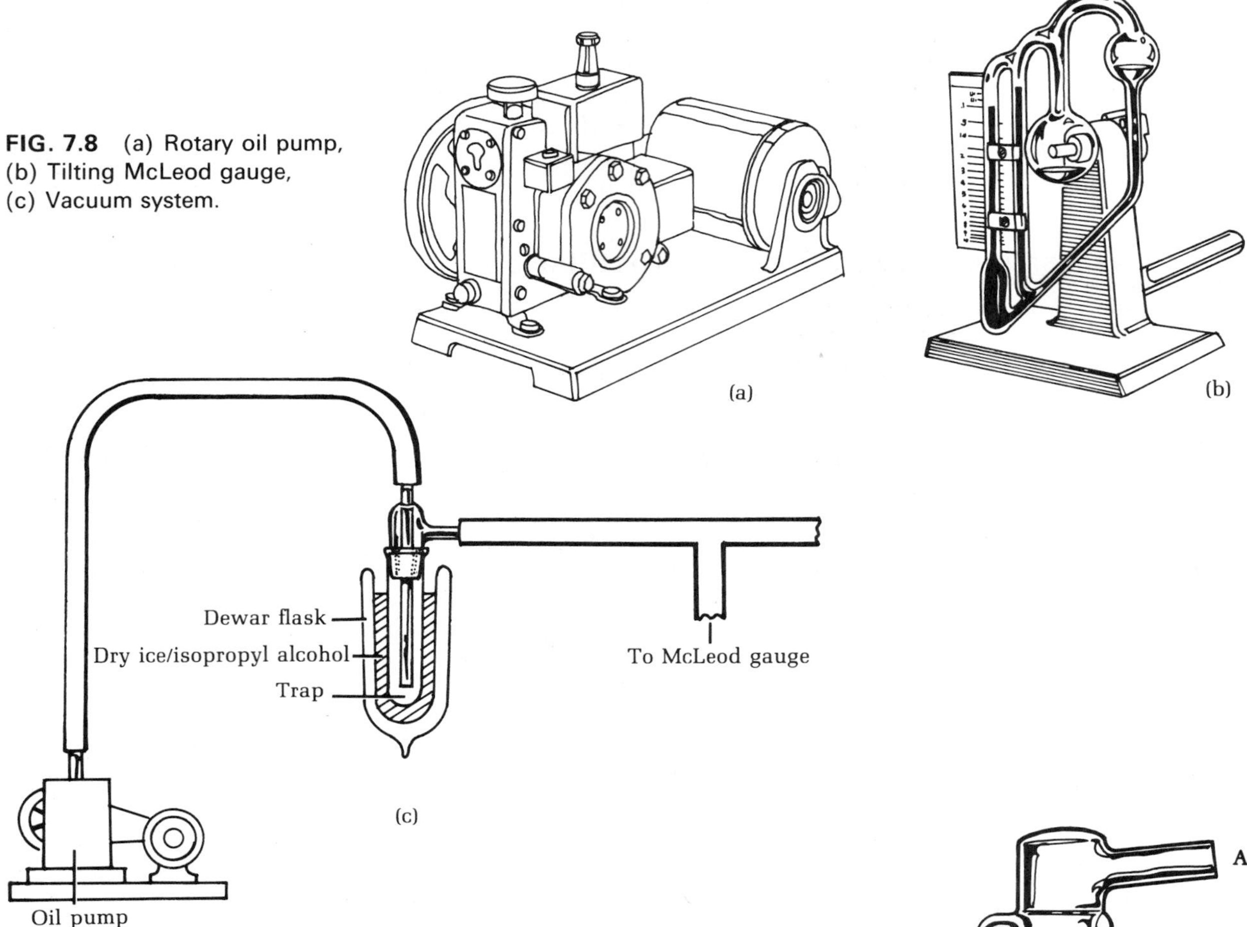

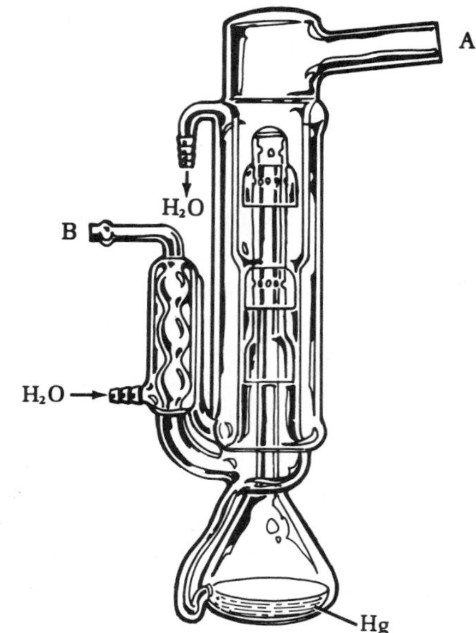

FIG. 7.9 High-speed three-stage mercury diffusion pump capable of producing a vacuum of 5×10^{-8} mm Hg. Mercury is boiled in the flask, the vapor rises in the center tube, is deflected downward in the inverted cups, and entrains gas molecules that diffuse in from the space to be evacuated, A. The mercury condenses to a liquid and is returned to the flask. The gas molecules are removed at B by an ordinary rotary vacuum pump.

4. Relationship between Boiling Point and Pressure

It is not possible to calculate the boiling point of a substance at some reduced pressure from a knowledge of the boiling temperature at 760 mm Hg, for the relationship between boiling point and pressure varies from compound to compound and is somewhat unpredictable. It is true, however, that boiling point curves for organic substances have much the same general disposition, as illustrated by the two lower curves in Fig. 7.10. These are similar and do not differ greatly from the curve for water. For substances boiling in the region 150–250°C at 760 mm Hg, the boiling point at 20 mm Hg is 100–120°C lower than at 760 mm Hg. Benzaldehyde, which is very sensitive to air oxidation at the normal boiling point of 178°C, distils at 76°C at 20 mm Hg, and the concentration of oxygen in the rarefied atmosphere is just $\frac{20}{760}$, or 3% of that in an ordinary distillation.

The curves all show a sharp upward inclination in the region of very low pressure. The lowering of the boiling point attending a reduction in pressure is much more pronounced at low than at high pressures. A drop in the atmospheric pressure of 10 mm Hg lowers the normal boiling point of an ordinary liquid by less than a degree, but a reduction of pressure from 20 to 10 mm Hg causes a drop of about 15°C in the boiling point. The effect at pressures below 1 mm is still more striking, and with development of practical forms of the highly efficient oil vapor or mercury vapor diffusion pump, distillation at a pressure of a few thousandths or ten thousandths of a millimeter has become a standard operation in many research laboratories. High-vacuum distillation, that is at a pressure below 1 mm Hg, affords a useful means of purifying extremely sensitive or very slightly volatile substances. Table 7.2 indicates the order of magnitude of the reduction in boiling

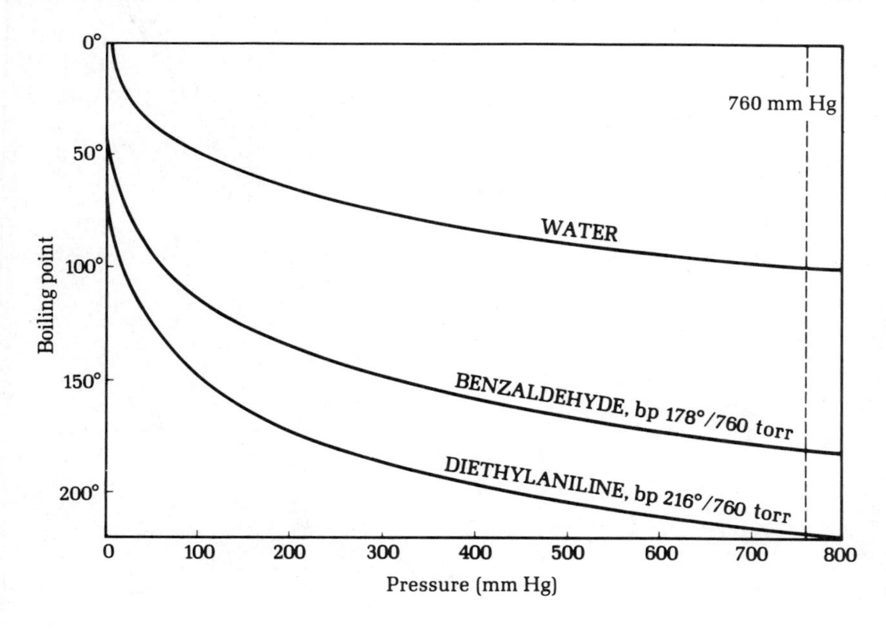

FIG. 7.10 Boiling point curves.

TABLE 7.2 Distillation of a (Hypothetical) Substance at Various Pressures

Method		Pressure (mm Hg)	bp (°C)
Ordinary distillation		760	250
Aspirator	summer	25	144
	winter	15	134
Rotary oil pump	poor condition	10	124
	good condition	3	99
	excellent condition	1	89
Mercury vapor pump		0.01	30

point attainable by operating in different ways and illustrates the importance of keeping vacuum pumps in good repair.

The boiling point of a substance at various pressures can be estimated from a pressure-temperature nomograph such as the one shown in Fig. 7.11. If the boiling point of a substance at 760 mm Hg is known, e.g., 300°C (column B), and the new pressure measured, e.g., 10 mm Hg (column C), then a straight line connecting these values on columns B and C when extended intersects column A to give the observed bp of 160°C. Conversely, a sub-

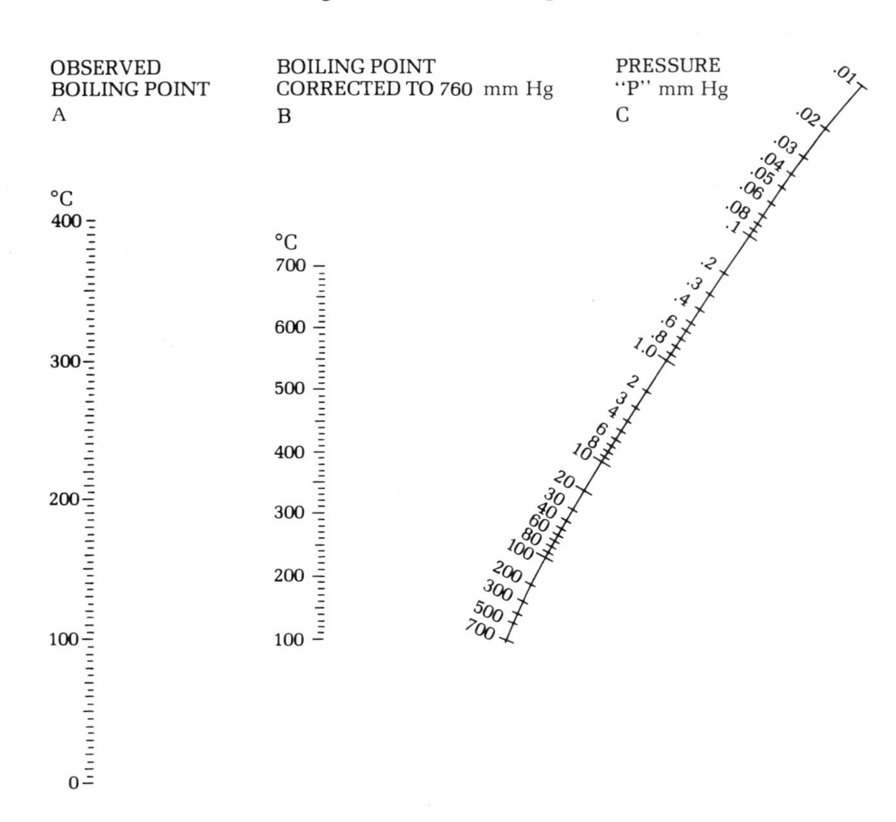

FIG. 7.11 Pressure-temperature nomograph.

stance observed to boil at 50°C (column A) at 1.0 mm Hg (column C) will boil at approximately 212°C at atmospheric pressure (column B).

Vacuum distillation is not confined to the purification of substances that are liquid at ordinary temperatures but often can be used to advantage for solid substances. The operation is conducted for a different purpose and by a different technique. A solid is seldom distilled to cause a separation of constituents of different degrees of volatility but rather to purify the solid. It is often possible in one vacuum distillation to remove foreign coloring matter and tar without appreciable loss of product, whereas several wasteful crystallizations might be required to attain the same purity. It is often good practice to distil a crude product and then to crystallize it. Time is saved in the latter operation because the hot solution usually requires neither filtration nor clarification. The solid must be dry and a test should be made to determine if it will distill without decomposition at the pressure of the pump available. That a compound lacks the required stability at high temperatures is sometimes indicated by the structure, but a high melting point should not be taken as an indication that distillation will fail. Substances melting as high as 300°C have been distilled with success at the pressure of an ordinary rotary vacuum pump. It is not necessary to observe the boiling point in distillations of this kind because the purity and identity of the distillate can be checked by melting point determinations. The omission of the customary thermometer simplifies the technique.

5. Molecular Distillation

At 1×10^{-3} mm Hg the mean free path of nitrogen molecules is 56 mm. This means that an average N_2 molecule can travel 56 mm before bumping into another N_2 molecule. In specially designed apparatus it is possible to distill almost any volatile molecule by operating at a very low pressure and having the condensing surface close to the material being distilled. The thermally energized molecule escapes from the liquid, moves about 10 mm without encountering any other molecules, and condenses on a cold surface. A simple apparatus for molecular distillation is illustrated in Fig. 7.12. The bottom of

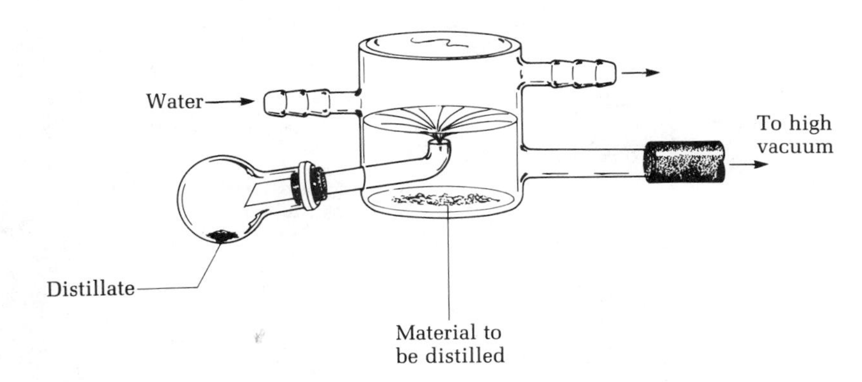

FIG. 7.12 Molecular distillation apparatus.

the apparatus is placed on a hot surface and the distillate moves a very short distance before condensing. If it is not too viscous it will drip from the point on the glass condensing surface and run down to the receiver.

The efficiency of molecular distillation apparatus is less than one theoretical plate (see Chap. 5). It is used to remove very volatile substances such as solvents and to separate volatile, high boiling substances at low temperatures from nonvolatile impurities.

In more sophisticated apparatus of quite different design the liquid to be distilled falls down a heated surface, which is again close to the condenser. This moving film prevents a buildup of nonvolatile material on the surface of the material to be distilled, which would cause the distillation to cease. Vitamin A is distilled commercially in this manner.

Part 2. Sublimation

Sublimation is the process whereby a solid evaporates from a warm surface and condenses on a cold surface, again as a solid (Fig. 7.13). This technique is particularly useful for the small-scale purification of solids because there is so little loss of material in transfer. If the substance has the correct properties, sublimation is preferred over crystallization when the amount of material to be purified weighs less than 100 mg.

As demonstrated in the first experiment, sublimation can occur readily at atmospheric pressure. For substances with lower vapor pressures vacuum sublimation is used. At very low pressure the sublimation becomes very similar to molecular distillation, where the molecule leaves the warm solid and passes unobstructed to a cold condensing surface and condenses in the form of a solid.

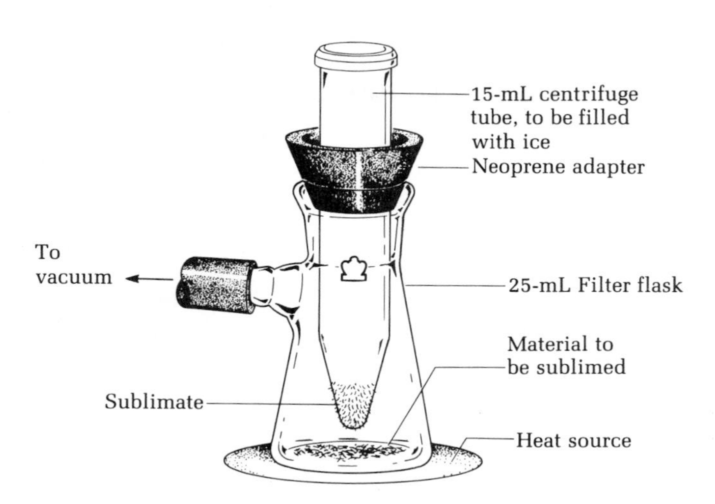

FIG. 7.13 Small-scale sublimation apparatus.

Since sublimation occurs from the surface of the warm solid, impurities can accumulate and slow down or even stop the sublimation, in which case it is necessary to open the apparatus, grind the impure solid to a fine powder, and restart the sublimation.

On a Larger Scale. Sublimation is a technique that is much easier to carry out on a small scale than on a large one. It would be unusual for a research chemist to sublime more than about 10 g of material, because the material tends to fall off the sublimer. In an apparatus such as that illustrated in Fig. 7.14(a), the solid to be sublimed is placed in the lower flask and connected via a lubricant-free rubber O-ring to the condenser, which in turn is connected to a vacuum pump. The lower flask is immersed in an oil bath at the appropriate temperature and the product sublimed and condensed onto the cool walls of the condenser. The parts of the apparatus are gently separated and the condenser inverted; the vacuum connection serves as a convenient funnel for product removal. For large-scale work the sublimator of Fig. 7.14(b) is used. The inner well is filled with a coolant (ice or dry ice). The sublimate clings to this cool surface, from which it can be removed by scraping and dissolving in an appropriate solvent.

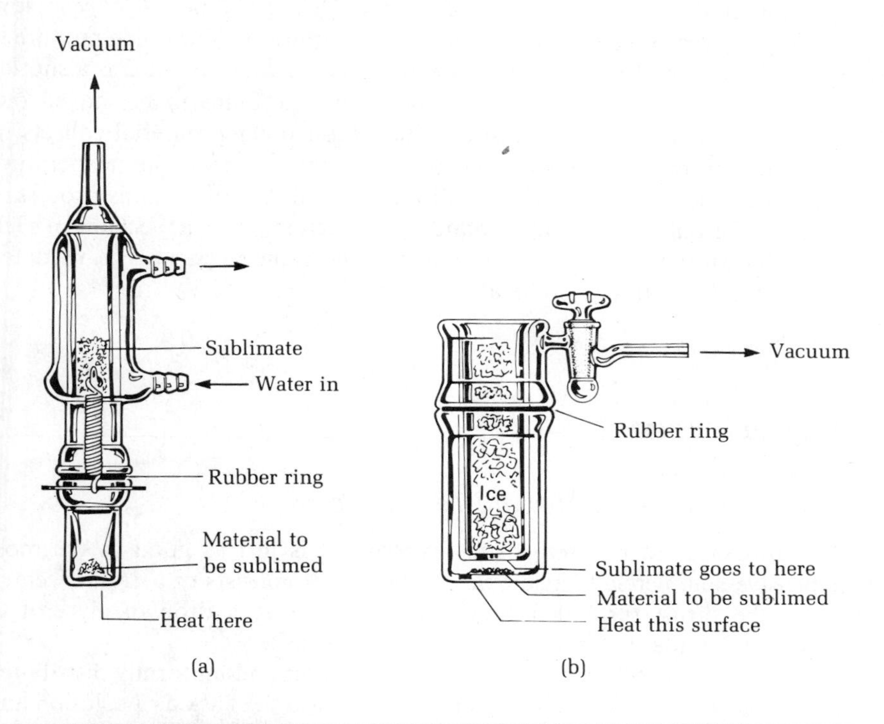

FIG. 7.14 (a) Mallory sublimator, (b) large vacuum sublimator.

Lyophilization

Lyophilization, also called freeze drying, is the process of subliming a solvent, usually water, with the object of recovering the solid that remains after the solvent is removed. This technique is extensively used to recover heat- and oxygen-sensitive substances of natural origin, such as proteins, vitamins, nucleic acids, and other biochemicals from dilute aqueous solution. The aqueous solution of the substance to be lyophilized is frozen and then subjected to a vacuum of about 1×10^{-3} mm Hg. The water sublimes and condenses as ice on the surface of a large and very cold condenser. The sample remains frozen during this entire process, without any external cooling being supplied, because of the very high heat of vaporization of water; thus the temperature of the sample never exceeds 0°C. Freeze drying on a large scale is employed to make "instant" coffee, tea, soup, rice, and all sorts of dehydrated foods for backpackers.

The Kugelrohr

The Kugelrohr (Ger., bulb-tube) is a widely used piece of research apparatus that consists of a series of bulbs that can be rotated and heated under vacuum. Bulb-to-bulb distillation frees the desired compound of very low boiling and very high boiling or nonvolatile impurities. The crude mixture is placed in bulb A (Fig. 7.15) in the heated glass chamber B. At C is a shutter mechanism that holds in the heat but allows the bulbs to be moved out one-by-one as distillation proceeds. The lowest boiling material collects in bulb D; then bulb C is moved out of the heated chamber, the temperature increased, and the next fraction collected in bulb C. Finally this process is repeated for bulb B. The bulbs rotate under vacuum using a mechanism such as the one used on rotary evaporators. The same apparatus is used for bulb-to-bulb fractional sublimation.

Experiment

Sublimation: Apparatus and Technique

The apparatus to be used in this experiment is just as good as the most expensive commercially available equipment. It consists of a 15-mL centrifuge tube thrust through a neoprene adapter (use a drop of glycerol to lubricate the adapter) fitted in a 25-mL filter flask.

The sample is either ground to a fine powder and uniformly distributed on the bottom of the flask or it is introduced into the flask as a solution and the solvent evaporated to deposit the substance on the bottom of the filter

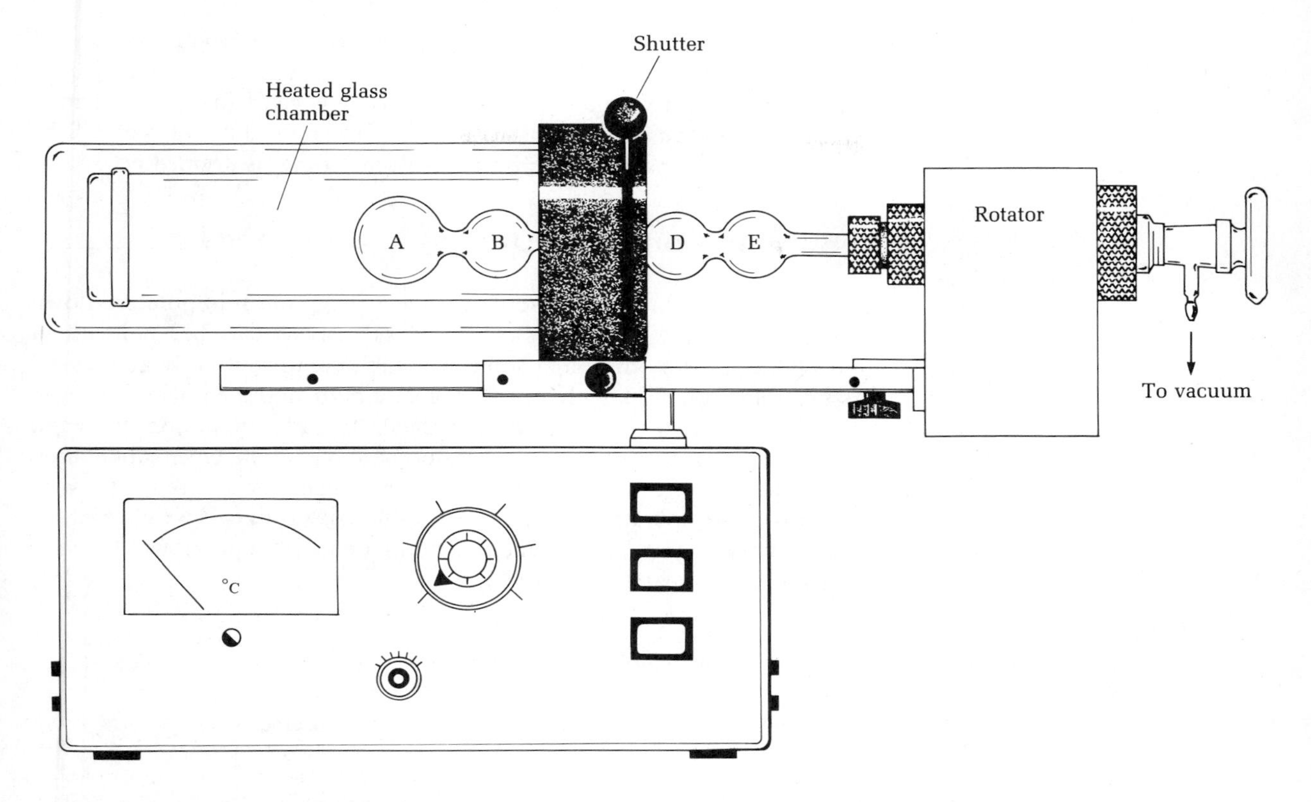

FIG. 7.15 Kugelrohr bulb-to-bulb distillation apparatus.

flask. If the compound sublimes easily, care must be exercised using the latter technique to ensure that the sample does not evaporate as the last of the solvent is being removed.

The 15-mL centrifuge tube and adapter are placed in the flask so that the tip of the centrifuge tube is about 3 to 8 mm above the bottom of the flask (Fig. 7.13). The flask is clamped with a three-prong clamp; the sidearm of the flask leads to an aspirator or vacuum pump. The vacuum source is turned on and the centrifuge tube filled with ice and water. The ice is not added before applying the vacuum so atmospheric moisture will not condense on the tube.

The filter flask is warmed cautiously on a hot sand bath until the product just begins to sublime. The heat is maintained at that temperature until sublimation is complete. Because some product will collect on the cool upper parts of the flask, it should be wrapped with a cone of aluminum foil to direct heat there and cause the material to collect on the surface of the centrifuge tube.

Once sublimation is judged complete the ice water is removed from the centrifuge tube with a pipette and replaced with water at room temperature. This will prevent moisture from condensing on the product once the vacuum is turned off and the tube removed from the flask. The product is scraped from the centrifuge tube with a metal spatula onto a piece of glazed paper. It is much easier to scrape the product from a centrifuge tube than a round-bottomed test tube. The last traces can be removed by washing off the tube with a few drops of an appropriate solvent, if that is deemed necessary.

1. Sublimation of an Unknown Substance

Into the bottom of a 25-mL filter flask place 50 mg of an impure unknown taken from the list in Table 7.3. These substances can be sublimed at atmospheric pressure although some will sublime more rapidly at reduced pressure. Close the flask with a rubber pipette bulb, and then place ice water in the centrifuge tube. Cautiously warm the flask until sublimation starts and then maintain that temperature throughout the sublimation. Once sublimation is complete remove the ice water from the centrifuge tube and replace it with water at room temperature. Collect the product, determine its weight and the percent recovery, and from the melting point identify the unknown. Hand in the product in a neatly labeled vial.

TABLE 7.3 Sublimation Unknowns

Substance	Mp (°C)	Substance	Mp (°C)
1,4-Dichlorobenzene	55	Benzoic acid	122
Naphthalene	82	Salicyclic acid	159
1-Naphthol	96	Camphor	177
Acetanilide	114	Caffeine	235

Extraction: Isolation of Caffeine from Tea and Cola Syrup

8

Extraction is one of man's oldest chemical operations. The preparation of a cup of coffee or tea involves the extraction of flavor and odor components from dried vegetable matter with hot water. Aqueous extracts of bay leaves, stick cinnamon, peppercorns, and cloves are used as food flavorings along with alcoholic extracts of vanilla and almond. For the last century and a half organic chemists have been extracting, isolating, purifying, and then characterizing the myriad compounds produced by plants that have been used for centuries as drugs and perfumes—substances such as quinine from cinchona bark, morphine from the opium poppy, cocaine from coca leaves, and menthol from peppermint oil. In research a Soxhlet extractor is often used (Fig. 8.1). The organic chemist commonly employs, in addition to solid/liquid extraction, two other types of extraction: liquid/liquid extraction and acid/base extraction.

Liquid/liquid Extraction

After a chemical reaction has been carried out the organic product is often separated from inorganic substances by liquid/liquid extraction. For example, in the synthesis of 1-bromobutane

$$2CH_3CH_2CH_2OH + 2NaBr + H_2SO_4 \rightarrow 2CH_3CH_2CH_2CH_2Br + 2H_2O + Na_2SO_4$$

1-butanol is heated with an aqueous solution of sodium bromide and sulfuric acid to produce the product and sodium sulfate. The 1-bromobutane is isolated from the reaction mixture by extraction with ether, a solvent in which 1-bromobutane is soluble and in which water and sodium sulfate are insoluble. The extraction is accomplished by simply adding ether to the aqueous mixture and shaking it. The ether is less dense than water and floats on top; it is removed and evaporated to leave the bromo compound, free of inorganic substances. The solvent used for extraction should have many properties of a satisfactory recrystallization solvent. It should readily dissolve the substance to be extracted, it should have a low boiling point so that it can readily be removed, it should not react with the solute or the other solvent, it should not be flammable or toxic, and it should be relatively inexpensive.

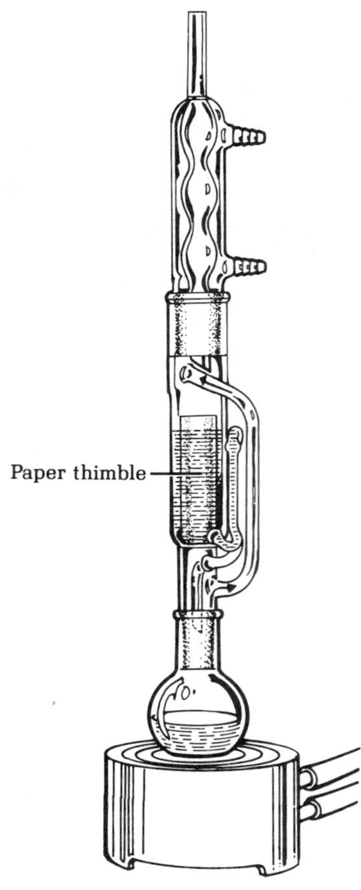

FIG. 8.1 Soxhlet extractor. For exhaustive extraction of solid mixtures, and even of dried leaves or seeds, the solid is packed into a filter paper thimble. Solvent vapor rises in the large diameter tube on the right, and condensed solvent drops onto the solid contained in the filter paper thimble, leaches out soluble material and, after initiating an automatic siphon, carries it to the boiling flask where nonvolatile extracted material accumulates. The same solvent is repeatedly used; substances of very slight solubility can be extracted by prolonged operation.

In addition it should not be miscible with water (the usual second phase). No solvent meets all these criteria, but several come close. Diethyl ether, usually referred to simply as ether, is probably the most common solvent used for extraction, but it is extremely flammable.

Ether has high solvent power for hydrocarbons and for oxygen-containing compounds and is so highly volatile (bp 34.6°) that it is easily removed from an extract at a temperature so low that even highly sensitive compounds are not likely to decompose.

Ether is useful for isolation of natural products that occur in animal and plant tissues having high water content. Although often preferred for research work because of these properties, ether is avoided in industrial processes because of the fire hazard, high solubility in water, losses in solvent recovery incident to volatility, and the oxidation on long exposure to air to a peroxide, which in a dry state may explode. Alternative water-immiscible solvents sometimes preferred, even though they do not match all the favorable properties of ether, are petroleum ether, ligroin, benzene, carbon tetrachloride, chloroform, dichloromethane, 1,2-dichloroethane, and 1-butanol. The chlorinated hydrocarbon solvents are heavier than water and hence, after equilibration of the aqueous and nonaqueous phases, the heavier lower layer is drawn off and the upper aqueous layer is extracted further and discarded. Chlorinated hydrocarbon solvents have the advantage of freedom from fire hazard, but their higher cost militates against their general use.

Distribution Coefficient

The extraction of a compound such as 1-butanol, which is slightly soluble in water as well as very soluble in ether, is an equilibrium process governed by the solubilities of the alcohol in the two solvents. The ratio of the solubilities is known as the distribution coefficient, also called the partition coefficient, k, and is an equilibrium constant with a certain value for a given substance, pair of solvents, and temperature.

To a good approximation the *concentration* of the solute in each solvent can be correlated with the *solubility* of the solute in the pure solvent, a figure that can be found readily in tables of solubility in reference books.

$$k = \frac{\text{concentration of C in ether}}{\text{concentration of C in water}} \cong \frac{\text{solubility of C in ether (g/100 mL)}}{\text{solubility of C in water (g/100 mL)}}$$

Consider a compound, A, which dissolves in ether to the extent of 12 g/100 mL and dissolves in water to the extent of 6 g/100 mL.

$$k = \frac{12\text{ g/100 mL ether}}{6\text{ g/100 mL water}} = 2$$

If a solution of 6 g of A in 100 mL of water is shaken with 100 mL of ether then

$$k = \frac{(x \text{ grams of A/100 mL ether})}{(6 - x \text{ grams of A/100 mL water})} = 2$$

from which

$$x = 4.0 \text{ g of A in the ether layer}$$

$$6 - x = 2.0 \text{ g left in the water layer}$$

It is, however, more efficient to extract the 100 mL of aqueous solution twice with 50-mL portions of ether rather than once with a 100-mL portion.

$$k = \frac{(x \text{ g of A/50 mL})}{(6 - x \text{ g of A/100 mL})} = 2$$

from which

$$x = 3.0 \text{ g in ether layer}$$

$$6 - x = 3.0 \text{ g in water layer}$$

If this 3.0 g/100 mL of water is extracted once more with 50 mL of ether we can calculate that 1.5 g will be in the ether layer, leaving 1.5 g in the water layer. So two extractions with 50-mL portions of ether will extract 3.0 g + 1.5 g = 4.5 g of A, while one extraction with a 100-mL portion of ether removes only 4.0 g of A. Three extractions with $33\frac{1}{3}$-mL portions of ether would extract 4.7 g. Obviously there is a point at which the increased amount of A extracted does not repay the effort of multiple extractions, but keep in mind that several small-scale extractions are more effective than one large-scale extraction.

Several small extractions are better than one large one.

Experiment

Distribution Coefficient for Benzoic Acid

In a reaction tube place about 50 mg of benzoic acid (weighed to the nearest milligram) and to this add exactly equal volumes of water followed by dichloromethane (about 0.8 mL each). While making this addition note which layer is the organic layer and which the aqueous. Cork the tube and shake the contents vigorously for at least 2 min. Allow the tube to stand undisturbed until the layers separate and then carefully draw off, using a Pasteur pipette, *all* of the aqueous layer without removing any of the organic layer. It may be of help to draw out the tip of the pipette to a fine point in a flame and, using this, to tilt the reaction tube on its side to make this separation as clean as possible.

Add about 600 mg of anhydrous sodium sulfate to the dichloromethane to remove traces of water, cork the tube, shake it and allow it to stand for

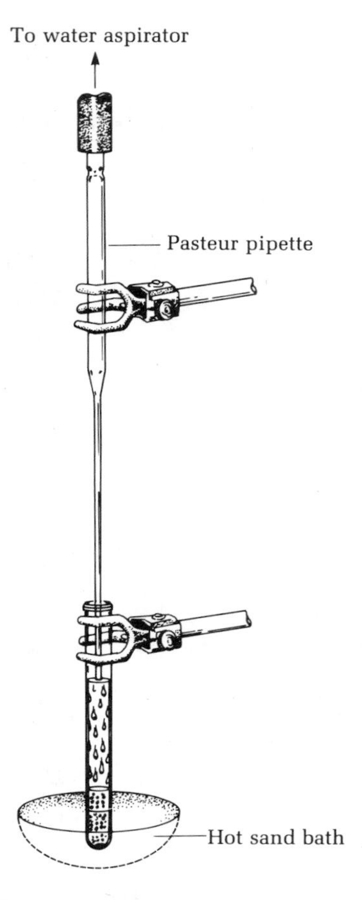

FIG. 8.2 Aspirator tube being used to remove solvent vapors.

about 5 min to complete the drying process. Using a dry Pasteur pipette transfer the dichloromethane to a tared (previously weighed) dry reaction tube or 10-mL Erlenmeyer flask containing a boiling chip. Complete the transfer by washing the drying agent with two more portions of solvent that are added to the original solution, and then evaporate the solvent. This can be done by boiling off the solvent while removing solvent vapors with an aspirator tube (Fig. 8.2) or by blowing a stream of air or nitrogen into the container while warming it. This operation should be performed in a hood.

From the weight of the benzoic acid in the dichloromethane layer the weight in the water layer can be obtained by difference. The ratio of the weight in dichloromethane to the weight in water is the distribution coefficient, because the volumes of the two solvents were equal. Report the value of the distribution coefficient.

Acid/Base Extraction

The third type of extraction, acid/base extraction, involves carrying out simple acid/base reactions in order to separate strong organic acids, weak organic acids, neutral organic compounds, and basic organic substances. The chemistry involved is given in the equations that follow, using benzoic acid, phenol, naphthalene, and aniline as examples of the four types of compounds.

$$C_6H_5COOH + NaHCO_3 \longrightarrow C_6H_5COO^-Na^+ + H_2O + CO_2$$

Benzoic acid
$pK_a = 4.17$
Covalent, sol. in org. solvents

Sodium benzoate
Ionic, sol. in water

$$C_6H_5COO^-Na^+ + HCl \longrightarrow C_6H_5COOH + NaCl$$

$$C_6H_5OH + NaOH \longrightarrow C_6H_5O^-Na^+ + H_2O$$

Phenol
$pK_a = 10$
Covalent, sol. in org. solvents

Sodium phenoxide
Ionic, sol. in water

$$C_6H_5O^-Na^+ + HCl \longrightarrow C_6H_5OH + NaCl$$

$$C_6H_5NH_2 + HCl \longrightarrow C_6H_5NH_3^+Cl^-$$

Aniline
$pK_b = 9.30$
Covalent, sol. in org. solvents

Anilinium chloride
Ionic, sol. in water

$$C_6H_5NH_3^+Cl^- + NaOH \longrightarrow C_6H_5NH_2 + H_2O + NaCl$$

Here is the strategy: The four organic compounds are dissolved in ether. The ether solution is shaken with a saturated aqueous solution of sodium bicarbonate, a weak base. This will react only with the strong acid, benzoic acid, to form the ionic salt, sodium benzoate, which dissolves in the aqueous layer and is removed. The ether solution now contains just phenol, naphthalene, and aniline. A 10% aqueous solution of sodium hydroxide is added and the mixture shaken. The hydroxide, a strong base, will react only with the phenol, a weak acid, to form sodium phenoxide, an ionic compound that dissolves in the aqueous layer and is removed. The ether now contains only naphthalene and aniline. Shaking it with dilute hydrochloric acid removes the aniline, a base, as the ionic anilinium chloride. The aqueous layer is removed. Evaporation of the ether now leaves naphthalene, the neutral compound. The other three compounds are recovered by adding acid to the sodium benzoate and sodium phenolate and base to the anilinium chloride to regenerate the covalent compounds benzoic acid, phenol, and aniline. These operations are conveniently represented in a flow diagram (Fig. 8.3).

The ability to separate strong from weak acids depends on the acidity constants of the acids and the basicity constants of the bases as follows. In the first equation consider the ionization of benzoic acid, which has an equilibrium constant, K_a, of 6.8×10^{-5}. The conversion of benzoic acid to

FIG. 8.3 Flow diagram for the separation of a strong acid, a weak acid, a neutral compound, and a base–benzoic acid, phenol, naphthalene, and aniline.

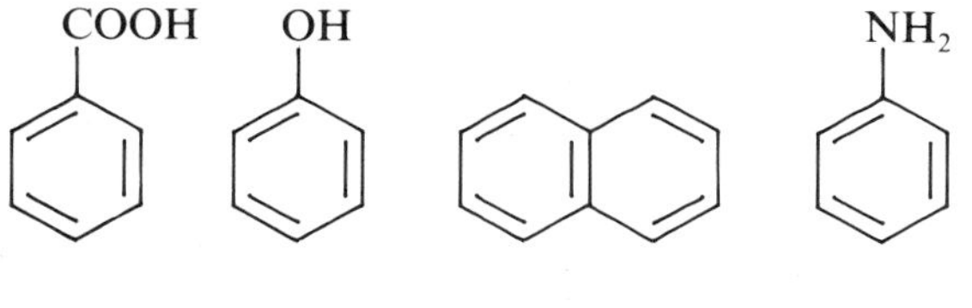

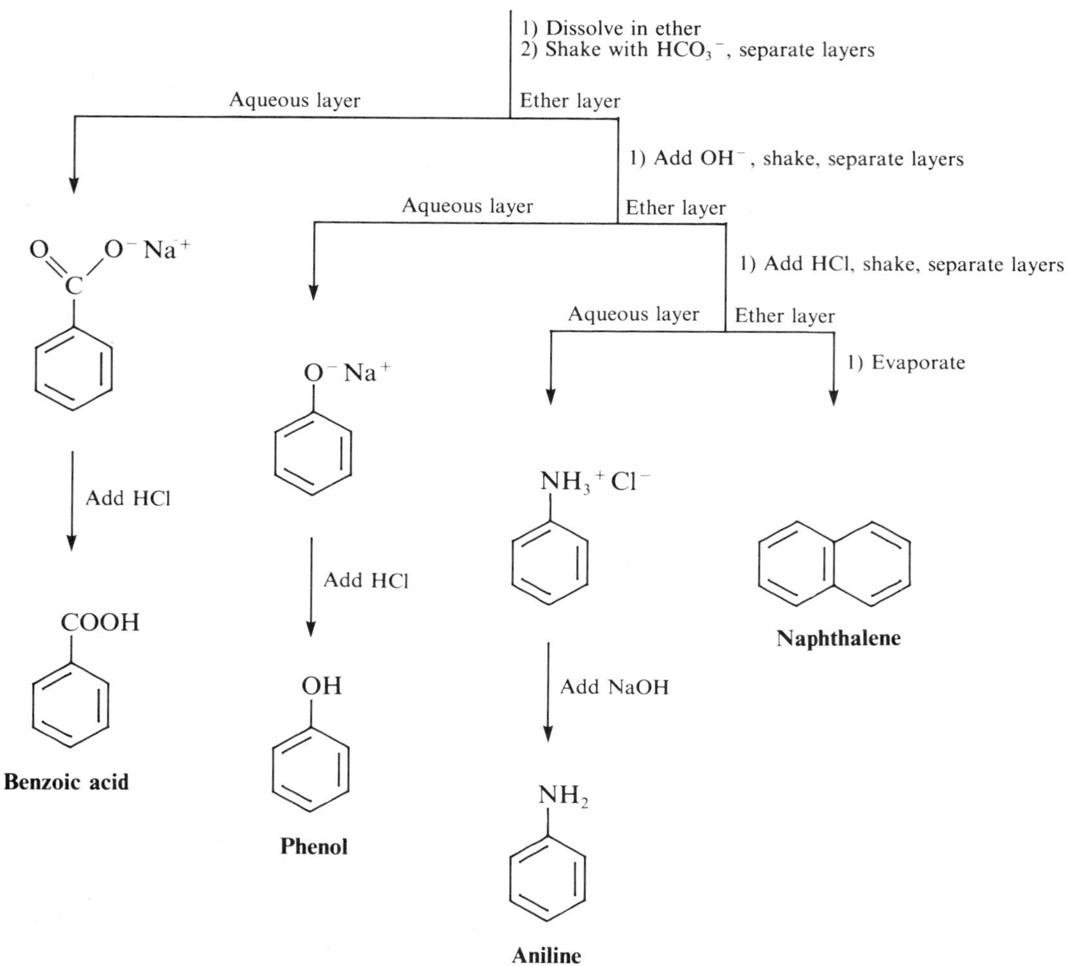

the benzoate anion in the fourth equation is governed by the equilibrium constant, K (Eq. 5), obtained by combining the third and fourth equations.

$$C_6H_5COOH + H_2O \rightleftharpoons C_6H_5COO^- + H_3O^+ \qquad (1)$$

$$K_a = \frac{[C_6H_5COO^-][H_3O^+]}{[C_6H_5COOH]} = 6.8 \times 10^{-5},\ pK_a = 4.17 \qquad (2)$$

$$K_w = [H_3O^+][OH^-] = 10^{-14} \quad (3)$$

$$C_6H_5COOH + OH^- \rightleftharpoons C_6H_5COO^- + H_2O \quad (4)$$

$$K = \frac{[C_6H_5COO^-]}{[C_6H_5COOH][OH^-]} = \frac{K_a}{K_w} = \frac{6.8 \times 10^{-5}}{10^{-14}} = 3.2 \times 10^8 \quad (5)$$

If 99% of the benzoic acid is converted to $C_6H_5COO^-$

$$\frac{[C_6H_5COO^-]}{[C_6H_5COOH]} = \frac{99}{1} \quad (6)$$

then from Eq. 5 the hydroxide ion concentration would need to be 3.2×10^{-7} M. Because saturated $NaHCO_3$ has $[OH^-] = 3 \times 10^{-4}$ M, the hydroxide ion concentration is high enough to convert benzoic acid completely to sodium benzoate.

For phenol with a K_a of 10^{-10} the minimum hydroxide ion concentration that will produce the phenoxide anion in 99% conversion is 10^{-2} M. The concentration of hydroxide in 10% sodium hydroxide solution is 10^{-1} M and so phenol in strong base is entirely converted to the water-soluble salt.

Liquid/liquid extraction and acid/base extraction are employed in the majority of organic reactions because it is unusual to have the product crystallize from the reaction mixture or to be able to distill the reaction product directly from the reaction mixture.

Practical Considerations

Emulsions. Imagine trying to extract a soap solution, e.g., a nonfoaming dishwasher detergent, into an organic solvent. A few shakes and you would have an absolutely intractable emulsion. An emulsion is a suspension of one liquid as droplets in another. Detergents stabilize emulsions, and so any time a detergent-like molecule happens to be in the material being extracted there is the danger of forming emulsions. Substances of this type are commonly found in nature, so one must be particularly wary of emulsion formation when making organic extracts of aqueous plant material, such as caffeine from tea. Emulsions, once formed, can be quite stable. You would be quite surprised to open your refrigerator one morning and see a layer of clarified butter floating on the top of a perfectly clear aqueous solution that had once been milk, but that is the classic example of an emulsion.

Prevention is the best cure for emulsions. This means shaking the solution to be extracted *very gently* until you see that the two layers will separate readily. If a bit of emulsion forms it may break simply on standing for a sufficient length of time. Making the aqueous layer highly ionic will

Shake gently to avoid emulsions

help. Add as much sodium chloride as will dissolve and shake the mixture gently. Vacuum filtration sometimes works and, when the organic layer is the lower layer, filtration through silicone-impregnated filter paper is an aid. Centrifugation works very well for breaking emulsions. This is easy on a small scale, but often the equipment is not available for large-scale centrifugation of organic liquids.

Pressure Buildup. The heat of the hand will cause pressure buildup in an extraction mixture that contains a very volatile solvent such as ether or dichloromethane. The extraction container, be it a reaction tube or a separatory funnel, must be opened carefully to vent this pressure.

Sodium bicarbonate solution is often used to neutralize acids when carrying out extractions. The result is the formation of carbon dioxide, which can cause foaming and high pressure buildup. Whenever bicarbonate is used add it very gradually with thorough mixing and frequent venting of the extraction device. If a large amount of acid is to be neutralized with bicarbonate the process should be carried out in a beaker.

Removal of Water. The organic solvents used for extraction dissolve not only the compound being extracted but also water. Evaporation of the solvent then leaves the desired compound contaminated with water. At room temperature water dissolves 7.5% of ether by weight and ether dissolves 1.5% of water. But ether is virtually insoluble in water saturated with sodium chloride (36.7 g/100 mL). If ether that contains dissolved water is shaken with a saturated aqueous solution of sodium chloride, water will be transferred from the ether to the aqueous layer. So, strange as it may seem, ethereal extracts routinely are dried by shaking them with saturated sodium chloride solution.

Saturated aqueous sodium chloride solution removes water from ether

Solvents such as dichloromethane do not dissolve nearly as much water and so are dried over a chemical drying agent. There are many choices of chemical drying agents for this purpose and the choice of which one to use is governed by four factors: the possibility of reaction with the substance being extracted, the speed with which it removes water from the solvent, the efficiency of the process, and the ease of recovery from the drying agent.

Some very good but specialized and reactive drying agents are potassium hydroxide, anhydrous potassium carbonate, sodium metal, calcium hydride, lithium aluminum hydride, and phosphorus pentoxide. Substances that are essentially neutral and unreactive and are widely used as drying agents include anhydrous calcium sulfate (Drierite), magnesium sulfate, molecular sieves, calcium chloride, and sodium sulfate.

Drierite, $CaSO_4$

Drierite, a specially prepared form of calcium sulfate, is a fast and effective drying agent. However it is difficult to ascertain whether enough has been used. An indicating type of Drierite is impregnated with cobalt chloride, which turns from blue to red when it is saturated with water. This works well when gases are being dried, but it should not be used for liquid extractions because the cobalt chloride dissolves in many protonic solvents.

Magnesium sulfate, $MgSO_4$

Magnesium sulfate is also a fast and fairly effective drying agent, but it is so finely powdered that it always requires careful filtration for removal.

Molecular sieves, zeolites

Molecular sieves are sodium alumino-silicates (zeolites) that have well-defined pore sizes. The 4A size adsorbs water to the exclusion of almost all organic substances and is a fast and effective drying agent, but like Drierite it is impossible to ascertain by appearance whether enough has been used. Molecular sieves in the form of 1/16-in. pellets are often used to dry solvents by simply adding them to the container.

Calcium chloride, $CaCl_2$

Calcium chloride is a very fast and effective drying agent, but may react with alcohols, phenols, amides, and carbonyl-containing compounds. Advantage is sometimes taken of this property to remove not only water from a solvent but, for example, a contaminating alcohol.

Sodium sulfate, Na_2SO_4. *The drying agent of choice for small-scale experiments.*

Sodium sulfate is the drying agent used almost exclusively in this book. It has a very high capacity for water, but is slow and not highly efficient in the removal of water. It has two advantages: first, it is granular and the solvent being dried can be poured off (decanted) from the drying agent or a Pasteur pipette can be used to draw off the drying agent without any filtration; and second, it has a tendency to clump together on the bottom of the container when excess water is present, but will fall freely through the solution when enough has been added. This latter property makes it easy to ascertain the correct quantity of drying agent to use. Complete drying of the solution is achieved by subsequent use of Drierite.

On a Larger Scale. A separatory funnel of capacity ranging from 30 mL to 6 L is used for larger-scale separations. Fitted with a Teflon stopcock it is gripped as shown in Fig. 8.4, shaken vigorously, and vented by opening the stopcock. It is placed in a ring and the layers allowed to separate as shown in Fig. 8.5. The lower layer is drawn off, after removing the stopper, by opening the stopcock.

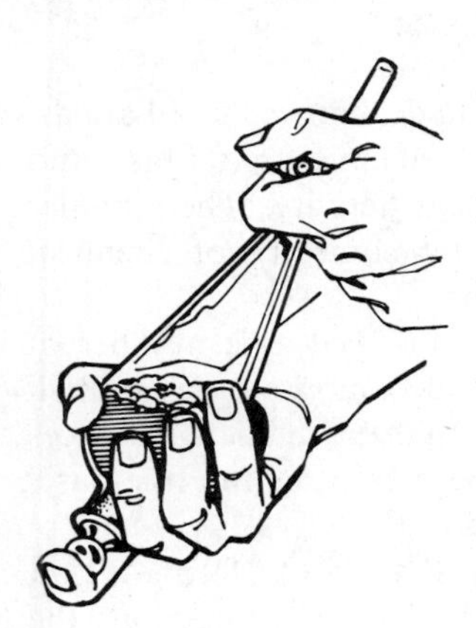

FIG. 8.4 Correct positions for holding a separatory funnel when shaking.

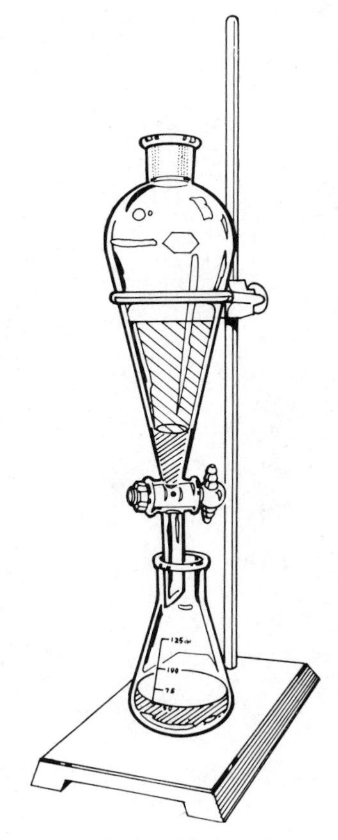

FIG. 8.5 Separatory funnel with Teflon stopcock.

Experiments

1. Extraction of Caffeine from Tea

Tea and coffee have been popular beverages for centuries, primarily because they contain the stimulant caffeine. It stimulates respiration, the heart, the central nervous system, and is a diuretic (promotes urination). It can cause nervousness and insomnia and, like many drugs, can be addicting, making it difficult to reduce the daily dose. A regular coffee drinker who consumes as little as four cups per day can experience headache, insomnia, and even nausea upon withdrawal of the drug.

Caffeine may be the most widely abused drug in the United States. During the course of a day an average person may unwittingly consume up to a gram of this substance. The caffeine content of some common foods and drugs is given in Table 8.1.

TABLE 8.1 Caffeine Content of Common Food and Drugs

Coffee	80 to 125 mg per cup
Coffee, decaffeinated	2 to 4 mg per cup
Tea	30 to 75 mg per cup
Cocoa	5 to 40 mg per cup
Milk chocolate	6 mg per oz
Baking chocolate	35 mg per oz
Coca-Cola	46 mg per 12 oz
Anacin, Bromo-Seltzer, Midol	32 mg per tablet
Excedrin, extra strength	65 mg per tablet
Dexatrim, Dietac, Vivarin	200 mg per tablet
Dristan	16 mg per tablet
No-Doz	100 mg per tablet

Caffeine belongs to a large class of compounds known as alkaloids. These are of plant origin, contain basic nitrogen, often have a bitter taste and complex structure, and usually have physiological activity. Their names usually end in "ine"; many are quite familiar by name if not chemical structure—nicotine, cocaine, morphine, strychnine.

Tea leaves contain tannins, which are acidic, as well as a number of colored compounds and a small amount of undecomposed chlorophyll (soluble in dichloromethane). In order to ensure that the acidic substances remain water soluble and that the caffeine will be present as the free base, sodium carbonate is added to the extraction medium.

The solubility of caffeine in water is 2.2 mg/mL at 25°C, 180 mg/mL at 80°C, and 670 mg/mL at 100°C. It is quite soluble in dichloromethane, the solvent used in this experiment to extract the caffeine from water.

Experiments

Procedure from Tea Leaves

In a 30-mL beaker place 10 mL of water, 2 g of sodium carbonate, and a wooden boiling stick. Bring the water to a boil on the sand bath, remove the boiling stick, and brew a very concentrated tea solution by immersing a tea bag in the very hot water for 5 min. Squeeze as much water from the bag as possible after it cools enough to handle. Be careful not to break the bag. Pour the extract into a 15-mL centrifuge tube and repeat the extraction process using 5 mL of water and 1 g of sodium carbonate. After about 5 min remove the tea bag, squeeze it to remove as much tea solution as possible, combine the aqueous extracts in the centrifuge tube and cool the solution in ice to below 40°C (the boiling point of dichloromethane).

Using three 1-mL portions of dichloromethane, extract the caffeine from the tea. Cork the tube and use a gentle rocking motion to carry out the extraction. Vigorous shaking will produce an intractable emulsion while extremely gentle mixing will fail to extract the caffeine. If you have ready access to a centrifuge the shaking can be more vigorous because any emulsions formed can be broken fairly well by centrifugation. After each extraction remove the lower organic layer into a reaction tube. Dry the combined extracts over anhydrous sodium sulfate for 5 or 10 min. Add the drying agent in portions with shaking until it no longer clumps together. Transfer the dry solution to a tared 25-mL filter flask and evaporate it to dryness. The residue will be crude caffeine (determine its weight) that is to be purified by sublimation.

Fit the filter flask with a No. 2 neoprene adapter through which is thrust a 15-mL centrifuge tube. If the centrifuge tube does not fit tightly, clamp it so it will not slip down into the flask. Connect the filter flask to the water aspirator through a trap and turn the water on full force. Clamp the flask with a large three-prong clamp, fill the centrifuge tube with ice and water, and heat the flask on a hot sand bath. Caffeine is reported to sublime at about 170°C under reduced pressure. Wrap the flask with a cone of aluminum foil to direct heat to the upper part of the flask and when sublimation ceases turn off the vacuum, remove the ice water, and allow the flask to cool somewhat before removing the centrifuge tube. Scrape the caffeine onto a tared weighing paper and, using the plastic funnel, transfer it to a small vial. At the discretion of the instructor determine the melting point using a sealed capillary. The melting point of caffeine is 238°C. Using the centrifugation technique to separate the extracts, 30 mg of crude caffeine can be obtained from one tea bag. This will give you 20 mg of sublimed material.

2. Extraction of Caffeine from Cola Syrup

Coca-Cola was originally flavored with extracts from the leaves of the coca plant and the kola nut. Coca is grown in northern South America; the Indians

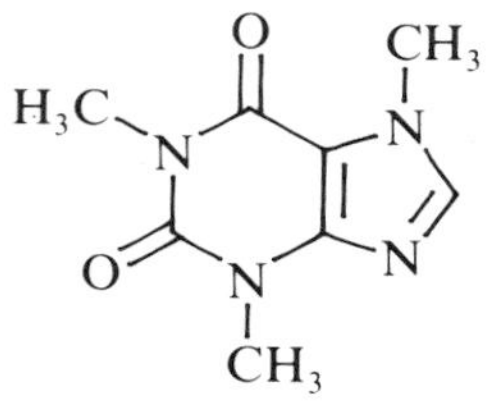

Caffeine
mp 238°C

of Peru and Bolivia have for centuries chewed the leaves to relieve the pangs of hunger and high mountain cold. The cocaine from the leaves causes local anesthesia of the stomach. It has limited use as a local anesthetic for surgery on the eye, nose, and throat. Unfortunately is is now a widely abused illicit drug. Kola nuts contain about 3% caffeine as well as a number of other alkaloids. The kola tree is in the same family as the cacao tree from which cocoa and chocolate are obtained. Modern cola drinks do not contain cocaine; however Coca-Cola contains 43 mg of caffeine per 12-oz bottle. The acidic taste of many soft drinks comes from citric, tartaric, phosphoric, and benzoic acids.

Automatic soft drink dispensing machines mix a syrup with carbonated water. In Experiment 2 caffeine is extracted from concentrated cola syrup.

Caffeine can also be extracted easily from tea bags. The procedure one would use to make a cup of tea, simply "steeping" the tea with very hot water for about 7 min, extracts most of the caffeine. There is no advantage to boiling the tea leaves with water for 20 min. Since caffeine is a white, slightly bitter, odorless, crystalline solid, it is obvious that water extracts more than just caffeine. When the brown aqueous solution is subsequently extracted with dichloromethane, primarily caffeine dissolves in the organic solvent. Evaporation of the solvent leaves crude caffeine, which on sublimation yields a relatively pure product. When the concentrated tea solution is extracted with dichloromethane, emulsions can form very easily. There are substances in tea that cause small droplets of the organic layer to remain suspended in the aqueous layer. This emulsion formation results from vigorous shaking. To avoid this problem, it might seem that one would boil the tea leaves with dichloromethane first and then extract the caffeine from the dichloromethane solution with water. In fact this does not work. Boiling 25 g of tea leaves with 50 mL of dichloromethane gives only 0.05 g of residue after evaporation of the solvent. Subsequent extractions give less material. Hot water causes the tea leaves to swell and is obviously a much more efficient extraction solvent. An attempt to sublime caffeine directly from tea leaves was also unsuccessful.

Procedure from Cola Syrup

Add 1 mL of concentrated ammonium hydroxide to a mixture of 5 mL of commercial cola syrup and 5 mL of water in a 15-mL centrifuge tube. Add 1 mL of dichloromethane and tip the tube gently back and forth for 5 min. Do not shake the mixture as in a normal extraction because an emulsion will form and the layers will not separate. After the layers have separated as much as possible remove the clear lower layer, leaving the emulsion behind. Using 1.5 mL of dichloromethane repeat the extraction in the same way twice more. At the final separation include the emulsion layer with the dichloromethane. Combine the extracts in a reaction tube and dry the solution with anhydrous sodium sulfate. Add the drying agent with shaking until it no longer clumps together. After 5 or 10 min remove the solution with a Pasteur pipette and place it in a tared filter flask. Wash off the drying agent with more dichloromethane and evaporate the mixture to dryness. Determine the crude weight of caffeine and then sublime it as described in the previous experiment.

3. Caffeine Salicylate

Preparation of a derivative of caffeine

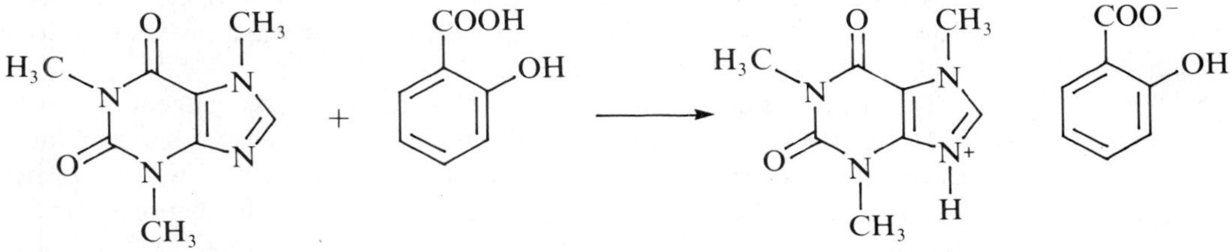

One way to confirm the identity of an organic compound is to prepare a derivative of it. Caffeine melts and sublimes at 238°C. It is an organic base and can therefore accept a proton from an acid to form a salt. The salt formed when caffeine combines with hydrochloric acid, like many amine salts, does not have a sharp melting point; it simply decomposes when heated. But the salt formed from salicylic acid, even though ionic, has a sharp melting point and can thus be used to help characterize caffeine.

Caffeine + **Salicylic acid** → **Caffeine salicylate**

Procedure

Crystallization from mixed solvents

To 10 mg of sublimed caffeine in a tared reaction tube add 7.5 mg of salicylic acid and 0.5 mL of dichloromethane. Heat the mixture to boiling and add petroleum ether (a poor solvent for caffeine) dropwise until the mixture just turns cloudy, indicating the solution is saturated. If too much petroleum ether is added then clarify it by adding a very small quantity of dichloromethane. Insulate the tube in order to allow it to cool slowly to room temperature, and then cool it in ice. The needle-like crystals are isolated by removing the solvent while the reaction tube is in the ice bath. Evaporate the last traces of solvent under vacuum, determine the weight of the derivative and its melting point. Caffeine salicylate is reported to melt at 137°C.

4. Separation of Acidic and Neutral Substances

A mixture of equal parts of benzoic acid, 2-naphthol, and 1,4-dimethoxybenzene (hydroquinone dimethyl ether) is to be separated by extraction from ether. Note the detailed directions for extraction carefully. In the next experiment you are to work out your own extraction procedure.

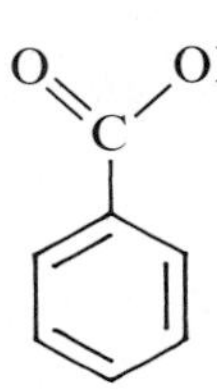

Benzoic acid
mp 123°C, pK_a 4.17

Procedure

Dissolve 0.15 g of the mixture in 2 mL of ether in a reaction tube (tube 1), then add 1.0 mL of a 10% aqueous solution of sodium bicarbonate to the

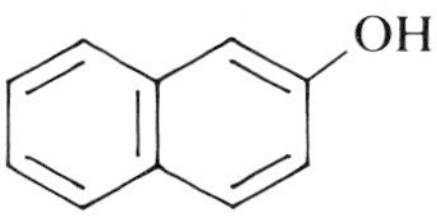

2-Naphthol
mp 123°C, pK_a 9.51

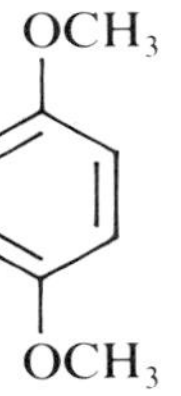

1,4-Dimethoxybenzene (Hydroquinone dimethyl ether)
mp 57°C

tube. Stopper the tube and shake it thoroughly for about three minutes. Vent the tube from time to time, although you will find very little pressure generated from carbon dioxide. Allow the layers to separate completely and then draw off the lower layer into another reaction tube (tube 2). Add another 0.15 mL of sodium bicarbonate solution to the tube, shake the contents as before and add the lower layer to tube 2. Exactly what chemical species is in tube 2? Add 0.2 mL of ether to tube 2, shake it thoroughly, remove the ether layer, and discard it. This is called *backwashing* and serves to remove any organic material that might contaminate the contents of tube 2.

Add 1.0 mL of 5% aqueous sodium hydroxide to tube 1, shake the mixture thoroughly, allow the layers to separate, draw off the lower layer using a clean Pasteur pipette and place it in tube 3. Extract tube 1 with two 0.15-mL portions of water and add these to tube 3. Backwash the contents of tube 3 with 0.15 mL of ether and discard the ether wash just as was done for tube 2. Exactly what chemical species is in tube 3?

To tube 1 add anhydrous sodium sulfate until the drying agent no longer clumps together when the contents are mixed. This can be a large quantity of drying agent because it does not react with the product. It will be washed off with ether after the drying process is finished. Allow 5 to 10 min for drying of the ether solution.

Add HCl *with care.* CO_2 *is released*

Using the concentration information given in the end papers of this book, calculate exactly how much concentrated hydrochloric acid is needed to neutralize the contents of tube 2. Then by dropwise addition of concentrated hydrochloric acid carry out this neutralization while testing the solution with litmus paper. An excess of hydrochloric acid does no harm. This reaction must be carried out with *extreme care* because much carbon dioxide is released in the neutralization. Add a boiling stick to the tube and very cautiously heat the tube to bring most of the solid benzoic acid into solution. Allow the tube to cool slowly to room temperature and then cool it in ice. Remove the solvent with a Pasteur pipette and recrystallize the residue from boiling water. Again allow the tube to cool slowly to room temperature then cool it in ice and at the appropriate time stir the crystals and collect them on the Hirsch funnel, using the procedures detailed in Chapter 3. The crystals can be transferred and washed on the funnel using a small quantity of ice water. The solubility of benzoic acid in water is 1.9 g/L at 0°C and 68 g/L at 95°C. Turn the crystals out onto a tared piece of paper, allow them to dry thoroughly, and determine the percent recovery of the benzoic acid. Assess the purity of the product by melting point.

In exactly the same way neutralize the contents of tube 3 with concentrated hydrochloric acid. This time, of course, there will be no carbon dioxide evolution. Again heat the tube to bring most of the material into solution, allow it to cool slowly, remove the solvent, and recrystallize the 2-naphthol from boiling water. At the appropriate time after the product has cooled slowly to room temperature and then in ice it is also collected on the Hirsch funnel, washed with a very small quantity of ice water, and allowed to dry. The percent recovery and melting point are determined.

The 1,4-dimethoxybenzene is recovered using the Pasteur pipette to remove the ether from the drying agent and to transfer it to a tared reaction tube. The drying agent is washed two or three times with additional ether to ensure complete transfer of the product. Put a boiling stick in the tube and evaporate the ether in the hood. An aspirator tube can also be used for this purpose (see Fig. 8.4), although the solvent is best removed by blowing a gentle stream of air into the tube using a Pasteur pipette while warming the tube in the hand or a beaker of warm water. Determine the weight of the crude 1,4-dimethoxybenzene and then recrystallize it from methanol/water. Reread Chapter 3 for detailed instructions on carrying out this process. The product is dissolved in about 0.5–1 mL of methanol and water is added until the solution gets cloudy, indicating the solution is saturated. This process is best carried out while heating the tube in a hot water bath at 50°C. Because the product melts at 58–60°C it is obviously impossible to have crystallization occur above 58°C. Allow the tube to cool slowly to room temperature and then cool it thoroughly in ice. The product is best isolated by removing the solvent using a Pasteur pipette. It can also be collected on the Hirsch funnel, using an ice-cold alcohol–water mixture to transfer and wash the product. Determine the percent recovery and the melting point.

5. Separation of a Neutral and Basic Substance

A mixture of equal parts of a neutral substance, naphthalene, and a basic substance, 4-chloroaniline, is to be separated by extraction from an ether solution. Naphthalene is completely insoluble in water. The base will dissolve in hydrochloric acid while the neutral naphthalene will remain in ether solution. 4-Chloroaniline is insoluble in cold water but will dissolve to some extent in hot water and it is soluble in ethanol. Naphthalene can be purified as described in Chapter 3.

Plan a procedure for separating 1.0 g of the mixture into its components and have the plan checked by the instructor before proceeding. A flow sheet is a convenient way to present the plan. Select the correct solvent or mixture of solvents for the recrystallization of 4-chloroaniline on the basis of solubility tests. Determine the weights and melting points of the isolated and purified products and calculate the percent recovery of each. Turn in the products in neatly labeled vials.

Questions

1. Suppose a reaction mixture, when diluted with water, afforded 300 mL of an aqueous solution of 30 g of the reaction product malononitrile, $CH_2(CN)_2$, which is to be isolated by extraction with ether. The solubility of malononitrile in ether at room temperature is 20.0 g per 100 mL and that in water is 13.3 g per 100 mL. What weight of malononitrile would be recovered by extraction with (a) three 100-mL portions

of ether; (b) one 300-mL portion of ether? *Suggestion:* For each extraction let x equal the weight extracted into the ether layer. In case (a) the concentration in the ether layer is $x/100$ and that in the water layer $(30 - x)/300$; the ratio of these quantities is equal to $k = 20/13.3$.

2. Why is it necessary to remove the stopper from a separatory funnel when liquid is being drained from it through the stopcock?

3. The pK_a of *p*-nitrophenol is 7.15. Would you expect this to dissolve in sodium bicarbonate solution? The pK_a of 3,6-dinitrophenol is 5.15. Will it dissolve in bicarbonate solution?

4. The distribution coefficient, k = (conc. in ligroin/conc. in water), between ligroin and water for solute A is 7.5. What weight of A would be removed from a solution of 10 g of A in 100 mL of water by a single extraction with 100 mL of ligroin? What weight of A would be removed by four successive extractions with 25-mL portions of ligroin? How much ligroin would be required to remove 98.5% of A in a single extraction?

5. In Experiment 1 how many moles of benzoic acid are present? How many moles of sodium bicarbonate are contained in 1 mL of a 10% aqueous solution? (A 10% solution has 1 g of solute in 10 mL of solvent.) Is the amount of sodium bicarbonate sufficient to react with all of the benzoic acid?

6. To isolate the benzoic acid from the bicarbonate solution, it is acidified with concentrated hydrochloric acid in Experiment 1. What volume of acid is needed to neutralize the bicarbonate? The concentration of hydrochloric acid is expressed in various ways on the inside back cover of this laboratory manual.

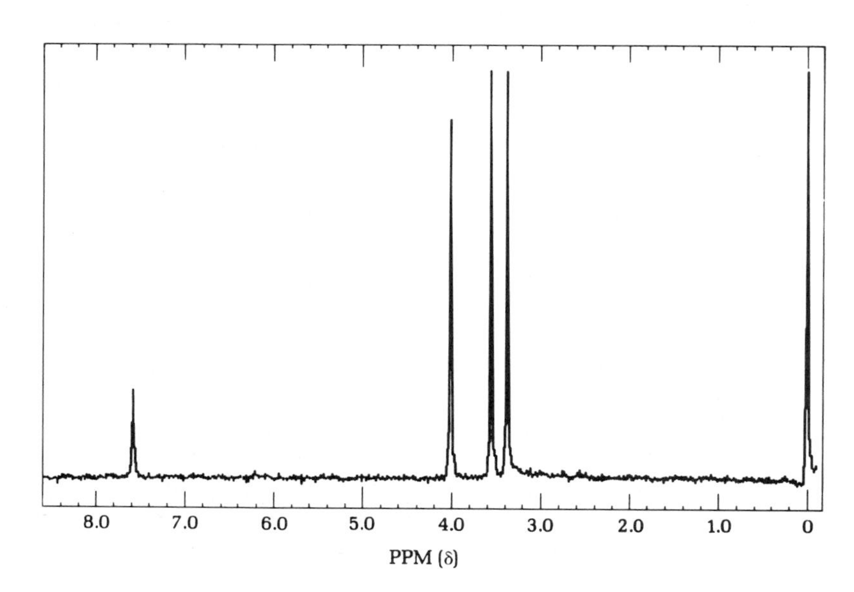

FIG. 8.6 ^{1}H nmr spectrum of caffeine.

7. How many moles of 4-*tert*-butyl phenol are in the mixture to be separated in Experiment 1? How many moles of sodium hydroxide are contained in 1 mL of 5% sodium hydroxide solution? What volume of concentrated hydrochloric acid is needed to neutralize this amount of sodium hydroxide solution?

9 Thin-Layer Chromatography: Analysis of Analgesics and Isolation of Lycopene from Tomato Paste

PRELAB EXERCISE: *Compare thin-layer chromatography (TLC) with column chromatography with regard to (1) quantity of material that can be separated, (2) the speed, (3) the solvent systems, and (4) the ability to separate compounds.*

Thin-layer chromatography (TLC) is a sensitive, fast, simple, and inexpensive analytical technique that will be used repeatedly in carrying out small-scale organic experiments. It is a true micro technique; as little as 10^{-9} g of material can be detected, although the usual sample size is from 1 to 100×10^{-6} g.

TLC requires micrograms of material

TLC involves spotting the sample to be analyzed near one end of a sheet of glass or plastic coated with a thin layer of an adsorbent. The sheet, which can be the size of a microscope slide, is placed on end in a covered jar containing a shallow layer of solvent. As the solvent rises by capillary action up through the adsorbent, differential partitioning occurs between the components of the mixture dissolved in the solvent and the stationary adsorbent phase. The more strongly a given component of the mixture is adsorbed onto the stationary phase, the less time it will spend in the mobile phase and the more slowly it will migrate up the TLC plate.

Uses of thin-layer chromatography

1. **To determine the number of components in a mixture.** TLC affords a quick and easy method for analyzing such things as a crude reaction mixture or an extract from some plant substance or a pain killer. Knowing the number and relative amounts of the components aids in planning further analytical and separation steps.
2. **To determine the identity of two substances.** If two substances spotted on the same TLC plate give spots in identical locations, they *may* be identical. If the spot positions are not the same the substances cannot be the same. It is possible for two closely related compounds that are not identical to have the same positions on a TLC plate.
3. **To monitor the progress of a reaction.** By sampling a reaction from time to time it is possible to watch the reactants disappear and the products appear using TLC. Thus the optimum time to halt the reaction can be

determined and the effect of changing such variables as temperature, concentrations, and solvents can be followed without the necessity of isolating the product.

4. **To determine the effectiveness of a purification.** The effectiveness of distillation, crystallization, extraction, and other separation and purification methods can be monitored using TLC, with the caveat that a single spot does not *guarantee* a single substance.
5. **To determine the appropriate conditions for a column chromatographic separation.** Thin-layer chromatography is generally unsatisfactory for purifying and isolating macroscopic quantities of material; however, the adsorbents most commonly used for TLC, silica gel and alumina, are used for column chromatography, discussed in a later chapter. Column chromatography is used to separate and purify up to about a gram of a solid mixture. The correct adsorbent and solvent used to carry out the chromatography can be determined rapidly by TLC.
6. **To monitor column chromatography.** As column chromatography is carried out the solvent is collected in a number of small flasks. Unless the desired compound is colored the various fractions must be analyzed in some way to determine which ones have the desired components of the mixture. TLC is a fast and effective method for doing this.

Adsorbents and Solvents

The two most common coatings for thin-layer chromatography plates are alumina, Al_2O_3, and silica gel, SiO_2. These are the same adsorbents most commonly used in column chromatography (Chapter 10) for the purification of macroscopic quantities of material. Of the two alumina, when anhydrous, is the more active, i.e., it will adsorb substances more strongly. It is thus the adsorbent of choice when the separation involves relatively nonpolar substrates such as hydrocarbons, alkyl halides, ethers, aldehydes, and ketones. To separate the more polar substrates such as alcohols, carboxylic acids, and amines, the less active adsorbent silica gel is used. In an extreme situation very polar substances on alumina do not migrate very far from the starting point (give low *Rf* values) and nonpolar compounds travel with the solvent front (give high *Rf* values) if chromatographed on silica gel. These extremes of behavior are markedly affected, however, by the solvents used to carry out the chromatography. A polar solvent will carry along with it polar substrates and nonpolar solvents will do the same with nonpolar compounds, another example of the generalization "like dissolves like."

In Table 9.1 are listed common solvents used in chromatography, both thin-layer and column. Only the environmentally safe solvents are listed; the polarities of such solvents as benzene, carbon tetrachloride, or chloroform can be matched by other less toxic solvents. In general these solvents are characterized by having low boiling points and low viscosities that allow them to migrate rapidly. They are listed in order of increasing polarity. A solvent more polar than methanol is seldom needed. Often just two solvents

TABLE 9.1 Chromatography Solvents

Solvent	Bp (°C)
Petroleum ether (pentanes)	35–60
Ligroin (hexanes)	60–80
Diethyl ether	35
Dichloromethane	40
Ethyl acetate	77
Acetone	56
2-Propanol	82
Ethanol	78
Methanol	65
Water	100
Acetic acid	118

TABLE 9.2 Order of Solute Migration on Chromatography

Solute	Solute
Fastest	
Alkanes	Ketones
Alkyl halides	Aldehydes
Alkenes	Amines
Dienes	Alcohols
Aromatic hydrocarbons	Phenols
Aromatic halides	Carboxylic acids
Ethers	Sulfonic acids
Esters	*Slowest*

are used in varying proportions; the polarity of the mixture is a weighted average of the two. Ligroin–ether mixtures are often employed in this way.

The order in which solutes migrate on thin-layer chromatography is the same as the order of solvent polarity. The largest *Rf* values are shown by the least polar solutes. In Table 9.2 the solutes are arranged in order of increasing polarity.

Apparatus and Procedure

TLC

A 1% solution of the substance to be examined is spotted onto the plate about 1 cm from the bottom end and the plate is inserted into a beaker or a 4-oz wide-mouth bottle containing 4 mL of an organic solvent. The bottle is lined with filter paper wet with solvent to saturate the atmosphere within the container. The top is put in place and the time noted. The solvent travels rapidly up in the thin layer by capillary action, and if the substance is a pure colored compound, one soon sees a spot either traveling along with the

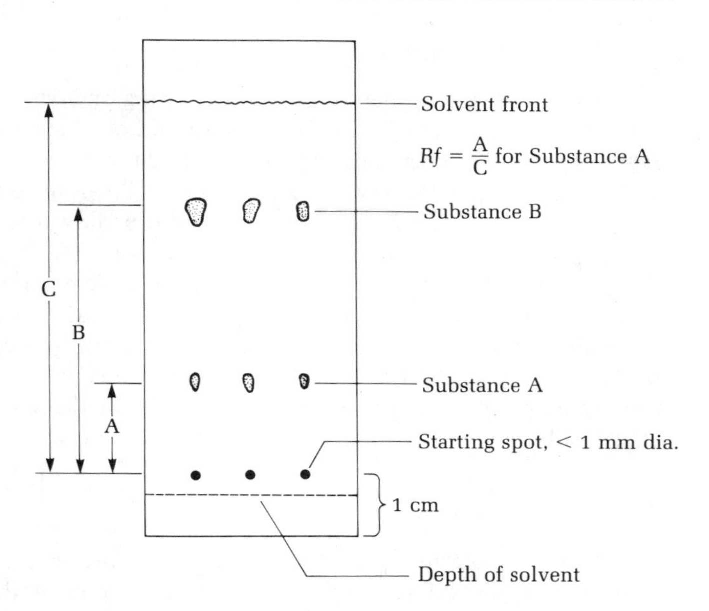

FIG. 9.1 Thin-layer chromatography plate.

solvent front or, more usually, at some distance behind the solvent front. One can remove the slide, quickly mark the front before the solvent evaporates, and calculate the *Rf* value. The *Rf* value is the ratio of the distance the spot travels from the point of origin to the distance the solvent travels (Fig. 9.1).

Colored compounds

If two colored compounds are present and an appropriate solvent is selected, two spots will appear.

Fortunately the method is not limited to colored substances. Any organic compound capable of being eluted from alumina will form a spot, which soon becomes visible when the solvent is let evaporate and the plate let stand in a stoppered 4-oz bottle containing a few crystals of iodine. Iodine vapor is adsorbed by the organic compound to form a brown spot. A spot should be outlined at once with a pencil because it will soon disappear as the iodine sublimes away; brief return to the iodine chamber will regenerate the spot. The order of elution and the elution power for solvents are the same as for column chromatography.

And colorless ones

The use of commercial TLC sheets such as Eastman silica gel with fluorescent indicator (No. 13181) is strongly recommended. These poly(ethylene terephthalate) sheets are coated with silica gel using polyacrylic acid as a binder. A fluorescent indicator has been added to the silica gel so that when the sheet is observed under 254-nm ultraviolet light, spots that either quench or enhance fluorescence can be seen. Iodine can also be used to visualize spots. The coating on these sheets is only 100 microns thick, so very small spots must be applied. Unlike student-prepared plates, these coated sheets (cut to 1 × 3-in. size with scissors) give very consistent results. A light pencil mark 1 cm from the end will guide spotting. A supply of these little precut sheets makes it a simple matter to examine most of the reactions in this book for completeness of reaction, purity of product, and side reactions.

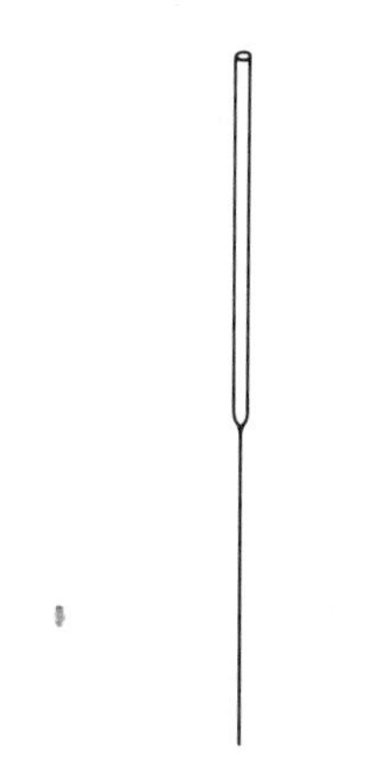

FIG. 9.2 Micropipette.

Spotting Test Solutions

This is done with micropipettes made by drawing open-end mp capillaries in a microburner flame (Fig. 9.2). The bore should be of such a size that, when the pipette is dipped deep into ligroin, the liquid flows in to form a tiny thread which, when the pipette is withdrawn, does not flow out to form a drop. To spot a test solution, let a 2–3 cm column of solution flow into the pipette, hold this vertically over a coated plate, aim it at a point on the right side of the plate and about 1 cm from the bottom, and lower the pipette until the tip just touches the adsorbent and liquid flows onto the plate; withdraw when the spot is about 1 mm in diamter. Make a second 1-mm spot on the left side of the plate, let it dry, and make two more applications of the same size (1-mm) at the same place. Determine whether the large or the small spot gives the better results.

Making TLC Plates

The adsorbent recommended for making TLC plates is a preparation of finely divided silica gel with gypsum binder and fluorescent indicator (Aldrich 28855-1) or alumina containing plaster of paris as binder (Fluka[1]). A slurry of this material in water can be applied to microscope slides by simple techniques of dipping (A) or coating (B); for a small class, or for occasional preparation of a few slides by an individual, the more economical coating method is recommended.

(A) By Dipping

FIG. 9.3 Preparation of TLC plates by dipping. The two microscope slides are grasped at one end, dipped into the silica gel slurry, removed and separated, and dried and activated by heating.

Place 15 g of silica gel or alumina formulated for TLC in a 40 × 80-mm weighing bottle[2] and stir with a glass rod or with a magnetic stirrer while gradually pouring in 75 mL of distilled water. Stir until lumps are eliminated and a completely homogenous slurry results. Grasp a pair of clean slides[3] at one end with the thumb and forefinger, dip them in the slurry of adsorbent (Fig. 9.3), withdraw with an even unhurried stroke, and touch a corner of the slides to the mouth of the weighing bottle to allow excess fluid to return to the container. Then dry the working surface by mounting the slides above a 70-watt hot plate, using as support two pairs of 1 × 7-cm strips of blotting paper (or a pair of applicator sticks). Drying takes about a minute and a half and is evident by inspection. Remove the plate with a forceps as soon as it is dry, for cooling takes longer. If you dip and dry 8–10 slides in succession,

1. Aluminumoxid Fluka für Dünnschicht Chromatographie, Typ D5, Fluka AG, Buchs, S. G. Switzerland; Fluka Alumina is distributed by Tridorn Chemical, Inc., 255 Oser Av., Hauppauge, N.Y. 11787.

2. Kimble Exax, 15146.

3. The slides should be washed with water and a detergent, rinsed with tap water and then with distilled water, and placed on a clean towel to dry. Dry slides should be touched only on the sides, for a touch on the working surface may spoil a chromatogram.

finished plates will be ready for use when you are through. Alternatively, dry the slides in an oven at 110°C. This heating activates the alumina. Coated slides dried at room temperature do not give as intense spots as those that have been heat-dried.

Keep the storage bottle or adsorbent closed when not in use. Note that this is not a stable emulsion but that the solid settles rapidly on standing. Stir thoroughly with a rod before each reuse.

Plates ready for use, as well as clean dry slides, are conveniently stored in a microscope slide box (25 slides).

(B) By Coating

Place 2 g of silica gel or alumina and 10 mL of distilled water in a 25-mL Erlenmeyer, stopper the flask and shake to produce an even slurry. Keep the flask stoppered when not in use. Place a clean, dry slide on a block of wood or box with the slide projecting about 1 cm on the left-hand side (if you are right-handed) so that it can be grasped easily on the two sides. Swirl the flask to mix the slurry and draw a portion into a medicine dropper. Hold the dropper vertically and, starting at the right end of the slide, apply emulsion until the entire upper face is covered; make further applications to repair pin holes and eliminate bubbles. Grasp the left end of the slide with a forceps and even the emulsion layer by tilting the slide to the left to cause a flow, and then to the right; tilt again to the front and to the rear. Dry the slide on a hot plate as you did when preparing slides by procedure A. The adsorbent should be about 0.25 mm thick.

Visualization of the Chromatogram

If the substance being chromatographed is colored then it is possible to detect the components visually. Colorless substances can be detected by placing the dry TLC plate in a jar containing a few crystals of iodine. The iodine vapor will be preferentially adsorbed by the substances on the plate and they will appear as brown spots on a lighter-colored background. The plate is removed from the jar and the outline of the spots traced lightly in pencil because the iodine will soon evaporate.

Iodine chamber

Plates that have been impregnated with a fluorescent indicator will show dark spots for the compounds under an ultraviolet light due to quenching of the fluorescence by the substance on the plate. Again trace the spots lightly in pencil while the plate is under the uv light. Don't look directly into the light; it will damage the eyes.

Don't look into lamp

A large number of specialized spray reagents have been developed that give specific colors for certain types of compounds, and there is a large literature on the solvents and adsorbents to use for the separation of given types of material.[4]

4. Egon Stahl, "Thin-Layer Chromatography," Springer Verlag, New York, 1969.

Experiments

1. Analgesics

Analgesics are substances that relieve pain. The most common of these is aspirin, a component of more than 100 nonprescription drugs. In Chapter 26, the history and background of this most popular drug is discussed. In the present experiment analgesic tablets will be analyzed by thin-layer chromatography to determine which analgesics they contain and whether they contain caffeine, which is often added to counteract the sedative effects of the analgesic.

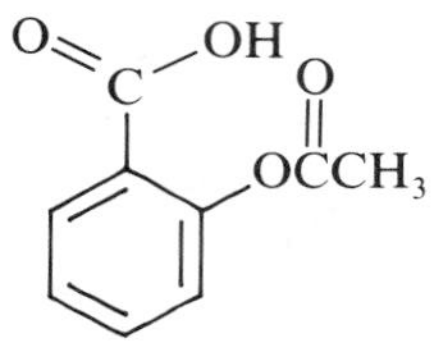

Aspirin
Acetylsalicyclic acid

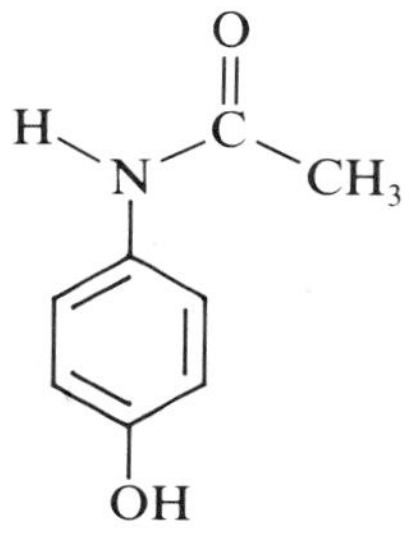

Acetaminophen
4-Acetamidophenol

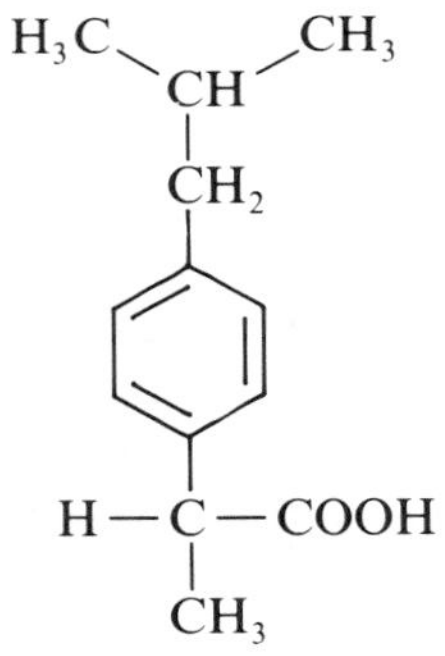

Ibuprofen
2-(4-Isobutylphenyl)propionic acid

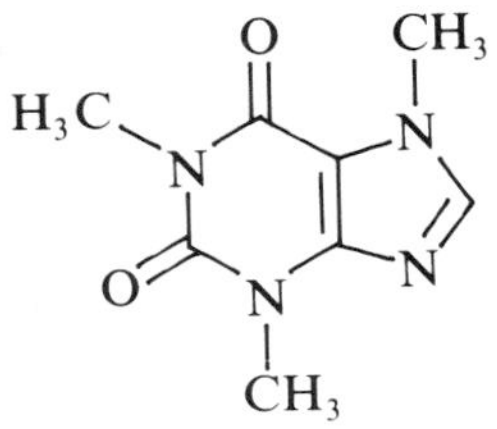

Caffeine

In addition to aspirin and caffeine the most common components of analgesics are, at present, acetaminophen and ibuprofen. Phenacetin, the P of the APC tablet and a former component of Empirin, has been removed from the market because of deleterious side effects. In addition to one or more of these substances each tablet contains a binder, often starch or silica gel. And to counteract the acidic properties of aspirin an inorganic buffering agent is added to some analgesics. Inspection of labels will reveal that most cold remedies and decongestants contain both aspirin and caffeine in addition to the primary ingredient.

To identify an unknown by TLC the usual strategy is to run chromatograms of known substances (the standards) and the unknown at the same time. If the unknown has one or more spots that correspond to spots with the same *Rf*'s as the standards, then those substances are probably present.

Proprietary drugs that contain one or more of the common analgesics and sometimes caffeine are sold under the names of Bayer Aspirin, Anacin, Datril, Advil, Excedrin, Extra Strength Excedrin, Tylenol, and Vanquish.

Procedure

Following the procedure outlined above, draw a light pencil line about 1 cm from the end of a chromatographic plate and on this line spot aspirin, acetaminophen, ibuprofen, and caffeine, which are available as reference standards. Make each spot as small as possible, preferably less than 0.5 mm in diameter. Examine the plate under the uv light to see that enough of the compound has been applied; if not, add more. On a separate plate run the unknown and one or more of the standards.

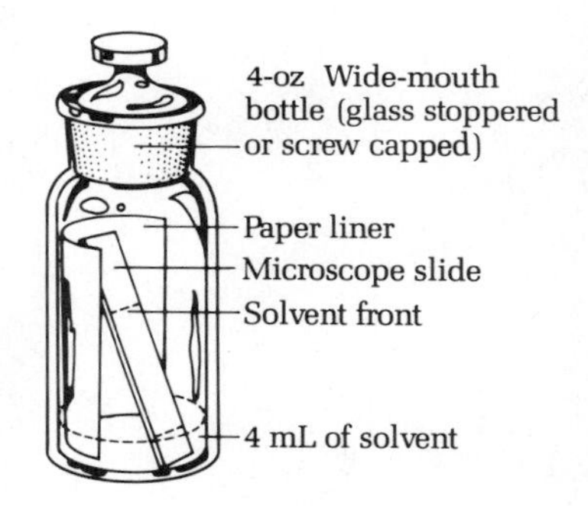

FIG. 9.4 A method of developing thin-layer chromatographic plates.

The unknown sample is prepared by crushing a part of a tablet, adding this powder to a reaction tube or small vial along with a few drops of ethanol, and then mixing the suspension. Not all of the tablet will dissolve, but enough will go into solution to spot the plate. The binder, starch or silica, will not dissolve.

Use as the solvent for the chromatogram a mixture of 95% ethyl acetate and 5% acetic acid (Fig. 9.4 or 9.5). After the solvent has risen to near the top of the plate, mark the solvent front with a pencil, remove the plate from the developing chamber, and allow the solvent to dry. Examine the plate under uv light to see the components as dark spots against a bright green-blue background. Outline the spots with a pencil. The spots can also be visualized by putting the plate in an iodine chamber made by placing a few crystals of iodine in the bottom of a capped 4-oz jar. Calculate the *Rf* values for the spots and identify the components in the unknown.

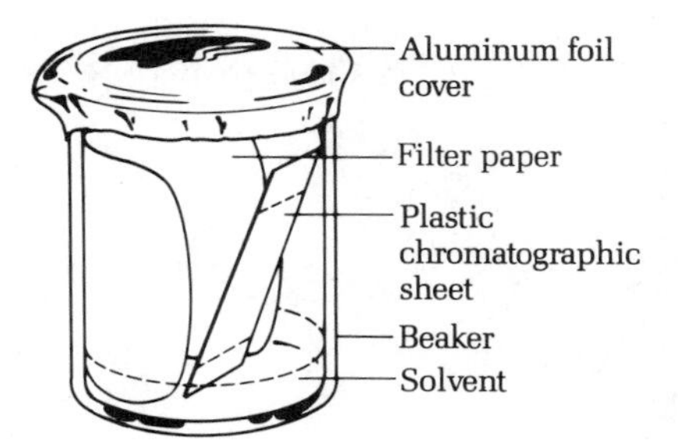

FIG. 9.5 An alternate method for developing TLC plates.

2. Plant Pigments

The botanist Michael Tswett discovered the technique of chromatography and applied it, as the name implies, to colored plant pigments. The leaves of plants contain, in addition to chlorophyll-a and -b, other pigments that are revealed in the fall when the leaf dies and the chlorophyll rapidly decomposes. Among the most abundant of the other pigments are the carotenoids, which include the carotenes and their oxygenated homologs, the xantho-

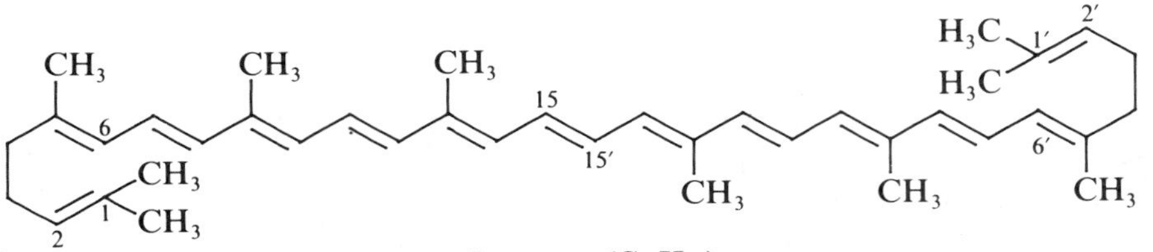

Lycopene ($C_{40}H_{56}$)
MW 536.85
mp 173°C, λ_{max}^{hexane} 475 nm

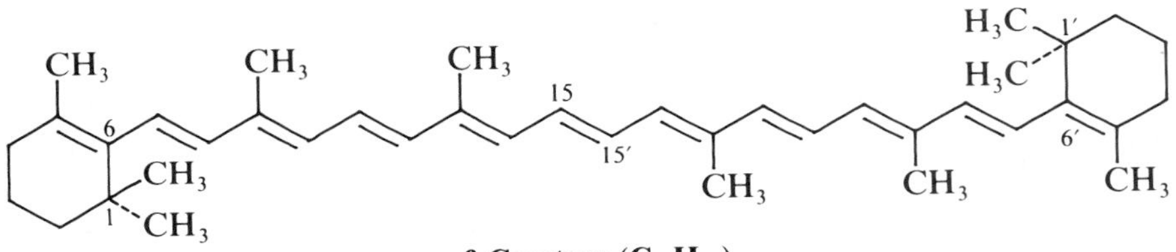

β-Carotene ($C_{40}H_{56}$)
mp 183°C, λ_{max}^{hexane} 451 nm

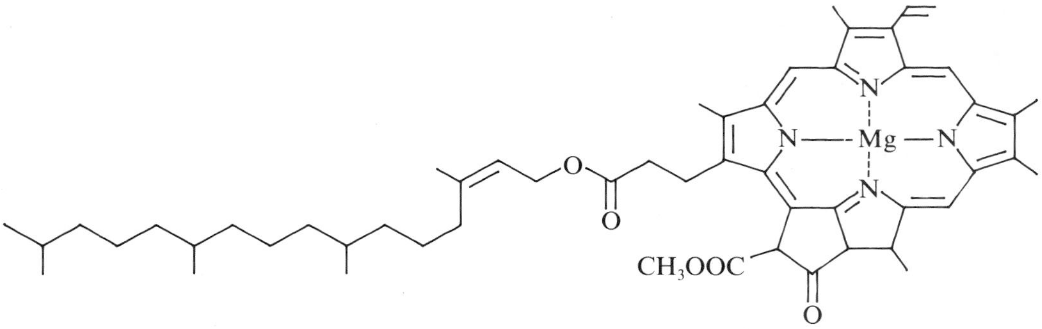

Chlorophyll-a

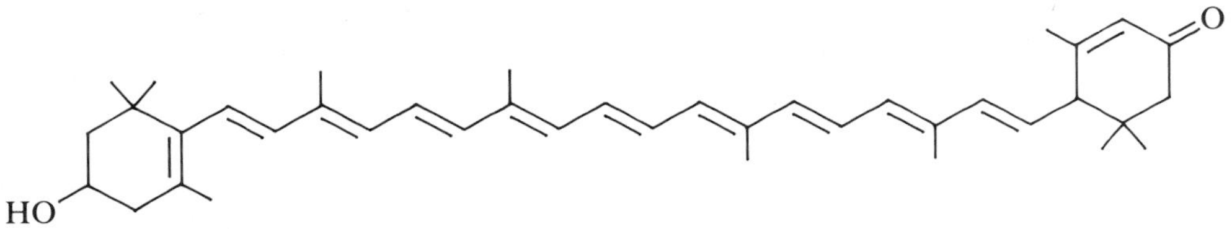

Lutein (a xanthophyll)

phylls. The bright orange β-carotene is the most important of these because it is transformed in the liver to vitamin A, which is required for night vision.

Cows eat fresh green grass that contains carotene, but they do not metabolize the carotene entirely and so it ends up in their milk. Butter made from this milk is therefore yellow. In the winter the silage cows eat does not contain carotene because it readily undergoes air oxidation and the butter made at that time is white. For some time an azo dye called Butter Yellow was added to winter butter to give it the accustomed color, but the dye was found to be a carcinogen. Now winter butter is colored with synthetic carotene, as is all margarine.

Lycopene, the red pigment of the tomato, is a C_{40}-carotenoid made up of eight isoprene units. β-Carotene, the yellow pigment of the carrot, is an isomer of lycopene in which the double bonds at $C_1—C_2$ and $C_1'—C_2'$ are replaced by bonds extending from C_1 to C_6 and from C_1' to C_6' to form rings. The chromophore in each case is a system of eleven all-*trans* conjugated double bonds; the closing of the two rings renders β-carotene less highly pigmented than lycopene.

Fresh tomato fruit contains about 96% of water, and R. Willstätter and H. R. Escher isolated from this source 20 mg of lycopene per kg of fruit. They then found a more convenient source in commercial tomato paste, from which seeds and skin have been eliminated and the water content reduced by evaporation in vacuum to a content of 26% solids, and isolated 150 mg of lycopene per kg of paste. The expected yield in the present experiment is 0.075 mg.

Lycopene and β-carotene from tomato paste and strained carrots

A jar of strained carrots sold as baby food serves as a convenient source of β-carotene. The German investigators isolated 1 g of β-carotene per kg of "dried" shredded carrots of unstated water content.

As an interesting variation, try extraction of lycopene from commercial catsup

The following procedure calls for dehydration of tomato or carrot paste with ethanol and extraction with dichloromethane, an efficient solvent for lipids.

Procedure

In a small mortar grind 2 g of green or brightly colored fall leaves (don't use ivy or waxy leaves) with 10 mL of ethanol, pour off the ethanol, which serves to break up and dehydrate the plant cells, and grind the leaves successively with three 1-mL portions of dichloromethane that are decanted or withdrawn with a Pasteur pipette and placed in a reaction tube. The pigments of interest are extracted by the dichloromethane. Alternatively place 0.5 g of carrot paste (baby food) or tomato paste in a reaction tube, stir and shake the paste with 3 mL of ethanol until the paste has a somewhat dry or fluffy appearance, remove the ethanol, and extract the dehydrated paste with three 1-mL portions of dichloromethane. Stir and shake the plant material with the solvent in order to extract as much of the pigments as possible.

These pigments are sensitive to light-catalyzed photochemical air oxidation. Work quickly, keep containers stoppered where possible, and protect solutions from undue exposure to light.

Fill the tube containing the dichloromethane extract from leaves or vegetable paste with a saturated sodium chloride solution and shake the mixture. Remove the aqueous layer and to the dichloromethane solution add anhydrous sodium sulfate until the drying agent no longer clumps together. Shake the mixture with the drying agent for about 5 min and then withdraw the solvent with a Pasteur pipette and place it in a reaction tube. Add to the solvent a few pieces of Drierite to complete the drying process. Gently stir the mixture for about 5 min, transfer the solvent to a tared reaction tube, wash off the drying agent with more solvent, and then evaporate the combined dichloromethane solutions under a stream of nitrogen or air while warming the tube in the hand or in a beaker of warm water. Carry out this evaporation in the hood.

Immediately cork the tube filled with nitrogen, determine its weight, and then add a drop or two of dichloromethane to dissolve the pigments for TLC analysis. Carry out the analysis without delay by spotting the mixture on a TLC plate about 1 cm from the bottom and 8 mm from the edge. Make one spot concentrated by repeatedly touching the plate, but ensure that the spot is as small as possible, less than 1.0 mm in diameter. The other spot can be of lower concentration. Develop the plate with 70 : 30 hexane : acetone. With other plates try cyclohexane and toluene as eluents and also hexane/ethanol mixtures of various compositions. The container in which the chromatography is carried out should be lined with filter paper that is wet with the solvent so the atmosphere in the container will be saturated with solvent vapor. On completion of elution mark the solvent front with a pencil and outline the colored spots. Examine the plate under the uv light. Are any new spots seen? Report colors and *Rf* values for all of your spots, and identify each as lycopene, carotene, chlorophyll, or xanthophyll.

Questions

1. Arrange the following in order of increasing *Rf* on thin-layer chromatography: acetic acid, acetaldehyde, 2-octanone, decane, and 1-butanol.
2. Why must the spot applied to a TLC plate be above the level of the developing solvent?
3. What will be the result of applying too much compound to a TLC plate?
4. Why is it necessary to run TLC in a closed container and to have the interior vapor saturated with the solvent?
5. What will be the appearance of a TLC plate if a solvent of too low polarity is used for the development? too high polarity?
6. A TLC plate showed two spots of *Rf* 0.25 and 0.26. The plate was removed from the developing chamber, dried carefully, and returned to the developing chamber. What would you expect to see after the second development was complete?

Column Chromatography: Cholesteryl Acetate and Fluorenone

10

Column chromatography is one of the most useful methods for the separation and purification of both solids and liquids when carrying out small-scale experiments. It becomes expensive and time-consuming, however, when more than about 10 g of material must be purified.

The application in the present experiment is typical: a reaction is carried out, it does not go to completion, and so column chromatography is used to separate the product from starting material, reagents, and by-products.

The theory of column chromatography is analogous to that of thin-layer chromatography. The most common adsorbents, silica gel and alumina, are the same ones used in TLC. The sample is dissolved in a small quantity of solvent (the eluent) and applied to the top of the column. The eluent, instead of rising by capillary action up a thin layer, flows down through the column filled with the adsorbent. Just as in TLC, there is an equilibrium established between the solute adsorbed on the silica gel or alumina and the eluting solvent flowing down through the column. Under some conditions the solute may be partitioning between an adsorbed solvent and the elution solvent; the partition coefficient, just as in the extraction process, determines the efficiency of separation in chromatography. The partition coefficient is determined by the solubility of the solute in the two phases, as was discussed in the extraction experiment (Chap. 8).

Three mutual interactions must be considered in column chromatography: the polarity of the sample, the polarity of the eluting solvent, and the activity of the adsorbent.

Adsorbent. A large number of adsorbents have been used for column chromatography—cellulose, sugar, starch, inorganic carbonates—but most separations employ alumina (Al_2O_3) and silica gel (SiO_2). Alumina comes in three forms: acidic, neutral, and basic. The neutral form of Brockmann Activity II or III, 150 mesh, is most commonly employed. The surface area of this alumina is about 150 m^2/g. Alumina as purchased will usually be Activity I, meaning it will strongly adsorb solutes. It must be deactivated by adding water, shaking and allowing the mixture to reach equilibrium over an hour or so. The amount of water needed to achieve certain activities is given in Table 10.1. The activity of the alumina on TLC plates is usually about III. Silica gel for column chromatography, 70–230 mesh, has a surface area of about 500 m^2/g and comes in only one activity.

TABLE 10.1 Alumina Activity

Brockmann Activity	I	II	III	IV	V
Percent by weight of water	0	3	6	10	15

Solvents. The elutropic series for a number of solvents is given in Table 10.2. The solvents are arranged in increasing polarity, with *n*-pentane the least polar. This is the order of ability of these solvents to dissolve polar organic compounds and to dislodge a polar substance adsorbed onto either silica gel or alumina, with *n*-pentane having the lowest solvent power.

TABLE 10.2 Elutropic Series for Solvents

n-Pentane (first)	Tetrahydrofuran
Petroleum ether	Dioxane
Cyclohexane	Ethyl acetate
Ligroin	2-Propanol
Carbon disulfide	Ethanol
Ethyl ether	Methanol
Dichloromethane	Acetic acid (last)

As a practical matter the following sequence of solvents is recommended in an investigation of unknown mixtures: elute first with petroleum ether; then ligroin, followed by ligroin containing 1%, 2%, 5%, 10%, 25%, and 50% ether; pure ether; ether and dichloromethane mixtures, followed by dichloromethane and methanol mixtures. A sudden change in solvent polarity will cause heat evolution as the alumina adsorbs the new solvent. This will cause undesirable vapor pockets and cracks in the column.

Solutes. The ease with which different classes of compounds elute from a column is indicated in Table 10.3. The order is similar to that of the eluting solvents—another application of "like dissolves like."

TABLE 10.3 Elution Order for Solutes

Alkanes (first)	Ketones
Alkenes	Aldehydes
Dienes	Amines
Aromatic hydrocarbons	Alcohols
Ethers	Phenols
Esters	Acids (last)

Sample and Column Size. In general the amount of alumina or silica gel used should weigh at least 30 times as much as the sample and the column when packed should have a height at least ten times the diameter. The density of silica gel is 0.4 g/mL and the density of alumina is 0.9 g/mL, so the optimum size for any column can be calculated.

On a Larger Scale. The appearance of the column, the method of packing, and the process of eluting and collecting fractions is exactly the same on a large scale. As noted, the process becomes expensive because of the quantities of adsorbent and eluting solvent used and it is slow because of the time necessary for the solvent to flow from the column. A relatively new process, flash chromatography, utilizes a very fine adsorbent and up to 50 psi pressure to force the solvent through the column "in a flash." Then one is faced with the problem of evaporating large volumes of solvent from many fractions. This can be done by simple distillation or more rapidly on a rotary evaporator (Fig. 10.1).

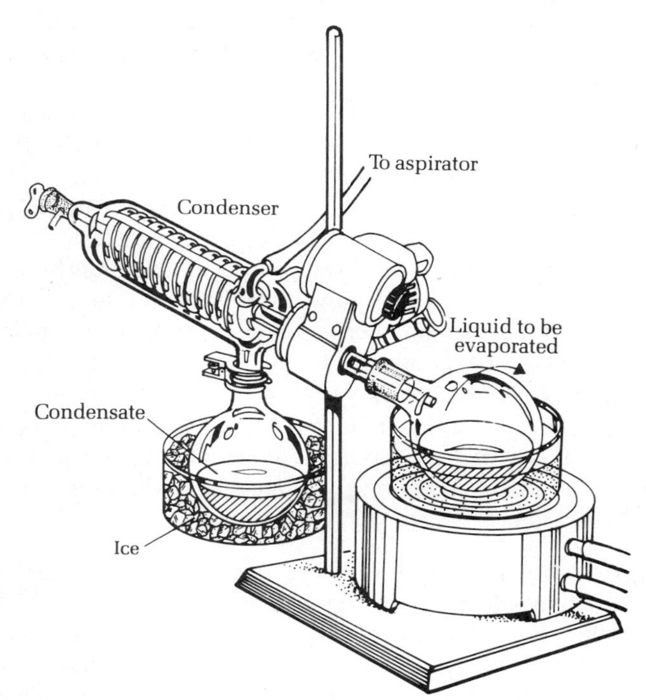

FIG. 10.1 Rotary evaporator. The rate of evaporation with this apparatus is very fast due to the thin film of liquid over the entire inner surface of the rotating flask that is heated under a vacuum. Foaming and bumping are also greatly reduced.

Packing the Chromatography Column

Uniform packing of the chromatography column is critical to the success of this technique. The sample is applied as a pure liquid or, if it is a solid, as a very concentrated solution in the solvent that will dissolve it best, regardless of polarity. As elution takes place, this narrow band of sample will separate into several bands corresponding to the number of components in the mixture and their relative polarities and molecular weights. It is essential that the components move through the column as a narrow horizontal band in order to come off the column in the least volume of solvent and not overlap with other components of the mixture. Therefore the column should be vertical and the packing should be perfectly uniform, without voids caused by air bubbles.

The preferred method for packing silica gel and alumina columns is the slurry method, whereby a slurry of the adsorbent and the first eluting solvent is made and poured into the column. When nothing is known about the mixture being separated, the column is prepared in petroleum ether, the least polar of the eluting solvents.

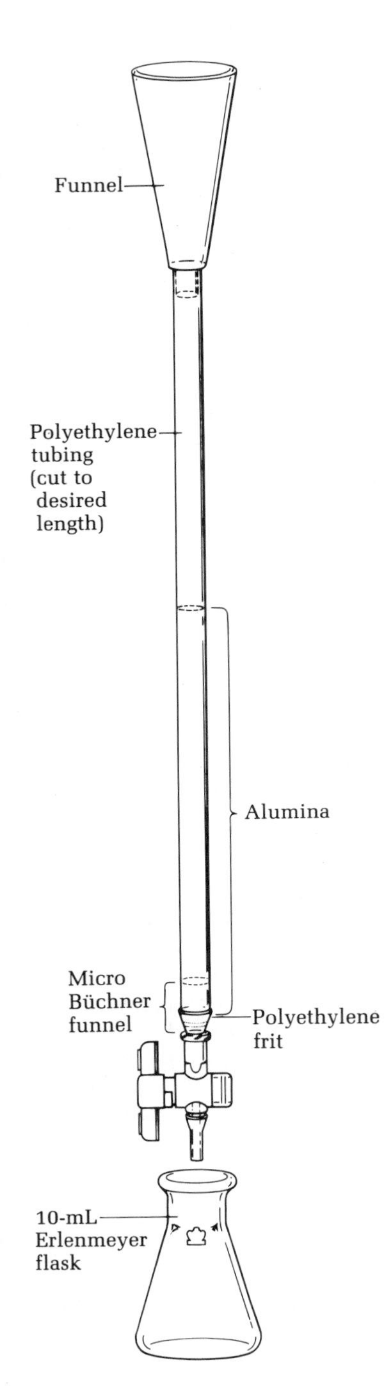

FIG. 10.2 Chromatographic column.

Procedure

Assemble the column as depicted in Fig. 10.2, being sure it is clamped in a vertical position. It is important to ascertain that the valve is made of polyethylene (translucent). If it is clear it may be a plastic such as polycarbonate, which will dissolve in organic solvents.

Close the valve and fill the column with petroleum ether to the bottom of the funnel. Prepare a slurry of 2.5 g of Activity III alumina in 8 mL of petroleum ether in a small beaker. This amount of alumina should fill the column to a height of 10 cm. Stir the slurry gently to get rid of air bubbles, and gently swirl, pour, and scrape the slurry into the funnel, which has a capacity of 10 mL. After some of the alumina has been added to the column, open the stopcock and allow solvent to drain slowly into an Erlenmeyer flask. Use this petroleum ether to rinse the beaker containing the alumina. As the alumina is being added tap the column with a glass rod or pencil so the alumina will pack tightly into the column. Continue to tap the column while cycling the petroleum ether through the column once more.

It is extremely important to **never let the column run dry at any time.** This will allow air to enter the column, which will result in uneven bands and poor separation.

Adding the Sample. The solvent is drained just to the surface of the alumina, which should be perfectly flat. The sample, in the most concentrated solution possible, is added to the column using a Pasteur pipette. It is allowed to run down the side of the column near the top of the alumina, without disturbing the alumina. The sample is run into the adsorbent, stopping when the solution reaches the top of the alumina, and then a small quantity of solvent is added without disturbing the surface. This is run down into the column and the process repeated until the sample is seen to be a band at the top of the column, at which point enough sand is poured onto the top of the column to form a layer about 3 mm thick. This will protect the surface of the alumina when more solvent is being added. Carefully add several milliliters of the first eluting solvent.

The eluent is collected in fractions of 5 mL each in ten 10-mL Erlenmeyer flasks or small vials.

Experiments

1. Acetylation of Cholesterol

Cholesterol + CH_3COCCH_3 (Acetic anhydride) ⟶ Cholesteryl acetate + CH_3COOH

Cholesterol is a solid alcohol; the average human body contains about 200 g distributed in brain, spinal cord, and nerve tissue and occasionally clogging the arteries and the gall bladder (see Chapter 22 for background and procedure for isolating cholesterol from human gallstones).

Alcohols combine with carboxylic acids in the presence of an acid catalyst to form esters and a mole of water. Because this is an equilibrium reaction, it does not go to completion unless some method, such as removal of the water, forces the equilibrium toward the side of the ester:

$$ROH + R'\overset{O}{\overset{\|}{C}}OH \rightleftharpoons RO\overset{O}{\overset{\|}{C}}R' + H_2O$$

There are a number of alternative methods for the preparation of esters. In the present experiment cholesterol is dissolved in acetic acid and allowed to react with acetic anhydride to form the ester cholesteryl acetate. The reaction does not take place rapidly and consequently does not go to completion under the conditions of this experiment. Thus when the reaction is over, both unreacted cholesterol and the product, cholesteryl acetate, are present. Separating these by fractional crystallization would be extremely difficult; but because they differ in polarity (the hydroxyl group of the cholesterol is the more strongly adsorbed on alumina), they are easily separated by column chromatography. Both molecules are colorless and hence cannot be detected visually. Each fraction should be sampled for thin-layer chromatography. In that way not only the presence but also the purity of each fraction can be accessed. It is also possible to put a drop of each fraction on a watch glass and evaporate it to see if the fraction contains product.

Procedure

In a reaction tube add 0.5 mL of acetic acid to 50 mg of cholesterol, which can be the material isolated from human gallstones. The initial thin slurry may set into a stiff paste of the molecular complex consisting of one molecule of cholesterol and one of acetic acid. Add 0.10 mL of acetic anhydride and a boiling chip and gently reflux the reaction mixture on a hot sand bath for no more than 30 min.

Cool the mixture, add 2 mL of water, and extract the product with three 2-mL portions of ether that are placed in a 4-in. test tube. Wash the ether extracts in the test tube with two 2-mL portions of water and one 2.5-mL portion of 10% sodium hydroxide (these three washes remove the acetic acid) and dry the ether by shaking it with 2.5 mL of saturated sodium chloride solution. Then complete the drying by adding enough anhydrous sodium sulfate to the solution so that it does not clump together.

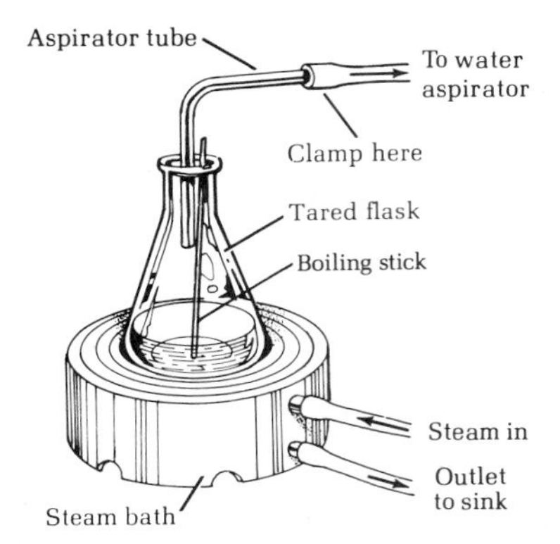

FIG. 10.3 Aspirator tube in use. A boiling stick may be necessary to promote even boiling.

Shake the ether solution with the drying agent for 10 min and then transfer it in portions to a tared reaction tube and evaporate to dryness. Use a drop of this ether solution to spot a TLC plate for later analysis. If the crude material weighs more than the theoretical weight, you will know it is not dry or it contains acetic acid, which can be detected by its odor. Dissolve this crude cholesteryl acetate in the minimum quantity of ether and apply it to the top of the chromatography column. Some skill is required to prevent the solution from dribbling from the pipette, a problem that can be solved by stuffing a tiny piece of cotton into the pipette tip. Push the cotton from the inside down to the tip of the pipette using a copper wire.

Elute the column with petroleum ether and collect two 5-mL fractions in tared 10-mL Erlenmeyer flasks or other suitable containers. Add a boiling stick to each flask and evaporate the solvent under an aspirator tube on a steam bath or sand bath. If the flask appears empty it can be used to collect later fractions. Lower the solvent layer to the top of the sand and elute with 25 mL of 70 : 30 ligroin : diethyl ether, collecting five 5-mL fractions. Evaporate the ligroin under an aspirator tube (Fig. 10.3). The last traces of ligroin can be removed using reduced pressure as shown in Fig. 10.4. Follow the 70 : 30 mixture with 20 mL of 50 : 50 ligroin : ether, collecting four 5-mL fractions. Save any flask that has any visible residue. Analyze the original mixture and each fraction by thin-layer chromatography on silica gel plates (Eastman #13181) using 1 : 1 ether : ligroin to develop the plates and either ultraviolet light or iodine vapor to visualize the spots.

FIG. 10.4 Drying a solid by reduced air pressure.

Cholesteryl acetate (mp 115°C) and cholesterol (mp 149°C) should appear, respectively, in early and late fractions with a few empty fractions (no residue) in between. If so, combine consecutive fractions of early and late material and determine the weights and melting points. Calculate the percentage of the acetylated material compared to the total recovered, and calculate the percentage yield from cholesterol.

2. Fluorene and Fluorenone

$Na_2Cr_2O_7$ / CH_3COOH

Fluorene
mp 114°C
MW 166.22

Fluorenone
mp 83°C
MW 180.21

The 9-position of fluorene is unusually reactive for a hydrocarbon. The protons on this carbon atom are acidic by virtue of being doubly benzylic and consequently this carbon can be oxidized by several reagents, including elemental oxygen. In the present experiment the very powerful and versatile oxidizing agent Cr(VI), in the form of chromium trioxide, is used to carry out this oxidation. Cr(VI) in a variety of other forms is used to carry out about a dozen oxidation reactions in this text. The *dust* of Cr(VI) salts is reported to be a carcinogen, so avoid breathing it.

Procedure

In a reaction tube dissolve 50 mg of fluorene in 0.25 mL of acetic acid by heating and add this hot solution to a solution of 0.15 g of sodium dichromate dihydrate in 0.5 mL of acetic acid. Heat the reaction mixture to 80°C for 15 min in a hot water bath, then cool it and add 1.5 mL of water. Stir the mixture for 2 min and then filter the mixture on the Hirsch funnel. Wash the product well with water and press out as much water as possible. Return the product to the reaction tube, add 2 mL of ether, and add anhydrous sodium sulfate until it no longer clumps together. Cork and shake the tube and allow the product to dry for 5 or 10 min before evaporating the ether in another tared reaction tube. Use ether to wash off the drying agent and to complete the transfer of product. Use this ether solution to spot a TLC plate. This crude mixture of fluorene and fluorenone will be separated by column chromatography.

Column Chromatography of the Fluorene–Fluorenone Mixture. Prepare a chromatographic column exactly as described above for the separation of cholesteryl acetate and cholesterol but use ligroin instead of petroleum ether as the initial solvent and to prepare the column. Dissolve the crude mixture of fluorene and fluorenone in 0.2 mL of warm toluene and add this to the surface of the alumina. Be sure to add the sample as a solution; should any sample crystallize, add a drop more toluene. Run the ligroin down to the surface of the alumina, add a few drops more ligroin, and repeat the process until the sample is seen as a narrow band at the top of the column. Carefully add a 3-mm layer of sand, then fill the column with ligroin, and collect 5-mL fractions in tared 10-mL Erlenmeyer flasks. Sample each flask for thin-layer chromatography and evaporate each to dryness. Final drying can be done under vacuum using the technique shown in Fig. 10.3. You are to decide when all of the product has been eluted from the column. The thin-layer plates can be developed using 20% dichloromethane in ligroin. Combine fractions that are identical and determine the melting points of the two substances.

Questions

1. What would be the effect of collecting larger fractions when carrying out either of the experiments described?
2. What would have been the result if a large quantity of petroleum ether alone were used as the eluent in either of the experiments described?
3. Once the chromatographic column has been prepared, why is it important to allow the level of the liquid in the column to drop to the level of the alumina before applying the solution of the compound to be separated?
4. A chemist started to carry out column chromatography on a Friday afternoon, got to the point at which the two compounds being separated were about three-fourths of the way down the column, and then returned on Monday to find the compounds had come off the column as a mixture. Speculate on the reason for this. The column had not run dry over the weekend.

Alkenes from Alcohols: Cyclohexene from Cyclohexanol

11

PRELAB EXERCISE: *Prepare a detailed flow sheet for the preparation of cyclohexene, indicating at each step which layer contains the desired product.*

$$\text{Cyclohexanol} \xrightarrow{H_3PO_4} \text{Cyclohexene} + H_2O$$

Cyclohexanol
mp 25°C, bp 161°C
den 0.96, MW 100.16

Cyclohexene
bp 83°C
den 0.81, MW 82.14

Dehydration of cyclohexanol to cyclohexene can be accomplished by pyrolysis of the cyclic secondary alcohol with an acid catalyst at a moderate temperature or by distillation over alumina or silica gel. The procedure selected for this experiment involves catalysis by phosphoric acid; sulfuric acid is no more efficient, causes charring, and gives rise to sulfur dioxide. When a mixture of cyclohexanol and phosphoric acid is heated in a flask equipped with a fractionating column, the formation of water is soon evident. On further heating, the water and the cyclohexene formed distil together by the principle of steam distillation, and any high-boiling cyclohexanol that may volatilize is returned to the flask. However, after dehydration is complete and the bulk of the product has distilled, the column remains saturated with water–cyclohexene that merely refluxes and does not distil. Hence, for recovery of otherwise lost reaction product, a chaser solvent is added and distillation is continued. A suitable chaser solvent is water-immiscible technical xylene, boiling point about 140°C; as it steam-distils it carries over the more volatile cyclohexene. When the total water-insoluble layer is separated, dried, and redistilled through the dried column the chaser again drives the cyclohexene from the column; the difference in boiling points is such that a sharp separation is possible. The holdup in the metal sponge–packed column is so great that if a chaser solvent is not used in the procedure the yield will be only about one-third that reported in the literature.

The mechanism of this reaction involves initial rapid protonation of the hydroxyl group by the phosphoric acid:

$$+ H_3PO_4 \rightleftharpoons + H_2PO_4^-$$

This is followed by loss of water to give the unstable secondary carbonium ion, which quickly loses a proton to water or the conjugate acid to give the alkene:

$$\longrightarrow \quad \longrightarrow \quad + H_3O^+$$

Experiment

Preparation of Cyclohexene

Handle phosphoric acid with care

Introduce 2.0 g of cyclohexanol, 0.5 mL of 85% phosphoric acid, and a boiling chip into a 5-mL round-bottomed long-necked flask. Add a stainless steel sponge to the neck of the flask and shake to mix the layers. Note the evolution of heat. Use the arrangement for fractional distillation as shown in Fig. 11.1. Note that the bulb of the thermometer must be *completely* below the side arm of the distilling head. Wrap the fractionating column and distilling head with glass wool. (See the experiment of fractional distillation for details of this technique.)

Use of a chaser

Heat the mixture gently on the sand bath and then distill until the residue in the flask has a volume of about 0.5–1.0 mL and very little distillate is being formed; note the temperature range. Let the assembly cool a little after removing it from the sand, remove the thermometer briefly, and add 2 mL of xylene (the chaser solvent) into the top of the column using a Pasteur pipette. Note the amount of the upper layer in the boiling flask and distill again until the volume of the layer has been reduced by about half. Transfer the contents of the vial into a reaction tube and rinse with a little xylene. Use xylene for rinsing in subsequent operations. Wash the mixture with an equal volume of saturated sodium chloride solution, remove the aqueous layer, and then add

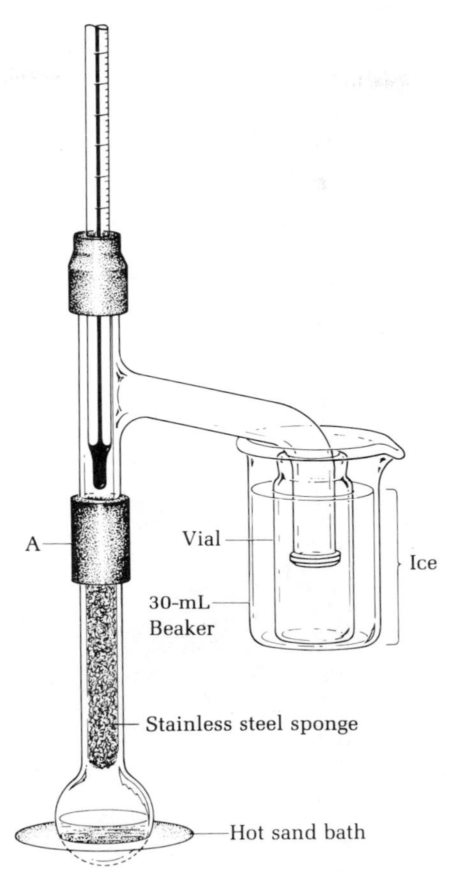

FIG. 11.1 Apparatus for the synthesis of cyclohexene from cyclohexanol. The column is packed with a stainless steel sponge.

sufficient anhydrous sodium sulfate to the reaction tube so that it does not clump together. Shake the solution with the drying agent and let it dry for at least 5 min. While this is taking place clean the distilling apparatus first with water, then ethanol and finally a little acetone. It is absolutely essential that the apparatus be completely dry; otherwise the product will be contaminated with whatever solvent is left in the apparatus. Transfer the dry cyclohexene solution to the distilling flask, add a boiling chip, note the atmospheric pressure, and distill the product. At the moment the temperature starts to rise above the plateau at which the product distills (83°C), stop the distillation to avoid contamination of the cyclohexene with xylene. A typical yield of this volatile alkene is about 1 g. Report your yield in grams and your percent yield. Run the infrared spectrum and interpret it for purity. Look for peaks due to starting material and xylene. Gas chromatography is especially useful in the analysis of this compound because the expected impurities differ markedly in boiling point from the product.

Note: Cyclohexene is extremely flammable: keep away from open flame

On a Larger Scale. The apparatus shown in Fig. 11.2 is used to collect cyclohexene. The distillation column is a larger version of the one used in this experiment and the condenser is water cooled. Even so the volatile product is collected in an ice-cooled receiver.

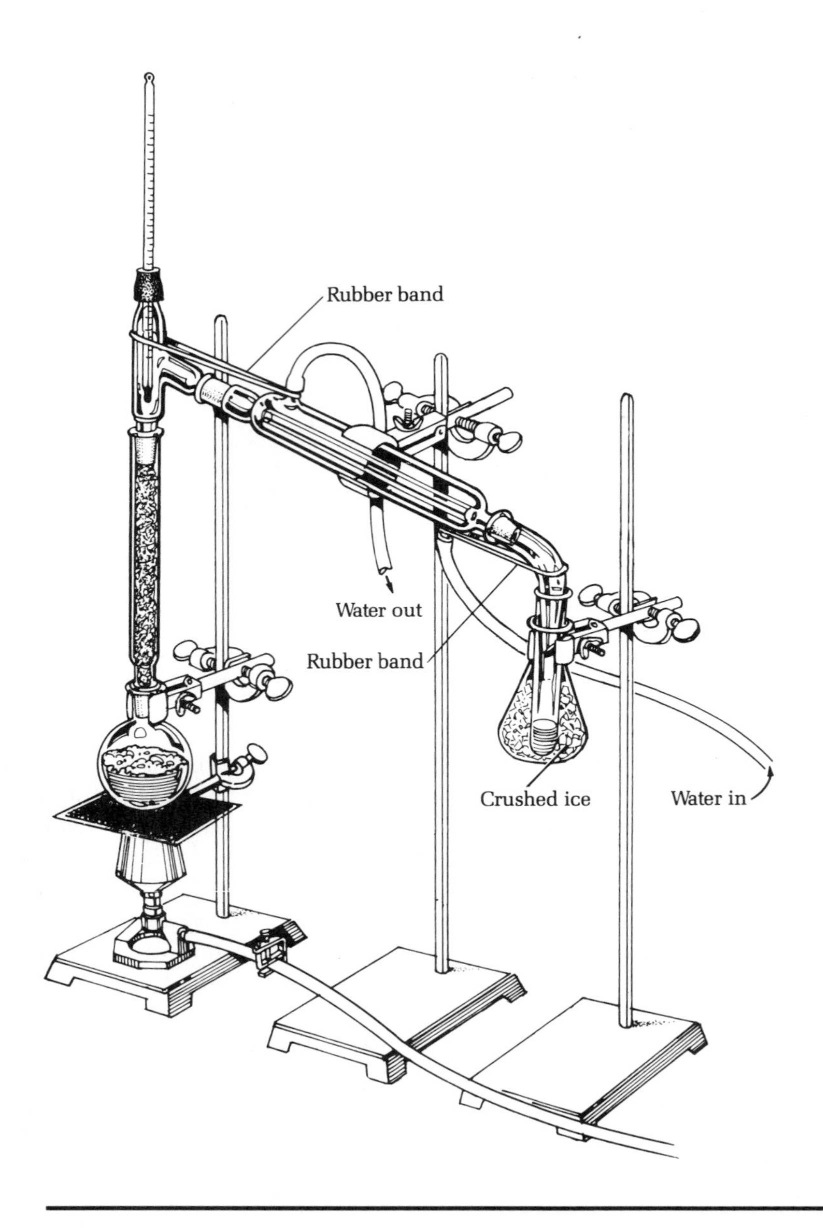

FIG. 11.2 Fractionation into an ice-cooled receiver.

Yield Calculations

Rarely do organic reactions give 100% yields of one pure product; an important objective of every experiment is a high yield of the desired product. The present experiment uses 2 g of starting cyclohexanol. This corresponds to 0.02 mole because the molecular weight of cyclohexanol is 100:

$$\frac{2.0\text{ g}}{100.16\text{ g/mole}} = 0.02\text{ mole} = 20\text{ millimoles}$$

If the reaction gave a 100% yield of cyclohexene then 0.02 mole of the alkene would be produced. The weight of 0.02 mole of cyclohexene is

$$0.02\text{ mole} \times 82.14\text{ g/mole} = 1.64\text{ g}$$

We call the 1.64 g the *theoretical yield*; it could be obtained if the reaction proceeded perfectly. If the actual yield is only 0.82 g, then the reaction would be said to give a 50% yield:

$$\frac{0.82\text{ g}}{1.64\text{ g}} \times 100 = 50\%$$

Typical student yields are included throughout this text. These are *not* theoretical yields; they suggest what an average or above-average student can expect to obtain for the experiment.

Questions

1. Assign the peaks in the ^{1}H nmr spectrum of cyclohexene (Fig. 11.3) to specific groups of protons on the molecule.

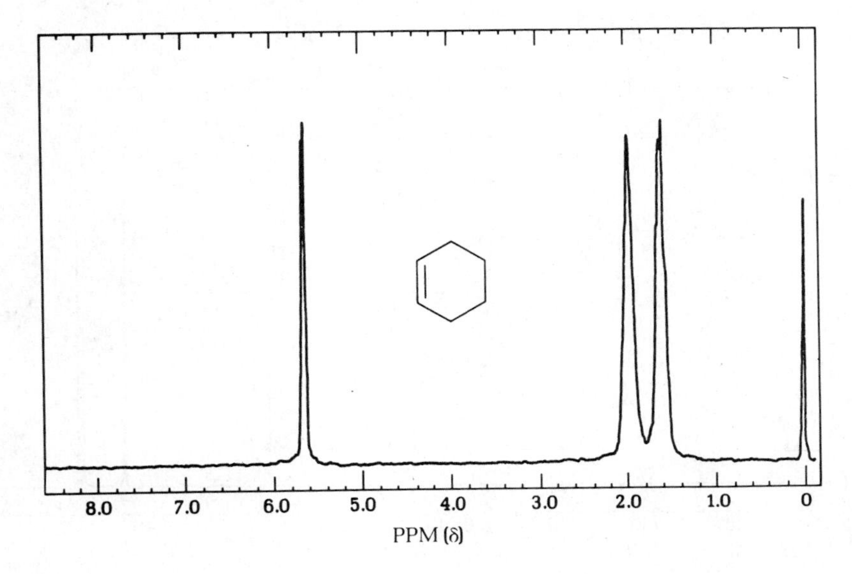

FIG. 11.3 ^{1}H nmr spectrum of cyclohexene.

2. What product(s) would be obtained by the dehydration of 1-methyl-1-hexanol? Of 2-methyl-1-hexanol?

3. Mixing cyclohexanol with phosphoric acid is an exothermic process while the production of cyclohexene is endothermic. Referring to the two chemical reactions on p. 128, construct an energy diagram showing the course of this reaction. Label the diagram with the starting alcohol, the oxonium ion (the protonated alcohol), the carbocation, and the product.

4. Is it reasonable that the ^{1}H spectrum of cyclohexene (Fig. 11.3) should closely resemble the ^{13}C spectrum (Fig. 11.4)?

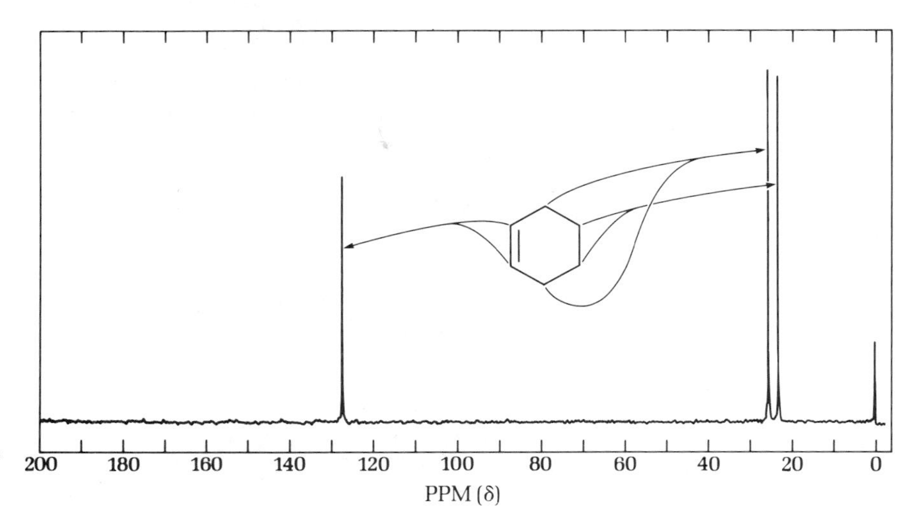

FIG. 11.4 ^{13}C nmr spectrum of cyclohexene.

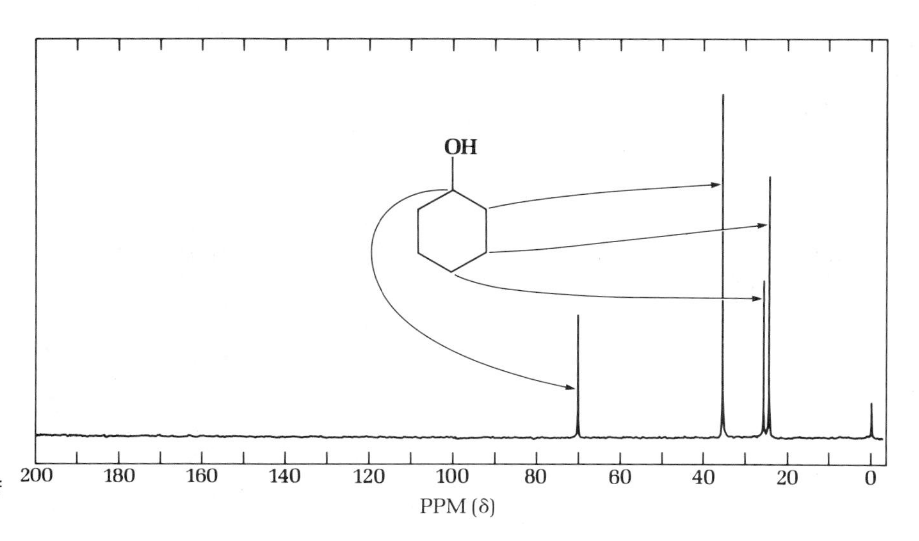

FIG. 11.5 ^{13}C nmr spectrum of cyclohexanol.

Butenes from Butanol: Analysis of a Mixture by Gas Chromatography

12

PRELAB EXERCISE: *If the dehydration of 2-methyl-2-butanol occurred on a purely statistical basis, what would be the relative proportions of 2-methyl-1-butene and 2-methyl-2-butene?*

Gas chromatography (gc), also called vapor phase chromatography (vpc) and gas-liquid chromatography (glc), is a means of separating volatile mixtures, the components of which may differ in boiling points by only a few tenths of a degree. The gc process is similar to fractional distillation, but instead of a glass column 25 cm long packed with a stainless-steel sponge, the gc column used is a 3–10 m long coiled metal tube (dia 6 mm), packed with ground firebrick. The firebrick serves as an inert support for a very high-boiling liquid (essentially nonvolatile), such as silicone oil and low-molecular-weight polymers like Carbowax. These are the *liquids* of gas-*liquid* chromatography and are referred to as the stationary phase. The sample (1–25 *micro*liters) is injected through a silicone rubber septum into the column, which is being swept with a current of helium (ca. 200 mL/min). The sample first dissolves in the high-boiling liquid phase and then the more volatile components of the sample evaporate from the liquid and pass into the gas phase. Helium, the carrier gas, carries these components along the column a short distance where they again dissolve in the liquid phase before reevaporation (Fig. 12.1)

Separation of mixtures of volatile samples

High-boiling liquid phase

Eventually the carrier gas, which is a very good thermal conductor, and the sample reach the detector, an electrically heated tungsten wire. As long as pure helium is flowing over the detector the temperature of the wire is rather low and the wire has a high resistance to the flow of electric current. Organic molecules have lower heat capacities than helium. Hence, when a mixture of helium and an organic sample flows over the detector wire it is cooled less efficiently and heats up. When the wire is hot its electrical resistance becomes lower and offers less resistance to the flow of current. The detector wire actually is one leg of a Wheatstone bridge, connected to a chart recorder that records, as a peak, the amount of current necessary to again balance the bridge. The record produced by the recorder is called a **chromatogram**.

The thermal conductivity detector

A gas chromatogram is simply a recording of current versus time (which is equivalent to a certain volume of helium) (Fig. 12.2). In the illustration the smaller peak from component A has the shorter retention time, T_1, and

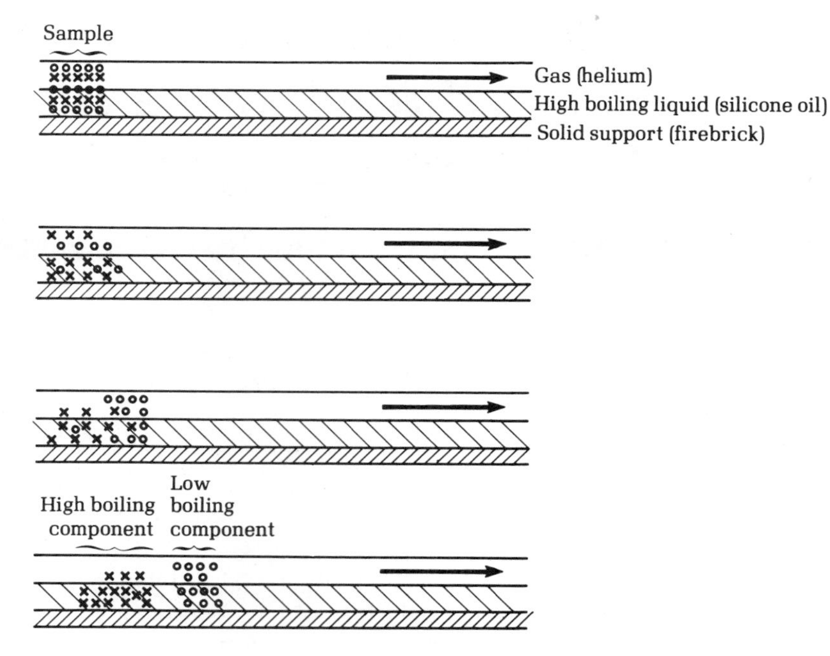

FIG. 12.1 Gas chromatography: Diagrammatic partition between gas and liquid phases.

The number and relative amounts of components in a mixture

so is a more volatile substance than component B (if the stationary phase is an inert liquid such as silicone oil). The areas under the two peaks are directly proportional to the amounts of A and B in the mixture. The retention time of a given component is a function of the column temperature, the helium flow rate, and the nature of the stationary phase. Hundreds of stationary phases are available; picking the correct one to carry out a given analysis is somewhat of an art, but widely used stationary phases are silicone oil and silicone rubber, both of which can be used at temperatures up to 300°C and separate mixtures on the basis of boiling point differences of the mixture's components. More specialized stationary phases will, for instance, allow alkanes to pass through readily (short retention time) while holding back (long retention time) alcohols by hydrogen bonding to the liquid phase.

The gas chromatograph

A diagram of a typical gas chromatograph is shown in Fig. 12.3. The carrier gas, usually helium, enters the chromatograph at ca. 60 lb/sq in. The

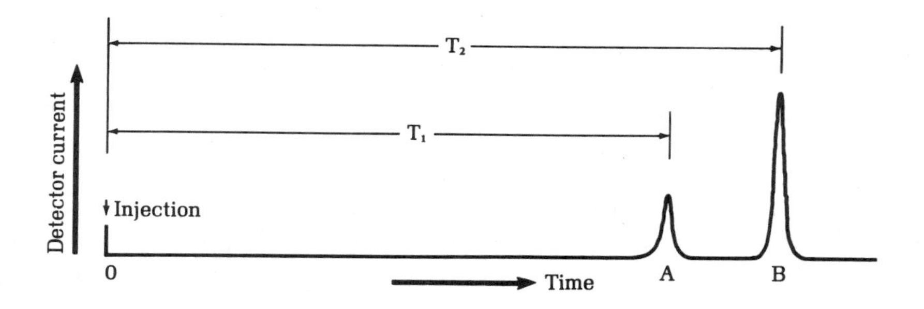

FIG. 12.2 Gas chromatogram.

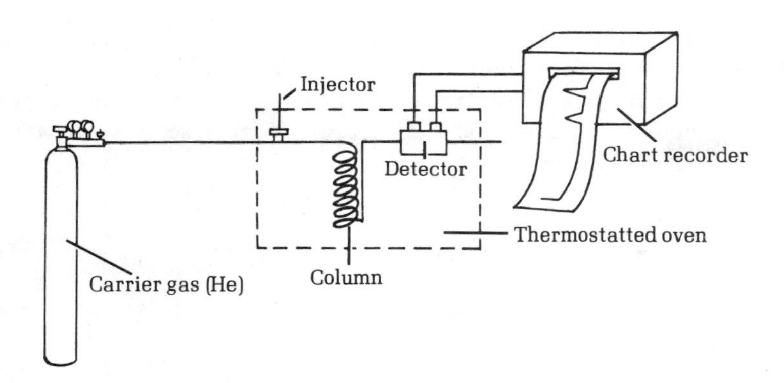

FIG. 12.3 Diagram of a gas chromatograph.

sample (1 to 25 microliters) is injected through a rubber septum using a small hypodermic syringe (Fig. 12.4). Handle the syringe with care. The syringe needles and plungers are thin and delicate; take care not to bend either one. Be wary of the injection port. It is very hot. The sample immediately passes through the column and then the detector. Injector, column, and detector are all enclosed in a thermostatted oven, which can be maintained at any temperature up to 300°C. In this way samples that would not volatilize enough at room temperature can be analyzed.

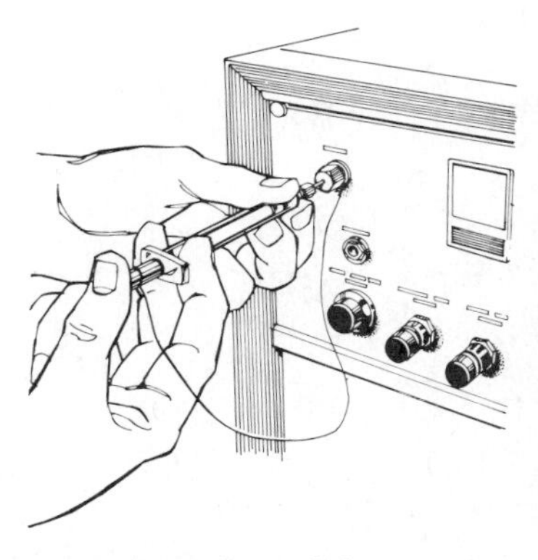

FIG. 12.4 One of the commercially available chromatographs.

Gas chromatography determines the number of components and their relative amounts in a very small sample. The small sample size is an advantage in many cases, but it precludes isolating the separated components. Some specialized chromatographs can separate samples as large as 0.5 mL per injection and automatically collect each fraction in a separate container. At the other extreme gas chromatographs equipped with flame ionization detectors can detect micrograms of sample, such as traces of pesticides in food or of drugs in blood and urine. Clearly a gas chromatogram gives little information about the chemical nature of the sample being detected. However, it is sometimes possible to collect enough sample at the exit port of the chromatograph to obtain an infrared spectrum. As the peak for the compound of interest appears on the chart paper, a 2-mm dia glass tube, 3 in. long and packed with glass wool, is inserted into the rubber septum at the exit port. The sample, if it is not too volatile, will condense in the cold glass tube. Subsequently, the sample is washed out with a drop or two of solvent and an infrared spectrum obtained.

Collecting a sample for an infrared spectrum

Dehydration

1. 2-Methyl-1-butene and 2-Methyl-2-butene[1]

The dilute sulfuric acid catalyzed dehydration of 2-methyl-2-butanol (*t*-amyl alcohol) proceeds readily to give a mixture of alkenes that can be analyzed

1. C. W. Schimelpfenig. *J. Chem. Ed.*, **39**, 310 (1962).

by gas chromatography. The mechanism of this reaction involves the intermediate formation of the relatively stable tertiary carbocation followed by loss of a proton either from a primary carbon atom to give the terminal olefin, 2-methyl-1-butene, or from a secondary carbon to give 2-methyl-2-butene.

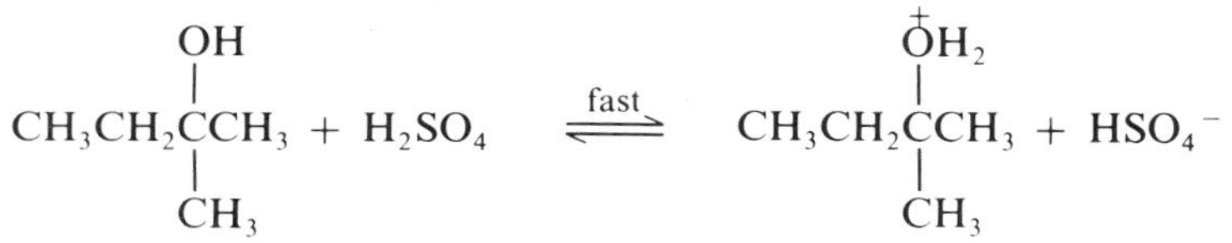

$$CH_3CH_2C(OH)(CH_3)CH_3 + H_2SO_4 \xrightleftharpoons{\text{fast}} CH_3CH_2C(\overset{+}{O}H_2)(CH_3)CH_3 + HSO_4^-$$

2-Methyl-2-butanol
bp 102°C
den 0.805
MW 88.15

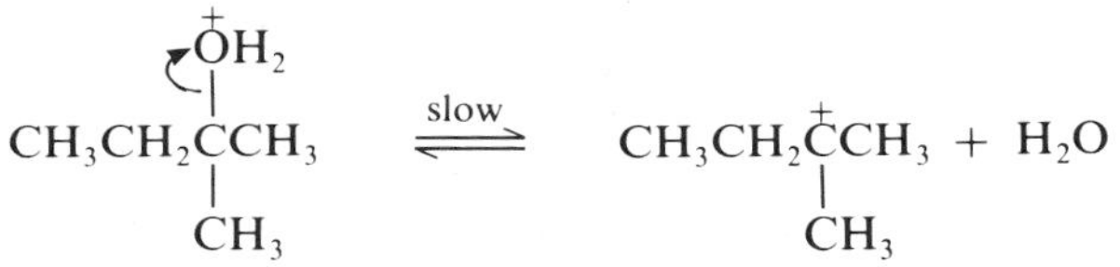

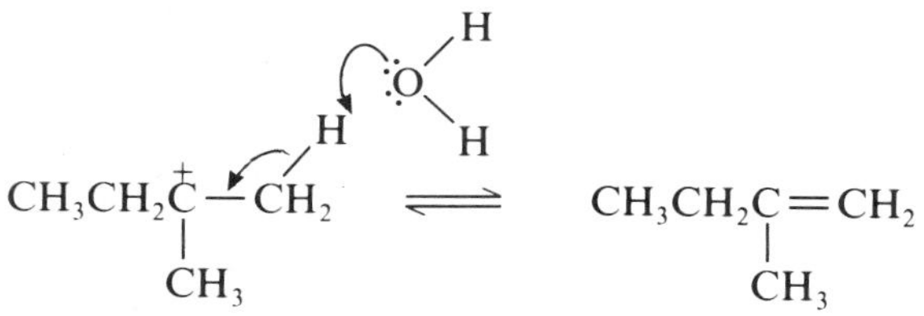

$$CH_3CH_2C(\overset{+}{O}H_2)(CH_3)CH_3 \xrightleftharpoons{\text{slow}} CH_3CH_2\overset{+}{C}(CH_3)CH_3 + H_2O$$

$$CH_3CH_2\overset{+}{C}(CH_3){-}CH_2{-}H \;\; + :\!OH_2 \rightleftharpoons CH_3CH_2C(CH_3){=}CH_2$$

2-Methyl-1-butene
bp 31.16°, den 0.662
MW 70.14

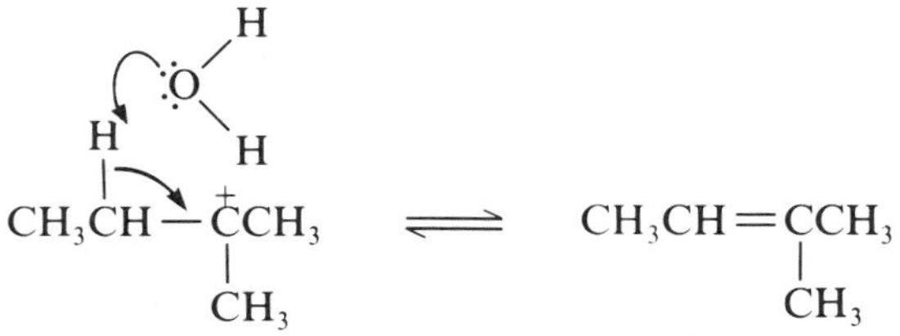

$$CH_3CH(H){-}\overset{+}{C}(CH_3)CH_3 \;\; + :\!OH_2 \rightleftharpoons CH_3CH{=}C(CH_3)CH_3$$

2-Methyl-2-butene
bp 38.57°C
den 0.662
MW 70.14

2-Methyl-2-butanol can also be dehydrated in high yield using iodine as a catalyst:

$$CH_3CH_2\underset{\displaystyle CH_3}{\overset{\displaystyle OH}{\overset{|}{\underset{|}{C}}}}CH_3 \xrightarrow{I_2\,(trace)} CH_3CH_2\underset{\displaystyle CH_3}{\overset{\displaystyle OI}{\overset{|}{\underset{|}{C}}}}CH_3 \xrightarrow{-HOI}$$

$$CH_3CH_2\underset{\displaystyle CH_3}{\underset{|}{C}}{=}CH_2 + CH_3CH{=}\underset{\displaystyle CH_3}{\underset{|}{C}}CH_3 + HI + HOI$$

$$HI + HOI \longrightarrow H_2O + I_2$$

Each step of this E_1 elimination reaction is reversible and thus the reaction is driven to completion by removing one of the products, the alkene. In these reactions several alkenes can be produced. The Saytzeff rule states that the more substituted alkene is the more stable and thus the one formed in larger amount. And the *trans* isomer is more stable than the *cis* isomer. With this information it should be possible to deduce which peaks on the gas chromatogram correspond to a given alkene and to predict the ratios of the products.

In analyzing your results from this experiment, consider the fact that the carbocation can lose any of six primary hydrogen atoms but only two secondary hydrogen atoms to give the product olefins.

Inject a few microliters of product into a gas chromatograph maintained at room temperature and equipped with a 6-mm dia × 3 m column packed with 10% SE-30 silicone rubber on Chromosorb-W or a similar inert packing. Mark the chart paper at the time of injection. In a few minutes two peaks should appear. After using the chromatograph, turn off the recorder and cap the pen. Make no unauthorized adjustments on the gas chromatograph. From your knowledge of the mechanisms of dehydration of secondary alcohols, which olefin should predominate? Does this agree with the boiling points? (In general, the compound with the shorter retention time has the lower boiling point.) Measure the relative areas under the peaks. One way to perform this integration is to cut out the peaks with scissors and weigh the two pieces of paper separately on an analytical balance. Although time-consuming, this method gives very precise results. If the peaks are symmetrical, their areas can be approximated by simply multiplying the height of the peak by its width at half-height.

Be sure the helium cylinder is attached to a bench or wall

Procedure

Into a 5-mL long-necked round-bottomed flask place 1 mL of water and then add dropwise with thorough mixing 0.5 mL of concentrated sulfuric acid.

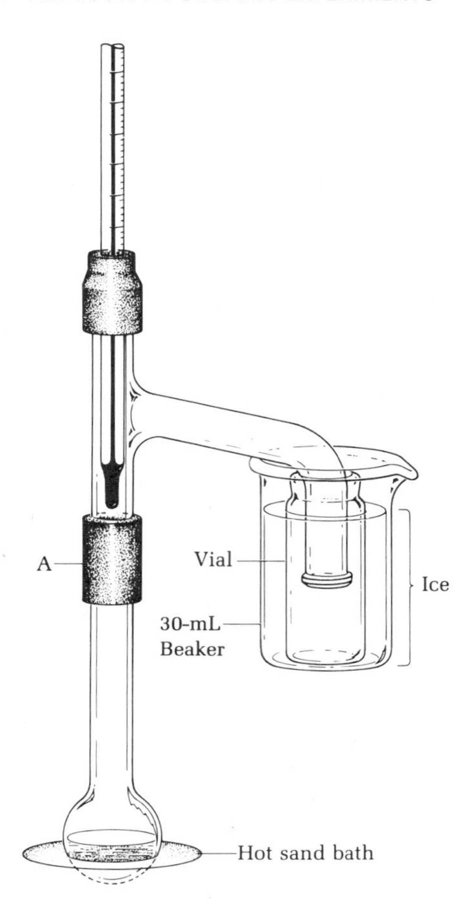

FIG. 12.5 Apparatus for dehydration of 2-methyl-2-butanol. Clamp at A with miniature 3-prong clamp. Beaker can be clamped with large 3-prong clamp.

Cool the hot solution in an ice bath and add to the cold solution 1.00 mL (0.80 g) of 2-methyl-2-butanol. Mix the reactants thoroughly, add a boiling stone, and set up the apparatus for simple distillation as shown in Fig. 12.5. Hold the flask in a towel while connecting the apparatus to guard against spills or flask breakage. Cool the receiver (a vial) in ice to avoid losing the very volatile products. Warm the flask on the sand bath to start the reaction and distill the products over the temperature range 30–45°C. After all of the products have distilled the rate of distillation will decrease markedly. Cease distillation at this point. After this the temperature registered on the thermometer will rise rapidly as water and sulfuric acid begin to distill.

To the distillate add 0.3 mL of *cold* 10% sodium hydroxide solution to neutralize any sulfurous acid present, draw off the aqueous layer, and dry the product over anhydrous sodium sulfate, adding the drying agent in small quantities until it no longer clumps together. Keep the vial cold. While the product is drying rinse out the distillation apparatus with water, ethanol, and then a small amount of acetone and *dry* it thoroughly, by drawing air through it using the aspirator. If this is not done carefully the products will be contaminated with acetone. Carefully transfer the dry mixture of butenes to the distilling apparatus using a Pasteur pipette, add a boiling chip to the now dry products, and distill into a tared, cold receiver. Collect the portion boiling up to 43°C. The yield reported in the literature is 84% when the reaction is carried out on a much larger scale. Weigh the product, calculate the yield, and analyze it by gas chromatography. The average yield on a small scale is 50%. 2-Methyl-1-butene boils at 31.2°C and 2-methyl-2-butene boils at 38.6°C.,

2. 1-Butene and *cis*- and *trans*-2-Butene[2]

In a 10 × 100-mm reaction tube place 0.10 mL (81 mg) of 2-butanol and, with cooling, 0.05 mL of concentrated sulfuric acid. Mix the reactants well and add a boiling chip. Stirring of the reaction mixture is not necessary because it is homogeneous. Insert a syringe needle through the bottom of a septum and place this on the reaction tube and connect the needle to a piece of polyethylene tubing that leads into the mouth of the distilling column, which is capped with a septum. The distilling column/collection tube is filled with water, capped with a finger, and inverted in the beaker of water and clamped for the collection of butene by the downward displacement of water (Fig. 12.6).

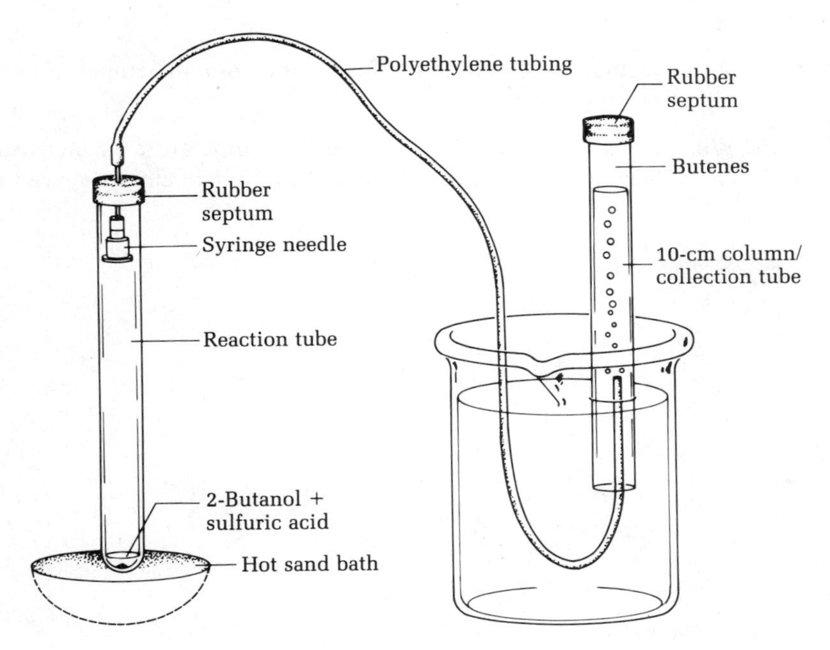

FIG. 12.6 Apparatus for dehydration of 2-butanol. Butenes are collected by the downward displacement of water.

2. G. K. Helmkamp and H. W. Johnson, Jr. *Selected Experiments in Organic Chemistry*, 3rd Ed., Freeman, New York, 1983, p. 99.

Lower the reaction tube into a hot (ca. 100°C) sand bath and increase the heat slowly to complete the reaction. Collect a few mL of butene, *remove the polyethylene tube from the water bath*, and cap the open end of the collection tube with another septum while it is under water. If the polyethylene tube is not removed from the water bath, water will be sucked back into the reaction tube when it cools. Use about 0.5 mL of the butene mixture to carry out the gas chromatographic analysis.

Questions

1. Write the structure of the three olefins produced by the dehydration of 3-methyl-3-pentanol.
2. When 2-methylpropene is bubbled into dilute sulfuric acid at room temperature, it appears to dissolve. What new substance has been formed?
3. A student wished to prepare ethylene gas by dehydration of ethanol at 140°C using sulfuric acid as the dehydrating agent. A low-boiling liquid was obtained instead of ethylene. What was the liquid and how might the reaction conditions be changed to give ethylene?
4. What would be the effect of increasing the carrier gas flow rate on the retention time?
5. What would be the effect of raising the column temperature on the retention time?
6. What would be the effect of raising the temperature or increasing the carrier gas flow rate on the ability to resolve two closely spaced peaks?

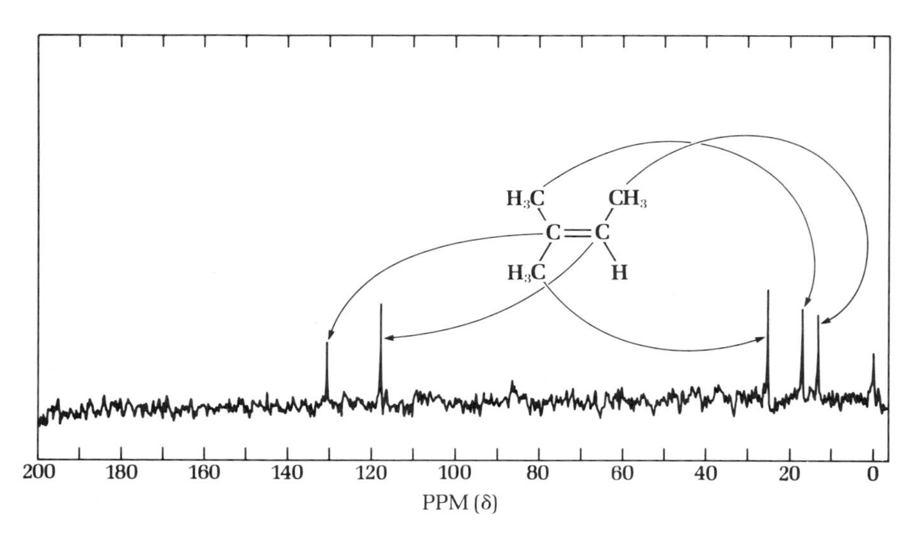

FIG. 12.7 ^{13}C nmr spectrum of 2-methyl-2-butene.

7. If you were to try a column one-half the length of the one you actually used, how do you think the retention times of the butenes would be affected? How do you think the separation of the peaks would be affected? How do you think the width of each peak would be affected?
8. From your knowledge of the dehydration of tertiary alcohols, which olefin should predominate in the dehydration of 2-methyl-2-butanol and why?
9. What is the maximum volume of the butene mixture that could be obtained by the dehydration of 81 mg of 2-butanol?
10. What other gases are in the collection tube besides the three butenes at the end of the second reaction?

13 Alkanes and Alkenes: Radical Initiated Chlorination of 1-Chlorobutane

Most of the alkanes from petroleum are used to produce energy by combustion, but a few percent are converted to industrially useful compounds by controlled reaction with oxygen or chlorine. The alkanes are inert to attack by most chemical reagents and will react with oxygen and halogens only under the special conditions of radical initiated reactions.

At room temperature an alkane such as butane will not react with chlorine. In order for a reaction to occur

$$CH_3CH_2CH_2CH_3 + Cl_2 \longrightarrow CH_3CH_2CH_2CH_2Cl + HCl$$

a precisely oriented four-center collision of the Cl_2 molecule with the butane molecule must occur with two bonds broken and two bonds formed simultaneously:

$$\begin{array}{cc} \text{H} & \text{Cl} \\ | & | \\ -\text{C}- & \text{Cl} \\ | & \end{array} \longrightarrow \begin{array}{c} \text{H}-\text{Cl} \\ \\ -\text{C}-\text{Cl} \\ | \end{array}$$

It is unlikely that the necessarily precise orientation of the two reacting molecules will be found. They must, of course, be within bonding distance of each other as well, so steric factors play a part.

If a concerted four-center reaction won't work, then an alternative possibility is a stepwise mechanism:

$$:\ddot{\underset{..}{Cl}}:\ddot{\underset{..}{Cl}}: \xrightarrow{\text{slow}} 2\ :\ddot{\underset{..}{Cl}}\cdot \qquad (1)$$

$$CH_3CH_2CH_2CH_3 \xrightarrow{\text{slow}} CH_3CH_2CH_2CH_2\cdot + H\cdot \qquad (2)$$

$$:\ddot{\underset{..}{Cl}}: + CH_3CH_2CH_2CH_2\cdot \xrightarrow{\text{fast}} CH_3CH_2CH_2CH_2Cl \qquad (3)$$

$$:\ddot{\underset{..}{Cl}}\cdot + H\cdot \xrightarrow{\text{fast}} H:\ddot{\underset{..}{Cl}}: \qquad (4)$$

For this series of reactions to occur, the chlorine molecule must dissociate into two chlorine atoms. Because chlorine has a bond energy of

58 kcal $mole^{-1}$, we would not expect any significant numbers of molecules to dissociate at room temperature. Thermal motion at 25°C can only break bonds having energies less than 30–35 kcal $mole^{-1}$. Although thermal dissociation would require a high temperature, the dissociation of chlorine into atoms (chlorine radicals) can be caused by violet and ultraviolet light:

$$Cl_2 \xrightarrow{h\nu} 2\ :\ddot{\underset{..}{Cl}}\cdot$$

The photon energy of red light is 48 kcal $mole^{-1}$ while light of 300 nm (ultraviolet) has a photon energy of 96 kcal $mole^{-1}$.

The chlorine radical can react with butane by abstraction of a hydrogen atom:

$$CH_3CH_2CH_2CH_3 + :\ddot{\underset{..}{Cl}}\cdot \rightleftharpoons CH_3CH_2CH_2CH_2\cdot + HCl$$

a reaction that is very slightly exothermic (and thus written as a reversible reaction). The butyl radical can react with chlorine:

$$CH_3CH_2CH_2CH_2\cdot + Cl_2 \rightarrow CH_3CH_2CH_2CH_2Cl + :\ddot{\underset{..}{Cl}}\cdot$$

a reaction that evolves 26 kcal $mole^{-1}$ of energy. The net result of these two reactions is a reaction of Cl_2 with butane "catalyzed" by $:\ddot{\underset{..}{Cl}}\cdot$, the chlorine radical. The whole process can be terminated by the reaction of radicals with each other:

$$CH_3CH_2CH_2CH_2\cdot + :\ddot{\underset{..}{Cl}}\cdot \rightarrow CH_3CH_2CH_2CH_2Cl$$

$$2\ CH_3CH_2CH_2CH_2\cdot \rightarrow CH_3(CH_2)_6CH_3$$

$$2\ :\ddot{\underset{..}{Cl}}\cdot \rightarrow Cl_2$$

To summarize, the process of light-induced radical chlorination involves three steps: chain initiation in which chlorine radicals are produced, chain propagation that involves no net consumption of chlorine radicals, and chain termination that destroys radicals.

$$Cl_2 \rightarrow 2\ :\ddot{\underset{..}{Cl}}\cdot \qquad \textit{Chain initiation}$$

$$\left.\begin{array}{l} CH_3CH_2CH_2CH_3 + :\ddot{\underset{..}{Cl}}\cdot \rightleftharpoons CH_3CH_2CH_2CH_2\cdot + HCl \\ CH_3CH_2CH_2CH_2\cdot + Cl_2 \rightarrow CH_3CH_2CH_2CH_2Cl + :\ddot{\underset{..}{Cl}}\cdot \end{array}\right\} \textit{Chain propagation}$$

$$\left.\begin{array}{r} CH_3CH_2CH_2CH_2\cdot + :\ddot{\underset{..}{Cl}}\cdot \rightarrow CH_3CH_2CH_2CH_2Cl \\ 2\ CH_3CH_2CH_2CH_2\cdot \rightarrow CH_3(CH_2)_6CH_3 \\ 2\ :\ddot{\underset{..}{Cl}}\cdot \rightarrow Cl_2 \end{array}\right\} \textit{Chain termination}$$

In the example the product is shown to be 1-chlorobutane, but in fact the reaction produces a mixture of 1-chlorobutane and 2-chlorobutane. If the reaction were to occur purely by chance, we would expect the ratio of products to be 6:4 because there are 6 primary hydrogens and 4 secondary hydrogens on butane. But because a secondary C—H bond is weaker than a primary C—H bond (98 vs 95 kcal mole^{-1}), we might expect more 2-chlorobutane than chance would dictate. In the chlorination of 2-methylbutane:

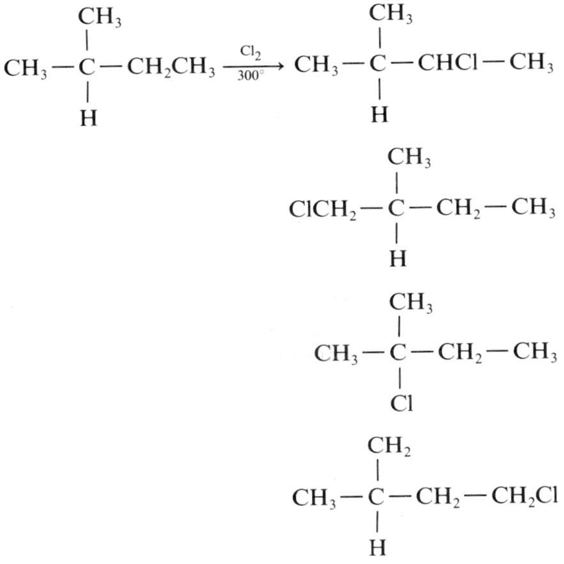

$$CH_3-\underset{\displaystyle H}{\overset{\displaystyle CH_3}{\underset{|}{\overset{|}{C}}}}-CH_2CH_3 \xrightarrow[300^\circ]{Cl_2}$$

Product	Found	Statistical Expectation Ratio	Statistical Expectation Percent
$CH_3-\underset{\displaystyle H}{\overset{\displaystyle CH_3}{\underset{\mid}{\overset{\mid}{C}}}}-CHCl-CH_3$	33%	2/12	17
$ClCH_2-\underset{\displaystyle H}{\overset{\displaystyle CH_3}{\underset{\mid}{\overset{\mid}{C}}}}-CH_2-CH_3$	30%	6/12	50
$CH_3-\underset{\displaystyle Cl}{\overset{\displaystyle CH_3}{\underset{\mid}{\overset{\mid}{C}}}}-CH_2-CH_3$	22%	1/12	8
$CH_3-\underset{\displaystyle H}{\overset{\displaystyle CH_2}{\underset{\mid}{\overset{\mid}{C}}}}-CH_2-CH_2Cl$	15%	3/12	25

As can be seen, the relatively weak tertiary C—H bond (92 kcal mole^{-1}) gives rise to 22% of product in contrast to the 8% expected on the basis of a random attack of $:\ddot{\underset{..}{Cl}}\cdot$ on the starting material.

The relative reactivities of the various hydrogens of 2-methylbutane on a per-hydrogen basis (referred to the primary hydrogens of C-4 as 1.0) can be calculated:

$$\frac{\text{C-2 tertiary}}{\text{C-4 primary}} = \frac{33/1}{15/3} = \frac{6.6}{1}$$

$$\frac{\text{C-3 secondary}}{\text{C-4 primary}} = \frac{33/2}{15/3} = \frac{3.3}{1}$$

A reaction of this type is of little use unless you happen to need the four products in the ratios found and can manage to separate them (their boiling points are very similar). But industrially the radical chlorination of methane and ethane are important reactions and the products can be separated easily:

$$CH_4 \xrightarrow[\Delta]{Cl_2} CH_3Cl + CH_2Cl_2 + CHCl_3 + CCl_4$$

	Chloromethane	**Dichloromethane**	**Trichloromethane**	**Tetrachloromethane**
Common names:	Methyl chloride	Methylene chloride	Chloroform	Carbon Tetrachloride
Boiling points:	−24°C	40°C	62°C	77°C

In the first experiment we will chlorinate 1-chlorobutane because it is easier to handle in the laboratory than gaseous butane and we will use sulfuryl chloride as our source of chlorine radicals because it is easier to handle than gaseous chlorine. Instead of using light to initiate the reaction we will use a chemical initiator, 2,2′-azobis-(2-methylpropionitrile). This azo compound (R—N=N—R) decomposes at moderate temperatures (80–100°C) to give two relatively stable radicals and nitrogen gas:

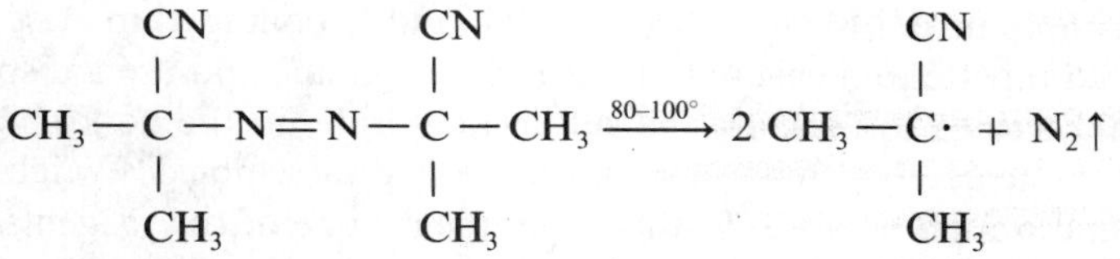

$$CH_3-\overset{CN}{\underset{CH_3}{C}}-N=N-\overset{CN}{\underset{CH_3}{C}}-CH_3 \xrightarrow{80\text{–}100^\circ} 2\ CH_3-\overset{CN}{\underset{CH_3}{C}}\cdot + N_2\uparrow$$

2,2′-Azobis-(2-methylpropionitrile)
MW 164.21, mp 102–103° (dec.)

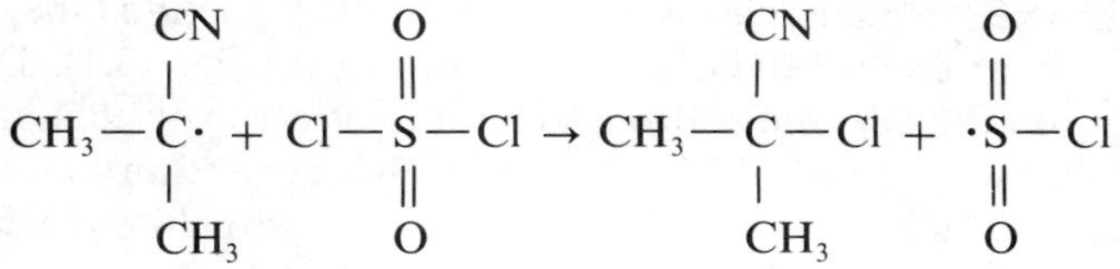

$$CH_3-\overset{CN}{\underset{CH_3}{C}}\cdot + Cl-\overset{O}{\underset{O}{\overset{\|}{\underset{\|}{S}}}}-Cl \rightarrow CH_3-\overset{CN}{\underset{CH_3}{C}}-Cl + \cdot\overset{O}{\underset{O}{\overset{\|}{\underset{\|}{S}}}}-Cl$$

Sulfuryl chloride

$$\cdot\overset{O}{\underset{O}{\overset{\|}{\underset{\|}{S}}}}-Cl \rightarrow \overset{O}{\underset{O}{\overset{\|}{\underset{\|}{S}}}}- + :\ddot{\underset{\cdot\cdot}{Cl}}\cdot$$

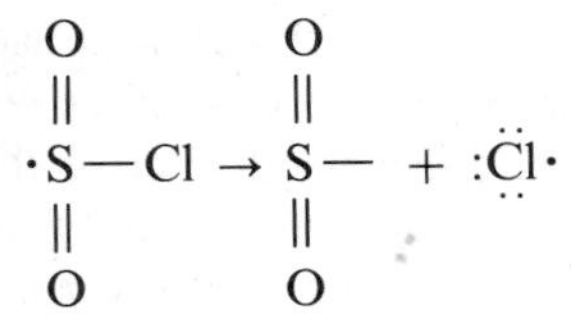

The radical monochlorination of 1-chlorobutane can give four products: 1,1-, 1,2-, 1,3- and 1,4-dichlorobutane. If the reaction occurred completely at random we would expect products in the ratios of the number of hydrogen atoms on each carbon, i.e., 3 : 2 : 2 : 2 respectively (33%, 22%, 22%, 22%). The object of the present experiment is to carry out the radical chlorination of 1-chlorobutane and then to determine the ratio of products using gas-liquid chromatography.

$$4\ CH_3CH_2CH_2CH_2Cl + 4\ SO_2Cl_2 \xrightarrow{R\cdot} \underset{\text{bp } 114^\circ C}{CH_3CH_2CH_2CHCl_2} + \underset{\text{bp } 124^\circ C}{CH_3CH_2CHClCH_2}$$

$$+ \underset{\text{bp } 134^\circ C}{CH_3CHClCH_2CH_2Cl} + \underset{\text{bp } 162^\circ C}{CH_2ClCH_2CH_2CH_2Cl}$$

$$+ 2\ SO_2 + 4\ HCl$$

1-Chlorobutane
MW 92.57
den 0.886
bp 77–78°C

Sulfuryl chloride
MW 134.97
den 1.67
bp 69°C

Experiments

1. Radical Chlorination of 1-Chlorobutane

To a 10 × 100-mm reaction tube add 1-chlorobutane (0.50 mL, 0.432 g, 4.6 mmole), sulfuryl chloride (0.16 mL, 0.27 g, 2.0 mmole), 2,2′-azobis-(2-methylpropionitrile) (4 mg, 0.025 mmole), and a boiling chip. Use a 1-mL graduated pipette to measure out the 1-chlorobutane and use a dispenser or a 0.5-mL syringe to measure the sulfuryl chloride (in the hood). Rinse the syringe immediately after using and leave it disassembled. Weigh the azo compound on the balance. In this experiment none of the reagents need be measured with great care; the 1-chlorobutane is in large excess and the azo compound is present in catalytic amounts.

Fit the reaction tube with a rubber septum, a syringe needle, and a piece of polyethylene tubing that leads down into another reaction tube, the mouth of which has a piece of damp cotton placed in it (Fig. 13.1). During this reaction sulfur dioxide and hydrogen choride gas are evolved and the damp cotton will absorb the gas. The amount of HCl evolved in this experiment (2 mmoles) is equal to 0.15 mL of concentrated hydrochloric acid. Be sure that the end of the polyethylene tube does not touch any water, because it would be sucked back into the reaction tube. Clamp the reaction tube with the reactants in a beaker of hot water maintained at 80°C so that just the tip of the tube is immersed in the water. This will cause the contents to boil gently and the vapors to condense on the cool upper walls of the reaction tube.

At the end of the reaction period remove the tube from the beaker of water, allow it to cool and then carefully add 0.5 mL of water dropwise to the tube from a Pasteur pipette. Note which layer is the aqueous one. Mix the contents thoroughly and then draw off and discard the water layer. Wash the organic phase in the same way with a 0.5-mL portion of 5% sodium bicarbonate solution and once with 0.5 mL of water. Carefully remove all of the water with a Pasteur pipette and then add anhydrous sodium sulfate to dry the product. Transfer the dry product (it should be perfectly clear, not cloudy) to a small tared (previously weighed) screw-capped sample vial or

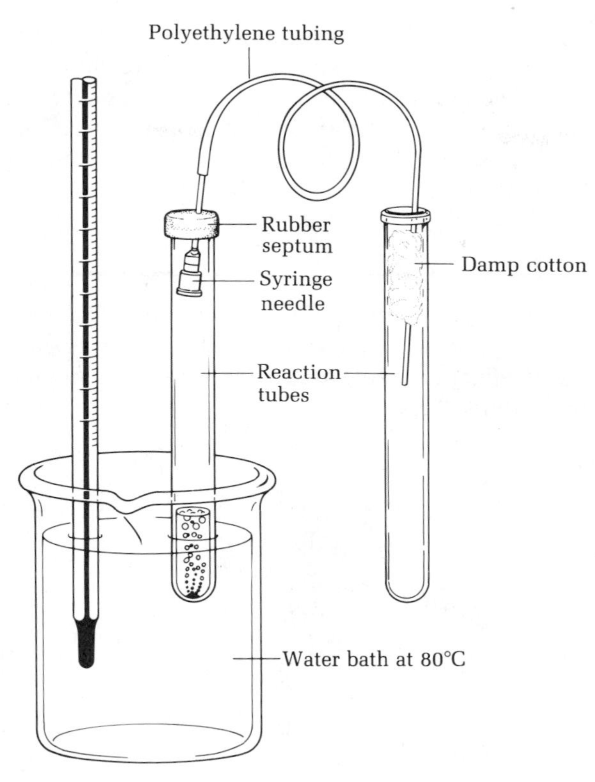

FIG. 13.1 Apparatus for chlorination of 1-chlorobutane with HCl and SO_2 trap. To be clamped appropriately.

corked reaction tube. Don't store the product in a polyethylene-capped container because it will dissolve in the polyethylene. It is not a good idea to store the product before analysis because the composition will change depending on which components evaporate or are absorbed by the cap on the container. Determine the weight of the product and then analyze it by gas–liquid chromatography.

Gas Chromatography

See Chapter 12 for information about gas chromatography. A Carbowax column works best, although any other nonpolar phase such as silicone rubber should work as well. With a nonpolar column packing the products are expected to come out in the order of their boiling points. A typical set of operating conditions would be column temperature 100°C, He flow rate 35 mL/min, column size 5-mm dia × 2 m, sample size 5 microliters, attenuation 16.

The amounts of each compound present in the reaction mixture are proportional to the areas under the peaks in the chromatogram. Because 1-chlorobutane is present in large excess, let this peak run off the paper, but be sure to keep the three product peaks on the paper. To determine relative peak areas simply cut out the peaks with a pair of scissors and weigh them.

This method of peak integration works very well and depends on the uniform thickness of paper, which results in its weight being proportional to its area. Calculate the relative percent of each product molecule and the partial rate factors relative to the primary hydrogens on carbon-4. Compare your results with those for the chlorination of 2-methylbutane, the data for which are given above.

On a Larger Scale. The chlorination of 25 mL of 1-chlorobutane is done in the apparatus depicted in Fig. 13.2. Heat is supplied by means of a steam bath, refluxing vapors are cooled with a water condenser, and HCl is trapped in the filter flask.

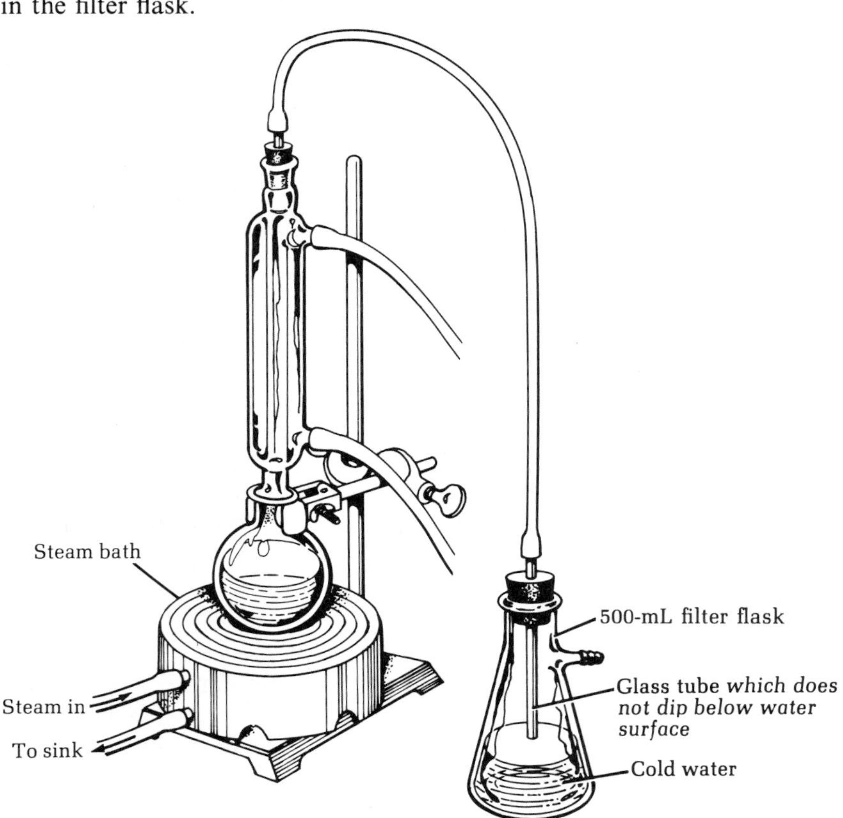

FIG. 13.2 Large-scale apparatus for chlorination of 1-chlorobutane with SO_2 and HCl gas trap.

Conduct this experiment under laboratory hood

Distinguishing between Alkanes and Alkenes

When an olefin such as cyclohexene is allowed to react with bromine at a very low concentration, substitution instead of addition occurs. Although addition of a halogen atom to the double bond occurs readily, the intermediate free radical (**1**) is not very stable and, if it does not encounter another halogen molecule soon, will revert to starting material. If, however, an allylic free radical (**2**) is formed initially, it is much more stable and will survive long enough to react with a halogen molecule, even when the halogen is at low concentration. This low halogen concentration favors allylic substitution, while high halogen concentration favors addition.

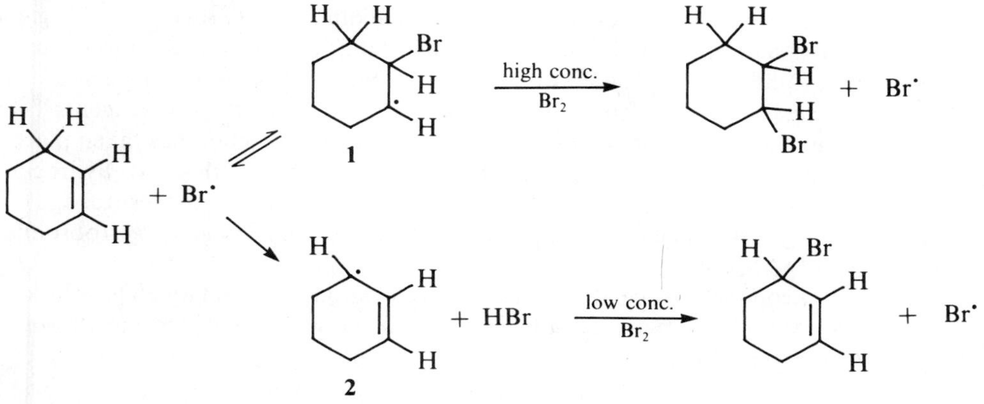

high conc. Br_2

$+ Br^{\cdot}$

1

$+ Br^{\cdot}$

$+ HBr$

low conc. Br_2

$+ Br^{\cdot}$

2

N-Bromosuccinimide (NBS) is a reagent that will continuously generate bromine molecules at a low concentration. This occurs when the HBr molecule from an allylic substitution reacts with NBS:

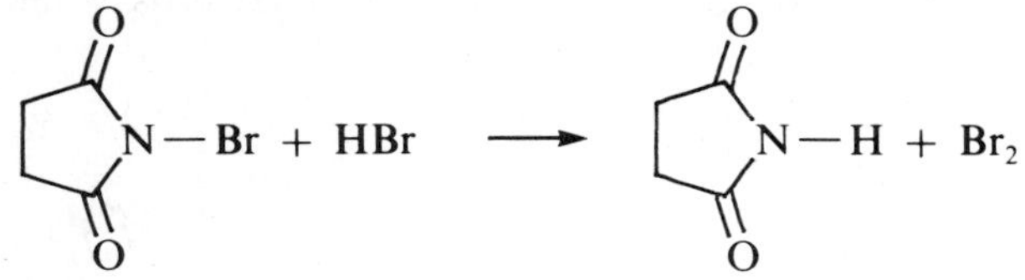

N—Br + HBr → N—H + Br_2

N-Bromosuccinimide

The net reaction is one of allylic bromination:

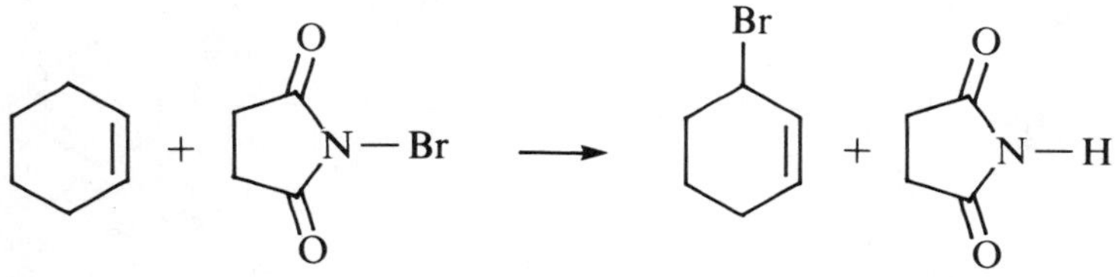

In aqueous solution NBS will react with cyclohexene to form the bromohydrin, a reaction that may or may not involve the intermediate formation of HOBr:

$$\text{cyclohexene} + \text{N-bromosuccinimide} + H_2O \longrightarrow \text{2-bromocyclohexanol (Br, OH)} + \text{succinimide (N–H)}$$

As seen in the preceding experiment, free radical halogenation of alkanes proceeds by substitution to give the chloroalkane and hydrogen chloride gas. Halogenation of an alkene, on the other hand, proceeds by addition across the double bond. These two different modes of reaction can be used to distinguish alkenes from alkanes. Bromine is used as the test reagent. The disappearance of the red bromine color indicates a reaction has taken place. Cyclohexene and dibromocyclohexane are both colorless. If hydrogen bromide is evolved, it can be detected by breathing across the test tube. The hydrogen bromide dissolves in the moist air to give a cloud of hydrobromic acid.

Bromine at a relatively high concentration in a nonaqueous solution such as dichloromethane can add to an alkene such as cyclohexene through a free radical process:

$$\text{cyclohexene} + Br^{\cdot} \longrightarrow \text{2-bromocyclohexyl radical} \xrightarrow{Br_2} \text{1,2-dibromocyclohexane} + Br^{\cdot}$$

Similarly, it will react with an alkane and will evolve a molecule of hydrogen bromide. The formation of bromine radicals is promoted by light:

$$Br_2 \xrightarrow{h\nu} 2\ Br^{\cdot}$$

$$CH_3CH_2CH_3 + Br^{\cdot} \longrightarrow HBr + CH_3\dot{C}HCH_3 \xrightarrow{Br_2} CH_3CH(Br)CH_3 + Br^{\cdot}$$

In aqueous solution a bromonium ion is formed as the intermediate when bromine reacts with an alkene. This intermediate bromonium ion can react with a molecule of bromine to form a dibromide, and it can react with water to form a bromohydrin. Both reactions give a mixture of products:

$$\text{cyclohexene} + Br_2 \longrightarrow \text{bromonium ion } (Br^{+}) + Br^{-} \longrightarrow \text{1,2-dibromocyclohexane}$$

$$\text{bromonium ion} \xrightarrow{H_2O} \text{2-bromocyclohexanol (Br, OH)}$$

A 10% solution of potassium permanganate is bright purple. When this reagent reacts with an olefin, the purple color disappears and a fine brown suspension of manganese dioxide may be seen. While we will use permanganate simply to distinguish between alkenes and alkanes, it is also an important preparative reagent:

$$\text{cyclohexene} + MnO_4^- \xrightarrow[KMnO_4]{\text{dilute neutral}} [\text{cyclic manganate ester (O, O, Mn, O, O)}] \xrightarrow{H_2O} \text{cyclohexane-1,2-diol (OH, OH)} + MnO_3^-$$

$$\text{cyclohexane-1,2-diol (OH, OH)} + MnO_4^- + H^+ \xrightarrow[KMnO_4]{\text{acidic}} \text{COOH, COOH} + MnO_2$$

Alkenes will react with sulfuric acid to form alkyl hydrogen sulfates. In the process the alkene will appear to dissolve in the concentrated sulfuric acid. Alkanes are completely unreactive toward sulfuric acid.

$$\text{cyclohexene} + H_2SO_4 \longrightarrow \text{cyclohexyl (H, } OSO_3^-\,H^+)$$

2. Tests for Alkanes and Alkenes

The following tests demonstrate properties characteristic of alkanes and alkenes, provide means of distinguishing between compounds of the two types, and distinguish between pure and impure alkanes. Use your own preparation of cyclohexene (dried over anhydrous sodium sulfate) as a typical alkene, purified 66–75°C ligroin (Eastman Organic Chemicals No. 513) as a typical alkane mixture (the bp of hexane is 69°C), and unpurified ligroin (Eastman No. P513) as an impure alkane. In the following experiments distinguish clearly, as you take notes, between *observations* and *conclusions* based on those observations. Write equations for all positive tests.

(a) Bromine in Nonaqueous Solution

Treat 0.5-mL samples of purified ligroin, unpurified ligroin, and cyclohexene with 2–3 drops of a 3% solution of bromine in dichloromethane. In case decolorization occurs, breathe across the mouth of the tube to see if hydrogen bromide can be detected. If the bromine color persists illuminate the solution and, if a reaction occurs, test as before for hydrogen bromide.

These tests can be carried out on a much smaller scale

(b) Bromine Water

Bromine (3%) in dichloromethane should be freshly prepared and stored in a brown bottle; test the reagent for decomposition by breathing across the bottle

Measure 1 mL of a 3% aqueous solution of bromine into each of three reaction tubes, then add 0.3-mL portions of purified ligroin to two of the tubes and 0.3 mL of cyclohexene to the third. Shake each tube and record the initial results. Put one of the ligroin-containing tubes in the desk out of the light and expose the other to bright sunlight or hold it close to a light bulb. When a change is noted compare the appearance with that of the mixture kept in the dark.

(c) Acid Permanganate Test

To 0.3-mL portions of purified ligroin, unpurified ligroin, and cyclohexene add a drop of an aqueous solution containing 1% potassium permanganate and 10% sulfuric acid and shake. If the initial portion of reagent is decolorized, add further portions.

(d) Sulfuric Acid

Caution: *student-prepared cyclohexene may be wet. Dry over anhydrous sodium sulfate before use.*

Cool 0.3-mL portions of purified ligroin and cyclohexene in ice, treat each with 1 mL of conc'd sulfuric acid, and shake. Observe and interpret the results. Is any reaction apparent? any warming? If the mixture separates into two layers, identify them.

(e) Bromination with Pyridinium Hydrobromide Perbromide ($C_5H_5\overset{+}{N}HBR_3$)[1]

This substance is a crystalline, nonvolatile, odorless complex of high molecular weight (319.84), which, in the presence of a bromine acceptor such as an alkene, dissociates to liberate one mole of bromine. For small-scale experiments it is much more convenient and agreeable to measure and use than free bromine.

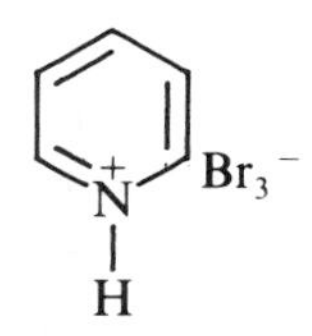

Pyridinium hydrobromide perbromide
MW 319.84

Note for the instructor

1. Crystalline material suitable for small-scale experiments is supplied by Aldrich Chemical Co. Massive crystals commercially available should be recrystallized from acetic acid (4 mL per g). Preparation: Mix 15 mL of pyridine with 30 mL of 48% hydrobromic acid and cool; add 25 g of bromine gradually with swirling, cool, collect the product with use of acetic acid for rinsing and washing. Without drying the solid, crystallize it from 100 mL of acetic acid. Yield of orange needles, 33 g (69%).

Add 80 mg of the reagent to a reaction tube and add 0.5 mL of acetic acid. Swirl the mixture and note that the solid is sparingly soluble. Add 20 mg (0.025 mL) of cyclohexene to the suspension of reagent. Swirl, crush any remaining crystals with a flattened stirring rod, and if after a time the amount of cyclohexene appears insufficient to exhaust the reagent, add a little more. When the solid is all dissolved, dilute with water and note the character of the product. By what property can you be sure that it is the reaction product and not starting material?

(f) Formation of a Bromohydrin

N-Bromosuccinimide in an aqueous solution will react with an olefin to form a bromohydrin:

$$\text{(cyclohexene)} + \text{N—Br} + H_2O \longrightarrow \text{(Br, OH)} + \text{N—H}$$

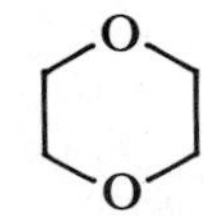

Dioxane
bp 101.5°C, den 1.04
(Dissolves organic compounds, miscible with water.)

***Caution!** Mild carcinogen*

Weigh 90 mg of N-bromosuccinimide, put it into reaction tube, and add 0.25 mL of dioxane and 0.5 millimole (40 mg) of cyclohexene. In another tube chill 0.1 mL of water and add to it 0.5 millimole of conc'd sulfuric acid. Transfer the cold dilute solution to the first tube with the capillary dropper. Note the result and the nature of the product that separates on dilution with water.

(g) Tests for Unsaturation

Determine which of the following hydrocarbons are saturated and which are unsaturated or contain unsaturated material. Use any of the above tests that seem appropriate.

Camphene
Pinene, the principal constituent of turpentine oil
Paraffin oil, a purified petroleum product
Gasoline produced by cracking
Cyclohexane
Rubber (The adhesive Grippit and other rubber cements are solutions of unvulcanized rubber. Squeeze a drop of it onto a stirring rod and dissolve it in toluene. For tests with permanganate or with bromine in dichloromethane use only a drop of the former and just enough of the latter to produce coloration.)

Questions

1. Draw the structure of the compound formed when cyclohexene dissolves in concentrated sulfuric acid.
2. The reaction of cyclohexene with cold dilute aqueous potassium permanganate gives a compound having the empirical formula $C_6H_{12}O_2$. What is the structure of this compound? What is the stereochemistry of the compound?
3. Cyclohexene reacts with bromine to give what compound? What is its stereochemistry?
4. 1-Hexene is brominated with pyridinium hydrobromide perbromide as in test (e). The reaction mixture is diluted with water. By what physical property can you be sure that a reaction product has been produced and not starting material?

Nucleophilic Substitution Reactions of Alkyl Halides 14

PRELAB EXERCISE: *Predict the outcomes of the two sets of experiments to be carried out with the eight halides used in the present experiment.*

The alkyl halides, R—X, where X = Cl, Br, I, and sometimes, F, play a central role in organic synthesis. They can easily be prepared from, among others, alcohols, alkenes, and industrially, alkanes. They in turn are the starting materials for the synthesis of a large number of new functional groups. These syntheses are often carried out by nucleophilic substitution reactions in which the halide is replaced by some nucleophile such as cyano, hydroxyl, ether, ester, alkyl—the list is long. As a consequence of the importance of this substitution reaction, it has been studied carefully by employing reactions such as the two used in this experiment. Some of the questions that can be asked include: how does the structure of the alkyl part of the alkyl halide affect the reaction; and what is the effect of changing the nature of the halide, the nature of the solvent, the relative concentrations of the reactants, the temperature of the reaction, or the nature of the nucleophile? In this experiment we shall explore the answers to a few of these questions.

In free radical reactions the covalent bond undergoes homolysis when it breaks

$$R{:}\ddot{\underset{..}{Cl}}{:} \rightarrow R\cdot + \cdot\ddot{\underset{..}{Cl}}{:}$$

whereas in ionic reactions it undergoes heterolysis

$$R{:}\ddot{\underset{..}{Cl}}{:} \rightarrow R^{+} + \ddot{\underset{..}{Cl}}{:}^{-}$$

A carbocation is often formed as a reactive intermediate in these reactions. This carbocation is sp^2 hybridized and trigonal-planar in structure with a

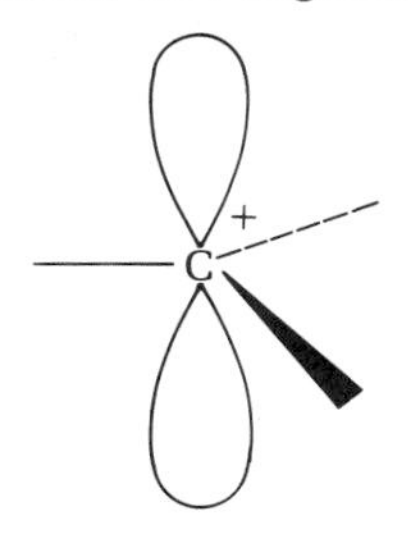

vacant *p*-orbital. Much experimental evidence of the type obtained in the present experiment indicates that the order of stability of carbocations is

$$\mathrm{R_3C^+} > \mathrm{R_2HC^+} > \mathrm{RH_2C^+} > \mathrm{H_3C^+}$$

The alkyl (R) groups stabilize the positive charge of the carbocation by displacing or releasing electrons toward the positive charge. Delocalization of the charge over several atoms stabilizes the charge.

Many organic reactions occur when a nucleophile, a species with an unshared pair of electrons, reacts with an alkyl halide to replace the halogen with the nucleophile.

$$\mathrm{Nu:^-} + \mathrm{R:\ddot{\underset{..}{X}}:} \rightarrow \mathrm{Nu:R} + \mathrm{:\ddot{\underset{..}{X}}:^-}$$

This substitution reaction can occur in one smooth step

$$\mathrm{Nu:^-} + \mathrm{R:\ddot{\underset{..}{X}}:} \rightarrow \left[\overset{\delta-}{\mathrm{Nu}} \cdots\cdots \mathrm{R} \cdots\cdots \overset{\delta-}{\mathrm{\ddot{\underset{..}{X}}:}}\right] \rightarrow \mathrm{Nu:R} + \mathrm{:\ddot{\underset{..}{X}}:^-}$$

or it can occur in two discrete steps depending primarily on the structure of

$$\mathrm{R:\ddot{\underset{..}{X}}:} \rightarrow \mathrm{R^+} + \mathrm{:\ddot{\underset{..}{X}}:^-}$$

$$\mathrm{Nu:^-} + \mathrm{R^+} \rightarrow \mathrm{Nu:R}$$

the R group. The nucleophile, $\mathrm{Nu:^-}$, can be a substance with a full negative charge, such as $\mathrm{:\ddot{\underset{..}{I}}:^-}$ or $\mathrm{H:\ddot{\underset{..}{O}}:^-}$, or a substance with an unshared pair of electrons such as exists on the oxygen atom in water, $\mathrm{H-\ddot{\underset{..}{O}}-H}$. Not all of the halides, $\mathrm{:\ddot{\underset{..}{X}}:^-}$, depart with equal ease in nucleophilic substitution reactions. In this experiment we shall investigate the ease with which the different halogens leave in one of the substitution reactions.

To distinguish between the reaction that occurs as one smooth step and the reaction that occurs as two discrete steps, it is necessary to study the kinetics of the reaction. If the reaction were carried out with several different concentrations of $\mathrm{R:\ddot{\underset{..}{X}}:}$ and $\mathrm{Nu:^-}$, we could determine if the reaction is bimolecular or unimolecular. In the case of the smooth, one-step reaction the nucleophile must collide with the alkyl halide. The kinetics of the reaction

$$\mathrm{Nu:^-} + \mathrm{R:\ddot{\underset{..}{X}}:} \rightarrow \left[\overset{\delta-}{\mathrm{Nu}} \cdots\cdots \mathrm{R} \cdots\cdots \overset{\delta-}{\mathrm{\ddot{\underset{..}{X}}:}}\right] \rightarrow \mathrm{Nu:R} + \mathrm{:\ddot{\underset{..}{X}}:^-}$$

$$\mathrm{Rate} = k\left[\mathrm{Nu:^-}\right]\left[\mathrm{R:\ddot{\underset{..}{X}}:}\right]$$

are found to depend on the concentration of both the nucleophile and the halide. Such a reaction is said to be a bimolecular nucleophilic substitution reaction, S_N2. If the reaction occurs as a two-step process

$$R{:}\ddot{\underset{..}{X}}{:} \xrightarrow{\text{slow}} R^+ + {:}\ddot{\underset{..}{X}}{:}^-$$

$$Nu{:}^- + R^+ \xrightarrow{\text{fast}} Nu{:}R$$

$$\text{Rate} = k[R{:}\ddot{\underset{..}{X}}{:}]$$

the rate of the first step, the slow step, depends only on the concentration of the halide and it is said to be a unimolecular nucleophilic substitution reaction, S_N1.

The S_N1 reaction proceeds through a planar carbocation. Even if the starting material were chiral, the product would be a mixture of enantiomers because the intermediate is planar.

Chiral alkyl chloride → Planar carbocation (Nu:⁻ attacks C⁺ from either side, *or*) → Enantiomers (1) and (2)

The S_N2 reaction occurs with inversion of configuration to give a product of the opposite chirality from the starting material.

$$Nu{:}^- + \text{(H)(D)(R)C—Cl} \longrightarrow \overset{\delta^-}{Nu}\text{------}C\text{------}\overset{\delta}{Cl} \longrightarrow Nu\text{—}C\text{(H)(D)(R)} + {:}\ddot{\underset{..}{Cl}}{:}^-$$

The order of reactivity for *simple* alkyl halides in the S_N2 reaction is

$$CH_3{-}X > R{-}CH_2{-}X > R{-}CH{-}X > (R{-}\underset{R}{\overset{R}{C}}{-}X)$$

The tertiary halide is in parentheses because it usually does not react by an S_N2 mechanism. The primary factor in this order of reactivity is steric hindrance, i.e., the ease with which the nucleophile can come within bonding distance of the alkyl halide. 2,2-Dimethyl-1-bromopropane

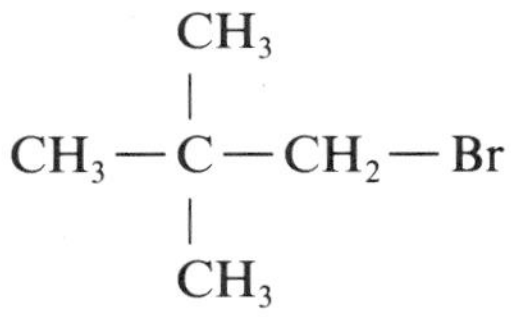

$$CH_3-\overset{\displaystyle CH_3}{\underset{\displaystyle CH_3}{\overset{|}{\underset{|}{C}}}}-CH_2-Br$$

even though it is a primary halide, reacts 100,000 times slower than ethyl bromide, CH_3CH_2Br, because of steric hindrance to attack on the bromine atom in the dimethyl compound.

The primary factor in S_N1 reactivity is the relative stability of the carbocation that is formed. For simple alkyl halides, this means that only tertiary halides react by this mechanism. The tertiary halide must be able to form a planar carbocation. Only slightly less reactive are the allyl carbocations, which derive their great stability from the delocalization of the charge on the carbon by resonance

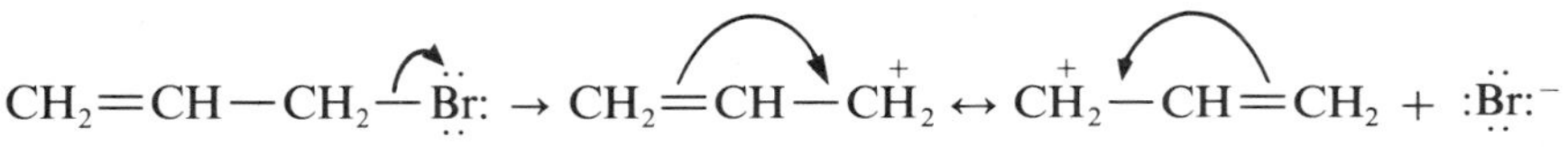

$$CH_2{=}CH-CH_2-\ddot{\underset{..}{Br}}: \rightarrow CH_2{=}CH-\overset{+}{C}H_2 \leftrightarrow \overset{+}{C}H_2-CH{=}CH_2 + :\ddot{\underset{..}{Br}}:^-$$

The nature of the solvent has a large effect on the rates of S_N2 reactions. In a solvent with a hydrogen atom attached to an electronegative atom such as oxygen, the protic solvent forms hydrogen bonds to the nucleophile.

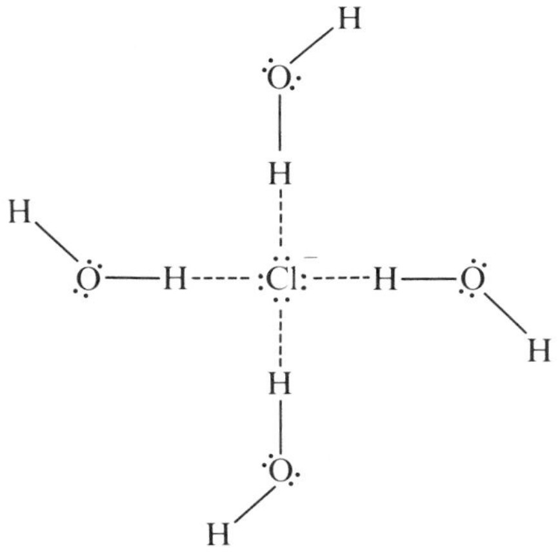

$$\begin{array}{ccccc} & & H-\ddot{O}-H & & \\ & & \vdots & & \\ H-\ddot{O}-H & \cdots & :\ddot{\underset{..}{Cl}}:^- & \cdots & H-\ddot{O}-H \\ & & \vdots & & \\ & & H-\ddot{O}-H & & \end{array}$$

These solvent molecules get in the way during an S_N2 reaction. If the solvent is polar and aprotic, solvation of the nucleophile cannot occur and the S_N2

reaction can occur up to a million times faster. Some common polar, aprotic solvents are

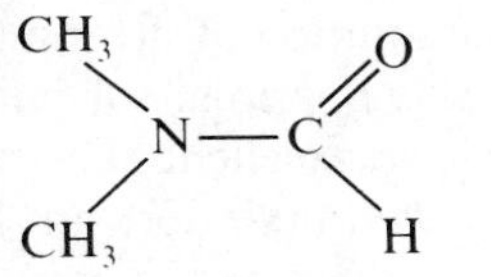

$(CH_3)_2N{-}CH{=}O$

N,N-Dimethylformamide

$CH_3{-}S(=O){-}CH_3$

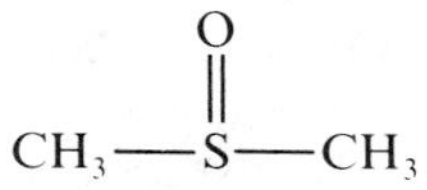

Dimethylsulfoxide

In the S_N1 reaction a polar protic solvent such as water stabilizes the transition state more than it does the reactants, lowering the energy of activation for the reaction and thus increasing the rate, relative to the rate in a nonpolar solvent. Acetic acid, ethanol, and acetone are relatively nonpolar solvents and have lower dielectric constants than the polar solvents water, dimethylsulfoxide, and N,N-dimethylformamide.

The rate of S_N1 and S_N2 reactions depends on the nature of the leaving group, the best leaving groups being the ones that form stable ions. Among the halogens we find that iodide ion, I^-, is the best leaving group as well as the best nucleophile in the S_N2 reaction.

Vinylic and aryl halides

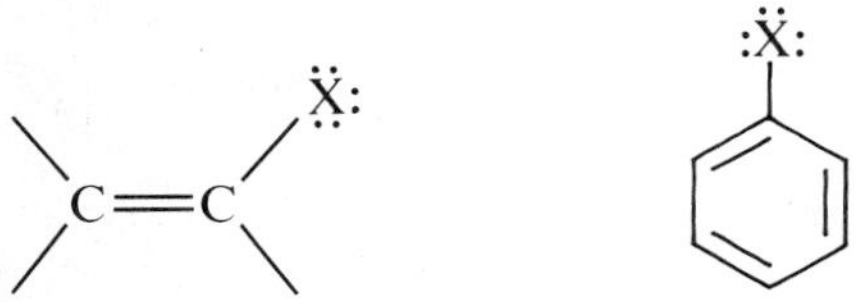

$C{=}C{-}\ddot{X}:$ $C_6H_5{-}\ddot{X}:$

do not normally react by S_N1 or S_N2 reactions because the resulting carbocations

$C{=}C^+$ $C_6H_5^+$

are relatively unstable. The electrons in the nearby double bonds repel the negatively charged nucleophile.

The rates of both S_N1 and S_N2 reactions depend on the temperature of the reaction. As the temperature increases the kinetic energy of the molecules increases, leading to a greater rate of reaction. The rate of many organic reactions will approximately double when the temperature increases about 10°C.

Experiments

In the experiments to follow, eight representative alkyl halides are treated with sodium iodide in acetone and with an ethanolic solution of silver nitrate. Acetone, with a dielectric constant of 21, is a relatively nonpolar solvent that will readily dissolve sodium iodide. The iodide ion is an excellent nucelophile and the nonpolar solvent, acetone, favors the S_N2 reaction; it does not favor ionization of the halide. The extent of reaction can be observed because sodium bromide and sodium chloride are not soluble in acetone and precipitate from solution if reaction occurs.

$$Na^+I^- + R{-}Cl \rightarrow R{-}I + NaCl\downarrow$$

$$Na^+I^- + R{-}Br \rightarrow R{-}I + NaBr\downarrow$$

When an alkyl halide is treated with an ethanolic solution of silver nitrate, the silver ion coordinates with an electron pair of the halogen. This weakens the carbon–halogen bond as a molecule of insoluble silver halide is formed, thus promoting an S_N1 reaction of the alkyl halide. The solvent, ethanol, favors ionization of the halide and the nitrate ion is a very poor nucleophile, so alkyl nitrates do not form by an S_N2 reaction.

$$R{-}\ddot{\underset{\cdot\cdot}{X}}: \xrightleftharpoons{Ag^+} \overset{\delta+}{R}\ddot{\underset{\cdot\cdot}{X}}\overset{\delta+}{Ag} \rightarrow R^+ + AgX\downarrow$$

On the basis of the foregoing discussion tertiary halides would be expected to react most rapidly and primary halides least rapidly.

Procedure

Label eight small containers (reaction tubes, 3-mL centrifuge tubes, 10 × 75-mm test tubes or 1-mL vials) and place 0.1 mL or 100 mg of each of the following halides in the tubes.

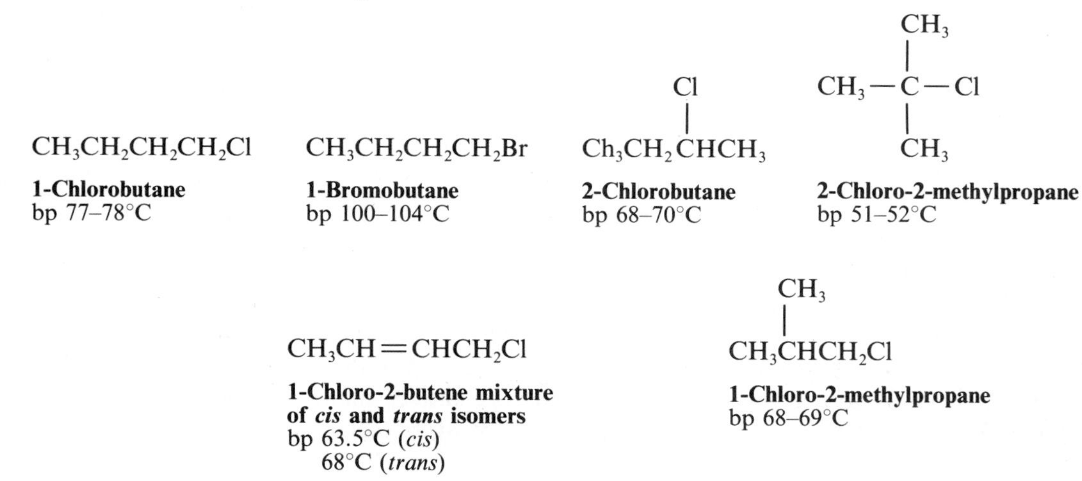

$(H_3C)_2C{=}CHCl$

1-Chloro-2-methylpropene
bp 68°C

1-Chloroadamantane
mp 165–166°C

To each tube then rapidly add 1 mL of an 18% solution of sodium iodide in acetone, stopper each tube, mix the contents thoroughly, and note the time. Note the time of first appearance of any precipitate. If no reaction occurs within about 5 min place those tubes in a 50°C water bath and watch for any reaction over the next 5 or 6 min.

Empty the tubes, rinse them with ethanol and place the same amount of each of the alkyl halides in each tube as in the first part of the experiment, add 1 mL of 1% ethanolic silver nitrate solution to each tube, mix the contents well, and note the time of addition as well as the time of appearance of the first traces of any precipitate. If a precipitate does not appear in 5 min heat those tubes in a 50° C water bath for 5 to 6 min and watch for any reaction.

To test the effect of solvent on the rate of S_N1 reactivity, compare the time needed for a precipitate to appear when 2-chlorobutane is treated with 1% ethanolic silver nitrate solution (above) and when treated with 1% silver nitrate in a mixture of 50% ethanol and 50% water.

In your analysis of the results from these experiments consider the following for both S_N1 and S_N2 conditions: The nature of the leaving group (Cl vs. Br) in the 1-halobutanes; the effect of structure, i.e., compare simple primary, secondary, and tertiary halides, unhindered primary vs. hindered primary halides, a simple tertiary halide vs. a complex tertiary halide, an allylic halide vs. a tertiary halide, and a vinylic halide vs. a primary halide; the effect of solvent polarity on the S_N1 reaction; and the effect of temperature on the reaction.

Questions

1. What would be the effect of carrying out the sodium iodide in acetone reaction with the alkyl halides using an iodide solution half as concentrated?

2. The addition of sodium or potassium iodide catalyzes many S_N2 reactions of alkyl chlorides or bromides. Explain.

15 The S_N2 Reaction: 1-Bromobutane

PRELAB EXERCISE: *Prepare a detailed flow sheet for the isolation and purification of n-butyl bromide. Indicate how each reaction by-product is removed and which layer is expected to contain the product in each separation step.*

$$CH_3CH_2CH_2CH_2OH \xrightarrow{NaBr,\ H_2SO_4} CH_3CH_2CH_2CH_2Br + NaHSO_4 + H_2O$$

1-Butanol	**1-Bromobutane**
bp 118°C	bp 101.6°C
den 0.810	den 1.275
MW 74.12	MW 137.03

Choice of reagents

A primary alkyl bromide can be prepared by heating the corresponding alcohol with (a) constant-boiling hydrobromic acid (47% HBr); (b) an aqueous solution of sodium bromide and excess sulfuric acid, which is an equilibrium mixture containing hydrobromic acid; or (c) with a solution of hydrobromic acid produced by bubbling sulfur dioxide into a suspension of bromine in water. Reagents (b) and (c) contain sulfuric acid at a concentration high enough to dehydrate secondary and tertiary alcohols to undesirable by-products (alkenes and ethers) and hence the HBr method (a) is preferred for preparation of halides of the types R_2CHBr and R_3CBr. Primary alcohols are more resistant to dehydration and can be converted efficiently to the bromides by the more economical methods (b) and (c), unless they are of such high molecular weight as to lack adequate solubility in the aqueous mixtures. The NaBr-H_2SO_4 method is preferred to the Br_2-SO_2 method because of the unpleasant, choking property of sulfur dioxide. The overall equation is given above, along with key properties of the starting material and principal product.

The procedure that follows specifies a certain proportion of 1-butanol, sodium bromide, sulfuric acid, and water; defines the reaction temperature and time; and describes operations to be performed in working up the reaction mixture. The prescription of quantities is based upon considerations of stoichiometry as modified by the results of experimentation. Before undertaking a preparative experiment you should analyze the procedure and calculate the molecular properties of the reagents. Construction of tables (see below) of properties of starting material, reagents, products, and by-products

Reagents

Reagent	MW	Den	Bp (°C)	Wt used (g)	Moles: *Theory*	Moles: *Used*
n-C_4H_9OH	74.12	0.810	118	0.80	0.011	0.011
NaBr	102.91	—	—	1.35	0.011	0.013
H_2SO_4	98.08	1.84	—	2.0	0.011	0.020

Product and By-Products

Compound	MW	Den	Bp (°C): *Given*	Bp (°C): *Found*	Yield (g): *Theory*	Yield (g): *Found*
n-C_4H_9Br	137.03	1.275	101.6		1.48 (100%)	__g__%
$CH_3CH_2CH{=}CH_2$			−6.3			
$C_4H_9OC_4H_9$			141			

provides guidance in regulation of temperature and in separation and purification of the product and should be entered in the laboratory notebook.

The laboratory notebook

$$2\,NaBr + H_2SO_4 \rightleftharpoons 2\,HBr + Na_2SO_4$$

One mole of 1-butanol theoretically requires one mole each of sodium bromide and sulfuric acid, but the procedure calls for use of a slight excess of bromide and twice the theoretical amount of acid. Excess acid is used to shift the equilibrium in favor of a high concentration of hydrobromic acid. The amount of sodium bromide taken, arbitrarily set at 1.2 times the theory as an insurance measure, is calculated as follows:

$$\frac{0.80\ (\text{g of } C_4H_9OH)}{74.12\ (\text{MW of } C_4H_9OH)} \times 102.91\ (\text{MW of NaBr}) \times 1.2 = 1.33\ \text{g of NaBr}$$

The theoretical yield is 0.011 mole of product, corresponding to the 0.011 mole of butyl alcohol taken; the maximal weight of product is calculated thus:

$$0.011\ (\text{mole of alcohol}) \times 137.03\ (\text{MW of product}) = 1.48\ \text{g 1-bromobutane}$$

The probable by-products are 1-butene, dibutyl ether, and the starting alcohol. The alkene is easily separable by distillation, but the other substances are in the same boiling point range as the product. However, all three possible by-products can be eliminated by extraction with concentrated sulfuric acid.

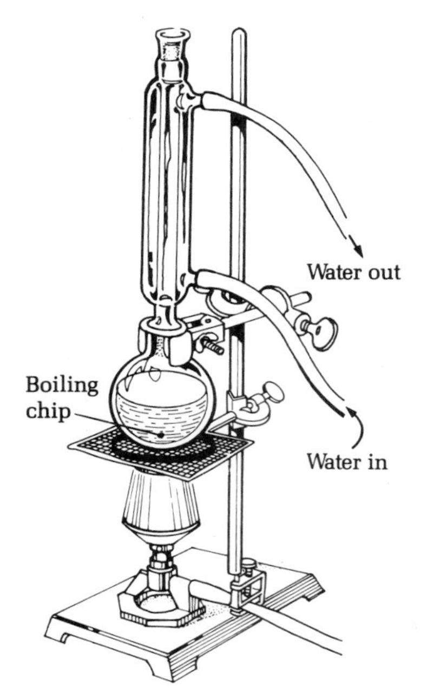

FIG. 15.1 Refluxing a reaction mixture.

Experiment

In a 5-mL round-bottomed long-necked flask dissolve 1.35 g of sodium bromide in 1.5 mL of water and 0.80 g of 1-butanol. Cautiously, with constant swirling, add 1.1 mL (2.0 g) of concentrated sulfuric acid dropwise to the solution. The NaBr will dissolve during heating. Fit the flask with a distillation head and ice-cooled receiver (Fig. 15.1) and reflux the reaction mixture on the sand bath for 30 min, taking care that none of the reactants distill during the reaction period. Wrap the upper end of the apparatus in wet cotton wool or a wet paper towel if escaping vapor is a problem. The upper layer that soon separates in the reaction flask is the alkyl bromide, because the aqueous solution of inorganic salts has the greater density. Remove the wet cotton wool, insulate the distilling column with glass wool, and distill the product into a collection vial until no more water-insoluble droplets come over, by which time the temperature of the distillate should have reached 115°C. If in doubt about whether all of the product has distilled, collect some of the distillate in a small tube and examine it carefully. The sample collected in the receiver is an azeotrope of 1-bromobutane and water containing some sulfuric acid, 1-butene, unreacted 1-butanol, and di-*n*-butyl ether. To ease cleanup wash the round-bottomed flask immediately. Rinse the distillation head with acetone so that it will be dry for use later in the experiment.

Transfer the distillate to a reaction tube, rinsing the vial with about 1 mL of water, which is then mixed with the sample in the reaction tube. Note that the 1-bromobutane now forms the lower layer. Remove the 1-bromobutane with a Pasteur pipette and place it in a dry reaction tube. Add 1 mL of concentrated sulfuric acid and mix the contents well by flicking the tube. The acid removes any unreacted starting material as well as any alkenyl or ethereal by-products. Allow the two layers to separate completely and then remove the sulfuric acid layer. The relative densities given above will help identify the two layers. An empirical method of distinguishing the layers is to remove a drop of the lower layer into a test tube of water to see whether the material is soluble (H_2SO_4) or not (1-bromobutane). Separate the layers and wash the 1-bromobutane layer with 1 mL of 10% sodium hydroxide solution to remove traces of acid, separate, and be careful to save the proper layer. In experiments of this type it is good practice to save all layers until the product is in hand.

Dry the cloudy 1-bromobutane by adding anhydrous sodium sulfate and mixing until the liquid clears and the sodium sulfate no longer clumps together. After 5 min decant the dried liquid into the dry 5-mL round-bottomed flask. This flask is conveniently dried by rinsing it with a milliliter of ethanol followed by a milliliter of acetone and then drawing air through it using a water aspirator. It is very important when drying apparatus of this type to remove all of the wash solvent, acetone in this case; otherwise it will contaminate the final product. Rinse the drying agent with two 1-mL portions of mesitylene (bp 162–164°C), which is then transferred to the round-bottomed flask. The high-boiling mesitylene is a "chaser"; it chases all of the bromobutane from the distilling flask. Otherwise about 0.3 mL would remain behind.

Add a boiling stone, pack the neck of the flask with a stainless steel sponge for fractional distillation, and fit it with a dry distilling head and thermometer. Wrap the column with glass wool and distill, collecting material boiling in the range 99–103°C. Stop collecting the moment the temperature begins to rise above 103°C. Most of the product will boil at 102°C. A typical yield is in the range of 1–1.2 g.

Put the sample in a vial of appropriate size; make a neatly printed label giving the name and formula of the product and your name. After all the time spent on the preparation, the final product should be worthy of a carefully executed and secured label.

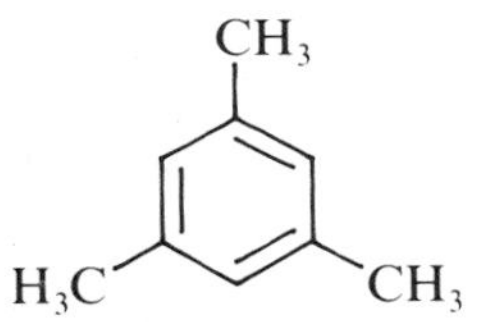

CH_3

H_3C CH_3

Mesitylene
1,3,5-Trimethylbenzene
MW 120.20
bp 162–164°C
den 0.864

On a Larger Scale. The apparatus depicted in Fig. 15.1 is used when preparing bromobutane from 20 mL of 1-butanol. The reaction mixture is heated with a flame and the refluxing vapors condensed in the water-cooled condenser and returned to the flask.

Questions

1. What experimental method would you recommend for the preparation of 1-bromooctane ? *t*-Butyl bromide?
2. Exlain why the crude product is apt to contain certain definite organic impurities.
3. How does each of these impurities react with sulfuric acid when the crude *n*-butyl bromide is shaken with this reagent?
4. How should the reaction conditions in the present experiment be changed to produce 1-chlorobutane?
5. Write a balanced equation for the reaction of sodium bisulfite with bromine.

6. What is the purpose of refluxing the reaction mixture for 30 min? Why not simply boil the mixture in an Erlenmeyer flask?
7. Write reaction mechanisms showing how 1-butene and di-*n*-butyl ether are formed.
8. Why is the resonance of the bromine-bearing carbon atom and the hydroxyl-bearing carbon atom farthest downfield in Figs. 15.2 and 15.3?

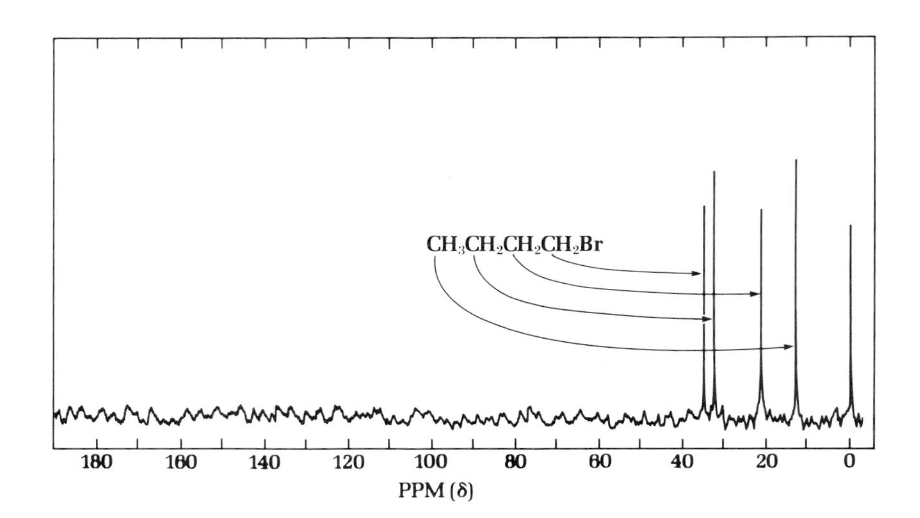

FIG. 15.2 ^{13}C nmr spectrum of 1-bromobutane.

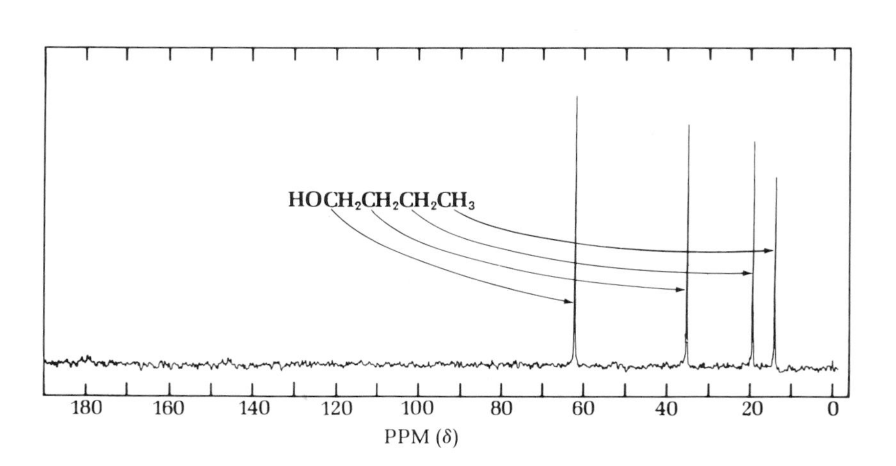

FIG. 15.3 ^{13}C nmr spectrum of 1-butanol.

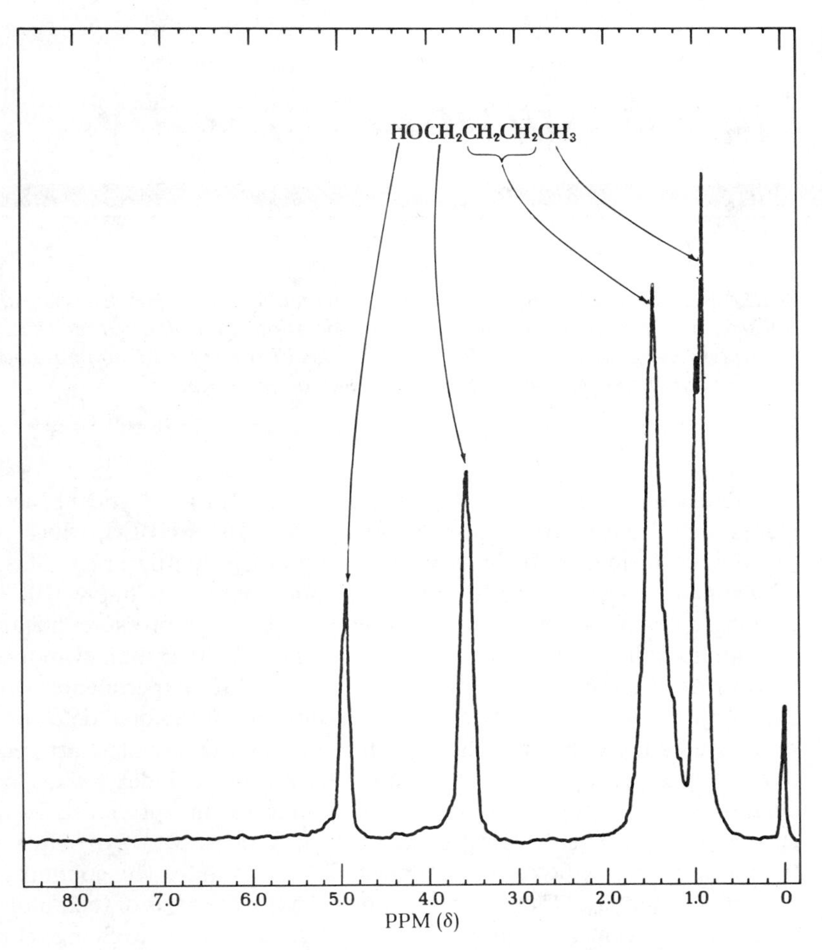

FIG. 15.4 ^{1}H nmr spectrum of 1-butanol.

16 Liquid Chromatography

PRELAB EXERCISE: *Read the instruction manual for the HPLC apparatus you will use, and acquaint yourself with the operating controls for the instrument. Consult reference works on HPLC for examples of solvents and column packings that will effect the separation you would like to carry out.*

The newest addition to the chromatographic family of analytical procedures is liquid chromatography, most commonly known as HPLC, which stands for high-performance (or high-pressure) liquid chromatography. This form of chromatography resembles column chromatography (Chapter 10) except that the process involves pumping liquid eluant at high pressure through the column instead of depending on gravity for flow. Rather than evaporating a number of fractions to detect separated products, the experimenter runs the eluant through a detector, most commonly an ultraviolet detector or a refractive index detector. The detector is connected to a chart recorder that gives a record of uv absorbance or refractive index versus mL of eluant—a chromatogram that is very similar to one produced by a gas chromatograph. The uv detector can detect as little as 10^{-10} g of solute, while the refractive index detector can detect 10^{-6} g of solute. The column can be packed with material that separates substances on the basis of their molecular sizes (exclusion chromatography), their ionic charges (ion exchange chromatography), their ability to adsorb to the packing as in ordinary column chromatography (adsorption chromatography), or their ability to partition between a stationary phase attached to the column and a mobile phase that is pumped through the column (partition chromatography). Partition column chromatography is most common in HPLC and will therefore be the focus of this discussion.

The partitioning of a solute between two immiscible solvents is what takes place in a separatory funnel during the process of extraction (Chapter 8). In partition chromatography this extraction process occurs repeatedly as the eluant, containing the mixture to be separated, flows past the fixed organic phase that is attached to beads of silica gel in the column. The advantage of this form of chromatography over gas chromatography is that substances of high molecular weight having very small chemical or structural differences can be separated. For example, insulins from a variety of animals can be separated from one another even though they differ by just one or two amino acids out of 51 and have molecular weights near 6000.

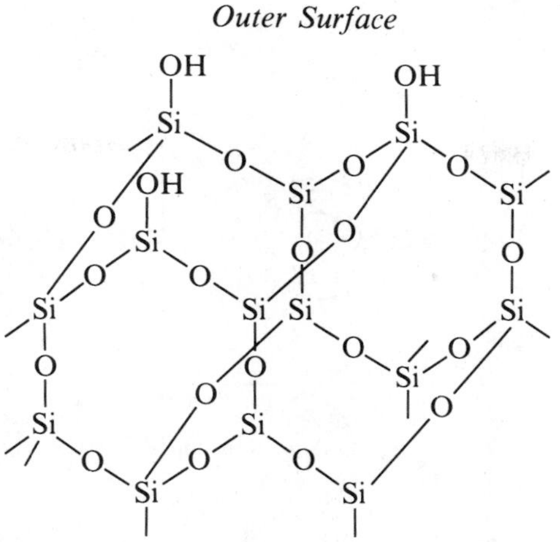

Outer Surface

FIG. 16.1 Silica gel.

The stationary phase in partition chromatography is usually a long hydrocarbon chain covalently bound to silica gel. Silica gel has the structure shown in Fig. 16.1. The siloxane bonds ($\sim$Si—O—Si$\sim$) are stable to water and the other solvents used in HPLC between pH 2 and 9. The silanol ($\sim$Si—OH) groups can be bound to long hydrocarbon chains through reactions such as the following:

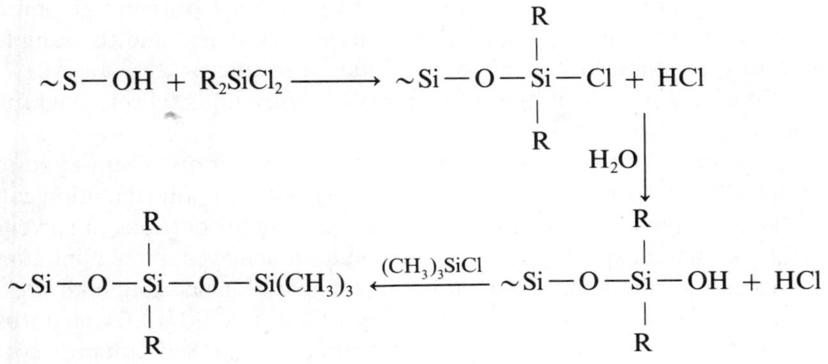

$$\sim\text{S—OH} + R_2SiCl_2 \longrightarrow \sim\text{Si—O—}\underset{R}{\overset{R}{\text{Si}}}\text{—Cl} + HCl$$

$$\downarrow H_2O$$

$$\sim\text{Si—O—}\underset{R}{\overset{R}{\text{Si}}}\text{—O—Si}(CH_3)_3 \xleftarrow{(CH_3)_3SiCl} \sim\text{Si—O—}\underset{R}{\overset{R}{\text{Si}}}\text{—OH} + HCl$$

$$R = (CH_2)_{17}CH_3 = \text{octadecyl}$$

The most common R group is the octadecyl group (C-18), which leaves the silica particles coated with hydrocarbon chains. The silica particles are very small ($\sim 40\,\mu$ dia $= 4 \times 10^{-3}$ cm dia) and very uniform in size. With very small, uniform particles, a molecule in the solvent can rapidly diffuse to the surface of the packing and undergo partitioning between the stationary and mobile phases. Such small-diameter particles present a large surface area to the eluant and offer considerable resistance to the flow of the eluant. Consequently, for a typical column 4 mm in diameter and 25 cm long, the inlet

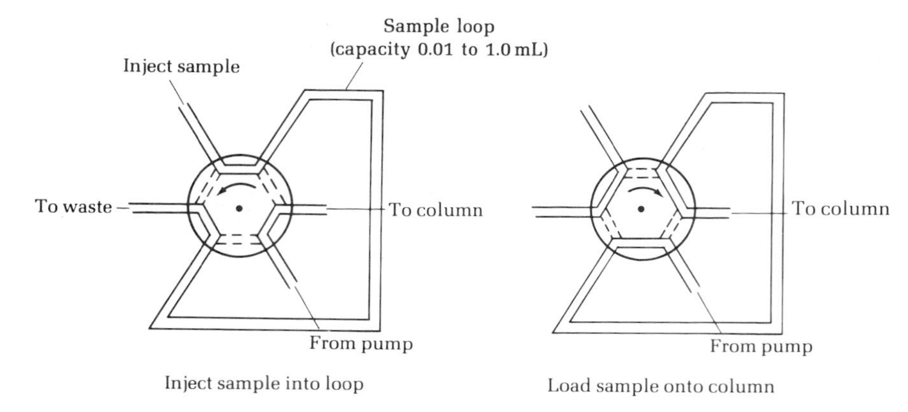

FIG. 16.2 Rotary loop injector for HPLC.

pressure necessary to pump the eluant at a flow rate of 1 mL/min is often in excess of 1000 psi and can reach as high as 5000 psi. Because the column is packed with such exceedingly fine particles, it is subject to clogging from particulate matter in the samples and solvent. Both must be filtered through very fine membrane filters or centrifuged before being applied to the column.

Because it would be difficult to inject a sample onto the column having a back pressure of > 1000 psi at the inlet, a loop injector is employed (Fig. 16.2). In the load position the sample is injected into the loop, which can be of any capacity, often 0.1 mL. During this time the pump is pushing pure solvent through the valve and column. When the valve is turned, the pump pushes the sample onto the column. Pressure is maintained and the sample goes onto the column in a small volume plug of solution.

A block diagram for a typical high-performance liquid chromatograph is shown in Fig. 16.3.

An isocratic HPLC is one in which only one solvent or solvent mixture is pumped through the column. An HPLC equipped for gradient elution can automatically mix two or more solvents in changing proportions (a solvent gradient) so that maximum separation of solutes is achieved in the minimum time. Because of the necessity for a pump capable of generating high pressures, a special injection valve, and a uv detector, the cost of HPLC apparatus is several times that for a gas chromatograph. The packed columns cost several hundred dollars each, and the carefully purified solvents cost several

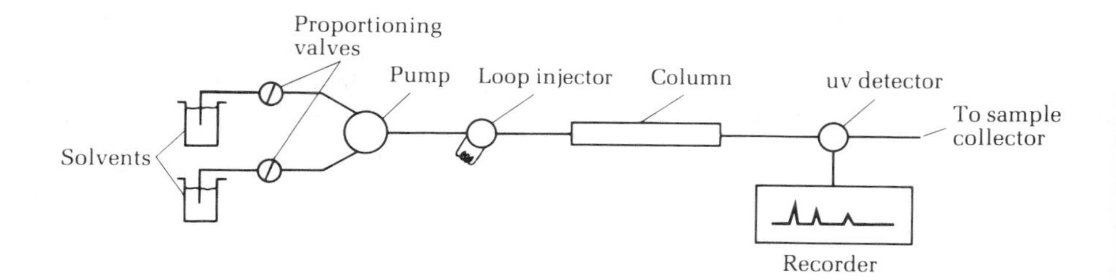

FIG. 16.3 Gradient elution high-performance liquid chromatograph.

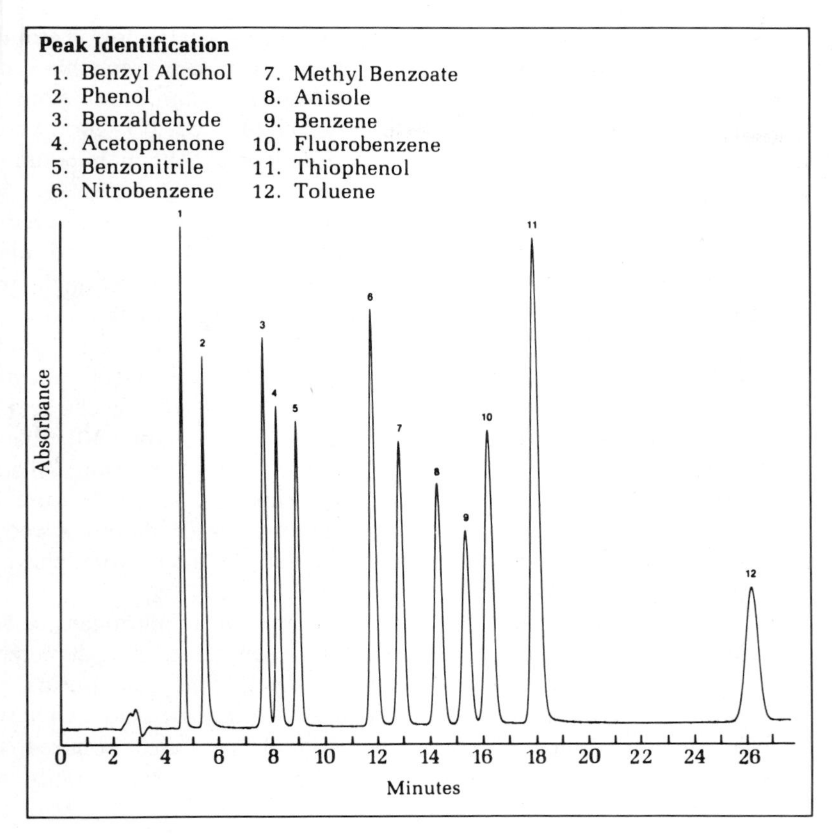

FIG. 16.4 Separation of 10 μl of a mixture of monosubstituted aromatic compounds by HPLC employing an octadecyl (C-18) column 4.5-mm dia × 25 cm long. The mobile phase is 40% acetonitrile/60% H_2O and the flow rate is 1 mL/min. A uv detector operating at 254 nm is used to detect the peaks.

times as much as reagent-grade solvents commonly found in the laboratory. In return for this investment, HPLC is proving to be an extremely versatile analytical tool of great sensitivity. HPLC is usually used for analytical purposes only; however, it is possible to collect the eluant, evaporate the solvent, and examine the residue by uv, ir, mass, or nmr spectroscopy. Up to 10 mg of material can be separated on one injection into an analytical HPLC by overloading the column. Special instruments are manufactured for preparative scale separations. Fig. 16.4 shows a typical separation of a variety of monosubstituted aromatic compounds.

Experiment

Many of the crude reaction mixtures and extracts encountered in the organic laboratory can be analyzed by HPLC. The most common apparatus is an isocratic (single eluant composition) system employing a C-18 column and an ultraviolet detector operating at 254 nm. Methanol–water mixtures or acetonitrile–water mixtures can be used to elute the compounds. It is absolutely necessary to employ HPLC-grade solvents—methanol, acetonitrile, and water—and to filter each sample through a membrane filter or to centrifuge it at high speed in a clinical centrifuge for 5 min. Employing these precautions ensures that the column will not be blocked. It is good practice to employ a short renewable guard column in front of the main column, as further insurance against blockage.

Use extreme caution to ensure that no insoluble material is injected onto the HPLC column. Do no exceed the recommended pump pressure.

Dissolve the sample in exactly the same solvent or solvent mixture being pumped through the column, filter or centrifuge the solution, and inject it into the chromatograph. If the retention time is not long enough or the components are not well resolved, increase the percent of the more polar solvent (water) in the eluant.

HPLC on an octadecyl column can be applied to the analysis of caffeine from tea or cola syrup (Chapter 8—use 20% methanol, 0.8% acetic acid, and 79.2% water as eluant), the acetylation of cholesterol (Chapter 10), pulegone and citronellal from citronellol (Chapter 25), cholesterol from gallstones (Chapter 22), the isolation of eugenol from cloves (Chapter 28—use 10% methanol, 5.4% acetic acid, and 84.6% water as eluant), isolation of lycopene and β-carotene (Chapter 9), and the product obtained from enzymatic reduction of ethylacetoacetate (Chapter 59).

The ingredients of common pain relievers (acetaminophen, caffeine, salicylamide, aspirin, and salicylic acid) can be separated using 20% methanol, 0.8% acetic acid, and 79.2% water until the caffeine peak appears. Then the eluant is changed to 20% methanol, 0.6% acetic acid, and 59.4% water using a flow rate of 3 mL/min. If commercial tablets are used, be sure to filter or centrifuge the solution before use in order to remove starch that is used as a binder in the tablets.

Questions

1. Which would you expect to elute first from a C-18 column, eluting with methanol: a C-10 or a C-20 saturated hydrocarbon?
2. Why would you have difficulty detecting saturated hydrocarbons using the uv detector?
3. Since saturated and monounsaturated fatty acids don't absorb uv light, how might they be modified for detection in an HPLC apparatus?

Separation and Purification of the Components of an Analgesic Tablet: Aspirin, Caffeine, and Acetaminophen

17

PRELAB EXERCISE: *Prepare a detailed flow sheet for the separation of aspirin, caffeine, and acetaminophen from an analgesic tablet.*

This experiment puts into practice the techniques learned in several previous chapters for separating and purifying, on a very small scale, the components in a common analgesic tablet. It is presumed that thin-layer or high-performance liquid chromatography has already shown that the tablet does indeed contain aspirin, caffeine, and acetaminophen.

Many pharmaceutical tablets are held together with a binder to prevent the components from crumbling on storage or while being swallowed. A close reading of the contents on the package will disclose the nature of the binder. Starch is commonly used, as is microcrystalline cellulose and silica gel. All of these have one property in common: they are insoluble in water and common organic solvents.

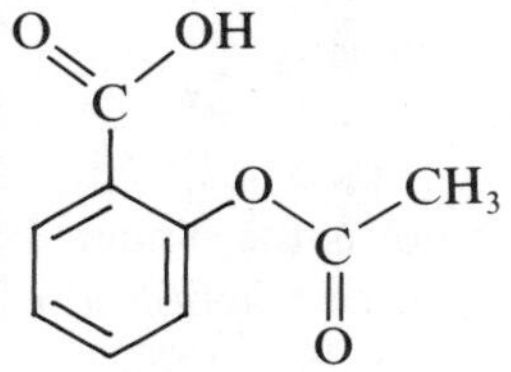

Aspirin
Acetylsalicylic acid
mp 135°C

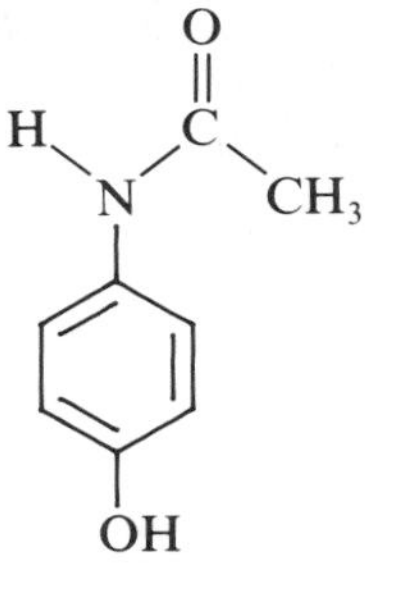

Acetaminophen
***p*-Hydroxyacetanilide**
mp 169–170.5°C

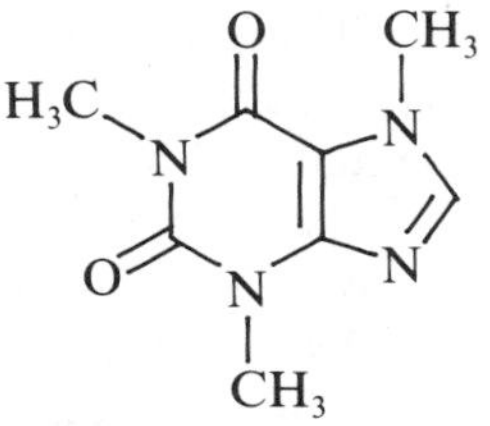

Caffeine
mp 238°C

Solubilities: Aspirin: 0.33 g/100 mL H_2O 25°C, 1 g/100 mL H_2O 37°C, 1 g/5 mL ethanol, 1 g/17 mL $CHCl_3$, 1 g/13 mL ether. Acetaminophen: v.sl.sol. cold H_2O, sol. hot H_2O, sol. methanol, ethanol, acetone, ethyl acetate, sl. sol. ether, ins. pet. ether, benzene. Caffeine: 1 g dissolves in 46 mL H_2O, 5.5 mL at 80°C, 1.5 mL at 100°C, 66 mL ethanol, 22 mL at 60°C, 50 mL acetone, 5.5 mL $CHCl_3$, and 530 mL ether.

Inspection of the structures of caffeine, acetylsalicylic acid, and acetaminophen reveals that one is a base, one a strong organic acid, and one a weak organic acid. One might be tempted to separate this mixture using exactly the same procedure as employed in Chapter 8 to separate benzoic acid, 2-naphthol, and 1,4-dimethoxybenzene, i.e., dissolve the mixture in dichloromethane; remove the benzoic acid (the strong acid) by reaction with bicarbonate ion, a weak base; then remove the naphthol, the weak acid, by reaction with hydroxide, a strong base. This process would leave the neutral compound, 1,4-dimethoxybenzene, in the dichloromethane solution.

In the present experiment the solubility data on the previous page reveals that the weak acid, acetaminophen, is not soluble in ether, chloroform, or dichloromethane, so it cannot be extracted by strong base. We can take advantage of this lack of solubility by dissolving the other two components, caffeine and aspirin, in dichloromethane, and removing the acetaminophen by filtration. The binder is also insoluble in dichloromethane, but treatment of the solid mixture with ethanol will dissolve the acetaminophen and not the binder. They can then be separated by filtration and the acetaminophen isolated by evaporation of the ethanol.

This experiment is a test of technique. It is not easy to separate and crystallize a few milligrams of a compound that occurs in a mixture.

Experiment

Procedure

In a mortar grind an Extra Strength Excedrin tablet to a very fine powder. The label states that this analgesic contains 250 mg of aspirin, 250 mg of acetaminophen, and 65 mg of caffeine per tablet. Place 300 mg of this powder in a reaction tube and add to it 2 mL of dichloromethane. Warm the mixture briefly and note that a large part of the material does not dissolve. Filter the mixture on the micro Büchner funnel (Fig. 17.1) into another reaction tube. Transfer the slurry to the funnel with a Pasteur pipette and complete the transfer with a small portion of dichloromethane. The rate of filtration can be increased by applying pressure to the top of the filter extension tube with a rubber bulb. Or use the pressure filtration apparatus (Fig. 17.2).

Transfer the powder on the filter to a reaction tube, add 1 mL of ethanol, and heat the mixture to boiling on the sand bath (boiling stick). Not all of the material will go into solution. That which does not is the binder. Filter the mixture on the same micro Büchner funnel into a tared reaction tube and complete the transfer and washing using a few drops of hot ethanol.

Acetaminophen

Evaporate about two-thirds of the filtrate by boiling off the ethanol, or better by warming the solution and blowing a stream of air into the reaction tube. Heat the residue to boiling (add a boiling stick to prevent bumping) and add more ethanol if necessary to bring the solid into solution. Allow the saturated solution to cool slowly to room temperature to deposit crystals of acetaminophen, which is reported to melt at 169–170.5°C. After the mixture has cooled to room temperature cool it in ice for several minutes, remove the solvent with a Pasteur pipette, wash the crystals once with two drops of ice-cold ethanol, remove the ethanol, and dry the crystals under aspirator vacuum while heating the tube on a steam bath.

Alternatively the original ethanol solution is evaporated to dryness and the residue recrystallized from boiling water. The crystals are best collected and dried on a Hirsch funnel. Once the crystals are dry determine their weight and melting point. Thin-layer chromatographic analysis and the melting

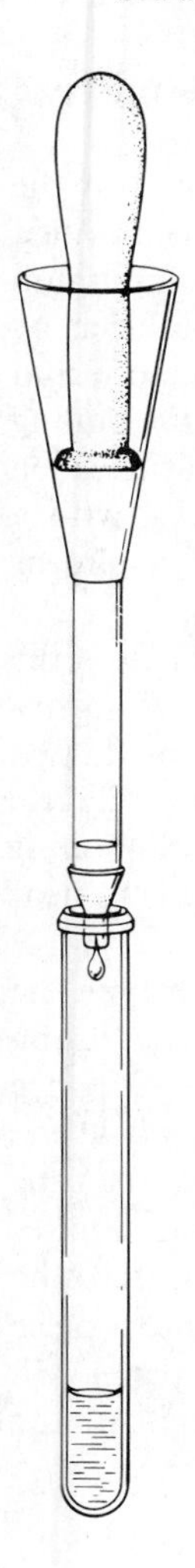

FIG. 17.1 Micro Büchner funnel assembly

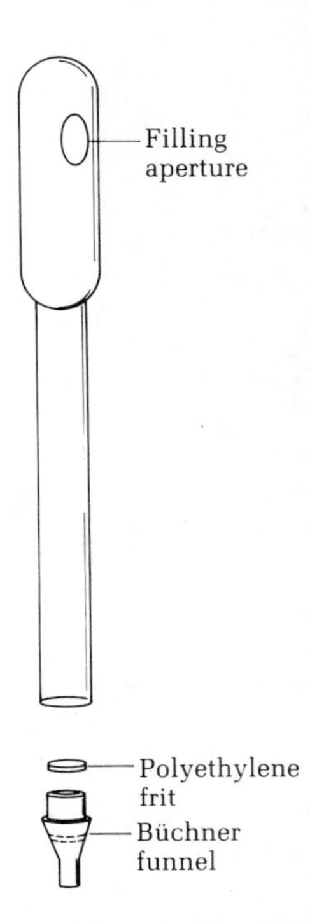

FIG. 17.2 Pressure filtration apparatus. Solution to be filtered is added through the aperture, which is closed by a finger as pressure is applied.

points of these crystals and the two other components of this mixture will indicate their purity.

The dichloromethane filtered from the binder and acetaminophen mixture should contain caffeine and aspirin. These can be separated by extraction either with acid, which will remove the caffeine as a water-soluble salt, or by extraction with base, which will remove the aspirin as a water-soluble salt. We shall use the latter procedure.

To the dichloromethane solution in a reaction tube add 1 mL of 10% sodium hydroxide solution and shake the mixture thoroughly. Remove the aqueous layer, add 0.2 mL more water, shake the mixture thoroughly and again remove the aqueous layer, which is combined with the first aqueous extract.

To the dichloromethane add anhydrous sodium sulfate until the drying agent no longer clumps together. Shake the mixture over a 5- or 10-min period to complete the drying process, then remove the solvent, wash the drying agent with more solvent, and evaporate the combined extracts to dryness under a stream of air to leave crude caffeine.

Caffeine

The caffeine can be purified by sublimation as has been done in the experiment in which it was extracted from tea (Chapter 8) or it can be purified by crystallization. Recrystallize the caffeine by dissolving it in the minimum quantity of 30% ethanol in tetrahydrofuran. It can also be crystallized by dissolving the product in a minimum quantity of hot toluene or acetone and adding to this solution ligroin until the solution is cloudy while at the boiling point. In any case allow the solution to cool slowly to room temperature, then cool the mixture in ice, and remove the solvent from the crystals with a Pasteur pipette. Remove the remainder of the solvent under aspirator vacuum, determine the weight of the caffeine and its melting point.

The aqueous hydroxide extract contains aspirin as the sodium salt of the carboxylic acid. To the aqueous solution add 10% hydrochloric acid dropwise until the solution tests strongly acid to indicator paper, and then add two more drops of acid. This will give a suspension of white acetylsalicylic acid in the aqueous solution. It could be filtered off and recrystallized from boiling water, but this would entail losses in transfer. An easier procedure is to heat the aqueous solution that contains the precipitated aspirin.

Aspirin

Add a boiling stick and heat the mixture to boiling, at which time the aspirin should dissolve completely. If it does not, add more water. Long boiling will hydrolyze the aspirin to salicylic acid, mp 157–159°C. Once completely dissolved the aspirin should be allowed to crystallize slowly as the solution cools to room temperature in an insulated container. Once the tube has reached room temperature it should be cooled in ice for several minutes and then the solvent removed with a Pasteur pipette. The crystals are to be washed with a few drops of ice-cold water and then scraped out onto a piece of filter paper. Squeezing the crystals between sheets of the filter paper will hasten drying. Once they are completely dry, determine the weight of the acetylsalicylic acid and its melting point.

Questions

1. Write equations showing how caffeine could be extracted from an organic solvent and subsequently regenerated.
2. Write equations showing how acetaminophen might be extracted from an organic solvent such as ether, if it were soluble.
3. Write detailed equations showing the mechanism by which aspirin is hydrolyzed in boiling, slightly acidic water.

Biosynthesis of Ethanol 18

PRELAB EXERCISE: *List the essential chemical substances, the solvent, and conditions for converting glucose to ethanol.*

Fermentation

Human beings have been preparing fermented beverages for more than 5000 years. Materials excavated from Egyptian tombs dating to the third millennium B.C. demonstrate the operations used in making beer and leavened bread. The history of fermentation, whereby sugar is converted to ethanol by the action of yeast, is also a history of chemistry. The word "gas" was coined by van Helmont in 1610 to describe the bubbles produced in fermentation. Leeuwenhoek observed and described the cells of yeast with his newly invented microscope in 1680. Joseph Black in 1754 discovered carbon dioxide and showed it to be a product of fermentation, the burning of charcoal, and respiration. Lavoisier in 1789 showed that sugar gives ethanol and carbon dioxide and made quantitative measurements of the amounts consumed and produced.

Glucose $\rightarrow 2\ C_2H_5OH + 2\ CO_2$

Once the mole concept was established, Gay-Lussac in 1815 could show that one mole of glucose gives exactly two moles of ethanol and two moles of carbon dioxide. But the process of fermentation stumped some great chemists. The little-known Kutzing wrote in 1837, on the basis of microscopic observation, "It is obvious that chemists must now strike yeast off the role of chemical compounds, since it is not a compound but an organized body, an organism." On the other side were chemists such as Berzelius, who believed that yeast had a catalytic action, and Liebig, who put forth a "theory of motion of the elements within a compound which caused a disturbance of equilibrium which was communicated to the elements of the substance with which it came in contact thus forming new compounds."

Pasteur's contribution

It remained for Pasteur to show that fermentation was a physiologic action associated with the life processes of yeast. Through his microscope he could see the yeast cells that grew naturally on the surface of grapes. He showed that grape juice carefully extracted from the center of a grape and exposed to clean air would not ferment. In his classic paper of 1857, he described fermentation as the action of a living organism; but because the conversion of glucose to ethanol and carbon dioxide is a balanced equation, other chemists disputed his findings. They searched for the substance in yeast that might cause the reaction. The search lasted for 40 years, eventually ended by a serendipitous experiment by Eduard Buchner. He made a cell-free extract of yeast that would still cause the conversion of sugar to alcohol.

Enzymes: fermentation catalysts

This cell-free extract contained the catalysts, which we now call enzymes, that were necessary for fermentation—a discovery that earned him the 1907 Nobel prize. In 1905 Harden discovered that inorganic phosphate added to the enzymes increased the rate of fermentation and was itself consumed. This result led him to eventually isolate fructose 1,6-diphosphate. Clearly the history of biochemistry is intimately associated with the study of alcoholic fermentation.

Sources of enzymes:
Grape skins
Malt
Saliva
Yeast

Ancient man discovered many of the essential reactions of alcoholic fermentation completely by accident. That crushed grapes would soon begin to froth and bubble and produce a pleasant beverage is a discovery lost in time. But what of those who lived in colder climates where the grape did not grow? How did they discover that the starch of wheat or barley could be converted to sugar by the enzymes in malt? When grain germinates, enzymes are produced that turn the starch into sugar. The process of malting involves letting the grain start to germinate and then heating and drying the sprouts to stop the process before the enzymes are used up. The color of the malt depends on the temperature of the drying. The darkest is used for stout and porter, the lighter for brown, amber, and pale ale. Because of a discovery some time ago that the resulting beverage did not spoil so rapidly if hops were added, we now also have beer.

Other sources exist for the amylases that catalyze the conversion of starch to glucose. The Peruvian campasinos (peasants) make a drink called "chicha" from masticated wheat, which is dried in small cakes. When water, yeast, and more ground wheat are added, the resulting mixture ferments to a beerlike beverage. The enzyme salivary amylase is the catalyst for this starch-to-glucose conversion.

The baker makes use of fermentation by taking advantage of the gas released to leaven his bread. In the present experiment baker's yeast is used to convert sucrose, ordinary table sugar, into ethanol and carbon dioxide

$$H_2O + \text{Sucrose} \xrightarrow{\text{Enzymes}} 4\ CH_3CH_2OH + 4\ CO_2$$

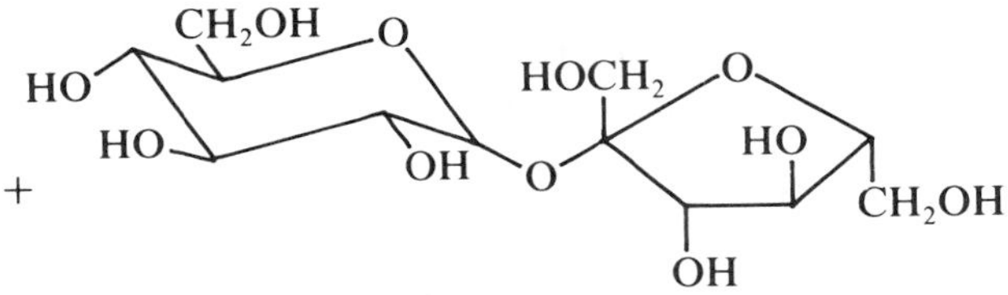

Sucrose

with the aid of some 14 enzymes as catalysts, in addition to adenosine triphosphate (ATP), phosphate ion, thiamine pyrophosphate, magnesium ion, and reduced nicotinamide adenine dinucleotide (NADH), all present in yeast. The fermentation process—known as the Emden-Meyerhof-Parnas scheme—involves the hydrolysis of sucrose to glucose and fructose which, as their phosphates, are cleaved to two three-carbon fragments. These fragments, as their phosphates, eventually are converted to pyruvic acid, which is decarboxylated to give acetaldehyde. Acetaldehyde, in turn, is reduced to ethanol in the final step. Each step requires a specific enzyme as a catalyst and often inorganic ions, such as magnesium and, of course, phosphate. Thirty-one kilocalories of heat are released per mole of glucose consumed in this sequence of anaerobic reactions.

This same sequence of reactions, up to the formation of pyruvic acid, occurs in the human body in times of stress when energy is needed, but not enough oxygen is available for normal aerobic oxidation. The pyruvic acid under these conditions is converted to lactic acid. It is the buildup of lactic acid in the muscles that is partly responsible for the feeling of fatigue.

1. Glucose → **2. Glucose-6-phosphate** ⇌ **3. Fructose-6-phosphate** → **4. Fructose-1,6-diphosphate** → **5. Dihydroxyacetone phosphate** + **6. Glyceraldehyde-3-phosphate**

6. → **7. 1,3-Diphosphoglyceric acid** → **8. 3-Phosphoglyceric acid** → **9. 2-Phosphoglyceric acid** → **10. Phosphoenolpyruvic acid**

10. → **11. Pyruvic acid** → (CO_2) **12. Acetaldehyde** → **13. Ethanol**

11. → (**14. Lactic acid**)

The Emden-Meyerhoff-Parnas scheme

The first step in the sequence is the formation of glucose-6-phosphate (**2**) from glucose (**1**). The reaction requires adenosine triphosphate (ATP), which is converted to the diphosphate (ADP) by catalysis with the enzyme glucokinase, which requires magnesium ion to function. This conversion is one of the reactions in which energy is released. In the living organism this energy can be used to do work; in fermentation it simply creates heat.

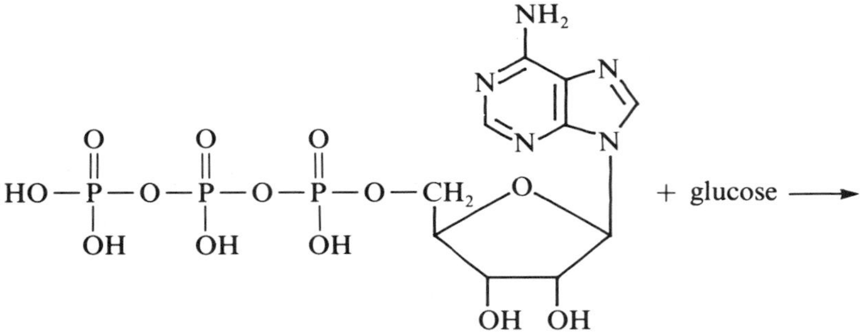

+ glucose ⟶

Adenosine triphosphate (ATP)

+ glucose-6-phosphate + 9 kcal/mole

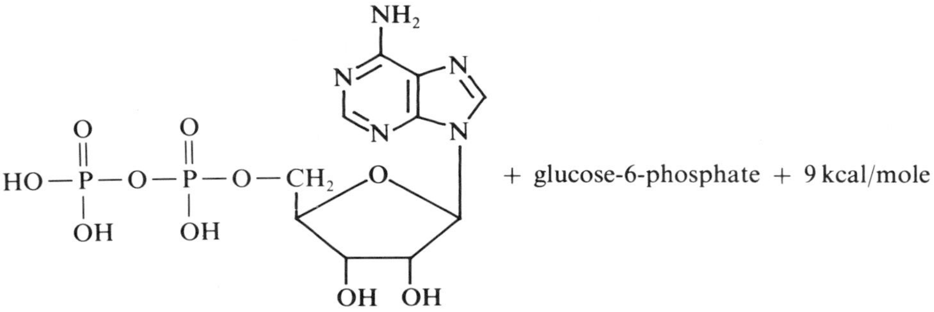

Adenosine diphosphate (ADP)

In the next step glucose-6-phosphate (**2**) is converted through the enol of the aldehyde to fructose-6-phosphate (**3**) by the enzyme phosphoglucoisomerase. The fructose monophosphate **3** is converted to the diphosphate **4** by the action of ATP under the influence of phosphofructokinase with the release of more energy. This diphosphate, **4**, undergoes a reverse aldol reaction catalyzed by aldolase to give dihydroxyacetone phosphate (**5**) and glyceraldehyde-3-phosphate (**6**). The latter two are interconverted by means of triosphosphate isomerase. The aldehyde of glyceraldehyde phosphate (**6**) is oxidized by nicotinamide adenine dinucleotide (NAD^+) in the presence of another enzyme to a carboxyl group that is phosphorylated with inorganic phosphate. In the next reaction ADP is converted to ATP as **7** loses phosphate to give **8**. A mutase converts the 3-phosphate, **8**, to the 2-phosphate, **9**. An enolase converts **9** to **10** and a kinase converts **10** to pyruvic acid (**11**).

A decarboxylase converts pyruvic acid (**11**) to acetaldehyde (**12**) in the fermentation process. Yeast alcohol dehydrogenase (YAD), a well-studied enzyme, catalyzes the reduction of acetaldehyde to ethanol. The reducing agent is reduced nicotinamide adenine dinucleotide, NADH.

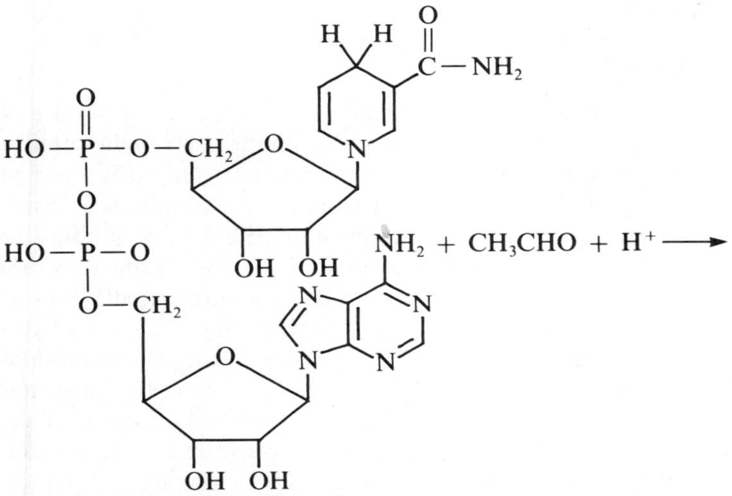

$+ CH_3CHO + H^+ \longrightarrow$

Reduced nicotinamide adenine dinucleotide, NADH

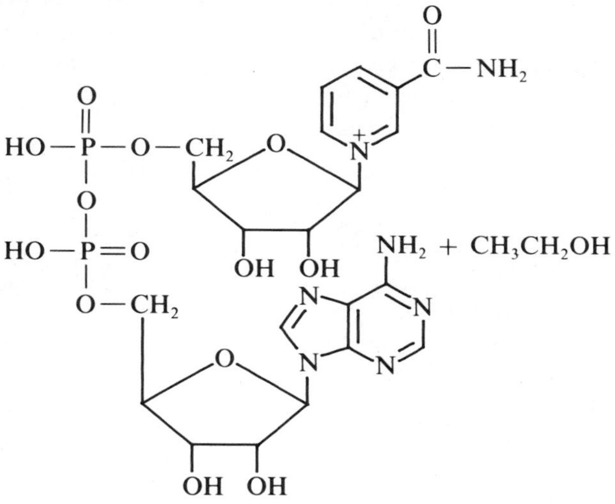

$+ CH_3CH_2OH$

Nicotinamide adenine dinucleotide, NAD$^+$

Enzymes are labile

Enzymes are remarkably efficient catalysts, but they are also labile (sensitive) to such factors as heat and cold, changes in pH, and various specific inhibitors. In the first experiment of this chapter you will have an opportunity to observe the biosynthesis of ethanol and to test the effects of various agents on the enzyme system.

This experiment involves the fermentation of 1.3 g of ordinary cane sugar using baker's yeast. The resulting dilute solution of ethanol, after removal of the yeast by filtration, can be distilled according to the procedures of Chapter 3.

Experiments

1. Fermentation of Glucose

Protect the fermenting reaction from exposure to oxygen and contaminating materials. Under aerobic conditions acetobacter bacteria can convert ethanol to acetic acid.

Reaction time: 1 week

To a 5-mL round-bottomed long-necked flask add 90 mg of dry yeast and 1.25 mL of warm (up to 50°C) water. Shake the mixture thoroughly until it is more or less homogeneous in appearance and then add to it 9 mg of disodium hydrogen phosphate, 1.30 g of sucrose, and an additional 3.75 mL of water, which should be warmed to about 45°C. Shake this mixture to ensure complete mixing and then fit the neck of the flask with a septum through which is inserted from the underside a syringe needle that is connected to an 8-in length of polyethylene tubing (Fig. 18.1). Lead this tubing beneath the surface of about 2 mL of a saturated aqueous solution of calcium hydroxide (limewater) in a reaction tube. The tube in the limewater will act as a seal to prevent air and unwanted enzymes from entering the flask but will allow gas to escape. Place the assembly in a warm spot (the optimum temperature for the reaction is 35°C) for a week, at which time the evolution of carbon dioxide will have ceased. What is the precipitate in the limewater?

On a small scale it will be necessary to provide external heat to maintain the fermentation or to group several reactions closely together in an insulated container. On a larger scale, since the fermentation is an exothermic reaction, there is enough heat evolved by the reaction to keep the mixture warm and promote the biosynthesis of ethanol.

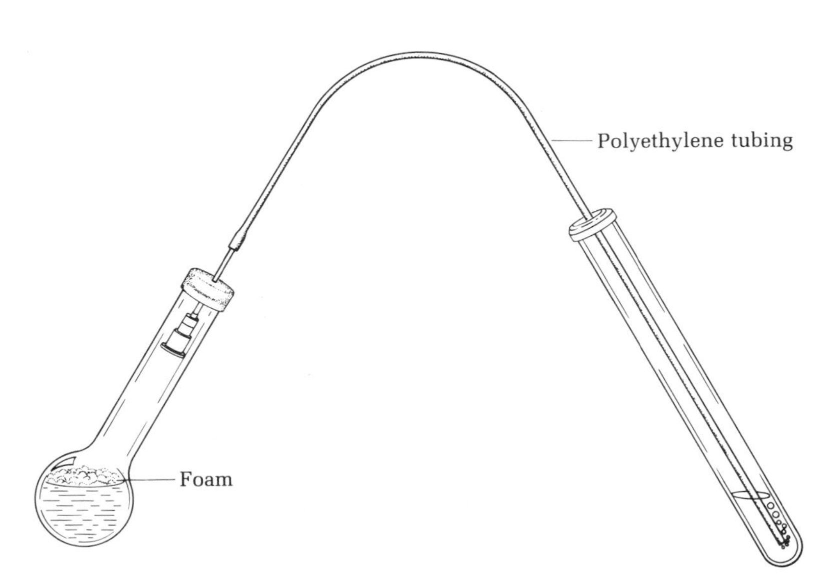

FIG. 18.1 Fermentation apparatus.

On completion of fermentation add about 0.25 g of Celite filter aid (diatomaceous earth, face powder) to the flask and shake it vigorously. Celite is added to make it possible to filter the solution. Otherwise the yeast cells will clog the filter paper. Filter the mixture on the Hirsch funnel into a 25-mL filter flask that is attached to the water aspirator through a trap by vacuum tubing (Figs 3.6–3.7). Because the small filter flask will easily tip over, clamp it to a ring stand. Moisten the filter paper with water and apply gentle suction (water supply to aspirator turned on full force, valve to trap partly open), and slowly pour the reaction mixture onto the filter. Wash out the flask with a milliliter of water and rinse the filter cake with this water. Full vacuum in this filtration will evaporate some of the desired ethanol.

The filtrate, which is a dilute solution of ethanol contaminated with a few bits of cellular material and other organic compounds (acetic acid if you are not careful), is saved in a stoppered container until it is distilled following the procedure outlined in Chapter 5.

2. Effect of Various Reagents and Conditions on Enzymatic Reactions

Prepare a fermentation mixture exactly as described in the first part of this experiment. After about 15 min, when the fermentation should be progressing nicely, split the mixture into five equal parts in reaction tubes. To one tube add 1.0 mL of water, to the next add 1.0 mL of 95% ethanol, and to the next add 1.0 mL of 0.5 M sodium fluoride solution. Heat the next tube for 5 min in a steam bath or boiling water bath and cool the next tube for 5 min in ice. Add to each tube two drops of mineral oil. The oil will float on the aqueous solution and prevent exposure to oxygen, which is necessary because fermentation is an anaerobic process. Place the tubes in a beaker of water at room temperature for 15 min, then connect them one at a time while in the beaker to an inverted 1.0-mL graduated pipette that has been plugged at one end as shown in Fig. 18.2 on the following page. Record the volume of gas in the pipette every minute for 5 min or count the bubbles evolved per minute. Plot a graph of volume of gas against time for each of the five reactions. What conclusions can you draw from the results of these five reactions?

***Caution**: Handle the sodium fluoride solution with great care—it is very poisonous if ingested*

Questions

1. Using yeast, can glucose be converted to ethanol? Can fructose be converted to ethanol?
2. Write the equation for the precipitate formed in the test tube containing calcium hydroxide.
3. In this experiment could 90% ethanol be made by adding more sugar to the fermentation flask?

FIG. 18.2 Apparatus for measuring carbon dioxide evolved.

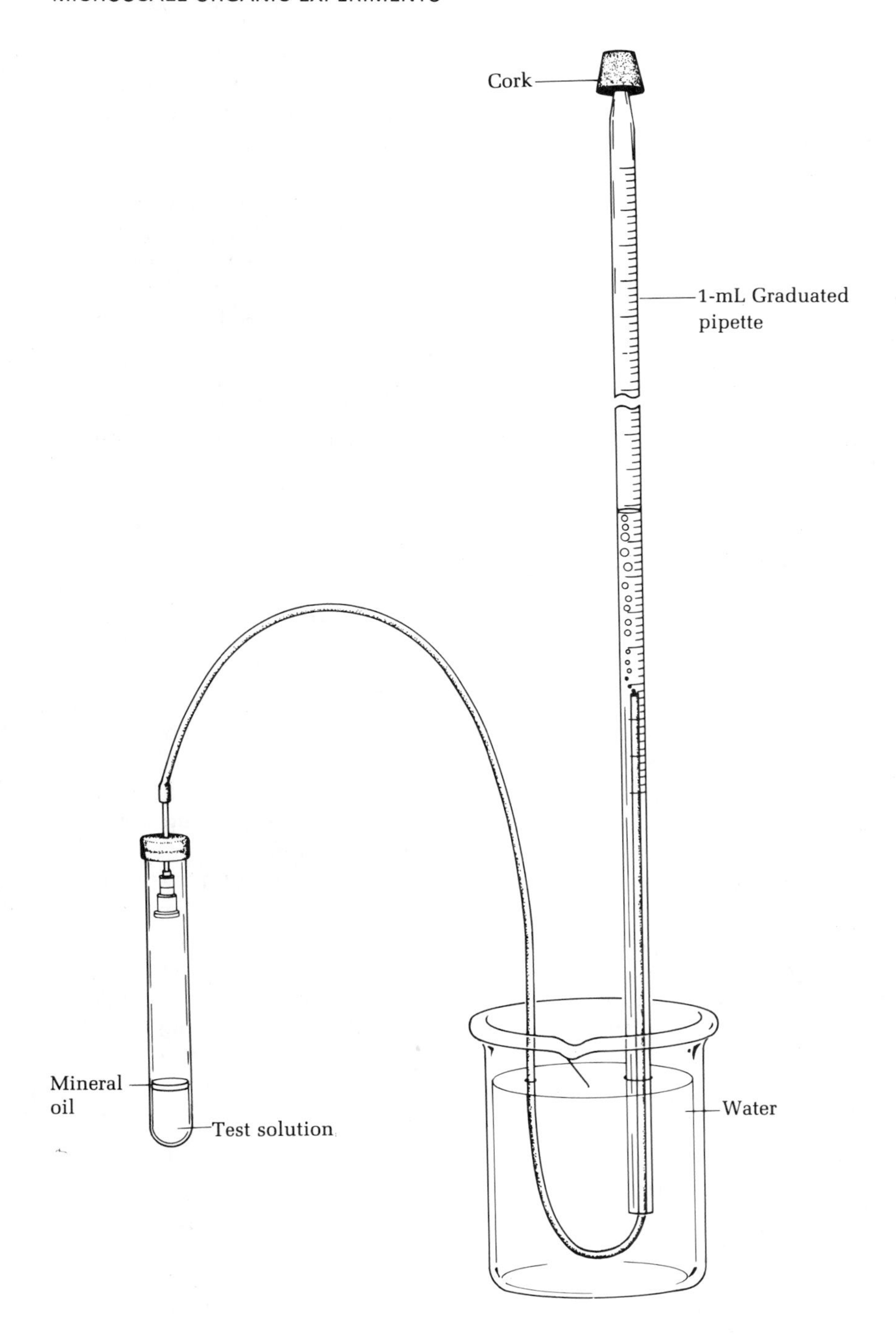

Infrared Spectroscopy 19

PRELAB EXERCISE: *When an infrared (ir) spectrum is run, there is a possibility that the chart paper is not properly placed or that the spectrometer is not mechanically adjusted. Describe how you could calibrate an infrared spectrum.*

ir—*To detect functional groups. A fast microanalytical technique.*

The presence and also the environment of functional groups in organic molecules can be identified by infrared spectroscopy. Like nuclear magnetic resonance and ultraviolet spectroscopy, infrared spectroscopy is nondestructive. Moreover, the small quantity of sample needed, the speed with which a spectrum can be obtained, the relatively small cost of the spectrometer, and the wide applicability of the method combine to make infrared spectroscopy one of the most useful tools available to the organic chemist.

Infrared radiation, which is electromagnetic radiation of longer wavelength than visible light, is detected not with the eyes but by a feeling of warmth on the skin. When absorbed by molecules, radiation of this wavelength (typically 2.5 to 15 microns), increases the amplitude of vibrations of the chemical bonds joining atoms.

Wavenumbers, cm^{-1}

Infrared spectra are measured in units of frequency or wavelength. The wavelength is measured in micrometers, *μm*, or microns, μ ($1\,\mu = 1 \times 10^{-4}$ cm). The positions of absorption bands are measured in frequency units by wavenumbers, $\bar{v}$, which are expressed in reciprocal centimeters, cm^{-1}, corresponding to the number of cycles of the wave in each centimeter.

$$\mathrm{cm}^{-1} = \frac{10{,}000}{\mu}$$

Various scales employed

Unlike ultraviolet and nmr spectra, infrared spectra are inverted and are not always presented on the same scale. Some spectrometers record the spectra on an ordinate linear in microns, but this compresses the low wavelength region. Other spectrometers present the spectra on a scale linear in reciprocal centimeters, but linear on two different scales, one between 4000 and 2000 cm^{-1}, which spreads out the low wavelength region, and the other a smaller one between 200 and 667 cm^{-1}.

To picture the molecular vibrations that interact with infrared light, imagine a molecule as being made up of balls (atoms) connected by springs (bonds). The vibration can be described by Hooke's law from classical mechanics, which says that the frequency of a stretching vibration is directly proportional to the strength of the spring (bond) and inversely proportional

C—H N—H Stretch at O—H short wavelengths

to the masses connected by the spring. Thus we find C—H, N—H and O—H bond-stretching vibrations are high frequency (short wavelength) compared to those of C—C and C—O, because of the low mass of hydrogen compared to that of carbon or oxygen. The bonds connecting carbon to bromine and iodine, atoms of large mass, vibrate so slowly that they are beyond the range of most common infrared spectrometers. A double bond can be regarded as a stiffer, stronger spring, so we find C=C and C=O vibrations at higher frequency than C—C and C—O stretching vibrations. And C≡C and C≡N stretch at even higher frequencies than C=C and C=O (but at lower frequencies than C—H, N—H, and O—H). These frequencies are in keeping with the bond strengths of single (~100 kcal/mole), double (~160 kcal/mole), and triple bonds (~220 kcal/mole).

The stretching vibrations noted above are intense and particularly easy to analyze. A nonlinear molecule of n atoms can undergo $3n - 6$ possible modes of vibration, which means cyclohexane with 18 atoms can undergo 48 possible modes of vibration. A single CH_2 group can vibrate in six different ways. Each vibrational mode produces a peak in the spectrum because it corresponds to the absorption of energy at a discrete frequency. These many modes of vibration create a complex spectrum that defies simple analysis, but even in very complex molecules certain functional groups have characteristic frequencies that can easily be recognized. Within these functional groups are the above-mentioned atoms and bonds, C—H, N—H, O—H, C=C, C=O, C≡C, and C≡N. Their absorption frequencies are given in Table 19.1.

Functional groups have characteristic frequencies

When the frequency of infrared light is the same as the natural vibrational frequency of an interatomic bond, light will be absorbed by the molecule and the amplitude of the bond vibration will increase. The intensity of infrared absorption bands is proportional to the change in dipole moment that a bond undergoes when it stretches. Thus the most intense bands (peaks) in an infrared spectrum are often from C=O and C—O stretching vibrations, while the C≡C stretching band for a symmetrical acetylene is almost nonexistent because the molecule undergoes no net change of dipole moment when it stretches:

Intensity ∝ to dipole moment change

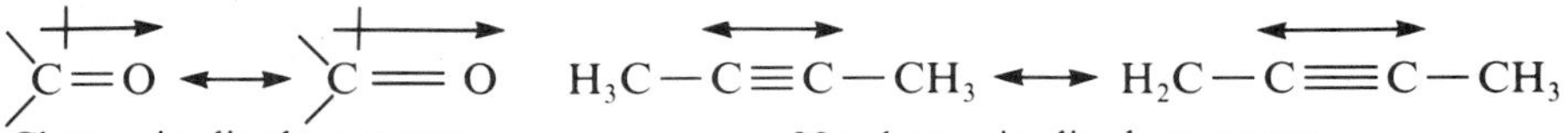

$$\rangle C=O \longleftrightarrow \rangle C=\!=O \qquad H_3C-C\equiv C-CH_3 \longleftrightarrow H_3C-C\equiv\!\equiv C-CH_3$$

Change in dipole moment — No change in dipole moment

Unlike proton nuclear magnetic resonance spectroscopy, where the area of the peaks is strictly proportional to the numbers of hydrogen atoms causing the peaks, the intensities of infrared peaks are not proportional to the numbers of atoms causing them. And where every peak or group of peaks in an nmr spectrum can be assigned to specific hydrogens in a molecule, the assignment of the majority of peaks in an infrared spectrum is usually not possible. Peaks to the right (longer wavelength) of 1250 cm^{-1} are the result of combinations of vibrations that are characteristic not of individual func-

TABLE 19.1 Characteristic Infrared Absorption Frequencies

	Wavenumber (cm^{-1})	Wavelength (μ)
O—H	3600–3400	2.78–2.94
N—H	3400–3200	2.94–3.12
C—H	3080–2760	3.25–3.62
C≡N	2260–2215	4.42–4.51
C≡C	2150–2100	4.65–4.76
C=O	1815–1650	5.51–6.06
C=C	1660–1600	6.02–6.25
C—O	1050–1200	9.52–8.33

tional groups, but of the molecule as a whole. This part of the spectrum is often referred to as the "fingerprint region," because it is uniquely characteristic of each molecule. While two organic compounds can have the same melting points or boiling points and can have identical ultraviolet and nmr spectra, they cannot have identical ir spectra. Infrared spectroscopy is thus the final arbiter in deciding whether two compounds are identical.

The "fingerprint region"

Analysis of Infrared Spectra

Three rules apply to all analyses: (1) pay most attention to the strongest absorptions, (2) pay more attention to peaks to the left (shorter wavelength) of 1250 cm^{-1}, and (3) pay as much attention to the absence of certain peaks as to the presence of others. The absence of characteristic peaks will definitely exclude certain functional groups. Be wary of weak O—H peaks because water is a common contaminant of many samples. Because KBr is hygroscopic it is often found in the spectra of KBr pellets.

Starting at the left-hand side of the spectrum:

O—H, *3600–3400 cm^{-1}*

The hydroxyl group gives a sharp peak at 3600 cm^{-1} for nonhydrogen-bonded groups and a broad peak at 3400 cm^{-1} for hydrogen-bonded groups. Depending on the concentration of the sample, the relative intensities of these two peaks will vary (see the spectrum of 1-naphthol, p. 446, where the sharp and broad peaks are just barely resolved). Carboxyl groups give a characteristic very intense and very broad wedge-shaped band, which extends to and obscures the C—H region [see the spectra of acetylsalicylic acid (aspirin), p. 243].

N—H, *3400–3200 cm^{-1}*

A sharp peak at 3400–3200 cm^{-1} is shown by N—H vibrations. The NH_2 group usually shows a doublet (see aniline, p. 340).

C—H, *3300–3000 cm^{-1}*

Unsaturated C—H bonds absorb in this region with alkynes at 3300 cm^{-1}, alkenes at 3080–3010 cm^{-1}, and aromatic compounds at 3050 cm^{-1}. Alkanes also absorb over this narrow range. All of these C—H peaks are strong except for the aromatic C—H. The complete absence of peaks to the left of

$3000\ cm^{-1}$ indicates no unsaturation while the complete absence of peaks to the right of $3000\ cm^{-1}$ indicates no aliphatic hydrogens. Nmr spectroscopy is the best method for identifying aromatic hydrogens.

C≡N, *2260–2215* cm^{-1} Although not common this functional group is very easily identified by this infrared peak because only one other peak, that at $2150–2100\ cm^{-1}$ from alkynes, appears near it. It is usually a strong and very sharp peak.

C≡C, *2150–2100* cm^{-1} Again alkynes are not common functional groups but, if terminal, they give strong and very sharp peaks in this range. If the alkyne is not terminal the peak can be weak or absent in the case of a symmetrical alkyne. No other functional groups give peaks near these for C≡C and C≡N.

C=O *1870–1825 and 1790–1765* cm^{-1} Acid anhydrides are also not common functional groups but are easily recognized because they always have two strong peaks in the indicated ranges. See Note below.

1815–1800 cm^{-1} Acid chlorides absorb at this frequency. The presence of halogen is, however, more easily determined by the Beilstein test.

1750–1735 cm^{-1} Esters and lactones give a very strong band in this frequency range. See Note.

1725–1705 cm^{-1} Peaks in this range are very strong and indicative of unsubstituted and unconjugated aldehydes and ketones. See Note.

1700 cm^{-1} Carboxylic acids produce a strong absorption at this frequency. See Note.

1690–1670 cm^{-1} Amides give strong peaks in this frequency range. See Note.

NOTE: The frequencies given for the C=O stretching vibrations for anhydrides, acid chlorides, esters, lactones, aldehydes, ketones, carboxylic acids and amides refer to the open chain or unstrained functional group in a nonconjugated system. If the carbonyl group is conjugated with a double bond or an aromatic ring, the frequency is $30\ cm^{-1}$ less. If it is conjugated to groups on both sides (cross-conjugated), the frequency is $50\ cm^{-1}$ less.

When the carbonyl group is in a ring smaller than six members the frequency is higher by about $25\ cm^{-1}$ and halogen or oxygen substitution on the carbon adjacent to an aldehyde or ketone carbonyl also increases the frequency by about $25\ cm^{-1}$. See the detailed list of carbonyl frequencies in Table 19.2.

C=C, *1660–1625* cm^{-1} Alkenes appear in this range, but the intensity of the peak is variable and cannot be relied upon. Conjugation shifts the peak to lower frequencies by about $30\ cm^{-1}$.

1600 cm^{-1} Aromatic rings give a medium to strong peak near this frequency and also a peak at $1450\ cm^{-1}$. Conjugated dienes also produce a peak at $1600\ cm^{-1}$.

—NO_2, *1520 and 1350* cm^{-1} Coupled stretching vibrations of the nitro group give rise to these two intense and easily recognizable bands.

—CH_3, *1375* cm^{-1} A methyl group displays a strong band at this frequency.

TABLE 19.2 Characteristic Infrared Carbonyl Stretching Frequencies (chloroform solutions)

	Wavenumber (cm^{-1})	Wavelength (μ)
Aliphatic ketones	1725–1705	5.80–5.87
Acid chlorides	1815–1785	5.51–5.60
α,β-Unsaturated ketones	1685–1666	5.93–6.00
Aryl ketones	1700–1680	5.88–5.95
Cyclobutanones	1775	5.64
Cyclopentanones	1750–1740	5.72–5.75
Cyclohexanones	1725–1705	5.80–5.87
β-Diketones	1640–1540	6.10–6.50
Aliphatic aldehydes	1740–1720	5.75–5.82
α,β-Unsaturated aldehydes	1705–1685	5.80–5.88
Aryl aldehydes	1715–1695	5.83–5.90
Aliphatic acids	1725–1700	5.80–5.88
α,β-Unsaturated acids	1700–1680	5.88–5.95
Aryl acids	1700–1680	5.88–5.95
Aliphatic esters	1740	5.75
α,β-Unsaturated esters	1730–1715	5.78–5.83
Aryl esters	1730–1715	5.78–5.83
Formate esters	1730–1715	5.78–5.83
Vinyl and phenyl acetate	1776	5.63
δ-Lactones	1740	5.75
γ-Lactones	1770	5.65
Acyclic anhydrides (two peaks)	1840–1800 1780–1740	5.44–5.56 5.62–5.75
Primary amides	1694–1650	5.90–6.06
Secondary amides	1700–1670	5.88–6.01
Tertiary amides	1670–1630	5.99–6.14

$>C(CH_3)_2$ *1385 and 1365* cm^{-1}

This grouping gives a strong doublet at these two frequencies.

$C_6H_5—O—$, ~ *1200* cm^{-1}
$—C—O—$, ~ *1150* cm^{-1}
$—CH—O—$, ~ *1150* cm^{-1}
$—CH_2—O—$, ~ *1050* cm^{-1}

These carbon-oxygen stretching or bending vibrations give rise to strong bands near the indicated positions, but these may vary. (See the spectrum of acetylsalicylic acid, p. 243). Esters often give two strong peaks some place in this range.

The infrared spectrometer

Figure 19.1 is a schematic representation of a typical double beam, optical null, infrared spectrometer. An electrically heated metal rod serves as the radiation source; the radiation passes through both the sample cell and reference cell, through combs and a beam chopper to the dispersion grating. In some instruments the radiation is dispersed by a prism made of sodium chloride, which is transparent to infrared radiation. Of the electromagnetic radiation frequencies spread out by the grating (or prism), only a small range of frequencies is allowed to pass through the slit to the detector, which is a thermocouple with a very rapid response time.

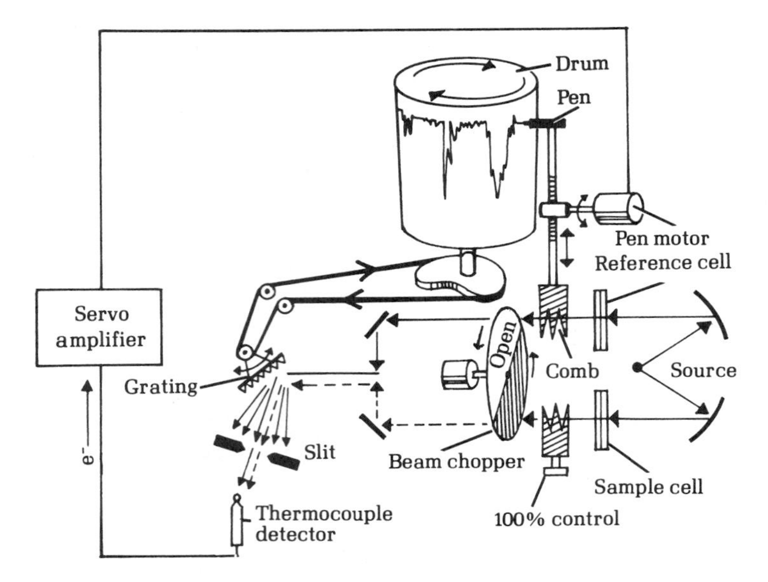

FIG. 19.1 Schematic diagram of a double-beam, optical null, infrared spectrometer.

The spectrometer works on the optical null principle. The detector senses infrared light coming alternately through the substances in the sample and reference cells. If the amount of light is the same from both beams the detector produces a direct current and nothing happens; but if less light comes through the sample beam than the reference beam (because of absorption of radiation by the sample molecules), then the detector senses an alternating current that alternates at the rate the chopper is turning. This current is amplified in the servo amplifier, which activates the pen motor. The pen motor moves the pen down the paper drawing an absorption band and at the same time drives a comb into the reference beam just far enough so that the detector will again sense a null, i.e., no alternating current. The motion of the drum holding the paper is linked to the grating so that as the drum moves the grating moves with it to scan the entire range of frequencies.

Correlation tables

Extensive correlation tables and discussions of characteristic group frequencies can be found in specialized references.[1] As one example, consider the band patterns of toluene, and of *o*-, *m*-, and *p*-xylene, which appear in the frequency range 2000 to 1650 cm^{-1} (Fig. 19.2). These band patterns are due to changes in the dipole moment accompanying changes in vibrational modes of the aromatic ring and are surprisingly similar to those for monosubstituted and other *o*-, *m*-, and *p*-disubstituted benzenes.

Experimental Aspects

*Sample, solvents, and equipment **must be dry***

Infrared spectra can be determined on neat (undiluted) liquids, on solutions with an appropriate solvent, and on solids as mulls and KBr pellets. Glass is opaque to infrared radiation; therefore, the sample and reference cells used

1. See list of references at end of chapter.

in infrared spectroscopy are sodium chloride plates. The sodium chloride plates are fragile and can be attacked by moisture. *Handle only by the edges.*

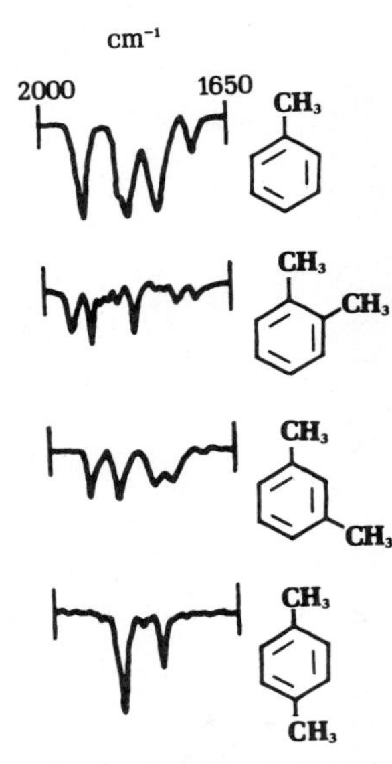

FIG. 19.2 Band patterns of toluene and *o*-, *m*-, and *p*-xylene. These peaks are very weak.

Spectra of Neat Liquids

To run a spectrum of a neat liquid (free of water!) remove a demountable cell (Fig. 19.3) from the desiccator and place a drop of the liquid between the salt plates, press the plates together to remove any air bubbles, and add the top rubber gasket and metal top plate. Next, put on all four of the nuts and *gently* tighten them to apply an even pressure to the top plate. Place the cell in the sample compartment (nearest the front of the spectrometer) and run the spectrum.

Although running a spectrum on a neat liquid is convenient and results in no extraneous bands to interpret, it is not possible to control the path length of the light through the liquid in a demountable cell. A low-viscosity liquid when squeezed between the salt plates may be so thin that the short path length gives peaks that are too weak. A viscous liquid, on the other

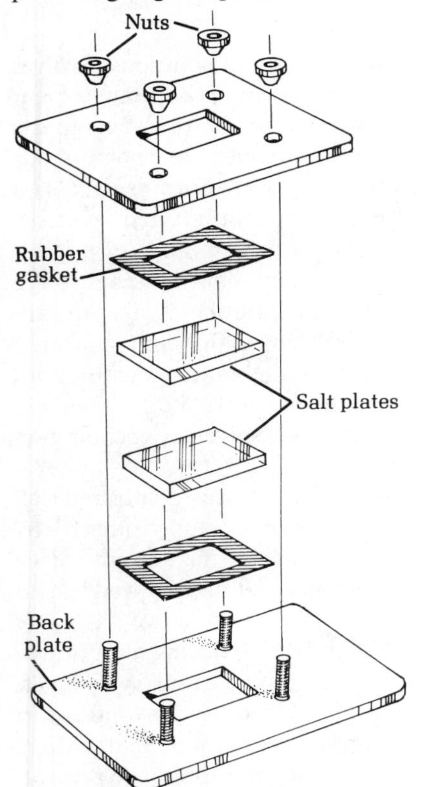

FIG. 19.3 Exploded view of a demountable salt cell for analyzing the infrared spectra of neat liquids.

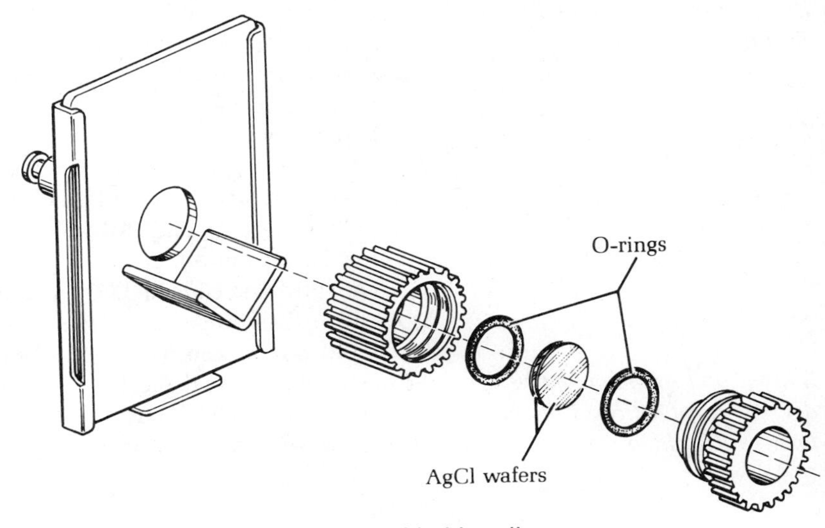

FIG. 19.4 Demountable silver chloride cell.

hand, may give peaks that are too intense. A properly run spectrum will have the most intense peak with an absorbance of about 1.0. Unlike the nmr spectrometer, the infrared spectrometer does not usually have a control to adjust the peak intensities; this control is possible only by adjusting the sample concentration.

Another demountable cell is pictured in Fig. 19.4 on page 191. The plates are thin wafers of silver chloride, which is transparent to infrared radiation. This cell has advantages over the salt cell in that the silver chloride disks are more resistant to breakage than NaCl plates, they are less expensive, and they are not affected by water. Because silver chloride is photosensitive the wafers must be stored in the dark to prevent them from turning black. Because one side of each wafer is recessed, the thickness of the sample can be varied according to the manner in which the cell is assembled. The disks are cleaned by rinsing them with an organic solvent such as acetone or ethanol and wiping dry with an absorbent paper towel.

Spectra of Solutions

Solvents: $CHCl_3$, CS_2, CCl_4

Caution! *Chloroform is toxic; use laboratory hood. Avoid the use of chloroform and carbon disulfide if possible.*

The most widely applicable method of running spectra of solutions involves dissolving an amount of the liquid or solid sample in an appropriate solvent to give a 10% solution. Just as in nmr spectroscopy, the best solvents to use are carbon disulfide and carbon tetrachloride; but because these compounds are not polar enough to dissolve many substances, chloroform is used as a compromise. Unlike nmr solvents, no solvent suitable in infrared spectroscopy is entirely free of absorption bands in the frequency range of interest (Figs. 19.5(a) and (b)). In chloroform, for instance, no light passes through the cell between 650 and 800 cm^{-1}. As can be seen from the figures, spectra obtained using carbon disulfide and chloroform cover the entire infrared frequency range. In practice, a base line is run with the same solvent in both cells to ascertain if the cells are clean and matched (Fig. 19.5(c)). Often it is necessary to obtain only one spectrum employing one solvent, depending on which region of the spectrum you need to use.

Three drops needed to fill cell

Three large drops of solution will fill the usual sealed infrared cell (Fig. 19.6). A 10% solution of a liquid sample can be approximated by dilution of one drop of the liquid sample with nine of the solvent. Since weights are more difficult to estimate, solid samples should be weighed to obtain a 10% solution.

Solvent and sample must be dry. Do not touch or breathe upon NaCl plates.

The infrared cell is filled by inclining it slightly and placing about three drops of the solution in the lower hypodermic port with a capillary dropper. The liquid can be seen rising between the salt plates through the window. In the most common sealed cell, the salt plates are spaced 0.1 mm apart. Make sure that the cell is filled past the window and that no air bubbles are present. Then place the Teflon stopper lightly but firmly in the hypodermic port. Be particularly careful not to spill any of the sample on the outside of the cell windows.

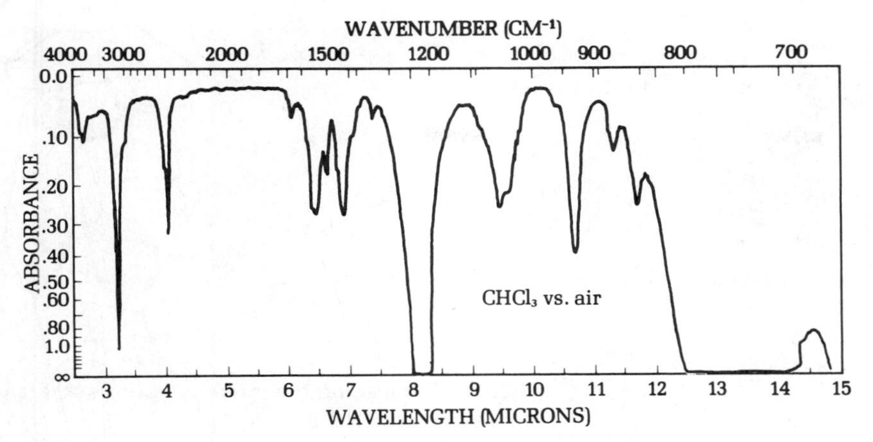

FIG. 19.5(a) Spectrum of chloroform in sample cell, air in reference cell. No infrared light passes through chloroform between 1200 and 1250 cm^{-1}, and between 650 and 800 cm^{-1}; therefore no information about sample absorption in those regions can be obtained.

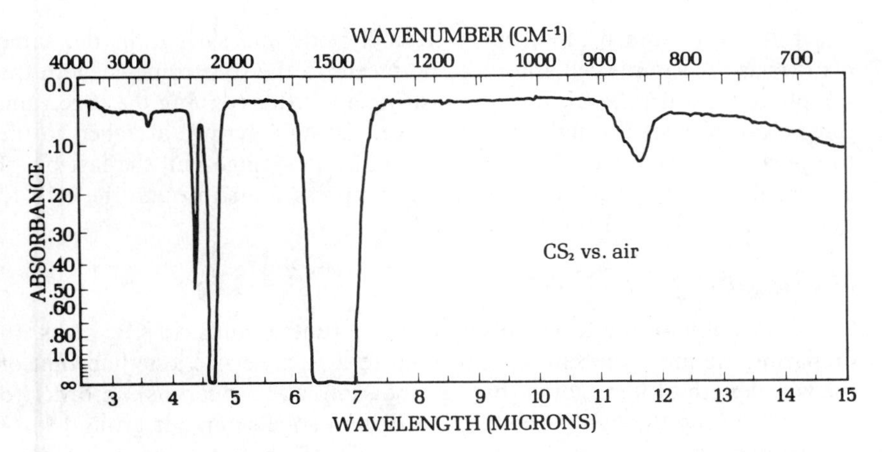

FIG. 19.5(b) Spectrum of carbon disulfide in sample cell, air in reference cell. No infrared light passes through carbon disulfide between 1430 and 1550 cm^{-1}.

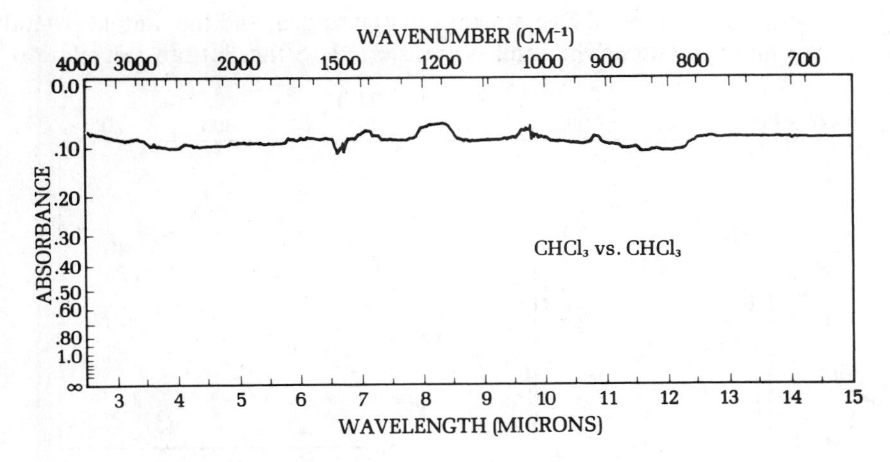

FIG. 19.5(c) Spectrum of chloroform in both sample and reference cells. A typical baseline.

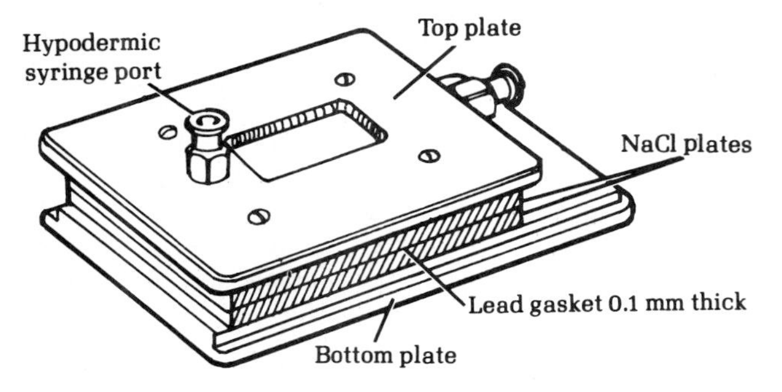

FIG. 19.6 Sealed infrared sample cell.

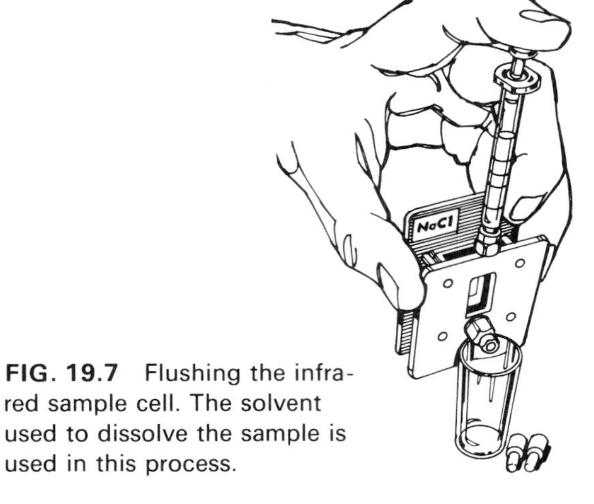

FIG. 19.7 Flushing the infrared sample cell. The solvent used to dissolve the sample is used in this process.

Dispose of waste solvents in the container provided

Fill the reference cell from a clean hypodermic syringe in the same manner as the sample cell and place both cells in the spectrometer, with the sample cell toward the front of the instrument. After running the spectrum, force clean solvent through the sample cell, using a syringe attached to the top port of the cell (Fig. 19.7). Finally, with the syringe pull the last bit of solvent from both cells, blow clean dry compressed air through the cells to dry them, and store them in a desiccator.

Mulls and KBr Disks

Solids insoluble in the usual solvents can be run as mulls or KBr disks. In preparing the mull, the sample is ground to a particle size less than that of the wavelength of light going through the sample (2.5 microns), in order to avoid scattering the light. About 15 to 20 mg of the sample is ground for 3 to 10 min in an agate mortar until it is spread over the entire inner surface of the mortar and has a caked and glassy appearance. Then, to make a mull, 1 or 2 drops of paraffin oil (Nujol) (Fig. 19.8) is added, and the sample ground 2 to 5 more minutes. The mull is transferred to the bottom salt plate of

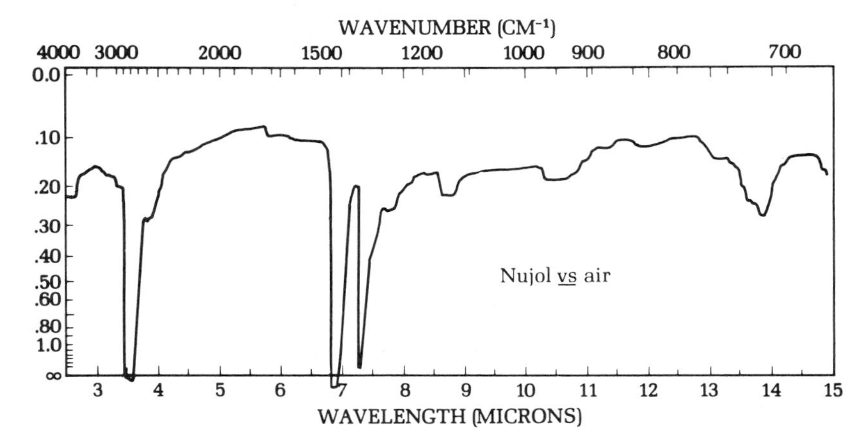

FIG. 19.8 Infrared spectrum of Nujol (paraffin oil).

a demountable cell (Fig. 19.3) using a rubber policeman, the top plate added and twisted to distribute the sample evenly and to eliminate all air pockets, and the spectrum run. Since the bands from Nujol obscure certain frequency regions, running another mull using Fluorolube as the mulling agent will allow the entire infrared spectral region to be covered. If the sample has not been ground sufficiently fine, there will be marked loss of transmittance at the short-wavelength end of the spectrum. After running the spectrum, the salt plates are wiped clean with a cloth saturated with an appropriate solvent.

Sample must be finely ground

Handle salt plates with care. Do not touch with fingers.

The spectrum of a solid sample can also be run by incorporating the sample in a KBr disk. This procedure needs only one disk to cover the entire spectral range, since KBr is completely transparent to infrared radiation. Although very little sample is required, making the disk calls for special equipment and time to prepare it. Since KBr is hygroscopic, water is a problem. The sample is first ground as for a mull and 1.5 mg of this is added to 300 mg of spectroscopic grade KBr (previously dried in an oven and stored in a desiccator). The two are gently mixed (not ground) and quickly placed in a 13-mm die and subjected to 14,000–16,000 lb/sq in. pressure for 3–6 min while under vacuum in a specially constructed hydraulic press. A transparent disk is produced, which is removed from the die with tweezers and placed in a special holder, prior to running the spectrum.

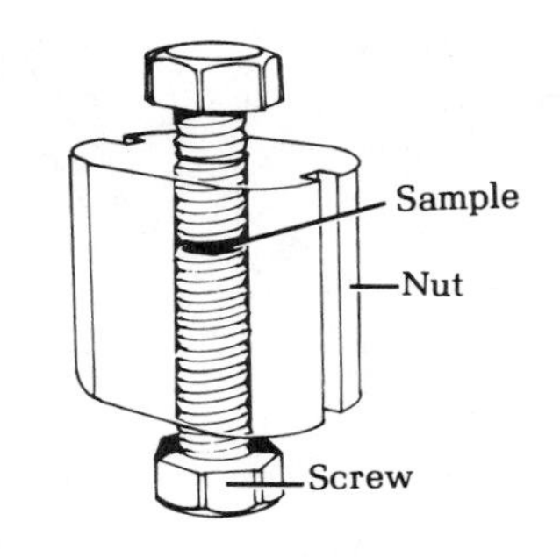

FIG. 19.9 KBr disk die. Pressure is applied by tightening the machine screws with a wrench.

A simple low-cost small press is illustrated in Fig. 19.9. The press consists of a large nut and two machine screws. The sample is placed between the two machine screws (which have polished faces), and the screws are tightened with a wrench with the nut held in a vise. The screws are then loosened and removed. The KBr disk is left in the nut, which is then mounted in the spectrometer to run the spectrum. An opaque area in the disk indicates insufficient pressure was applied. Too much pressure can result in crushed disks.

Running the Spectrum

Satisfactory spectra are easily obtained with the lower cost spectrometers, even those which have only a few controls and require few adjustments. To run a spectrum the paper must be positioned accurately, the pen set between 90% and 100% transmittance with the 100% control (0.0 and 0.05 absorbance), and the speed control set for a fast scan (usually one of about three minutes). The calibration of a given spectrum can be checked by backing up the drum and superimposing a spectrum of a thin polystyrene film. This film, mounted in a cardboard holder that has the frequencies of important peaks printed on it, will be found near most spectrometers. The film is held in the sample beam and parts of the spectrum to be calibrated are rerun. The spectrometer gain (amplification) should be checked frequently and adjusted when necessary. To check the gain, put the pen on the 90% transmittance line with the 100% control. Place your finger in the sample beam so that the pen goes down to 70% *T*. Then quickly remove your finger. The pen should overshoot the 90% *T* line by 2%.

Calibration with polystyrene film

Throughout the remainder of this book representative infrared spectra of starting materials and products will be presented and the important bands in each spectrum identified.

Experiment

Unknown Carbonyl Compound

Sample, solvents, and apparatus must be dry

Run the infrared spectrum of an unknown carbonyl compound obtained from the laboratory instructor. Be particularly careful that all apparatus and solvents are completely free of water, which will damage the sodium chloride cell plates. The spectrum can be calibrated by positioning the spectrometer pen at a wavelength of about 6.2 μ without disturbing the paper, and rerunning the spectrum in the region from 6.2 to 6.4 μ while holding the polystyrene calibration film in the sample beam. This will superimpose a sharp calibration peak at 6.246 μ (1601 cm^{-1}) and a less intense peak at 6.317 μ (1583 cm^{-1}) on the spectrum. Determine the frequency of the carbonyl peak and list the possible types of compounds that could correspond to this frequency (Table 19.2).

References

1. L. J. Bellamy, *The Infrared Spectra of Complex Organic Molecules*, Chapman and Hall, New York, 2nd Ed., 1980.
2. R. T. Conley, *Infrared Spectroscopy*, Allyn and Bacon, Boston, 2nd Ed., 1972.
3. K. Nakanishi and P. H. Solomon, *Infrared Absorption Spectroscopy*, Holden-Day, San Francisco, 1977.
4. R. M. Silverstein, C. G. Bassler, and T. C. Morrill, *Spectrometric Identification of Organic Compounds*, John Wiley, New York, 4th Ed., 1981. (Includes ir, uv, and nmr.)
5. H. A. Szymanski and R. E. Erickson, *Infrared Band Handbook*, Plenum, New York, 2nd Ed., 1970.
6. C. J. Pouchert, *The Aldrich Library of Infrared Spectra*, Aldrich Chemical Co., Milwaukee, 1970.
7. J. W. Cooper, *Spectroscopy Techniques for Organic Chemists*, John Wiley, New York, 1980.
8. G. Socrates, *Infrared Characteristic Group Frequencies*, John Wiley, New York, 1980.
9. J. R. Ferraro and L. J. Basile, eds., *Fourier Transform Infrared Spectroscopy*, Vol. 1, 1973; Vol. 2, 1979; Vol. 3, 1982, Techniques; Vol. 4, 1985, Applications to Chemical Systems, Academic Press, New York.

Nuclear Magnetic Resonance Spectroscopy 20

PRELAB EXERCISE: *Outline the preliminary solubility experiments you would carry out on an unknown using inexpensive solvents, before preparing a solution of the compound for nmr spectroscopy.*

nmr—Determination of the number, kind, and relative locations of hydrogen atoms (protons) in a molecule

Nuclear magnetic resonance (nmr) spectroscopy is a means of determining the number, kind, and relative locations of certain atoms, principally hydrogen, in molecules. Experimentally, the sample, 0.3 mL of a 20% solution in a 5-mm o.d. glass tube, is placed in the probe of the spectrometer between the faces of a powerful (14,000 gauss) permanent or electromagnet and irradiated with radiofrequency energy (60 MHz; 60,000,000 cps for protons). The absorption of radiofrequency energy versus magnetic field strength is plotted by the spectrometer to give a spectrum.

Integration

In a typical ^{1}H spectrum (Fig. 20.1, ethyl iodide) the relative numbers of hydrogen atoms (protons) in the molecule are determined from the *integral*, the stair-step line over the peaks. The height of the step is proportional to the area under the nmr peak, and in nmr spectroscopy (contrasted with infrared, for instance) the area of each group of peaks is directly proportional to the number of hydrogen atoms causing the peaks. Integrators are part of all nmr spectrometers, and running the integral takes no more time than running the spectrum. The different kinds of protons are indicated by their *chemical shifts*.[1] For ethyl iodide the two protons adjacent to the electronegative iodine atom are *downfield* (at lower magnetic field strength) (δ = 3.20 ppm) from the three methyl protons at δ = 1.83 ppm. Tables of chemical shifts for protons in various environments can be found in reference books[2] and are given graphically in Fig. 20.2 on p. 199.

Chemical shifts, δ

The relative locations of the five protons in ethyl iodide are indicated by the pattern of peaks on the spectrum. The three peaks indicate methyl

1. Chemical shift is a measure of a peak's position relative to the peak of a standard substance (e.g., TMS, tetramethylsilane), which is assigned a chemical shift value of 0.0.

2. Suggested references are L. M. Jackman and S. Sternhell, *Applications of Nuclear Magnetic Resonance Spectroscopy in Organic Chemistry*, 2nd ed., Pergamon Press, New York, 1969, and R. J. Abraham and P. Loftus, *Proton and Carbon-13 Spectroscopy, An Integrated Approach*, Heyden and Son, Philadelphia, 1978. See also K. L. Williamson, *Basic NMR Spectroscopy*, Vocational Media Associates/Prentice-Hall Media, Box 1050, Mount Kisco, NY 10549. This is an audio-visual program in six parts (The NMR phenomenon, Chemical shifts, Coupling constants, Preparing the sample, Running the spectrum, and Applications).

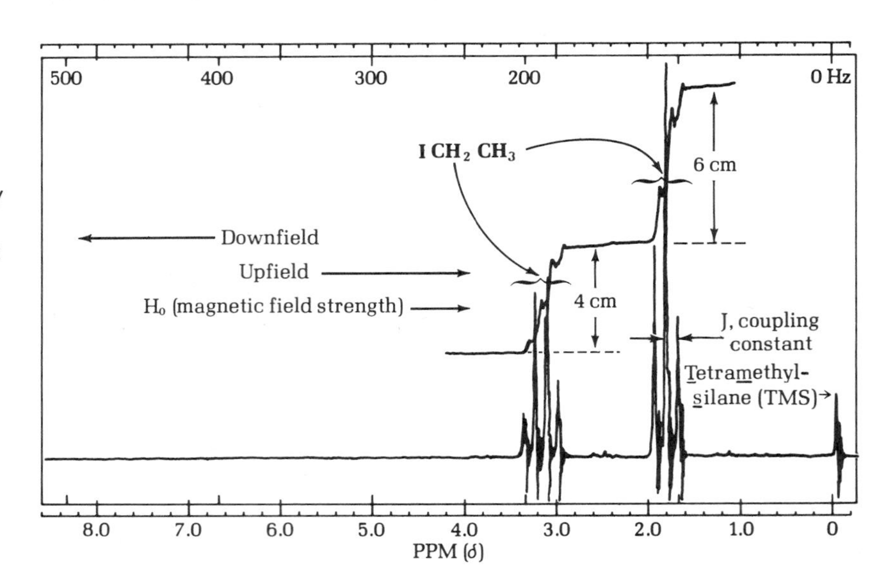

FIG. 20.1 Proton nmr spectrum of ethyl iodide. The staircase-like line is the integral. In the integral mode of operation the recorder pen moves from left to right and moves vertically a distance proportional to the areas of the peaks over which it passes. Hence, the relative area of the quartet of peaks at 3.20 ppm to the triplet of peaks at 1.83 ppm is given by the relative heights of the integral (4 cm is to 6 cm as 2 is to 3). The relative numbers of hydrogen atoms are proportional to the peak areas (2H and 3H).

protons adjacent to two protons; four peaks indicate methylene protons adjacent to three methyl protons. In general, in molecules of this type, a given set of protons will appear as $n + 1$ peaks if they are adjacent to n protons. The distance between adjacent peaks in the quartet and triplet is the *coupling constant*, J.

Coupling constant, J

Not all nmr spectra are so easily analyzed as the spectrum for ethyl iodide. Consider the one for 3-hexanol (Fig. 20.3). Twelve protons give rise to an unintelligible group of peaks between 1.0 and 2.2 ppm. It is not clear from a 3-hexanol spectrum which of the two low-field peaks (on the left-hand side of the spectrum) should be assigned to the hydroxyl proton and which should be assigned to the proton on C-3.

Deuterium exchange

Deuterium atoms give no peaks in the nmr spectrum; thus the peak in the 3-hexanol spectrum for the proton on C-3 will be evident if the spectrum of the alcohol is one in which the hydroxyl proton of the alcohol has been replaced by deuterium. The hydroxyl proton of 3-hexanol is acidic and will exchange rapidly with the deuterium of D_2O.

Shift reagents

Addition of a few milligrams of a hexacoordinate complex of europium {tris(dipivaloylmethanato)europium(III),[Eu(dpm)$_3$]} to an nmr sample, which contains a Lewis base center (an amine or basic oxygen, such as a hydroxyl group), has a dramatic effect on the spectrum. This lanthanide (soluble in CS_2 and $CDCl_3$) causes large shifts in the positions of peaks arising from the protons near the metal atom in this molecule and is therefore referred to as a shift reagent. It produces the shifts by complexing with the unshared electrons of the hydroxyl oxygen, the amine nitrogen, or other Lewis base centers. As part of the complex the europium atom exerts a large dipolar effect on nearby hydrogens, with resulting changes in the nmr spectrum.

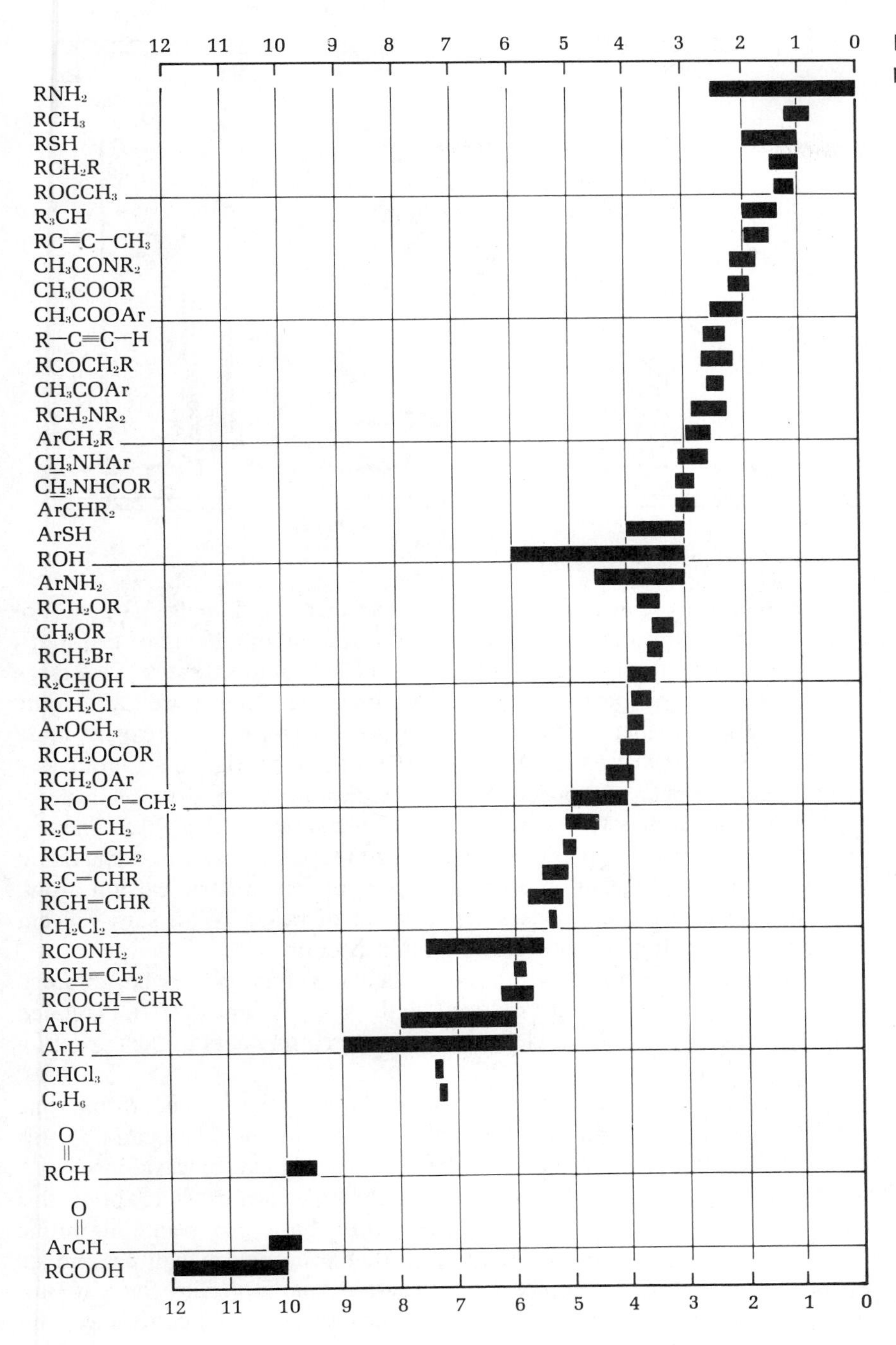

FIG. 20.2 ^{1}H Chemical shifts, ppm from TMS.

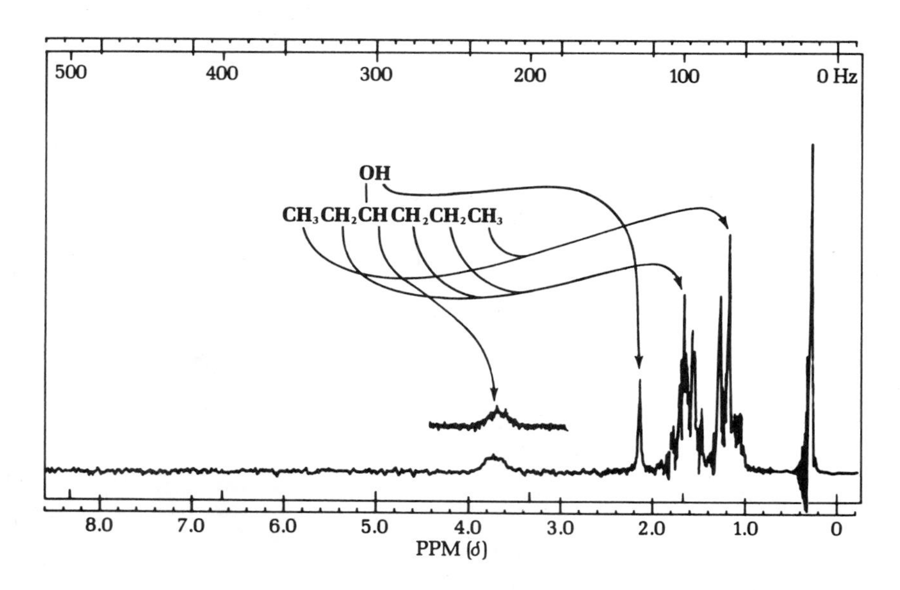

FIG. 20.3 Proton nmr spectrum of 3-hexanol.

With no shift reagent present, the nmr spectrum of 2-methyl-3-pentanol (Fig. 20.4(A)) is not readily analyzed. Addition of about 10 mg of Eu(dpm)$_3$ (the shift reagent) to the sample (0.5 mL of a 0.4 M solution) causes very large downfield shifts of peaks owing to protons near the coordination site (Fig. 20.4(B)).[3] Further additions of 10-mg portions of the shift reagent cause further downfield shifts (Figs. 20.4(D)–(F)), so that in Fig. 20.4(G) only the methyl peaks appear within 500 Hz of TMS. Finally, when the mole ratio of Eu(dpm)$_3$ to alcohol is 1 : 1 we find the spectrum shown in Fig. 20.5. The two protons on C-4 and the two methyls on C-2 are magnetically nonequivalent because they are adjacent to a chiral center (an asymmetric carbon atom, C-3), and therefore each gives a separate set of peaks. With shift reagent added, this spectrum can be analyzed by inspection.

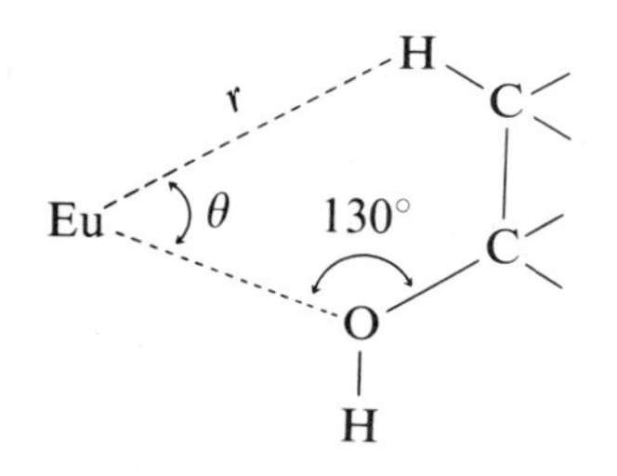

$$\frac{\Delta H}{H} = \frac{3\cos^2\theta - 1}{r^3}$$

Quantitative information about molecular geometry can be obtained from shifted spectra. The shift induced by the shift reagent, $\Delta H/H$, is related to the distance (r) and the angle (θ), which the proton bears to the europium atom.

Eu(dpm)$_3$ is a metal chelate (*chele*, Greek, claw). The β-diketone dipivaloylmethane (2,2,6,6-tetramethylheptane-3,5-dione) is a ligand, in this case a bidentate ligand, attached to the europium at two places. Since Eu^{3+} is hexacoordinate, three dipivaloylmethane molecules cluster about this metal atom. However, when a molecule with a basic group like an amine or an alcohol is in solution with Eu(dpm)$_3$ the europium will expand its coordination sphere to complex with this additional molecule. Such a complex is weak and so its nmr spectrum is an average of the complexed and uncomplexed molecule.

3. See also K. L. Williamson, D. R. Clutter, R. Emch, M. Alexander, A. E. Burroughs, C. Chua, and M. E. Bogel, *J. Am. Chem. Soc.*, **96**, 1471 (1974).

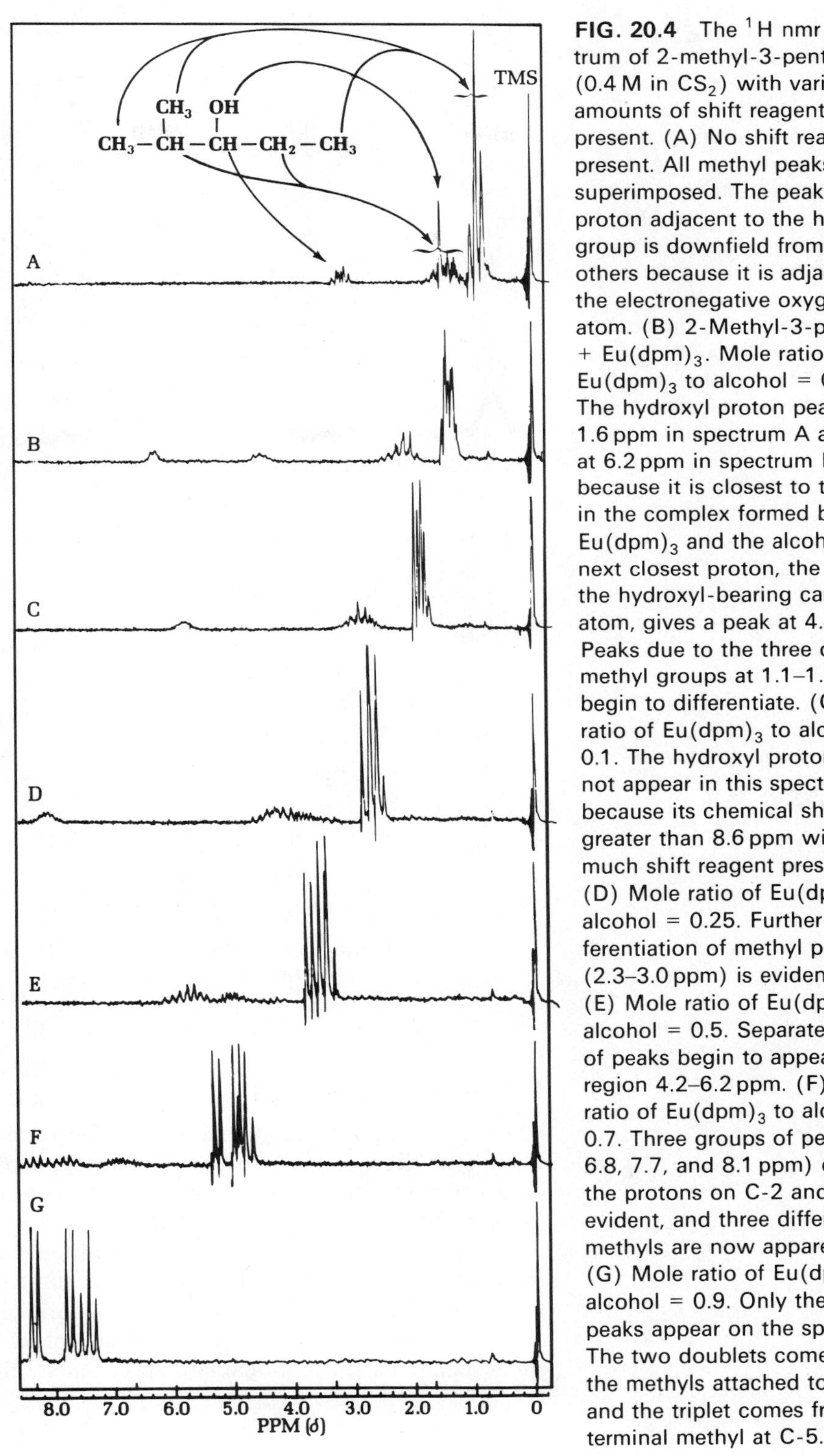

FIG. 20.4 The ^{1}H nmr spectrum of 2-methyl-3-pentanol (0.4 M in CS_2) with various amounts of shift reagent present. (A) No shift reagent present. All methyl peaks are superimposed. The peak for the proton adjacent to the hydroxyl group is downfield from the others because it is adjacent to the electronegative oxygen atom. (B) 2-Methyl-3-pentanol + $Eu(dpm)_3$. Mole ratio of $Eu(dpm)_3$ to alcohol = 0.05. The hydroxyl proton peak at 1.6 ppm in spectrum A appears at 6.2 ppm in spectrum B because it is closest to the Eu in the complex formed between $Eu(dpm)_3$ and the alcohol. The next closest proton, the one on the hydroxyl-bearing carbon atom, gives a peak at 4.4 ppm. Peaks due to the three different methyl groups at 1.1–1.5 ppm begin to differentiate. (C) Mole ratio of $Eu(dpm)_3$ to alcohol = 0.1. The hydroxyl proton does not appear in this spectrum because its chemical shift is greater than 8.6 ppm with this much shift reagent present. (D) Mole ratio of $Eu(dpm)_3$ to alcohol = 0.25. Further differentiation of methyl peaks (2.3–3.0 ppm) is evident. (E) Mole ratio of $Eu(dpm)_3$ to alcohol = 0.5. Separate groups of peaks begin to appear in the region 4.2–6.2 ppm. (F) Mole ratio of $Eu(dpm)_3$ to alcohol = 0.7. Three groups of peaks (at 6.8, 7.7, and 8.1 ppm) due to the protons on C-2 and C-4 are evident, and three different methyls are now apparent. (G) Mole ratio of $Eu(dpm)_3$ to alcohol = 0.9. Only the methyl peaks appear on the spectrum. The two doublets come from the methyls attached to C-2 and the triplet comes from the terminal methyl at C-5.

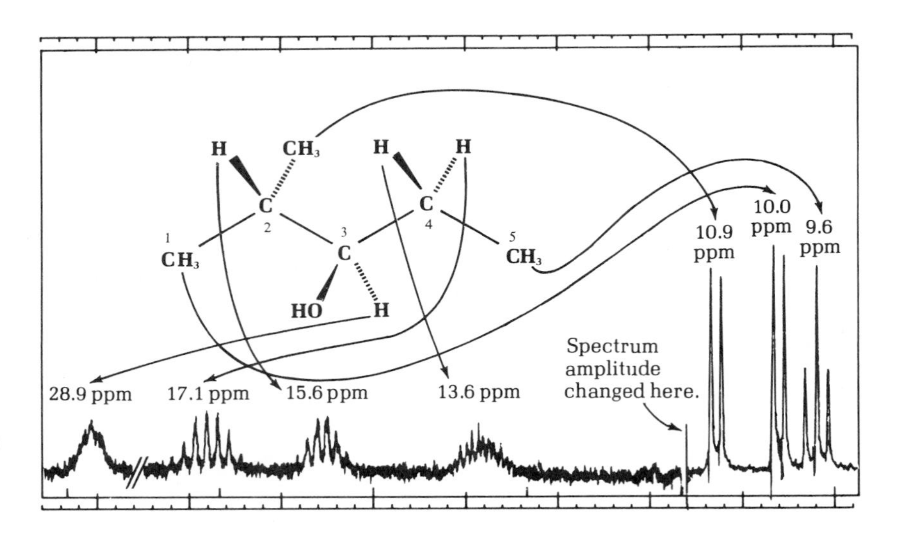

FIG. 20.5 Proton nmr spectrum of 2-methyl-3-pentanol containing $Eu(dpm)_3$. Mole ratio of $Eu(dpm)_3$ to alcohol = 1.0. Compare this spectrum to those shown in Fig. 20.4. Protons nearest the hydroxyl group are shifted most. Methyl groups are recorded at reduced spectrum amplitude. Note the large chemical shift difference between the two protons on C-4. The average conformation of the molecule is the one shown and was calculated from the equation on p. 200.

Tris(dipivaloylmethanato)europiumIII, $Eu(dpm)_3$
MW 701.78, mp 188–189°C

Carbon-13 Spectra

Although the most common and least expensive nmr spetrometers are those capable of observing protons, spectrometers capable of observing carbon atoms of mass 13 are becoming prevalent. In many cases ^{13}C spectra are much simpler and easier to interpret than ^{1}H spectra because the spectrum, as usually presented, consists of a single line for each chemically and magnetically distinct carbon atom. The element carbon consists of 98.9% of carbon with mass 12 and spin 0 (nmr inactive) and only 1.1% of ^{13}C with spin $\frac{1}{2}$ (nmr active). Carbon, with such a low concentration of spin $\frac{1}{2}$ nuclei, gives such a small signal in a conventional nmr spectrometer that special means must be employed to obtain an observable spectrum. In a CW (continuous wave) spectrometer, proton spectra are produced by sweeping the radiofrequency through the spectrum while holding the magnetic field constant, a process that requires anywhere from one to five minutes. If one tries to obtain a ^{13}C spectrum under these conditions the signal is so small it cannot be distinguished from the random noise in the background. In order to increase the signal-to-noise ratio a number of spectra are averaged in a small computer built into the spectrometer. Since noise is random and the signal coherent, the signal will increase in size and the noise decrease as many spectra are averaged together. Spectra are accumulated rapidly by applying a very short pulse of radiofrequency energy to the sample and then storing the resulting free induction decay signal in digital form in the computer. The free induction decay signal (FID), which takes just 0.6 s to acquire, contains frequency information about all the signals in the spectrum. In a few minutes several hundred FID's can be obtained and averaged in the computer. The FID is converted to a spectrum of conventional appearance by carrying out a Fourier transform computation on the signal using the spectrometer's computer. These principles are illustrated for very dilute proton spectra in Fig. 20.6.

CW spectrometer

Fourier transform spectrometer

Because only one in a hundred carbon atoms have mass 13 the chances of a molecule having two ^{13}C atoms adjacent to one another are small. Consequently, coupling of one carbon with another is not observed. Coupling of the ^{13}C atoms with ^{1}H atoms leads to excessively complex spectra, so this is ordinarily eliminated by noise decoupling the protons. This decoupling distorts the peaks such that peak areas of carbon spectra are not proportional to the number of carbon atoms present. Unlike proton spectra, which are observed over a range of approximately 15 ppm, carbon spectra occur over a 200-ppm range and peaks from magnetically distinct carbons rarely overlap. For example, each of the twelve carbons of sucrose gives a separate line (Fig. 20.7). The same factors that control proton chemical shifts are operative for carbon atoms. Carbons of high electron density (e.g., methyl groups) appear upfield near the carbon atoms of tetramethylsilane, the zero of reference, while carbon atoms bearing electron-withdrawing groups or atoms appear downfield. It should not be surprising that carbonyl carbons are found furthest downfield, between 160 and 220 ppm. Figure 20.8 on p. 205 gives the chemical shift ranges for carbon atoms.

Noise decoupling

Experiment

Procedure

A typical ^{1}H nmr sample is 0.3 to 0.5 mL of a 10–20% solution of a nonviscous liquid or a solid in a proton-free solvent contained in a 5-mm dia glass tube. The sample tube must be of uniform outside and inside diameter with uniform wall thickness. Test a sample tube by rolling it down a very slightly inclined piece of plate glass. Reject all tubes that roll unevenly.

Sample size: 0.3 mL of 10–20% solution plus TMS

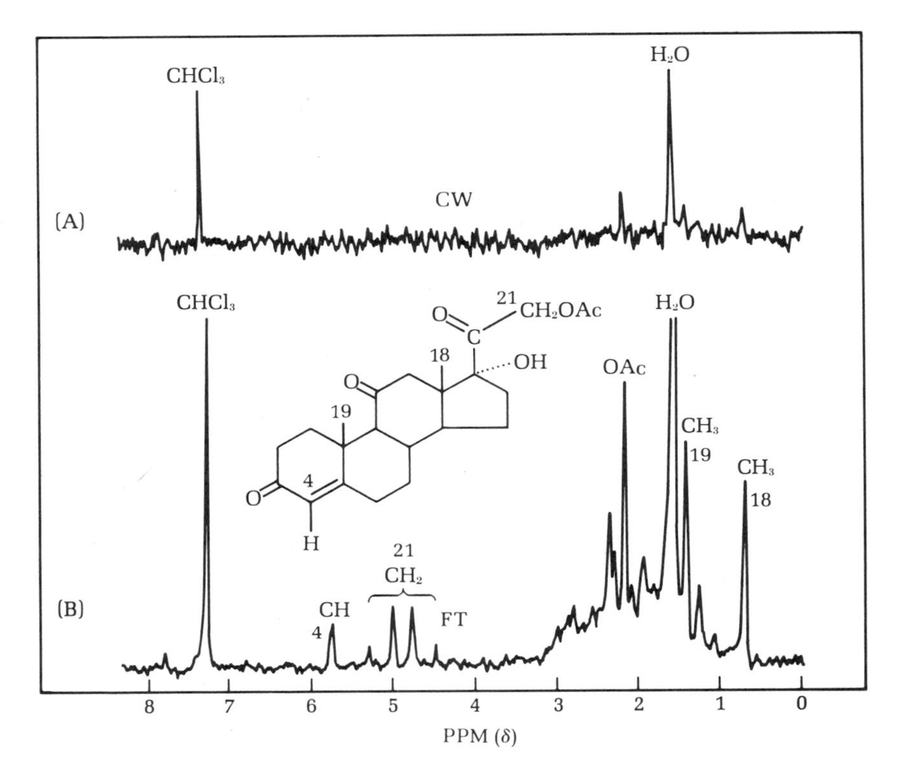

FIG. 20.6 ^{1}H nmr spectra of cortisone acetate, 300 μg/0.3 mL, in $CDCl_3$. (A) Continuous wave (CW) spectrum, 500-s scan time. (B) Fourier transform (FT) spectrum of the same sample, 250 scans (500 s). The H_2O and $CHCl_3$ are contaminants.

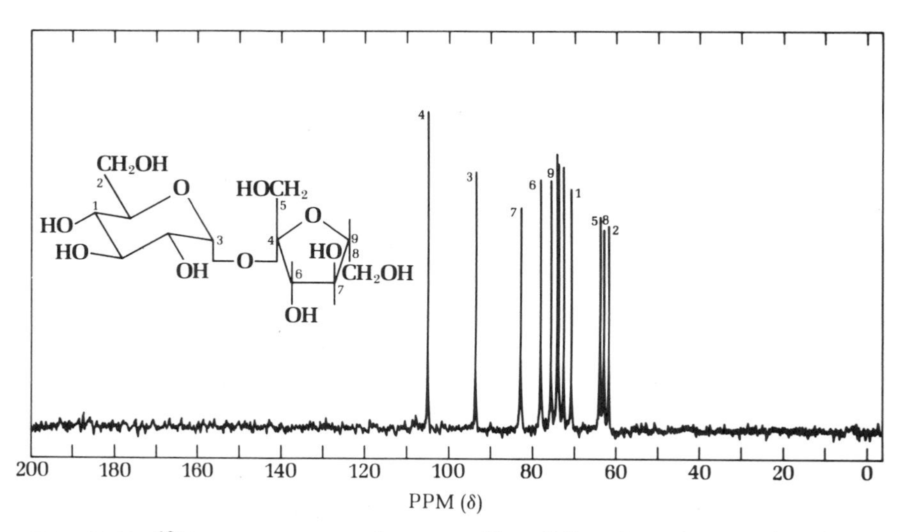

FIG. 20.7 ^{13}C nmr spectrum of sucrose. Not all lines have been assigned to individual carbon atoms.

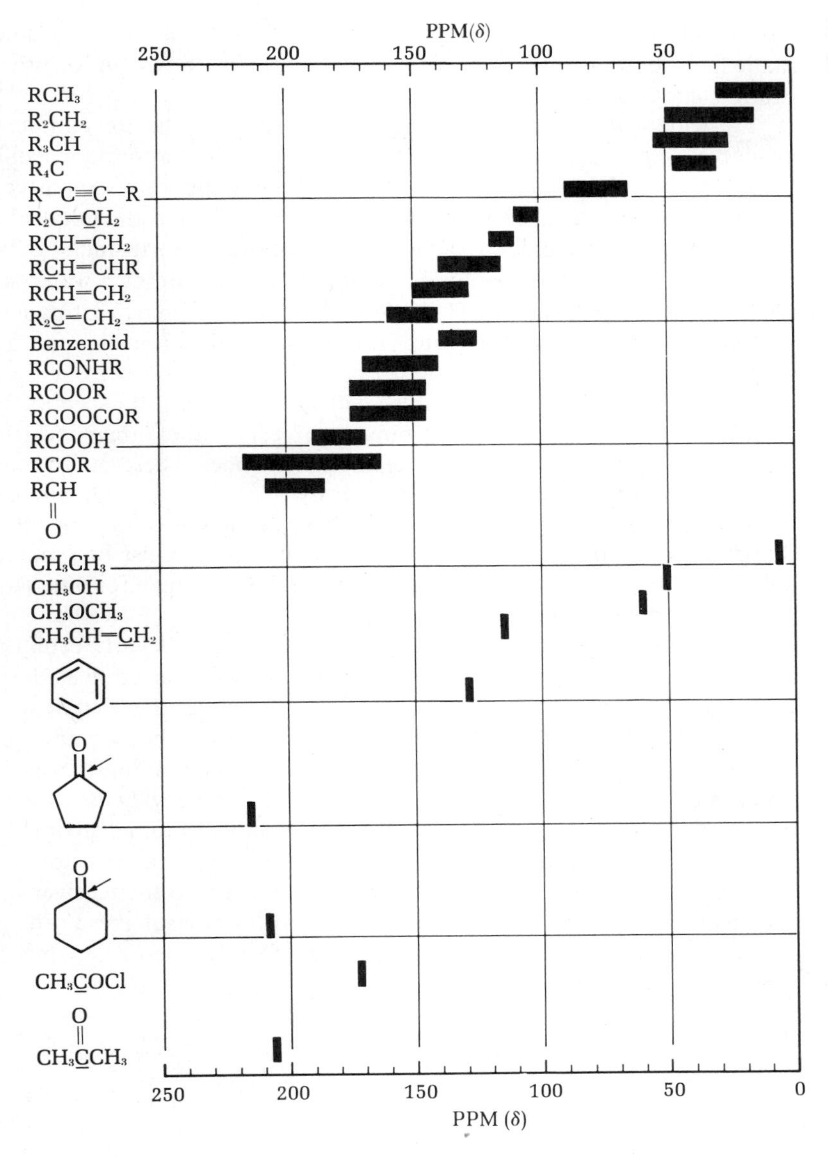

FIG. 20.8 ^{13}C chemical shifts, ppm from TMS.

The ideal solvent, from the nmr standpoint, is carbon tetrachloride. It is proton-free and nonpolar but unfortunately a poor solvent. Carbon disulfide is an excellent compromise. It will, however, react with amines.

Deuterochloroform ($CDCl_3$) is one of the most widely used nmr solvents. Although more expensive than nondeuterated solvents, it will dissolve a wider range of samples than carbon disulfide or carbon tetrachloride. Residual protons in the $CDCl_3$ will always give a peak at 7.27 ppm. Chemical shifts of protons are measured relative to the sharp peak of the protons in

CS_2, the solvent of choice; use only under laboratory hood: toxic and flammable

tetramethylsilane (taken as 0.0 ppm). Stock solutions of 3–5% tetramethylsilane in carbon disulfide and in deuterochloroform are useful for preparing routine samples.

TMS boils at room temperature. It is expensive and flammable. Handle with care.

A wide variety of completely deuterated solvents are commercially available, e.g., deuteroacetone (CD_3COCD_3), deuterodimethylsulfoxide (CD_3SOCD_3), deuterobenzene (C_6D_6), although they are expensive. For highly polar samples a mixture of the expensive deuterodimethylsulfoxide with the less expensive deuterochloroform will often be satisfactory. Water-soluble samples are dissolved in deuterated water containing a water-soluble salt [DSS, $(CH_3)_3SiCH_2CH_2CH_2SO_3SO_3^-Na^+$] as a reference substance. The protons on the three methyl groups bound to the silicon in this salt absorb at 0.0 ppm.

Erratic spectra from ferromagnetic impurities; remove by filtration

Solid impurities in nmr samples will cause very erratic spectra. If two successive spectra taken within minutes of each other are not identical, suspect solid impurities, especially ferromagnetic ones. These can be removed by filtration of the sample through a tightly packed wad of glass wool in a capillary pipette (Fig. 20.9). If very high resolution spectra (all lines very sharp) are desired, oxygen, a paramagnetic impurity, must be removed by bubbling a fine stream of pure nitrogen through the sample for 60 s. Routine samples do not require this treatment.

The usual nmr sample has a volume of 0.3 mL to 0.5 mL, even though the volume sensed by the spectrometer receiver coils (referred to as the active volume) is much smaller (Fig. 20.10). To average the magnetic fields produced by the spectrometer within the sample, the tube is spun by an air turbine at thirty to forty revolutions per second while taking the spectrum. Too rapid spinning or an insufficient amount of sample will cause the vortex produced by the spinning to penetrate the active volume, giving erratic nonreproducible spectra. A variety of microcells are available for holding and proper positioning of small samples with respect to the receiver coils of the spectrometer (Fig. 20.11). The vertical positioning of these cells in the spectrometer is critical. If microcells are used, only one or two mg of the

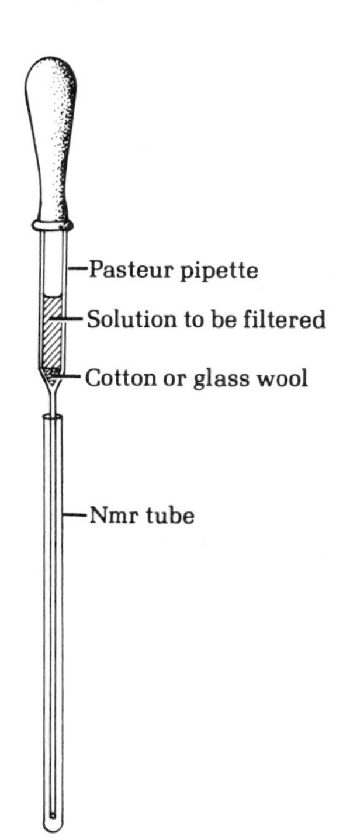

FIG. 20.9 Micro filter for nmr samples. Solution to be filtered is placed in the top of the Pasteur pipette, rubber bulb put in place, and pressure applied to force sample through the cotton or glass wool into an nmr sample tube.

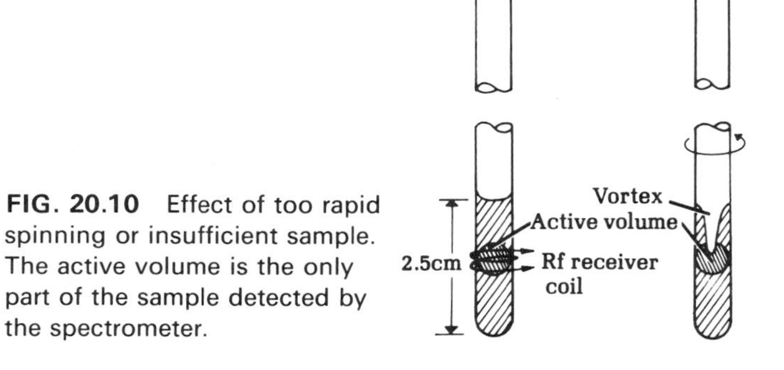

FIG. 20.10 Effect of too rapid spinning or insufficient sample. The active volume is the only part of the sample detected by the spectrometer.

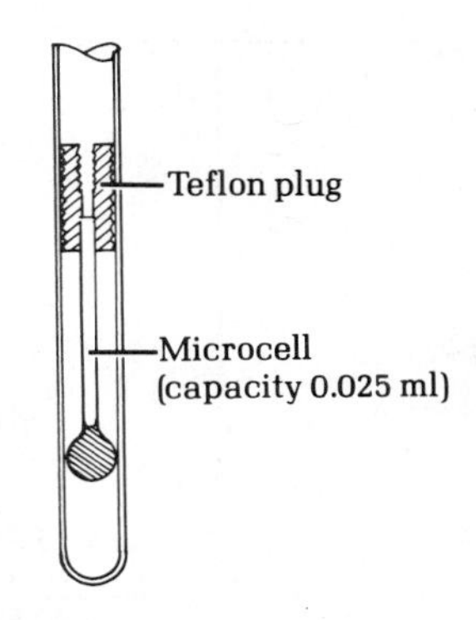

FIG. 20.11 Nmr microcell positioned in an nmr tube. The Teflon plug supports the microcell and allows the sample to be centered in the active volume of the spectrometer.

sample are needed to give satisfactory spectra, in contrast to the 20–30 mg usually needed for a CW spectrum.

Adjusting the Spectrometer

Sweep zero

To be certain the spectrometer is correctly adjusted and working properly, record the spectrum of the standard sample of chloroform and tetramethylsilane (TMS) usually found with the spectrometer. While recording the spectrum from left to right, the $CHCl_3$ peak should be brought to 7.27 ppm and the TMS peak to 0.0 ppm with the sweep zero control. The most important adjustment, the resolution control (also called homogeneity, or Y-control), should be adjusted for each sample so the TMS peak is as high and narrow as possible, with good ringing (Fig. 20.12). The signal (the peak traced by the spectrometer) should also be properly phased; it will then have the same appearance in both forward and backward scans (Fig. 20.12).

Resolution: sharp, narrow peaks with good ringing = high resolution

Use scrap paper for all but the final, best spectrum

Small peaks symmetrically placed on each side of a principal peak are artifacts called spinning side bands (Fig. 20.13). They are recognized as such by changing the spin speed (see again Fig. 20.13), which causes the spinning side bands to change positions. Two controls on the spectrometer determine the height of a signal as it is recorded on the paper. One, the spectrum amplitude control, increases the size of the signal as well as the baseline noise (the jitter of the pen when no signal is present). The other, the radiofrequency (rf) power control, increases the size of the signal alone, but only to a point,

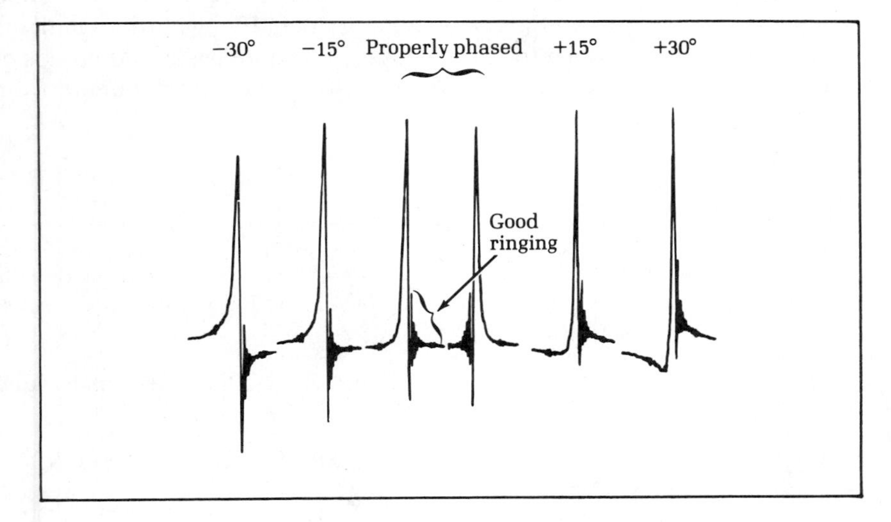

FIG. 20.12 Effect of phasing on signal shape. The TMS peaks on both forward and backward scans are quite high and narrow, with good ringing and perfect symmetry.

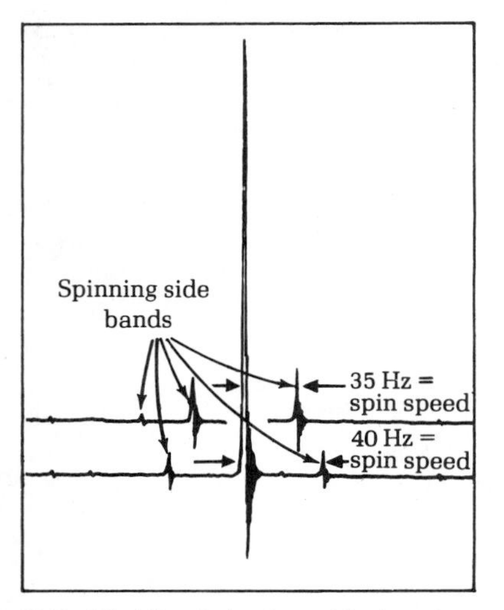

FIG. 20.13 Spinning side bands.

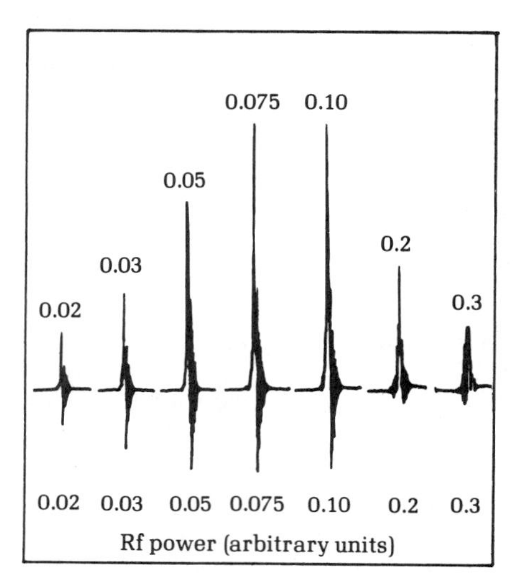

FIG. 20.14 Saturation of the nmr signal for chloroform. At the optimum rf power level (0.075–0.10) the signal reaches maximum intensity. Further increase of power causes the signal to become smaller and broader, with poor ringing–saturation.

after which saturation occurs (Fig. 20.14). Applying more than the optimum rf power will cause the peak to become distorted and of low intensity.

Identification of Unknown Alcohol or Amine by ^{1}H nmr

Shift reagents are expensive

Dispose of used solvents in the container provided

Using a stock solution of 4% TMS in carbon disulfide, prepare 0.5 mL of a 0.4 M (or 10%) solution of an unknown alcohol or amine. Filter the solution, if necessary, into a clean, *dry* nmr tube. Set the TMS peak at 0.0 ppm, check the phasing, maximize the resolution, and run a spectrum over a 500-Hz range using a 250-s sweep time. To the unknown solution add about 5 mg of Eu(dpm)$_3$ (the shift reagent), shake thoroughly to dissolve, and run another spectrum. Continue adding Eu(dpm)$_3$ in 10-mg portions until the spectrum is shifted enough for easy analysis. Integrate peaks and groups of peaks if in doubt about their relative areas. To protect the Eu(dpm)$_3$ from moisture store in a desiccator.

References

1. L. M. Jackman and S. Sternhell, *Applications of Nuclear Magnetic Resonance Spectroscopy in Organic Chemistry*, Pergamon Press, Elmsford, N.Y., 2nd Ed., 1969.

2. D. Shaw, *Fourier Transform NMR Spectroscopy*, Elsevier, Amsterdam, 1976.

3. T. C. Farrer and E. D. Becker, *Pulse and Fourier Transform NMR*, Academic Press, New York, 1971.

4. M. L. Martin, J.-J. Delpuech, and G. J. Martin, *Practical NMR Spectroscopy*, Heyden & Son, Philadelphia, 1980.

5. E. D. Becker, *High Resolution NMR; Theory and Chemical Applications*, 2nd Ed., Academic Press, New York, 1980.

6. *High Resolution NMR Spectra Catalogs*, Varian Associates, Palo Alto, Vol. 1, 1962, Vol. 2, 1963.

7. J. W. Cooper, *Spectroscopic Techniques for Organic Chemists*, John Wiley, New York, 1980.

8. J. B. Stothers, *Carbon-13 NMR Spectroscopy*, Academic Press, New York, 1972.

9. G. C. Levy, R. Lichter, and G. L. Nelson, *Carbon-13 NMR Spectroscopy*, 2nd Ed., Wiley Interscience, New York, 1980.

10. L. F. Johnson and W. C. Jankowski, *Carbon-13 NMR Spectra*, Wiley Interscience, New York, 1972.

11. R. J. Abraham and P. Loftus, *Proton and Carbon-13 NMR Spectroscopy*. Heyden & Son, Philadelphia, 1978.

12. J. W. Akitt, *NMR and Chemistry*, 2nd Ed., Chapman and Hall, London, 1983.

13. O. Jardetzky and G. C. K. Roberts, *NMR in Molecular Biology*, Academic Press, New York, 1981.

14. A. E. Derome, *Modern NMR Techniques for Chemistry Research*, Pergamon Press, New York, 1986.

Questions

1. Propose a structure or structures consistent with the proton nmr spectrum of Fig. 20.15. Numbers adjacent to groups of peaks refer to relative peak areas. Account for missing lines.

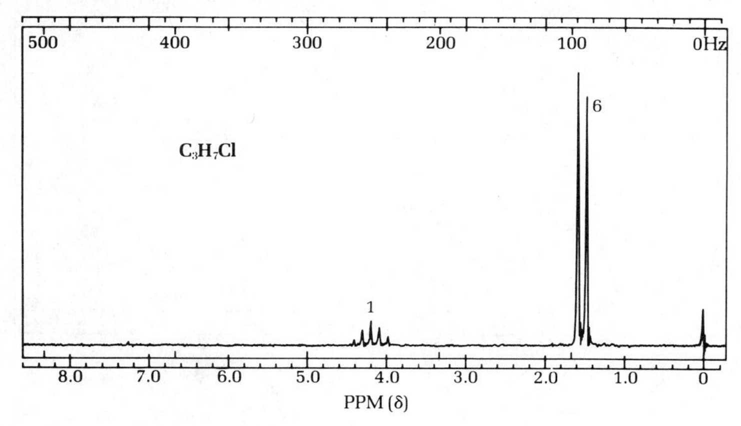

FIG. 20.15 Proton nmr spectrum, Question 1.

2. Propose a structure or structures consistent with the proton nmr spectrum of Fig. 20.16. Numbers adjacent to groups of peaks refer to relative peak areas.

3. Propose a structure or structures consistent with the proton nmr spectrum of Fig. 20.17.

4. Propose structures for a, b, and c consistent with the carbon-13 nmr spectra of Figs. 20.18, 20.19, and 20.20. These are isomeric alcohols with the empirical formula $C_4H_{10}O$.

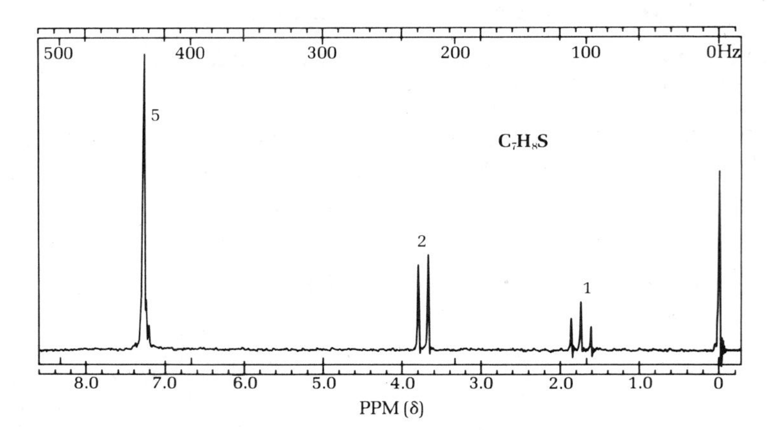

FIG. 20.16 Proton nmr spectrum, Question 2.

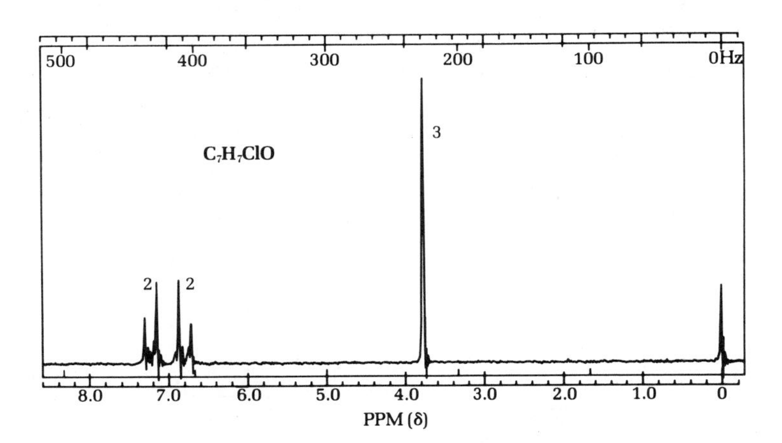

FIG. 20.17 Proton nmr spectrum, Question 3.

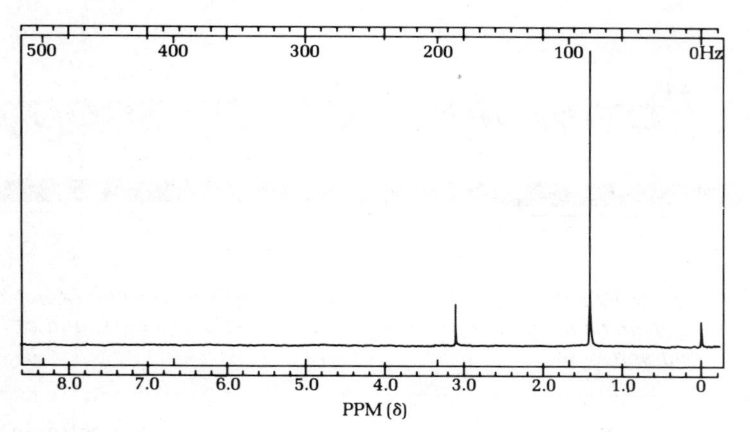

FIG. 20.18 ^1H nmr spectrum of $C_4H_{10}O$, Question 4(a).

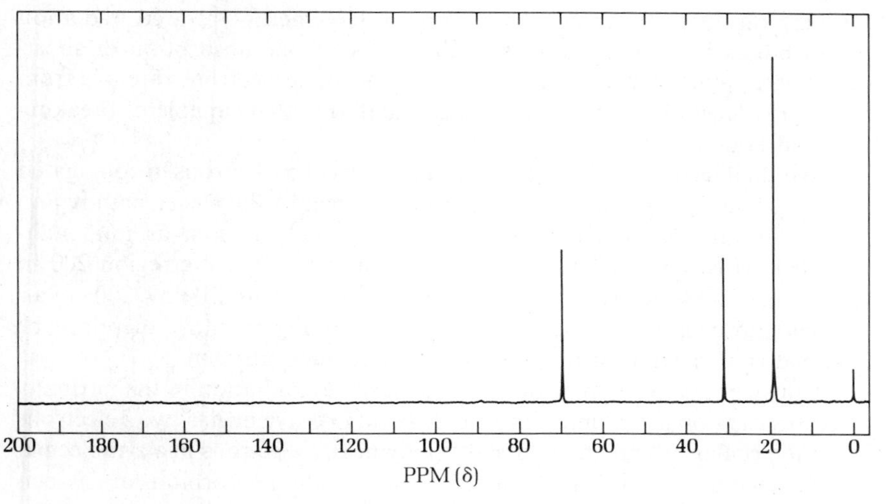

FIG. 20.19 ^{13}C nmr spectrum of $C_4H_{10}O$, Question 4(b).

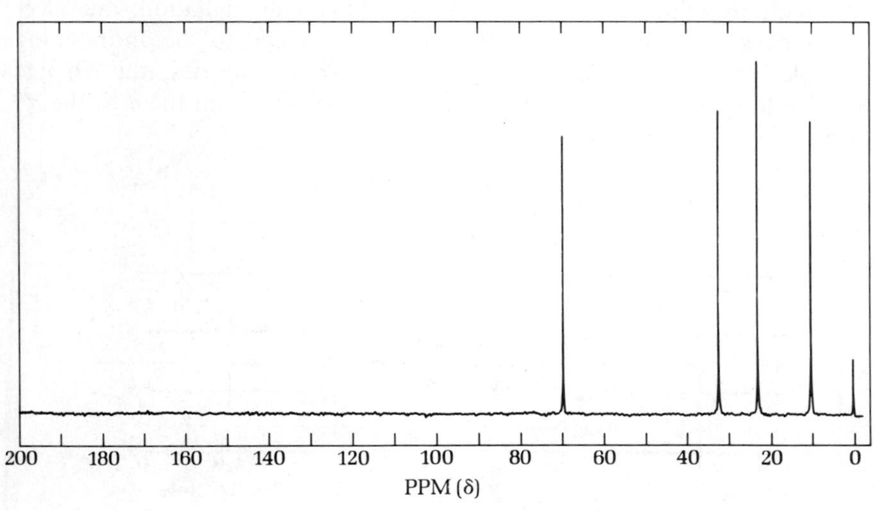

FIG. 20.20 ^{13}C nmr spectrum of $C_4H_{10}O$, Question 4(c).

21 Ultraviolet Spectroscopy

PRELAB EXERCISE: *Predict the appearance of the ultraviolet spectrum of n-butyl amine in acid, of methoxybenzene in acid and base, and of benzoic acid in acid and base.*

uv—*Electronic transitions within molecules*

Ultraviolet spectroscopy gives information about electronic transitions within molecules. Whereas absorption of low-energy infrared radiation causes bonds in a molecule to stretch and bend, the absorption of short-wavelength, high-energy ultraviolet radiation causes electrons to move from one energy level to another with energies that are often capable of breaking chemical bonds.

We shall be most concerned with transitions of π-electrons in conjugated and aromatic ring systems. These transitions occur in the wavelength region 200 to 800 nm (nanometers, 10^{-9} meters, formerly known as mμ, millimicrons). Most common ultraviolet spectrometers cover the region 200 to 400 nm as well as the visible spectral region 400 to 800 nm. Below 200 nm air (oxygen) absorbs uv radiation; spectra in that region must therefore be obtained in a vacuum or in an atmosphere of pure nitrogen.

Consider ethylene, even though it absorbs uv radiation in the normally inacessible region at 163 nm. The double bond in ethylene has two *s* electrons in a σ-molecular orbital and two, less tightly held, *p* electrons in a π-molecular orbital. Two unoccupied, high-energy-level, antibonding orbitals are associated with these orbitals. When ethylene absorbs uv radiation, one electron moves up from the bonding π-molecular orbital to the antibonding π^*-molecular orbital (Fig. 21.1). As the diagram indicates, this change requires less energy than the excitation of an electron from the σ to the σ^* orbital.

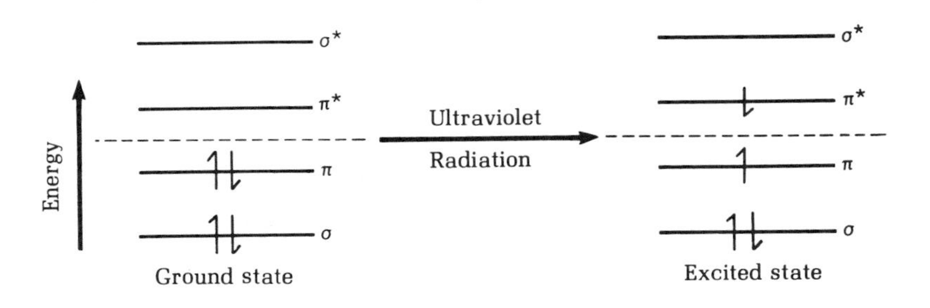

FIG. 21.1 Electronic energy levels of ethylene.

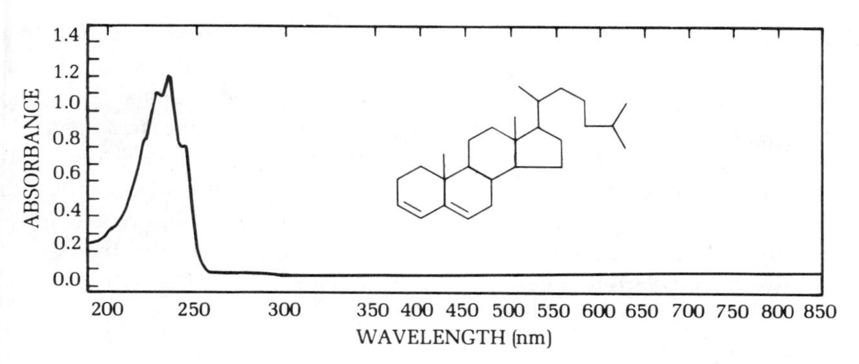

FIG. 21.2 The ultraviolet spectrum of cholesta-3,5-diene in ethanol.

By comparison with infrared spectra and nmr spectra, uv spectra are fairly featureless (Fig. 21.2). This condition results as molecules in a number of different vibrational states undergo the same electronic transition, to produce a band spectrum instead of a line spectrum.

Band spectra

Unlike ir spectroscopy, ultraviolet spectroscopy lends itself to precise quantitative analysis of substances. The intensity of an absorption band is usually given by the molar extinction coefficient, ϵ, which according to the Beer-Lambert Law is equal to the absorbance, A, divided by the product of the molar concentration, c, and the path length, l, in centimeters.

Beer-Lambert Law

$$\epsilon = \frac{A}{cl}$$

The wavelength of maximum absorption (the tip of the peak) is given by λ_{max}. Because uv spectra are so featureless it is common practice to describe a spectrum like that of cholesta-3,5-diene (Fig. 21.2) as λ_{max} 234 nm (ϵ = 20,000), and not bother to reproduce the actual spectrum.

λ_{max}, wavelength of maximum absorption

The extinction coefficients of conjugated dienes and enones are in the range 10,000–20,000, so only very dilute solutions are needed for spectra. In the example of Fig. 21.2 the absorbance at the tip of the peak, A, is 1.2, and the path length is the usual 1 cm; so the molar concentration needed for this spectrum is 6×10^{-5} mole per liter.

ϵ, extinction coefficient

$$c = \frac{A}{l\epsilon} = \frac{1.2}{20{,}000} = 6 \times 10^{-5} \text{ mole per liter}$$

which is 0.221 mg per 10 mL of solvent.

The usual solvents for uv spectroscopy are 95% ethanol, methanol, water, and saturated hydrocarbons such as hexane, trimethylpentane, and isooctane; the three hydrocarbons are often specially purified to remove impurities that absorb in the uv region. Any transparent solvent can be used for spectra in the visible region.

Spectro grade solvents

Sample cells for spectra in the visible region are made of glass, but uv cells must be of the more expensive fused quartz, since glass absorbs uv

UV cells are expensive; handle with care

radiation. The cells and solvents must be clean and pure, since very little of a substance produces a uv spectrum. A single fingerprint will give a spectrum!

Ethylene has λ_{max} 163 nm (ϵ = 15,000) and butadiene has λ_{max} 217 nm (ϵ = 20,900). As the conjugated system is extended, the wavelength of maximum absorption moves to longer wavelengths (toward the visible region): for example, lycopene with 11 conjugated double bonds has λ_{max} 470 nm (ϵ = 185,000), Fig. 21.3. Since lycopene absorbs blue visible light at 470 nm the substance appears bright red. It is responsible for the color of tomatoes; its isolation is described in Chapter 9.

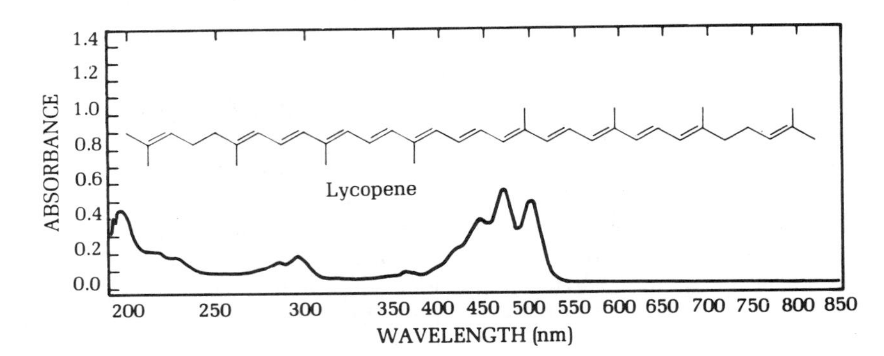

FIG. 21.3 The ultraviolet-visible spectrum of lycopene in isooctane.

Woodward and Fieser Rules for dienes and dienones

The wavelengths of maximum absorption of conjugated dienes and polyenes and conjugated enones and dienones are given by the Woodward and Fieser Rules, Tables 21.1 and 21.2.

The application of the rules in the above tables is demonstrated by the spectra of pulegone (1) and carvone (2), Fig. 21.4, with the calculations given in Tables 21.4 and 21.5.

These rules will be applied in a later experiment in which cholesterol is converted into an α,β-unsaturated ketone.

TABLE 21.1 Rules for the Prediction of λ_{max} for Conjugated Dienes and Polyenes

	Increment (nm)
Parent acyclic diene (butadiene)	217
Parent heteroannular diene	214
Double bond extending the conjugation	30
Alkyl substituent or ring residue	5
Exocyclic location of double bond to any ring	5
Groups: OAc, OR	0
Solvent correction, see Table 21.3	0
λ_{max} = Total	

TABLE 21.2 Rules for the Prediction of λ_{max} for Conjugated Enones and Dienones

β—C(β)=C(α)—C(R)=O and δ—C(δ)=C(γ)—C(β)=C(α)—C(R)=O		Increment (nm)
Parent α,β-unsaturated system		215
Double bond extending the conjugation		30
R (alkyl or ring residue), OR, $OCOCH_3$	α	10
	β	12
	δ, and higher	18
α-Hydroxyl, enolic		35
α-Cl		15
α-Br		23
exo-Location of double bond to any ring		5
Homoannular diene component		39
Solvent correction see Table 21.3		
	λ_{max}^{EtOH} =	Total

TABLE 21.3 Solvent Correction

Solvent	Factor for Correction to Ethanol
Hexane	+11
Ether	+7
Dioxane	+5
Chloroform	+1
Methanol	0
Ethanol	0
Water	−8

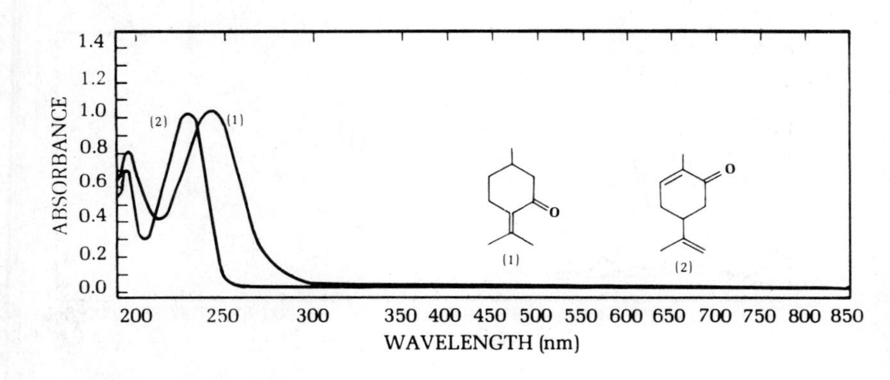

FIG. 21.4 Ultraviolet spectra of (1) pulegone and (2) carvone in hexane.

TABLE 21.4 Calculation of λ_{max} for Pulegone

Parent α,β-unsaturated system	215 nm
α-Ring residue, R	10
β-Alkyl group (two methyls)	24
Exocyclic double bond	5
Solvent correction (hexane)	−11
	Calcd λ_{max} 245 nm; found 244 nm

TABLE 21.5 Calculation of λ_{max} for Carvone

Parent α,β-unsaturated system	215 nm
α-Alkyl group (methyl)	10
β-Ring residue	12
Solvent correction (hexane)	−11
	Calcd λ_{max} 226 nm; found 229 nm

No simple rules exist for calculation of aromatic ring spectra, but several generalizations can be made. From Fig. 21.5 it is obvious that as polynuclear aromatic rings are extended linearly, λ_{max} shifts to longer wavelengths.

As alkyl groups are added to benzene, λ_{max} shifts from 255 nm for benzene to 261 nm for toluene to 272 nm for hexamethylbenzene. Substituents bearing nonbonding electrons also cause shifts of λ_{max} to longer wavelengths, e.g., from 255 nm for benzene to 257 nm for chlorobenzene, 270 nm for phenol, and 280 nm for aniline (ϵ = 6,200–8,600). That these effects are the result of interaction of the π-electron system with the nonbonded electrons is seen dramatically in the spectra of vanillin and the derived anion (Fig. 21.6). Addition of two more nonbonding electrons in the anion causes λ_{max} to shift from 279 nm to 351 nm and ϵ to increase. Removing the electrons from the nitrogen of aniline by making the anilinium cation causes λ_{max} to decrease from 280 nm to 254 nm (Fig. 21.7). These changes of λ_{max} as a function of pH have obvious analytical applications.

Effect of acid and base on λ_{max}

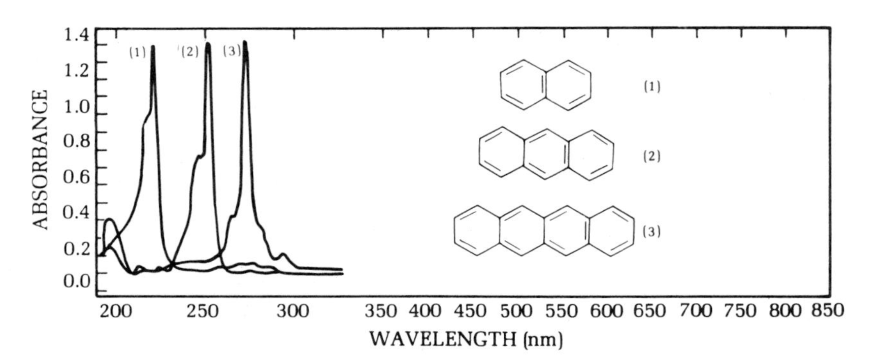

FIG. 21.5 The ultraviolet spectra of (1) naphthalene, (2) anthracene, and (3) tetracene.

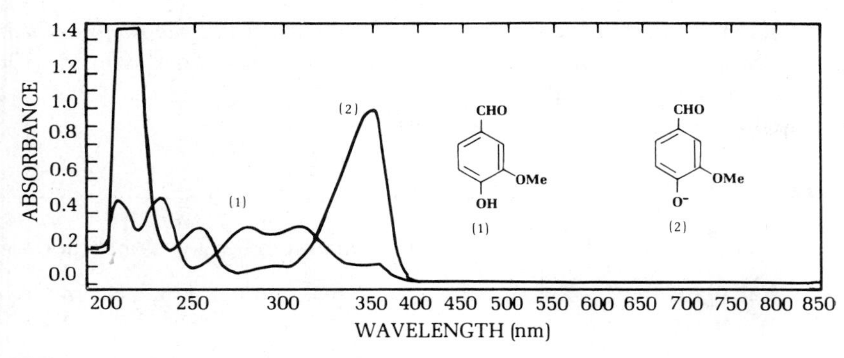

FIG. 21.6 Ultraviolet spectrum of (1) neutral vanillin and (2) the anion of vanillin.

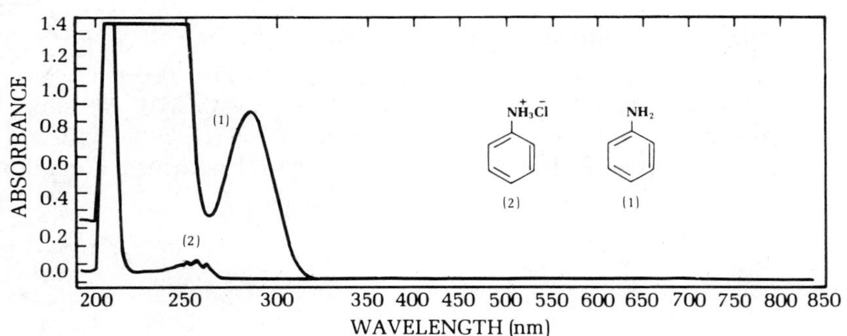

FIG. 21.7 Ultraviolet spectrum of (1) aniline and (2) aniline hydrochloride

Intense bands result from π-π conjugation of double bonds and carbonyl groups with the aromatic ring. Styrene, for example, has λ_{max} 244 nm (ϵ = 12,000) and benzaldehyde λ_{max} 244 nm (ϵ = 15,000).

Experiment

Ultraviolet Spectrum of Unknown Acid, Base, or Neutral Compound

Determine whether an unknown compound obtained from the instructor is acidic, basic, or neutral from the ultraviolet spectra in the presence of acid and base as well as in neutral media.

References

1. A. E. Gillam and E. S. Stern, An *Introduction to Electronic Absorption Spectroscopy in Organic Chemistry*, Edward Arnold, London, 1967.
2. C. N. R. Rao, *Ultraviolet and Visible Spectroscopy*, Butterworths, London, 2nd Ed., 1967.
3. H. H. Jaffe and M. Orchin, *Theory and Application of Ultraviolet Spectroscopy*, John Wiley, New York, 1962.

4. R. M. Silverstein, C. G. Bassler, and T. C. Morrill, *Spectrometric Identification of Organic Compounds*, John Wiley, New York, 4th Ed., 1981. (Includes ir, uv and nmr.)

Questions

1. Calculate the ultraviolet absorption maximum for 2-cyclohexene-1-one.
2. Calculate the ultraviolet absorption maximum for 3,4,4-trimethyl-2-cyclohexene-1-one.
3. Calculate the ultraviolet absorption maximum for 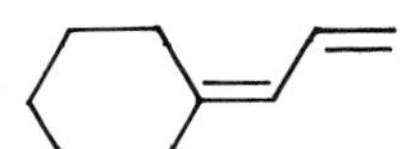
4. What concentration, in g/mL, of a substance with MW 200 should be prepared in order to give an absorbance value, A, equal to 0.8 if the substance has $\epsilon = 16{,}000$ and a cell with path length 1 cm is employed?

Cholesterol from Human Gallstones 22

***PRELAB EXERCISE:** Explain how the process of bromination and debromination of cholesterol can free it from impurities. Write a detailed mechanism for the bromination of cholesterol, taking care to consider the stereochemistry of the product.*

In this experiment cholesterol will be isolated from human gallstones. Cholesterol is an unsaturated alcohol containing 27 carbon atoms and 46 hydrogen atoms:

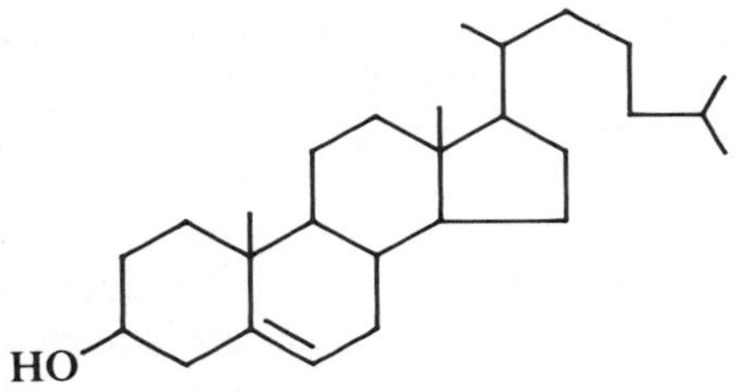

HO

It is a solid, mp 148.5°C, and is insoluble in water but soluble in boiling ethanol and dioxane.

Gallstones

The gall bladder is attached to the undersurface of the liver just below the rib cage. It retains bile produced by the liver and feeds it into the upper part of the small intestine as needed for digestion. Bile consists primarily of bile acids, which are carboxylic acids closely resembling cholesterol and which aid in the digestion of fats by functioning as emulsifying agents. The gall bladder also harbors free cholesterol. If the concentration of cholesterol in the bile exceeds a certain critical level, it will come out of solution and agglomerate into particles that grow to form gallstones. An amateur geologist given a bottle of gallstones to identify once labeled them a "riverbed conglomerate"—and indeed they do resemble stones in color, texture, and hardness. They come in a variety of shapes and colors and can be up to an inch in diameter.

As gallstones collect they irritate the lining of the gall bladder causing severe pain, nausea, and vomiting. The stones can block the bile duct and lead to fatal complications. The remedy is surgery although recently gallstones have been dissolved right in the gall bladder, not with boiling alcohol but with methyl *tert*-butyl ether. The ether is injected directly into the gall bladder through a 1.7-mm tube, which pierces the skin and liver on its way

Cholesterol: soluble in methyl tert-butyl ether in the gall bladder

to the gall bladder. Depending on the number and size of the gallstones the dissolution process can take from 7 to 18 h.

In the average human, approximately 200 g of cholesterol is concentrated primarily in the spinal cord, brain, and nerve tissue. Insoluble in water and plasma, it is transported in the bloodstream bound to lipoproteins, which are proteins attached to lipids (fats). Recent research has divided these lipoproteins, when centrifuged, into two broad classes—high-density (HDL) and low-density (LDL) lipoproteins. A relatively high concentration of HDL bound to cholesterol seems to cause no problems and in fact is beneficial; but a high ratio of LDL-cholesterol leads to the deposition of cholesterol both in the gall bladder (resulting in gallstones) and also on the walls of the arteries (causing a plaque that cuts off blood flow and hastens hardening of the arteries or atherosclerosis).

HDL: high-density lipoproteins (good)
LDL: low-density lipoproteins (bad)

Mounting evidence points to unsaturated fats such as those found in vegetable oils as favoring the HDL-cholesterol bond, while LDL-cholesterol formation is speeded by saturated fats such as those found in animals. The HDL-cholesterol level goes down with smoking or eating large amounts of sugar. It goes up with regular exercise and with the consumption of *moderate* amounts of alcohol (a glass of wine per day). The 1985 Nobel prize in physiology or medicine went to Michael Brown and Joseph Goldstein for their pioneering work on LDL- and HDL-cholesterol.

1985 Nobel prize: Brown and Goldstein

The average American woman at age 75 has a 50% chance of developing gallstones while for a man of the same age the chance is only half as great. Gallstones and coronary heart disease are also much more common in fat people. Almost 70% of the women in certain American Indian tribes get gallstones before the age of 30, whereas only 10% of black women are afflicted. Swedes and Finns have gallstones more often than Americans; the problem is almost unknown among the Masai people of East Africa.

The most common treatment at present for gallstones is surgical removal, an operation performed 500,000 times each year in the United States. This is the source of the gallstones used in the present experiment.

Experiment

Cholesterol from Gallstones[1]

In a 10 × 100 mm reaction tube containing a boiling stick dissolve 200 mg of crushed human gallstones in 2.0 mL of 2-butanone by gentle heating on a hot sand bath. Note the height of the solution in the reaction tube. In another reaction tube also containing a boiling stick, heat 0.5 mL of 2-butanone to boiling. As soon as the gallstones have disintegrated and the cholesterol has

Note for the instructor

1. Obtainable from the department of surgery of a hospital. Wrap the stones in a towel and crush by light pounding with a hammer. Do not breathe the dust as the gallstones may not be sterile. (See also *Instructor's Manual for Organic Experiments*, 6th ed. for a procedure for making "artificial gallstones.")

Bilirubin
$\lambda_{max}^{CHCl_3}$ 450 nm

dissolved, filter the hot solution on the micro Büchner funnel into another reaction tube. The pressure filtration apparatus or that depicted in Fig. 22.1 is used to give a water-clear filtrate.

Transfer the hot solution to the Büchner funnel using a warm Pasteur pipette. The pipette is warmed by immersing it in the hot vapors of the boiling 2-butanone. Add the solution through the side hole of the pressure filtration apparatus, close the hole with a finger and apply pressure to filter the solution or apply pressure using a rubber bulb as in Fig. 22.1. Add a 6 mm filter paper to the funnel if the filtrate is not clear. Wash out the reaction tube and pipette with the 0.5 mL of 2-butanone and rinse the Büchner funnel as well. The brown residue removed by filtration is primarily the bile pigment bilirubin, a metabolite of hemoglobin.

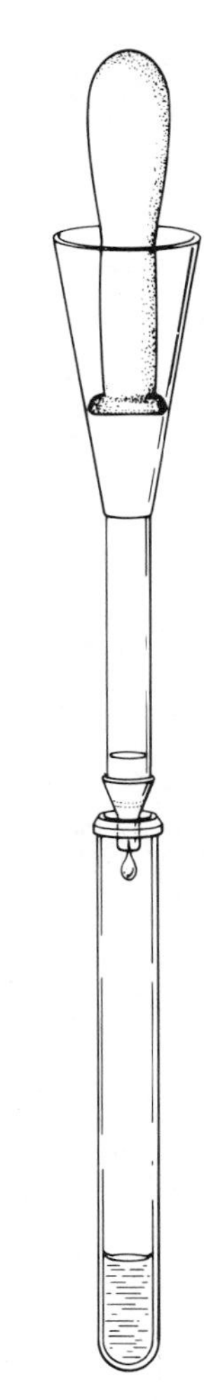
FIG. 22.1 Micro Büchner filtration assembly.

Under a stream of nitrogen or air (in the hood) evaporate the solution to 1.5 mL, dilute the hot solution with 1 mL of methanol, and then, while the solution is boiling (boiling stick), add water dropwise until a very faint cloudiness persists. Cholesterol is completely insoluble in water and not very soluble in methanol. Addition of water will produce a solution saturated with cholesterol at the boiling point. This is the critical part of this experiment. Add a drop of water, redissolve any precipitated cholesterol, and continue adding water until the solution is saturated while hot.

Cork the tube, wrap it with some insulating material, and allow the solution to cool slowly without disturbance to room temperature; then cool it in ice. Slow cooling produces large crystals. Collect the product on the Hirsch funnel or by withdrawing the solvent using the Pasteur pipette. Turn the crystals out onto a piece of filter paper, squeeze out excess solvent and allow them to dry. Determine the weight and melting point of the cholesterol and calculate the yield from gallstones. This material can be purified by bromination and debromination (Chapter 23) and it can be acetylated.

Question

1. Cholesterol is an alcohol. Why is it more soluble in organic solvents than in water?

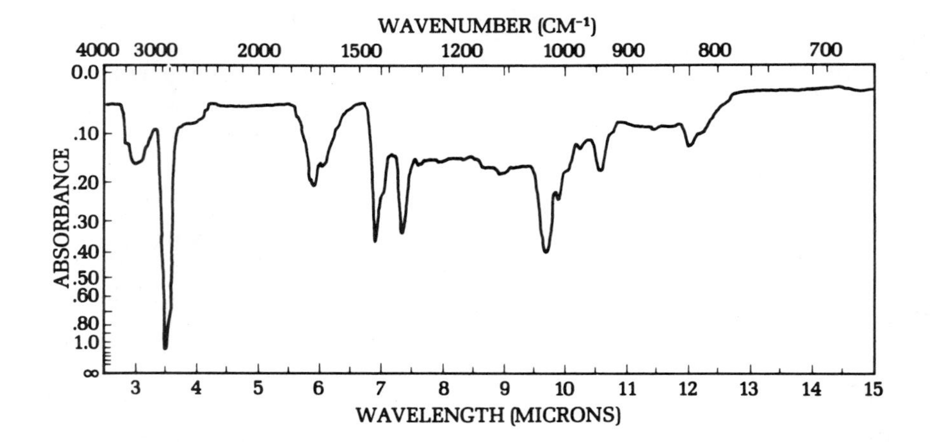

FIG. 22.2 Infrared spectrum of cholesterol.

Bromination and Debromination: Purification of Cholesterol

23

PRELAB EXERCISE: *Make a molecular model of 5α,6β-dibromocholestan-3β-ol and convince yourself that the bromine atoms in this molecule are* trans *and diaxial.*

The bromination of a double bond is an important and well-understood organic reaction. In the present experiment it is employed for the very practical purpose of purifying crude cholesterol through the process of bromination, crystallization, and then zinc dust debromination.

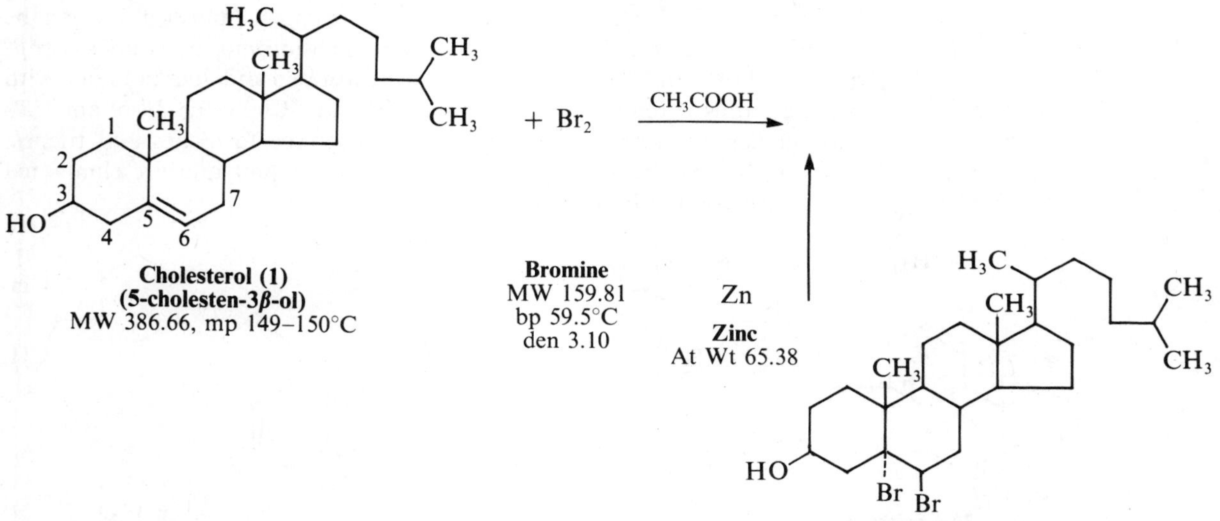

Cholesterol (1)
(5-cholesten-3β-ol)
MW 386.66, mp 149–150°C

+ Br_2 —CH_3COOH→

Bromine
MW 159.81
bp 59.5°C
den 3.10

Zn

Zinc
At Wt 65.38

5α,6β-Dibromocholestan-3β-ol (3)

The reaction involves nucleophilic attack by the alkene on bromine with the formation of a tertiary carbocation that probably has some bromonium ion character resulting from sharing of the nonbonding electrons on bromine with the electron-deficient C-5 carbon. This ion is attacked from the backside by bromide ion to form dibromocholesterol with the bromine atoms in the *trans* and diaxial configuration, the usual result when brominating a cyclohexene.

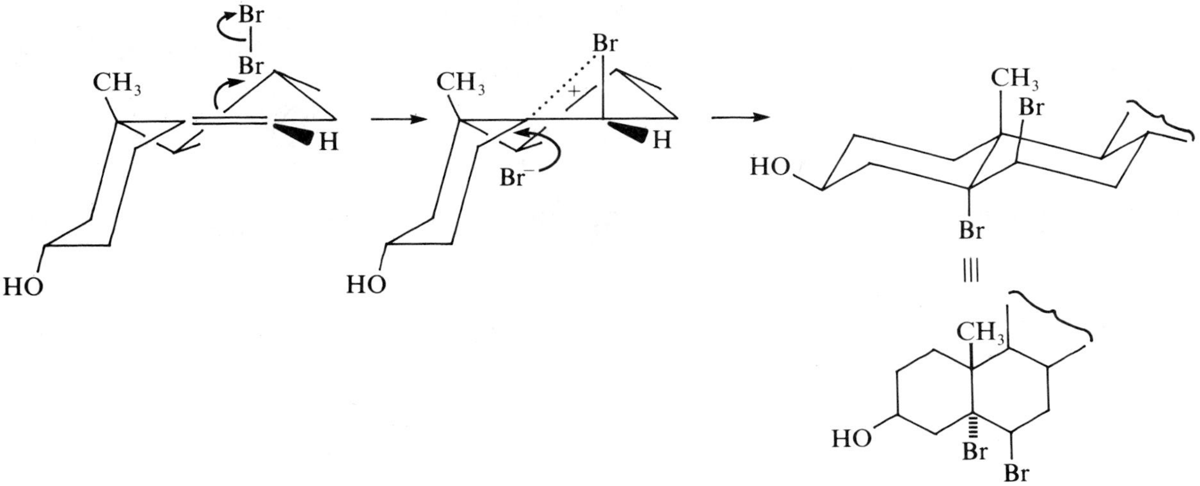

Cholesterol isolated from natural sources contains small amounts (0.1–3%) of 3β-cholestanol, 7-cholesten-3β-ol, and 5,7-cholestadien-3β-ol.[1] These are so very similar to cholesterol in solubility that their removal by crystallization is not feasible. However, complete purification can be accomplished through the sparingly soluble dibromo derivative 5α,6β-dibromocholestan-3β-ol. 3β-Cholestanol is saturated and does not react with bromine; thus it remains in the mother liquor. 7-Cholesten-3β-ol and 5,7-cholestadien-3β-ol are dehydrogenated by bromine to dienes and trienes, respectively, that likewise remain in the mother liquor and are eliminated along with colored by-products.

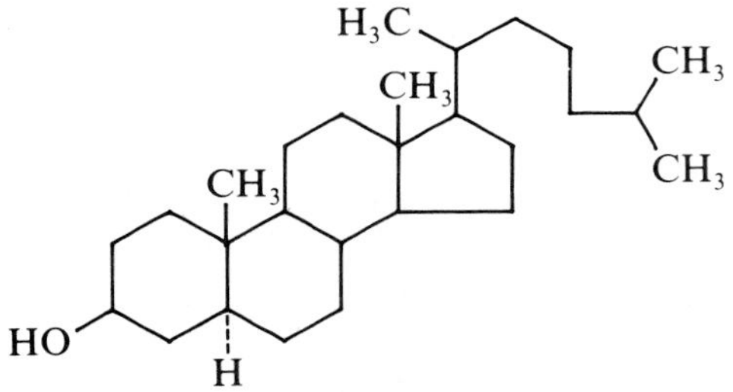

3β-Cholestanol

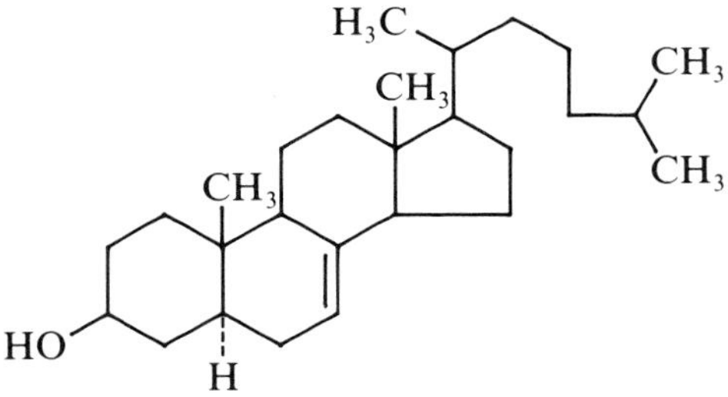

7-Cholesten-3β-ol

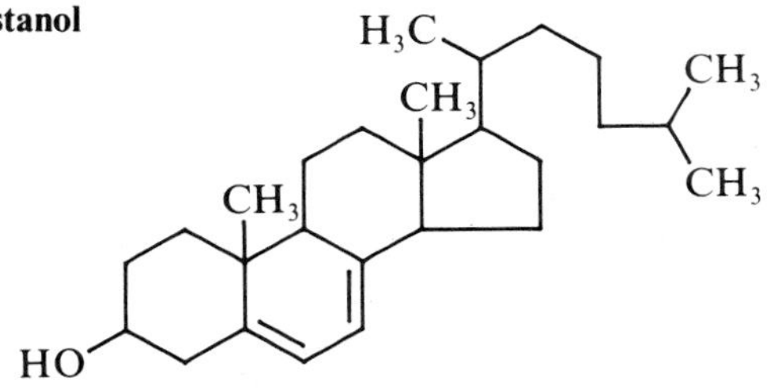

5,7-Cholestadien-3β-ol

1. A fourth companion, cerebrosterol, or 24-hydroxycholesterol, is easily eliminated by crystallization from alcohol.

The cholesterol dibromide that crystallizes from the reaction solution is collected, washed free of the impurities or their dehydrogenation products, and debrominated with zinc dust, with regeneration of cholesterol in pure form. Specific color tests can differentiate between pure cholesterol and tissue cholesterol purified by ordinary methods.

Experiments

1. Bromination of Cholesterol

In a 10 × 75 mm test tube dissolve 100 mg of gallstone cholesterol or of commercial cholesterol (content of 7-cholesten-3β-ol about 0.6%) in 0.7 mL of ether by gentle warming and then add 0.5 mL of a solution of bromine and sodium acetate in acetic acid.[2] Cholesterol dibromide begins to crystallize in a minute or two. Cool in an ice bath and stir the crystalline paste with a stirring rod for about 10 min to ensure complete crystallization, and at the same time cool a mixture of 0.3 mL of ether and 0.7 mL of acetic acid in ice. Then collect the crystals on the Hirsch funnel and wash with the iced ether–acetic acid solution to remove the yellow mother liquor. The short test tube is used to facilitate this transfer of the crystalline paste. This material is not easy to filter on a very small scale because the crystals are very small. Finally wash the crystals on the filter with a few drops of ethanol, continuing to apply suction, and transfer the white solid without drying it (dry weight 120 mg) to a reaction tube.

2. Zinc Dust Debromination

Add 2 mL of ether, 0.5 mL of acetic acid, and 20 mg of Zn dust[3] and mix the contents. In about 3 min the dibromide dissolves; after 5–10 min of mixing, zinc acetate usually separates to form a white precipitate (the dilution sometimes is such that no separation occurs). Stir for 5 min more and then add water (no more than 50 mg) until any solid present (zinc acetate) dissolves to make a clear solution. Decant the solution from the zinc into a reaction tube, wash the ethereal solution twice with 2 mL of water and then with 1 mL of 10% sodium hydroxide (to remove traces of acetic acid). Then shake the ether solution with anhydrous sodium sulfate, pipette off the dry solution, wash the drying agent with a little ether, add 1 mL of methanol (and a boiling stick), and evaporate the solution to the point where most of the ether is removed and the purified cholesterol begins to crystallize. Remove the solution from the heat, let crystallization proceed at room temperature

Notes for the instructor

2. Weigh a 125-mL Erlenmeyer flask on a balance placed in the hood, add 4.5 g of bromine by a capillary dropping tube (avoid breathing the vapor), and add 50 mL of acetic acid and 0.4 g of sodium acetate (anhydrous).

3. If the reaction is slow add more zinc dust. The amount specified is adequate if material is taken from a freshly opened bottle, but zinc dust deteriorates on exposure to air.

and then in an ice bath, collect the crystals, and wash them with cold methanol; you should obtain 0.6–0.7 g, mp 149–150°C.

Questions

1. What is the purpose of the acetic acid in this reaction?
2. Why might old zinc dust not react in the debromination reaction?
3. Why does not acetate ion attack the intermediate bromonium ion to give the 5-acetoxy-6-bromo compound in the bromination of cholesterol?

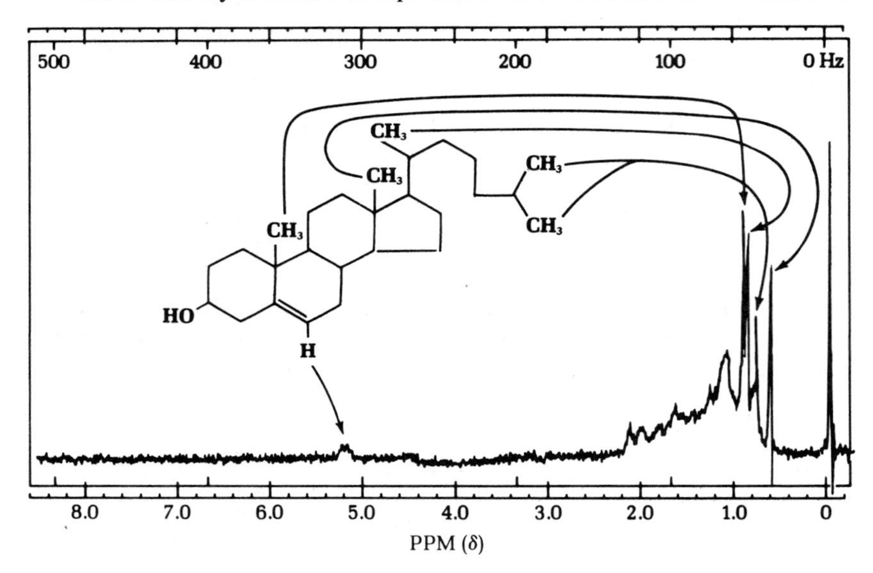

FIG. 23.1 ^{1}H nmr spectrum of cholesterol.

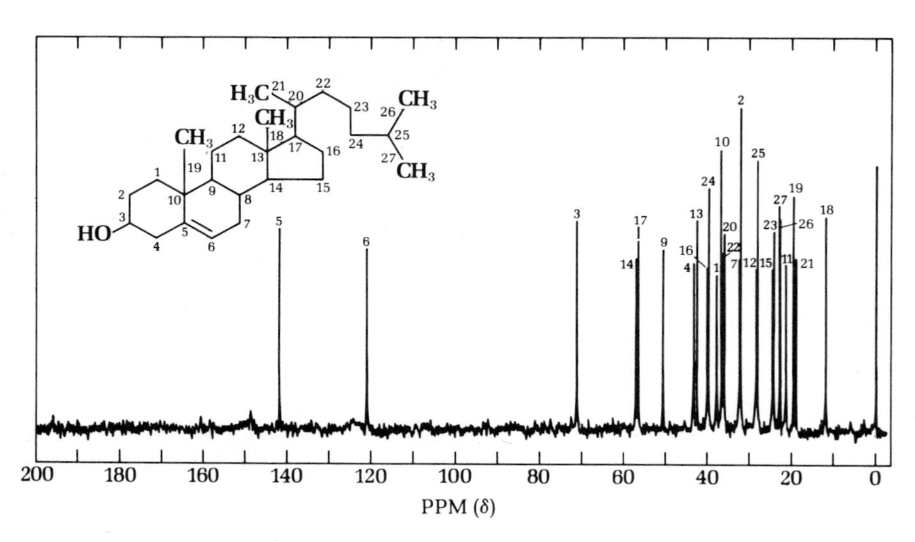

FIG. 23.2 ^{13}C nmr spectrum of cholesterol.

Pulegone from Citronellol: Oxidation with Pyridinium Chlorochromate

24

PRELAB EXERCISE: *Write a balanced equation for the oxidation of citronellol to citronellal.*

A variety of reagents and conditions can be employed to oxidize alcohols to carbonyl compounds. The choice of which reagents and set of conditions to use depends on such factors as the scale of the reaction, the speed of the reaction, anticipated yield, and ease of isolation of the products. Cyclohexanol can be oxidized to cyclohexanone by dichromate in acetic acid. This is much faster than permanganate oxidation of the same alcohol. Cholesterol can be oxidized to 5-cholesten-3-one by either the Jones reagent (chromium trioxide in sulfuric acid and water) or the Collins reagent, $(C_5H_5N)_2CrO_3$, prepared from chromium trioxide in anhydrous pyridine. Both of these reagents have the advantage that they will not cause isomerization of the labile Δ^5 double bond to the Δ^4 position. Recently E. J. Corey has added two new Cr^{6+} reagents to the repertory. Although they might seem similar to the other dichromate reagents, they each have unique advantages. While not nearly as fast as the Collins or Jones reagents, the Corey reagents are characterized by extraordinary ease of product isolation, because one merely removes the reagent by filtration and evaporates the solvent to obtain the product. Both reagents are easily prepared or can be purchased commercially (Aldrich). The reagent used in this experiment, pyridinium chlorochromate, $C_5H_5NH^+ClCrO_3^-$, will oxidize primary alcohols to aldehydes in high yield without oxidizing the aldehyde further to a carboxylic acid.[1] It will oxidize secondary alcohols to ketones, but being somewhat acidic will cause rearrangement of 5-cholesten-3-one to 4-cholesten-3-one.

Oxidizing agents:
$Cr_2O_7^{2-}$ in HOAc, $KMnO_4$,
CrO_3 in aq H_2SO_4, Jones reagent,
CrO_3 + pyridine, Collins reagent,
Corey reagents, PCC

Pyridine + CrO_3 + HCl ⟶ $C_5H_5NH^+$ $ClCrO_3^-$

Pyridine

Pyridinium chlorochromate, PCC
MW 215.56

1. E. J. Corey and J. W. Suggs, *Tetrahedron Letters*, 2647 (1975).

$$(C_6H_5)_2CHOH \xrightarrow[1.5\,h,\ 25^\circ C]{PCC-CH_2Cl_2} (C_6H_5)_2C{=}O \quad 100\%$$

$$CH_3(CH_2)_8CH_2OH \xrightarrow{PCC-CH_2Cl_2} CH_3(CH_2)_8\overset{O}{\overset{\|}{C}}H \quad 92\%$$

The other Corey reagent, pyridinium dichromate (PDC), $(C_5H_5NH^+)_2Cr_2O_7^{2-}$, dissolved in dimethylformamide (DMF) will oxidize allylic alcohols to α,β-unsaturated aldehydes without oxidizing the aldehyde to the carboxylic acid.[2]

$(C_5H_5NH^+)_2\,Cr_2O_7^{2-}$

Pyridinium dichromate, PDC

$$\text{cyclohex-2-en-1-ol} \xrightarrow[4\text{–}5\,h,\ 0^\circ C]{PDC-DMF} \text{cyclohex-2-en-1-one} \quad 86\%$$

$$C_6H_5CH{=}CH{-}CH_2OH \xrightarrow[4\text{–}5\,h,\ 0^\circ C]{PDC-DMF} C_6H_5CH{=}CH{-}\overset{O}{\overset{\|}{C}}H \quad 97\%$$

If the primary alcohol is not conjugated, then it is oxidized to the carboxylic acid:

$$C_6H_{11}CH_2OH \xrightarrow[7\text{–}9\,h,\ 25^\circ C]{PDC-DMF} C_6H_{11}COOH \quad 84\%$$

$$\text{1-methylcyclohex-3-enyl-}CH_2OH \xrightarrow[7\text{–}9\,h,\ 25^\circ C]{PDC-DMF} \text{1-methylcyclohex-3-enyl-}\overset{O}{\overset{\|}{C}}OH \quad 90\%$$

When suspended in CH_2Cl_2, pyridinium dichromate will oxidize primary alcohols to the corresponding aldehydes and no further:

2. E. J. Corey and G. Schmidt, *Tetrahedron Letters*, 399 (1979).

$$C_6H_5CH_2OH \xrightarrow[10\,h,\ 25^\circ C]{PDC-CH_2Cl_2} C_6H_5CHO \quad 83\%$$

$$CH_3(CH_2)_8CH_2OH \xrightarrow[20\,h,\ 25^\circ C]{PDC-CH_2Cl_2} CH_3(CH_2)_8CHO$$

$$\xrightarrow[24\,h,\ 25^\circ C]{PDC-CH_2Cl_2}$$

Experiments

In the present experiment pyridinium chlorochromate (PCC) is used to oxidize citronellol, a constituent of rose and geranium oil, to the corresponding aldehyde. If PDC in CH_2Cl_2 were used, the reaction would stop at this point and citronellal could be isolated in 92% yield. Or if sodium acetate (a buffer) were added to the PCC reaction, the oxidation would also stop at the aldehyde stage, in 82% yield. Recently it has been found that PCC on alumina will effect the same oxidation in less time and even higher yield.[3]

$$RCH_2OH \xrightarrow{PCC\ on\ alumina} RCHO$$

If the buffer is omitted, the pyridinium chlorochromate is acidic enough to cause the intermediate citronellal to cyclize to isopulegols (four possible isomers). These secondary alcohols are oxidized to two isomeric isopulegones. On treatment with base, the double bond will migrate to give pulegone, as shown on the following page.

The course of the reaction can easily be followed by thin layer chromatography.[4] The isolation of the oxidation products, citronellal and isopulegone, could not be simpler: the reduced reagent is removed by filtration and the solvent is evaporated. In the case of the second experiment the reaction can be stopped once isopulegone is formed, or the isopulegone can be isomerized to pulegone as a separate reaction.

3. Y.-S. Cheng, W.-L. Liu and S. Chen, *Synthesis*, 223 (1980).

4. E. J. Corey and J. W. Suggs, *J. Org. Chem.*, **41**, 380 (1976). TLC R_f values: citronellol, 0.17; citronellal, 0.65; isopulegols, 0.27 and 0.35; isopulegone, 0.41; pulegone, 0.36, CH_2Cl_2 developer.

Citronellol
MW 156.27
bp 222°C

PCC—CH_2Cl_2

Citronellal
MW 154.24

Isopulegols
(four possible isomers)

PCC—CH_2Cl_2

Isopulegone
(two possible isomers)

OH^-
Ethanol

Pulegone
MW 152.23
bp 224°C

Preparation of PCC

$$C_5H_5N + CrO_3^- + HCl \longrightarrow C_5H_5NH^+ \; ClCrO_3^-$$

Pyridinium chlorochromate, PCC
MW 215.56

Dissolve 12 g of chromium trioxide in 22 mL of 6 N hydrochloric acid, and add 9.5 g of pyridine during a 10-min period while maintaining the temperature at 45°. Cool the mixture to 0° and collect the crystalline yellow-orange pyridinium chlorochromate on a sintered glass funnel and dry it for 1 h *in vacuo*. The compound is not hygroscopic and is stable at room temperature.

Caution! *The dust, not the solutions, of Cr^{6+} is carcinogenic.*

Preparation of PCC-Alumina Reagent

To prepare PCC on alumina, cool the reaction mixture described above to 10°C until the product crystallizes; then reheat the solution to 40°C to dissolve the solid. To the resulting solution add 100 g of alumina with stirring at 40°C. The solvent is removed on a rotary evaporator, and the orange solid is dried in vacuum for 2 h at room temperature. If stored under vacuum in the dark, the reagent will retain its activity for several weeks.

1. Citronellal from Citronellol Using Pyridinium Chlorochromate on Alumina

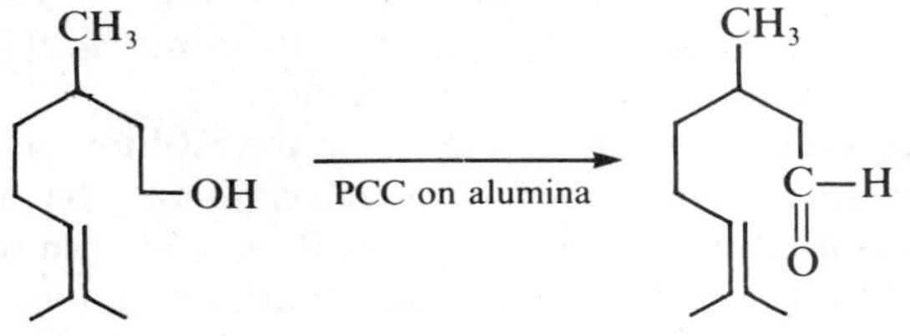

CH_3 CH_3

OH PCC on alumina C—H O

Citronellol **Citronellal**

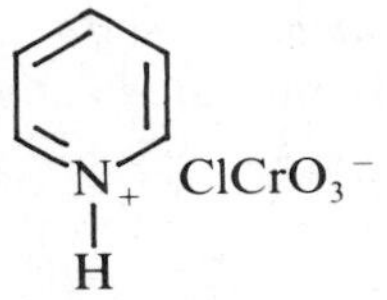

To a 10-mL Erlenmeyer flask add 1.5 g (1.22 mmoles) of PCC on alumina, 120 mg of citronellol, and 2 mL of hexane. Stir the mixture using a magnetic stirrer or by shaking for up to 3 h (follow the course of the reaction by TLC), remove the stirring bar, and then remove the solid by filtration on the Hirsch funnel under vacuum. Wash the alumina on the filter with three 2 mL-portions of ether, and remove the solvents from the filtrate by evaporation under a gentle stream of air, while warming the flask in a beaker of water. The last trace of solvent can be removed under vacuum. The residue should be pure citronellal, bp 90°C at 14 mm. Check its purity by TLC on silica gel using dichloromethane as eluant. Using infrared spectroscopy as a thin film between sodium chloride plates look for the presence of unoxidized hydroxyl at 3600–3400 cm^{-1}.

N^+ $ClCrO_3^-$ H

Pyridinium chlorochromate
MW 215.56

2. Isopulegone from Citronellol Using Pyridinium Chlorochromate in CH_2Cl_2

To 1 g of pyridinium chlorochromate suspended in 6.25 mL of dichloromethane in a 10-mL Erlenmeyer flask, add 0.25 g of citronellol. The best yields are obtained if the mixture is stirred at room temperature for 36 h or more using a magnetic stirrer. Alternatively, let the mixture sit at room temperature for a week with occasional shaking. The chromium salts are removed by vacuum filtration and washed on the filter with 1 mL of dichloromethane. Remove the solvent by evaporation with a gentle stream of nitrogen or air to give isopulegone. Follow the course of the reaction and analyze the product by TLC on silica gel plastic sheets, developing with dichloromethane. Save 1 mg of the product for the ultraviolet spectrum (see the next experiment).

Dispose of Cr salts in heavy metal waste container

3. Pulegone from Isopulegone

Add 250 mg of crude isopulegone isomers to a reaction tube followed by 2 mL of a 2% solution of sodium hydroxide in ethanol. Warm the mixture on a hot water bath or on a steam bath for 1 hr. Evaporate the ethanol under a stream of air or nitrogen and then dissolve the residue in 1 mL of dichloromethane. Prepare a chromatographic column from 2.5 g of silica gel in dichloromethane and pour the solution of pulegone through the column followed by 2 mL of dichloromethane. This very crude chromatography is to remove the very small quantity of sodium hydroxide present in the pulegone. Evaporate the solvent to leave pure pulegone.

The pulegone can also be isolated from the reaction mixture by evaporating the ethanol, adding 1 mL of water, and extracting the product with two 1-mL portions of ether. The ether extract is washed with 0.5 mL of water followed by 0.5 mL of saturated sodium chloride solution, and then dried over anhydrous sodium sulfate. The ether is removed from the drying agent with a Pasteur pipette, the drying agent is washed with more ether, and the ether evaporated under a stream of air or nitrogen to leave pure pulegone.

Ultraviolet spectra α,β-unsaturated ketone

Pulegone occurs naturally in pennyroyal oil and has a pleasant odor between those of peppermint and camphor. It is an α,β-unsaturated ketone and therefore should have λ_{max} 244 nm as calculated by application of the Fieser and Woodward Rules. (See p. 214.) The starting material is not a conjugated ketone and so should not have intense uv absorption. Compare the two compounds by dissolving 1 mg of each in 50 mL of ethanol and determine their ultraviolet spectra.

Questions

1. Draw the chair conformations of the four possible isopulegols.
2. Assign as many peaks as possible to specific protons in the ^{1}H nmr spectra of citronellol (Fig. 24.1) and citronellal (Fig. 24.2).
3. Give the mechanism for the base-catalyzed isomerization of isopulegone to pulegone.

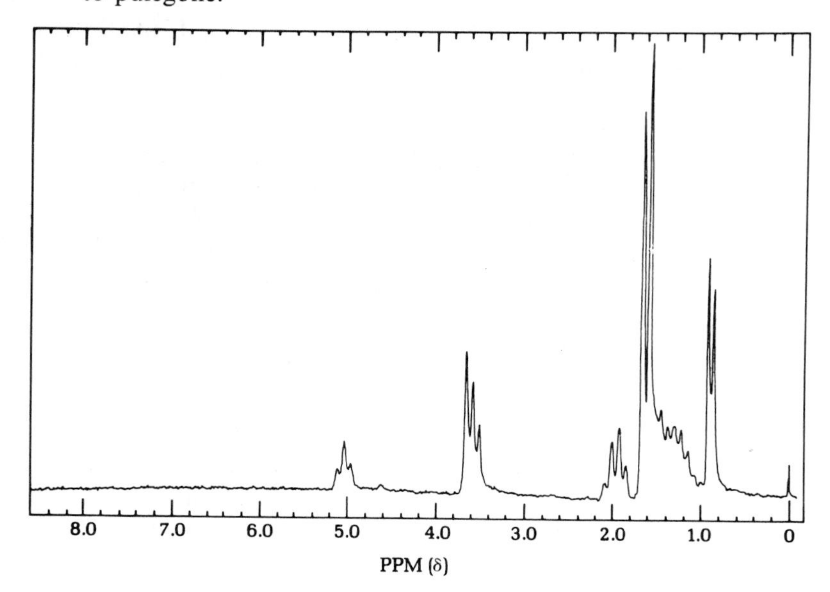

FIG. 24.1 ^{1}H nmr spectrum of citronellol.

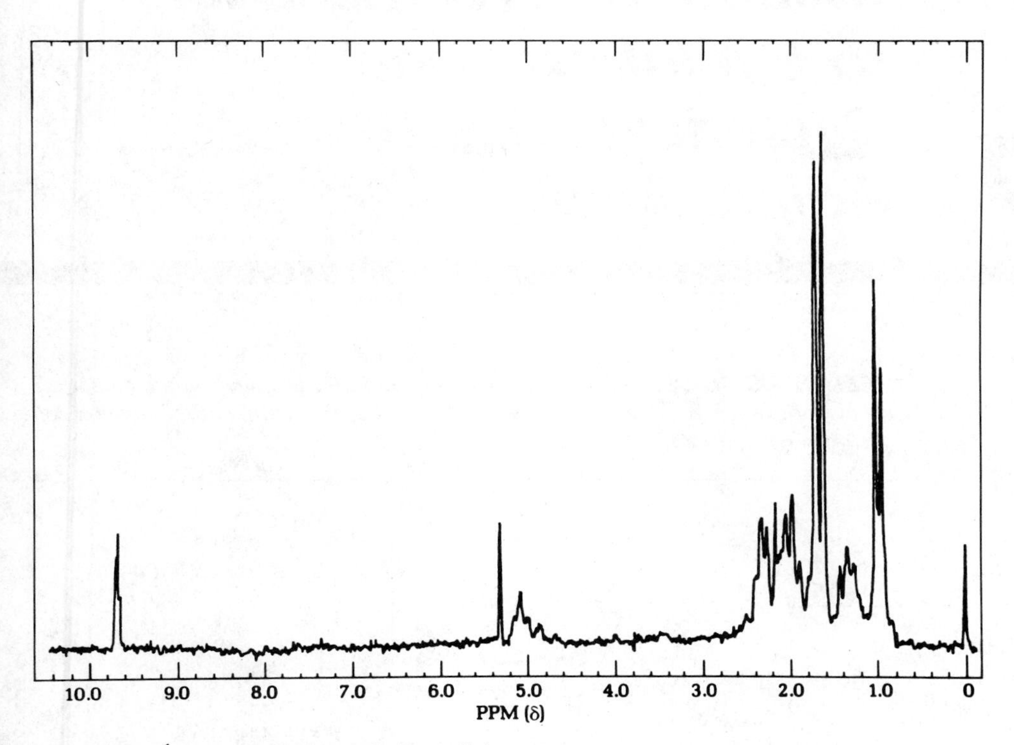

FIG. 24.2 ^{1}H nmr spectrum of citronellal.

25 Oxidation: Cyclohexanol to Cyclohexanone; Cyclohexanone to Adipic Acid

PRELAB EXERCISE: *Write a balanced equation for the dichromate oxidation of cyclohexanol to cyclohexanone and for the permanganate oxidation of cyclohexanone to adipic acid.*

$$\text{Cyclohexanol} \xrightarrow[\text{HOAc}]{\text{PCC or } Na_2Cr_2O_7} \text{Cyclohexanone} \underset{}{\overset{OH^-}{\rightleftharpoons}} [\text{enolate}] \xrightarrow{KMnO_4} \text{Adipic acid}$$

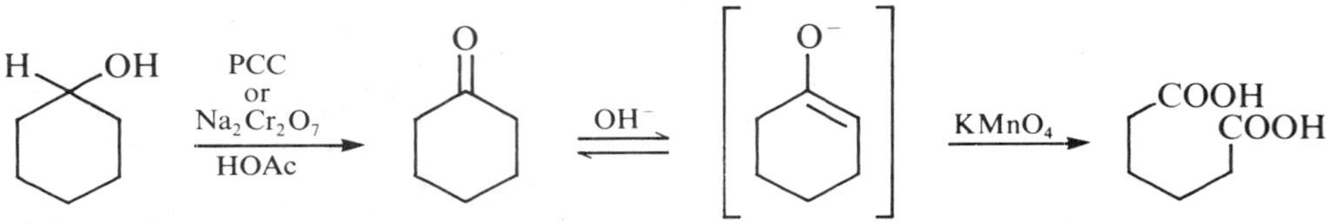

Cyclohexanol
bp 161.5°C, den 0.96
MW 100.16

Cyclohexanone
bp 157°C, den 0.95
MW 98.14
solubility 1.5 g/100 mL $H_2O^{10°}$

Adipic acid
mp 153°C, MW 146.14
solubility 1.4 g/100 g $H_2O^{15°}$

The oxidation of a secondary alcohol to a ketone can be accomplished by a very large number of oxidizing agents, including sodium dichromate in acetic acid and pyridinium chlorochromate. The ketone can be oxidized further to the dicarboxylic acid giving adipic acid. Both of these oxidations can be carried out by permanganate ion to give the diacid. Nitric acid is a powerful oxidizing agent that can oxidize cyclohexane, cyclohexene, cyclohexanol, or cyclohexanone to adipic acid.

The mechanism of oxidation of an alcohol to a ketone by dichromate appears to be the following:

$$H_2O + Cr_2O_7^{2-} \rightleftharpoons 2\ HCrO_4^-$$

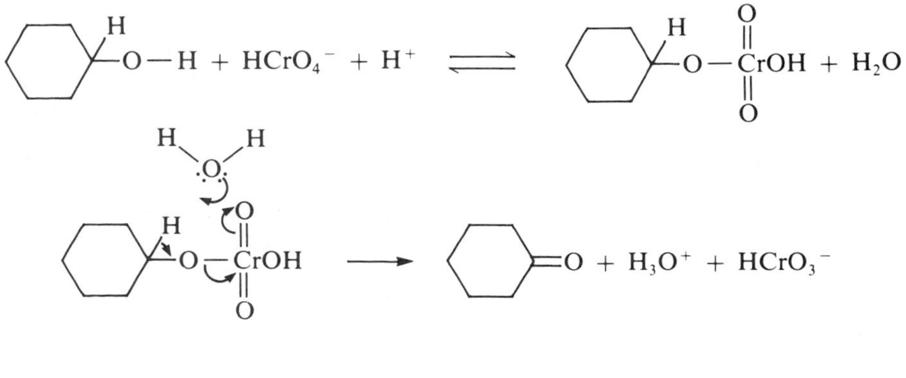

$$C_6H_{10}(H){-}O{-}H + HCrO_4^- + H^+ \rightleftharpoons C_6H_{10}(H){-}O{-}CrO_2OH + H_2O$$

$$C_6H_{10}(H){-}O{-}CrO_2OH \xrightarrow{H_2O} C_6H_{10}{=}O + H_3O^+ + HCrO_3^-$$

A number of intermediate valence states of chromium are involved in this reaction as orange Cr^{6+} ultimately is reduced to green Cr^{3+}. The course of the oxidation can be followed by these color changes.

The oxidation of a ketone by permanganate to the dicarboxylic acid takes place through the enol form of the ketone. The reaction can be followed as the bright purple permanganate solution reacts to give a brown precipitate of manganese dioxide. A possible mechanism for this reaction is the following:

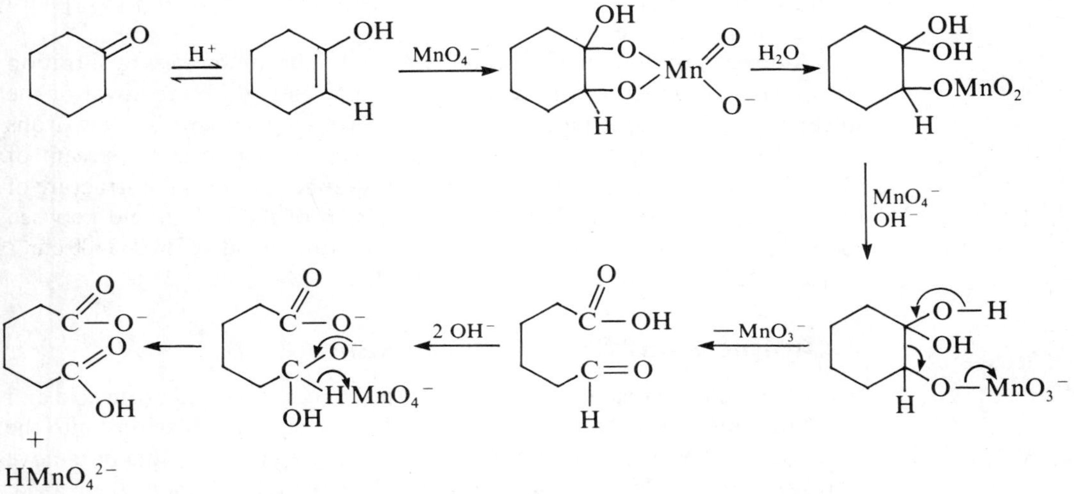

$$3\ HMnO_4^{2-} + H_2O \longrightarrow 2\ MnO_2 + MnO_4^- + 5\ OH^-$$

The balanced equation for the oxidation of cyclohexanone to adipic acid is:

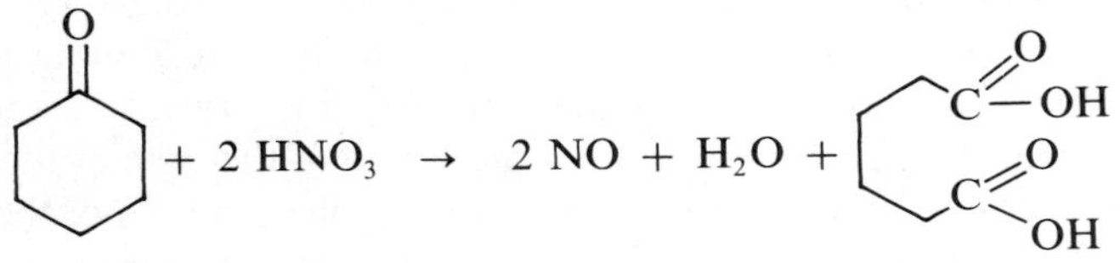

$$\text{cyclohexanone} + 2\ HNO_3 \rightarrow 2\ NO + H_2O + \text{adipic acid}$$

In this reaction nitric acid is reduced to nitric oxide.

In the following experiments cyclohexanol is oxidized to cyclohexanone using pyridinium chlorochromate in dichloromethane. The progress of the reaction can be followed by thin-layer chromatography. On a larger scale this reaction would be carried out using sodium dichromate in acetic acid because the reagents are less expensive, the reaction is faster, and much less solvent is required. It is interesting to contrast the micro- and macroscale reactions, and for this reason the large-scale procedure is presented in full in the section On a Larger Scale.

Experiments

1. Cyclohexanone from Cyclohexanol

To a 10-mL Erlenmeyer flask add 0.62 g of finely powdered pyridinium chlorochromate (see previous chapter for preparation), 4 mL of dichloromethane, and 250 mg of cyclohexanol. The mixture is stirred magnetically or shaken over a period of days (a week does no harm) until thin-layer chromatography on silica gel indicates the reaction is complete. Elute the TLC plates with dichloromethane.

At the end of the reaction period remove the chromium salts by filtration on the Hirsch funnel, or, if the salts are coarse enough, by removal of the solvent using a Pasteur pipette. Wash the chromium salts with a few drops of dichloromethane, and evaporate the solvent under a gentle stream of nitrogen or air to leave the product, cyclohexanone. Confirm the structure of this product by obtaining an infrared spectrum of the pure liquid between sodium chloride or silver chloride plates. Look for a band at 3600–3400 cm^{-1} indicative of unoxidized alcohol (or water).

2. Adipic Acid from Cyclohexanone

In a 10 × 100 mm reaction tube place 1.0 mL of concentrated nitric acid and a boiling chip. Clamp the tube and also a Pasteur pipette, which dips into the tube and is connected to a water aspirator pulling a gentle vacuum to remove nitrogen oxides. Add to the tube one small drop of cyclohexanone from a vial containing 150 mg of the ketone. Warm the tube until brown oxides of nitrogen are seen emanating from the nitric acid and an exothermic reaction begins. Do not add more cyclohexanone until it is quite clear (evolution of brown oxides of nitrogen, generation of heat) that the reaction has started. Remove the tube from the heat. Add the remainder of the cyclohexanone to the hot nitric acid over a period of about three minutes. The reaction is extremely exothermic; no external heating will be necessary. After the completion of cyclohexanone addition, heat the flask to boiling for 1 min.

As the tube cools to room temperature, fine crystals of adipic acid should appear. If they do not, scratch the inside of the tube at the liquid-air interface with a small glass rod to initiate crystallization. Cool the tube in a mixture of ice and water for at least 3 min, then stir the crystals with the tip of a glass Pasteur pipette and remove the solvent by forcing the pipette to the bottom of the tube. Rap the tube on a hard surface and remove more solvent from the crystals, taking care to keep the tube cold. Add 0.2 mL of ice water to the crystals, cool thoroughly in ice, and remove as much of the wash liquid as possible using the Pasteur pipette, bearing in mind that the product is *very* soluble in water. To remove the last bit of water force a piece of rolled filter paper down onto the surface of the crystals. Remove the filter paper, then the product. Press the crystals between sheets of filter paper to effect further drying, then allow the crystals to dry in air. Determine the mp, the infrared spectrum, and the homogeneity of the product by thin-layer chromatography. For TLC, use as the eluent three parts 95% ethanol and one part ammonium hydroxide.

On a Larger Scale. The following procedure is used to prepare about 11 g of cyclohexanone.

In a 125-mL Erlenmeyer flask dissolve 15 g of sodium dichromate dihydrate in 25 mL of acetic acid by swirling the mixture on the hot plate and then cool the solution with ice to 15°C. In a second Erlenmeyer flask chill a mixture of 15.0 g of cyclohexanol and 10 mL of acetic acid in ice. After the first solution is cooled to 15°C, transfer the thermometer and adjust the temperature in the second flask to 15°C. Wipe the flask containing the dichromate solution, pour the solution into the cyclohexanol-acetic acid mixture, rinse the flask with a little solvent (acetic acid), note the time, and take the initially light orange solution from the ice bath, but keep the ice bath ready for use when required.[1] The exothermic reaction that is soon evident can get out of hand unless controlled. When the temperature rises to 60°C cool in ice just enough to prevent a further rise and then, by intermittent brief cooling, keep the temperature close to 60°C for 15 min. No further cooling is needed, but the flask should be swirled occasionally and the temperature watched. The usual maximal temperature is 65°C (25–30 min). When the temperature begins to drop and the solution becomes pure green, the reaction is over. Allow 5–10 min more reaction time and then pour the green solution into a 250-mL round-bottomed flask, rinse the Erlenmeyer flask with 100 mL of water, and add the solution to the flask for steam distillation (Chapter 6) of the product. Distil as long as any oil passes over with the water and, because cyclohexanone is appreciably soluble in water, continue somewhat beyond this point (about 80 mL will be collected).

Cr^{6+} salts are toxic; weigh in hood

Reaction time 45 min

Alternatively, instead of setting up the apparatus for steam distillation, simply add a boiling chip to the 250-mL flask and distil 40 mL of liquid, cool the flask slightly, add 40 mL of water to it and distil 40 mL more. Note the temperature during the distillation. This is a steam distillation in which steam is generated *in situ* rather than from an outside source.

Cyclohexanone is fairly soluble in water. Dissolving inorganic salts such as potassium carbonate or sodium chloride in the aqueous layer will decrease the solubility of cyclohexanone such that it can be completely extracted with ether. This process is known as "salting out."

Dispose of Cr salts in heavy metal waste container

1. When the acetic acid solutions of cyclohexanol and dichromate were mixed at 25°C rather than at 15°C the yield of crude cyclohexanone was only 6.9 g. A clue to the evident importance of the initial temperature is suggested by an experiment in which the cyclohexanol was dissolved in 12.5 mL of benzene instead of 10 mL of acetic acid and the two solutions were mixed at 15°C. Within a few minutes orange-yellow crystals separated and soon filled the flask; the substance probably is the chromate ester, $(C_6H_{11}O)_2CrO_2$. When the crystal magma was let stand at room temperature the crystals soon dissolved, exothermic oxidation proceeded, and cyclohexanone was formed in high yield. Perhaps a low initial temperature ensures complete conversion of the alcohol into the chromate ester before side reactions set in.

To salt out the cyclohexanone add to the distillate 0.2 g of sodium chloride per milliliter of water present and swirl to dissolve the salt. Then pour the mixture into a separatory funnel, rinse the flask with ether, add more ether to a total volume of 25–30 mL, shake, and draw off the water layer. Then wash the ether layer with 25 mL of 10% sodium hydroxide solution to remove acetic acid, test a drop of the wash liquor to make sure it contains excess alkali, and draw off the aqueous layer.

To dry the ether, which contains dissolved water, shake the ether layer with an equal volume of saturated aqueous sodium chloride solution. Draw off the aqueous layer, pour the ether out of the neck of the separatory funnel into an Erlenmeyer flask, add about 5 g of anhydrous sodium sulfate, and complete final drying of the ether solution by occasional swirling of the solution over a 5-min period. Remove the drying agent by decantation or gravity filtration into a tared Erlenmeyer flask and rinse the flask that contained the drying agent, the sodium sulfate, and the funnel with ether. Add a boiling chip to the ether solution and evaporate the ether on the steam bath under an aspirator tube. Cool the contents of the flask to room temperature, evacuate the crude cyclohexanone under aspirator vacuum to remove final traces of ether, and weigh the product. Yield is 11–12.5 g.

The crude cyclohexanone can be purified by simple distillation or used directly in the following experiment.

Questions

1. In the oxidation of cyclohexanol to cyclohexanone, what purpose does the acetic acid serve?

2. Explain the order of the chemical shifts of the carbon atoms in the ^{13}C spectra of cyclohexanone (Fig. 25.1) and adipic acid (Fig. 25.2).

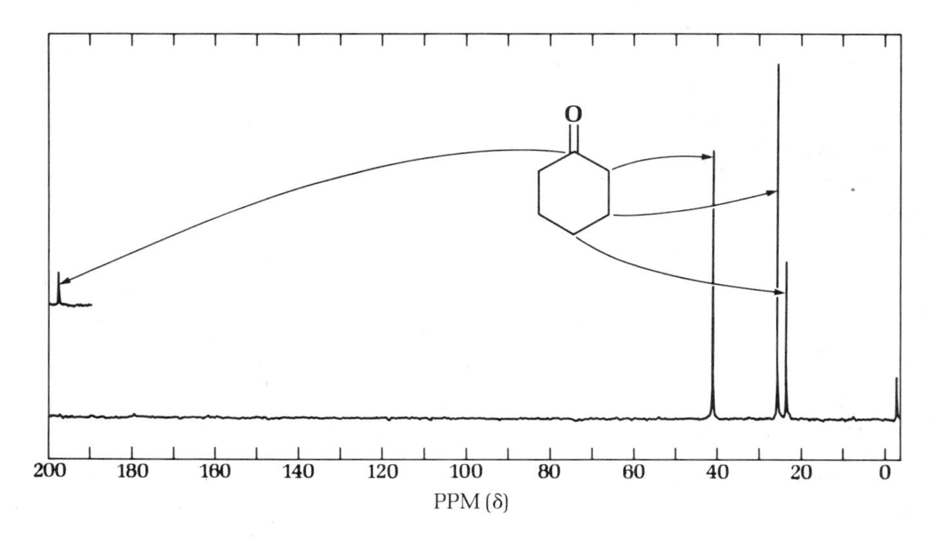

FIG. 25.1 ^{13}C nmr spectrum of cyclohexanone.

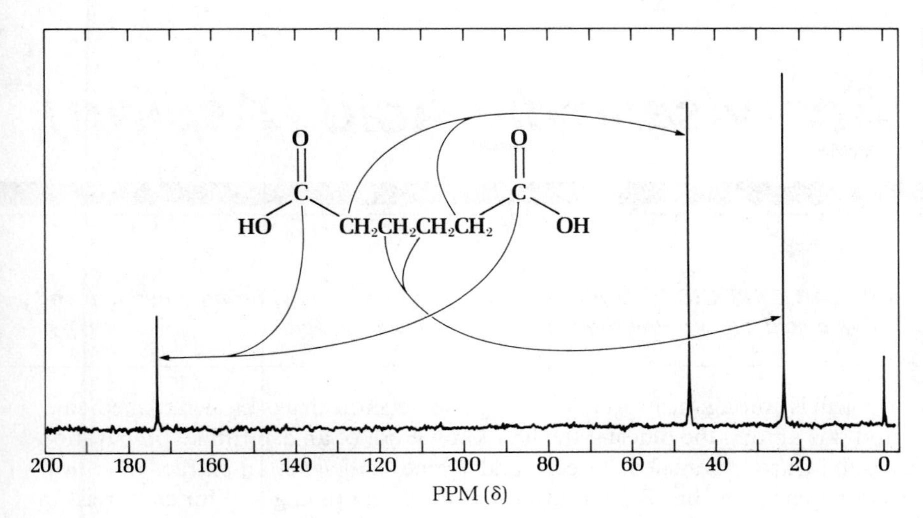

FIG. 25.2 ^{13}C nmr spectrum of adipic acid.

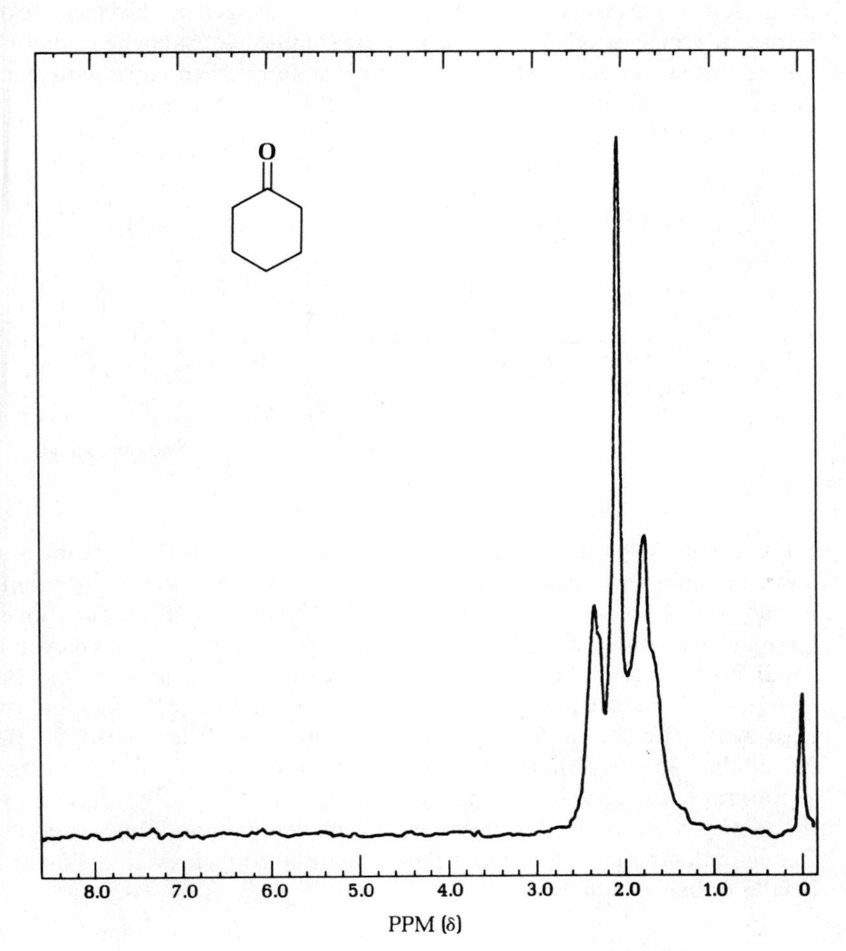

FIG. 25.3 ^{1}H nmr spectrum of cyclohexanone.

26 Acetylsalicylic Acid (Aspirin)

PRELAB EXERCISE: *Write detailed mechanisms showing how pyridine and sulfuric acid catalyze the formation of acetylsalicylic acid.*

Aspirin is among the most fascinating and versatile drugs known to medicine, and it is among the oldest—the first known use of an aspirinlike preparation can be traced to ancient Greece and Rome. Salicigen, an extract of willow and poplar bark, has been used as a pain reliever (analgesic) for centuries. In the middle of the last century it was found that salicigen is a glycoside formed from a molecule of salicylic acid and a sugar molecule. Salicylic acid is easily synthesized on a large scale by heating sodium phenoxide with carbon dioxide at 150°C under slight pressure (the Kolbe synthesis):

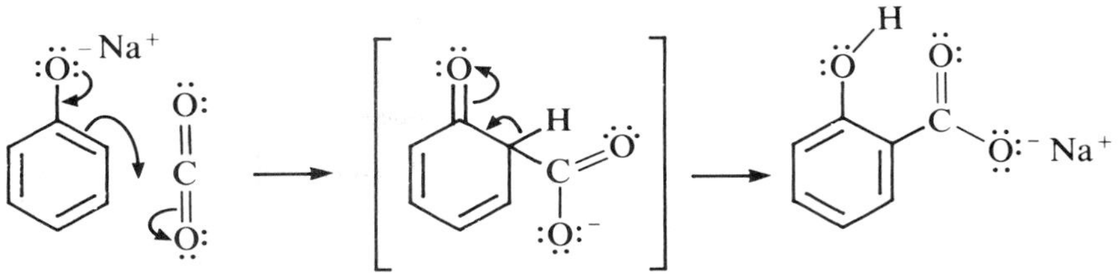

Sodium salicylate

But unfortunately salicylic acid attacks the mucous membranes of the mouth and esophagus and causes gastric pain that may be worse than the discomfort it was meant to cure. Felix Hoffmann, a chemist for Friedrich Bayer, a German dye company, reasoned that the corrosive nature of salicylic acid could be altered by addition of an acetyl group; and in 1893 the Bayer Company obtained a patent on acetylsalicylic acid, despite the fact that it had been synthesized some forty years previously by Charles Gerhardt. Bayer coined the name Aspirin for their new product to reflect its acetyl nature and its natural occurrence in the Spiraea plant. Over the years they have allowed the term aspirin to fall into the public domain so it is no longer capitalized. The manufacturers of Coke and Sanka work hard to prevent a similar fate befalling their products.

In 1904 the head of Bayer, Carl Duisberg, decided to emulate John D. Rockefeller's Standard Oil Company and formed an "interessen gemeinshaft" (I.G.) of the dye industry (Farbenindustrie). This cartel completely dominated the world dye industry before World War I and it continued to prosper between the wars even though some of their assets were seized and sold after World War I. After World War I an American company, Sterling Drug, bought the rights to aspirin. The company's Glenbrook Laboratories division still is the major manufacturer of aspirin in the United States (Bayer Aspirin).

Because of their involvement at Auschwitz the top management of IG Farbenindustrie was tried and convicted at the Nuremberg trials after World War II and the cartel broken into three large branches, Bayer, Hoechst, and BASF (Badische Anilin and Sodafabrik), each of which do more business than DuPont, the largest American chemical company.

By law all drugs sold in the United States must meet purity standards set by the Food and Drug Administration, and so all aspirin is essentially the same. Each 5-grain tablet contains 0.325 g of acetylsalicylic acid held together with a binder. The remarkable difference in price for aspirin is primarily a reflection of the advertising budget of the company that sells it.

Aspirin is an analgesic (painkiller), an antipyretic (fever reducer), and an anti-inflammatory agent. It is the premier drug for reducing fever, a role for which it is uniquely suited. As an anti-inflammatory, it has become the most widely effective treatment for arthritis. Patients suffering from arthritis must take so much aspirin (several grams per day) that gastric problems may result. For this reason aspirin is often combined with a buffering agent. Bufferin is an example of such a preparation.

The ability of aspirin to diminish inflammation is apparently due to its inhibition of the synthesis of prostaglandins, a group of C-20 molecules that enhance inflammation. Aspirin transfers the oxygenase activity of prostaglandin synthetase by moving the acetyl group to a terminal amine group of the enzyme.

If aspirin were a new invention, the U.S. Food and Drug Administration (FDA) would place many hurdles in the path of its approval. It has been implicated, for example, in Reyes syndrome, a brain disorder that strikes children and young people under 18. It has an effect on platelets, which play a vital role in blood clotting. In newborn babies and their mothers, aspirin can lead to uncontrolled bleeding and problems of circulation for the baby —even brain hemorrhage in extreme cases. This same effect can be turned into an advantage, however. Heart specialists urge potential stroke victims to take aspirin regularly to inhibit clotting in their arteries.

Aspirin is found in more than 100 common medications, including Alka-Seltzer, Anacin ("contains the pain reliever doctors recommend most"), APC, Coricidin, Excedrin, Midol, and Vanquish. Despite its side-effects, aspirin remains the safest, cheapest, and most effective nonprescription drug. It is made commercially employing the same synthesis used here.

The mechanism for the acetylation of salicylic acid is as follows:

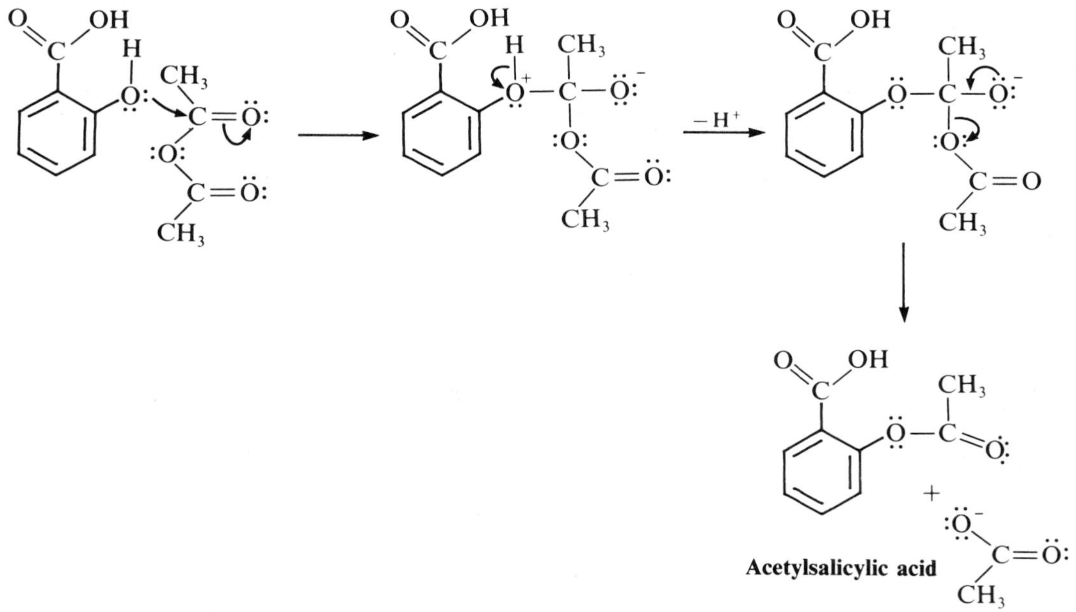

Acetylsalicylic acid

Experiment

Synthesis of Acetylsalicylic Acid (Aspirin)

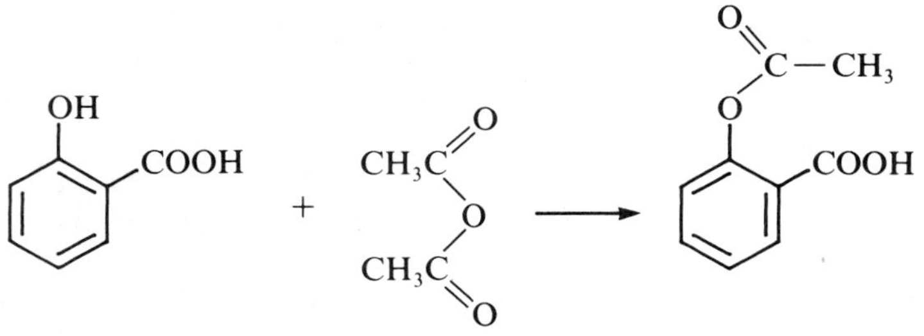

Salicylic acid	Acetic anhydride	Acetylsalicylic acid
MW 138.12, mp 159°C	MW 102.09, bp 140°C	MW 180.15, mp 128–137°C

To a reaction tube add 138 mg of salicylic acid, a boiling chip, and one small drop of 85% phosphoric acid followed by 0.3 mL of acetic anhydride, which will serve to wash the reactants to the bottom of the tube. Mix the reactants thoroughly and then heat the reaction tube on the steam bath or in a beaker of 90°C water for 5 min. Cautiously add 0.2 mL of water to the reaction mixture to decompose excess acetic anhydride. This will be an exothermic reaction. When the reaction is over, add 0.3 mL more water and allow the tube to cool slowly to room temperature. If crystallization of the product does not occur during the cooling process, add a seed crystal or scratch the inside of the tube with a glass stirring rod. Cool the tube in ice until crystallization is complete (at least 10 min), then remove the solvent with a Pasteur pipette. If the crystals are too fine for this procedure, then collect the product by vacuum filtration on the Hirsch funnel. Complete the transfer of the product to the funnel using a very small quantity of ice water. In either case turn the product out onto a piece of filter paper and squeeze the crystals between sheets of filter paper to absorb excess water. Allow the product to dry thoroughly in air before determining the weight and calculating the yield. Determine the melting point and the infrared spectrum in chloroform solution. Aspirin is hydrolyzed by boiling water, but the reaction is not rapid; if desired, therefore, the product may be recrystallized from a small quantity of very hot water.

Compare a tablet of commercial aspirin with your sample. Test the solubility of the tablet in water and in toluene and observe if it dissolves completely. Compare its behavior when heated in a melting point capillary with the behavior of your sample. If an impurity is found, it is probably some substance used as binder for the tablets. Is it organic or inorganic?

Questions

1. Hydrochloric acid is about as strong a mineral acid as sulfuric acid. Why would it not be a satisfactory catalyst in this reaction?
2. How do you account for the smell of vinegar when an old bottle of aspirin is opened?

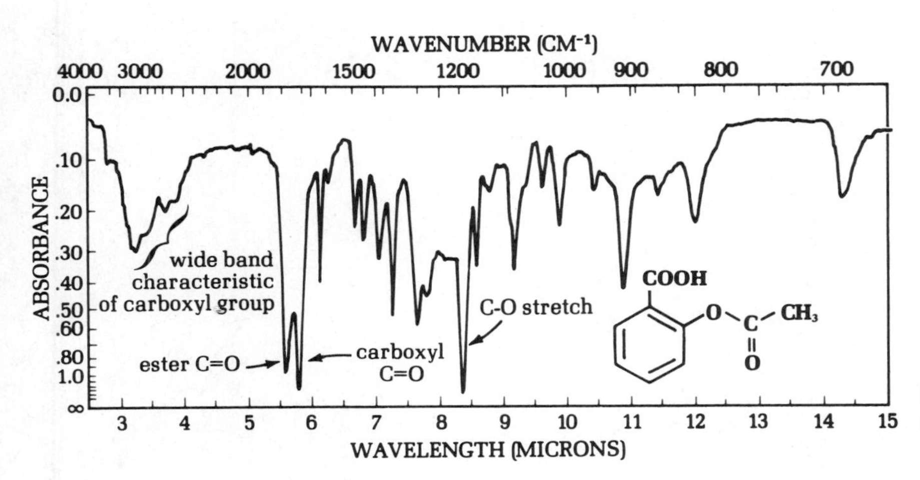

FIG. 26.1 Infrared spectrum of acetylsalicylic acid (aspirin) in $CHCl_3$.

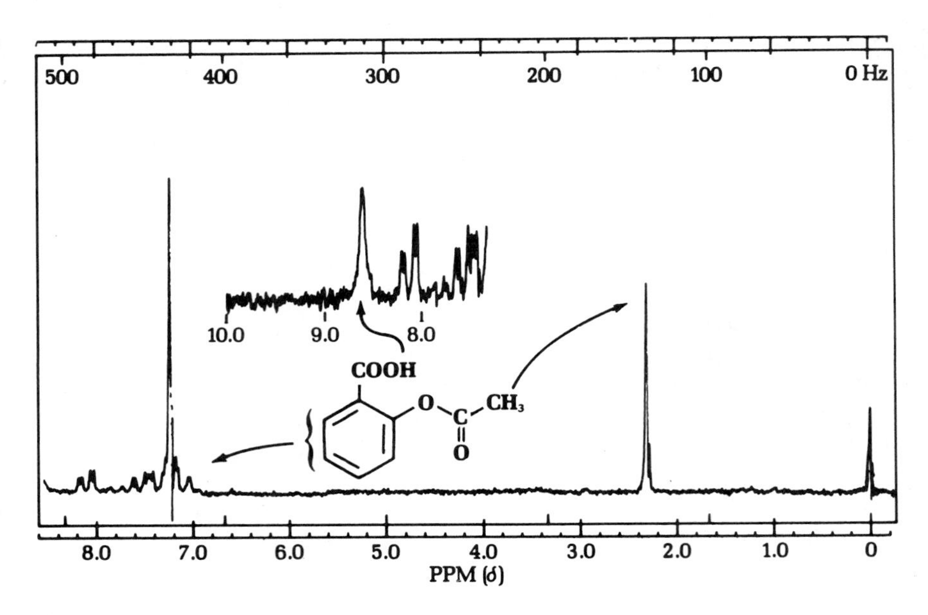

FIG. 26.2 ^{1}H nmr spectrum of acetylsalicylic acid (aspirin).

Esterification 27

PRELAB EXERCISE: *Give the detailed mechanism for the acid catalyzed* hydrolysis *of methyl benzoate.*

The ester group

$$R-\overset{\overset{\displaystyle O}{\|}}{C}-O-R'$$

is an important functional group that can be synthesized in a number of different ways. The low molecular weight esters have very pleasant odors and indeed are the major components of the flavor and odor components of a number of fruits. Although the natural flavor may contain nearly a hundred different compounds, single esters approximate the natural odors and are often used in the food industry for artificial flavors and fragrances (Table 27.1).

Flavors and fragrances

Esters can be prepared by the reaction of a carboxylic acid with an alcohol in the presence of a catalyst such as concentrated sulfuric acid, hydrogen chloride, *p*-toluenesulfonic acid, or the acid form of an ion exchange resin:

$$\underset{\textbf{Acetic acid}}{CH_3\overset{\overset{\displaystyle O}{\|}}{C}-OH} + \underset{\textbf{Methanol}}{CH_3OH} \overset{H^+}{\rightleftharpoons} \underset{\textbf{Methyl acetate}}{CH_3\overset{\overset{\displaystyle O}{\|}}{C}-OCH_3} + H_2O$$

This Fischer esterification reaction reaches equilibrium after a few hours of refluxing. The position of the equilibrium can be shifted by adding more of the acid or of the alcohol, depending on cost or availability. The mechanism of the reaction involves initial protonation of the carboxyl group, attack by the nucleophilic hydroxyl, a proton transfer, and loss of water followed by loss of the catalyzing proton to give the ester. Because each of these steps is completely reversible, this process is also, in reverse, the mechanism for the hydrolysis of an ester:

Fischer esterification

$$R-C(=\ddot{O}:)-\ddot{O}-H \underset{-H^+}{\overset{+H^+}{\rightleftharpoons}} R-C(=\overset{+}{\ddot{O}}-H)-\ddot{O}-H \underset{-R'-\ddot{O}H}{\overset{+R'-\ddot{O}H}{\rightleftharpoons}} R-C(H-\ddot{O}:)(\ddot{O}H)(H-\overset{+}{O}-R') \rightleftharpoons$$

$$R-C(H-\ddot{O}:)(\overset{+}{\ddot{O}}H-H)(:\ddot{O}-R') \underset{+H_2O}{\overset{-H_2O}{\rightleftharpoons}} R-C(=\overset{+}{\ddot{O}}-H)-\ddot{O}-R' \underset{+H^+}{\overset{-H^+}{\rightleftharpoons}} R-C(=\ddot{O}:)-\ddot{O}-R'$$

TABLE 27.1 Fragrances and Boiling Points of Esters

Ester	Formula	Bp (°C)	Flavor
Isobutyl formate	$HC(=O)-OCH_2CH(CH_3)CH_3$	98.4	Raspberry
n-Propyl acetate	$CH_3C(=O)-OCH_2CH_2CH_3$	101.7	Pear
Methyl butyrate	$CH_3CH_2CH_2C(=O)-OCH_3$	102.3	Apple
Ethyl butyrate	$CH_3CH_2CH_2C(=O)-OCH_2CH_3$	121	Pineapple
Isobutyl propionate	$CH_3CH_2C(=O)-OCH_2CH(CH_3)CH_3$	136.8	Rum
Isoamyl acetate	$CH_3C(=O)-OCH_2CH_2CH(CH_3)CH_3$	142	Banana
Benzyl acetate	$CH_3C(=O)-OCH_2-C_6H_5$	206	Peach
Octyl acetate	$CH_3C(=O)-OCH_2(CH_2)_6CH_3$	210	Orange
Methyl salicylate	$o\text{-}HOC_6H_4-C(=O)-OCH_3$	222	Wintergreen

Other methods are available for the synthesis of esters, most of them more expensive but readily carried out on a small scale. For example, alcohols react with anhydrides and with acid chlorides:

$$\underset{\textbf{Ethanol}}{CH_3CH_2OH} + \underset{\textbf{Acetic anhydride}}{CH_3\overset{O}{\overset{\|}{C}}-O-\overset{O}{\overset{\|}{C}}CH_3} \longrightarrow \underset{\textbf{Ethyl acetate}}{CH_3\overset{O}{\overset{\|}{C}}-OCH_2CH_3} + \underset{\textbf{Acetic acid}}{CH_3\overset{O}{\overset{\|}{C}}-OH}$$

$$\underset{\textbf{1-Propanol}}{CH_3CH_2CH_2OH} + \underset{\textbf{Acetyl chloride}}{CH_3\overset{O}{\overset{\|}{C}}-Cl} \longrightarrow \underset{\textbf{\textit{n}-Propyl acetate}}{CH_3\overset{O}{\overset{\|}{C}}-OCH_2CH_2CH_3} + HCl$$

In the latter reaction an organic base such as pyridine is usually added to react with the hydrogen chloride.

A number of other methods can be used to synthesize the ester group. Among these are the addition of 2-butene to an acid to form *t*-butyl esters, the addition of ketene to make acetates, and the reaction of a silver salt with an alkyl halide:

Other ester syntheses

$$\underset{\textbf{2-Butene}}{CH_2=\overset{CH_3}{\overset{|}{C}}CH_3} + \underset{\textbf{Propionic acid}}{CH_3CH_2\overset{O}{\overset{\|}{C}}-OH} \xrightarrow{H^+} \underset{\textbf{\textit{t}-Butyl propionate}}{CH_3CH_2\overset{O}{\overset{\|}{C}}-O\overset{CH_3}{\overset{|}{\underset{CH_3}{\underset{|}{C}}}}CH_3}$$

$$\underset{\textbf{Ketene}}{CH_2=C=O} + \underset{\textbf{Benzyl alcohol}}{HOCH_2-C_6H_5} \longrightarrow \underset{\textbf{Benzyl acetate}}{CH_3\overset{O}{\overset{\|}{C}}-OCH_2-C_6H_5}$$

$$\underset{\textbf{Silver acetate}}{CH_3\overset{O}{\overset{\|}{C}}-OAg} + \underset{\textbf{1-Bromo-3-methylpropane}}{BrCH_2CH_2\overset{CH_3}{\overset{|}{C}}HCH_3} \longrightarrow \underset{\textbf{Isoamyl acetate}}{CH_3\overset{O}{\overset{\|}{C}}-OCH_2CH_2\overset{CH_3}{\overset{|}{C}}HCH_3}$$

As noted above, esterification is an equilibrium process. Consider the reaction of acetic acid with 1-butanol to give *n*-butyl acetate:

$$CH_3\overset{O}{\overset{\|}{C}}-OH + HOCH_2CH_2CH_2CH_3 \overset{H^+}{\rightleftharpoons} CH_3\overset{O}{\overset{\|}{C}}-OCH_2CH_2CH_2CH_3 + H_2O$$

Acetic acid	1-Butanol	*n*-Butyl acetate	
MW 61.06	MW 74.12	MW 116.16	MW 18
bp 117.9°C	bp 117.7°C	bp 126.5°C	

The equilibrium constant is:

$$K_{eq} = \frac{[n\text{-BuAc}][H_2O]}{[n\text{-BuOH}][HOAc]}$$

For primary alcohols reacting with unhindered carboxylic acids, $K_{eq} \approx 4$. If equal quantities of 1-butanol and acetic acid are allowed to react, at equilibrium the theoretical yield of ester is only 67%. To upset the equilibrium we can, by Le Chatelier's principle, increase the concentration of either the alcohol or acid, as noted above. If either one is doubled the theoretical yield increases to 85%. When one is tripled it goes to 90%. But note that in the example cited the boiling point of the relatively nonpolar ester is only about 8°C higher than the boiling points of the polar acetic acid and 1-butanol, so a difficult separation problem exists if either starting material is increased in concentration and the product isolated by distillation.

Another way to upset the equilibrium is to remove water. This can be done by adding to the reaction mixture molecular sieves, an artificial zeolite, which preferentially adsorb water. Most other drying agents, such as anhydrous sodium sulfate, will not remove water at the temperatures used to make esters.

Azeotropic distillation

A third way to upset the equilibrium is to preferentially remove the water as an azeotrope. The following can be found in any chemistry handbook table of ternary (three-component) azeotropes:

TABLE 27.2 A Ternary Azeotrope

		Azeotrope			
			Percent Composition		
	Bp (°C)	Bp (°C)	In azeotrope	Upper layer	Lower layer
1-Butanol	117.7	90.7	8.0	11.0	2.0
n-Butyl acetate	126.7		63.0	86.0	1.0
Water	100.0		29.0	3.0	97.0

These data tell us that the vapor that distils from a mixture of 1-butanol, *n*-butyl acetate, and water will boil at 90.7°C and the vapor contains 8% alcohol, 63% ester, and 29% water. The vapor is homogeneous, but when it condenses it separates into two layers. The upper layer is composed of 11% alcohol, 86% ester, and 3% water; but the lower layer consists of 97% water with only traces of alcohol and ester. If some ingenious way to remove the lower layer from the condensate and still return the upper layer to the reaction mixture can be devised, then the equilibrium can be upset and nearly 100% of the ester can be produced in the reaction flask.

Microscale Dean-Stark apparatus

The apparatus shown in Fig. 27.1, modeled after that of Dean and Stark, achieves the desired separation of the two layers. The mixture of equimolar quantities of 1-butanol and acetic acid is placed in the flask along with an acid

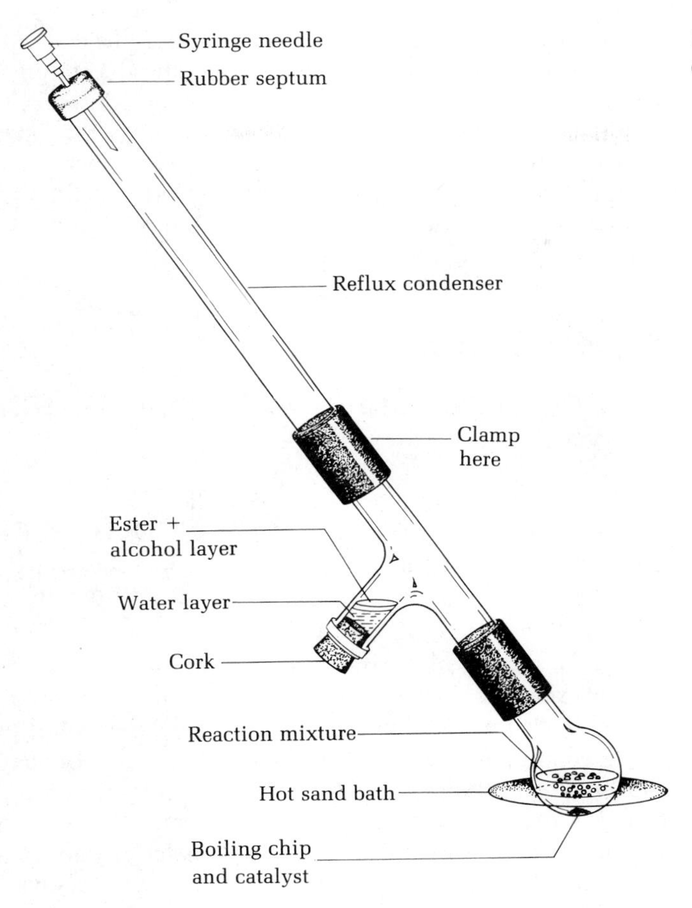

FIG. 27.1 Azeotropic esterification apparatus.

catalyst. The vapor, the temperature of which is 90.7°C, condenses and runs down to the sidearm, which is closed with a cork. The layers separate, with the denser water layer remaining in the sidearm while the lighter ester plus alcohol layer runs down into the reaction flask. As soon as the theoretical quantity of water has collected, the reaction is over and the product in the flask should be ester of at least 97% purity.

In this, and the other esterification reactions in this experiment that require an acid catalyst, the acid form of an ion exchange resin is used. This resin is a cross-linked polystyrene that bears sulfonic acid groups. Essentially it is an immobilized form of *p*-toluenesulfonic acid, an organic substituted sulfuric acid.

An ion exchange catalyst

$$\text{-(CH}-\text{CH}_2-\text{CH}-\text{CH}_2-\text{CH)}_n\text{-}$$

$$\text{SO}_3^-\,\text{H}^+ \qquad \text{SO}_3^-\,\text{H}^+ \qquad \text{SO}_3^-\,\text{H}^+$$

This catalyst has the distinct advantage that at the end of the reaction it can be removed simply by filtration. Immobilized catalysts of this type are becoming more and more common in organic synthesis.

If concentrated sulfuric acid were used as the catalyst, it would be necessary to dilute the reaction mixture with ether; wash the ether layer successively with water, sodium carbonate solution, and saturated sodium chloride solution; and then dry the ether layer with anhydrous sodium sulfate before evaporating the ether.

Experiments

1. *n*-Butyl Acetate by Azeotropic Distillation of Water

$$CH_3\overset{O}{\overset{\|}{C}}-OH + HOCH_2CH_2CH_2CH_3 \overset{H^+}{\rightleftharpoons} CH_3\overset{O}{\overset{\|}{C}}-OCH_2CH_2CH_2CH_3 + H_2O$$

Acetic acid	**1-Butanol**	***n*-Butyl acetate**	
MW 61.06	MW 74.12	MW 116.16	MW 18
bp 117.9°C	bp 117.7°C	bp 126.5°C	

Ion exchange resin catalyst

In a 5-mL short-necked round-bottomed flask, place 0.2 g of Dowex 50X2-100 ion exchange resin, 0.61 g of acetic acid, 0.74 g of 1-butanol, and a boiling chip. Attach the addition port with the sidearm corked and an empty distilling column as shown in Fig. 27.1 and clamp the apparatus at the angle shown. The septum and empty syringe needle prevent moisture from diffusing into the apparatus. Heat the flask on a hot sand bath and boil the reaction mixture while holding a thermometer just above the boiling liquid. You should note a temperature of about 91°C. Remove the thermometer, replace the septum and needle, and allow the reaction mixture to reflux in such a manner that the vapors condense about one-third of the way up the empty distilling column, which is functioning as an air condenser. Note that the material that condenses is not homogeneous as droplets of water begin to collect in the upper part of the apparatus. As the sidearm fills with condensate, it is cloudy at first and then two layers separate. When the volume of the lower aqueous layer does not appear to increase the reaction is over. This will take about 1–1.5 h. Carefully remove the apparatus from the heat, allow it to cool, and then tip the apparatus very carefully to allow all of the upper layer in the sidearm to run back into the reaction flask. Disconnect the apparatus. Determine how much water has collected in the sidearm by either removing the water with a syringe, the needle of which pierces the cork, or gravimetrically.

Remove the product from the reaction flask with a Pasteur pipette, determine its weight and boiling point, and assess its purity by thin-layer chromatography and/or gas chromatography and infrared spectroscopy. Look for the presence of hydroxyl and carboxyl absorption bands in the infrared spectrum.

2. Isobutyl Propionate by Fischer Esterification

$$CH_3CH_2\overset{O}{\overset{\|}{C}}-OH + HOCH_2\overset{CH_3}{\overset{|}{C}}HCH_3 \overset{H^+}{\rightleftharpoons} CH_3CH_2\overset{O}{\overset{\|}{C}}-OCH_2\overset{CH_3}{\overset{|}{C}}HCH_3 + H_2O$$

Propanoic acid MW 74.08, bp 141°C — **2-Methyl-1-propanol** MW 74.12, bp 108°C — **Isobutyl propionate** MW 130.19, bp 136.8°C

To a reaction tube add 112 mg of 2-methyl-1-propanol (isobutyl alcohol), 148 mg of propanoic acid (propionic acid), 50 mg of Dowex 50X2-100 ion exchange resin and a boiling chip. Cap the tube with a septum and empty syringe needle to prevent moisture from entering the apparatus. Reflux the resulting mixture for 1 h or more, cool it to room temperature, remove the product mixture from the resin with a Pasteur pipette, and chromatograph the product on a silica gel column.

Chromatography Procedure

Assemble the column as depicted in Fig. 27.2, being sure it is clamped in a vertical position. Close the valve and fill the column with dichloromethane to the bottom of the funnel. Prepare a slurry of 1 g of silica gel in 4 mL of dichloromethane in a small beaker. Stir the slurry gently to get rid of air bubbles, and gently swirl, pour, and scrape the slurry into the funnel, which has a capacity of 10 mL. After some of the silica gel has been added to the column, open the stopcock and allow solvent to drain slowly into an Erlenmeyer flask. Use this dichloromethane to rinse the beaker containing the silica gel. As the silica gel is being added, tap the column with a glass rod or pencil so the adsorbent will pack tightly into the column. Continue to tap the column while cycling the dichloromethane through the column once more, then add 1 g of anhydrous potassium carbonate to the top of the silica gel. The potassium carbonate will remove water from the esterification mixture as well as react with any carboxylic acid present. Run the solvent down to the top of the potassium carbonate layer.

It is extremely important to *never let the column run dry at any time*. Otherwise air will enter the column, resulting in uneven bands and poor separation.

Adding the Sample. The solvent is drained just to the surface of the potassium carbonate. Using a Pasteur pipette, add the sample to the column and let it run into the adsorbent, stopping when the solution reaches the top of the potassium carbonate. The flask and ion exchange resin are rinsed twice with 0.5-mL portions of dichloromethane that are run into the column, with the eluent being collected in a tared reaction tube. The elution is completed with 1 mL more dichloromethane.

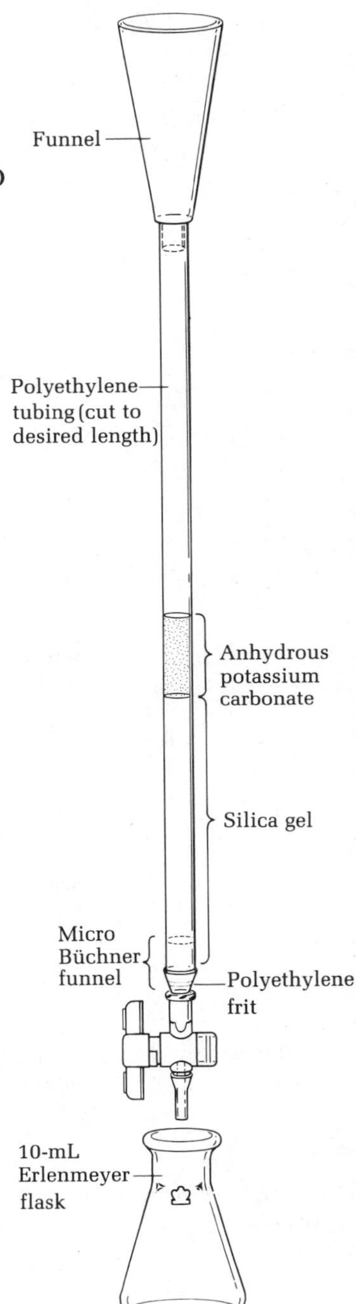

FIG. 27.2 Chromatographic column for esters.

Analysis of the Product. Evaporate the dichloromethane under a stream of air or nitrogen in the hood and remove the last traces by connecting the reaction tube to the water aspirator. Since the dichloromethane boils at 40°C and the product at 137°C, the separation of the two is easily accomplished. Determine the weight of the product and its boiling point and calculate the yield. The ester should be a perfectly clear, homogeneous liquid. Obtain an infrared spectrum and analyze it for the presence of unreacted alcohol and carboxylic acid. Check the purity of the product by thin-layer and/or gas chromatography.

3. Benzyl Acetate from Acetic Anhydride

$$C_6H_5CH_2OH + CH_3C(=O)-O-C(=O)CH_3 \longrightarrow C_6H_5CH_2O-C(=O)CH_3 + CH_3C(=O)-OH$$

Benzyl alcohol	**Acetic anhydride**	**Benzyl acetate**	**Acetic acid**
MW 108.14	MW 102.09	MW 150.18	MW 60.06
bp 205°C	bp 138–140°C	bp 206°C	bp 117–118°C

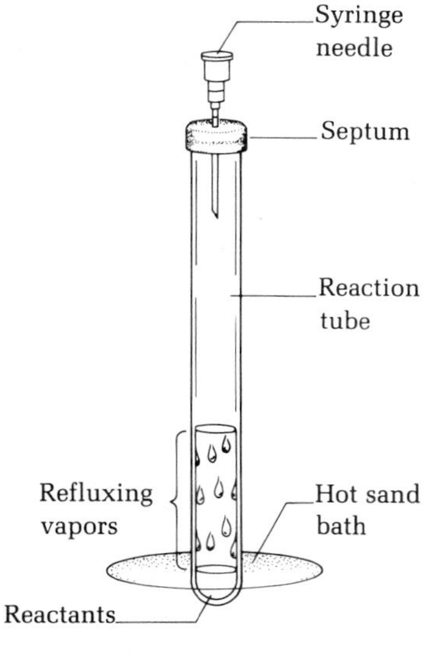

FIG. 27.3 Esterification apparatus.

To a reaction tube add 108 mg of benzyl alcohol and 102 mg of acetic anhydride and a boiling chip. Cap the tube with a septum and empty syringe needle to prevent moisture from entering the apparatus. Reflux the mixture for at least an hour, cool the mixture to room temperature, and chromatograph the liquid in exactly the same manner described immediately above. Analyze the product by thin-layer chromatography and by infrared spectroscopy as a thin film between salt or silver chloride plates. Note the presence or absence of hydroxyl and carboxyl bands.

4. Other Esterifications

This experiment lends itself to wide-ranging experimentation. All three methods of esterification described above can, in principle, be applied to any unhindered primary or secondary alcohol. These methods work well for any of the esters in Table 27.1 and for hundreds of others as well.

Questions

1. Throughout this chapter the esters have been given their more common trivial names. Name the esters in Table 27.1 according to the IUPAC system of nomenclature.

2. Why might it be difficult to purify methyl acetate by column chromatography as described in experiments 2 and 3?

Diels-Alder Reaction 28

PRELAB EXERCISE: *Describe in detail the laboratory operations, reagents, and solvents you would employ to prepare:*

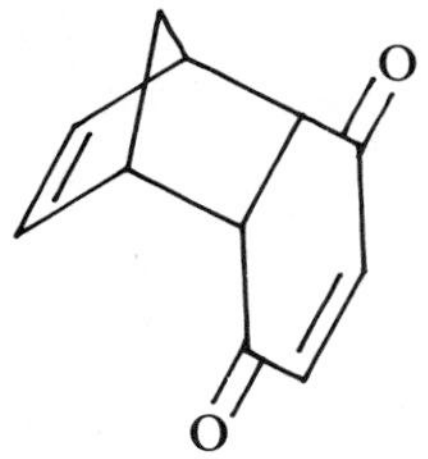

Otto Diels and his pupil Kurt Alder received the Nobel Prize in 1950 for their discovery and work on the reaction that bears their names. Its great usefulness lies in its high yield and high stereospecificity. A cycloaddition reaction, it involves the 1,4-addition of a conjugated diene in the s-*cis*-conformation to an alkene in which two new sigma bonds are formed from two pi bonds.

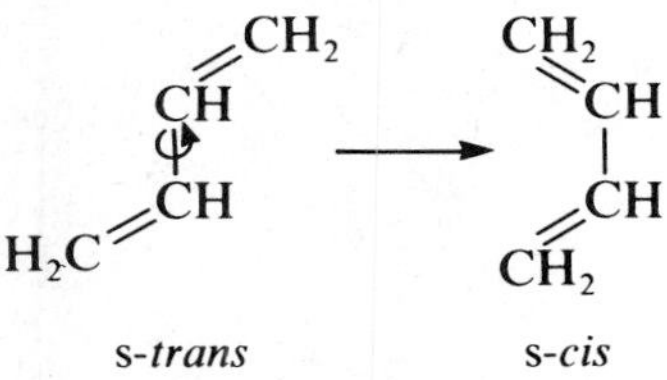

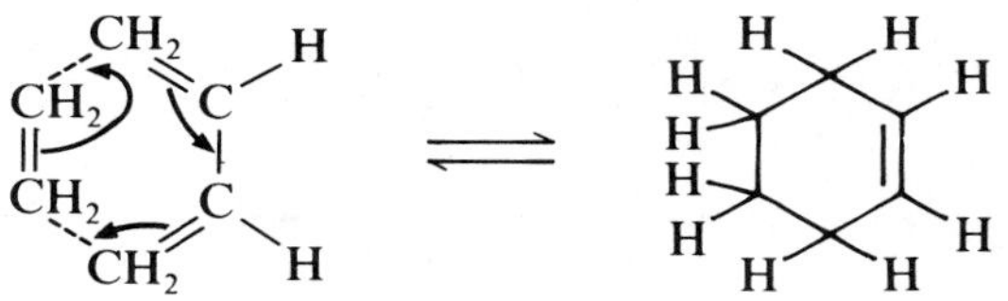

s-*trans* s-*cis*

The adduct is a six-membered ring alkene. The diene can have the two conjugated bonds contained within a ring system as with cyclopentadiene or cyclohexadiene, or the molecule can be an acyclic diene that must be in the *cis* conformation about the single bond before reaction can occur.

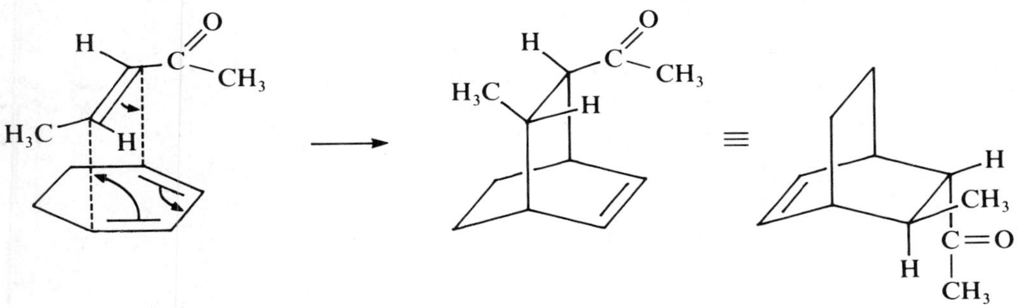

The reaction works best when there is a marked difference between the electron densities in the diene and the alkene with which it reacts, the dienophile. Usually the dienophile has electron-attracting groups attached to it while the diene is electron rich, e.g., as in the reaction of methyl vinyl ketone with 1,3-butadiene.

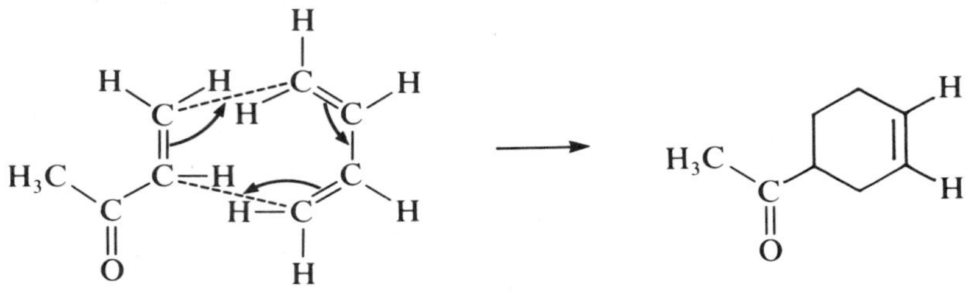

Methyl vinyl ketone **1,3-Butadiene**

Retention of the configurations of the reactants in the products implies that both new sigma bonds are formed almost simultaneously. If not, then the intermediate with a single new bond could rotate about that bond before the second sigma bond is formed, thus destroying the stereospecificity of the reaction.

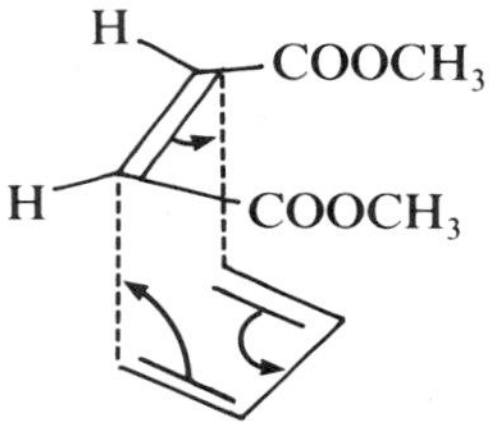

Dimethyl maleate
+
1,3
Butadiene

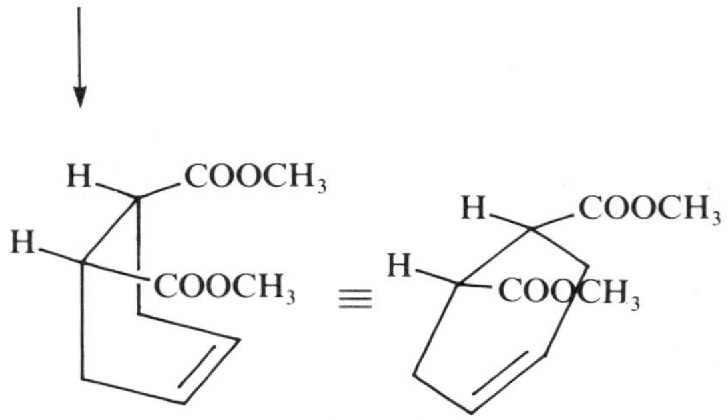

cis-isomer

This does not happen:

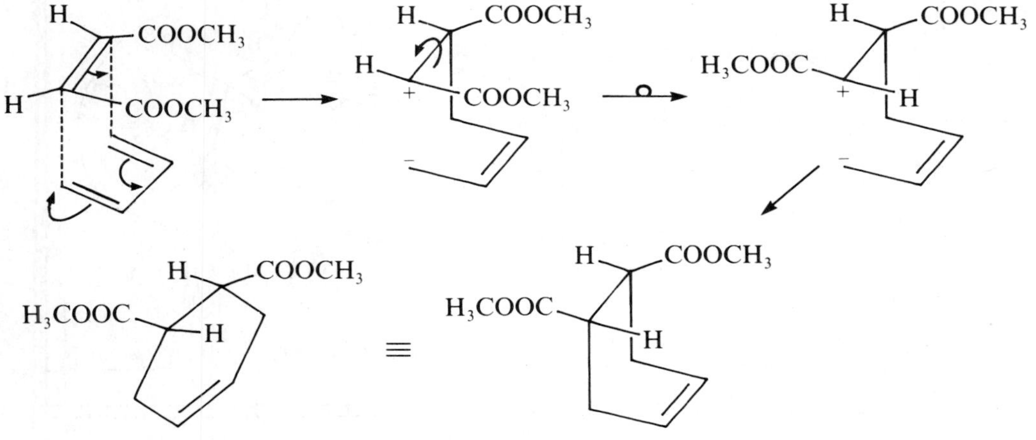

trans-isomer

A highly stereospecific reaction

This reaction is not polar in that no charged intermediates are formed. Neither is it radical because no unpaired electrons are involved. It is instead known as a concerted reaction, or one in which several bonds in the transition state are simultaneously made and broken. When a cyclic diene and a cyclic dienophile react with each other as in the present reaction, more than one stereoisomer may be formed. The isomer that predominates is the one which involves maximum overlap of pi electrons in the transition state. The transition state for the formation of the *endo*-isomer in the present reaction involves a sandwich with the diene directly above the dienophile. To form the *exo*-isomer the diene and dienophile would need to be arranged in a stair-step fashion.

The Diels-Alder reaction has been used extensively in the synthesis of complex natural products because it is possible to exploit the formation of a number of chiral centers in one reaction and also the regioselectivity of the reaction. For example, the first step in R. B. Woodward's synthesis of cortisone was the formation of Diels-Alder adduct.

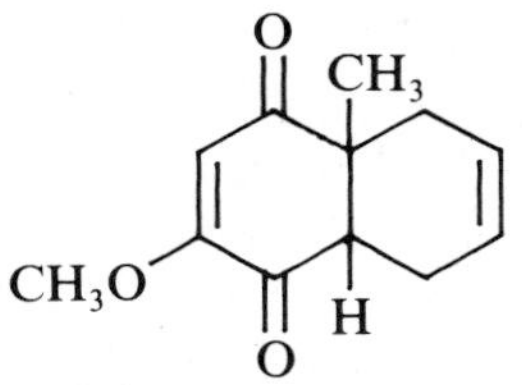

Woodward's Diels-Alder adduct

But the reaction is also subject to steric hindrance, especially when the difference between the electron-withdrawing and donating characters of the two reactants is not great. When Woodward tried to synthesize cantharidin, the active ingredient in Spanish fly, by the Diels-Alder condensation of furan with dimethylmaleic anhydride, the reaction did not work. Cantharidin is a powerful vesicant (blister-former).

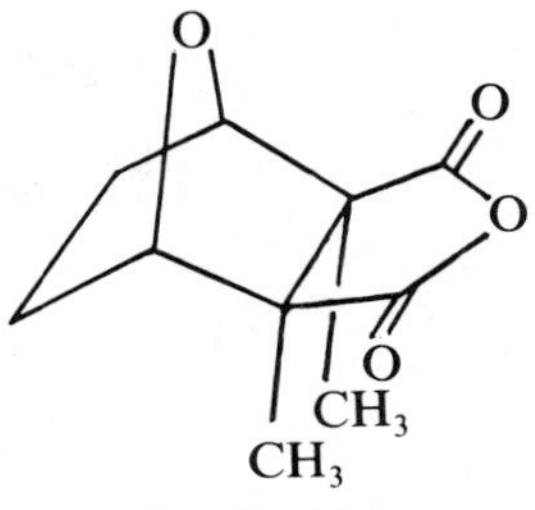

Cantharidin

Woodward and Roald Hoffmann, who received a Nobel prize for their work, formulated the theoretical rules involving the correlation of orbital symmetry, which govern the Diels-Alder and other electrocyclic reactions.

Cyclopentadiene is obtained from the light oil from coal tar distillation but exists as the stable dimer, dicyclopentadiene, which is the Diels-Alder adduct from two molecules of the diene. Thus, generation of cyclopentadiene

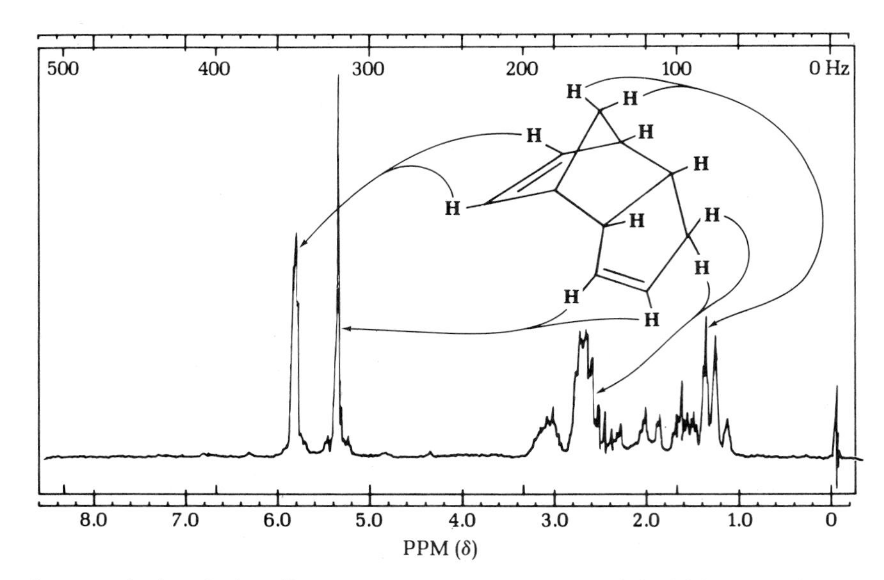

FIG. 28.1 ^{1}H nmr spectrum of dicyclopentadiene.

by pyrolysis of the dimer represents a reverse Diels-Alder reaction. See Figs. 28.1 and 28.2 for nmr and infrared spectra of dicyclopentadiene. In the Diels-Alder addition of cyclopentadiene and maleic anhydride the two molecules approach each other in the orientation shown in the drawing (see below), as this orientation provides maximal overlap of π-bonds of the two reactants and favors formation of an initial π-complex and then the final *endo*-product. Dicyclopentadiene also has the *endo*-configuration.

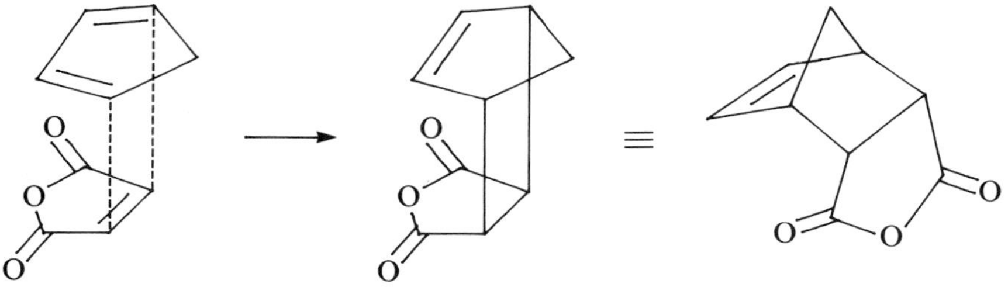

Experiments

In experiment 1 unique apparatus for the microscale cracking of dicyclopentadiene is used. The dicyclopentadiene is dispensed from a septum-capped container into a syringe so the vile odor of dicyclopentadiene is not detected in the laboratory.

Gas chromatography reveals that cyclopentadiene is 8% dimerized in 4 h and 50% dimerized in 24 h at room temperature. It should be kept on ice and used as soon as possible after being prepared.

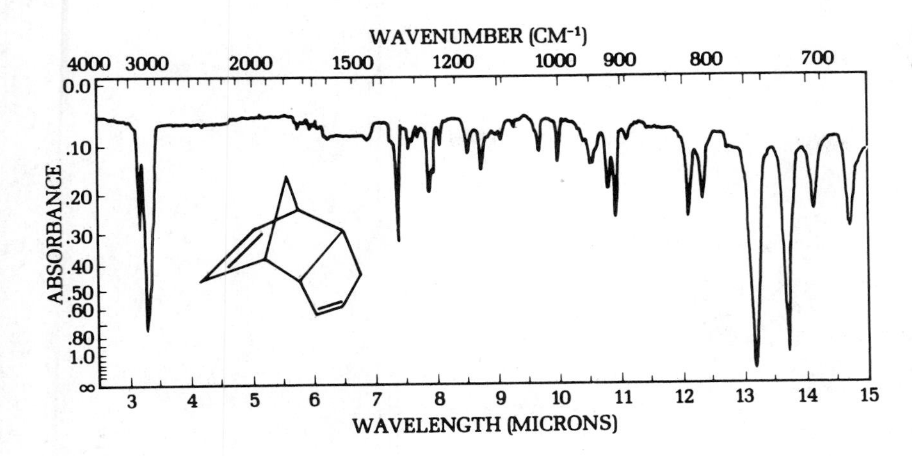

FIG. 28.2 Infrared spectrum of dicyclopentadiene.

1. Microscale Cracking of Dicyclopentadione

H H

~160°C ⇌ room temperature

+

Dicyclopentadiene
den 0.98
MW 132.20

Cyclopentadiene
bp 41°C, den 0.80
MW 66.10

Place approximately 2.5 mL of mineral oil in a short-necked 5-mL round-bottomed flask equipped with an addition port bearing a septum on the sidearm and topped with a distillation head and thermometer (Fig. 28.3). Heat the flask on a sand bath that has been heated to 200–250°C (do not leave a thermometer in the hot sand bath) and is supported on a ring, so that it may be lowered if necessary.

Place a small tared collection vial in an ice-filled 50-mL beaker at the end of the distilling head, taking care to keep water out of the vial. Using a tared 2-mL syringe, draw 0.5 mL of dicyclopentadiene from the septum-capped storage container after first injecting some air into the container to overcome the vacuum. Stick the needle of the filled syringe into a rubber stopper or cork to avoid loss of the contents until they are used.

When the mineral oil is hot, inject the dicyclopentadiene through the septum on the addition port. Add it dropwise at such a rate that the

FIG. 28.3 Apparatus for microscale cracking of dicyclopentadiene. Clamp at A. Add dicyclopentadiene dropwise via syringe so that distillate temperature does not exceed 42°C.

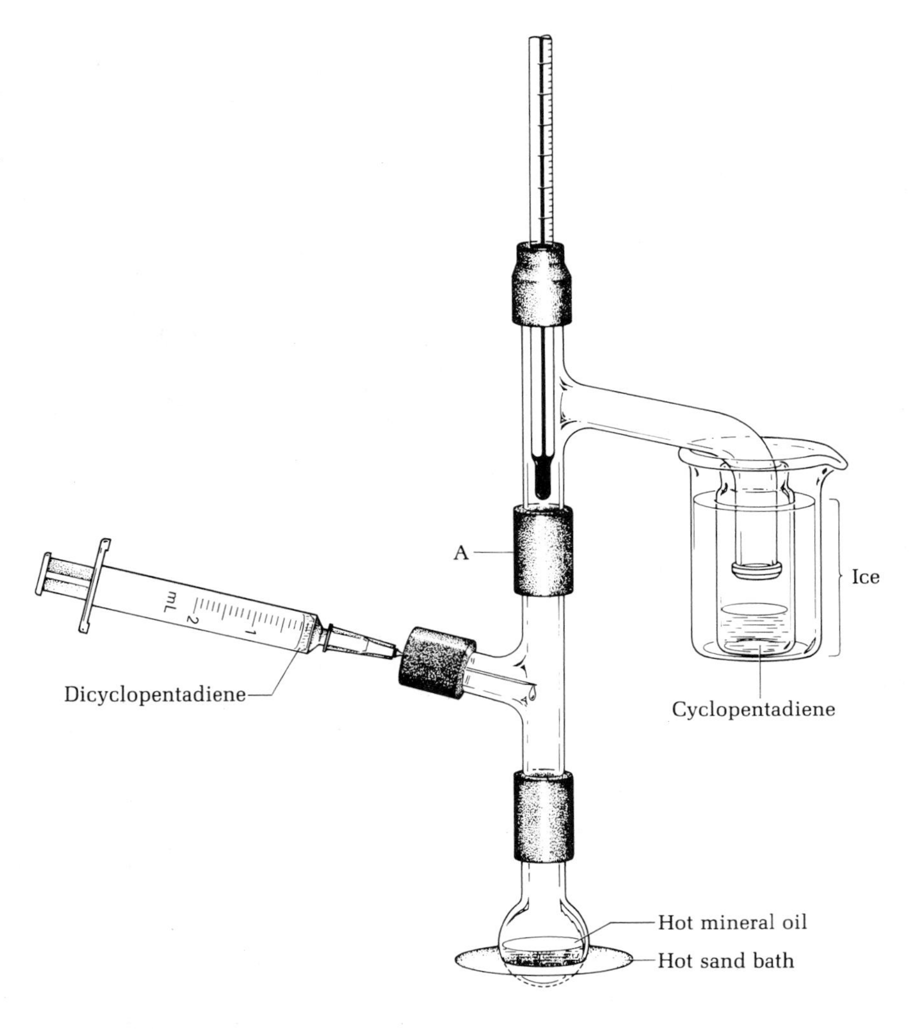

Reverse Diels-Alder reaction

temperature of the thermometer never exceeds 42°C, the boiling point of cyclopentadiene. The process will require about 20 min. Once the dicyclopentadiene has all been added, lower the sand bath and turn it off, remove the syringe, weigh it and weigh the product after closing the vial with a foil-lined cap. Rinse the syringe with a few drops of acetone in the hood. Calculate the percent yield of cyclopentadiene. If the product is cloudy, add a small quantity of anhydrous sodium sulfate to dry it. Keep this cyclopentadiene on ice and use it the same day it is prepared.

2. *cis*-Norbornene-5,6-*endo*-dicarboxylic anhydride

Maleic anhydride
mp 53°C, MW 98.06

***cis*-Norbornene-5,6-*endo*-dicarboxylic anhydride**
mp 165°C, MW 164.16

Dissolve 0.20 g of powdered maleic anhydride in 1 mL of ethyl acetate in a tared 10 × 100 mm reaction tube and then add 1 mL of bp 60°–80°C ligroin. This combination of solvents is used because the product is too soluble in pure ethyl acetate and not soluble enough in pure ligroin. To the solution of maleic anhydride add 0.20 mL (0.160 g) of dry cyclopentadiene, mix the reactants, and observe the reaction. Allow the tube to cool to room temperature, during which time crystallization of the product should occur. If crystallization does not occur, scratch the inside of the test tube with a stirring rod at the liquid-air interface. The scratch marks on the inside of the tube often form the nuclei on which crystallization can occur. Should crystallization occur very rapidly at room temperature, the crystals will be very small. If so, save a seed crystal, heat the mixture until the product dissolves, seed it, and allow it to cool slowly to room temperature. You will be rewarded with large platelike crystals. Remove the solvent from the crystals with a

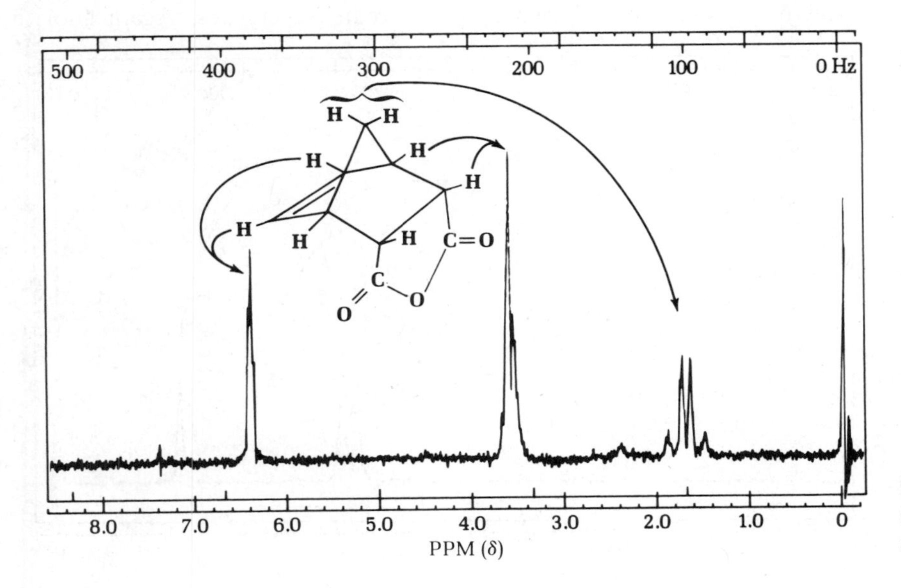

FIG. 28.4 ^{1}H nmr spectrum of *cis*-norbornene-5,6-*endo*-dicarboxylic anhydride.

Pasteur pipette that is forced to the bottom of the tube, wash the crystals with one portion of cold ligroin, and remove the solvent. Scrape the product onto a piece of filter paper, allow the crystals to dry in air, determine their weight and calculate the yield of the product. Determine the melting point of the product and turn in any material not used in the next experiment. Thin-layer chromatography of the product is hardly necessary; it is quite pure.

3. *cis*-Norbornene-5,6-*endo*-dicarboxylic acid

H_2O + →

***endo,cis*-Diacid**

To 0.2 g (200 mg) of the anhydride from the previous experiment, add 2.5 mL of water and a boiling stick in a 10 × 100 mm reaction tube. Heat the mixture to boiling by immersing the tube in a hot sand bath. The anhydride may appear to melt and form globules on the bottom of the tube. As the reaction proceeds the anhydride will react with the water, and the diacid, which is soluble in boiling water, will be formed. Remove the boiling stick from the hot solution and allow the mixture to cool to room temperature.

If crystallization of the diacid does not occur, follow exactly the same procedure used for the anhydride. On slow cooling with simultaneous crystal growth the solution will deposit long needlelike crystals. Again cool the

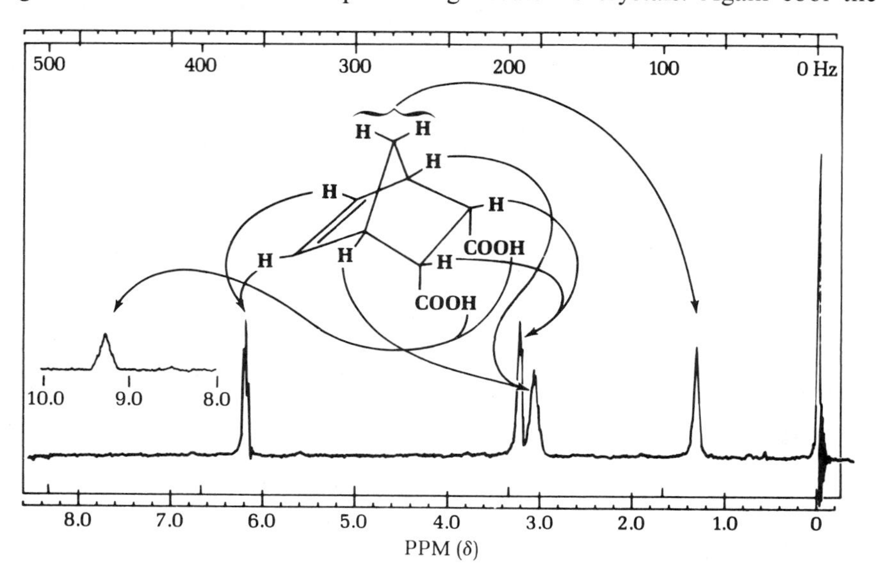

FIG. 28.5 ^{1}H nmr spectrum of *cis*-norbornene-5,6-*endo*-dicarboxylic acid.

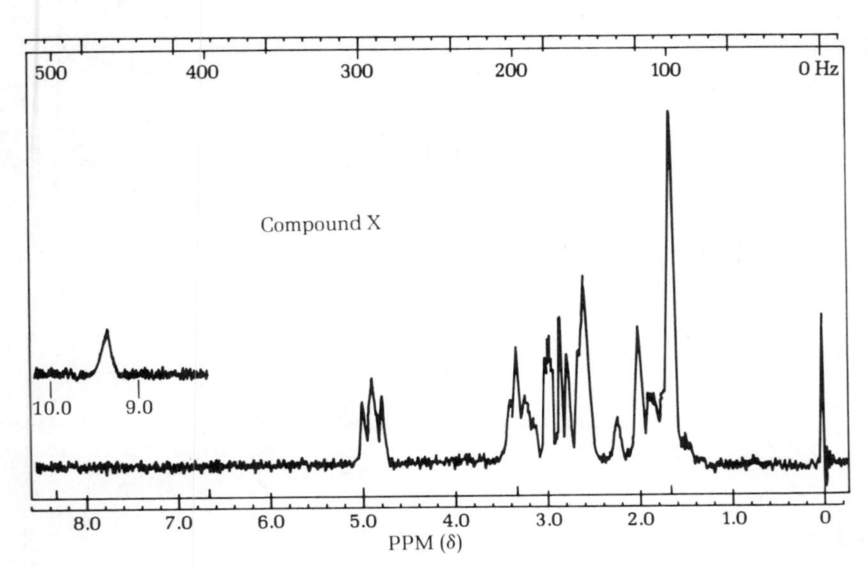

FIG. 28.6 ^{1}H nmr spectrum of compound X, prepared from *endo,cis*-diacid.

mixture in ice, allow sufficient time for crystal growth to occur, and then remove the solvent with a Pasteur pipette. Wash the crystals once with a small quantity of ice water, remove the solvent, and scrape the product out onto a piece of filter paper to dry. Weigh the diacid and determine the mp and percent yield. The melting point depends on the rate of heating as the anhydride reforms and water splits out.[1]

The temperature of decomposition is variable

4. Synthesis of Compound X[2]

To a tared 10 × 100 mm reaction tube add 0.15 g of the *endo, cis*-diacid from the previous experiment followed by 0.25 mL of concentrated sulfuric acid. Warm the mixture on the steam bath or in a beaker of boiling water for about 2 min, cool, and then *cautiously* add 0.70 mL of water to the test tube. This should be done dropwise with vigorous mixing of the contents after the addition of each drop. The product will crystallize as a fine powder.

Save a seed crystal and heat the tube on a hot sand bath until the crystals redissolve. Seed the solution and allow it to cool slowly to room temperature. Compound X will crystallize in platelike crystals. The crystallization process for this compound is fairly slow; allow at least 10 min for the solution to come

1. The *endo,cis*-diacid is stable to alkali but can be isomerized to the *trans*-diacid (mp 192°C) by conversion to the dimethyl ester (3 g of acid, 10 mL methanol, 0.5 mL concentrated H_2SO_4; reflux 1 h). This ester is equilibrated with sodium ethoxide in refluxing ethanol for 3 days and saponified. For an account of a related epimerization and discussion of the mechanism, see J. Meinwald and P. G. Gassman, *J. Am. Chem. Soc.*, **82**, 5445 (1960). See also K. L. Williamson, Y.-F. Li, R. Lacko and C. H. Youn, *J. Am. Chem. Soc.*, **91**, 6129 (1969) and K. L. Williamson and Y.-F. Li, *J. Am. Chem. Soc.*, **92**, 7654 (1970).

2. Introduced by James A. Deyrup.

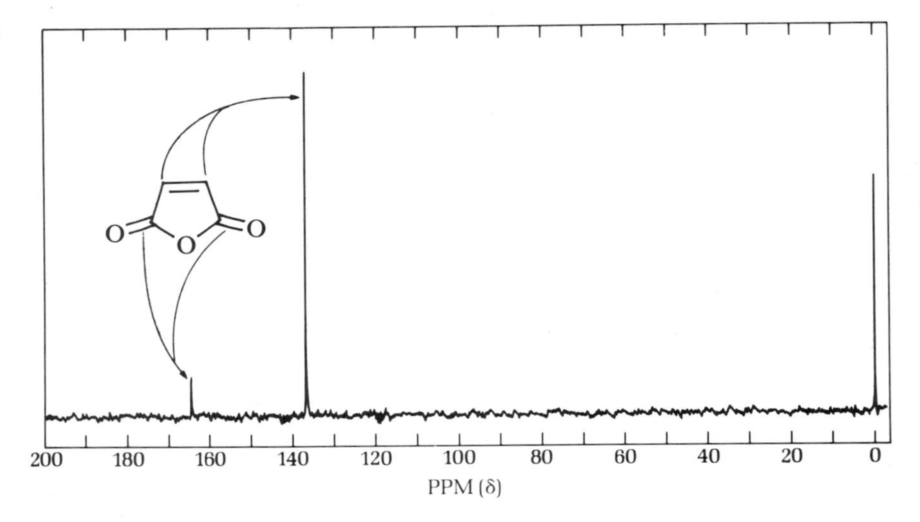

FIG. 28.7 ^{13}C nmr spectrum of maleic anhydride.

to room temperature and a further 10 min in the ice bath before removing the solvent with a Pasteur pipette. Wash the crystals in the tube with one small portion of ice water, remove the liquid, and scrape the product onto a piece of filter paper. Squeeze the crystals between sheets of filter paper to complete the drying process and then determine the weight, yield, and mp of the product. Speculate upon its structure.

What is X?

Study the nmr spectrum of Compound X and compare it with that of the starting material. Note that the spectrum of X is much more complex than that of the starting material, yet X is an isomer of the starting material. What functional group is missing from X that is present in the starting material? What functional group is unchanged in both? What intermediate is formed when X dissolves in concentrated sulfuric acid?

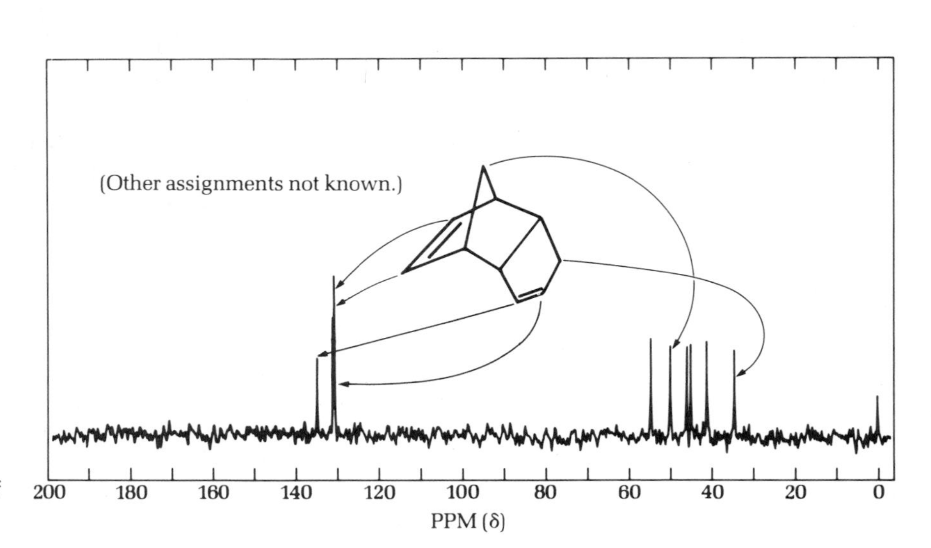

FIG. 28.8 ^{13}C nmr spectrum of dicyclopentadiene.

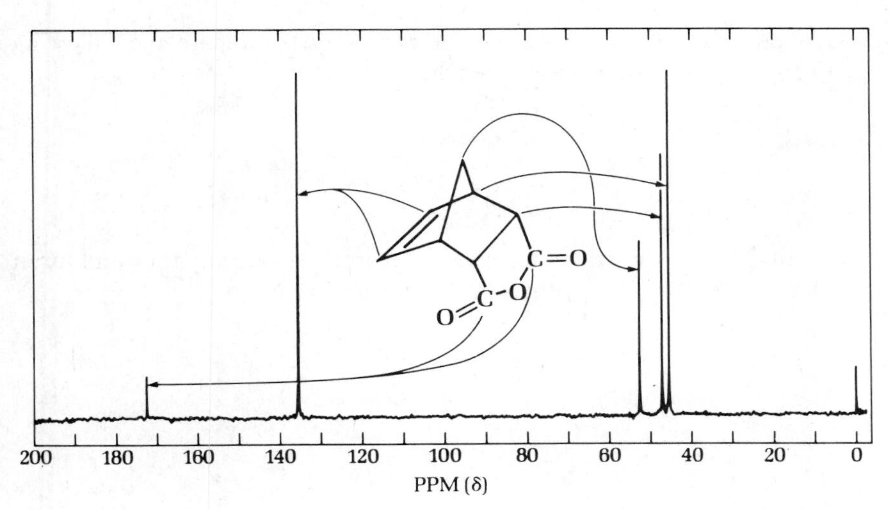

FIG. 28.9 ^{13}C nmr spectrum of *cis*-norbornene-5,6-*endo*-dicarboxylic anhydride.

Questions

1. In the cracking of dicyclopentadiene, why is it necessary to distil the product very slowly?

2. Draw the products of the following reactions:

(a)

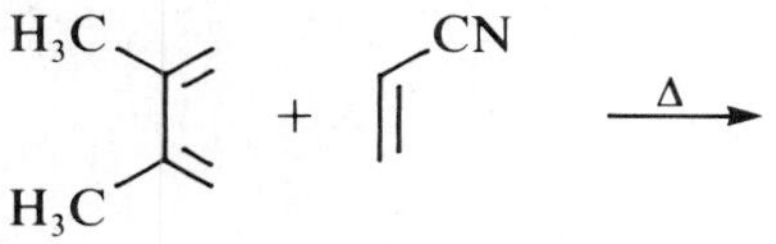

H_3C, H_3C + CN $\xrightarrow{\Delta}$

(b)

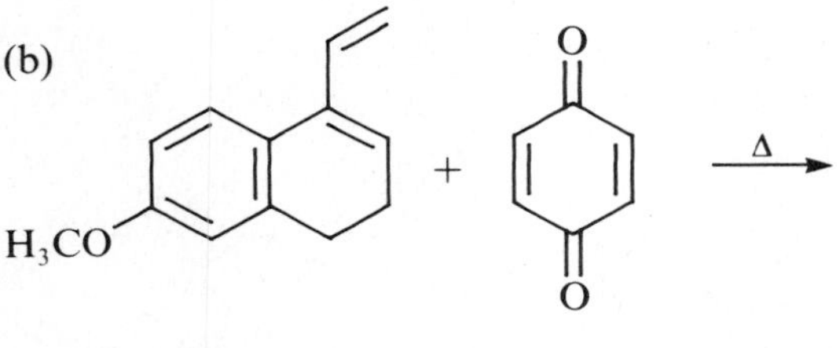

H_3CO + O, O $\xrightarrow{\Delta}$

(c)

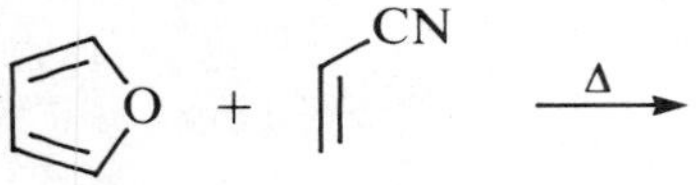

O + CN $\xrightarrow{\Delta}$

3. What starting material would be necessary to prepare the following compound by the Diels-Alder reaction?

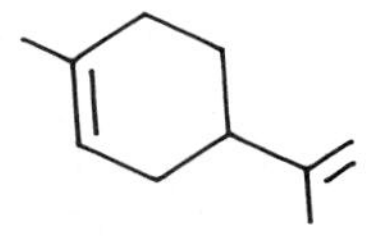

4. If the Diels-Alder reaction between dimethylmaleic anhydride and furan had worked would cantharidin have been formed?

Ferrocene [Bis(cyclopentadienyl)iron] 29

PRELAB EXERCISE: *Propose a detailed outline of the procedure for the synthesis of ferrocene, paying particular attention to the time required for each step.*

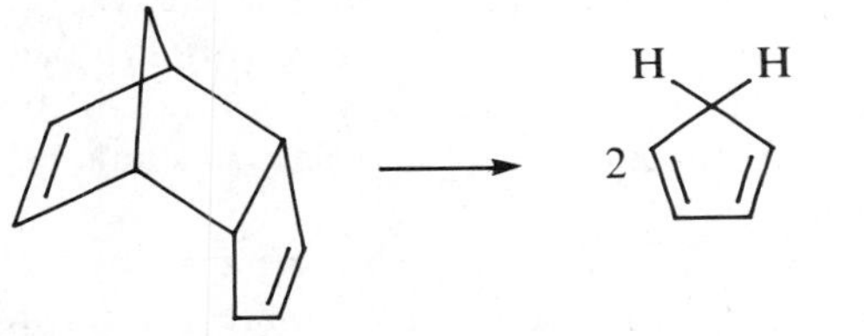

Dicyclopentadiene → 2 Cyclopentadiene

Dicyclopentadiene
den 0.98
MW 132.20

Cyclopentadiene
bp 41°C, den 0.80
MW 66.10

Cyclopentadiene + KOH → Potassium cyclopentadienide (K^+) + H_2O

Potassium cyclopentadienide

2 $C_5H_5^- K^+$ + $FeCl_2 \cdot 4\ H_2O$ → Fe(C_5H_5)$_2$ + 2 KCl + 4 H_2O

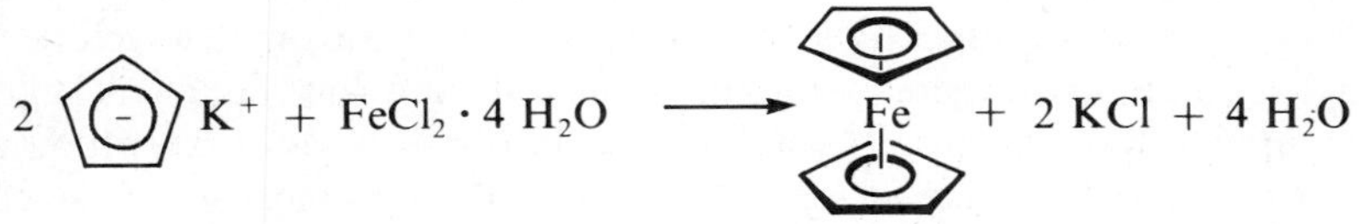

Iron(II) chloride tetrahydrate
MW 198.81

Ferrocene
Bis(cyclopentadienyl)iron
MW 186.04, mp 172–174°C

The Grignard reagent is a classical organometallic compound. The magnesium ion in Group IIA of the periodic table needs to lose two and only two electrons to achieve the inert gas configuration. This metal has a strong tendency to form ionic bonds by electron transfer:

$$RBr + Mg \longrightarrow \overset{\delta-}{R}-\overset{\delta+}{Mg}Br$$

In the transition metals the situation is not so simple. Consider the bonding between iron and carbon monoxide in $Fe(CO)_5$:

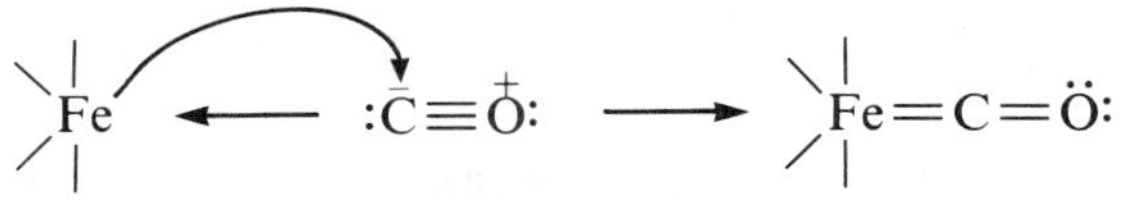

$$\text{Fe} \longleftarrow :\bar{C}{\equiv}\overset{+}{O}: \longrightarrow \text{Fe}{=}C{=}\ddot{O}:$$

The pair of electrons on the carbon atom is shared with iron to form a σ bond between the carbon and iron. The π bond between iron and carbon is formed from a pair of electrons in the *d*-orbital of iron. The π bond is thus formed by the overlap of a *d* orbital of iron with the p-π bond of the carbonyl group. This mutual sharing of electrons results in a relatively nonpolar bond.

Iron has 6 electrons in the $3d$ orbital, 2 in the $4s$, and none in the $4p$ orbital. The inert gas configuration requires 18 electrons—ten $3d$, two $4s$, and six $4p$ electrons. Iron pentacarbonyl enters this configuration by accepting two electrons from each of the five carbonyl groups, a total of 18 electrons. Back-bonding of the d-π type distributes the excess electrons among the five carbon monoxide molecules.

Early attempts to form σ-bonded derivatives linking alkyl carbon atoms to iron were unsuccessful, but P. L. Pauson in 1951 succeeded in preparing a very stable substance, ferrocene, $C_{10}H_{10}Fe$, by reacting two moles of cyclopentadienylmagnesium bromide with anhydrous ferrous chloride. Another group of chemists, Wilkinson, Rosenblum, Whiting, and Woodward, recognized that the properties of ferrocene (remarkable stability to water, acids, and air and its ease of sublimation) could only be explained if it had the structure depicted and that the bonding of the ferrous iron with its six electrons must involve all twelve of the π-electrons on the two cyclopentadiene rings, with a stable 18-electron inert gas structure as the result.

In the present experiment ferrocene is prepared by reaction of the anion of cyclopentadiene with iron(II) chloride. Abstraction of one of the acidic allylic protons of cyclopentadiene with base gives the aromatic cyclopentadienyl anion. It is considered aromatic because it conforms to the Hückel rule in having $4n + 2\pi$ electrons (where n is 1). Two molecules of this anion will react with iron(II) to give ferrocene, the most common member of the class of metal-organic compounds referred to as metallocenes. In this centrosymmetric sandwich-type π complex, all carbon atoms are equidistant from the iron atom, and the two cyclopentadienyl rings rotate more or less freely with respect to each other. The extraordinary stability of ferrocene (stable to 500°C) can be attributed to the sharing of the 12 π electrons of the two cyclopentadienyl rings with the six outer shell electrons of iron(II) to give the iron a stable 18-electron inert gas configuration. Ferrocene is soluble in organic solvents, can be dissolved in concentrated sulfuric acid and recovered unchanged, and is resistant to other acids and bases as well (in the absence of oxygen). This behavior is consistent with that of an aromatic compound; ferrocene is found to undergo electrophilic aromatic substitution reactions with ease.

Ferrocene: soluble in organic solvents, stable to 500°C

Cyclopentadiene readily dimerizes at room temperature by a Diels-Alder reaction to give dicyclopentadiene. This dimer can be "cracked" by heating (an example of the reversibility of the Diels-Alder reaction) to give low-boiling cyclopentadiene. In most syntheses of ferrocene the anion of cyclopentadiene is prepared by reaction of the diene with metallic sodium. Subsequently, this anion is allowed to react with anhydrous iron(II) chloride. In the present experiment the anion is generated using powdered potassium hydroxide, which functions as both a base and a dehydrating agent.

The anion of cyclopentadiene rapidly decomposes in air, and iron(II) chloride, although reasonably stable in the solid state, is readily oxidized to the iron(III) (ferric) state in solution. Consequently this reaction must be carried out in the absence of oxygen, accomplished by bubbling nitrogen gas through the solutions to displace dissolved oxygen and to flush air from the apparatus. In research laboratories rather elaborate apparatus is used to carry out an experiment in the absence of oxygen. In the present experiment, because no gases are evolved, no heating is necessary, and the reaction is only mildly exothermic, very simple apparatus is used.

Experiment

To a 5-mL short-necked, round-bottomed flask quickly add 0.75 g of finely powdered potassium hydroxide,[1] followed by 1.25 mL of dimethoxyethane. Cap the flask with a septum and pass nitrogen into the flask or, better, through the solution for about 2 min. This is done by connecting a tank of nitrogen via a rubber tube to a 22 gauge needle and adjusting the nitrogen flow to a few milliliters per minute by bubbling it under a liquid such as acetone. With the nitrogen flow adjusted insert an empty 18 gauge needle through the septum of the flask as an outlet and then insert the 22 gauge nitrogen inlet (Fig. 29.1). Shake the flask vigorously to dislodge the solid potassium hydroxide from the bottom of the flask while passing in nitrogen. This shaking will help to dissolve the solid and will also serve to saturate the solution with nitrogen.

To a 10 × 100 mm reaction tube add 0.35 g of finely powdered iron(II) chloride tetrahydrate and 1.5 mL of dimethyl sulfoxide. Cap the tube with a rubber septum, insert an empty 18 gauge needle through the septum and pass nitrogen into the tube for about 2 min to displace the oxygen present. Remove the needles and then shake the vial vigorously to dissolve all of the iron chloride.

Using an accurate glass syringe, inject 0.300 mL of freshly prepared cyclopentadiene (see Chapter 28 for the preparation of cyclopentadiene) into the flask containing the potassium hydroxide. Don't grasp the body of the

$CH_3OCH_2CH_2OCH_3$
1,2-Dimethoxyethane (Ethylene glycol dimethyl ether, monoglyme), bp 85°C
Completely miscible with water

Potassium hydroxide is extremely corrosive and hygroscopic. Immediately wash any spilled powder or solutions from the skin and wipe up all spills. Keep containers tightly closed. Work in the hood.

$CH_3S(=O)CH_3$
Dimethyl sulfoxide, DMSO (Methyl sulfoxide), bp 189°C
Completely miscible with water

1. Potassium hydroxide is easily ground to a fine powder in 25-g batches in one minute employing an ordinary food blender (e.g. Waring, Osterizer). The finely powdered base is transferred in a hood to a bottle with a tightly fitting cap. Alternatively grind about a gram of potassium hydroxide in a mortar and transfer it rapidly to the reaction flask.

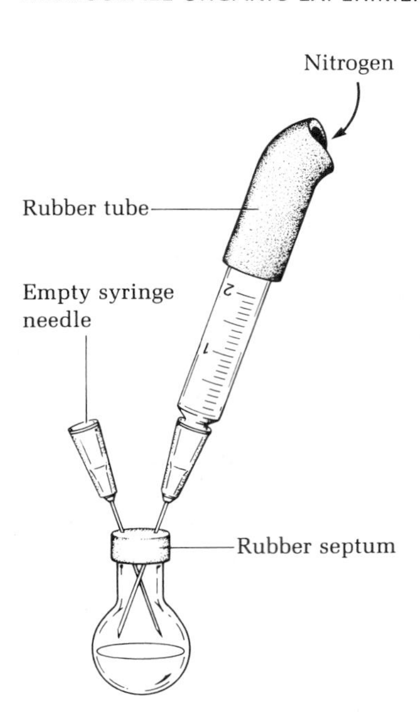

FIG. 29.1 Apparatus for flushing air from reaction flask.

Dimethyl sulfoxide is rapidly absorbed through the skin. Wash off spills with water. Wear disposable gloves when shaking the apparatus.

syringe because the heat of your hand will cause the cyclopentadiene to volatilize. Shake the flask vigorously and note the color change as the potassium cyclopentadienide is formed. After waiting about 5 min for the anion to form, pierce the septum with an empty needle for pressure relief and inject the iron(II) chloride solution in six 0.25-mL portions over a 10-min period. Between injections remove both needles from the septum and shake the flask vigorously. After all of the iron(II) chloride solution has been added, rinse the reaction tube with 0.25 mL more dimethyl sulfoxide and add this to the flask. Continue to shake the solution for about 15 min to complete the reaction.

Isolation of ferrocene

To isolate the ferrocene, pour the dark slurry onto a mixture of 4.5 mL of 6 M hydrochloric acid and 5 g of ice in a 30-mL beaker. Stir the contents of the beaker throughly to dissolve and neutralize all the potassium hydroxide. Collect the crystalline orange ferrocene on a Hirsch funnel, wash the crystals well with water, press out excess water, squeeze the product between sheets of filter paper to complete the drying, and then purify the ferrocene by sublimation.

Purification by sublimation

To sublime the ferrocene, add the crude product to a 50-mL filter flask equipped with a neoprene filter adapter and a 15-mL centrifuge tube that is pushed to within 5 mm of the bottom of the flask (Fig. 29.2). Apply a vacuum to the flask and warm it gently over a hot sand bath to remove the last traces of moisture from the ferrocene; then add ice to the centrifuge tube and heat the flask more strongly to sublime the product. Wrap the flask with a cone of aluminum foil to funnel heat to the upper portion of the flask. In this way all of the ferrocene will collect on the centrifuge tube.

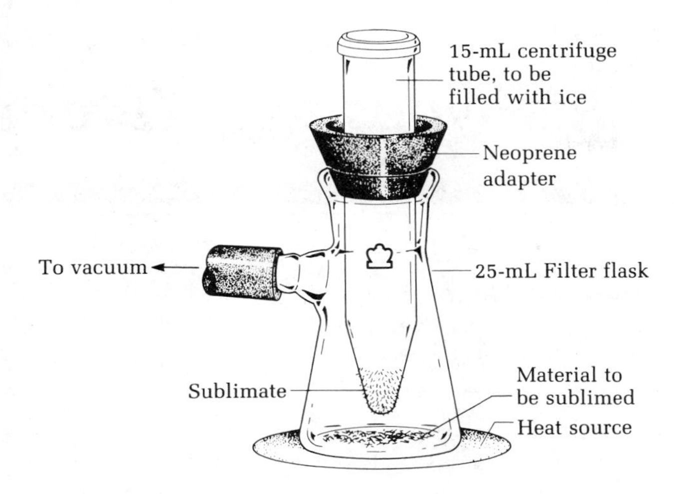

FIG. 29.2 Apparatus for the sublimation of ferrocene.

When sublimation is complete, cool the flask, remove the ice water from the centrifuge tube, and replace it with room temperature water (to prevent moisture from collecting on the tube) and then disconnect the vacuum. Transfer the product to a tared stoppered vial, determine the weight, and calculate the percent yield. Determine the melting point in an evacuated capillary since the product sublimes at the melting point (Fig. 29.3). See Chapter 4 for this technique.

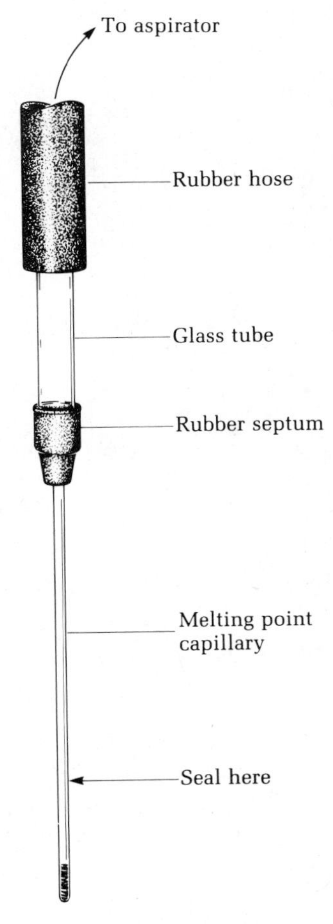

FIG. 29.3 Evacuation of a melting point capillary prior to sealing.

Questions

1. If ferrocene is stable to air and all of the reagents are stable to air before the reaction begins, why must air be so carefully excluded from this reaction?
2. What special properties do the solvents dimethoxyethane and dimethyl sulfoxide have compared to diethyl ether, for example, that make them particularly suited for this reaction?
3. What is there about ferrocene that allows it to sublime easily where many other compounds do not?

30 Aldehydes and Ketones

PRELAB EXERCISE: *Outline a logical series of experiments designed to identify an unknown aldehyde or ketone with the least effort. Consider the time required to complete each identification reaction.*

The carbonyl group occupies a central place in organic chemistry. Aldehydes and ketones—compounds such as formaldehyde, acetaldehyde, acetone and 2-butanone—are very important industrial chemicals used by themselves and as starting materials for a host of other substances. For example, in 1985 8.2 *billion* lb of formaldehyde-containing plastics were produced in the United States.

The carbonyl carbon is sp^2 hybridized, the bond angles between adjacent groups are 120°, and the four atoms R, R′, C, and O lie in one plane:

$$R-\overset{\overset{\displaystyle \ddot{O}:}{\|}}{C}-R'$$

The electronegative oxygen polarizes the carbon-oxygen bond, rendering the carbon electron deficient and hence subject to nucleophilic substitution.

$$\rangle C{=}\ddot{O}: \longleftrightarrow \rangle \overset{+}{C}-\ddot{\underset{\cdot\cdot}{O}}:^{-} \quad \text{or} \quad \rangle \overset{\delta+}{C}{\overset{\cdots}{=}}\overset{\delta-}{\underset{\cdot\cdot}{O}}:$$

Geometry of the carbonyl group

Attack on the sp^2 hybridized carbon occurs via the π-electron cloud above or below the plane of the carbonyl group:

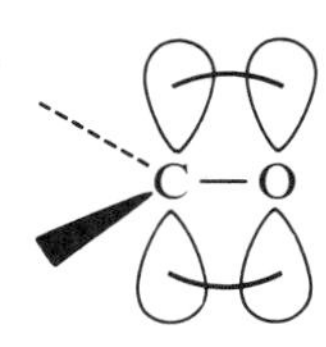

Many reactions of carbonyl groups are acid-catalyzed. The acid attacks the electronegative oxygen, which bears a partial negative charge, to create a carbocation that subsequently reacts with the nucleophile:

$$R_2C{=}\ddot{O}{:} + H^+ \rightleftharpoons R_2C{=}\overset{+}{\ddot{O}}{-}H \longleftrightarrow R_2\overset{+}{C}{-}\ddot{O}{-}H$$

$$R_2C{=}\overset{+}{\ddot{O}}{-}H + \ddot{N}u{-}H \rightleftharpoons R_2C(Nu){-}\ddot{O}{-}H + H^+$$

The strength of the nucleophile and the structure of the carbonyl compound determine whether the equilibrium lies on the side of the carbonyl compound or the tetrahedral adduct. Water, a weak nucleophile, does not usually add to the carbonyl group to form a stable compound:

$$RR'C{=}\ddot{O}{:} + H_2O \rightleftharpoons RR'C(\ddot{O}H)_2$$

A stable hydrate: chloral hydrate

but in the special case of trichloroacetaldehyde the electron-withdrawing trichloromethyl group allows a stable hydrate to form:

$$Cl_3C(H)C{=}\ddot{O}{:} + H_2O \rightleftharpoons Cl_3C(H)C(\ddot{O}{-}H)_2$$

The compound so formed, chloral hydrate, was discovered by Liebig in 1832 and was introduced as one of the first sedatives and hypnotics (sleep-inducing substances) in 1869. It is now most commonly encountered in detective fiction as a "Mickey Finn" or "knockout drops."

Reaction with an alcohol. Hemiacetals

In an analogous manner, an aldehyde or ketone can react with an alcohol. The product, a hemiacetal or hemiketal, is usually not stable, but in the case of certain cyclic hemiacetals the product can be isolated. Glucose is an example of a stable hemiacetal.

$$R{-}CH{=}\ddot{O}{:} + H{-}\ddot{O}{-}R' \rightleftharpoons R{-}CH(\ddot{O}{:}^-){-}\overset{+}{\ddot{O}}(H){-}R' \rightleftharpoons R{-}CH(\ddot{O}{-}H){-}\ddot{O}{-}R'$$

Hemiacetal
usually not stable

Glucose
a stable cyclic hemiacetal

Bisulfite addition products

The bisulfite ion is a strong nucleophile but a weak acid. It will attack the unhindered carbonyl group of an aldehyde or methyl ketone to form an addition product:

$$RHC{=}O + {:}SO_3H^- \, Na^+ \rightleftharpoons R{-}CH(O^- Na^+){-}SO_3H \longrightarrow R{-}CH(OH){-}SO_3^- Na^+$$

Since these bisulfite addition compounds are ionic water-soluble compounds and can be formed in up to 90% yield, they serve as a useful means of separating aldehydes and methyl ketones from mixtures of organic compounds. At high sodium bisulfite concentrations these adducts crystallize and can be isolated by filtration. The aldehyde or ketone can be regenerated by adding either a strong acid or base:

$$C_6H_5{-}CH(OH){-}SO_3^- Na^+ + HCl \longrightarrow C_6H_5{-}CHO + SO_2 + H_2O + NaCl$$

$$CH_3CH_2{-}C(OH)(CH_3){-}SO_3^- Na^+ + NaOH \longrightarrow CH_3CH_2{-}C(=O){-}CH_3 + Na_2SO_3 + H_2O$$

A similar reaction occurs between aldehydes and ketones and hydrogen cyanide, which, like bisulfite, is a weak acid but a strong nucleophile. The reaction is hazardous to carry out because of the toxicity of cyanide, but the cyanohydrins are useful synthetic intermediates:

$$CH_3CH_2-\overset{\overset{\displaystyle :\ddot{O}:}{\|}}{C}-CH_3 + HCN \longrightarrow CH_3CH_2\underset{\underset{\displaystyle CN}{|}}{\overset{\overset{\displaystyle :\ddot{O}-H}{|}}{C}}CH_3 \xrightarrow{H_2SO_4} CH_3CH=\underset{\underset{\displaystyle COOH}{|}}{C}CH_3$$

$$CN \xrightarrow{HCl,\ H_2O} CH_3CH_2\underset{\underset{\displaystyle COOH}{|}}{\overset{\overset{\displaystyle :\ddot{O}-H}{|}}{C}}CH_3 \qquad CN \xrightarrow{LiAlH_4} CH_3CH_2\underset{\underset{\displaystyle CH_2NH_2}{|}}{\overset{\overset{\displaystyle :\ddot{O}-H}{|}}{C}}CH_3$$

Cyanohydrin formation and reactions

Amines are good nucleophiles and readily add to the carbonyl group:

$$R-\underset{\displaystyle H}{C}{=}\ddot{O}: + H_2\ddot{N}R' \rightleftharpoons R-\underset{\underset{\displaystyle H}{|}}{\overset{\overset{\displaystyle :\ddot{O}:^-}{|}}{C}}-\overset{+}{N}H_2R' \rightleftharpoons R-\underset{\underset{\displaystyle H}{|}}{\overset{\overset{\displaystyle :\ddot{O}-H}{|}}{C}}-\ddot{N}HR'$$

The reaction is strongly dependent on the pH. In acid the amine is protonated ($R\overset{+}{N}H_3$) and is no longer a nucleophile. In strong base there are no protons available to catalyze the reaction. But in weakly acid solution (pH 4–6) the equilibrium between acid and base (a) is such that protons are available to protonate the carbonyl (b) and yet there is free amine present to react with the protonated carbonyl (c):

$$CH_3\ddot{N}H_2 + HCl \rightleftharpoons CH_3\overset{+}{N}H_3 + Cl^- \qquad \textbf{(a)}$$

$$CH_3-\underset{\displaystyle H}{C}{=}\ddot{O}: + HCl \rightleftharpoons \left[CH_3-\underset{\displaystyle H}{C}{=}\overset{+}{\ddot{O}}-H \longleftrightarrow CH_3-\underset{\displaystyle H}{C^+}-\ddot{O}-H \right] + Cl^- \qquad \textbf{(b)}$$

$$CH_3-\underset{\displaystyle H}{C}{=}\overset{+}{\ddot{O}}-H + CH_3\ddot{N}H_2 \rightleftharpoons CH_3-\underset{\underset{\displaystyle H}{|}}{\overset{\overset{\displaystyle :\ddot{O}-H}{|}}{C}}-\overset{+}{N}H_2CH_3 \xrightleftharpoons{-H^+} CH_3-\underset{\underset{\displaystyle H}{|}}{\overset{\overset{\displaystyle :\ddot{O}-H}{|}}{C}}-\underset{\underset{\displaystyle H}{|}}{\ddot{N}}-CH_3 \qquad \textbf{(c)}$$

The intermediate hydroxyamino form of the adduct is not stable and spontaneously dehydrates under the mildly acidic conditions of the reaction to give an imine, commonly referred to as a Schiff base:

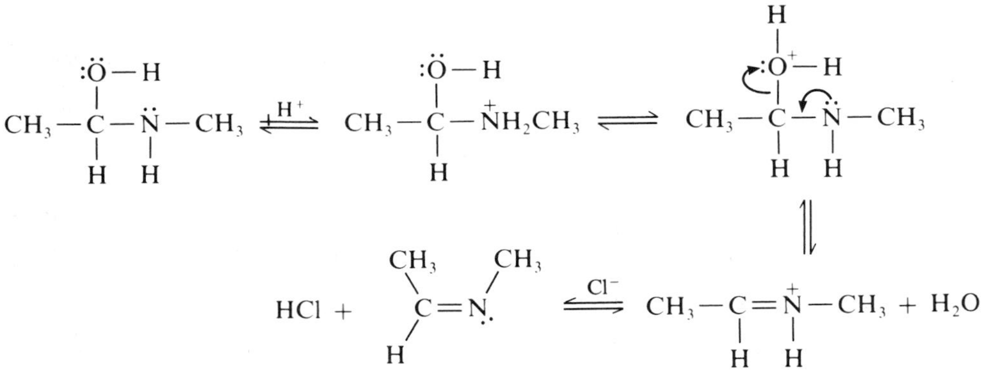

$$CH_3-\underset{H}{\underset{|}{C}}(OH)-\underset{H}{\underset{|}{\ddot{N}}}-CH_3 \overset{H^+}{\rightleftharpoons} CH_3-\underset{H}{\underset{|}{C}}(OH)-\overset{+}{N}H_2CH_3 \rightleftharpoons CH_3-\underset{H}{\underset{|}{C}}(\overset{+}{O}H_2)-\underset{H}{\underset{|}{N}}-CH_3$$

$$\rightleftharpoons CH_3-\underset{H}{\underset{|}{C}}=\underset{H}{\underset{|}{\overset{+}{N}}}-CH_3 + H_2O \overset{Cl^-}{\rightleftharpoons} HCl + (CH_3)(H)C=\ddot{N}-CH_3$$

Schiff base

Imine or Schiff base formation

The biosynthesis of most amino acids proceeds through Schiff base intermediates.

Three rather special amines form useful stable imines:

Solid derivatives of aldehydes and ketones

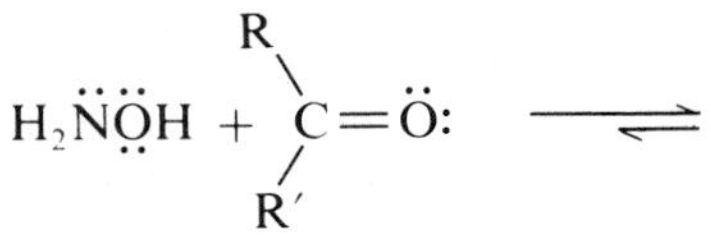

$$H_2\ddot{N}\ddot{O}H + RR'C=\ddot{O}: \rightleftharpoons RR'C=\ddot{N}-\ddot{O}-H$$

Hydroxylamine **Oxime**

$$H_2NNH\overset{O}{\overset{\|}{C}}NH_2 + RR'C=O \rightleftharpoons RR'C=N-\underset{H}{\underset{|}{N}}-\overset{O}{\overset{\|}{C}}-NH_2$$

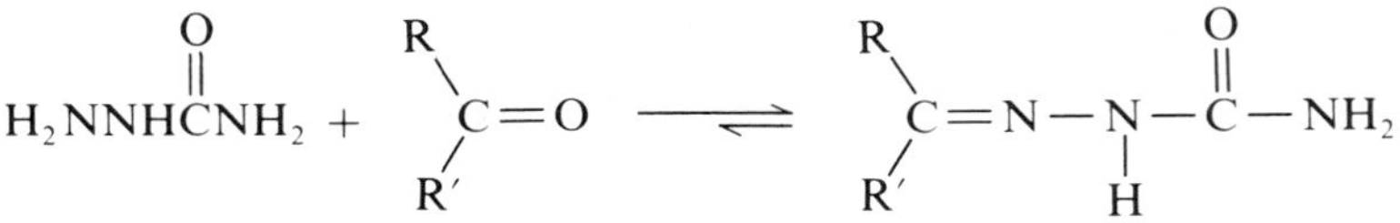

Semicarbazide **Semicarbazone**

Oximes
Semicarbazones
2,4-Dinitrophenylhydrazones

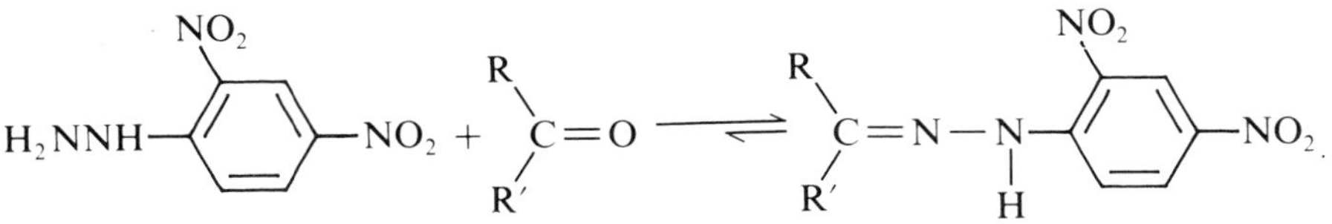

$$H_2NNH-C_6H_3(NO_2)_2 + RR'C=O \rightleftharpoons RR'C=N-\underset{H}{\underset{|}{N}}-C_6H_3(NO_2)_2$$

2,4-Dinitrophenylhydrazine **2,4-Dinitrophenylhydrazone**

These imines are solids and are useful for the characterization of aldehydes and ketones. For example, infrared and nmr spectroscopy may indicate that a certain unknown is acetaldehyde. It is difficult to determine the boiling point of a few milligrams of a liquid, but if it can be converted to a solid derivative the mp *can* be determined with that amount. The 2,4-dinitrophenylhydrazones are usually the derivatives of choice because they are nicely crystalline compounds with well-defined melting or decomposition points and they increase the molecular weight by 180. Ten milligrams of acetaldehyde will give 51 mg of 2,4-DNP:

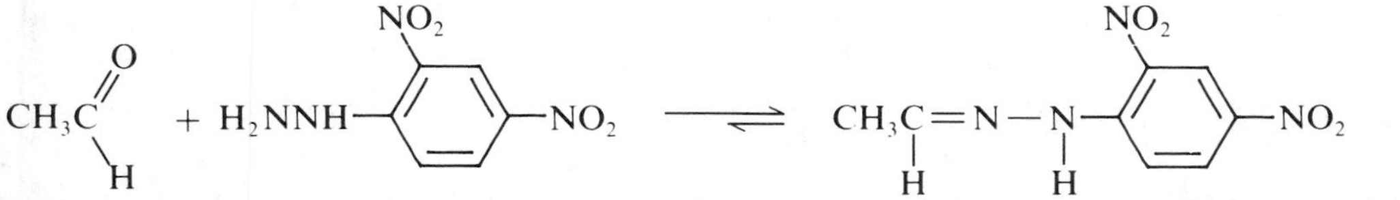

$$CH_3CHO + H_2NNH{-}C_6H_3(NO_2)_2 \rightleftharpoons CH_3CH{=}N{-}NH{-}C_6H_3(NO_2)_2$$

Acetaldehyde MW 44.05 bp 20.8°C

2,4-Dinitrophenylhydrazine MW 198.14 mp 196°C

Acetaldehyde 2,4-dinitrophenylhydrazone MW 224.19 mp 168.5°C

Before the advent of nmr and ir spectroscopy the chemist was often called upon to identify aldehydes and ketones by purely chemical means. Aldehydes can be distinguished chemically from ketones by their ease of oxidation to carboxylic acids. The oxidizing agent, an ammoniacal solution of silver nitrate, Tollens' reagent, is reduced to metallic silver, which is deposited on the inside of a test tube as a silver mirror.

$$2\ Ag(NH_3)_2OH + R{-}CHO \longrightarrow 2\ Ag + R{-}COO^- \quad NH_4^+ + H_2O + 3\ NH_3$$

Another way to distinguish aldehydes from ketones is to use Schiff's reagent. This is a solution of the red dye, Basic Fuchsin, which is rendered colorless on treatment with sulfur dioxide. In the presence of an aldehyde the colorless solution turns magenta.

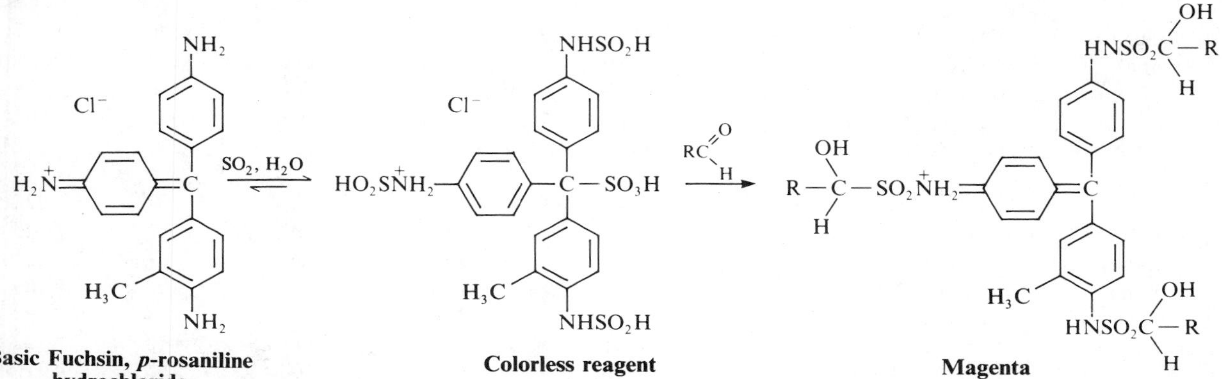

Cl⁻ $H_2\overset{+}{N}{=}$ ... SO_2, H_2O ⇌ Cl⁻ $HO_2S\overset{+}{N}H_2{-}$ ... $C{-}SO_3H$ $\xrightarrow{RCHO}$ $R{-}CH(OH){-}SO_2\overset{+}{N}H_2{=}$ ...

Basic Fuchsin, *p*-rosaniline hydrochloride

Colorless reagent

Magenta

Methyl ketones can be distinguished from other ketones by the iodoform test. The methyl ketone is treated with iodine in a basic solution. Introduction of the first iodine atom increases the acidity of the remaining methyl protons, so halogenation stops only when the triiodo compound has been produced.

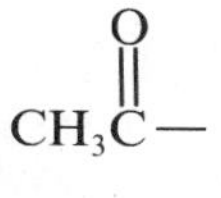

$CH_3\overset{O}{\overset{\|}{C}}{-}$

A methyl ketone

The base then allows the relatively stable triiodomethyl carbanion to leave and a subsequent proton transfer gives iodoform, a yellow crystalline solid of mp 119–123°C. The test is also positive for compounds easily oxidized to methyl ketones, such as CH_3CHOH— and ethanol. Acetaldehyde also gives a positive test because it is both a methyl ketone and an aldehyde.

$$R-\overset{\ddot{O}:}{\overset{\|}{C}}-CH_3 + OH^- \rightleftharpoons R-\overset{\ddot{O}:}{\overset{\|}{C}}-\overset{-}{\ddot{C}}H_2 + H_2O$$

$$R-\overset{\ddot{O}:}{\overset{\|}{C}}-\overset{-}{\ddot{C}}H_2 + I_2 \quad R-\overset{O}{\overset{\|}{C}}-CH_2I + I^- \xrightarrow{OH^-} \xrightarrow{I_2} \xrightarrow{OH^-} \xrightarrow{I_2} R-\overset{\ddot{O}:}{\overset{\|}{C}}-CI_3$$

$$R-\overset{\ddot{O}:}{\overset{\|}{C}}-CI_3 + OH^- \longrightarrow R-\underset{:\ddot{O}H}{\overset{:\ddot{O}:^-}{C}}-CI_3 \longrightarrow R-C\begin{matrix}{/\!\!/ :\ddot{O}} \\ {\backslash :\ddot{O}-H}\end{matrix} + :\overset{-}{C}I_3 \longrightarrow R-C\begin{matrix}{/\!\!/ :\ddot{O}} \\ {\backslash \ddot{O}:^-}\end{matrix} + \underset{\substack{\textbf{Iodoform} \\ \text{mp } 123^\circ C}}{CHI_3}$$

TABLE 30.1 Melting Points of Derivatives of Some Aldehydes and Ketones

Compound[c]	Formula	MW	Den	Water solubility	Melting Points (°C) Phenyl-hydrazone	2,4-DNP	Semi-carbazone
Acetone	CH_3COCH_3	58.08	0.79		42	126	187
n-Butanal	$CH_3CH_2CH_2CHO$	72.10	0.82	4 g/100 g	Oil	123	95(106)[a]
Diethyl ketone	$CH_3CH_2COCH_2CH_3$	86.13	0.81	4.7 g/100 g	Oil	156	138
Furfural	$C_4H_3O \cdot CHO$	96.08	1.16	9 g/100 g	97	212(230)[a]	202
Benzaldehyde	C_6H_5CHO	106.12	1.05	Insol.	158	237	222
Heptane-2,6-dione	$CH_3COCH_2CH_2CH_2COCH_3$	114.14	0.97	∞	120[b]	257[b]	224[b]
2-Heptanone	$CH_3(CH_2)_4COCH_3$	114.18	0.83	Insol.	Oil	89	123
3-Heptanone	$CH_3(CH_2)_3COCH_2CH_3$	114.18		Insol.	Oil	81	101
n-Heptanal	$n\text{-}C_6H_{13}CHO$	114.18	0.82	Insol.	Oil	108	109
Acetophenone	$C_6H_5COCH_3$	120.66	1.03	Insol.	105	238	198
2-Octanone	$CH_3(CH_2)_5COCH_3$	128.21	0.82	Insol.	Oil	58	122
Cinnamaldehyde	$C_6H_5CH{=}CHCHO$	132.15	1.10	Insol.	168	255	215
Propiophenone	$C_6H_5COCH_2CH_3$	134.17	1.01	Insol.	about 48°	191	182

a. Both melting points have been found, depending on crystalline form of derivative.
b. Monoderivative or diderivative.
c. See Tables 70.3 and 70.14 for additional aldehydes and ketones.

Experiments

1. Unknowns

You will be given an unknown that may be any of the aldehydes or ketones listed in Table 30.1 or a noncarbonyl compound. Hence, the first test should be that with 2,4-dinitrophenylhydrazine reagent; if this test is negative, report accordingly and proceed to another unknown. At least one derivative of the unknown is to be submitted to the instructor; but if you first do the bisulfite and iodoform characterizing tests, the results may suggest derivatives whose melting points will be particularly revealing.

In conducting the following tests you should perform three tests simultaneously: on a compound known to give a positive test, on a compound known to give a negative test, and on the unknown. In this way you will be able to determine whether the reagents are working as they should as well as interpret a positive or a negative test.

2. 2,4-Dinitrophenylhydrazones

$$RR'C{=}\ddot{O}: + H_2\ddot{N}{-}\ddot{N}H{-}C_6H_3(NO_2)_2 \xrightarrow{H^+} RR'C{=}\ddot{N}{-}\ddot{N}H{-}C_6H_3(NO_2)_2$$

To 2 mL of the stock solution[1] of 2,4-dinitrophenylhydrazine in phosphoric acid add about 20 mg of the compound to be tested. Two milliliters of the 0.1 M solution contains 0.2 millimole (0.0002 mole) of the reagent. If the compound to be tested has a molecular weight of 100 then 20 mg is 0.2 millimole. Warm the reaction mixture for a few minutes in a water bath and then let crystallization proceed. Collect the product by suction filtration, wash the crystals with a large amount of water to remove all phosphoric acid, press a piece of moist litmus paper onto the crystals and if they are acidic wash them with more water. Press the product as dry as possible between sheets of filter paper and recrystallize from boiling ethanol. Occasionally a high molecular weight derivative won't dissolve in a reasonable quantity (3 mL) of ethanol. In that case cool the hot suspension and isolate the crystals by suction filtration. The boiling ethanol treatment removes impurities so that an accurate melting point can be obtained on the isolated material.

1. Dissolve 2.0 g of 2,4-dinitrophenylhydrazine in 50 mL of 85% phosphoric acid by heating, cool, add 50 mL of 95% ethanol, cool again, and clarify by suction filtration from a trace of residue.

An alternative procedure is applicable when the 2,4-dinitrophenylhydrazone is known to be sparingly soluble in ethanol. Measure 0.2 millimole (40 mg) of crystalline 2,4-dinitrophenylhydrazine into a 10-mL Erlenmeyer flask, add 6 mL of 95% ethanol, digest on the steam bath until all particles of solid are dissolved, and then add 0.2 millimole of the compound to be tested and continue warming. If there is no immediate change, add, from a Pasteur pipette, 1–2 drops of concentrated hydrochloric acid as catalyst and note the result. Warm for a few minutes, then cool and collect the product. This procedure would be used for an aldehyde like cinnamaldehyde ($C_6H_5CH{=}CHCHO$).

The alternative procedure strikingly demonstrates the catalytic effect of hydrochloric acid, but it is not applicable to a substance like diethyl ketone, whose 2,4-dinitrophenylhydrazone is much too soluble to crystallize from the large volume of ethanol. The first procedure is obviously the one to use for an unknown.

3. Semicarbazones

$$RR'C{=}\ddot{O}: + Cl^- \; H_3\overset{+}{N}{-}\ddot{N}H{-}C({=}O){-}NH_2 + C_5H_5N \longrightarrow RR'C{=}\ddot{N}{-}\ddot{N}H{-}C({=}O){-}NH_2 + C_5H_5\overset{+}{N}H \; Cl^- + H_2O$$

Semicarbazide hydrochloride **Pyridine** **Semicarbazone** **Pyridine hydrochloride**

Semicarbazide (mp 96°C) is not very stable in the free form and is used as the crystalline hydrochloride (mp 173°C). Since this salt is insoluble in methanol or ethanol and does not react readily with typical carbonyl compounds in alcohol-water mixtures, a basic reagent, pyridine, is added to liberate free semicarbazide.

To 0.25 mL of the stock solution[2] of semicarbazide hydrochloride, which contains 0.5 millimole of the reagent, add 0.5 millimole of the compound to be tested and enough methanol (0.5 mL) to produce a clear solution; then add 5 drops of pyridine (a twofold excess) and warm the solution gently on the steam bath for a few minutes. Cool the solution slowly to room temperature. It may be necessary to scratch the inside of the test tube in order to induce crystallization. Cool the tube in ice, collect the product by suction filtration, and wash it with water followed by a small amount of cold methanol. Recrystallize the product from methanol, ethanol, or ethanol/water.

2. Prepare a stock solution by dissolving 1.11 g of semicarbazide hydrochloride in 5 mL of water; 0.5 mL of this solution contains 1 millimole of reagent.

4. Tollens' Test

Test for aldehydes

$$R-C(=O)H + 2\ Ag(NH_3)_2OH \longrightarrow 2\ Ag + RCOO^-NH_4^+ + H_2O + 3\ NH_3$$

Clean 4 or 5 reaction tubes by adding a few milliliters of 10% sodium hydroxide solution to each and heating them in a water bath while preparing Tollens' reagent.

To 1.0 mL of 5% silver nitrate solution in a 4″ test tube, add 0.5 mL of 10% sodium hydroxide in a test tube. To the gray precipitate of silver oxide, Ag_2O, add 0.25 mL of 2.8% ammonia solution (0.1 mL of concentrated ammonium hydroxide diluted to 1 mL). Stopper the tube and shake it. Repeat the process until *almost* all of the precipitate dissolves (1.5 mL of ammonia at most); then dilute the solution to 5 mL. Empty the reaction tubes of sodium hydroxide solution, rinse them, and add 0.5 mL of Tollens' reagent to each. Add one drop (no more) of the substance to be tested by allowing it to run down the inside of the inclined reaction tube. Set the tubes aside for a few minutes without agitating the contents. If no reaction occurs, warm the mixture briefly on a water bath. As a known aldehyde try one drop of a 0.1 M solution of glucose. A more typical aldehyde to test is benzaldehyde.

At the end of the reaction destroy excess Tollens' reagent with nitric acid: It can form an explosive fulminate on standing. Nitric acid can also be used to remove silver mirrors from the test tubes.

5. Schiff's Test

Very sensitive test for aldehydes

Add 1 drop of the unknown to 0.7 mL of Schiff's reagent.[3] A magenta color will appear within 10 min with aldehydes. As in all of these tests compare the colors produced by a known aldehyde, a known ketone, and the unknown compound.

6. Iodoform Test

$$R-\overset{\overset{\displaystyle :\ddot{O}}{\|}}{C}-CH_3 + 3\ I_2 + 4\ OH^- \longrightarrow R-C(=\ddot{O}:)(-\ddot{O}:^-) + CHI_3 + 3\ :\ddot{I}:^- + 3\ H_2O$$

Test for methyl ketones

The reagent contains iodine in potassium iodide solution[4] at a concentration such that 0.5 mL of solution, on reaction with excess methyl ketone, will yield 43 mg of iodoform. If the substance to be tested is water soluble, dissolve

3. Schiff's reagent is prepared by dissolving 0.05 g Basic Fuchsin (*p*-rosaniline hydrochloride) in 50 mL of water and then adding 2 mL of a saturated aqueous solution of sodium bisulfite. After 1 h add 1 mL of concentrated hydrochloric acid.

4. Dissolve 5 g of iodine in a solution of 10 g of potassium iodide in 40 mL of water.

1 drop of a liquid or an estimated 15 mg of a solid in 0.5 mL of water in a reaction tube; add 0.5 mL of 10% sodium hydroxide and then slowly add 0.75 mL of the iodine solution. In a positive test the brown color of the reagent disappears and yellow iodoform separates. If the substance to be tested is insoluble in water, dissolve it in 0.5 mL of 1,2-dimethoxyethane, proceed as above, and at the end dilute with 2.5 mL of water.

Suggested test substances are hexane-2,5-dione (water soluble), *n*-butyraldehyde (water soluble), and acetophenone (water insoluble).

Iodoform can be recognized by its odor and yellow color and, more securely, from the melting point (119–123°C). The substance can be isolated by suction filtration of the test suspension or by adding 0.5 mL of dichloromethane, shaking the stoppered test tube to extract the iodoform into the small lower layer, withdrawing the clear part of this layer with a capillary dropping tube, and evaporating it in a small tube on the steam bath. The crude solid is crystallized from methanol-water.

7. Bisulfite Test

Forms with unhindered carbonyls

$$\mathrm{R(H)C{=}\ddot{O}{:}} + \mathrm{Na^{+}SO_3H^{-}} \rightleftharpoons \mathrm{R{-}C(H)(\ddot{O}H){-}SO_3^{-}Na^{+}}$$

Put 0.4 mL of the stock solution[5] into a reaction tube and add 2 drops of the substance to be tested. Shake each tube during the next 10 min and note the results. A positive test will result from aldehydes, unhindered cyclic ketones such as cyclohexanone, and unhindered methyl ketones.

If the bisulfite test is applied to a liquid or solid that is very sparingly soluble in water, formation of the addition product is facilitated by adding a small amount of methanol before the addition to the bisulfite solution.

8. nmr and ir Spectroscopy

A peak at 9.6–10 ppm is highly characteristic of aldehydes because almost no other peaks appear in this region. On some spectrometers an offset will be needed to detect this region. Similarly a sharp singlet at 2.2 ppm is very characteristic of methyl ketones; but beware of contamination of the sample by acetone, which is often used to clean glassware.

Infrared spectroscopy is extremely useful in analyzing all carbonyl-containing compounds, including aldehydes and ketones. See the extensive discussion in Chapter 19.

5. Prepare a stock solution from 25 g of sodium bisulfite dissolved in 100 mL of water with brief swirling.

Questions

1. What is the purpose of making derivatives of unknowns?
2. Why are 2,4-dinitrophenylhydrazones better derivatives then phenylhydrazones?
3. Using chemical tests how would you distinguish among 2-pentanone, 3-pentanone, and pentanal?
4. Draw the structure of a compound, C_5H_8O, which gives a positive iodoform test and does not decolorize permanganate.
5. Draw the structure of a compound, C_5H_8O, which gives a positive Tollens' test and does not react with bromine in dichloromethane.
6. Draw the structure of a compound, C_5H_8O, which reacts with phenylhydrazine, decolorizes bromine in dichloromethane, and does not give a positive iodoform test.
7. Draw the structure of two geometric isomers, C_5H_8O, which give a positive iodoform test.
8. Assign the various peaks in the ^1H nmr spectrum of 2-butanone to specific protons in the molecule (Fig. 30.4).
9. Assign the various peaks in the ^1H nmr spectrum of crotonaldehyde to specific protons in the molecule (Fig. 30.5).
10. What vibrations cause the peaks at about $3.4\,\mu$ ($3000\,cm^{-1}$) in the infrared spectrum of fluorene (Fig. 30.2)?

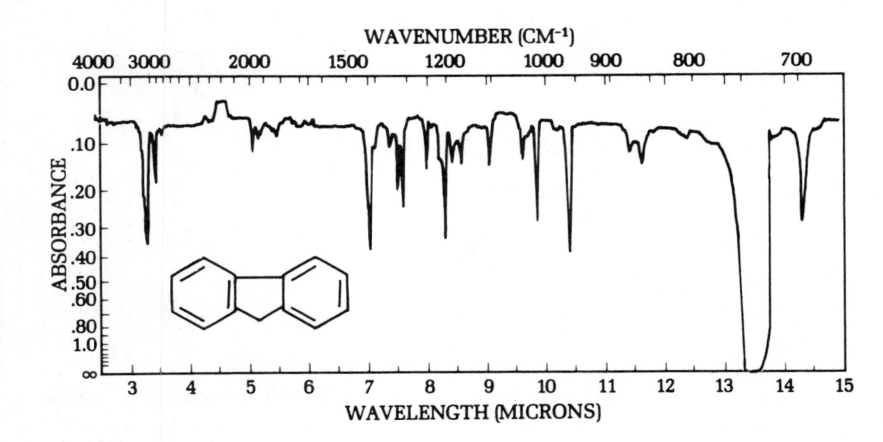

FIG. 30.1 Infrared spectrum of fluorene in CS_2.

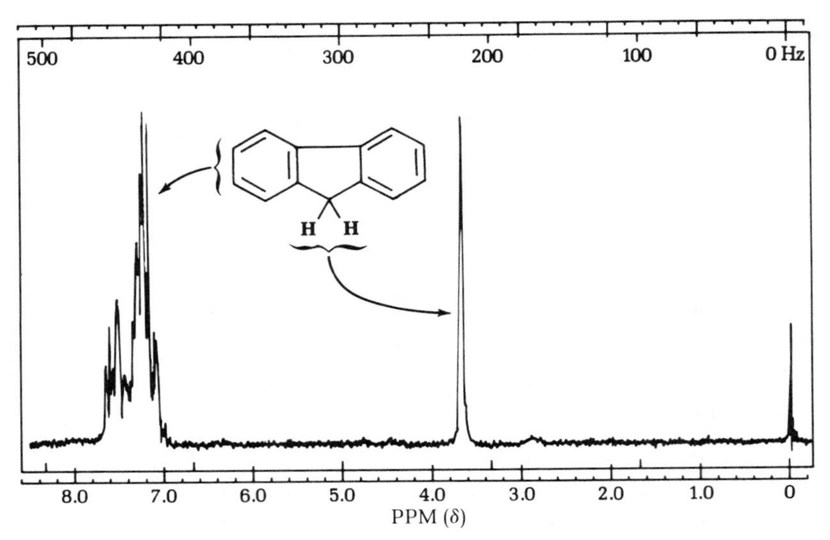

FIG. 30.2 ^1H nmr spectrum of fluorene.

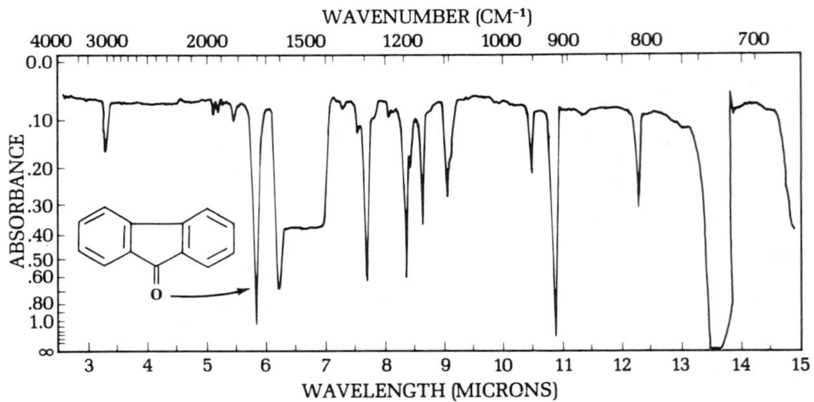

FIG. 30.3 Infrared spectrum of fluorenone.

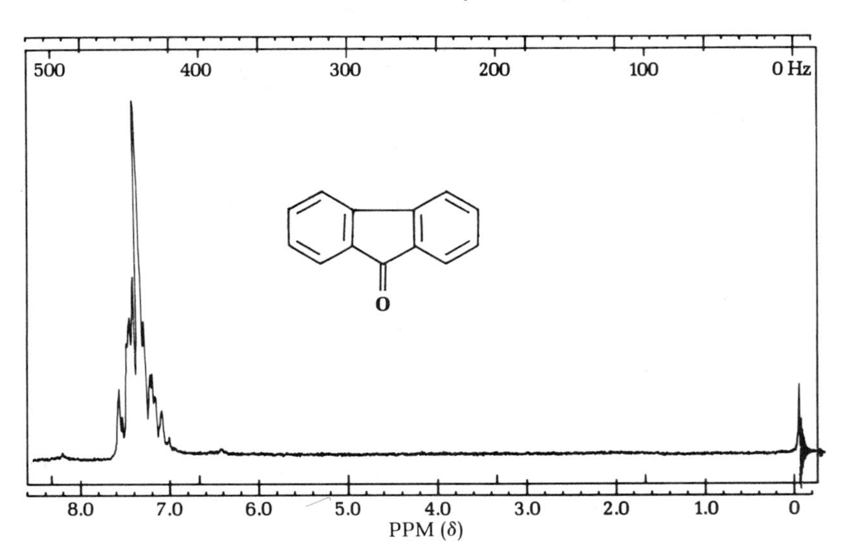

FIG. 30.4 ^1H nmr spectrum of fluorenone.

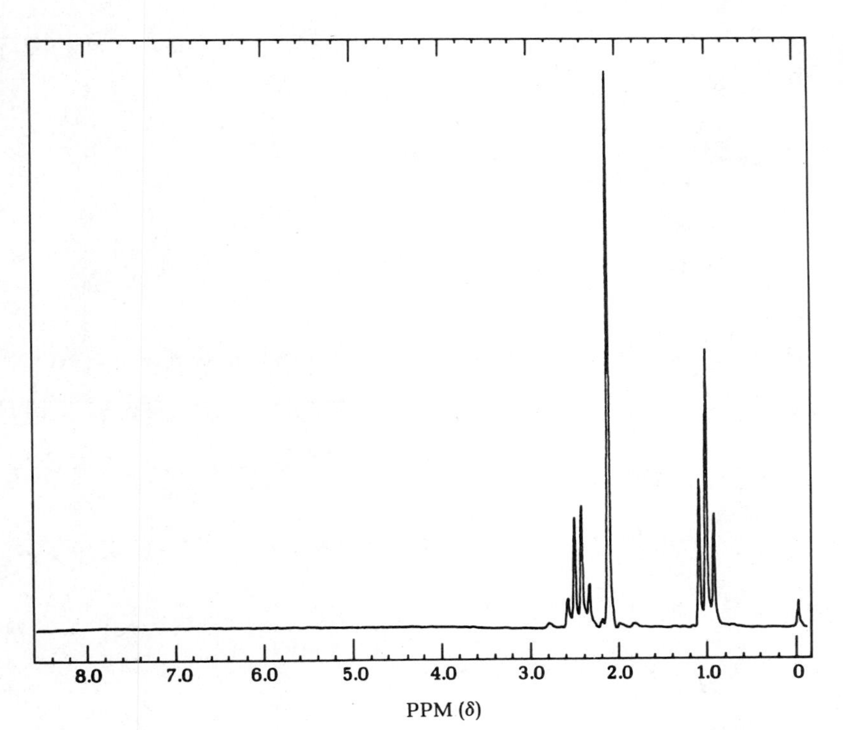

FIG. 30.5 ^{1}H nmr spectrum of 2-butanone, $CH_3COCH_2CH_3$.

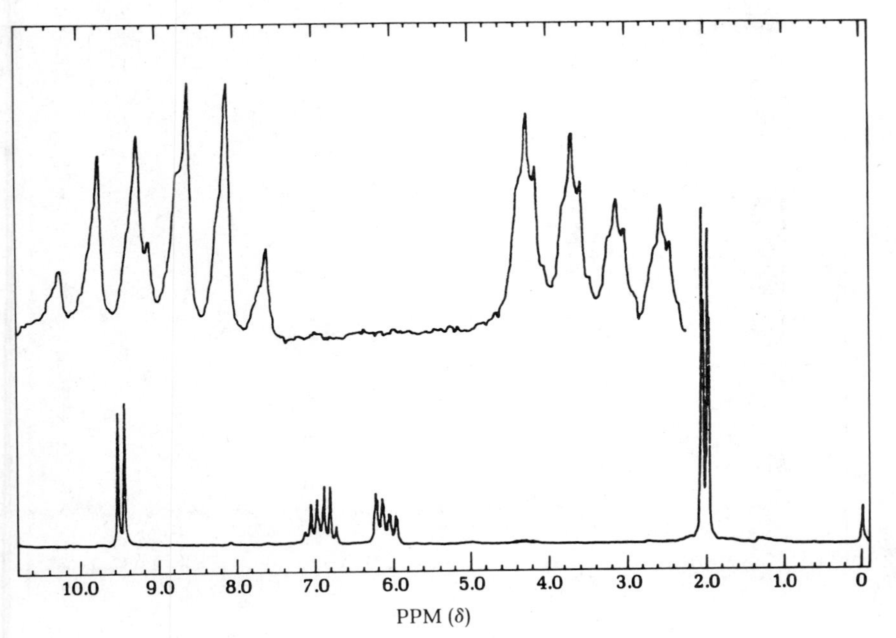

FIG. 30.6 ^{1}H nmr spectrum of crotonaldehyde, $CH_3CH{=}CHCHO$.

FIG. 30.7 ^{13}C nmr spectrum of 2-butanone, $CH_3COCH_2CH_3$.

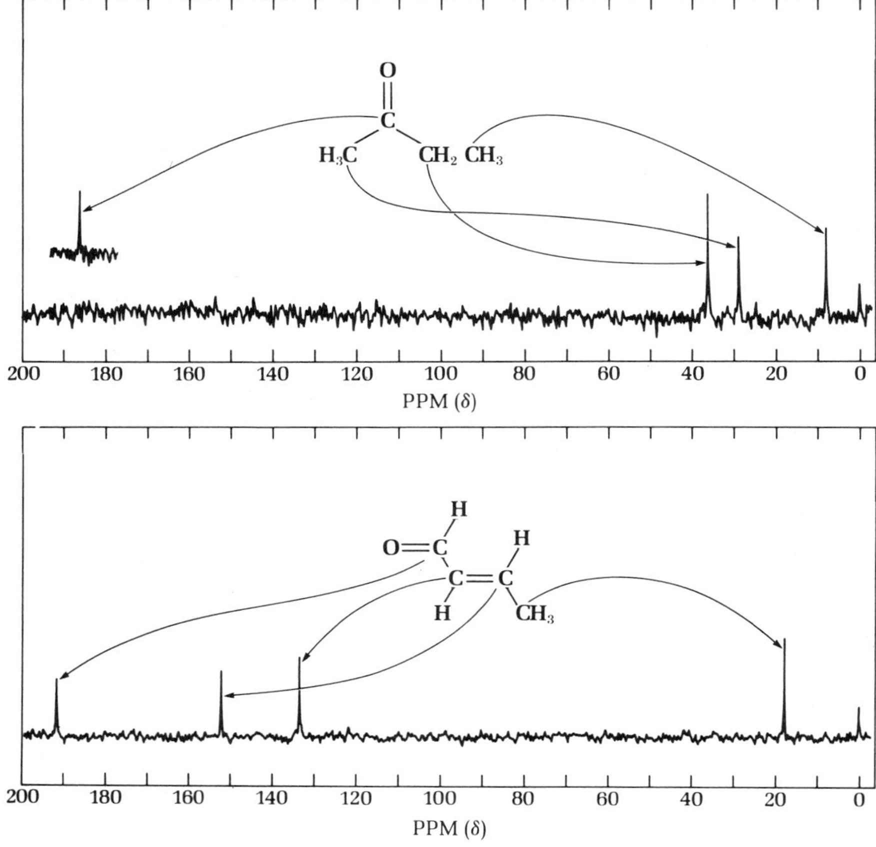

FIG. 30.8 ^{13}C nmr spectrum of crotonaldehyde.

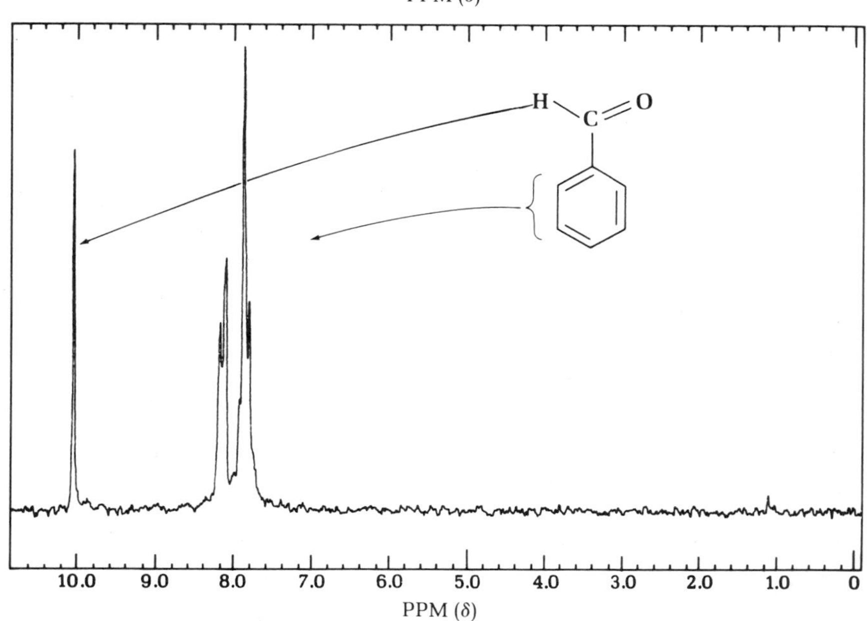

FIG. 30.9 ^{1}H nmr spectrum of benzaldehyde.

Grignard Synthesis of Triphenylmethanol and Benzoic Acid

31

PRELAB EXERCISE: *Prepare a flow sheet for the preparation of triphenylmethanol. Through a knowledge of the physical properties of the solvents, reactants, and products, show how the products are purified. Indicate which layer in separations should contain the product.*

In 1912 Victor Grignard received the Nobel prize in chemistry for his work on the reaction that bears his name, a carbon-carbon bond-forming reaction by which almost any alcohol may be formed from appropriate alkyl halides and carbonyl compounds. The Grignard reagent is easily formed by reaction of an alkyl halide, in particular a bromide, with magnesium metal in anhydrous ether. Although the reaction can be written and thought of as simply

$$R{-}Br + Mg \longrightarrow R{-}Mg{-}Br$$

it appears that the structure of the material in solution is rather more complex. There is evidence that dialkylmagnesium is present

$$2\ R{-}Mg{-}Br \rightleftharpoons R{-}Mg{-}R + MgBr_2$$

Structure of the Grignard reagent

and that the magnesium atoms, which have the capacity to accept two electron pairs from donor molecules to achieve a four-coordinated state, are solvated by the unshared pairs of electrons on ether:

$$\begin{array}{c} Et\diagdown\ \ \diagup Et \\ \ddot{O} \\ R{-}\overset{\cdot\cdot}{\underset{\cdot\cdot}{Mg}}{-}Br \\ \ddot{O} \\ Et\diagup\ \ \diagdown Et \end{array}$$

A strong base and strong nucleophile

The Grignard reagent is a strong base and a strong nucleophile. As a base it will react with all protons that are more acidic than those found on alkenes and alkanes. Thus Grignard reagents react readily with water, alcohols, amines, thiols, etc., to regenerate the alkane:

$$R-Mg-Br + H_2O \longrightarrow R-H + MgBrOH$$
$$R-Mg-Br + R'OH \longrightarrow R-H + MgBrOR'$$
$$R-Mg-Br + R'NH_2 \longrightarrow R-H + MgBrNHR'$$

The starting material for preparing the Grignard reagent can contain no acidic protons. The reactants and apparatus must all be completely dry; otherwise the reaction will not start. If proper precautions are taken, however, the reaction proceeds smoothly.

The magnesium metal, in the form of turnings, has a coat of oxide on the outside. A fresh surface can be exposed by crushing a piece under the absolutely dry ether in the presence of the organic halide. Reaction will begin at the exposed surface, as evidenced by a slight turbidity in the solution and evolution of bubbles. Once the exothermic reaction starts it proceeds easily, the magnesium dissolves, and a solution of the Grignard reagent is formed. The solution is often turbid and gray due to impurities in the magnesium. The reagent is not isolated but reacted immediately with, most often, an appropriate carbonyl compound:

$$R-Mg-Br + R'-\overset{\overset{\ddot{O}:}{\|}}{C}-R'' \longrightarrow R'-\underset{\underset{R}{|}}{\overset{\overset{:\ddot{O}:^{-}\,MgBr^{+}}{|}}{C}}-R''$$

to give, in another exothermic reaction, the magnesium alkoxide. In a simple acid-base reaction this alkoxide is reacted with acidified ice water to give the covalent, ether-soluble alcohol and the ionic water-soluble magnesium salt:

$$R'-\underset{\underset{R}{|}}{\overset{\overset{:\ddot{O}:^{-}\,MgBr^{+}}{|}}{C}}-R'' + H^{+}Cl^{-} \longrightarrow R'-\underset{\underset{R}{|}}{\overset{\overset{:\ddot{O}-H}{|}}{C}}-R'' + MgBrCl$$

The great versatility of this reaction lies in the wide range of reactants that undergo the reaction. Thirteen representative reactions are shown on the following page.

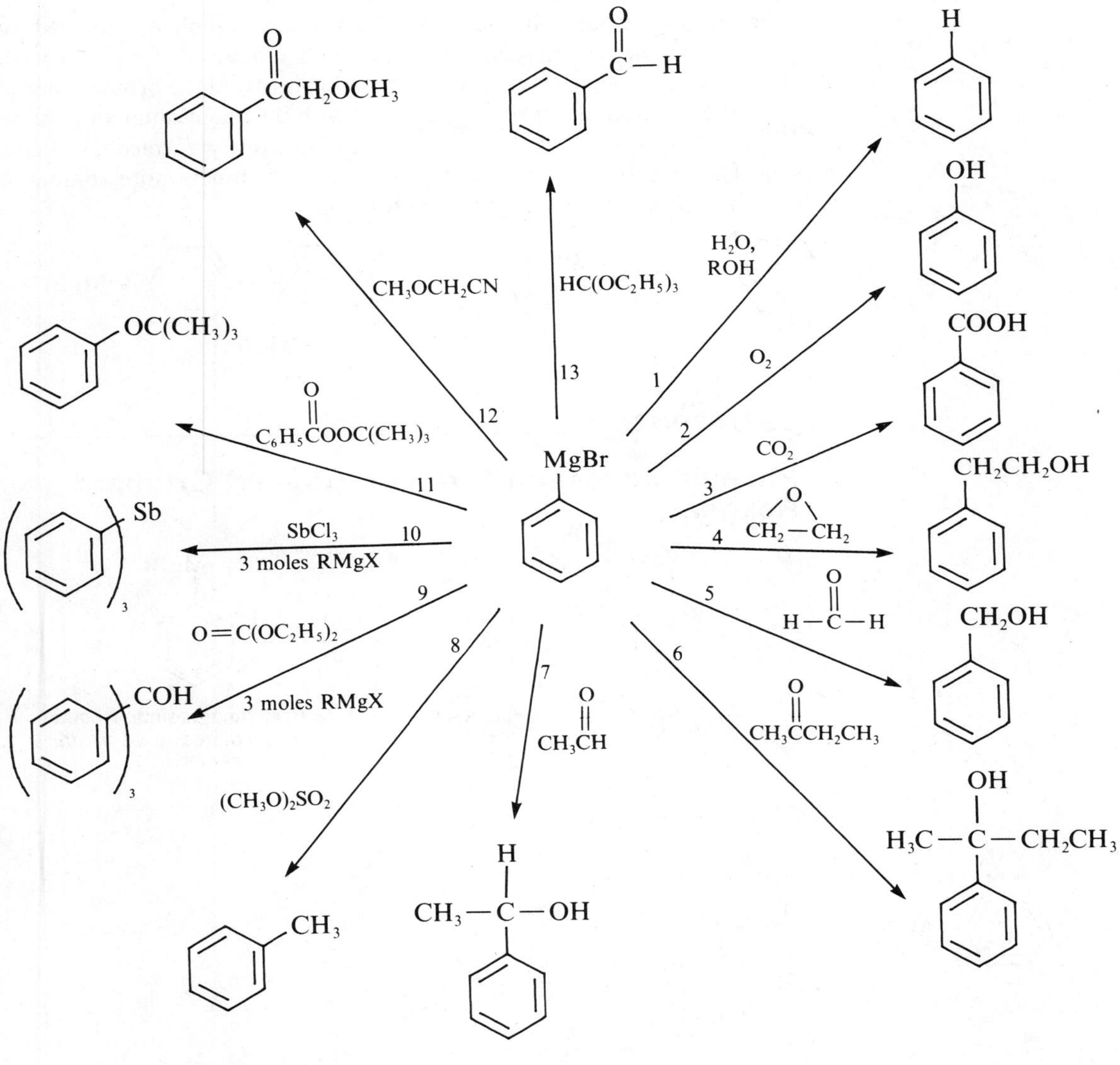

In every case except reaction 1 the intermediate alkoxide must be hydrolyzed to give the product. The reaction with oxygen (reaction 2) is usually not a problem because the ether vapor over the reagent protects it from attack by oxygen, but this reaction is one reason why the reagent cannot usually be stored without special precautions. Reactions 6 and 7 with ketones and aldehydes giving respectively tertiary and secondary alcohols are among the most common. Reactions 8–12 are not nearly so common.

A versatile reagent

In the present experiment we shall carry out another common type of Grignard reaction, the formation of a tertiary alcohol from a mole of the reagent and one of a ketone, benzophenone.

The primary impurity in the present experiment is biphenyl, formed by the reaction of phenylmagnesium bromide with unreacted bromobenzene. The most effective way to lessen this side reaction is to add the bromobenzene slowly to the reaction mixture so it will react with the magnesium and not be present in high concentration to react with previously formed Grignard reagent. The impurity is easily eliminated since it is much more soluble in hydrocarbon solvents than triphenylmethanol.

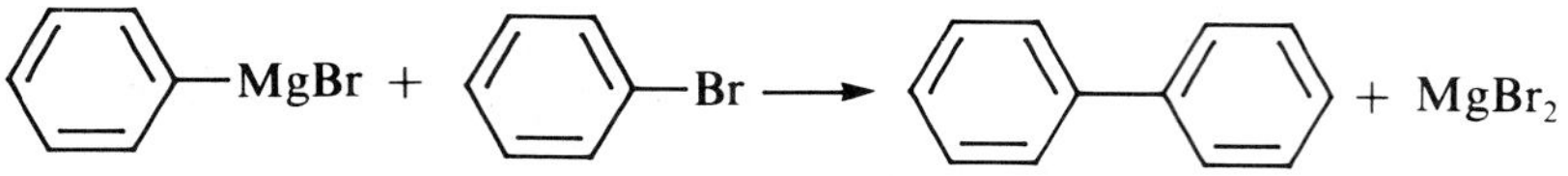

$$C_6H_5{-}MgBr + C_6H_5{-}Br \longrightarrow C_6H_5{-}C_6H_5 + MgBr_2$$

Biphenyl
mp 72°C

Experiments

1. Phenylmagnesium Bromide (Phenyl Grignard Reagent)

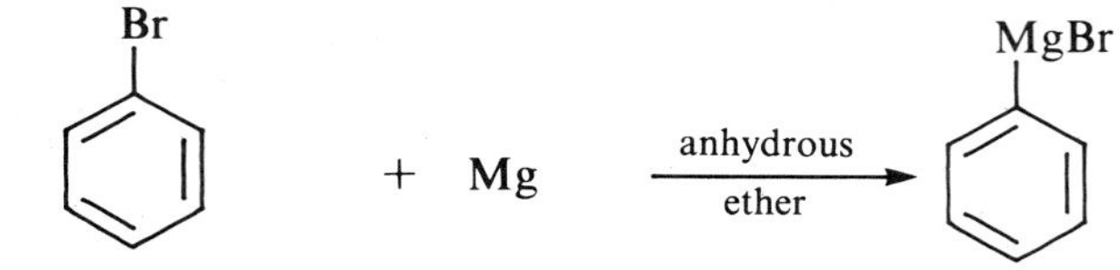

$$C_6H_5Br + Mg \xrightarrow[\text{ether}]{\text{anhydrous}} C_6H_5MgBr$$

Bromobenzene
MW 157.02
bp 156°C
den 1.491

Magnesium
At Wt
24.31

Phenylmagnesium bromide
not isolated, used *in situ*

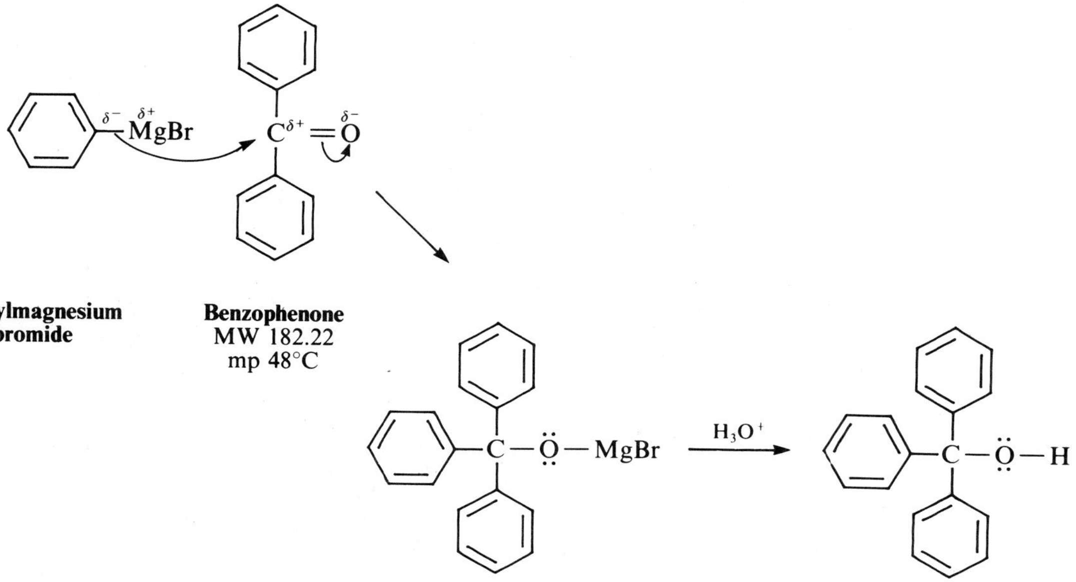

$$C_6H_5^{\delta-}{-}\overset{\delta+}{MgBr} + (C_6H_5)_2C^{\delta+}{=}\overset{\delta-}{O} \longrightarrow (C_6H_5)_3C{-}\ddot{O}{-}MgBr \xrightarrow{H_3O^+} (C_6H_5)_3C{-}\ddot{O}{-}H$$

Phenylmagnesium bromide

Benzophenone
MW 182.22
mp 48°C

Triphenylmethanol
MW 260.34
mp 164.2°C

Advance Preparation. It is absolutely imperative that all equipment and reagents be absolutely dry. The glassware to be used—two reaction tubes, two one-dram vials, and a stirring rod—should be dried in a 110°C oven for at least 30 min, as should the magnesium. The plastic and rubber ware should be rinsed with acetone, if it appears to be either dirty or wet with water, and then placed in a desiccator for at least 12 h. Do not place plastic ware in the oven. New sealed packages of syringes can be used without prior drying.

Procedure

Remove a reaction tube from the oven and immediately cap it with a septum. In the operations that follow, keep the tube capped except when it is necessary to open it. After it cools to room temperature, add 2.00 millimoles (48.6 mg) of magnesium. Since the magnesium comes in rather large pieces, it may not be possible to use exactly this quantity of magnesium. If not, be sure to make the magnesium the limiting reagent by adding a 5% excess of bromobenzene.

Using a dry syringe add to the magnesium by injection through the septum 0.5 mL of anhydrous ether. This is not the common laboratory solvent, but specially dried and packaged ether that is completely dry. Your laboratory instructor will demonstrate transfer from the storage container used in your laboratory. A very convenient container is a 50-mL septum-capped bottle. This method of dispensing the solvent has two advantages: the ether is kept anhydrous and there is little possibility of its catching fire. Ether is extremely flammable; do not work with this solvent near flames.

Ether can be kept anhydrous by storing over Linde 5 A molecular sieves. Discard after 90 days because of peroxide formation.

Into an oven-dried vial weigh 2.1 mmoles (340 mg) of dry (stored over molecular sieves) bromobenzene. Using a syringe, add to this vial 0.7 mL of anhydrous ether, cap the vial, mix the contents, and with the same syringe remove all of the solution from the vial. This can be done virtually quantitatively so you do not need to rinse the vial. Don't worry about air bubbles in the syringe. Immediately cap the empty vial to keep it dry for later use. Inject about 6–8 drops of the bromobenzene-ether mixture into the reaction tube and mix the contents. Pierce the septum with another syringe needle for pressure relief (Fig. 31.1).

If the reaction does not start within 2–3 min, remove the septum, syringe, and empty syringe needle and crush the magnesium with a dry stirring rod. You can do this easily in the confines of the 10-mm dia. reaction tube; there is little danger of poking the stirring rod through the bottom of the tube. Immediately replace the septum, syringe, and syringe needle. The reaction should start within seconds. The formerly clear solution becomes cloudy and soon begins to boil as the magnesium metal reacts with the bromobenzene to form the Grignard reagent, phenylmagnesium bromide.

Once the Grignard reaction starts it will continue; any further difficulty in initiating the reaction can be dealt with by trying the following expedients in succession.

Starting the Grignard reaction

1. Warm the reaction tube with your hands or a beaker of warm water. Then see if boiling continues when the tube is removed from the warmth.

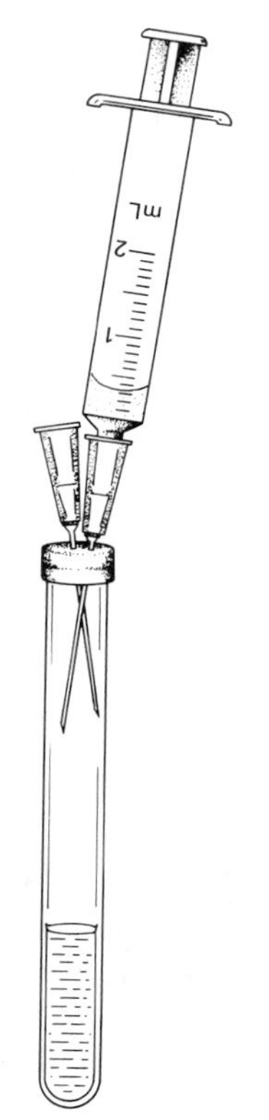

FIG. 31.1 Reaction tube with addition syringe and pressure relief needle.

2. Try further mashing of the metal with a stirring rod.
3. Add a tiny crystal of iodine as a starter (in this case the ethereal solution of the final reaction product should be washed with sodium bisulfite solution to remove the yellow color).
4. Add a few drops of a preformed Grignard reagent from some other source, such as some phenylmagnesium bromide or perhaps methyl magnesium iodide.
5. Start afresh, taking greater care with the dryness of apparatus and reagents, and sublime a crystal or two of iodine on the surface of the magnesium to generate Gattermann's "activated magnesium" before beginning the reaction again.

To the refluxing mixture add slowly and dropwise over a period of several minutes the remainder of the solution of bromobenzene in ether at such a rate that the reaction remains under control at all times. Should it get out of control, cool the tube in a beaker of water. You will have to warm the solution to start the reaction once more.

Once the reaction begins, spontaneous boiling of the diluted mixture may be slow or become slow. If so, mount the tube so the bottom is just immersed (2 mm) in a beaker of warm (40°C) water and reflux the solution gently until the magnesium has disintegrated and the solution has acquired a cloudy or brownish appearance. The reaction is complete when none or a very small quantity of the metal remains. Check to see that the volume of ether has not decreased. If it has, add more anhydrous ether. A damp piece of paper tissue wrapped about the upper part of the tube will help condense the ether. Since the solution of the Grignard reagent deteriorates on standing, the next step should be started at once. The phenylmagnesium bromide can be converted to triphenylmethanol or to benzoic acid.

2. Triphenylmethanol

In an oven-dried vial dissolve 2.0 mmoles (0.364 g) of benzophenone in 1.0 mL of anhydrous ether by capping the vial and mixing the contents thoroughly. With a dry syringe remove all of the solution from the vial and add it dropwise with thorough shaking after each drop to the solution of the Grignard reagent. Add the benzophenone at such a rate as to maintain the ether at a gentle reflux. The exothermic reaction can be kept under control by cooling the reaction tube if necessary in a beaker of 20°C water. Rinse the vial with a few drops of anhydrous ether after all of the first solution has been added and add this rinse to the reaction tube.

After all of the benzophenone has been added the mixture should be homogeneous. If not mix it thoroughly, using a stirring rod if necessary. The syringe can be removed, but leave the pressure relief needle in place. Reflux the reaction mixture very gently using a beaker of warm water as the heat source for 30 min to complete the reaction. Should ether evaporate replace it.

At the end of the reaction period, cool the tube in ice and add to it dropwise with stirring (use a glass rod or a spatula) 2 mL of 3 N hydrochloric

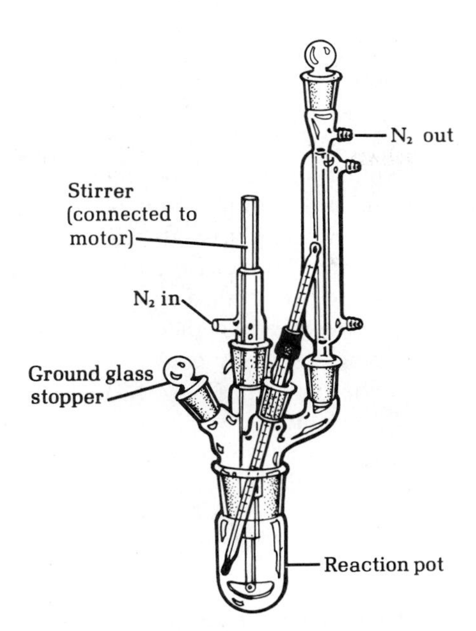

FIG. 31.2 Semimicro-scale, research-type apparatus for Grignard reaction, with provision for a motor-driven stirrer and an inlet and outlet for dry nitrogen.

acid. A creamy white precipitate of triphenylmethanol will separate. Add more ether (it need not be anhydrous) to the reaction tube and shake the contents to dissolve all of the triphenylmethanol. The result should be two perfectly clear layers. Remove a drop of the ether layer for TLC analysis. Any bubbling seen at the interface or in the lower layer is leftover magnesium reacting with the hydrochloric acid. Remove the aqueous layer and shake the ether layer with an equal volume of saturated aqueous sodium chloride solution in order to remove water and any remaining acid. Carefully remove all of the aqueous layer and then dry the ether layer by adding anhydrous sodium sulfate to the reaction tube until the drying agent no longer clumps together. Cork the tube and shake it from time to time over a 5- or 10-min period to complete the drying.

Using a Pasteur pipette, remove the ether from the drying agent and place it in another tared, dry reaction tube. Use more ether to wash off the drying agent and combine these ether extracts. Evaporate the ether in a hood by blowing a gentle stream of nitrogen or air onto the surface of the solution while warming the tube in a beaker of water or in the hand.

After all of the solvent has been removed, determine the weight of the crude product. Note the odor of biphenyl, the product of the side reaction that takes place between bromobenzene and phenylmagnesium bromide during the first reaction.

Trituration (grinding) of the crude product with petroleum ether will remove the biphenyl. Stir the crystals with 0.5 mL of petroleum ether in the ice bath, remove the solvent as thoroughly as possible, and recrystallize the residue from boiling 2-propanol (about 2 mL). Allow the solution to cool slowly to room temperature and then cool it thoroughly in ice. Triphenylmethanol crystallizes slowly, so allow the mixture to remain in the ice as long

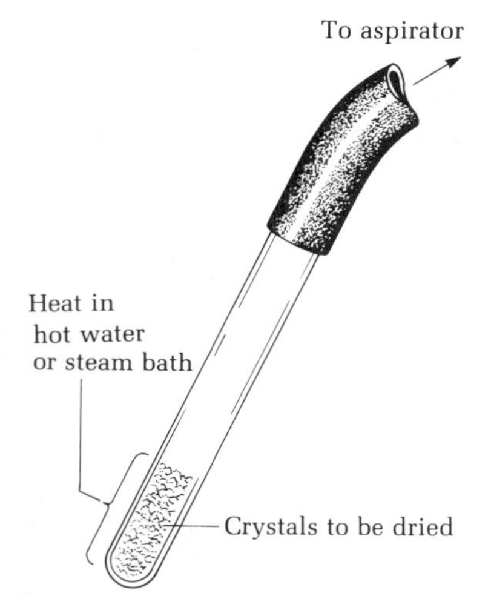

FIG. 31.3 Apparatus for drying crystals in reaction tube under vacuum.

as possible. Stir the ice-cold mixture well and collect the product by vacuum filtration on the Hirsch funnel.

An alternative method for purification of the triphenylmethanol utilizes a mixed solvent. Dissolve the crystals in the smallest possible quantity of warm ether and add to the solution 1.5 mL of ligroin. Add a boiling stick to the solution and boil off some of the ether until the solution becomes slightly cloudy, indicating it is saturated. Allow the solution to cool slowly to room temperature. Triphenylmethanol is deposited slowly as large thick prisms. Cool the solution in ice and after allowing time for complete crystallization to occur, remove the ether with a Pasteur pipette and wash the crystals once with a few drops of a cold 1 : 4 ether/ligroin mixture. Dry the crystals in the tube under vacuum (Fig. 31.3).

Determine the weight, melting point, and percent yield of the triphenylmethanol. Analyze the crude and recrystallized product by thin-layer chromatography on silica gel (see Chapter 9), developing the plate with dichloromethane. An infrared spectrum can be determined in chloroform solution or by preparing a mull or KBr disk (see Chapter 19).

3. Benzoic Acid

$$C_6H_5MgBr + CO_2 \longrightarrow C_6H_5C(=O)O^-MgBr^+ \xrightarrow{HCl} C_6H_5C(=O)OH + MgBr$$

Phenylmagnesium bromide

Carbon dioxide
MW 44.01
mp −78.5°C (sublimes)

Benzoic acid
MW 122.12
mp 123°C

Caution! Handle dry ice with a towel or gloves. Contact with the skin can cause frostbite since dry ice sublimes at −78.5°C.

Prepare 2 mmoles of phenylmagnesium bromide exactly as described in part 1 of this experiment. In a dry 30-mL beaker place a small piece of dry ice (solid carbon dioxide). Wipe off the surface of the piece of dry ice with a dry towel to remove frost and replace it in the beaker. Remove the pressure relief needle from the reaction tube and then insert a syringe through the septum, turn the tube upside down, and draw into the syringe as much of the reagent solution as possible. Squirt this solution onto the piece of dry ice and then rinse out the reaction tube with a milliliter of anhydrous ether and squirt this onto the dry ice. Allow excess dry ice to sublime and then hydrolyze the salt by the addition of 2 mL of 3 N hydrochloric acid.

Transfer the mixture from the beaker to a reaction tube and shake it thoroughly. Two homogeneous layers should result. Add 1–2 mL of acid or of ether if necessary. Remove the aqueous layer, and shake the ether layer with 1 mL of water, which is removed and discarded. Then extract the benzoic acid by adding to the ether layer 0.7 mL of 10% sodium hydroxide solution; shaking the mixture thoroughly; and withdrawing the aqueous layer, which is placed in a very small beaker or vial. The extraction is repeated with another 0.5-mL portion of base and finally 0.5 mL of water. Now that the extraction is complete, the ether, which can be discarded, contains primarily diphenyl, the by-product formed during the preparation of the phenylmagnesium bromide.

The combined aqueous extracts are heated briefly to about 50°C to drive off dissolved ether from the aqueous solution and then made acidic by the addition of concentrated hydrochloric acid (test with indicator paper). Cool the mixture thoroughly in an ice bath. Collect the benzoic acid on the Hirsch funnel and wash it with about 1 mL of ice water while on the funnel. A few crystals of this crude material are saved for a mp determination and the remainder of the product is recrystallized from boiling water.

The solubility of benzoic acid is 68 g/L at 95°C and 1.7 g/L at 0°C. Allow the hot solution to cool slowly to room temperature; then cool it in ice for several minutes before collecting the product by vacuum filtration on the Hirsch funnel. Turn the product out onto a piece of filter paper, squeeze out excess water, and allow it to dry thoroughly. Once dry, weigh it, calculate the percent yield, and determine the mp along with the mp of the crude material. The infrared spectrum may be determined as a solution in chloroform (1 g of benzoic acid dissolves in 4.5 mL of chloroform) or as a mull or KBr disk (see Chapter 19).

Questions

1. Why does rapid addition of bromobenzene to magnesium favor the formation of the undesirable by-product, biphenyl, over phenylmagnesium bromide?
2. Triphenylmethanol can also be prepared from the reaction of ethyl benzoate with phenylmagnesium bromide and by the reaction of diethylcarbonate ($C_2H_5O\overset{\overset{O}{\|}}{C}OC_2H_5$) with phenylmagnesium bromide. Write stepwise reaction mechanisms for these two reactions.
3. If the ethyl benzoate used to prepare triphenylmethanol is wet, what by-product is formed?
4. Exactly what weight of dry ice is needed to react with 2 mmoles of phenylmagnesium bromide?
5. In the synthesis of benzoic acid, benzene is often detected as an impurity. How does this come about?
6. The benzoic acid could have been extracted from the ether layer using sodium bicarbonate solution. Give equations showing how this might be done and how the benzoic acid would be regenerated. What practical reason makes this extraction less desirable than sodium hydroxide extraction?

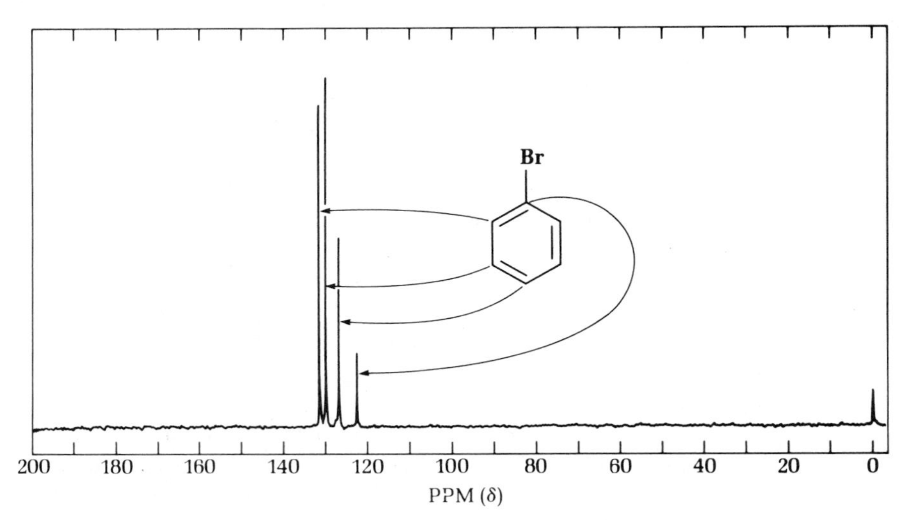

FIG. 31.4 ^{13}C nmr spectrum of bromobenzene.

Reactions of Triphenylmethyl Carbocation, Carbanion, and Radical

32

PRELAB EXERCISE: *Write all of the resonance structures of the triphenylmethyl carbocation.*

Triphenylmethanol, prepared in the experiment in Chapter 31, has played an interesting part in the history of organic chemistry. It was converted to the first stable carbocation and the first stable free radical. In this experiment triphenylmethanol is easily converted to the triphenylmethyl (trityl) carbocation, carbanion, and radical. Each of these is stabilized by ten contributing resonance forms and consequently is unusually stable. Because of their long conjugated systems, these forms absorb radiation in the visible region of the spectrum and thus can be detected visually.

Stable carbocation carbanion and radical

Experiments

1. The Triphenylmethyl Carbocation, the Trityl Carbocation

The reactions of triphenylmethanol are dominated by the ease with which it dissociates to form the relatively stable triphenylmethyl carbocation. When colorless triphenylmethanol is dissolved in concentrated sulfuric acid, an orange-yellow solution results that gives a fourfold depression of the melting point of sulfuric acid, meaning that four moles of ions are produced. If the triphenylmethanol simply were protonated only two moles of ions would result.

$$(C_6H_5)_3COH + 2\ H_2SO_4 \rightleftharpoons (C_6H_5)_3C^+ + H_3O^+ + 2\ HSO_4^-$$

$$(C_6H_5)_3COH + H_2SO_4 \not\rightleftharpoons (C_6H_5)_3COH_2^+ + HSO_4^-$$

The central carbon atom in the carbocation is sp^2 hybridized and thus the three carbons attached to it are coplanar and disposed at angles of 120°:

$$\begin{array}{ccc} C & & C \\ & \overset{+}{C} & \\ & | & \\ & C & \end{array}$$

However the three phenyl groups, because of steric hindrance, cannot lie in one plane. Therefore the carbocation is propeller-shaped:

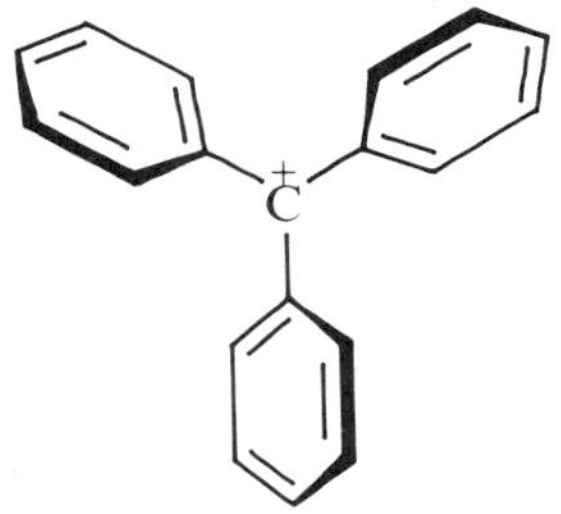

Triphenylmethanol is a tertiary alcohol and undergoes, as expected, S_N1 reactions. The intermediate cation however, is stable enough to be seen in sulfuric acid solution as a red-brown to yellow solution. Upon dissolution in concentrated sulfuric acid, the hydroxyl is protonated and loses water, which is itself protonated to drive the reaction toward the carbocation. The bisulfate ion is a very weak nucleophile and does not compete with methanol in the formation of the product, trityl methyl ether.

Trityl Methyl Ether

$$(C_6H_5)_3C{-}OH \underset{}{\overset{H^+}{\rightleftharpoons}} (C_6H_5)_3C{-}\overset{+}{O}H_2 \rightleftharpoons (C_6H_5)_3C^+ + H_2O \overset{H^+}{\rightleftharpoons} H_3O^+$$

Yellow

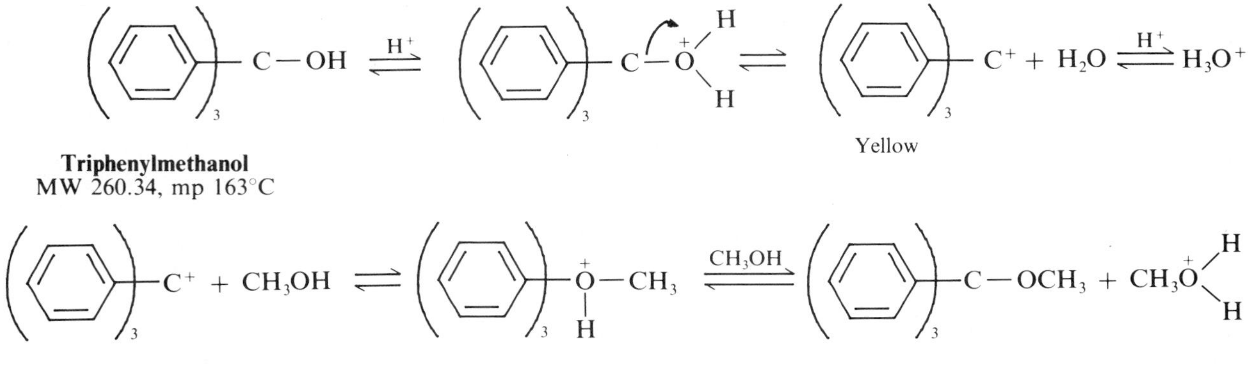

Triphenylmethanol
MW 260.34, mp 163°C

$$(C_6H_5)_3C^+ + CH_3OH \rightleftharpoons (C_6H_5)_3C{-}\overset{+}{O}(H){-}CH_3 \overset{CH_3OH}{\rightleftharpoons} (C_6H_5)_3C{-}OCH_3 + CH_3\overset{+}{O}H_2$$

Triphenylmethyl methyl ether
Trityl methyl ether
MW 274.37, mp 96°C

In a 10 × 100 mm reaction tube, place 50 mg of triphenylmethanol, grind the crystals to a fine powder with a glass stirring rod, and add 0.5 mL of concentrated sulfuric acid. Continue to stir the mixture to dissolve all of the alcohol. Using a Pasteur pipette, transfer the sulfuric acid solution to 3 mL of ice-cold methanol in another reaction tube. Use some of the cold methanol to rinse out the first tube. Pour the methanol solution into 12 mL of cold 10% sodium hydroxide solution in a 25-mL Erlenmeyer flask. Induce crystallization if necessary by cooling the solution, adding a seed crystal, or scratching the solution with a glass stirring rod at the liquid-air interface.

Warm the solution to room temperature to allow inorganic salts to dissolve, and then collect the product by filtration on the Hirsch funnel. Wash the crystals well with water and squeeze them between sheets of filter paper to aid drying. Determine the weight of the crude material, calculate the percent yield, and save a sample for melting point determination. Recrystallize the product from boiling methanol and determine the melting point of the purified trityl methyl ether.

Triphenylmethyl Bromide, Trityl Bromide

In this experiment the triphenylmethanol is dissolved in a good ionizing solvent, acetic acid, and allowed to react with a strong acid and nucleophile, hydrobromic acid. The intermediate carbocation reacts immediately with bromide ion.

Reactions that proceed through the carbocation

$$(C_6H_5)_3C{-}OH + HBr \rightleftharpoons (C_6H_5)_3C{-}\overset{+}{O}H_2\ Br^- \rightleftharpoons (C_6H_5)_3C^+ + H_2O \xrightarrow{Br^-} (C_6H_5)_3C{-}Br$$

Triphenylmethyl bromide
Trityl bromide
MW 323.24, mp 154°C

Dissolve 100 mg of triphenylmethanol in 2 mL of warm acetic acid, add 0.2 mL of 47% hydrobromic acid, and heat the mixture for 5 min on the steam bath or in a beaker of boiling water. Cool the mixture in ice, collect the product on the Hirsch funnel, wash the product with ligroin, allow it to dry, determine the weight, and calculate the percent yield. Recrystallize the product from ligroin and compare the melting points of the recrystallized and crude material. The compound crystallizes slowly; allow adequate time for crystals to form.

Triphenylmethyl Iodide, Trityl Iodide?

In a reaction very similar to the preparation of the bromide, the iodide might be prepared. Add sodium bisulfite to react with any iodine formed.

$$(C_6H_5)_3C{-}OH \underset{H_2O,\ H^+}{\overset{HI,\ -H_2O}{\rightleftharpoons}} (C_6H_5)_3C{-}I$$

Triphenylmethyl iodide
Trityl iodide
MW 370.22, mp 183°C

Dissolve 100 mg of triphenylmethanol in 2 mL of warm acetic acid, add 0.2 mL of 47% hydriodic acid, heat the mixture for 1 hr on the steam bath, cool it, and add it to a solution of 0.1 g of sodium bisulfite dissolved in 2 mL of water in a 10-mL Erlenmeyer flask. Collect the product on the Hirsch funnel, wash it with water, press out as much water as possible, and recrystallize the crude, moist product from methanol (about 3 or 4 mL). Determine the weight of the dry product, calculate the percent yield, and determine the melting point. Run the Beilstein test. What has been produced and why?

Triphenylmethyl Fluoroborate and the Tropilium Ion

Tropilium—an aromatic ion

Reaction of triphenylmethanol with fluoroboric acid in the presence of acetic anhydride generates the stable salt, trityl fluoroborate. The fluoroboric acid protonates the triphenylmethanol, which loses the elements of water in an equilibrium reaction. The water reacts with the acetic anhydride to form acetic acid and thus drives the reaction to completion.

The salt so formed is the fluoroborate of the triphenyl carbocation; it is a powerful base. It can be isolated, but in this case will be used *in situ* to react with the hydrocarbon, cycloheptatriene.

This hydrocarbon is more basic than most hydrocarbons and will lose a proton to the triphenylmethyl carbocation to give triphenylmethane and the cycloheptatrienide carbocation, the tropilium ion. To demonstrate that hydride ion transfer has taken place, isolate triphenylmethane.

The tropilium ion has a planar structure, each carbon bears a single proton, and the ion contains 6π electrons—it is aromatic, a characteristic that can be confirmed by nuclear magnetic resonance spectroscopy.

The Hückel rule: $4n + 2\pi$ electrons

The fluoroborate group can be displaced by the iodide ion to prepare tropilium iodide. You can see that the iodide ion is ionic by watching the reaction of the aqueous solution of this ion with silver ion.

$(C_6H_5)_3C{-}OH \xrightleftharpoons{HBF_4} (C_6H_5)_3C{-}\overset{+}{O}H_2 \rightleftharpoons (C_6H_5)_3C^+ + H_2O$

Triphenylmethanol
MW 260.34
mp 163°C

$\xrightarrow{BF_4^-,\ CH_3COOCOCH_3}$ $(C_6H_5)_3C^+BF_4^- + 2CH_3COOH$

Triphenylmethyl fluoroborate

$(C_6H_5)_3C^+BF_4^-$ + cycloheptatriene → $(C_6H_5)_3CH$ + tropilium BF_4^- $\xrightarrow[MW\ 149.89]{NaI}$ tropilium I^-

Cycloheptatriene
MW 92.14
bp 117°C

Triphenylmethane
MW 244.34
mp 94°C

Tropilium fluoroborate
MW 177.94

Tropilium iodide
MW 218.04

In a 10 × 100 mm reaction tube place 1.75 mL of acetic anhydride, cool the tube and add 88 mg of fluoroboric acid. Add 195 mg of triphenylmethanol with thorough stirring. Warm the mixture to give a homogeneous dark solution of the triphenylmethyl fluoroborate, then add 78 mg of cycloheptatriene. The color of the trityl cation should fade during this reaction and the tropilium fluoroborate begin to precipitate. Add 2 mL of anhydrous ether to the reaction tube, stir the contents well while cooling on ice, and collect the product by filtration on the Hirsch funnel. Wash the product with 2 mL of dry ether and then dry the product between sheets of filter paper.

Caution! *Handle fluoroboric acid and acetic anhydride with great care. Both of these reagents are toxic and corrosive. Avoid breathing the vapors or any contact with the skin. In case of skin contact, rinse the affected part under running water for at least 10 min. Carry out this experiment in the hood.*

To the filtrate add 10% sodium hydroxide solution and shake the flask to allow all of the acetic anhydride and fluoroboric acid to react with the base. Test the aqueous layer with indicator paper to ascertain that neutralization is complete and then transfer the mixture to a reaction tube and draw off the aqueous layer. Wash the ether once with 2 mL of water, once with 2 mL of saturated sodium chloride solution, and dry the ether over anhydrous sodium sulfate, adding the drying agent until it no longer clumps together.

Transfer the ether to a tared reaction tube and evaporate the solvent to leave crude triphenylmethane. Remove a sample for melting point determination and recrystallize the residue from an appropriate solvent, determined by experimentation. Prove to yourself that the compound isolated is indeed triphenylmethane. Obtain an infrared spectrum and a nuclear magnetic resonance spectrum.

Structure proof: nmr

To determine the nmr spectrum of the tropilium fluoroborate dissolve about 50 mg of the product in 0.3 mL of deuterated dimethyl sulfoxide that contains 1% tetramethylsilane as a reference. Compare the nmr spectrum obtained with that of the starting material, cycloheptatriene. The spectrum of the latter can be obtained in deuterochloroform, again using tetramethylsilane as the reference compound.

The tropilium fluoroborate can be converted to tropilium iodide by dissolving the borate in the minimum amount of hot, but not boiling, water (a few drops) and adding to this solution 0.25 mL of a saturated solution of sodium iodide. Cool the mixture in ice, remove the solvent with a pipette, and wash the crystals with 0.5 mL of ice-cold methanol. Scrape most of the crystals onto a piece of filter paper and allow them to dry before determining the weight. To the crystalline residue in the reaction tube add a few drops of water, warm the tube if necessary to dissolve the tropilium iodide, and then add a drop of 2% aqueous silver nitrate solution and note the result.

2. The Trityl Carbanion

The triphenylmethyl carbanion, the trityl anion, can be generated by the reaction of triphenylmethane with the very powerful base, *n*-butyllithium. The reaction generates the blood-red lithium triphenylmethide and butane. The triphenylmethyl anion reacts much as a Grignard reagent does. In the present experiment it reacts with carbon dioxide to give triphenylacetic acid after acidification. Avoid an excess of *n*-butyllithium; on reaction with carbon dioxide, it gives the vile-smelling pentanoic acid.

$$(C_6H_5)_3C{-}H + Li^+ \, \bar{C}H_2CH_2CH_2CH_3 \longrightarrow (C_6H_5)_3C{:}^- \, Li^+ + CH_3CH_2CH_2CH_3$$

Triphenylmethane
MW 244.34
mp 94°C

***n*-Butyllithium**
MW 64.06

$$\downarrow CO_2, H^+$$

$$(C_6H_5)_3C{-}COOH$$

Triphenylacetic acid
MW 288.35, mp 270–273°C

Procedure

To a reaction tube (Fig. 32.1) that has been dried for at least 30 min in a 110°C oven and capped with a rubber septum, add 100 mg of triphenylmethane and 3 mL of anhydrous ether. Add an empty syringe needle to the septum; flush out the air in the tube by passing a slow current of nitrogen into the tube for about 30 s. Remove the nitrogen inlet needle, and with thorough mixing of the solution add to the reaction mixture 0.4 mL of a 1.0 M solution of *n*-butyllithium in hexane (a commercial product). Use great care in handling *n*-butyllithium. It reacts avidly with air and violently with water.

Reaction with dry ice

In a 30-mL beaker place one or two small pieces of dry ice (solid carbon dioxide) that have been wiped free of any adhering frost. Handle the dry ice with a towel or gloves. Contact with the skin can cause frostbite because the sublimation temperature of dry ice is −78.5°C. Calculate how much carbon dioxide is needed to react with the anion prepared from 100 mg of triphenylmethane; use at least a ten-fold excess. Using a Pasteur pipette transfer the solution of the anion to the dry ice; reaction is immediate. Allow the unreacted dry ice to sublime; then add 3 mL of 5% aqueous sodium hydroxide solution, dissolve as much solid as possible, and transfer the liquid to a reaction tube. Shake the mixture, draw off the ether layer, and wash the aqueous layer with two 2.5-mL portions of ether, which are also discarded. Using 10% hydrochloric acid, acidify the aqueous layer to pH < 4. Collect the precipitate, wash it with water, and allow it to dry in the air. Take the melting point. When pure, triphenylacetic acid melts at 267°C. The purity and identity of the product can be assessed using thin layer chromatography and infrared spectroscopy (using a potassium bromide disk).

3. Triphenylmethyl, a Stable Free Radical

In the early part of the nineteenth century many attempts were made to prepare methyl, ethyl, and similar radicals in a free state, as sodium had been

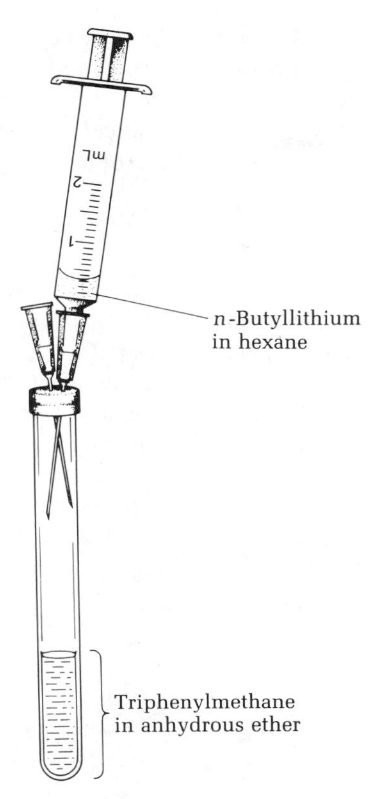

FIG. 32.1 Apparatus for the synthesis of the trityl carbanion.

prepared from sodium chloride. Many well-known chemists tried: Gay-Lussac's CN turned out to be cyanogen, $(CN)_2$; Bunsen's cacodyl from $(CH_3)_2AsCl$, proved to be $(CH_3)_2As-As(CH_3)_2$; and the Kolbe electrolysis gave $CH_3CH_2CH_2CH_3$ instead of CH_3CH_2. Moses Gomberg at the University of Michigan prepared the first free radical in 1900. He had prepared triphenylmethane and chlorinated it to give triphenylmethyl chloride, which he hoped to couple to form hexaphenylethane:

Gomberg: The first free radical

$$2\,(C_6H_5)_3C-Cl + 2Ag \not\longrightarrow (C_6H_5)_3C-C(C_6H_5)_3 + 2\,AgCl$$

Triphenylmethyl chloride **Hexaphenylethane**

The solid that he obtained instead was a high-melting, sparingly soluble, white solid which turned out on analysis to be not a hydrocarbon but instead an oxygen-containing compound $(C_{38}H_{30}O_2)$. Repeating the experiment in the absence of air, he obtained a yellow solution that, on evaporation in the absence of air, deposited crystals of a colorless hydrocarbon that was remarkably reactive. It readily reacted with oxygen, bromine, chlorine, and iodine. On dissolution the white hydrocarbon gave the yellow solution in the absence of air; the hydrocarbon was deposited when he evaporated the solution once more. Gomberg interpreted these results as follows: "The experimental evidence . . . forces me to the conclusion that we have to deal here with a free radical, triphenylmethyl, $(C_6H_5)_3C\cdot$. The action of zinc results, as it seems to me, in a mere abstraction of the halogen:

$$2(C_6H_5)_3C-Cl + Zn \longrightarrow 2(C_6H_5)_3C\cdot + ZnCl_2$$

Now as a result of the removal of the halogen atom from triphenylchloromethane, the fourth valence of the methane is bound either to take up the complicated group $(C_6H_5)_3C\cdot$ or remain as such, with carbon as trivalent. Apparently the latter is what happens."

For a long time after Gomberg first carried out this reaction in 1900 it was assumed the radical dimerized to form hexaphenylethane:

$$2\,(C_6H_5)_3C\cdot \longrightarrow (C_6H_5)_3C-C(C_6H_5)_3$$

But in 1968 nmr and uv evidence showed the radical is in equilibrium with a different substance:

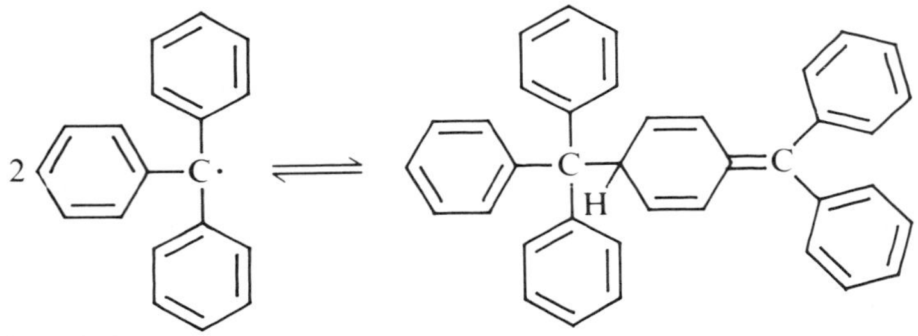

Hexaphenylethane cannot be synthesized because of steric hindrance.

In this experiment the trityl radical is prepared in much the same fashion as Gomberg used, by the reaction of trityl bromide with zinc in the absence of oxygen. The yellow solution will then deliberately be exposed to air to give the peroxide.

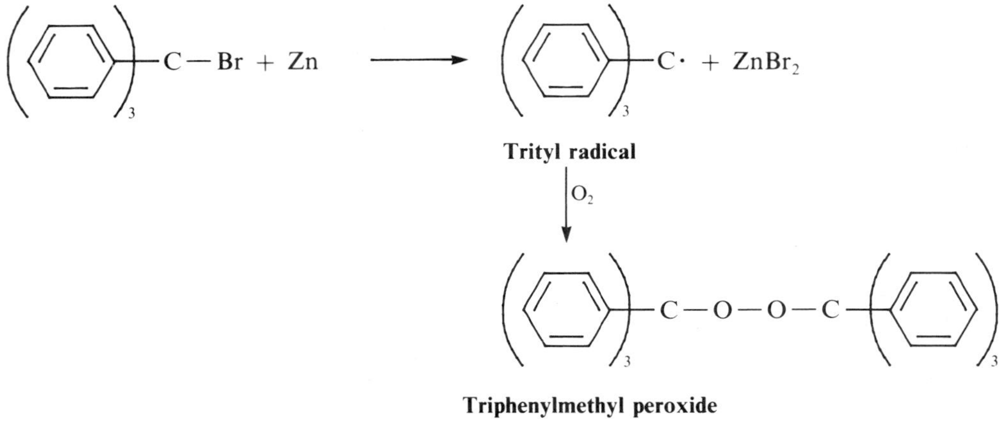

$$(C_6H_5)_3C-Br + Zn \longrightarrow (C_6H_5)_3C\cdot + ZnBr_2$$

Trityl radical

$\downarrow O_2$

$$(C_6H_5)_3C-O-O-C(C_6H_5)_3$$

Triphenylmethyl peroxide

Procedure

The trityl radical

In a reaction tube dissolve 100 mg of trityl bromide or chloride in 0.5 mL of toluene. Material prepared in part 1.2 above may be used; it need not be recrystallized. Cap the tube with a septum, insert an empty needle, and flush the tube with nitrogen while shaking the contents. Add 0.2 g of fresh zinc dust as quickly as possible, then flush the tube of oxygen once more. Shake the tube vigorously for about 10 min, and note the appearance of the reaction mixture. Using a Pasteur pipette transfer the solution, but not the zinc, to a 30-mL beaker. Roll the solution around the inside of the beaker to give it maximum exposure to the air and after a few minutes collect the peroxide on the Hirsch funnel. Wash the solid with a little cold toluene, allow it to dry in the air, and determine the melting point. It is reported to melt at 186°C.

4. A Puzzle

Carry out both of the following reactions. Compare the products formed. Run necessary solubility and simple qualitative tests. Interpret your results and propose structures for the compounds produced in reactions 1 and 2.

Compound 1

Malonic acid: $HOOCCH_2COOH$

In a reaction tube place 100 mg of triphenylmethanol and 250 mg of malonic acid. Grind the two solids together with a glass stirring rod and then heat the reaction tube at 160°C for 10 min. Use an aspirator tube mounted at the mouth of the tube to carry away undesirable fumes. Allow the tube to cool and then dissolve the contents in about 0.25 mL of toluene. Dilute the solution with 1 mL of 60–80°C ligroin, and after cooling the tube in ice isolate the crystals, wash them with a little cold ligroin, and once they are dry determine their weight and melting point.

Compound 2

Chloroacetic acid $ClCH_2COOH$

In a reaction tube dissolve 100 mg of triphenylmethanol in 2 mL of acetic acid and then add to this solution 0.4 mL of an acetic acid solution containing 5% of chloroacetic acid and 1% of sulfuric acid. Heat the mixture for 5 min, add about 1 mL of water to produce a hot saturated solution, and let the tube cool slowly to room temperature. Collect the product by filtration, and wash the crystals with 1 : 1 methanol–water. Once they are dry, determine the weight and melting point and then perform tests on Compounds 1 and 2 to deduce their identities.

Questions

1. Give the mechanism for the free radical chlorination of triphenylmethane.
2. Is the propeller-shaped triphenylmethyl carbocation a chiral species?

3. Without carrying out the experiments, speculate on the structures of Compounds 1 and 2 made in Experiment 4.
4. Is the production of triphenylmethyl peroxide a chain reaction?
5. What product would you expect from the reaction of carbon tetrachloride, benzene, and aluminum chloride?

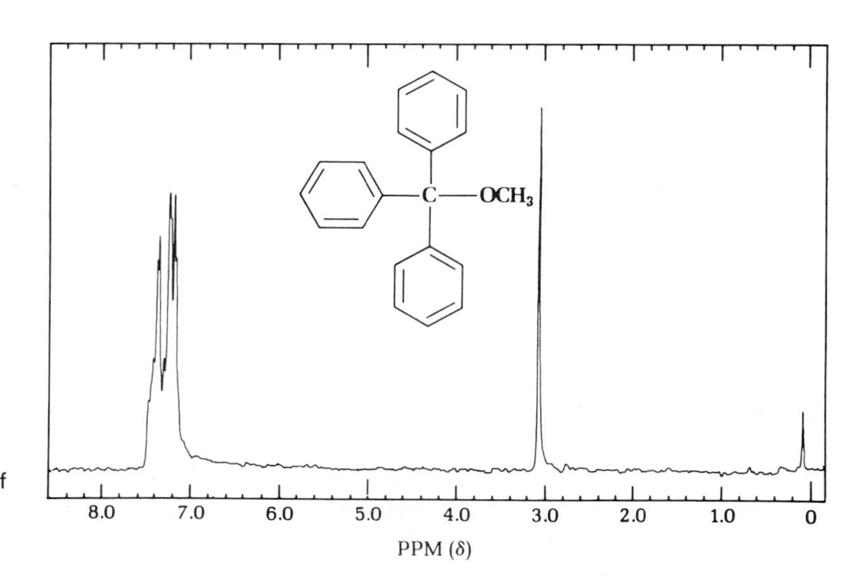

FIG. 32.2 ^{1}H nmr spectrum of triphenylmethyl methyl ether (trityl methyl ether).

Dibenzalacetone by the Aldol Condensation

33

PRELAB EXERCISE: *Calculate the volumes of benzaldehyde and acetone needed for this reaction, taking into account the densities of the liquids and the number of moles of each required.*

$$2\,C_6H_5CHO + CH_3\overset{O}{\overset{\|}{C}}CH_3 \xrightarrow[\text{MW 40.01}]{\text{NaOH}} C_6H_5\text{–}CH{=}CH\overset{O}{\overset{\|}{C}}CH{=}CH\text{–}C_6H_5$$

Benzaldehyde
MW 106.13, bp 178°C
den 1.04

Acetone
(2-Propanone)
MW 58.08, bp 56°C
den 0.790

Dibenzalacetone
(1,5-Diphenyl-1,4-pentadien-3-one)
MW 234.30, mp 110–111°C

The reaction of an aldehyde with a ketone employing sodium hydroxide as the base is an example of a mixed aldol condensation reaction, the Claisen-Schmidt reaction. Dibenzalacetone is readily prepared by condensation of acetone with two equivalents of benzaldehyde. The aldehyde carbonyl is more reactive than that of the ketone and therefore reacts rapidly with the anion of the ketone to give a β-hydroxyketone, which easily undergoes base-catalyzed dehydration. Depending on the relative quantities of the reactants, the reaction can give either mono- or dibenzalacetone.

In the present experiment sufficient ethanol is present as solvent to readily dissolve the starting material, benzaldehyde, and also the intermediate, benzalacetone. The benzalacetone, once formed, can then easily react with another mole of benzaldehyde to give the product, dibenzalacetone. The detailed mechanism for the formation of benzalacetone is:

$$CH_3\overset{O}{\overset{\|}{C}}CH_3 + OH^- \rightleftharpoons \left[CH_3\overset{O}{\overset{\|}{C}}\text{–}CH_2^- \leftrightarrow CH_3\overset{O^-}{\overset{|}{C}}{=}CH_2 \right] + H_2O$$

$$CH_3\overset{O}{\overset{\|}{C}}CH_2^- + \overset{H}{C}(=O)\text{–}C_6H_5 \rightleftharpoons CH_3\overset{O}{\overset{\|}{C}}CH_2\overset{H}{\underset{O^-}{C}}\text{–}C_6H_5 \xrightleftharpoons{H_2O} CH_3\overset{O}{\overset{\|}{C}}CH_2\overset{H}{\underset{OH}{C}}\text{–}C_6H_5 + OH^-$$

$$\mathrm{CH_3\overset{O}{\overset{\|}{C}}-\underset{H}{\overset{H}{C}}-\underset{OH}{\overset{H}{C}}-C_6H_5} \xrightarrow{^{-}OH} \mathrm{CH_3\overset{O}{\overset{\|}{C}}-\underset{H}{C}=\overset{H}{C}-C_6H_5} + H_2O + OH^-$$

Benzalacetone
4-Phenyl-3-butene-2-one
mp 42°C

Experiment

If sodium hydroxide gets on the skin, wash until the skin no longer has a "soapy" feeling

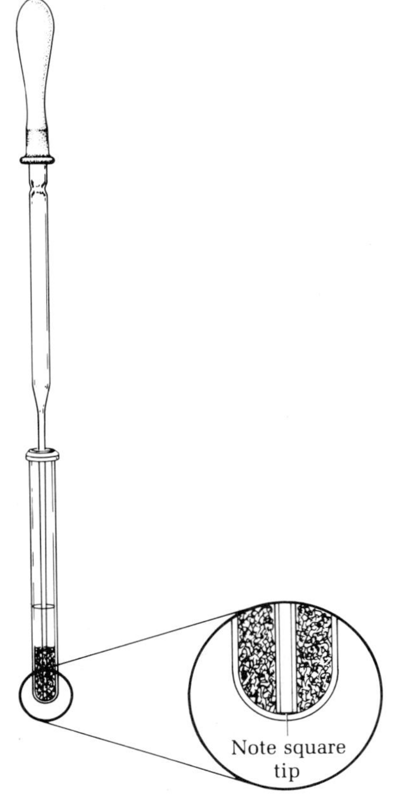

FIG. 33.1 Pasteur pipette filtration technique. Solvent is removed between pipette tip and bottom of tube, leaving crystals in reaction tube.

Into a 10 × 100 mm reaction tube place 2 mL of 10% sodium hydroxide solution. To this solution add 1.6 mL of 95% ethanol and then, from a 1-mL syringe, 0.204 mL of benzaldehyde. Rinse the syringe with a small quantity of acetone and then add 0.073 mL of acetone to the reaction mixture. Alternatively the benzaldehyde (0.212 g) and the acetone (0.058 g) can be weighed out. Cap the tube immediately and shake the mixture vigorously. The benzaldehyde, initially insoluble, goes into solution and a water-clear, pale yellow solution results. After a minute or so it suddenly becomes cloudy and a yellow precipitate of the product forms. Continue to shake the tube from time to time for the next 30 min. If the product fails to crystallize, open the tube and scratch the inside of the tube with a glass rod. Remove the liquid from the tube using a Pasteur pipette by squeezing the bulb of the pipette, pressing the tip against the bottom of the tube, and bringing the liquid into the pipette, leaving the crystals in the tube (Fig. 33.1). Add 3 mL of water, cap, and shake the tube vigorously. Remove the wash liquid as before and wash the crystals twice more with water. After the final washing rap the tube sharply on the desk to drive the crystals to the bottom of the tube; remove remaining water with the pipette. It is important to remove as much of the water as possible. Some might be absorbed into a roll of filter paper forced into the crystals. Weigh the tube to determine the weight of crude wet product. Add 2 mL of 95% ethanol to the damp crystals and warm the tube on a hot plate. Insert a wooden boiling stick to promote even boiling and heat the mixture until the crystals dissolve.

Remove the tube from the hot sand bath and place it in an insulated container to cool slowly to room temperature. Should the product separate as an oil, try to obtain a seed crystal, heat the solution to dissolve the oil, and add the seed crystal as the solution cools. If it continues to oil out add a bit more ethanol. Collect the product by removing the solvent with a pipette after cooling the tube for several minutes in ice. Wash the crystals once with about 0.5 mL of ice-cold ethanol while the tube is in ice. Dry the product under vacuum by attaching the tube to an aspirator for a few minutes. Determine the weight of the dibenzalacetone, its mp, and calculate the percent yield. In a typical experiment the yield will be 0.10 g, mp 110.5–112°C.

Questions

1. Why is it important to maintain equivalent proportions of reagents in this reaction?
2. What side products do you expect in this reaction? How are they removed?
3. What do the melting points of the crude and recrystallized products tell you about purity?
4. Exactly how would you prepare benzalacetone?

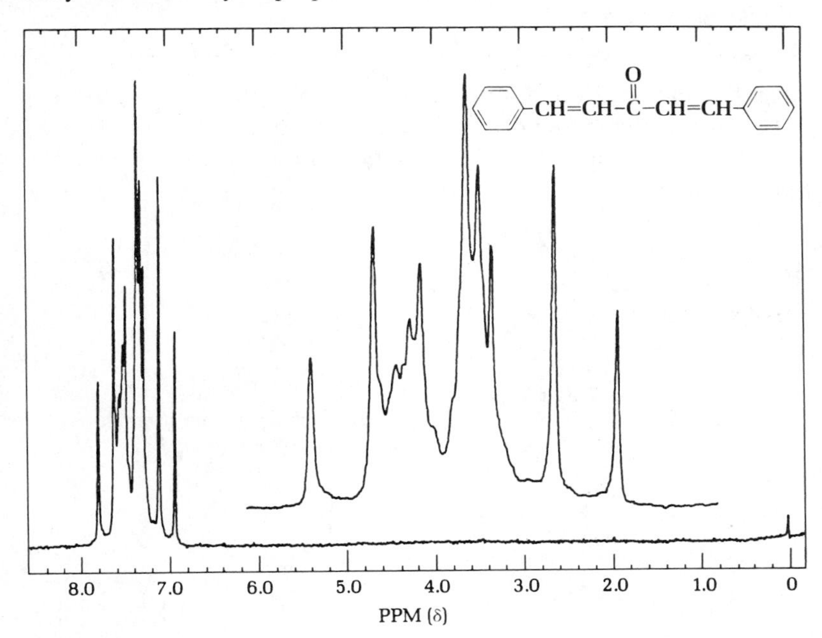

FIG. 33.2 ^1H nmr spectrum of dibenzalacetone. (An expanded spectrum appears above the normal spectrum.)

34 Wittig-Horner Reaction

PRELAB EXERCISE: *Account for the fact that the Wittig-Horner reaction of cinnamaldehyde gives almost exclusively the E,E-butadiene with very little contaminating E,Z-product.*

The Wittig reaction affords an invaluable method for the conversion of a carbonyl compound to an olefin, for example the conversion of benzaldehyde and *n*-butyl bromide to a mixture of *cis*- and *trans*-1-phenyl-2-*n*-propylethylene.

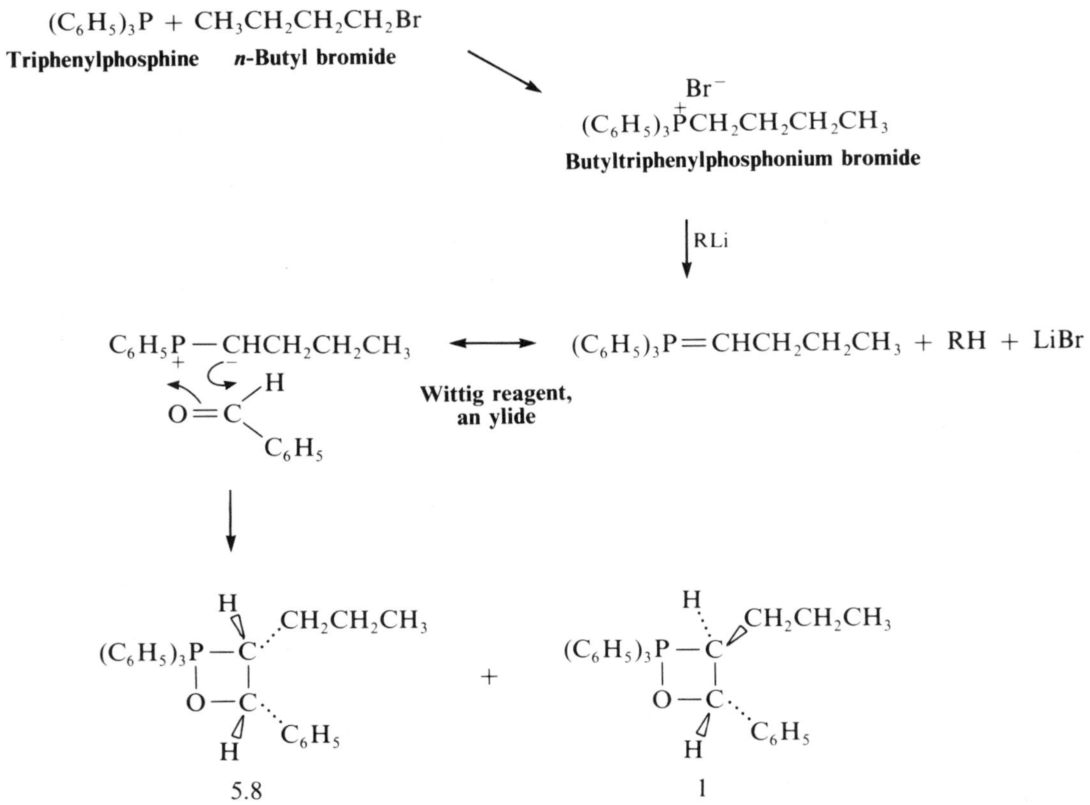

$$(C_6H_5)_3P + CH_3CH_2CH_2CH_2Br \longrightarrow (C_6H_5)_3\overset{+}{P}CH_2CH_2CH_2CH_3 \; Br^-$$

Triphenylphosphine **n-Butyl bromide** → **Butyltriphenylphosphonium bromide**

$$\xrightarrow{RLi}$$

$$C_6H_5\overset{+}{P}-\overset{-}{C}HCH_2CH_2CH_3 \longleftrightarrow (C_6H_5)_3P{=}CHCH_2CH_2CH_3 + RH + LiBr$$

Wittig reagent, an ylide

$$O{=}C(H)(C_6H_5)$$

$$\downarrow$$

$$(C_6H_5)_3P-C(H)(CH_2CH_2CH_3) \;|\; O-C(H)(C_6H_5) \quad + \quad (C_6H_5)_3P-C(H)(CH_2CH_2CH_3) \;|\; O-C(H)(C_6H_5)$$

5.8 : 1

3-Propyl-2,2,2,4-tetraphenyl-1,2-oxaphosphetane

$$(C_6H_5)_3P{-}C(H)(CH_2CH_2CH_3){-}O{-}C(H)(C_6H_5) \longrightarrow C_6H_5CH{=}CHCH_2CH_2CH_3\ (cis) + (C_6H_5)_3P{=}O + C_6H_5CH{=}CHCH_2CH_2CH_3\ (trans)$$

***cis*-1-Phenyl-2-*n*-propylethylene** **Triphenyl phosphine oxide** ***trans*-1-phenyl-2-*n*-propylethylene**

Because the active reagent, an ylide, is unstable, it is generated in the presence of the carbonyl compound by dehydrohalogenation of an alkyltriphenylphosphonium bromide with phenyllithium in dry ether in a nitrogen atmosphere. The existence of the four-membered ring intermediate was proved by nmr in 1984.[1]

When the halogen compound employed in the first step has an activated halogen atom ($RCH{=}CHCH_2X$, $C_6H_5CH_2X$, $XCH_2CO_2H_5$) a simpler procedure known as the Horner phosphonate modification of the Wittig reaction is applicable. When benzyl chloride is heated with triethyl phosphite, ethyl chloride is eliminated from the initially formed phosphonium chloride

$$C_6H_5{-}CH_2Cl + (CH_3CH_2O)_3P \longrightarrow C_6H_5{-}CH_2P^+(OCH_2CH_3)_3\ Cl^- \longrightarrow C_6H_5{-}CH_2P^+(O^-)(OCH_2CH_3)_2 + ClCH_2CH_3$$

Benzyl chloride bp 179°C, MW 126.59 den 1.10

Triethyl phosphite bp 156°C, MW 166.16 den 0.94

Benzyltriethylphosphonium chloride

Diethyl benzylphosphonate bp 156°C/9 torr

1. B. E. Maryanoff, A. B. Reitz, and M. S. Mutter, *J. Am. Chem. Soc.*, **106**, 1873 (1984).

Diethyl benzylphosphonate + $CH_3O^-Na^+$ (MW 52.02) ⟶ Ylide ($C_6H_5\bar{C}H-P(=O)(OCH_2CH_3)(OCH_2CH_3)$, Na^+) + CH_3OH

Ylide + Cinnamaldehyde ⟶ *E,E*-1,4-Diphenyl-1,3-butadiene + $Na^+\bar{O}-P(=O)(OCH_2CH_3)-OCH_2CH_3$

Cinnamaldehyde
MW 132.16, den 1.11
bp 248°C

***E,E*-1,4-Diphenyl-1,3-butadiene**
mp 153°C, MW 206.27

with the production of diethyl benzylphosphonate. This phosphonate is stable, but in the presence of a strong base such as the sodium methoxide used here it condenses with a carbonyl component in the same way that a Wittig ylide condenses. Thus, it reacts with benzaldehyde to give *E*-stilbene and with cinnamaldehyde to give *E,E*-1,4-diphenyl-1,3-butadiene.

Simplified Wittig reaction

***Caution:** Take care to keep organophosphorus compounds off the skin*

Experiment

The success of this reaction is strongly dependent on the purity of the starting materials. Benzyl chloride and triethylphosphite are usually pure enough as received from the supplier. Cinnamaldehyde from a new, previously unopened bottle should be satisfactory, but since it is oxidized in air extremely rapidly it should be distilled, preferably under nitrogen or at reduced pressure, if there is any doubt about its quality. Sodium methoxide, direct from reputable suppliers, has been found on occasion to be completely inactive. No easy method exists for determining its activity; therefore it is best prepared following the procedure of *Organic Syntheses*, Col. Vol. IV, 651 (1963).[2]

To a 10 mm × 100 mm reaction tube add 316 mg (0.0025 mole) of benzyl chloride (α-chlorotoluene) and 415 mg (0.44 mL, 0.0025 mole) of triethylphosphite and a boiling chip (Fig. 34.1). Because the triethyl phosphite has an offensive odor obtain this material from the dispenser that has been calibrated to deliver the correct quantity. Place the reaction tube to

Handle benzyl chloride in the hood. Lachrymator and cancer-suspect agent.

2. For enough sodium methoxide for 50 reactions, add 3.5 g of sodium spheres to 50 mL of anhydrous methanol in a 100-mL flask equipped with a condenser. Add the sodium a few pieces at a time through the top of the condenser at such a rate as to keep the reaction under control. It is safest to wait for complete reaction of one portion of sodium before adding the next. After all of the sodium has reacted, remove the methanol on a rotary evaporator, first over a steam bath and then over a bath heated to 150°C. The sodium methoxide will be a free-flowing white powder that should keep for several weeks in a desiccator. Yield 8.2 g.

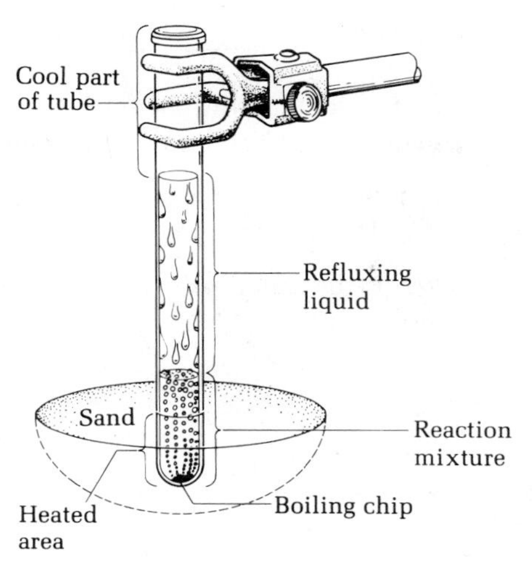

FIG. 34.1 Reaction tube for preparation of phosphonate.

a depth of about 1 cm in a sand bath which has been preheated to 210–220°C. Reflux the reaction mixture for 1 h. The vapors condense on the cool upper portion of the reaction tube so a condenser is not necessary. At an internal temperature of about 140°C (which you need not monitor), ethyl chloride is evolved and by the end of the reflux period the temperature of the reaction mixture will be 200–220°C. At the end of the reaction period remove the tube from the sand bath, cool it to room temperature, and then add the contents to 160 mg (0.003 mole) of sodium methoxide in a 10-mL Erlenmeyer flask using 1 mL of dry dimethylformamide to complete the transfer. Cool the mixture in ice, mix the contents of the flask thoroughly, then add 330 mg (0.0025 mole) of freshly distilled cinnamaldehyde in 1 mL of dry dimethylformamide dropwise with thorough mixing in the ice bath. Allow the mixture to come to room temperature. Note the changes that take place over the next few minutes.

$O{=}C(H)-N(CH_3)_2$

Dimethylformamide
MW 73.10, bp 153°C

A highly polar solvent capable of dissolving ionic compounds such as sodium methoxide yet miscible with water—avoid skin contact

After a *minimum* of 10 min add 2 mL of methanol to the Erlenmeyer with thorough stirring. Then almost fill the flask with water and stir the mixture until it is homogeneous in color. The hydrocarbon precipitates from the reaction mixture upon the addition of the methanol and water. Collect the product by vacuum filtration on the Hirsch funnel, wash the crystals first with water to remove the red color, and then with ice-cold methanol to remove the yellow color. The hydrocarbon is completely insoluble in water and only sparingly soluble in methanol. Weigh the dry product, which should be faint yellow in color; determine its melting point; and calculate the crude yield. If the melting point is below 150°C, recrystallize the product from cyclohexane (10 mL per g) and again determine the mp and yield. Hand in the product and give crude and recrystallized weights, melting points, and yields.

Questions

1. Show how 1,4-diphenyl-1,3-butadiene might be synthesized from benzaldehyde and an appropriate halogenated compound.
2. Explain why the methyl groups of trimethylphosphite give two peaks in the ^{1}H nmr spectrum (Figure 34.2).
3. Write the equation for the reaction between sodium methoxide and moist air.

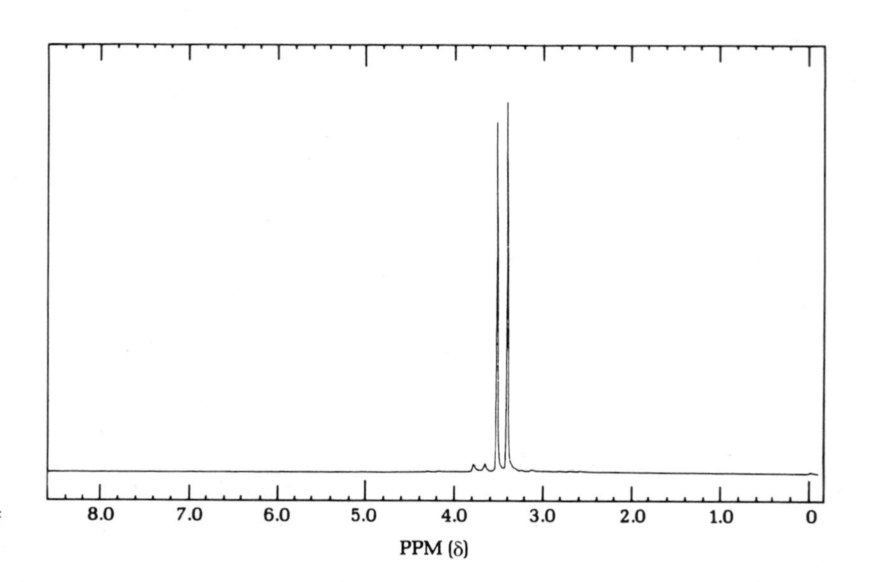

FIG. 34.2 ^{1}H nmr spectrum of trimethyl phosphite.

p-Terphenyl by the Diels-Alder Reaction 35

PRELAB EXERCISE: *Why does the dicarboxylate **5** undergo double decarboxylation to give terphenyl, while the diacid formed from the reaction of cyclopentadiene with maleic anhydride (Chapter 24) does not undergo decarboxylation under the same reaction conditions?*

E-E-1,4-Diphenyl-1,3-butadiene (**1**) is most stable in the transoid form, but at a suitably elevated temperature the cisoid form present in the equilibrium adds to dimethyl acetylenedicarboxylate (**2**) to give dimethyl-1,4-diphenyl-1,4-dihydrophthalate (**3**). This low-melting ester is obtained as an oil and when warmed briefly with methanolic potassium hydroxide is isomerized to the high-melting *E*-ester (**4**). The free *E*-acid can be obtained in 86% yield by refluxing the suspension of (**3**) in methanol for 4 h; but in the recommended procedure the isomerized ester is collected, washed to remove dark mother liquor, and hydrolyzed by brief heating with potassium hydroxide in a high-boiling solvent. The final step, an oxidative decarboxylation, is rapid and nearly quantitative. It probably involves reaction of the oxidant with the dianion (**5**) with removal of two electrons and formation of a diradical, which loses carbon dioxide with formation of *p*-terphenyl (**6**).

Diels-Alder reaction

Experiment

In a 10 × 100-mm reaction tube place 100 mg of *E,E*-1,4-diphenyl-1,3-butadiene prepared by the Wittig reaction in Chapter 34. Add 74 mg of dimethyl acetylenedicarboxylate (*caution, skin irritant*[1]), followed by 0.35 mL of triethylene glycol dimethyl ether (triglyme, bp 222°C) and a boiling chip. Reflux the mixture on a hot sand bath for 30 min.

$CH_3OCH_2CH_2OCH_2$
$|$
$CH_3OCH_2CH_2OCH_2$

Triethylene glycol dimethyl ether (triglyme). Miscible with water. bp 222°C

Cool the yellowish solution in cold water, add 2.5 mL of ether, and extract the solution three times with 1.5-mL portions of water to remove the high-boiling solvent, shake the ethereal solution once with saturated sodium chloride solution, and dry the solution over anhydrous sodium sulfate. Add the drying agent until it no longer clumps together. Remove the ether solution with a Pasteur pipette and place it in a tared reaction tube. Evaporate

Reaction time: 30 min; reflux over a flame

1. This ester is a powerful lachrymator (tear producer) and vesicant (blistering agent) and should be dispensed from a bottle provided with a pipette and a pipetter. Even a trace of ester on the skin should be washed off with methanol, followed by soap and water.

Transoid form ⇌ Cisoid form + $COOCH_3$–C≡C–$COOCH_3$ → ... →

(1) ***E,E*-1,4-Diphenyl-1,3-butadiene** MW 206.27

(2) MW 142.11

(3) MW 348.38, mp 98°C

$\xrightarrow{OH^-}$ $\xrightarrow{K_3Fe(CN)_6}$

(4) mp 170°C

(5)

(6) ***p*-Terphenyl** MW 230.29, mp 211°C

the solvent under a stream of nitrogen or air, removing the last traces under aspirator vacuum.

While the evaporation is in progress, dissolve 35 mg of potassium hydroxide in 0.35 mL of methanol by heating and stirring. Crystallization of the yellow oil containing **3** can be initiated by cooling and scratching; this provides assurance that the reaction has proceeded properly. Transfer the methanolic potassium hydroxide and heat with stirring on the hot sand bath for about 1 min until a stiff paste of crystals of the isomerized ester, **4**, appears. Cool, thin the mixture with methanol, collect the product and wash it free of dark mother liquor, and spread it thinly on a paper for rapid drying. The yield of pure, white ester, **4**, is about 120 mg. Solutions in methanol are strongly fluorescent.

*Rapid hydrolysis of the hindered ester (**4**)*

Place the ester **4** in a reaction tube, add 50 mg of potassium hydroxide, and pour in 0.35 mL of triethylene glycol. Stir the mixture with a thermometer and heat, raising the temperature to 140°C in the course of about 5 min. By intermittent heating, keep the temperature close to 140°C for 5 min longer and then cool the mixture under the tap. Pour into a 4-in. test tube and rinse the reaction tube with about 3.3 mL of water. Heat to boiling and, in case there is a small precipitate or the solution is cloudy, add a little pelletized Norit decolorizing charcoal, swirl, and filter the alkaline solution. Then add 225 mg of potassium ferricyanide and heat on the hot sand bath for about 5 min to dissolve the oxidant and to coagulate the white precipitate, which soon separates and is collected by filtration on the Hirsch funnel. The product can be air-dried overnight or dried to constant weight by heating in an evacuated reaction tube on the steam bath. The yield of colorless *p*-terphenyl, mp 209–210°C, is 50 to 60 mg. The *p*-terphenyl crystallizes well from dioxane; however, dioxane is a mild carcinogen and must be handled with care.

Questions

1. What is the driving force for the isomerization of **3** to **4**?
2. Why is the *trans* and not the *cis* diester **4** formed in the isomerization of **3** to **4**?
3. Why does hydrolysis of **4** in methanol require 4 h whereas hydrolysis in triethylene glycol requires only 10 min?

36 Nitration of Methyl Benzoate

PRELAB EXERCISE: *Show what ionic form 3-aminobenzoic acid is in at pH 8, pH 5, pH 2, and comment on its solubility in water at these three pH values.*

Benzene and somewhat less reactive aromatic compounds such as methyl benzoate can be nitrated with a mixture of nitric and sulfuric acids that ionizes completely to generate the nitronium and hydronium ions:

$$HNO_3 + 2H_2SO_4 \rightleftharpoons NO_2^+ + 2HSO_4^- + H_3O^+$$

Hot concentrated nitric acid is also a good oxidizing agent. For example, benzoin is oxidized easily to benzil (Chapter 50). Activated aromatic compounds such as amines and phenols can be nitrated using just concentrated nitric acid:

$$\underset{95\%}{2HNO_3} \rightleftharpoons \underset{5\%}{NO_2^+ + NO_3^- + H_2O}$$

A strongly deactivated benzene ring such as is found in nitrobenzene requires the use of heat, concentrated sulfuric acid, and fuming nitric acid to prepare *m*-dinitrobenzene. 1,3,5-Trinitrobenzene cannot be prepared by nitration of *m*-dinitrobenzene.

In the present experiment sulfuric acid serves as the solvent:

$$C_6H_5C(=O)OCH_3 + H_2SO_4 \rightleftharpoons \left[C_6H_5\overset{+}{C}(OH)OCH_3 \longleftrightarrow C_6H_5C(=\overset{+}{O}H)OCH_3 \right] + HSO_4^-$$

and nitration occurs at the *meta* position because of the partial positive charges residing at the *ortho* and *para* positions:

Experiment

Nitration of Methyl Benzoate

$$HNO_3 + 2H_2SO_4 \rightleftharpoons NO_2^+ + 2HSO_4^- + H_3O^+$$

Nitronium ion Hydronium ion

$COOCH_3$ $\xrightarrow[H_2SO_4]{HNO_3}$ $COOCH_3$ / NO_2

Methyl benzoate
MW 136.16
bp 199.6°C
den 1.09

Methyl 3-nitrobenzoate
MW 181.15
mp 78°C

Cool 0.6 mL of concentrated sulfuric acid to 0°C in a 10 × 100 mm reaction tube and then add to it 0.30 g of methyl benzoate. Again cool the mixture to 0°C and add dropwise, using a Pasteur pipette, a mixture of 0.2 mL of concentrated sulfuric acid and 0.2 mL of concentrated nitric acid. Keep the reaction mixture in ice. Using a stirring rod keep the reaction well mixed during the addition of the acids and do not allow the temperature of the mixture to rise above about 15°C as judged by touching the reaction tube.

Use care in handling concentrated sulfuric and nitric acids

After all the nitric acid has been added, warm the mixture to room temperature and after 15 min pour it onto 2.5 g of ice in a small beaker. Isolate the solid product by suction filtration using the Hirsch funnel and a 25-mL filter flask. Wash the product well with water, then with one 0.2-mL portion of ice-cold methanol. If the methanol is not ice-cold, product can be lost in this washing step. Save a small sample for melting point determination and analysis by thin-layer chromatography and infrared spectroscopy.

The remainder is weighed and crystallized from an equal weight of methanol in a reaction tube. Alternatively the sample can be dissolved in a *slightly* larger quantity of methanol and water added dropwise to make the hot solution saturated with the product. Slow cooling should produce large crystals, with a mp of 78°C. The crude material can be obtained in about 80% yield with a mp of 74–76°C.

Questions

1. Why does methyl benzoate dissolve in concentrated sulfuric acid?
2. What would you expect the structure of the dinitro ester to be?

37 Friedel-Crafts Alkylation of Benzene and Dimethoxybenzene; Host-Guest Chemistry

PRELAB EXERCISE: *Prepare a flow sheet for each reaction, indicating how the catalysts and unreacted starting materials are removed from the reaction mixture.*

Friedel-Crafts alkylation of aromatic rings most often employs an alkyl halide and a strong Lewis acid catalyst. Some of the catalysts that can be used, in order of decreasing activity, are the halides of Al, Sb, Fe, Ti, Sn, Bi, and Zn. Although useful, the reaction has several limitations. The aromatic ring must be unsubstituted or bear activating groups and because the product, an alkylated aromatic molecule, is more reactive than the starting material, multiple substitution usually occurs. Furthermore, primary halides will rearrange under the reaction conditions:

$$C_6H_6 + CH_3CH_2CH_2Cl \xrightarrow{AlCl_3} C_6H_5CH_2CH_2CH_3 + C_6H_5CH(CH_3)_2$$

	$C_6H_5CH_2CH_2CH_3$	$C_6H_5CH(CH_3)_2$
−6°C:	60%	40%
+35°C:	40%	60%

In the present reaction a tertiary halide and the most powerful Friedel-Crafts catalyst, $AlCl_3$, are allowed to react with benzene. The initially formed *t*-butylbenzene is a liquid while the product, 1,4-di-*t*-butylbenzene, which has a symmetrical structure, is a beautifully crystalline solid. The alkylation reaction probably proceeds through the carbocation under the conditions of the present experiment:

$$\text{C}_6\text{H}_6 + 2\ (CH_3)_3CCl \xrightarrow{AlCl_3} \left[(CH_3)_3C^+ \ AlCl_4^- \right] \longrightarrow p\text{-}[(CH_3)_3C]_2C_6H_4$$

Benzene
MW 78.11, den 0.88
bp 80°C

2-Chloro-2-methylpropane
(*t*-Butyl chloride)
MW 92.57, den 0.85
bp 51°C

1,4-Di-*t*-butylbenzene
MW 190.32, mp 77–79°C
bp 167°C

The reaction is reversible. If 1,4-di-*t*-butylbenzene is allowed to react with *t*-butyl chloride and aluminum chloride (1.3 moles) at 0–5°C, 1,3-di-*t*-butylbenzene, 1,3,5-tri-*t*-butylbenzene, and unchanged starting material are found in the reaction mixture. Thus, the mother liquor from crystallization of 1,4-di-*t*-butylbenzene in the present experiment probably contains *t*-butylbenzene, the desired 1,4-di-product, the 1,3-di-isomer, and 1,3,5-tri-*t*-butylbenzene.

Although the mother liquor probably contains a mixture of several components, the 1,4-di-*t*-butylbenzene can be isolated easily as an inclusion complex. Inclusion complexes are examples of host-guest chemistry. The molecule thiourea, NH_2CSNH_2, the host, has the interesting property of crystallizing in a helical crystal lattice that has a cylindrical hole in it. The guest molecule can reside in this hole if it is the correct size. It is not bound to the host, and there are often a nonintegral number (on the average) of host molecules per guest. The inclusion complex of thiourea and 1,4-di-*t*-butylbenzene crystallizes very nicely from a mixture of the other hydrocarbons and thus more of the product can be obtained. Because thiourea is very soluble in water the product is recovered from the complex by shaking it with a mixture of ether and water. The complex immediately decomposes and the product dissolves in the ether layer, from which it can be recovered.

Compare the length of the 1,4-di-*t*-butylbenzene molecule with the length of *n*-alkanes and predict the host/guest ratio. You can then check your prediction experimentally.

Experiments

1. 1,4-Di-*t*-butylbenzene

Measure, using a 0.5-mL plastic syringe, 0.40 mL of dry 2-chloro-2-methylpropane (*t*-butyl chloride) and 0.20 mL of dry benzene into a dry 10 × 100

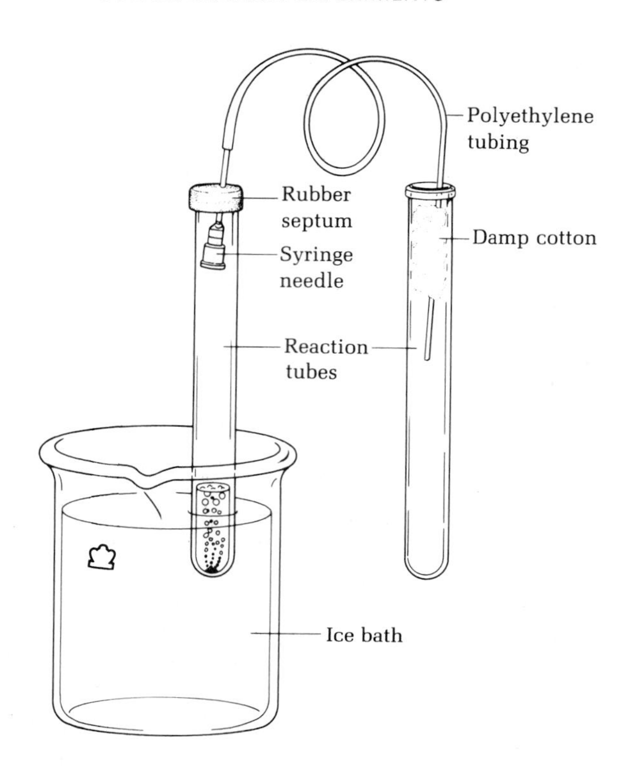

FIG. 37.1 Hydrogen chloride gas trap for Friedel-Crafts reaction.

Caution! *Benzene is a mild carcinogen. Handle in the hood, do not breathe vapors or allow liquid to come in contact with the skin.*

Aluminum chloride dust is extremely hygroscopic and irritating. It hydrolyzes to hydrogen chloride on contact with moisture. Clean up spilled material immediately.

Add powdered anhydrous sodium sulfate until it no longer clumps together

mm reaction tube equipped with a septum, inverted needle, and polyethylene tubing as seen in Fig. 37.1. The benzene and the alkyl chloride will be found in septum-stoppered containers. Cool the tube in ice and then add to it 40 mg of aluminum chloride. Weighing and transferring this small quantity is difficult because aluminum chloride reacts with great rapidity with moist air. Keep the reagent bottle closed as much of the time as possible while weighing the reagent into a very small dry, capped vial. Since the aluminum chloride is a catalyst, the amount need not be exactly 20 mg.

Mix the contents of the reaction tube by flicking the tube with the finger. After an induction period of about 2 min, a vigorous reaction sets in, with bubbling and liberation of hydrogen chloride. The hydrogen chloride is trapped using the apparatus depicted in Fig. 37.1. The wet cotton in the empty reaction tube will dissolve the hydrogen chloride. Near the end of the reaction the product separates as a white solid. When this occurs, remove the tube from the ice and let it stand at room temperature for 5 min.

Add about 1.0 mL of ice water to the reaction mixture, mix the contents thoroughly, and extract the product with three 0.8-mL portions of ether. Wash the combined ether extracts with about 1.5 mL of saturated sodium chloride solution and dry the ether over anhydrous sodium sulfate. Add sufficient drying agent so that it does not clump together. After 5 min transfer the ether solution to a dry, tared reaction tube using more ether to wash the drying agent and evaporate the ether under a stream of air in the hood.

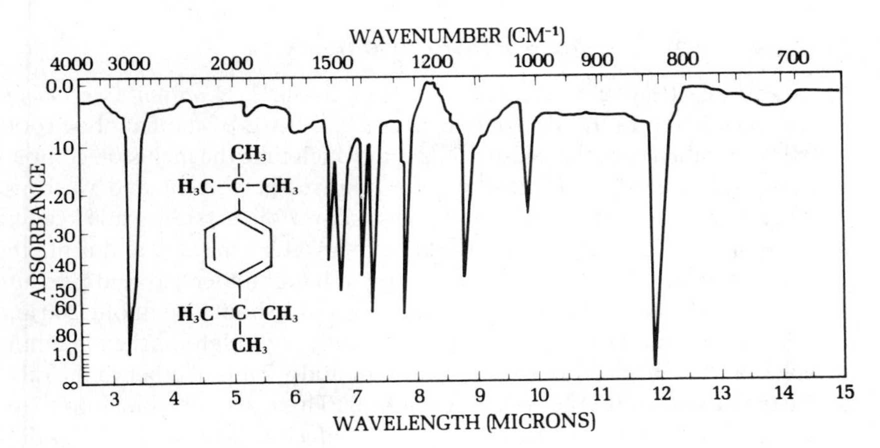

FIG. 37.2 Infrared spectrum of 1,4-di-*t*-butylbenzene.

Remove the last traces of ether under water aspirator vacuum. The oily product should solidify on cooling and weigh about 300 mg.

Spontaneous crystallization gives beautiful needles or plates

For crystallization, dissolve the product in 0.40 mL of methanol and let the solution come to room temperature without disturbance. After thorough cooling at 0°C, remove the methanol with a Pasteur pipette and rinse the crystals with a drop of ice-cold methanol while keeping the reaction tube in ice. Save this methanol solution for analysis by thin-layer chromatography. The yield of recrystallized material after drying under aspirator vacuum should be about 160 mg. Remove a sample of crystals for analysis by infrared spectroscopy, thin-layer chromatography, and melting point determination. Using thin-layer chromatography compare the pure crystalline product with the residue left after evaporation of the methanol.

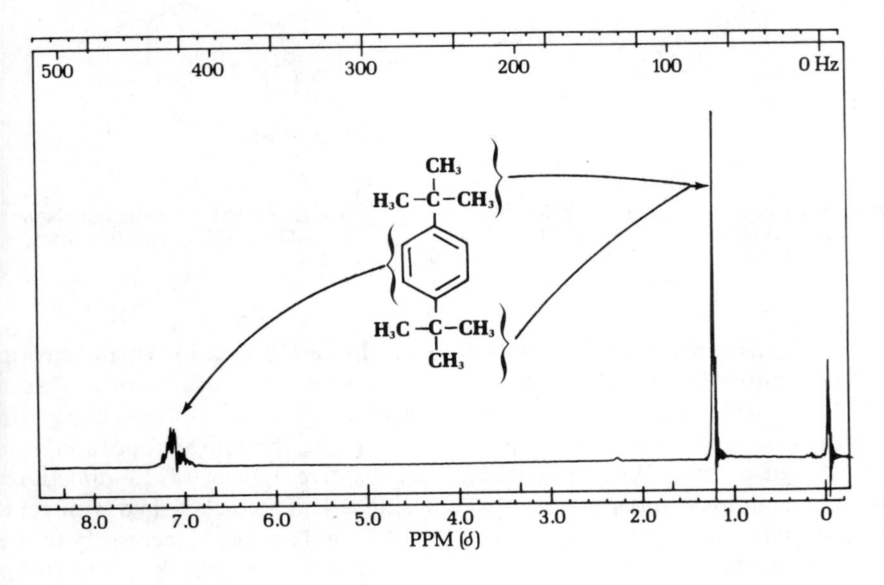

FIG. 37.3 ^{1}H nmr spectrum of 1,4-di-*t*-butylbenzene.

Preparation of Thiourea Inclusion Complex

***Caution!** Thiourea is a mild carcinogen. Handle the solid in a hood. Do not breathe dust.*

$$H_2N\overset{\overset{S}{\|}}{C}NH_2$$

Thiourea
MW 76.12

Inclusion complex starts to crystallize in 10 min

In a tared reaction tube dissolve 200 mg of thiourea (*Caution! See margin note.*) and 120 mg of 1,4-di-*t*-butylbenzene in 2.0 mL of methanol at room temperature; then cool the mixture in ice, at which time the inclusion complex will crystallize. Using a Pasteur pipette remove the solvent and wash the product twice with just enough methanol to cover the crystals while keeping the tube on ice. Connect the reaction tube to a water aspirator and using the heat of the hand evaporate the remaining methanol under reduced pressure until the weight of the tube is constant. The yield should be about 200 mg.

Remove a small sample, determine carefully the weight of the remaining complex, and then add about 1.2 mL of water and 1.2 mL of ether to the tube. Shake the mixture until the crystals disappear. This causes the breakup of the complex with the thiourea remaining in the aqueous layer and the 1,4-di-*t*-butylbenzene passing into the ether layer. Draw off the aqueous layer and dry the ether layer with anhydrous sodium sulfate. Add sufficient drying agent so that it does not clump together. More ether can be added if necessary. Transfer the ether to a tared reaction tube and wash the drying agent twice with fresh portions of ether. The object is to make a quantitative transfer of the butylbenzene. Evaporate the ether and remove the last traces under aspirator vacuum as before. After the weight of the tube is constant, record the weight of the hydrocarbon. Calculate the number of molecules of thiourea per molecule of hydrocarbon (probably *not* an integral number).

2. 1,4-Di-*t*-butyl-2,5-dimethoxybenzene

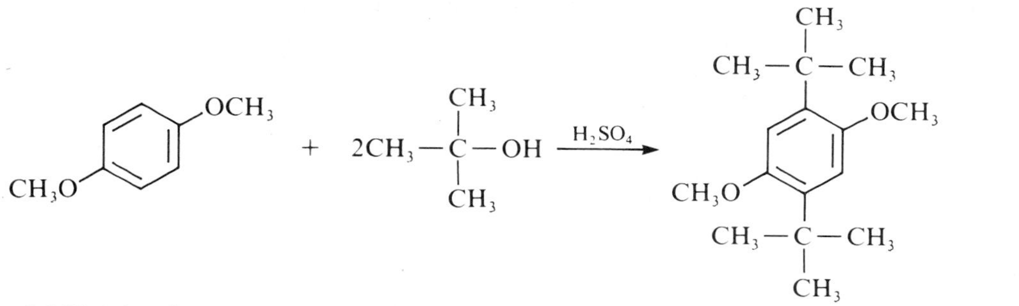

$$CH_3O\text{-}C_6H_4\text{-}OCH_3 + 2CH_3-\overset{\overset{CH_3}{|}}{\underset{\underset{CH_3}{|}}{C}}-OH \xrightarrow{H_2SO_4} (CH_3)_3C\text{-}C_6H_2(OCH_3)_2\text{-}C(CH_3)_3$$

1,4-Dimethoxybenzene (Hydroquinone dimethyl ether)
MW 138.16, mp 57°C

2-Methyl-2-propanol (*t*-Butyl alcohol)
MW 74.12, den 0.79
mp 25.5°C, bp 82.8°C

1,4-Di-*t*-butyl-2,5-dimethoxybenzene
MW 250.37, mp 104–105°C

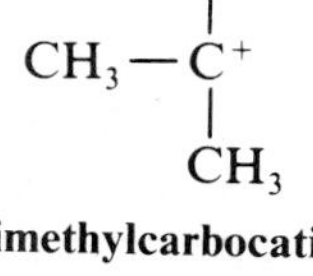

$$CH_3-\overset{\overset{CH_3}{|}}{\underset{\underset{CH_3}{|}}{C^+}}$$

Trimethylcarbocation

This experiment illustrates the Friedel-Crafts alkylation of an activated benzene molecule with a tertiary alcohol in the presence of sulfuric acid as the Lewis acid catalyst. As in the reaction of benzene and *t*-butyl chloride the substitution involves attack by the electrophilic trimethylcarbocation.

In a 10 × 100 mm reaction tube dissolve 120 mg of 1,4-dimethoxybenzene (hydroquinone dimethyl ether) in 0.4 mL of acetic acid with gentle warming and add to it 0.2 mL of *t*-butyl alcohol (it may be necessary to melt this alcohol). Cool the mixture in ice and then add to it dropwise from a

Pasteur pipette a mixture of 0.2 mL concentrated sulfuric acid and 0.2 mL of 30% fuming sulfuric acid. After each drop of acid is added, mix the solution thoroughly. At the end of this addition considerable solid reaction product should have separated. Stir the mixture thoroughly with a glass stirring rod and then remove the reaction tube from the ice and allow it to warm to room temperature and remain at 20–25°C for at least 10 min to complete the reaction. Then cool the mixture in ice to cause crystallization to occur. *Very carefully* add a drop of water to the mixture, stir it with the glass rod, and continue to add water dropwise with cooling and mixing until 2.5 mL have been added. Remove the solvent from the cold solution with a Pasteur pipette, wash the crystals *thoroughly* with water, and finally with three 0.2-mL portions of ice-cold methanol while cooling the reaction tube on ice. Recrystallize the product from hot methanol. Remove the solvent using a Pasteur pipette after allowing the mixture to cool to room temperature and then to 0°C in ice. Remove the last traces of methanol under aspirator vacuum while warming the tube in the hand. The yield of large plates of 1,4-di-*t*-butyl-2,5-dimethoxybenzene should be about 80 to 100 mg. Analyze the product by thin-layer chromatography and infrared spectroscopy. Determine the melting point and the percent yield.

$H_2SO_4 + SO_3$
Fuming sulfuric acid
Caution! *Highly corrosive. Reacts violently with water. Measure these reagents in the hood.*

Antics of growing crystals

R. D. Stolow of Tufts University reported[1] that growing crystals of the di-*t*-butyldimethoxy compound change shape in a dramatic manner; thin plates curl and roll up and then uncurl so suddenly that they propel themselves for a distance of several centimeters. If you do not observe this phenomenon during crystallization of a small sample, you may be interested in consulting the papers cited and pooling your sample with others for trial on a large scale. The solvent mixture recommended by the Tufts workers for observation of the phenomenon is 9.7 mL of acetic acid and 1.4 mL of water per gram of product.

Figures 37.4 and 37.5 present the infrared and nmr spectra of the starting hydroquinone dimethyl ether. Can you predict the appearance of the nmr spectrum of the product?

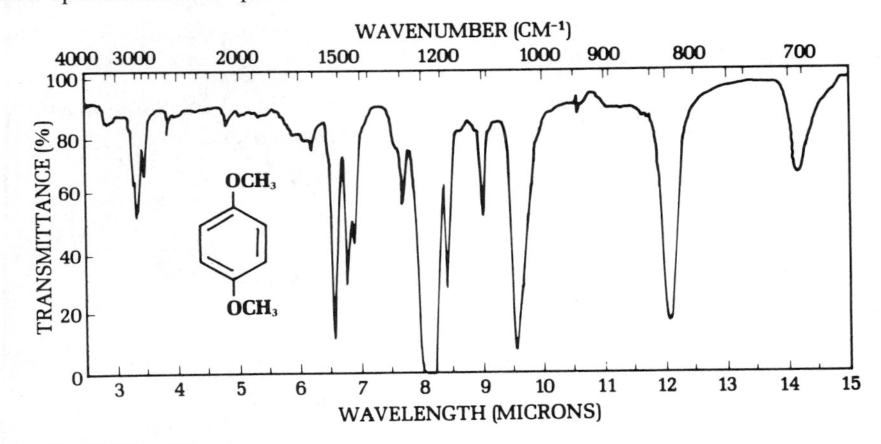

FIG. 37.4 Infrared spectrum of 1,4-dimethoxybenzene.

1. R. D. Stolow and J. W. Larsen, *Chemistry and Industry*, 449 (1963). See also J. M. Blatchly and N. H. Hartshorne, *Transactions of the Faraday Society*, **62**, 512 (1966).

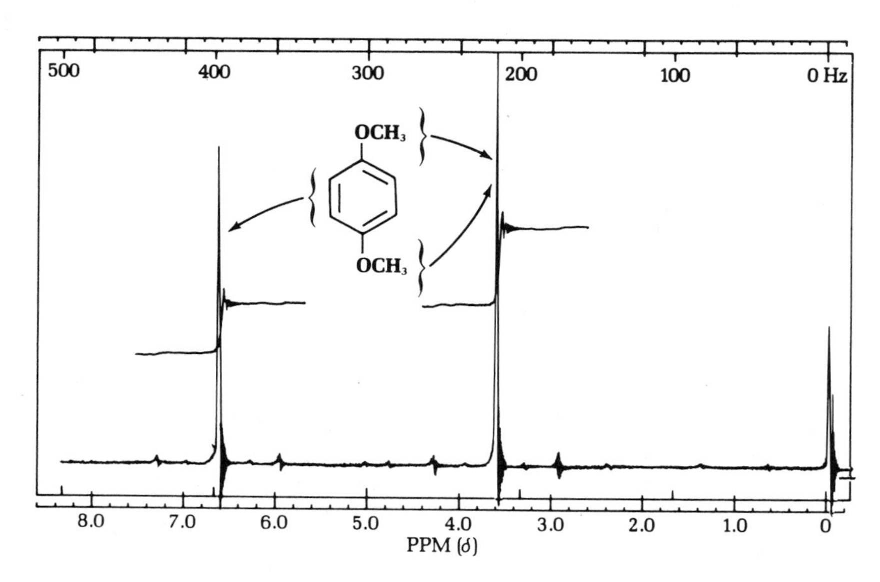

FIG. 37.5 ^{1}H nmr spectrum of 1,4-dimethoxybenzene.

Questions

1. Explain why the reaction of 1,4-di-*t*-butylbenzene with *t*-butyl chloride and aluminum chloride gives 1,3,5-tri-*t*-butylbenzene.
2. Why must aluminum chloride be protected from exposure to the air?
3. Why does fuming sulfuric acid react violently with water?
4. Draw a detailed mechanism for the formation of *t*-butyl-2,5-dimethoxybenzene.
5. Why is the 1,4 isomer, 1,4-di-*t*-butyl-2,5-dimethoxybenzene, the major product in the alkylation of dimethoxybenzene? Would you expect either of the following compounds to be formed as side products: 1,3-di-*t*-butyl-2,5-dimethoxybenzene; 1,4-dimethoxy-2,3-di-*t*-butylbenzene? Why or why not?
6. Suggest two other compounds that might be used in place of *t*-butyl alcohol to form 1,4-di-*t*-butyl-2,5-dimethoxybenzene.

Friedel-Crafts Acylation of Ferrocene: Acetylferrocene

38

PRELAB EXERCISE: *How many possible isomers could exist for diacetylferrocene?*

Ferrocene
Bis(cyclopentadienyl)iron
MW 186.04

+

Acetic Anhydride
MW 102.09
bp 139.5°C, den 1.08

$\xrightarrow{85\%\ H_3PO_4}$

Acetylferrocene
(Acetylcyclopentadienyl)cyclopentadienyliron
MW 228.08, mp 85–86°C

The Friedel-Crafts acylation of benzene requires aluminum chloride as the catalyst, but ferrocene, which has been referred to as a "superaromatic" compound, can be acylated under much milder conditions with phosphoric acid as catalyst. Since the acetyl group is a deactivating substituent the addition of a second acetyl group, which requires more vigorous conditions, will occur in the nonacetylated cyclopentadienyl ring to give 1,1′-diacetylferrocene. Because ferrocene gives just one monoacetyl derivative and just one diacetyl derivative, it was assigned an unusual sandwich structure.

Experiment

To 93 mg of dry sublimed ferrocene (material from Chapter 29) in a 10 × 100 mm reaction tube add 0.35 mL (0.32 g) of acetic anhydride followed by 0.1 mL (60 mg) of 85% phosphoric acid. Cap the tube with a septum bearing an empty syringe needle and warm it on a steam bath or in a beaker of boiling water while agitating the mixture to dissolve the ferrocene. Heat the mixture for 10 min more and then cool the tube thoroughly in ice. Carefully add to the solution 0.5 mL of ice water dropwise with thorough mixing, followed by the dropwise addition of 10% aqueous sodium hydroxide solution until the mixture is neutral (test with indicator paper and avoid an excess of base). Collect the product on the Hirsch funnel, wash it thoroughly with water, and press it as dry as possible between sheets of filter paper. Save

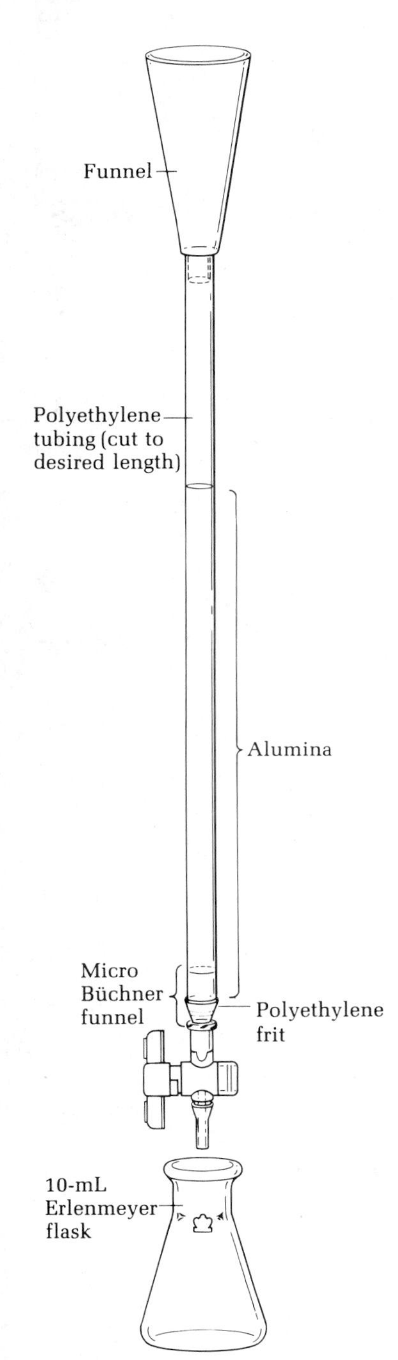

FIG. 38.1 Chromatographic column.

a sample of this material for melting point and thin-layer chromatographic analysis and purify the remainder by column chromatography.

Column chromatography. Assemble the column as depicted in Figure 38.1, being sure it is clamped in a vertical position. Close the valve and fill the column with petroleum ether to the bottom of the funnel. Prepare a slurry of 2.5 g of Activity III alumina in 8 mL of petroleum ether in a small beaker. This amount of alumina should fill the column to a height of 10 cm. Stir the slurry gently to get rid of air bubbles, and gently swirl, pour, and scrape the slurry into the funnel, which has a capacity of 10 mL. After some of the alumina has been added to the column open the stopcock and allow solvent to drain slowly into an Erlenmeyer flask. Use this petroleum ether to rinse the beaker containing the alumina. As the alumina is being added tap the column with a glass rod or pencil so the alumina will pack tightly into the column. Continue to tap the column while cycling the petroleum ether through the column once more.

It is extremely important to **never let the column run dry at any time.** Air can enter a dry column, which results in uneven bands and poor separation.

Adding the Sample. The solvent is drained just to the surface of the alumina, which should be perfectly flat. The sample of crude acetylferrocene is dissolved in the minimum quantity of dichloromethane and added to the column using a Pasteur pipette. It is allowed to run down the side of the column near the top of the alumina, without disturbing the alumina. The sample is run into the adsorbent, stopping when the solution reaches the top of the alumina. The transfer is completed with a few drops more of dichloromethane and then a small quantity of petroleum ether is added, without disturbing the surface. This is run down into the column and the process repeated until the sample is seen to be a band at the top of the column, at which point enough sand is poured onto the top of the column to form a layer about 3 mm thick. This will protect the surface of the alumina when more solvent is being added.

Elution. Carefully add petroleum ether to the column and begin to elute the product from the column. A band of unreacted ferrocene will come down the column as an orange-yellow band. Collect this in a tared 10-mL flask. Then elute the column with a 50 : 50 mixture of petroleum ether and diethyl ether. This will bring down the acetylferrocene as an orange-red band. Collect this in a separate tared 10-mL Erlenmeyer flask. Spot a thin-layer plate with these two solutions as well as the crude acetylferrocene. Any diacetylferrocene will be seen as a dark band at the top of the column. Evaporate the solvents from the two Erlenmeyer flasks and determine the weights of the residues.

Recrystallize the acetylferrocene from hot ligroin or hexane. Dissolve the residue in the flask in the minimum quantity of boiling solvent (about 1 mL) and transfer this hot solution with a Pasteur pipette to a reaction tube.

Allow the tube to cool slowly to room temperature, and then cool it in ice for at least 10 min. Acetylferrocene crystallizes as dark red rosettes of needles. Remove the solvent, scrape out the product onto filter paper, determine the weight, calculate the percent yield, and determine the melting point. The eluent is collected in fractions of 5 mL each in ten 10-mL Erlenmeyer flasks or small vials.

Analyze your product by thin-layer chromatography. Dissolve very small samples of pure ferrocene, the crude reaction mixture, and recrystallized acetylferrocene, each in a few drops of toluene; spot the three solutions with microcapillaries on silica gel plates; and develop the chromatogram with 30 : 1 toluene–absolute ethanol. Visualize the spots under a uv lamp if the silica gel has a fluorescent indicator or by adsorption of iodine vapor. Do you detect unreacted ferrocene in the reaction mixture and/or a spot that might be attributed to diacetylferrocene?

Question

What is the structure of the intermediate species that attacks ferrocene to form acetylferrocene? What other organic molecule is formed?

39 Alkylation of Mesitylene

PRELAB EXERCISE: *How much formic acid is consumed in this reaction?*

The reaction of an alkyl halide with an aromatic molecule in the presence of aluminum chloride, the Friedel-Crafts reaction, proceeds through the formation of an intermediate carbocation:

$$(CH_3)_2CHCl + AlCl_3 \rightleftharpoons (CH_3)_2CH^{+} + AlCl_4^{-}$$

$$(CH_3)_2CH^{+} + C_6H_6 \rightleftharpoons (CH_3)_2CH{-}C_6H_6^{+}$$

$$(CH_3)_2CH{-}C_6H_6^{+} \xrightleftharpoons{AlCl_4^-} (CH_3)_2CH{-}C_6H_5 + HCl + AlCl_3$$

Friedel and Crafts (who later became the president of MIT) discovered this reaction in 1879, but seven years before Baeyer and his colleagues carried out very similar reactions using aldehydes as the alkylating agent and strong acids as catalysts. These reactions, like the Friedel-Crafts reaction, proceed through carbocation intermediates. Consider the synthesis of DDT (1,1,1-trichloro-2,2-di(*p*-chlorophenylethane)):

$$CCl_3{-}C({=}O){-}H + H_2SO_4 \rightleftharpoons CCl_3{-}\overset{+}{C}(OH){-}H + HSO_4^-$$

$CCl_3{-}\overset{+}{C}(OH){-}H$ + chlorobenzene $\rightleftharpoons$ (arenium ion) $\rightleftharpoons$ $CCl_3{-}CH(OH){-}C_6H_4Cl$ + H^+

$\rightleftharpoons$ $CCl_3{-}CH(\overset{+}{O}H_2){-}C_6H_4Cl$ $\underset{-H_2O}{\rightleftharpoons}$ $CCl_3{-}\overset{+}{C}H{-}C_6H_4Cl$ + chlorobenzene $\rightleftharpoons$ (arenium ion) $\underset{-H^+}{\rightleftharpoons}$ $CCl_3{-}CH(C_6H_4Cl)_2$

DDT

Trichloroacetaldehyde (chloral) forms a carbocation on reaction with concentrated sulfuric acid. This reacts primarily at the *para*-position of chlorobenzene; and the intermediate alcohol, being benzylic, in the presence of acid readily forms a new carbocation, which in turn attacks another molecule of chlorobenzene. Even though synthesized in 1872, the remarkable insecticidal properties of DDT were not recognized until about 1940. It took another 25 years to realize that this compound, which is resistant to normal biochemical degradation, was building up in rivers, lakes, and streams and causing long-term environmental damage to wildlife. It is now outlawed in this country.

In the present experiment, discovered by Baeyer in 1872, formaldehyde is allowed to react with mesitylene in the presence of formic acid. The sequence of reactions is very similar to those that form DDT:

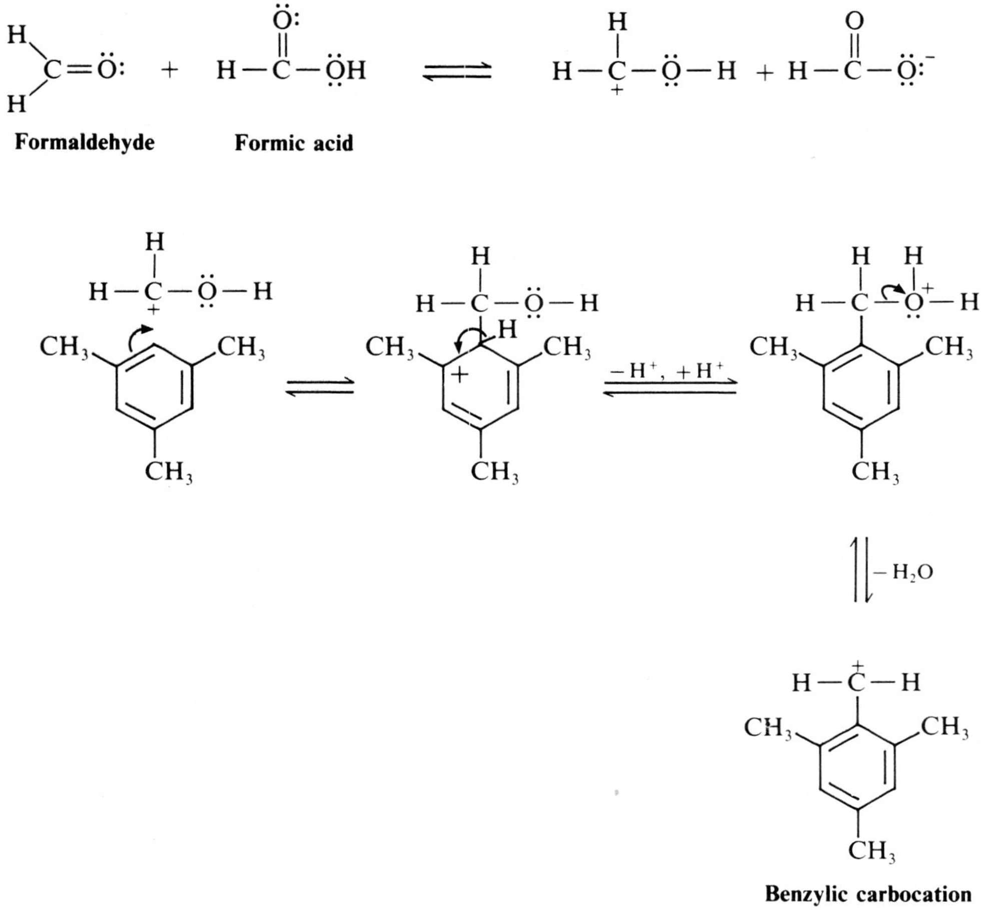

$$H_2C{=}O + H{-}C({=}O){-}OH \rightleftharpoons H_2\overset{+}{C}{-}O{-}H + H{-}C({=}O){-}O^-$$

Formaldehyde   Formic acid

$$H_2\overset{+}{C}{-}OH + \text{mesitylene} \rightleftharpoons \text{arenium ion} \xrightleftharpoons{-H^+,\ +H^+} ArCH_2{-}\overset{+}{O}H_2 \xrightleftharpoons{-H_2O} ArCH_2^+$$

Benzylic carbocation

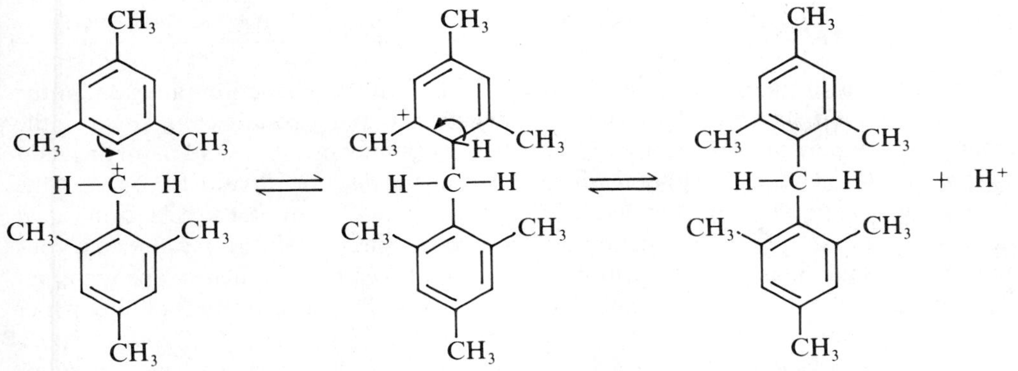

Dimesitylmethane

When aluminum chloride, a much more powerful catalyst, is used in this reaction, the methyl groups on the mesitylene rearrange and disproportionate to form a number of products, including polymeric material. The strongly activated ring of phenol reacts with formaldehyde at the *ortho*- and *para*-positions to form a polymer, Bakelite (see Chapter 67, Polymers).

A convenient form of formaldehyde to use in a reaction of this type is paraldehyde, a polymer that readily decomposes to formaldehyde:

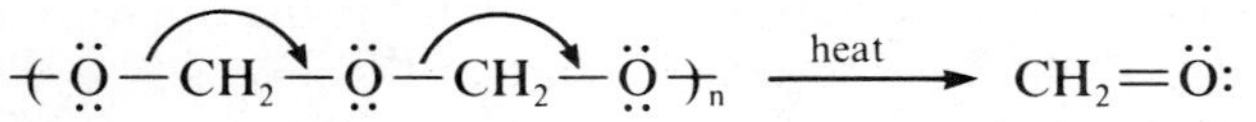

$$\text{+}\ddot{O}-CH_2-\ddot{O}-CH_2-\ddot{O}\text{+}_n \xrightarrow{\text{heat}} CH_2=\ddot{O}:$$

Paraldehyde — Formaldehyde

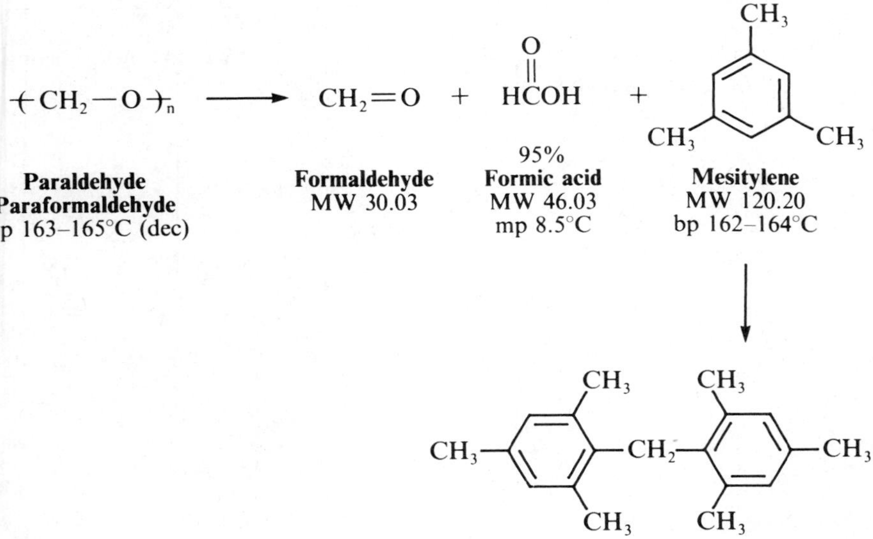

$$\text{+}CH_2-O\text{+}_n \longrightarrow CH_2=O + HCOH + \text{mesitylene}$$

Paraldehyde
Paraformaldehyde
mp 163–165°C (dec)

Formaldehyde
MW 30.03

95%
Formic acid
MW 46.03
mp 8.5°C

Mesitylene
MW 120.20
bp 162–164°C

Dimesitylmethane
MW 252.41

Experiment

Caution! Avoid all contact with formic acid or its vapors. Should any come in contact with your skin, wash it off immediately with a large quantity of water.

To 10 mg of paraformaldehyde in a 10 × 100 mm reaction tube add, in the hood, 0.06 mL (7.3 mg) of 95% formic acid and a boiling chip. Dissolve the paraformaldehyde in the formic acid by boiling on a hot sand bath, then add 0.133 mL (0.115 g) of mesitylene. Reflux the reaction mixture for 2 h, cool the mixture to room temperature and then in ice. Remove the excess formic acid using a Pasteur pipette and then wash the crystals with water, aqueous sodium carbonate solution and again with water, and then scrape them out onto a piece of filter paper. Squeeze the crystals between sheets of filter paper to dry.

Determine the weight of the crude product and save a few crystals for a mp determination. Recrystallize the product by dissolving it in the minimum quantity of boiling ligroin. Allow the solution to cool to room temperature, add a seed crystal if necessary, and then cool the mixture for at least 15 min in ice before removing the solvent using a Pasteur pipette. Another solvent for crystallization is a mixture of 0.75 mL of toluene and 0.1 mL of methanol, which is adequate for 0.5 g of product. Obtain the infrared and nuclear magnetic resonance spectra of the pure material and the melting points of the crude and recrystallized product. Turn in the pure product along with a card giving relevant data on the substance including the number of moles of each of the starting materials. From the spectra and any tests you may wish to run, confirm the structure of the product.

Question

The intermediate benzylic carbocation is stabilized by resonance. Draw the contributing resonance structures.

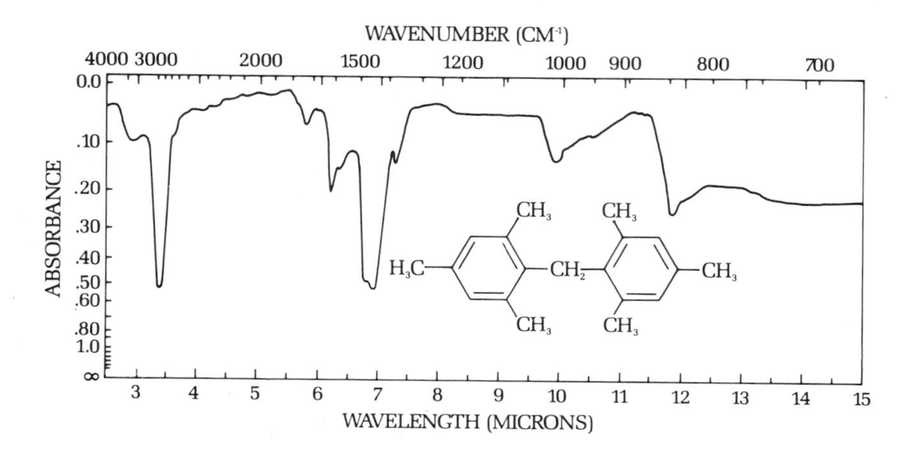

FIG. 39.1 Infrared spectrum of mesitylene.

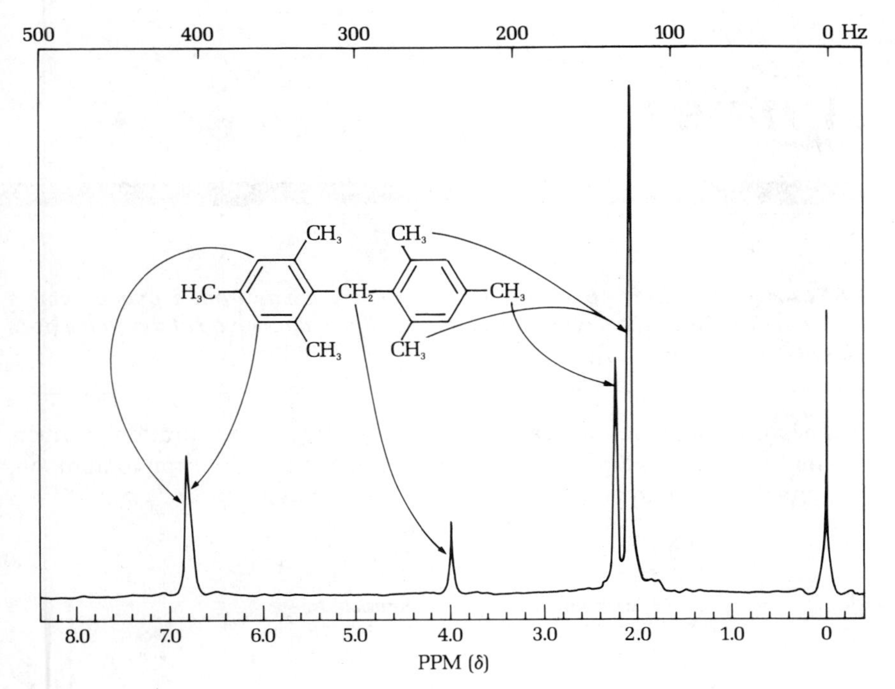

FIG. 39.2 ^{1}H nmr spectrum of mesitylene.

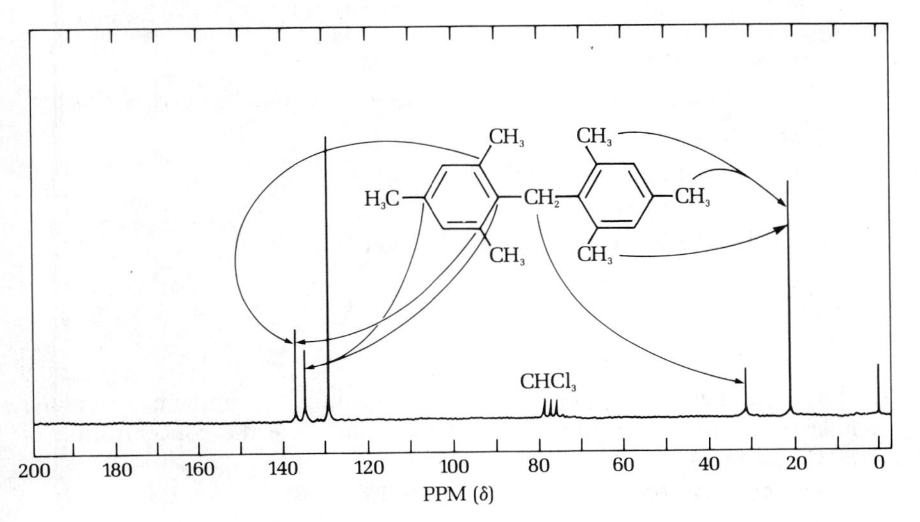

FIG. 39.3 ^{13}C nmr spectrum of dimesitylmethane.

40 Amines

PRELAB EXERCISE: *Explain why an ordinary amide from a primary amine does not dissolve in alkali while the corresponding sulfonamide of the amine does dissolve in alkali.*

Amines are weak bases and can be regarded as organic substitution products of ammonia. Just as ammonia reacts with acids to form the ammonium ion, so amines react with acid to form the organoammonium ion:

$$\ddot{N}H_3 + H_3\ddot{O}^+ \rightleftharpoons \overset{+}{N}H_4 + H_2\ddot{O}:$$

Ammonia **Ammonium ion**

$$CH_3\ddot{N}H_2 + H_3\ddot{O}^+ \rightleftharpoons CH_3\overset{+}{N}H_3 + H_2\ddot{O}:$$

Methylamine **Methylammonium ion**

When an amine dissolves in water the following equilibrium is established:

$$R\ddot{N}H_2 + H_2O \rightleftharpoons R\overset{+}{N}H_3 + OH^-$$

and from this the basicity constant can be defined as

$$K_b = \frac{[RNH_3^+][OH^-]}{[RNH_2]}$$

Strong bases have the larger values of K_b, meaning the amine has a greater tendency to accept a proton from water to increase the concentration of RNH_3^+ and OH^-.

The basicity constant for ammonia is 1.8×10^{-5}.

$$\ddot{N}H_3 + H_2O \rightleftharpoons NH_4^+ + OH^-$$

$$K_b = 1.8 \times 10^{-5} = \frac{[NH_4^+][OH^-]}{[NH_3]}$$

Alkyl amines such as methyl amine, CH_3NH_2, or *n*-propylamine, $CH_3CH_2CH_2NH_2$, are stronger bases than ammonia because the alkyl groups are electron donors and increase the effective electron density on nitrogen. Aromatic amines on the other hand are weaker bases than ammonia. Delocalization of the unshared pair of electrons on nitrogen onto the aromatic ring means the electrons are not available to be shared with an acidic proton.

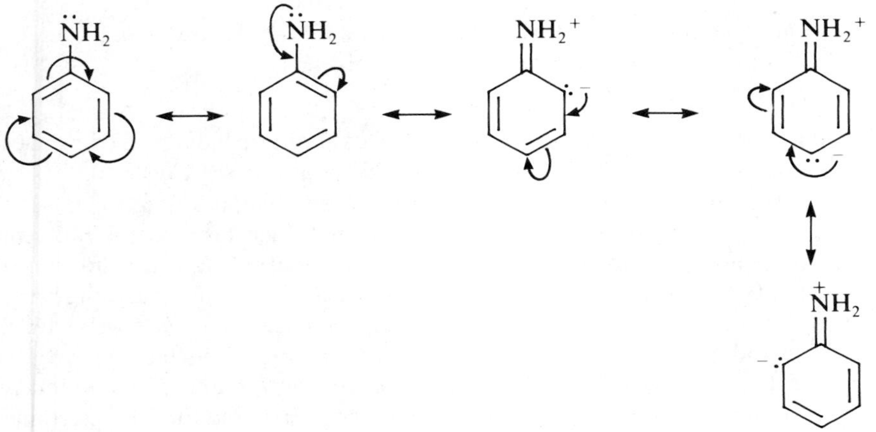

Low-molecular-weight amines have a powerful fishy odor. Slightly higher-molecular-weight diamines have names suggestive of their odors: $H_2N(CH_2)_5NH$, cadaverine and $H_2N(CH_2)_4NH_2$, putrescine. The lower-molecular-weight amines with up to about five carbon atoms are soluble in water. Many amines, especially the liquid aromatic amines, undergo light-catalyzed free radical air oxidation reactions to give a large variety of highly colored decomposition products. The higher-molecular-weight amines that are insoluble in water will dissolve in acid to form ionic amine salts. This reaction is useful for both characterization and for separation purposes. The ionic amine salt on treatment with base will regenerate the amine.

Nitriles, $R-C\equiv N$, are not basic and neither are amides or substituted amides. The adjacent electron-withdrawing carbonyl oxygen effectively removes the unshared pair of electrons on nitrogen:

Amides are not basic

$$R-C(=\ddot{O}:)-\ddot{N}H_2 \longleftrightarrow R-C(-\ddot{O}:^-)=\overset{+}{N}H_2$$

The object of the present experiment is to identify an unknown amine or amine salt. Procedures for solubility tests are given in Section 1.

Section 2 presents the Hinsberg test, a test for distinguishing between primary, secondary, and tertiary amines, Section 3 gives procedures for preparation of solid derivatives for melting point characterizations, and Section 4 gives spectral characteristics. Apply the procedures to known substances along with the unknown.

Experiments

1. Solubility

Substances to be tested:

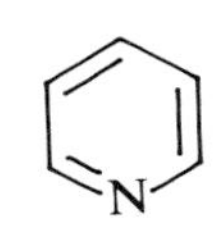

Pyridine

Aniline, $C_6H_5NH_2$, by 184°C
p-Toluidine, $CH_3C_6H_4NH_2$, mp 43°C
Pyridine, C_5H_5N, bp 115°C (a tertiary amine)
Methylamine hydrochloride, dec. (salt of CH_3NH_2, bp − 6.7°C)
Aniline hydrochloride, dec. (salt of $C_6H_5NH_2$)
Aniline sulfate, dec.

CH_3

NH_2

***p*-Toluidine**

Amines are often contaminated with dark oxidation products. The amines can easily be purified by an instant micro distillation. See Fig. 5.10.

First, see if the substance has a fishy, ammonialike odor, for if so it probably is an amine of low molecular weight. Then test the solubility in water by putting 1 drop if a liquid or an estimated 10 mg if a solid into a reaction tube, adding 0.1 mL of water (a 5-mm column) and first seeing if the substance dissolves in the cold. If the substance is a solid, rub it well with a stirring rod and break up any lumps before drawing a conclusion.

If the substance is *readily soluble in cold water* and if the odor is suggestive of an amine, test the solution with pH paper and further determine if the odor disappears on addition of a few drops of 10% hydrochloric acid. If the properties are more like those of a salt, add a few drops of 10% sodium hydroxide solution. If the solution remains clear, addition of a little sodium chloride may cause separation of a liquid or solid amine.

If the substance is not soluble in cold water, see if it will dissolve on heating; be careful not to mistake the melting of a substance for dissolving. If it *dissolves in hot water*, add a few drops of 10% alkali and see if an amine precipitates. (If you are in doubt as to whether a salt has dissolved partially or not at all, pour off the supernatant liquid and make it basic.)

If the substance is *insoluble in hot water*, add 10% hydrochloric acid, heat if necessary, and see if it dissolves. If so, make the solution basic and see if an amine precipitates.

2. Hinsberg Test

The procedure for distinguishing amines with benzenesulfonyl chloride is to be run in parallel on the following substances:

Use caution in handling amines. Many are toxic.

Aniline, $C_6H_5NH_2$ (bp 184° C)
N-Methylaniline, $C_6H_5NHCH_3$ (bp 194°C)
Triethylamine, $(CH_3CH_2)_3N$ (bp 90°C)

Primary and secondary amines react in the presence of alkali with benzenesulfonyl chloride, $C_6H_5SO_2Cl$, to give sulfonamides.

$$C_6H_5SO_2Cl + H_2\ddot{N}R + NaOH \longrightarrow C_6H_5SO_2\ddot{N}HR + NaCl + H_2O \quad 1$$

Primary amine — (Soluble in alkali)

$$C_6H_5SO_2Cl + H\ddot{N}R_2 + NaOH \longrightarrow C_6H_5SO_2\ddot{N}R_2 + NaCl + H_2O \quad 2$$

Secondary amine — (Insoluble in alkali)

The sulfonamides are distinguishable because the derivative from a primary amine has an acidic hydrogen, which renders the product soluble in alkali (reaction 1); whereas the sulfonamide from a secondary amine is insoluble (reaction 2). Tertiary amines lack the necessary acidic hydrogen for formation of benzenesulfonyl derivatives.

Low-molecular-weight amines have very bad odors; work in the hood

$$C_6H_5SO_2\ddot{N}HR + NaOH \longrightarrow C_6H_5SO_2\ddot{N}{:}^{-}R\ Na^{+} + H_2O$$

In a reaction tube add 50 mg of the amine, 200 mg of benzenesulfonyl chloride, and 1 mL of methanol. Over the hot sand bath or a steam bath heat the mixture to just below the boiling point, cool, and add 2 mL of 20% sodium hydroxide. Shake the mixture for 5 min and then allow the tube to stand for 10 more min with occasional shaking. If the odor of benzenesulfonyl chloride is detected, warm the mixture to hydrolyze it. Cool the mixture and acidify it by adding 6 M hydrochloric acid, dropwise, and with stirring. If a precipitate is seen at this point the amine is either primary or secondary. If no precipitate is seen, the amine is tertiary.

Distinguishing among primary (1°), secondary (2°), and tertiary (3°) amines

Do not allow benzenesulfonyl chloride to come in contact with the skin

If a precipitate is present remove it by filtration on the Hirsch funnel, wash it with 2 mL of water, and transfer it to a reaction tube. Add 2.5 mL of 5% sodium hydroxide solution and warm the mixture to 50°C. Shake the tube vigorously for 2 min. If the precipitate dissolves, the amine is primary. If it does not dissolve it is secondary. The sulfonamide of a primary amine can be recovered by acidifying the alkaline solution. Once dry these sulfonamides can be used to characterize the amine by their melting points.

3. Solid Derivatives

Acetyl derivatives of primary and secondary amines are usually solids suitable for melting point characterization and are readily prepared by reaction with acetic anhydride, even in the presence of water. Benzoyl and benzenesulfonyl derivatives are made by reaction of the amine with the appropriate acid chloride in the presence of alkali, as in Section 2 (the benzenesulfonamides of aniline and of *N*-methylaniline melt at 110°C and 79°C, respectively).

Solid derivatives suitable for characterization of tertiary amines are the methiodides and picrates:

$$R_3N + CH_3I \longrightarrow R_3\overset{+}{N}CH_3 \; I^-$$

$$\text{2,4,6-}(NO_2)_3C_6H_2OH + NR_3 \longrightarrow \text{2,4,6-}(NO_2)_3C_6H_2O^- \; \overset{+}{N}HR_3$$

2,4,6-Trinitrophenol (Picric Acid) **Amine picrate**

Typical derivatives are to be prepared; and although determination of melting points is not necessary because the values are given, the products should be saved for possible identification of unknowns.

A. Acetylation of Aniline with Acetic Anhydride

$$C_6H_5NH_2 + CH_3{-}\overset{O}{\overset{\|}{C}}{-}O{-}\overset{O}{\overset{\|}{C}}{-}CH_3 \longrightarrow C_6H_5NH{-}\overset{O}{\overset{\|}{C}}{-}CH_3 + CH_3COOH$$

Aniline MW 93.13 bp 184°C

Acetic anhydride MW 102.09 bp 138–140°C

Acetanilide MW 135.17 mp 114°C

In a dry reaction tube place 9.3 mg of freshly distilled aniline, 10.2 mg of pure acetic anhydride and a small boiling chip. Reflux the mixture for about 8 min, then add water to the mixture dropwise with heating until all of the product is in solution. This will require very little water. Allow the mixture to cool spontaneously to room temperature and then cool the mixture in ice. Remove the solvent and recrystallize the crude material again from boiling water. Isolate the crystals in the same way, cool the tube in ice water, and wash the crystals with a drop or two of ice-cold acetone. Remove the acetone with a Pasteur pipette and dry the product in the reaction tube. Once dry determine the weight, calculate the percent yield and determine the melting point. With patience this reaction and the recrystallizations can be carried out in a melting point tube on one-tenth the indicated quantities of material using essentially the same techniques.

B. Acetylation of Aniline Hydrochloride with Acetic Anhydride

Dissolve 26 mg of aniline hydrochloride in 0.25 mL of water, add 21 mg of acetic anhydride followed immediately by 25 mg of sodium acetate. Warm the mixture, cool it to room temperature, and then in ice. Recrystallize the product from water and wash it with acetone as described above. See p. 344 for mechanism.

C. Formation of an Amine Picrate from Picric Acid and Triethylamine

Picric acid + $(CH_3CH_2)_3N$ ⟶ Triethylamine picrate

Picric acid
MW 222.11, mp 120–122°C

Triethylamine
MW 101.19
bp 88.8°C

Triethylamine picrate
MW 323.30, mp 171°C

Dissolve 30 mg of moist (35% water) picric acid in 0.25 mL of methanol and to the warm solution add 10.1 mg of triethylamine and let the solution stand to deposit crystals of the picrate, an amine salt that has a characteristic melting point. This picrate melts at 171°C.

Picric acid can explode if allowed to become dry. Use only the moist reagent (35% water) and do not allow the reagent to dry out.

4. The nmr and ir Spectra of Amines

The proton bound to nitrogen can appear between 0.6 and 7.0 ppm on the nmr spectrum, the position depending upon solvent, concentration, and structure of the amine. The peak is sometimes extremely broad owing to slow exchange and interaction of the proton with the electric quadrupole of the nitrogen. If addition of a drop of D_2O to the sample causes the peak to disappear, this is evidence for an amine hydrogen; but alcohols, phenols, and enols will also exhibit this exchange behavior. See Fig. 40.1 for the nmr spectrum of aniline, in which the amine hydrogens appear as a sharp peak at 3.3 ppm. Infrared spectroscopy can also be very useful for identification purposes. Primary amines, both aromatic and aliphatic, show a weak doublet between 3300 and 3500 cm^{-1} and a strong absorption between 1560 and 1640 cm^{-1} due to NH bending (Fig. 40.2). Secondary amines show a single peak between 3310 and 3450 cm^{-1}. Tertiary amines have no useful infrared absorptions. In Chapter 21 the characteristic ultraviolet absorption shifts of aromatic amines in the presence and absence of acids were discussed.

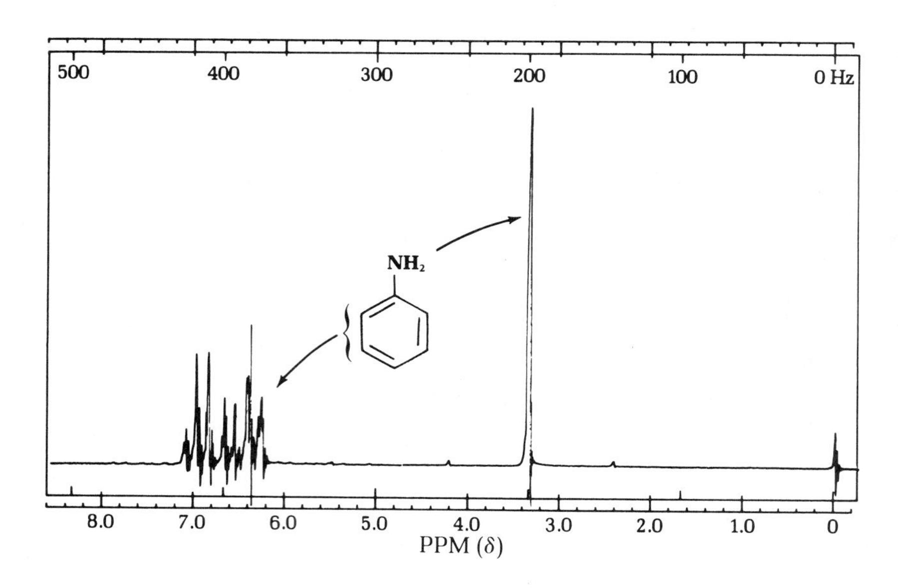

FIG. 40.1 The nmr spectrum of aniline.

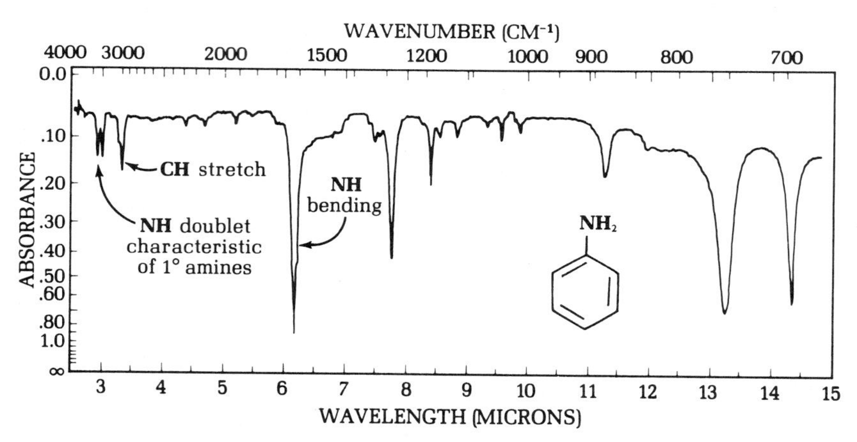

FIG. 40.2 Infrared spectrum of aniline in CS_2.

5. Unknowns (see Tables 70.5 and 70.6)

Determine first if the unknown is an amine or an amine salt and then determine whether the amine is primary, secondary, or tertiary. Complete identification of your unknown may be required.

Questions

1. How could you most easily distinguish between samples of 2-aminonaphthalene and of acetanilide?
2. Would you expect the reaction product from benzenesulfonyl chloride and ammonia to be soluble or insoluble in alkali?

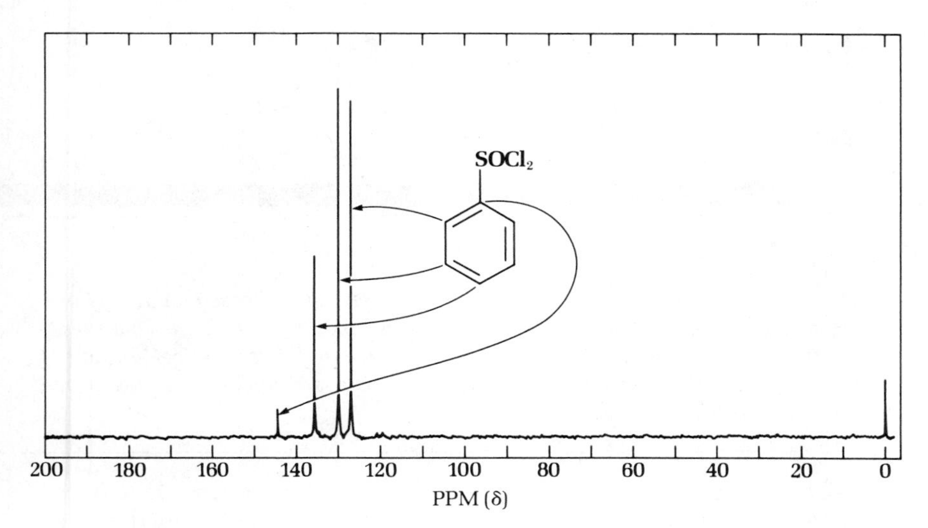

FIG. 40.3 ^{13}C nmr spectrum of benzenesulfonyl chloride.

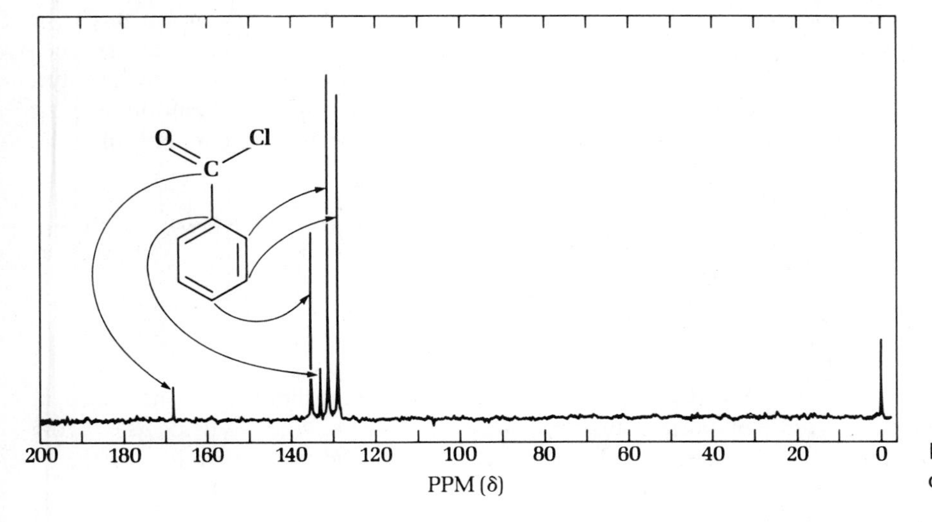

FIG. 40.4 ^{14}C nmr spectrum of benzoyl chloride.

3. Is it safe to conclude that a substance is a tertiary amine because it forms a picrate?
4. Why is it usually true that amines that are insoluble in water are odorless?
5. Technical dimethylaniline contains traces of aniline and of methylaniline. Suggest a method for elimination of these impurities.
6. How would you prepare aniline from aniline hydrochloride?
7. Write a balanced equation for the reaction of benzenesulfonyl chloride with sodium hydroxide solution in the absence of an amine. What solubility would you expect the product to have in acid and base?

41 Sulfanilamide from Nitrobenzene

PRELAB EXERCISE: *Prepare detailed flow sheets for the three reactions. Look up the solubilities of all reagents and indicate in separation steps in which layer you would expect to find the desired product. Calculate the theoretical amount of ammonium hydroxide needed to react with 5 g of p-acetaminobenzenesulfonyl chloride to form sulfanilamide.*

Paul Ehrlich, the father of immunology and chemotherapy, discovered Salvarsan, an arsenical "magic bullet" (a favorite phrase of his) used to treat syphilis. He hypothesized at the beginning of this century that it might be possible to find a dye that would selectively stain, or dye, a bacterial cell and thus destroy it. In 1932 I.G. Farbenindustrie patented a new azo dye, Prontosil, which they put through routine testing for chemotherapeutic activity when it was noted it had particular affinity for protein fibers like silk.

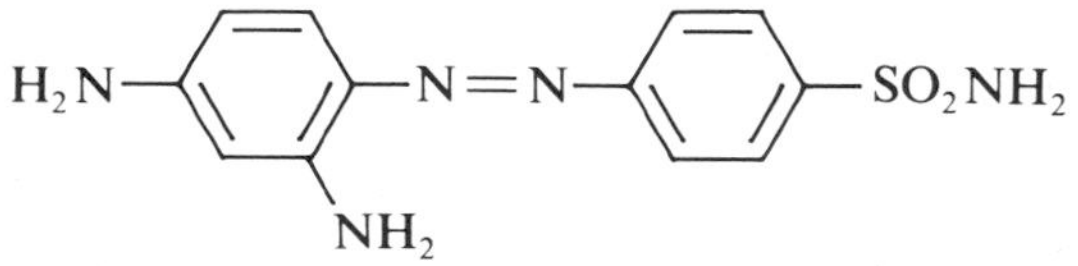

$H_2N-C_6H_3(NH_2)-N{=}N-C_6H_4-SO_2NH_2$

Prontosil

The dye was found to be effective against streptococcal infections in mice, but somewhat surprisingly, ineffective *in vitro* (outside the living animal). A number of other dyes were tested, but only those having the

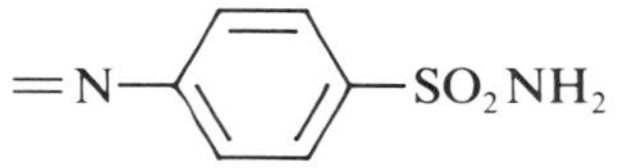

$=N-C_6H_4-SO_2NH_2$

group were effective. French workers hypothesized that the antibacterial activity had nothing to do with the identity of the compounds as dyes, but rather with the reduction of the dyes in the body to *p*-aminobenzenesulfonamide, known commonly as sulfanilamide.

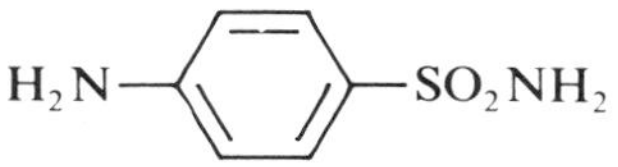

$H_2N-C_6H_4-SO_2NH_2$

Sulfanilamide R=H

On the basis of this hypothesis sulfanilamide was tested and found to be the active substance.

Because sulfanilamide had been synthesized in 1908, its manufacture could not be protected by patents, so the new drug and thousands of its derivatives were rapidly synthesized and tested. When the R group in sulfanilamide is replaced with a heterocyclic ring system—pyridine, thiazole, diazine, merazine, etc.—the sulfa drug so produced is often faster-acting or less toxic than sulfanilamide. Although they have been supplanted for the most part by antibiotics of microbial origin, these drugs still find wide application in chemotherapy.

Unlike that of most drugs, the mode of action of the sulfa drugs is now completely understood. Bacteria must synthesize folic acid in order to grow. Higher animals, like man, do not synthesize this vitamin and hence must acquire it in their food. Sulfanilamide inhibits the formation of folic acid, stopping the growth of bacteria; and because the synthesis of folic acid does not occur in man, only the bacteria are affected.

A closer look at these events reveals that bacteria synthesize folic acid using several enzymes, including one called dihydropteroate synthetase, which catalyzes the attachment of *p*-aminobenzoic acid to a pteridine ring system. When sulfanilamide is present it competes with the *p*-aminobenzoic acid (note the structural similarity) for the active site on the enzyme. This activity makes it a competitive inhibitor. Once this site is occupied on the enzyme, folic acid synthesis stops and bacterial growth stops. Folic acid can also be synthesized in the laboratory.[1]

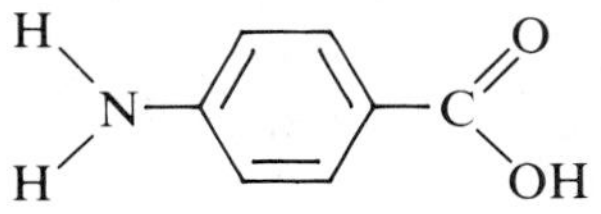

p-Aminobenzoic acid

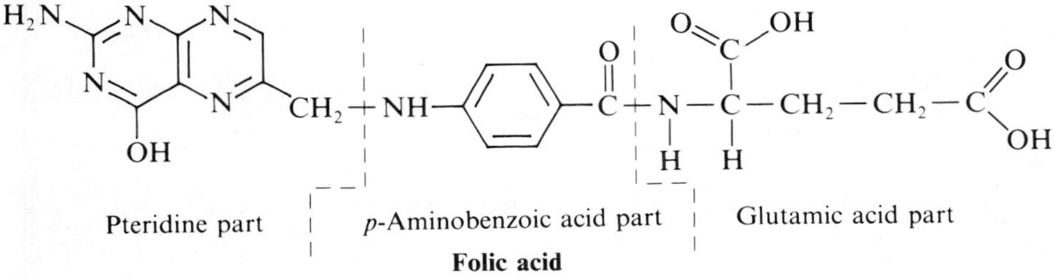

Pteridine part | *p*-Aminobenzoic acid part | Glutamic acid part

Folic acid

This experiment is a multistep synthesis of sulfanilamide starting with nitrobenzene. Nitrobenzene is reduced with tin and hydrochloric acid to give

1. L. T. Plante, K. L. Williamson, and E. J. Pastore, "Methods in Enzymology," Vol. 66, D. B. McCormick and L. D. Wright, eds., Academic Press, New York, 1980, p. 533.

the anilinium ion, which is converted to aniline with base:

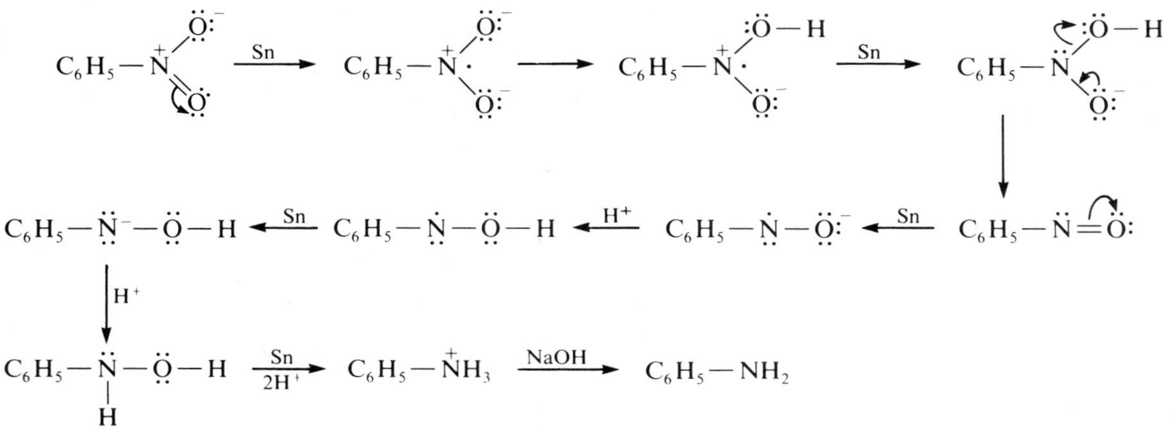

$$C_6H_5-\overset{+}{N}(=O)O^- \xrightarrow{Sn} C_6H_5-\overset{+}{N}^{\cdot}(O^-)_2 \longrightarrow C_6H_5-\overset{+}{N}^{\cdot}(OH)(O^-) \xrightarrow{Sn} C_6H_5-\overset{+}{N}(OH)(O^-)$$

$$\downarrow$$

$$C_6H_5-N^--O-H \xleftarrow{Sn} C_6H_5-\dot{N}-O-H \xleftarrow{H^+} C_6H_5-\dot{N}-O^- \xleftarrow{Sn} C_6H_5-N=O$$

$$\downarrow H^+$$

$$C_6H_5-N(H)-O-H \xrightarrow[2H^+]{Sn} C_6H_5-\overset{+}{N}H_3 \xrightarrow{NaOH} C_6H_5-NH_2$$

Anilinium ion **Aniline**

The acetylation of aniline in aqueous solution to give acetanilide serves to protect the amine group from reaction with chlorosulfonic acid. The acetylation takes place readily in aqueous solution. Aniline reacts with acid to give the water-soluble anilinium ion:

$$C_6H_5-\ddot{N}H_2 + H^+ \longrightarrow C_6H_5-\overset{+}{N}H_3$$

Aniline **Anilinium ion**

The anilinium ion reacts with acetate ion to set up an equilibrium that liberates a small quantity of aniline:

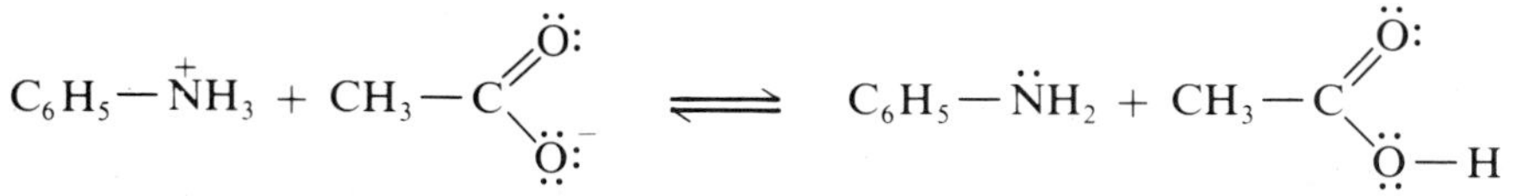

$$C_6H_5-\overset{+}{N}H_3 + CH_3-C(=O)O^- \rightleftharpoons C_6H_5-\ddot{N}H_2 + CH_3-C(=O)O-H$$

This aniline reacts with acetic anhydride to give water-insoluble acetanilide:

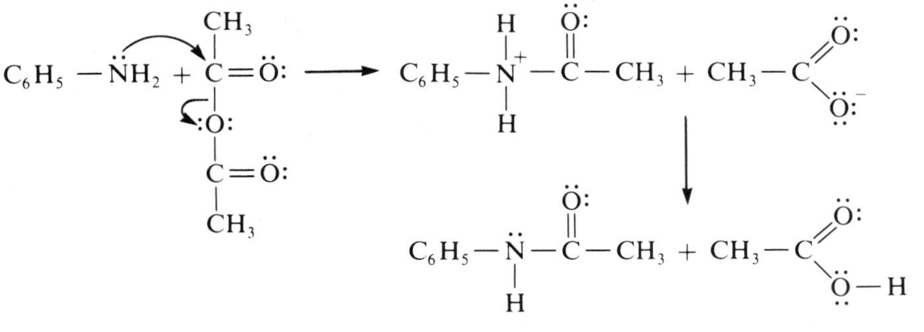

$$C_6H_5-\ddot{N}H_2 + CH_3-C(=O)-O-C(=O)-CH_3 \longrightarrow C_6H_5-\overset{+}{N}H_2-C(=O)-CH_3 + CH_3-C(=O)O^-$$

$$\downarrow$$

$$C_6H_5-\ddot{N}(H)-C(=O)-CH_3 + CH_3-C(=O)O-H$$

Acetanilide

This upsets the equilibrium, releasing more aniline that can then react with acetic anhydride to give more acetanilide.

Acetanilide reacts with chlorosulfonic acid in an electrophilic aromatic substitution:

$$CH_3-C(=\ddot{O}:)-\ddot{N}H-C_6H_5 + :\ddot{O}^-{-}\overset{++}{S}(:\ddot{O}:^-)(\ddot{O}-H)-\ddot{C}l: \longrightarrow CH_3-C(=\ddot{O}:)-\ddot{N}H-C_6H_5^{+}(H)-\overset{++}{S}(:\ddot{O}:^-)_2-\ddot{O}-H$$

Chlorosulfonic acid

$$\longrightarrow CH_3-C(=\ddot{O}:)-\ddot{N}H-C_6H_4-\overset{++}{S}(:\ddot{O}:^-)_2-\ddot{O}-H$$

$$CH_3-C(=\ddot{O}:)-\ddot{N}H-C_6H_4-\overset{++}{S}(:\ddot{O}:^-)_2-\ddot{O}-H \xrightarrow{H-\ddot{O}-\overset{++}{S}(:\ddot{O}:^-)_2-\ddot{C}l:} H-\ddot{O}-\overset{++}{S}(:\ddot{O}:^-)_2-\ddot{O}-H + CH_3-C(=\ddot{O}:)-\ddot{N}H-C_6H_4-\overset{++}{S}(:\ddot{O}:^-)_2-\ddot{C}l:$$

p-Acetamidobenzenesulfonyl chloride

$$CH_3-C(=\ddot{O}:)-\ddot{N}H-C_6H_4-\overset{++}{S}(:\ddot{O}:^-)_2-\ddot{C}l: + 2\,R\ddot{N}H_2 \longrightarrow CH_3-C(=\ddot{O}:)-\ddot{N}H-C_6H_4-\overset{++}{S}(:\ddot{O}:^-)_2-\ddot{N}H-R + R\overset{+}{N}H_3Cl$$

p-Acetamidobenzenesulfonamide

The protecting amide group is removed from the *p*-acetamidobenzenesulfonamide by acid hydrolysis. The amide group is more easily hydrolyzed than the sulfonamide group:

$$CH_3-C(=\ddot{O}:)-\ddot{N}H-C_6H_4-SO_2NHR \xrightarrow{H^+} CH_3-\overset{+}{C}(:\ddot{O}-H)-\ddot{N}H-C_6H_4-SO_2NHR \;\; (+\; H-\ddot{O}-H)$$

$$\longrightarrow CH_3-C(:\ddot{O}-H)(H-\overset{+}{\ddot{O}}-H)-\ddot{N}H-C_6H_4-SO_2NHR \;\; (H^+)$$

$$\longrightarrow CH_3-C(:\ddot{O}-H)(H-\ddot{O}:)-\overset{+}{N}H_2-C_6H_4-SO_2NHR$$

$$\longrightarrow CH_3-C(=\ddot{O}:)-\ddot{O}-H + H_2N-C_6H_4-SO_2NHR$$

Sulfanilamide

$C_6H_5NO_2$ $\xrightarrow[(2)\ OH^-]{(1)\ Sn,\ HCl}$ $C_6H_5NH_2$ $\xrightarrow[HCl,\ NaOAc]{(CH_3CO)_2O}$ $C_6H_5NHCOCH_3$ $\xrightarrow{HSO_3Cl}$

Nitrobenzene
MW 123.11
bp 210.9°C, den 1.12

Aniline
MW 93.12
bp 184.4°C, den 1.02

Acetanilide
MW 135.16
mp 114°C

$p\text{-}CH_3CONHC_6H_4SO_2Cl$ $\xrightarrow{NH_3}$ $p\text{-}CH_3CONHC_6H_4SO_2NH_2$ $\xrightarrow{dil.\ HCl}$ $p\text{-}H_2NC_6H_4SO_2NH_2$

***p*-Acetaminobenzene-sulfonyl chloride**
MW 233.68

***p*-Acetaminobenzene-sulfonamide**
MW 214.25

Sulfanilamide (*p*-Aminobenzene-sulfonamide)
MW 172.20, mp 163–164°C

In the present experiment nitrobenzene is reduced to aniline by tin and hydrochloric acid. A double salt with tin having the formula $(C_6H_5NH_3)_2SnCl_4$ separates partially during the reaction, and at the end it is decomposed by addition of excess alkali, which converts the tin into water-soluble stannite or stannate (Na_2SnO_2 or Na_2SnO_3). The aniline liberated is separated from inorganic salts and the insoluble impurities derived from the tin by steam distillation and is then dried, distilled, and acetylated with acetic anhydride in aqueous solution. Treatment of the resulting acetanilide with excess chlorosulfonic acid effects substitution of the chlorosulfonyl group and affords *p*-acetaminobenzenesulfonyl chloride. The alternative route to this intermediate via sulfanilic acid is unsatisfactory, because sulfanilic acid being dipolar is difficult to acetylate. In both processes the amino group must be protected by acetylation to permit formation of the acid chloride group. The next step in the synthesis is ammonolysis of the sulfonyl chloride and the terminal step is removal of the protective acetyl group.

Use the total product obtained at each step as starting material for the next step and adjust the amounts of reagents accordingly. Keep a record of your working time. Aim for a high overall yield of pure final product in the shortest possible time. Study the procedures carefully before coming to the laboratory so that your work will be efficient. A combination of consecutive steps that avoids a needless isolation saves time and increases the yield.

Experiments

1. Aniline

In a 5-mL round-bottomed short-necked flask place 1.25 g of granulated tin, 600 mg of nitrobenzene, and 2.75 mL of concentrated hydrochloric acid. Have an ice bath ready to control the reaction, insert a thermometer, and let the mixture react until the temperature reaches 60°C and then cool it briefly, if necessary, so the temperature does not rise above 62°C. Swirl the reaction mixture and maintain the temperature in the range of 55–60°C for 15 min. Remove the thermometer and rinse it with water, fit the flask with an empty fractionating column that will function as an air condenser, and heat the reaction mixture on the steam bath with frequent swirling until droplets of nitrobenzene are absent from the condenser, and the color due to an intermediate reduction product is gone (about 15 min). During this period dissolve 2 g of sodium hydroxide in 5 mL of water and cool to room temperature.

Measure nitrobenzene (toxic) in the hood. Do not breathe the vapors.

Handle aniline with care. It may be a carcinogen.

Reaction time: 0.5 h

Handle the hot solution of sodium hydroxide with care; it is very corrosive

At the end of the reduction reaction, cool the acid solution in ice (to prevent volatilization of the aniline) during the gradual addition of the solution of alkali. This alkali neutralizes the aniline hydrochloride, releasing aniline, which will now be volatile in steam. Remove the air condenser and attach the addition adapter and distillation head to the flask along with an ice-cooled vial as a receiver (Fig. 41.1). Add a boiling chip and heat the contents of the flask to boiling. A mixture of water and aniline will steam distill (see Chapter 6) and the distillate will be cloudy as the two layers, aniline and water, separate. As the distillation proceeds, add water through the addition port, using a syringe, to keep the volume in the distilling flask constant. Since aniline is fairly soluble in water (3.6 g/100 g of water at 18°C), distillation should be continued somewhat beyond the point at which the distillate has lost its original turbidity (2.5–3 mL more). Make an accurate estimate of the volume of the distillate, e.g., by pouring it into a 10-mL graduated cylinder.

Dispose of tin salts and solutions in the container provided

At this point aniline can be isolated. You could reduce the solubility of aniline by dissolving in the steam distillate 0.2 g of sodium chloride per milliliter. Extract the aniline with 2–3 portions of dichloromethane, dry the extract, distill the dichloromethane (bp 41°C), and then distill the aniline (bp 184°C). Or the aniline can be converted directly to acetanilide. The procedure calls for pure aniline, but note that the first step is to dissolve the aniline in water and hydrochloric acid. Your steam distillate is a mixture of aniline and water, both of which have been distilled. Are they not both water-white and presumably pure? Hence, an attractive procedure would be to assume that the steam distillate contains the theoretical amount of aniline and to add to it, in turn, appropriate amounts of hydrochloric acid, acetic anhydride, and sodium acetate, calculated from the quantities given in the next experiment.

Alternative choices

Salting out aniline

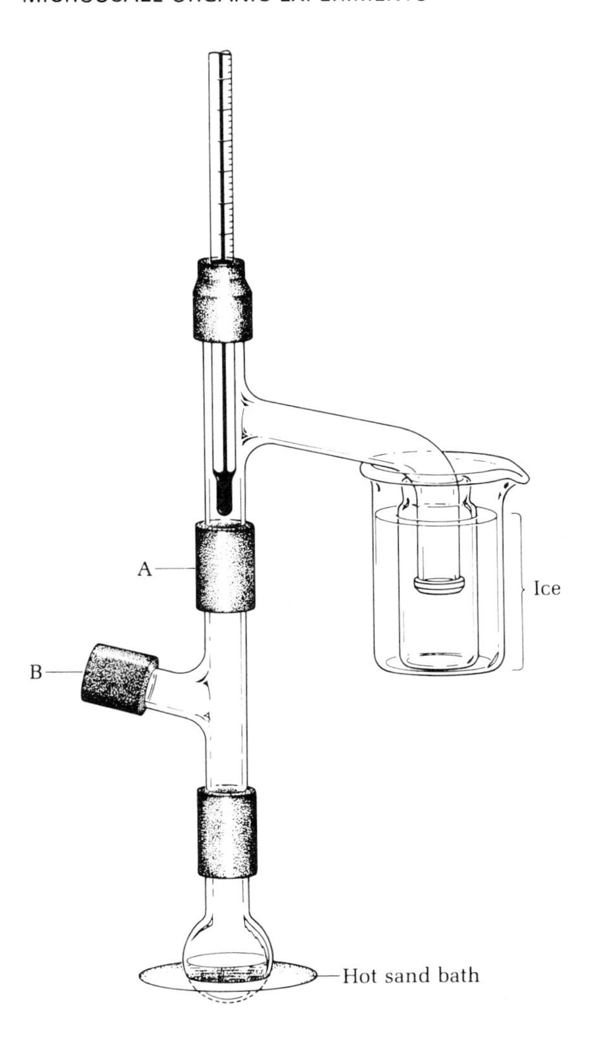

FIG. 41.1 Small-scale steam distillation apparatus. Clamp at A with three-prong micro clamp. Add water via syringe at B.

2. Acetanilide

Acetylation in Aqueous Solution

Handle aniline and acetic anhydride under hood; avoid contact with skin

Add to the steam distillate, which consists of a mixture of aniline and water, 0.43 mL of concentrated hydrochloric acid and make up the volume to 13 mL. Add 0.620 mL of acetic anhydride, and also prepare a solution of 0.53 g of anhydrous sodium acetate in 3 mL of water. Add the acetic anhydride to the solution of aniline hydrochloride with stirring and at once add the sodium acetate solution. Stir, cool in ice, and collect the product by suction filtration on the Hirsch funnel. It should be colorless and the mp close to 114°C. Since the acetanilide *must be completely dry* for use in the next step, it is advisable to put the material in a tared 25-mL Erlenmeyer flask and to heat it on the steam bath under evacuation until the weight is constant. (See Fig. 41.5.)

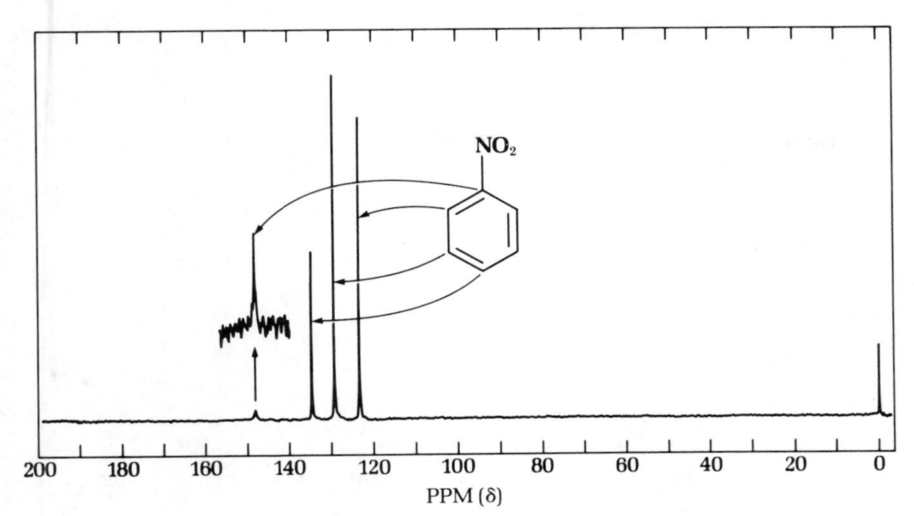

FIG. 41.2 ^{13}C nmr spectrum of nitrobenzene.

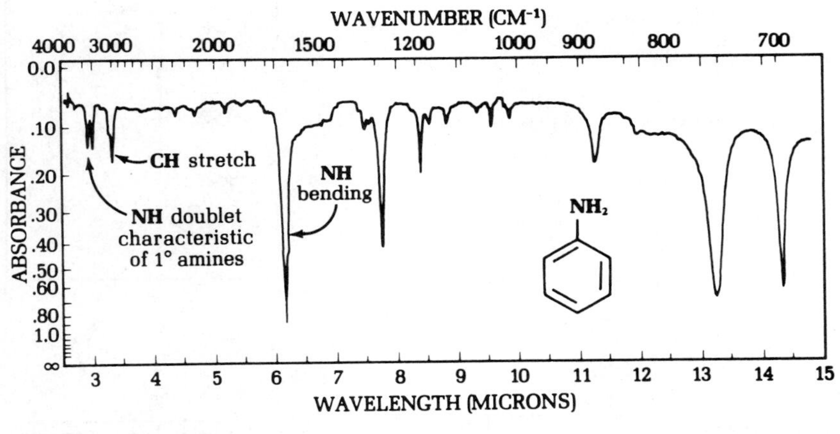

FIG. 41.3 Infrared spectrum of aniline.

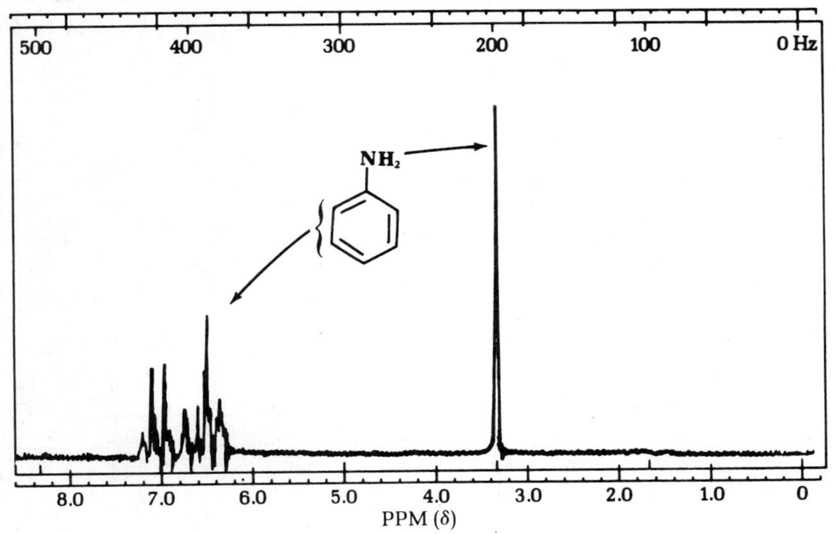

FIG. 41.4 ^{1}H nmr spectrum of aniline.

FIG. 41.5 Drying of acetanilide under reduced pressure. Heat flask on steam bath.

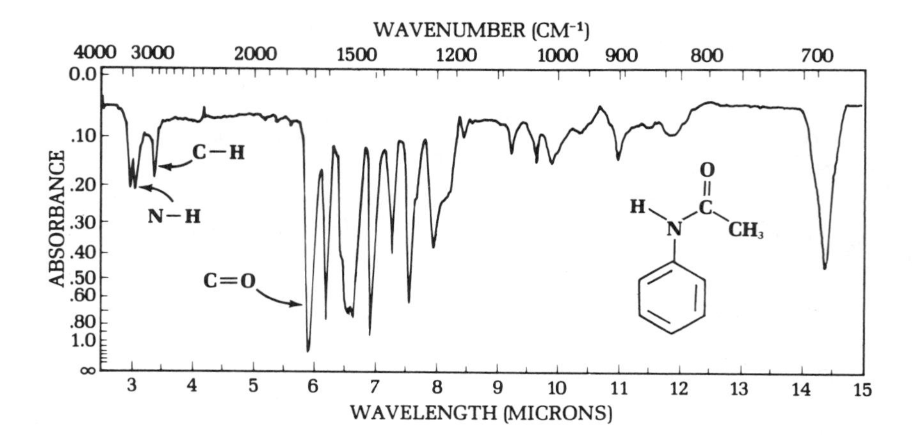

FIG. 41.6 Infrared spectrum of acetanilide in $CHCl_3$.

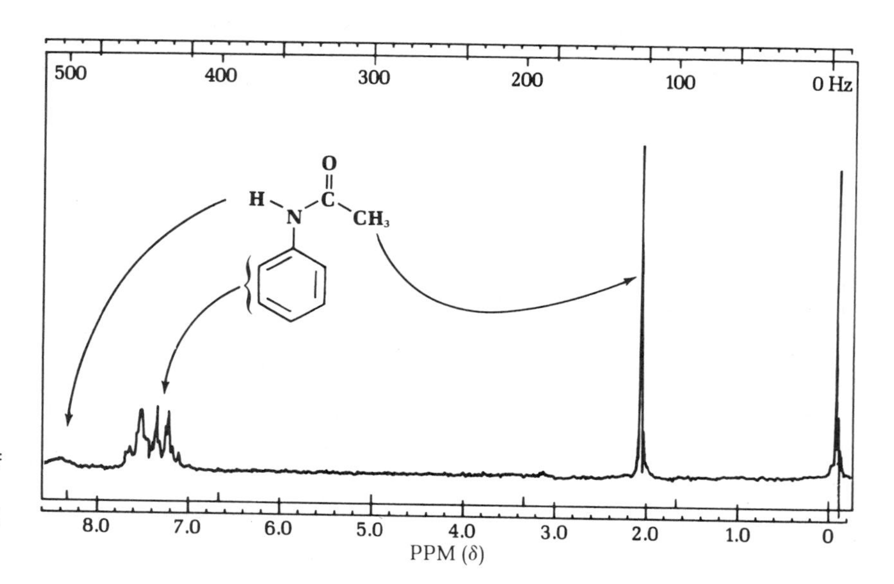

FIG. 41.7 ^1H nmr spectrum of acetanilide. The amide proton shows a characteristically broad peak.

3. Sulfanilamide

The chlorosulfonation of acetanilide in the preparation of sulfanilamide is conducted without solvent in the 25-mL Erlenmeyer flask used for drying the precipitated acetanilide from procedure 2. Because the reaction is most easily controlled when the acetanilide is in the form of a hard cake, the dried solid is melted by heating the flask over a hot sand bath; as the melt cools, the flask is swirled to distribute the material as it solidifies over the lower walls of the flask. Let the flask cool while making provision for trapping the hydrogen chloride evolved in the chlorosulfonation.

Fit the Erlenmeyer with a stopper connected by an inverted syringe needle to a short length of polyethylene tubing leading into a reaction tube that contains a small piece of damp cotton to trap hydrogen chloride vapors (Fig. 41.8). Cool the flask containing the acetanilide (0.250 g) in an ice bath, and add 0.625 mL of chlorosulfonic acid dropwise from a capped vial. Connect the flask to the gas trap between additions. The flask is now removed from the ice bath and swirled until a part of the solid has dissolved and the evolution of hydrogen chloride is proceeding rapidly. Occasional cooling in ice may be required to prevent too brisk a reaction. In 5–10 min the reaction subsides and only a few lumps of acetanilide remain undissolved.

Caution! *Chlorosulfonic acid is a corrosive chemical and reacts violently with water. Withdraw with pipette and pipetter (see Figs. 1.24 and 1.25). Neutralize any spills and drips immediately.*

When this point has been reached, heat the mixture on the steam bath for 10 min to complete the reaction, cool the flask in ice, and deliver the oily product by drops with a capillary dropper while stirring it into 3.5 mL of ice water contained in a 10-mL Erlenmeyer flask in the hood. Use extreme caution when adding the oil to ice water and when rinsing out any containers that have held chlorosulfonic acid. Rinse the flask with cold water and stir the precipitated *p*-acetaminobenzenesulfonyl chloride for a few minutes until an even suspension of granular white solid is obtained. Collect and wash the

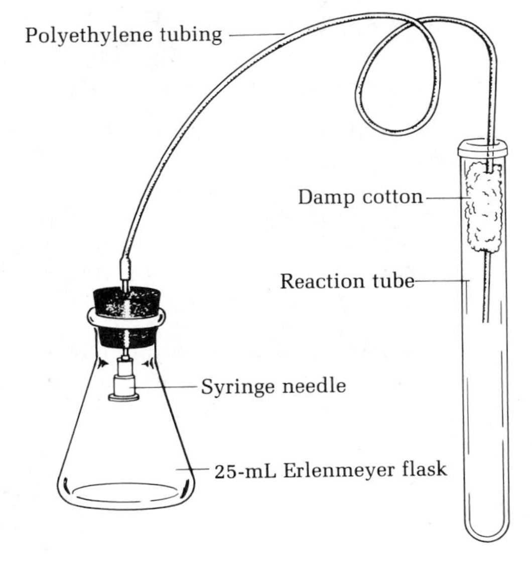

FIG. 41.8 Hydrogen chloride trap.

Do not let the mixture stand before addition of ammonia

solid on a Hirsch funnel. After pressing and draining the filter cake, transfer the solid to the rinsed reaction flask, add 0.75 mL of concentrated aqueous ammonia solution and 0.75 mL of water, and heat the mixture over a hot sand bath to just below the boiling point with occasional swirling in the hood. Heat the mixture in this manner for 5 min. During this treatment a change can be noted as the sulfonyl chloride undergoes transformation to a more pasty suspension of the amide. Cool the suspension well in an ice bath, collect the *p*-acetaminobenzenesulfonamide by suction filtration, press the cake on the Hirsch funnel, and allow it to drain thoroughly. Any excess water will unduly dilute the acid used in the next step.

Transfer the still-moist amide to the well-drained reaction flask, add 0.25 mL of concentrated hydrochloric acid and 0.5 mL of water, boil the mixture gently until the solid has all dissolved (5–10 min), and then continue the heating at the boiling point for 10 min longer (do not evaporate to dryness). The solution when cooled to room temperature should deposit no solid amide, but, if it is deposited, heating should be continued for a further period. The cooled solution of sulfanilamide hydrochloride is shaken with granulated decolorizing charcoal and filtered by removal of the solution with a Pasteur pipette. Place the solution in a 30-mL beaker and cautiously add an aqueous solution of 0.25 g of sodium bicarbonate while stirring to neutralize the hydrochloride. After the foam has subsided, test the suspension with indicator paper, and, if it is still acidic, add more bicarbonate until the neutral point is reached. Cool thoroughly in ice and collect the granular, white precipitate of sulfanilamide. The crude product (mp 161–163°C) on crystallization from alcohol or water affords pure sulfanilamide, mp 163–164°C, with about 90% recovery.

Questions

1. Why is an acetyl group added to aniline (making acetanilide) and then removed to regenerate the amine group in sulfanilamide?
2. What happens when chlorosulfonic acid comes in contact with water? Write balanced equations.
3. Acetic anhydride, like any anhydride, reacts with water to form a carboxylic acid. How then is it possible to carry out an acetylation in aqueous solution? What is the purpose of the hydrochloric acid and the sodium acetate in this reaction?
4. What happens when *p*-acetaminobenzenesulfonyl chloride is allowed to stand for some time in contact with water?

The Sandmeyer Reaction: 1-Bromo-4-Chlorobenzene and 2-Iodobenzoic Acid

42

PRELAB EXERCISE: *Outline the steps necessary to prepare 4-bromotoluene, 4-iodotoluene, and 4-fluorotoluene from benzene.*

The Sandmeyer reaction is a versatile means of replacing the amine group of a primary aromatic amine with a number of different substituents:

$$C_6H_5NH_2 \longrightarrow C_6H_5\overset{+}{N}{\equiv}N\ Cl^-$$

$$C_6H_5\overset{+}{N}{\equiv}N\ Cl^- \xrightarrow{HBF_4} C_6H_5\overset{+}{N}{\equiv}N\ BF_4^- \xrightarrow{\Delta} C_6H_5F$$

$$C_6H_5\overset{+}{N}{\equiv}N\ Cl^- \xrightarrow{CuCl} C_6H_5Cl$$

$$C_6H_5\overset{+}{N}{\equiv}N\ Cl^- \xrightarrow{CuBr} C_6H_5Br$$

$$C_6H_5\overset{+}{N}{\equiv}N\ Cl^- \xrightarrow{KI} C_6H_5I$$

$$C_6H_5\overset{+}{N}{\equiv}N\ Cl^- \xrightarrow{H_3O^+,\ \Delta} C_6H_5OH$$

$$C_6H_5\overset{+}{N}{\equiv}N\ Cl^- \xrightarrow{H_3PO_2} C_6H_5H$$

$$C_6H_5\overset{+}{N}{\equiv}N\ Cl^- \xrightarrow{CuCN} C_6H_5C{\equiv}N$$

The diazonium salt is formed by the reaction of nitrous acid with the amine in acid solution. Nitrous acid is not stable and must be prepared *in situ*; in strong acid it dissociates to form nitroso ions, ^+NO, which attack the nitrogen of the amine. The intermediate so formed loses a proton, rearranges, and finally loses water to form the resonance-stabilized diazonium ion.

$$NaNO_2 + HCl \rightleftharpoons NO\ddot{N}O + Na^+Cl^-$$

Sodium nitrite — Nitrous acid

$$H_3O^+ + HO\ddot{N}O \rightleftharpoons H_2O + H_2\overset{+}{\ddot{O}}\ddot{N}O \rightleftharpoons 2\,H_2O + :\overset{+}{N}{=}O$$

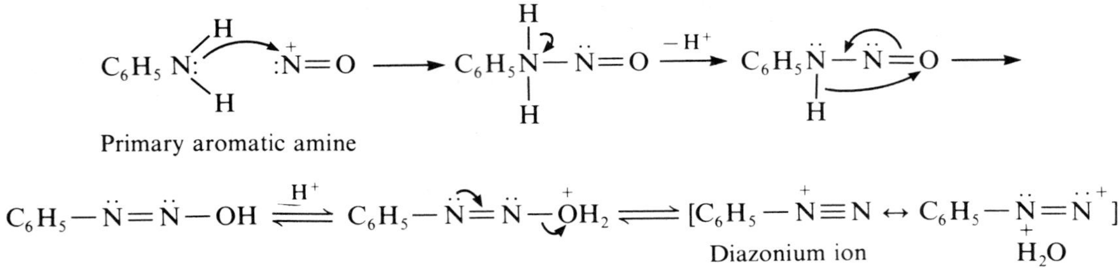

$$C_6H_5\ddot{N}H_2 + :\overset{+}{N}{=}O \longrightarrow C_6H_5\overset{+}{N}H_2{-}\ddot{N}{=}O \xrightarrow{-H^+} C_6H_5\ddot{N}H{-}\ddot{N}{=}O \longrightarrow$$

Primary aromatic amine

$$C_6H_5{-}\ddot{N}{=}\ddot{N}{-}OH \underset{}{\overset{H^+}{\rightleftharpoons}} C_6H_5{-}\ddot{N}{=}\ddot{N}{-}\overset{+}{O}H_2 \rightleftharpoons [C_6H_5{-}\overset{+}{N}{\equiv}N \leftrightarrow C_6H_5{-}\ddot{N}{=}\overset{+}{\ddot{N}}] + H_2O$$

Diazonium ion

The diazonium ion is reasonably stable in aqueous solution at 0°C; on warming up it will form the phenol, as seen above. A versatile functional group, it will undergo all of the reactions depicted above as well as couple to aromatic rings activated with substituents such as amino and hydroxyl groups to form the huge class of azo dyes (see Chapter 66).

$$C_6H_5\overset{+}{N}{\equiv}N\ Cl^- + C_6H_5N(CH_3)_2 \longrightarrow C_6H_5{-}N{=}N{-}C_6H_4{-}N(CH_3)_2$$

Benzene diazonium chloride — **N,N-Dimethylaniline** — **A diazo compound**

Diazonium salts are not ordinarily isolated because the dry solid is explosive.

1-Bromo-4-chlorobenzene

$$NaNO_2 + HCl \longrightarrow Na^+Cl^- + HONO$$

Sodium nitrite MW 69.01 — **Nitrous acid**

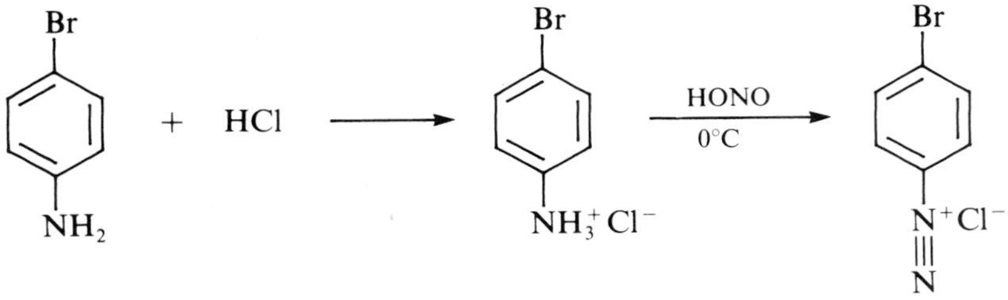

$$4\text{-}BrC_6H_4NH_2 + HCl \longrightarrow 4\text{-}BrC_6H_4NH_3^+Cl^- \xrightarrow[0^\circ C]{HONO} 4\text{-}BrC_6H_4N^+{\equiv}N\ Cl^-$$

4-Bromoaniline MW 172.03, mp 62–64°C — **4-Bromoaniline hydrochloride** — **4-Bromobenzenediazonium chloride**

$$\text{4-}BrC_6H_4N^+{\equiv}N\,Cl^- + CuCl \longrightarrow \text{4-}BrC_6H_4Cl$$

1-Bromo-4-chlorobenzene
MW 191.46, mp 68°C

In Experiment 2, 4-bromoaniline is dissolved in the required amount of hydrochloric acid, two more equivalents of acid are added, and the mixture cooled in ice to produce a paste of the crystalline amine hydrochloride. When this salt is treated at 0–5°C with one equivalent of sodium nitrite, nitrous acid is generated and reacts to produce the diazonium salt. The excess hydrochloric acid (beyond the two equivalents required to form the amine hydrochloride and react with sodium nitrite) maintains acidity sufficient to prevent formation of the diazoamino compound and rearrangement of the diazonium salt.

Experiments

1. Copper(I) Chloride Solution

$$2\,CuSO_4 \cdot 5\,H_2O + 4\,NaCl + NaHSO_3 + NaOH \longrightarrow 2\,CuCl + 3\,Na_2SO_4 + 2\,HCl + 10\,H_2O$$

$CuSO_4 \cdot 5H_2O$ MW 249.71; NaCl MW 58.45; $NaHSO_3$ MW 104.97; NaOH MW 40.01; CuCl MW 99.02

In a reaction tube dissolve 0.15 g of copper(II) sulfate crystals ($CuSO_4 \cdot 5\,H_2O$) in 0.5 mL of water by boiling and then add 50 mg of sodium chloride, which may give a small precipitate of basic copper(II) chloride. Prepare a solution of sodium sulfite from 35 mg of sodium bisulfite and 0.2 mL of 10% sodium hydroxide solution and add this dropwise to the hot copper(II) sulfate solution. Rinse the tube that contained the sulfite with a drop of water and add it to the copper(I) chloride mixture. Shake the reaction mixture well and then allow it to cool in ice while diazotizing the amine. When you are ready to use the copper(I) chloride, remove the water above the solid, wash the white solid once with water, remove the water, and dissolve the solid in 0.225 mL of concentrated hydrochloric acid. The solution is susceptible to air oxidation and should not stand for an appreciable time before use.

$NaHSO_3$, sodium bisulfite, sodium hydrogen sulfite not *$Na_2S_2O_2$, sodium hydrosulfite*

2. Diazotization

In a 10 × 100 mm reaction tube place 86 mg of 4-bromoaniline, and 0.25 mL of 10% hydrochloric acid. Warm the mixture on the sand bath to dissolve the amine and ensure transformation into hydrochloride. Cool the tube in ice (the hydrochloride will crystallize) and add an ice-cold solution of 35 mg of sodium nitrite in 0.1 mL of water. Use a drop of water to complete the transfer of the nitrite solution. Mix the solution thoroughly; after about 5 min the solid should have dissolved to give a clear solution of the diazonium salt.

3. Sandmeyer Reaction

Cool the copper(I) chloride solution in ice and add the diazonium chloride solution dropwise with thorough mixing using a Pasteur pipette. Rinse the tube that contained the diazonium chloride solution with a drop of water. Allow the reaction mixture to warm to room temperature and observe the reaction mixture closely. After 10 min extract the product with three 0.5-mL portions of ether; wash the ether extract with 0.2 mL of 10% sodium hydroxide solution, which will remove any 4-bromophenol; and then with 0.5 mL of saturated sodium chloride solution. Dry the ether solution over anhydrous sodium sulfate. Add sufficient drying agent so that it no longer clumps together. After 5 or 10 min remove the ether, wash the drying agent with ether, and evaporate the ether solutions in a small, tared filter flask. Use care in this evaporation because the product is volatile. Determine the weight of the crude product and then purify the product by sublimation at atmospheric pressure (see Chapter 7).

The 15-mL centrifuge tube (Fig. 42.1) is filled with ice water while the filter flask is heated on the sand bath. Wrap the filter flask in aluminum foil to direct heat to the upper surfaces and close the sidearm with a rubber pipette bulb. Replace the ice water in the centrifuge tube with room temperature water before removing it so that moisture will not condense on the tube. Scrape the product onto a piece of glazed paper, determine the weight and the melting point and analyze it for purity by thin-layer chromatography and infrared spectroscopy in a chloroform solution.

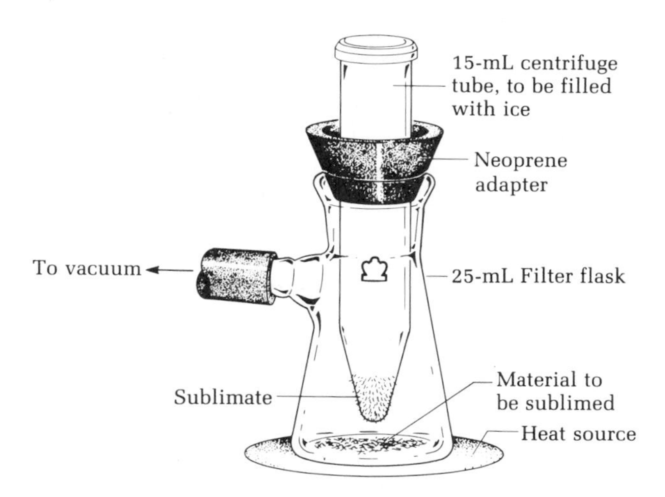

FIG. 42.1 Sublimation apparatus.

4. 2-Iodobenzoic Acid

2-Aminobenzoic acid + $NaNO_2$ + 2 HCl ⟶ 2-carboxybenzenediazonium chloride ($COOH$, $\overset{+}{N}{\equiv}N$ Cl^-) + Na^+Cl^- + 2 H_2O

2-Aminobenzoic acid
MW 137.14, mp 144–148°C

2-carboxybenzenediazonium chloride —KI→ **2-Iodobenzoic acid** + N_2 + K^+Cl^-

2-Iodobenzoic acid
MW 248.02
mp 126–163°C

To a reaction tube containing 137 mg of 2-aminobenzoic acid (anthranilic acid) add 1 mL of water and 0.25 mL of concentrated hydrochloric acid. Heat to dissolve the amino acid and form the hydrochloride, then cool the tube in ice while bubbling nitrogen gas through the solution to displace air from the tube. Cap the tube with a septum and then, with the tube in an ice bath, add a solution of 75 mg of sodium nitrite dissolved in 0.3 mL of water using a syringe. This addition should be made dropwise with thorough agitation of the reaction tube. Rinse the syringe with 0.1 mL of water, which is also injected into the reaction mixture. After 5 min a solution of 0.17 g of potassium iodide in 0.25 mL of water is added, when a brown complex partially separates.

The mixture is let stand without cooling for 5 min and then cautiously warmed to 40°C. At this point a vigorous reaction begins with nitrogen gas evolution, foaming, and separation of a tan solid. After reacting for 10 min the mixture is heated in a beaker of boiling water for 10 min and then cooled in ice. A few milligrams of sodium sulfite are added to destroy any iodine present and the granular tan product collected and washed with water.

The still-moist product is dissolved in 0.7 mL of ethanol, 0.35 mL of water is added, and the hot, brown solution treated with enough granular Norit to remove most of the color. The solution is transferred to another reaction tube to remove the decolorizing charcoal and diluted at the boiling point with 0.35–0.40 mL of water and let stand. 2-Iodobenzoic acid separates in large, slightly yellow needles. The yield should be about 150 mg with a mp near 164°C. 2-Iodobenzoic acid is the precursor for benzyne (Chaps. 48 and 49).

Questions

1. Nitric acid is generated by the action of sulfuric acid on sodium nitrate. Nitrous acid is prepared by the action of hydrochloric acid on sodium nitrite. Why is nitrous acid prepared *in situ*, rather than obtained from the reagent shelf?

2. How would 4-bromoaniline be prepared from benzene?

Malonic Ester Synthesis: Synthesis of a Barbiturate 43

PRELAB EXERCISE: *Which barbiturate discussed in this chapter cannot be synthesized by the acetoacetic ester reaction?*

Barbiturates are central nervous system depressants used as hypnotic drugs and anesthetics. They are all derivatives of barbituric acid (R=R′=H), which has no sedative properties. It is called an acid because the carbonyl groups render the imide hydrogens acidic:

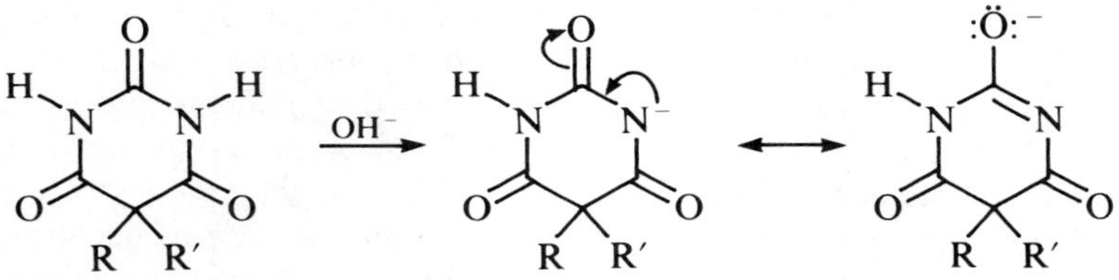

Barbituric acid R = R′ = H

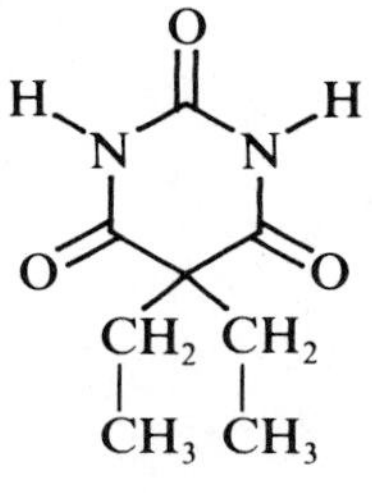

Veronal

Barbituric acid was first synthesized in 1864 by Adolph von Baeyer. It apparently was named at a tavern on St. Barbara's day and is derived from urea. At the turn of the century the great chemist Emil Fischer synthesized the first hypnotic (sleep-inducing) barbiturate, the 5,5-diethyl derivative, at the direction of von Mering. von Mering, who made the seminal discovery that removal of the pancreas causes diabetes, named the new derivative of barbituric acid Veronal because he regarded Verona as the most restful city on earth.

Barbiturates are the most widely used sleeping pills and are classified according to their duration of action. Because it is quite easy to put almost any conceivable R group on to diethyl malonate, several thousand derivatives of barbituric acid have been synthesized. Studies of these derivatives have shown that as the alkyl chain, R, gets longer or as double bonds are introduced into the chain, the duration of action and the time of onset of action decreases. Maximum sedation occurs when the alkyl chains contain five or six carbons, as found in amobarbital (R=ethyl, R′=3-methylbutyl) and pentobarbital (R=ethyl, R′=1-methylbutyl).

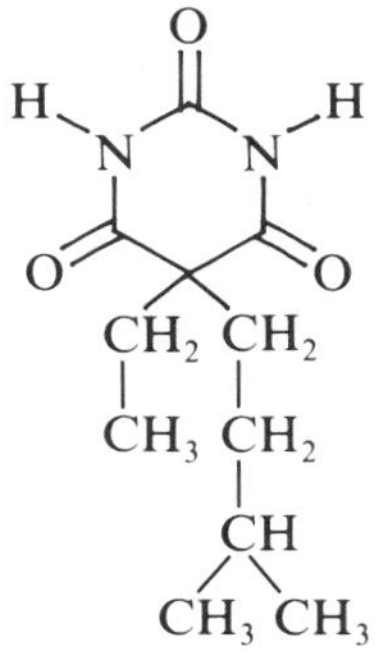

Amobarbital 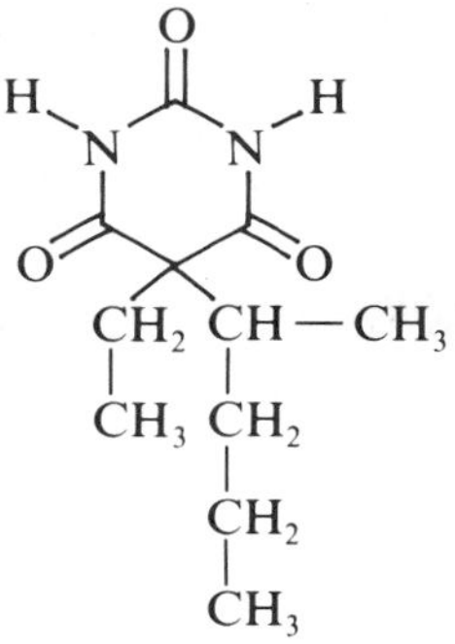 **Pentobarbital** 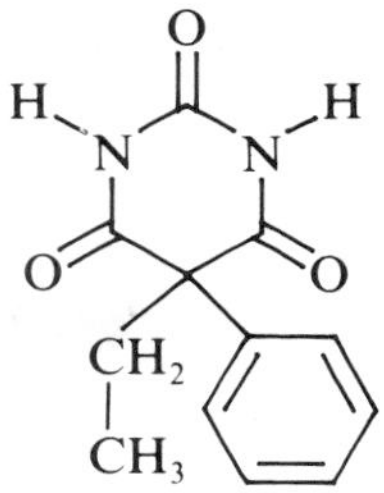 **Phenobarbital**

These two molecules illustrate how subtle changes in molecular structure can affect action. Amobarbital requires 30 min to take effect and sedation lasts for 5–6 h, while pentobarbital takes effect in 15 min and sedation lasts only 2–3 h. Phenobarbital (R=ethyl, R′=phenyl), on the other hand, requires over an hour to take effect, but sedation lasts for 6–10 h. When the alkyl chains are made much longer the sedative properties decrease and the substances become anticonvulsants, which are used to treat epileptic seizures. If the alkyl group is too long or is substituted at one of the two nitrogens, convulsants are produced.

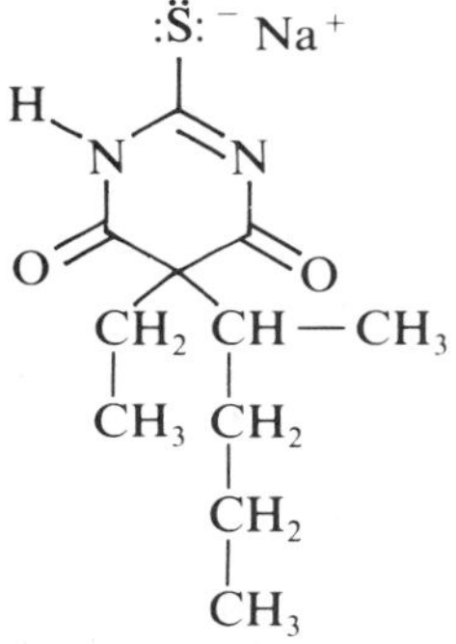

Sodium pentothal

The usual dosage is 10–50 mg per pill. Continuous use of barbiturates leads to physiologic dependence (addiction) and withdrawal symptoms are just like those experienced by a heroin addict. Replacing the oxygen atom at carbon-2 with sulfur results in the compound pentothal. Given intravenously as the sodium salt, it is a fast-acting general anesthetic. Like all general anesthetics, the details of its mode of action are unknown. In low, subanesthetic doses sodium pentothal reduces inhibitions and the will to resist and thus functions as the so-called "truth serum," apparently because it takes less mental effort to tell the truth than to prevaricate.

Most barbiturates are made from diethylmalonate. The methylene protons between the two carbonyl groups are acidic and will give a highly stabilized enolate anion.

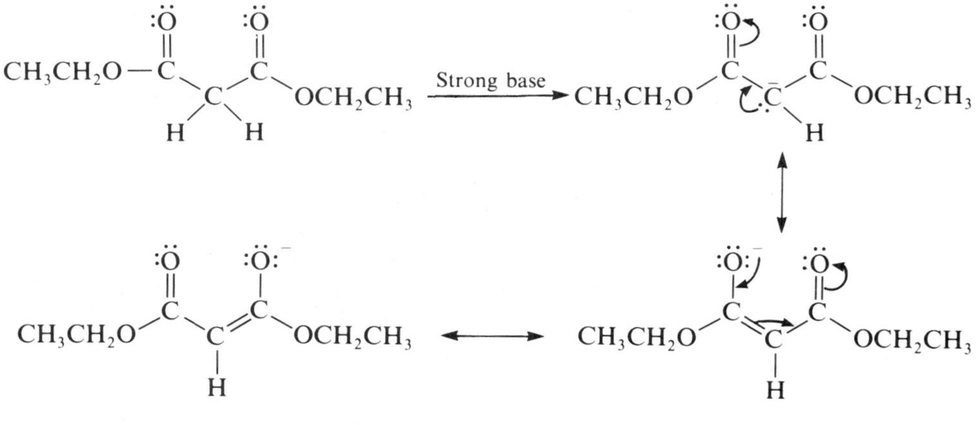

$$CH_3CH_2O-C(=O)-CH_2-C(=O)-OCH_2CH_3 \xrightarrow{\text{Strong base}} CH_3CH_2O-C(=O)-\bar{C}H-C(=O)-OCH_2CH_3$$

The acidic protons can be removed with a strong base, most often sodium ethoxide in dry ethanol. In the present experiment carbonate functions as the strong base because in the absence of water it is not solvated. In association with the tricaprylmethylammonium ion, it is soluble in the organic phase and can react with the diethyl malonate.

$$K_2CO_{3(s)} + 2\ H_3C{-}\overset{+}{N}[(CH_2)_7CH_3]_3\ Cl^- \rightleftharpoons \left(H_3C{-}\overset{+}{N}[(CH_2)_7CH_3]_3\right)_2 CO_3^{=} + 2\ KCl$$

Tricaprylmethylammonium chloride
Phase transfer catalyst

Soluble in organic phase

The enolate anion of diethylmalonate can be alkylated by an S_N2 displacement of bromide:

$$CH_3CH_2{-}C(=O){-}\overset{-}{C}H{-}C(=O){-}OCH_2CH_3 + CH_3CH_2CH_2CH_2{-}Br \longrightarrow CH_3CH_2O{-}C(=O){-}CH(CH_2CH_2CH_2CH_3){-}C(=O){-}OCH_2CH_3 + Br^-$$

and because the product still contains one acidic proton, the process can be repeated using the same or a different alkyl halide. The product is often hydrolyzed and decarboxylated to give a carboxylic acid:

$$CH_3CH_2O{-}C(=O){-}CH(CH_2CH_2CH_2CH_3){-}C(=O){-}OCH_2CH_3 \xrightarrow[\Delta]{H^+,\ H_2O} O{=}C(OH){-}CH(CH_2CH_2CH_2CH_2){-}C(=O){-}OH \xrightarrow{-CO_2} HO{-}C(=O){-}CH_2CH_2CH_2CH_2CH_3$$

In this experiment the *n*-butyldiethylmalonate is allowed to react with urea in the presence of a strong base, sodium ethoxide, to give the barbiturate.

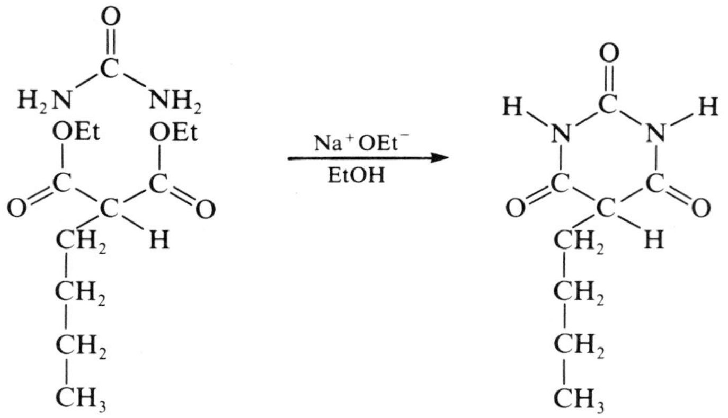

$$H_2NC(=O)NH_2 + EtOC(=O)CH(CH_2CH_2CH_2CH_3)C(=O)OEt \xrightarrow[\text{EtOH}]{Na^+OEt^-} \text{n-Butylbarbituric acid}$$

n-Butylbarbituric acid

Barbiturates with only one alkyl substituent are rapidly degraded by the body and therefore are physiologically inactive.

Experiments

1. *n*-Butyldiethylmalonate

$$CH_3CH_2OCCH_2COCH_2CH_3 + K_2CO_3 \xrightarrow{CH_3\overset{+}{N}[(CH_2)_7CH_3]_3Cl^-} [CH_3CH_2OCC\overset{-}{H}COCH_2CH_3]K^+ + H_2O + CO_2$$

Diethylmalonate
MW 160.17
bp 199.3°C

Tricaprylmethylammonium chloride
MW 404.17

$$\xrightarrow{CH_3CH_2CH_2CH_2Br} CH_3CH_2OCCH(CH_2CH_2CH_2CH_3)COCH_2CH_3$$

1-Bromobutane
MW 137.03
bp 101.6°C

***n*-Butyldiethylmalonate**
MW 216.28
bp 235–240°C

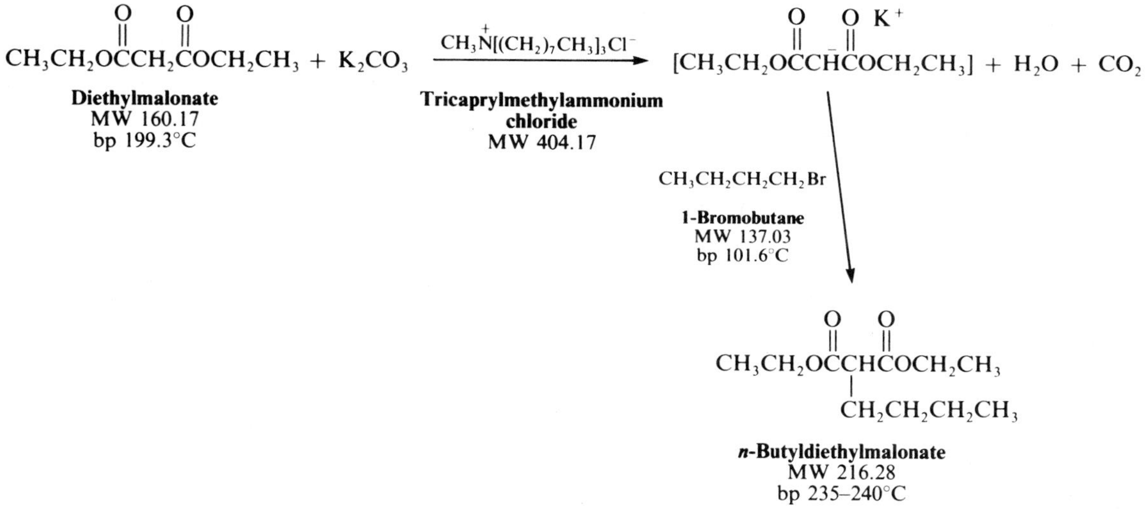

Into a 5-mL long-necked round-bottomed flask weigh 40 mg of tricaprylmethylammonium chloride, 400 mg of diethylmalonate, 350 mg of

1-bromobutane, and 415 mg of anhydrous potassium carbonate. Attach the empty distillation column as an air condenser and reflux the mixture for 1.5 h. Allow the mixture to cool somewhat and then cool it in ice. Using 2.5 mL of water transfer the reaction mixture to a reaction tube, shake the mixture, draw off the organic layer, and extract the aqueous layer with two 1.5-mL portions of ether. Dry the combined organic extracts over anhydrous sodium sulfate, remove the ether solution from the sodium sulfate, and complete the drying of the ether over Drierite (commercial anhydrous calcium sulfate). Place the dry ether solutions in a *dry*, tared 5-mL round-bottomed long-necked flask and remove the ether by distillation or evaporation in the hood. Complete the removal of the ether by connecting the flask to the aspirator. Determine the weight and calculate the yield of product. Analyze it by thin-layer chromatography on silica gel using dichloromethane as the eluent. Compare the product with a sample of authentic, pure *n*-butyldiethylmalonate and with the starting materials diethylmalonate and bromobutane. If the sample is primarily the desired product, it may be used in the next reaction. It must be perfectly clear in appearance and absolutely dry.

2. *n*-Butylbarbituric acid

$$H_2NC(=O)NH_2 + CH_3CH_2OC(=O)CH(CH_2CH_2CH_2CH_3)C(=O)OCH_2CH_3 \xrightarrow{Na^+\,{}^-OCH_2CH_3} n\text{-Butylbarbituric acid}$$

Urea
MW 60.06
mp 133–135°C

***n*-Butyldiethylmalonate**
MW 216.28
bp 235–240°C

***n*-Butylbarbituric acid**
MW 184.18
mp 209–210°C

To a perfectly dry 5-mL round-bottomed long-necked flask add 2.5 mL of absolute (100%, anhydrous) ethanol. Add to the flask the empty distillation column, which will function as an air condenser. Fit the top of the column with a rubber septum bearing an empty syringe needle, which will serve to prevent moisture from diffusing into the reaction mixture (Fig. 43.1). To the ethanol add 23 mg of sodium by removing the septum and dropping the metal down the air condenser. The sodium metal is most easily handled in the form of small spheres, which are stored in mineral spirits. They should have a grey appearance and not have a yellowish crust on the outside of each piece. Using a pair of tweezers remove a sphere, rinse it in ligroin, blot off the solvent on a piece of filter paper, and weigh it into a dry closed vial. Transfer the sodium piece by piece to the flask containing the ethanol.

***Caution!** Handle sodium metal with care. It will react violently with water and can cause fire or an explosion. Avoid all contact with moisture in any form. Dispose of scrap material in the designated beaker that contains 1-propanol.*

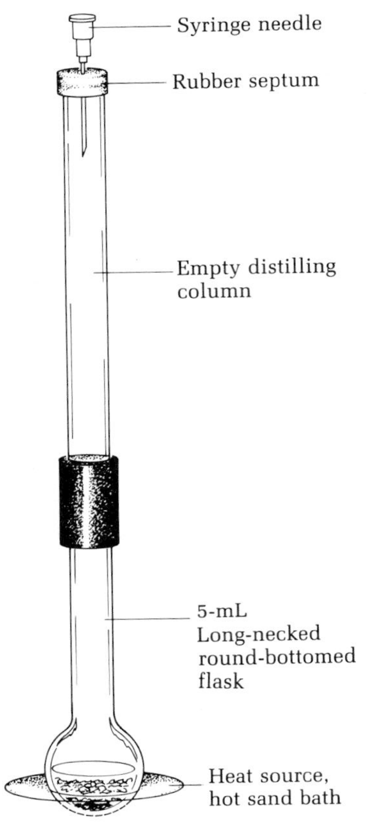

FIG. 43.1 Apparatus for synthesis of *n*-butylbarbituric acid.

Be extremely careful to have no water near the sodium. If a piece of sodium is accidentally dropped place it in the receptacle provided in the hood. Do not dispose of it in a trash can as it can easily start a fire due to oxidation. If necessary warm the flask on the sand bath to dissolve the last traces of sodium. Do not proceed until all traces of sodium have disappeared. To the sodium ethoxide solution add 216 mg of *dry n*-butylmalonate (synthesized in the previous reaction or a commercial sample (Aldrich)). Immediately add 60 mg of urea dissolved in 1.1 mL of absolute ethanol. Reflux the resulting mixture on a warm sand bath or a steam bath for 2 h. What solid separates during this reaction? It may cause the mixture to bump during the first part of the refluxing period.

At the completion of the reaction acidify the solution with 1 mL of 10% hydrochloric acid (check pH with indicator paper) and then reduce the volume in the flask to one-half by distillation of the ethanol or by removal of the solvent on the rotary evaporator. Cool the solution on ice until no more product crystallizes; then collect it by vacuum filtration on the Hirsch funnel. Wash unreacted *n*-butyldiethylmalonate (detected by its odor) from the crystals with ligroin, press it dry, and then recrystallize the product from boiling water (20 mL of water per gram of product). The acid crystallizes slowly. Cool the solution for at least 30 min before collecting the product by vacuum filtration. Dry the barbituric acid on filter paper until the next laboratory period, then weigh it, calculate the percent yield, and determine the mp. Pure *n*-butylbarbituric acid melts at 209–210°C.

Questions

1. The anion of diethylmalonate is usually made by reacting the diester with sodium methoxide. What weight of sodium would be required in the present experiment if that method were employed?

2. Outline all steps in the synthesis of pentobarbital.

3. What problem might be encountered if $BrCH_2C(CH_3)_3$ were used as the halide in this synthesis?

Photochemistry: The Synthesis of Benzopinacol

44

PRELAB EXERCISE: Draw a mechanism for the base-catalyzed cleavage of benzopinacol.

Photochemistry, as the name implies, is the chemistry of reactions initiated by light. A molecule can absorb light energy and then undergo isomerization, fragmentation, rearrangement, dimerization, or hydrogen atom abstraction. The last reaction is the subject of this experiment. On irradiation with sunlight, benzophenone abstracts a proton from the solvent 2-propanol and becomes reduced to benzopinacol.

$$C_6H_5\overset{\overset{\displaystyle O}{\|}}{C}C_6H_5 \quad + \quad CH_3\overset{\overset{\displaystyle OH}{|}}{C}HCH_3 \quad \xrightarrow{h\nu} \quad C_6H_5-\underset{\underset{\displaystyle C_6H_5}{|}}{\overset{\overset{\displaystyle OH}{|}}{C}}-\underset{\underset{\displaystyle C_6H_5}{|}}{\overset{\overset{\displaystyle OH}{|}}{C}}-C_6H_5$$

Benzophenone
MW 182.21
mp 48°C

2-Propanol
MW 60.09
bp 82°C

Benzopinacol
MW 366.44 mp 189°C

The energy of light varies with its frequency where h is Planck's constant, ν is the frequency of the light, λ is its wavelength, and c is the speed of light.

$$\Delta E = h\nu = h\frac{c}{\lambda}$$

The fact that benzophenone is colorless means it does not absorb visible light, yet irradiation of benzophenone in alcohol results in a chemical change. Pyrex glass is not transparent to ultraviolet light with a wavelength shorter than 290 nm, so some wavelength between 290 nm and 400 nm (the edge of the visible region) must be responsible. The uv spectrum of benzophenone indicates an absorption band centered at about 355 nm. Light of that wavelength is absorbed by benzophenone.

What, in a general sense, occurs when a molecule absorbs light? A photon is absorbed only if its energy (wavelength) corresponds exactly to the difference between two electronic energy levels in the molecule. In benzophenone the electrons most loosely held and thus most easily excited are the two pairs of nonbonded electrons on the carbonyl oxygen, called the *n*-electrons.

$$\rangle C{=}\ddot{O}\text{:} \leftarrow n\text{-electrons}$$

These electrons have paired spins $\rangle C{=}O$ ⇅ and only one is excited by the photon of light. The electron goes into the lowest unoccupied excited energy level, the π^*. Each electronic energy level of benzophenone has within it many vibrational and rotational energy levels. An electron can reside in many of these in the lower electronic energy level, the ground state, S_0, and it can be promoted to many vibrational and rotational energy levels in the first excited state (S_1). Hence the ultraviolet spectrum does not appear as a single sharp peak, but as a band made up of many peaks that arise from transitions between these many energy levels. See Fig 44.1.

The spin of the electron cannot change in going from the ground state, S_0, to the excited singlet state, S_1 (conservation of angular momentum). Once in a higher vibrational or rotational state it can drop to S_1 by vibrational relaxation, losing some of its energy as heat. It then undergoes either fluorescence or intersystem crossing. As seen in Fig 44.1, the rate for fluorescence, a light-emitting process in which the electron returns to the ground state, is 10^4 times slower than intersystem crossing; so benzophenone does not fluoresce. The electron flips its orientation during intersystem crossing so

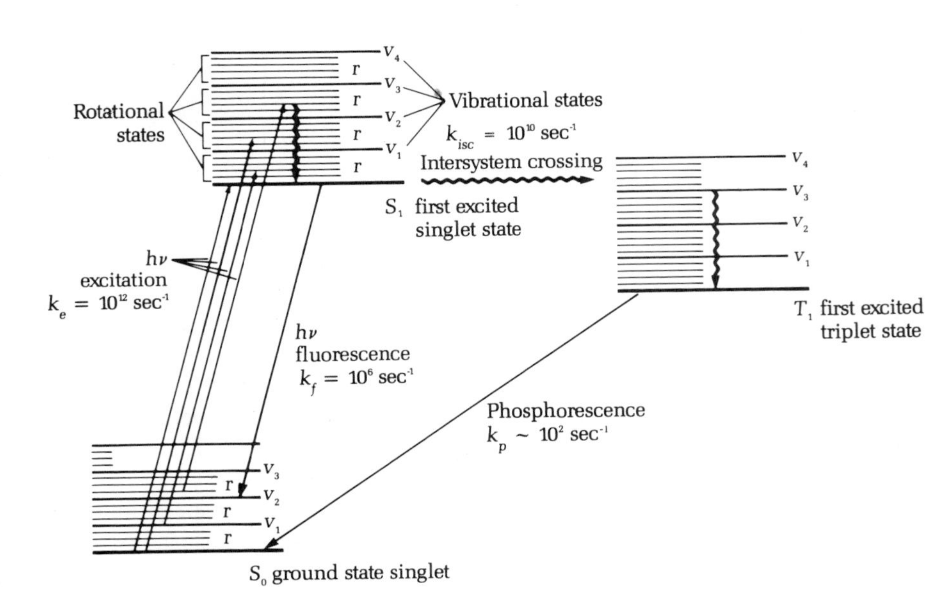

FIG. 44.1 Electronic energy levels and possible transitions.

that it has the same orientation as the electron with which it was paired in the ground state. In this new state, called the *triplet*, it can lose energy as light; but this process, phosphorescence, is a slow one. The lifetime of the triplet state is long enough for chemical reactions to take place. The triplet can also lose energy as heat in a radiationless transition but the probability of this happening is relatively low. This situation can be represented diagrammatically:

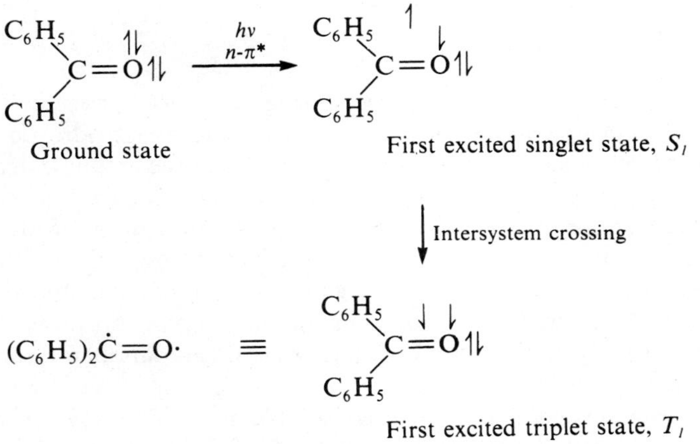

$$(C_6H_5)_2C{=}O \xrightarrow[n\text{-}\pi^*]{h\nu} (C_6H_5)_2C{=}O$$

Ground state — First excited singlet state, S_1

Intersystem crossing

$$(C_6H_5)_2\dot{C}{=}O\cdot \equiv (C_6H_5)_2C{=}O$$

First excited triplet state, T_1

This T_1 state is a diradical and can abstract a methine proton from the solvent, then a hydroxyl proton to give acetone and diphenyl hydroxy radical, which then dimerizes to give the product.

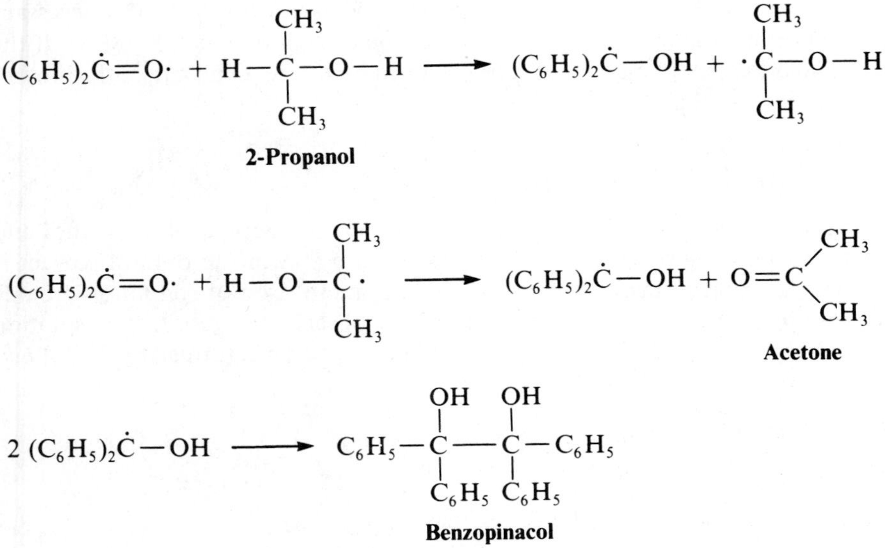

$$(C_6H_5)_2\dot{C}{=}O\cdot + H{-}C(CH_3)_2{-}O{-}H \longrightarrow (C_6H_5)_2\dot{C}{-}OH + \cdot C(CH_3)_2{-}O{-}H$$

2-Propanol

$$(C_6H_5)_2\dot{C}{=}O\cdot + H{-}O{-}C(CH_3)_2\cdot \longrightarrow (C_6H_5)_2\dot{C}{-}OH + O{=}C(CH_3)_2$$

Acetone

$$2\,(C_6H_5)_2\dot{C}{-}OH \longrightarrow C_6H_5{-}C(OH)(C_6H_5){-}C(OH)(C_6H_5){-}C_6H_5$$

Benzopinacol

Experiments

1. Benzopinacol

This experiment should be done when there is good prospect for several hours of bright sun. The benzopinacol is cleaved by alkali to benzhydrol and benzophenone (experiment 2) and it is rearranged in acid to benzopinacolone (experiment 3).

Procedure

In a 2-mL or l-dram ampoule dissolve 100 mg of benzophenone in 1 mL of isopropyl alcohol by warming on a sand bath. Add a microdrop of acetic acid, cool the ampoule in dry ice, and seal the neck of the ampoule in a flame (Fig. 44.2). Set the ampoule outside in bright sunlight. On a larger scale this experiment can take as long as a week because of absorption of ultraviolet radiation by the thick glass walls of a flask and by self-absorption of the thick layers of solution. On a microscale the glass of an ampoule is very thin and the thickness of the entire solution is small, so the reaction proceeds very rapidly. A rough parabolic reflector fashioned from aluminum foil can speed the reaction even more.

Reaction time: one day to one week

Since benzopinacol is but sparingly soluble in the isopropyl alcohol, its formation is followed by the separation of small, colorless crystals (benzophenone forms large, thick prisms) from around the walls of the ampoule. If the reaction mixture is exposed to direct sunlight throughout the daylight hours, the first crystals separate in less than an hour and the reaction is complete in a day. In winter the reaction will take longer and any benzophenone which crystallizes must be brought into solution by warming on the steam bath. When the reaction appears to be over, chill the tube if necessary and collect the product. The material should be pure, mp 188–189°C. If the yield is low, more material can be obtained by further exposure of the mother liquor to sunlight.

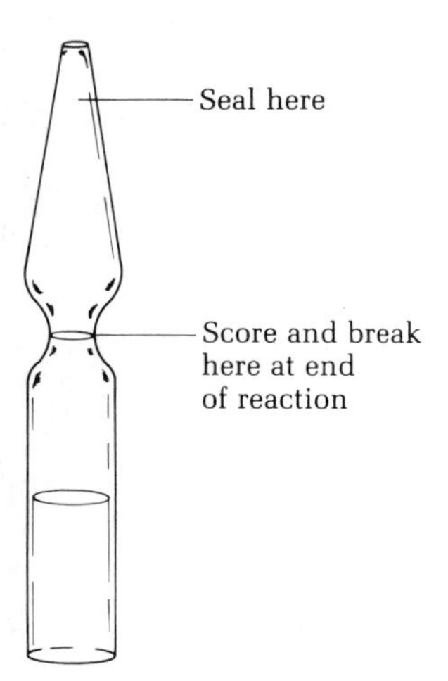

FIG. 44.2 Microscale photolysis in 1-dram (1.77-mL) ampoule.

2. Alkaline Cleavage

Suspend a small test sample (about 15 mg) of benzopinacol in alcohol and heat to boiling on a steam bath, making sure that the amount of solvent is not sufficient to dissolve the solid. Add a micro drop of sodium hydroxide solution, heat for a minute or two, and observe the result. The solution contains equal parts of benzhydrol and benzophenone, formed by the following reaction:

$$(C_6H_5)_2C(OH)-C(OH)(C_6H_5)_2 \xrightarrow{\overset{-}{R}O\overset{+}{N}a} (C_6H_5)_2CH(OH) + O{=}C(C_6H_5)_2$$

Benzopinacol mp 189°C — **Benzhydrol** mp 68°C — **Benzophenone** mp 48°C

The low-melting products resulting from the cleavage are much more soluble than the starting material. Analyze the mixture by TLC.

Benzophenone can be converted into benzhydrol in nearly quantitative yield by following the procedure outlined above for the preparation of benzopinacol, modified by addition of a *very* small piece of sodium (5 mg) instead of the acetic acid. The reaction is complete when, after exposure to sunlight, the greenish-blue color disappears. To obtain the benzhydrol the solution is diluted with water, acidified, and evaporated. Benzopinacol is produced as before by photochemical reduction, but it is at once cleaved by the sodium alkoxide, the benzophenone formed by cleavage is converted into more benzopinacol, cleaved, and eventually consumed.

3. Pinacolone Rearrangement

This acid-catalyzed, carbonium ion rearrangement is characterized by rapidity and by the high yield.

In a reaction tube place 50 mg of benzopinacol, 0.25 mL of acetic acid, and one *very* small crystal of iodine (0.0005 g). Heat to boiling for a minute or two on the sand bath until the crystals are dissolved, and then reflux the red solution for 5 min. On cooling, the pinacolone separates as a stiff paste. Thin the paste with alcohol, collect the product by vacuum filtration on the Hirsch funnel, and wash it free of iodine with cold 95% ethanol. The material should be pure; the expected yield is 95%.

$$(C_6H_5)_2C(OH)-C(OH)(C_6H_5)_2 \underset{}{\overset{H^+}{\rightleftharpoons}} \left[(C_6H_5)_2C(\overset{+}{O}H_2)-C(OH)(C_6H_5)_2\right]$$

Benzopinacol

$$\xrightarrow{-H_2O} \left[(C_6H_5)_2\overset{+}{C}-C(:OH)(C_6H_5)_2\right]$$

(Carbocation)

$$\longrightarrow \left[(C_6H_5)_3C-C(=\overset{+}{O}H)-C_6H_5\right] \overset{-H^+}{\rightleftharpoons} (C_6H_5)_3C-C(=O)-C_6H_5$$

Benzopinacolone
mp 179–180°C

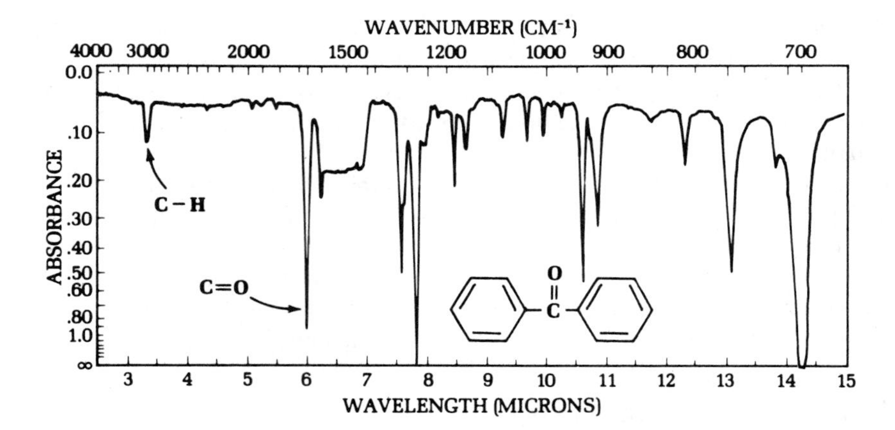

FIG. 44.3 Infrared spectrum of benzophenone in CS_2.

Question

Would the desired reaction occur if ethanol or *t*-butyl alcohol were used instead of isopropyl alcohol in the attempted photochemical dimerization of benzophenone?

Luminol: Synthesis of a Chemiluminescent Substance

45

PRELAB EXERCISE: *Write a balanced equation for the reduction of nitrophthalhydrazide to aminophthalhydrazide using sodium hydrosulfite.*

The oxidation of 3-aminophthalhydrazide, **1**, commonly known as luminol, is attended with a striking emission of blue-green light. Most exothermic chemical reactions produce energy in the form of heat, but a few produce light and release little or no heat. This phenomenon, chemiluminescence, is usually an oxidation reaction. In the case of luminol an alkaline solution of the compound is allowed to react with a mixture of hydrogen peroxide and potassium ferricyanide. The dianion **2** is oxidized to the singlet excited state (two unpaired electrons of like spin) of the amino phthalate ion, **3**. This slowly undergoes intersystem crossing to the triplet excited state (two unpaired electrons of opposite spin) **4**, which decays to the ground state ion, **5**, with the emission of one quantum of light (a photon) per molecule. Very few molecules are more efficient in chemiluminescence than luminol.

Luminol, **1**, is made by reduction of the nitro derivative, **8**, formed on thermal dehydration of a mixture of 3-nitrophthalic acid, **6**, and hydrazine, **7**. An earlier procedure for effecting the first step called for addition of hydrazine sulfate to an alkaline solution of the acid, evaporation to dryness, and baking the resulting mixture of the hydrazine salt and sodium sulfate at 165°C, and it required a total of 4.5 h for completion. This working time can be drastically reduced by adding high-boiling (bp 290°C) triethylene glycol to an aqueous solution of the hydrazine salt, distilling the excess water, and raising the temperature to a point where dehydration to **8** is complete within a few minutes. Nitrophthalhydrazide, **8**, is insoluble in dilute acid but soluble in alkali, by virtue of enolization; and it is conveniently reduced to luminol, **1**, by sodium hydrosulfite (sodium dithionite) in alkaline solution. In dilute, weakly acidic or neutral solution luminol exists largely as the dipolar ion **9**, which exhibits beautiful blue fluorescence.[1]

1. Several methods of demonstrating the chemiluminscence of luminol are described by E. H. Huntress, L. N. Stanley, and A. S. Parker, *J. Chem. Ed.*, **11**, 142 (1934). The mechanism of the reaction has been investigated by Emil H. White and co-workers (*J. Am. Chem. Soc.*, **86**, 940 and 942 (1964)).

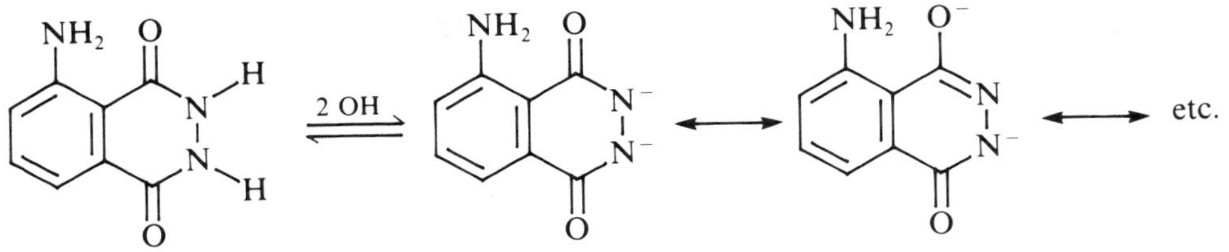

NH_2 O N H N H O

1

Luminol
(3-Aminophthalhydrazide)

2 OH

NH_2 O N^- N^- O ⟷ NH_2 O^- N N^- O ⟷ etc.

2

H_2O_2, $K_3Fe(CN)_6$

NH_2 COO^- COO^-

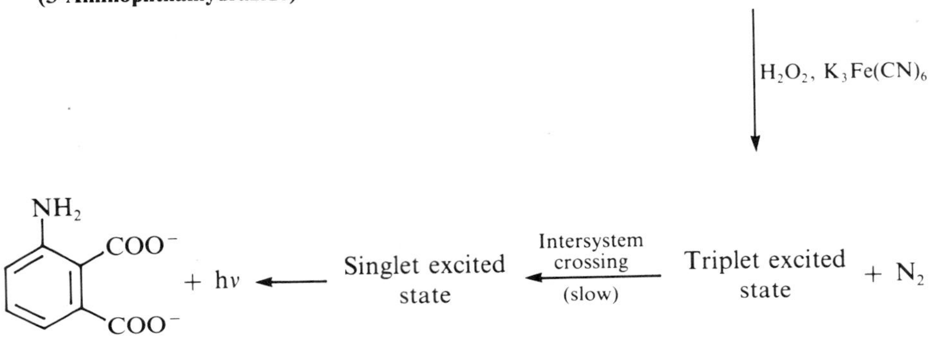

Ground state

5

+ hν ⟵ Singlet excited state

4

Intersystem crossing (slow)

Triplet excited state + N_2

3

NO_2 COOH COOH

6

3-Nitrophthalic acid
MW 211.13
mp 222°C

+ NH_2 NH_2

7

Hydrazine

Δ

NO_2 O C N H N H O

8

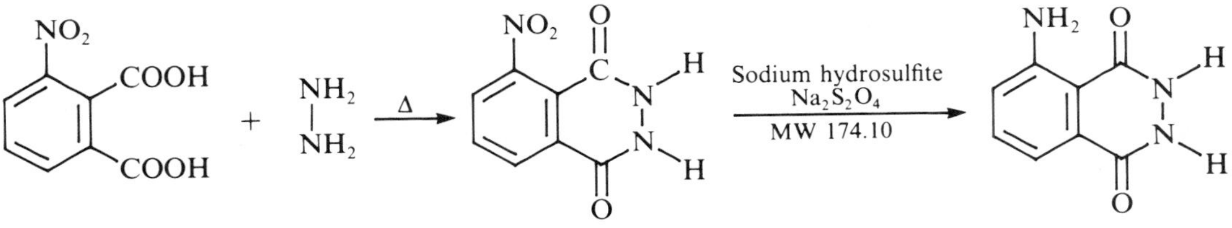

Sodium hydrosulfite
$Na_2S_2O_4$
MW 174.10

NH_2 O N H N H O

1

Luminol
MW 177.16
mp 332°C

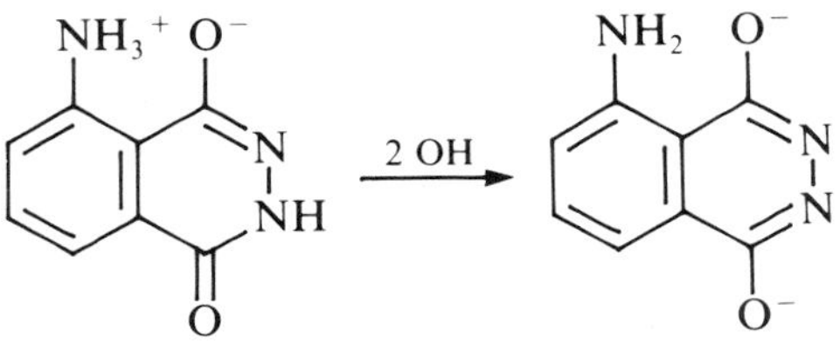

NH_3^+ O^- N NH O

9

2 OH

NH_2 O^- N N O^-

10

An alkaline solution contains the doubly enolized anion **10** and displays particularly marked chemiluminescence when oxidized with a combination of hydrogen peroxide and potassium ferricyanide.

Experiment

Procedure

First heat a tube containing 3 mL of water on the steam bath. Then heat a mixture of 200 mg of 3-nitrophthalic acid and 0.4 mL of an 8% aqueous solution of hydrazine[2] (*caution*) in a 10 × 100 mm reaction tube over a hot sand bath until the solid is dissolved. Add 0.6 mL of triethylene glycol and clamp the tube in a vertical position above the hot sand bath. Insert a boiling chip, a thermometer, and an aspirator tube connected to an aspirator, and boil the solution vigorously to distill the excess water. There will be a period during which the solution will boil at 110°C and then over a 3- or 4-min period it will rise to 215°C. Lift the tube from the hot sand and by intermittent gentle heating maintain a temperature of 215–220°C for 2 min. Remove the tube, cool to about 100°C (crystals of the product often appear), add the 3 mL of hot water, cool the tube in cold water, and collect the light yellow granular nitro compound (**8**) by vacuum filtration on the Hirsch funnel.[3] The dry weight should be about 140 mg.

Hydrazine is a carcinogen. Handle with care. Wear gloves and carry out this experiment in the hood.

The nitro compound need not be dried and can be transferred at once, for reduction, to the uncleaned tube in which it was prepared. Add 1.0 mL of 10% sodium hydroxide solution, stir with a rod, and to the resulting deep brown-red solution add 0.6 g of sodium hydrosulfite dihydrate (*not* sodium hydrogen sulfite or sodium bisulfite). Wash the solid down the walls with a little water. Heat to the boiling point, stir, and keep the mixture hot for 5 min, during which time some of the reduction product may separate. Then add 0.4 mL of acetic acid, cool the tube in a beaker of cold water, and stir; collect the resulting precipitate of light-yellow luminol (**1**) by vacuum filtration on the Hirsch funnel. The filtrate on standing overnight usually deposits a further crop of luminol (20–40 mg).

The two-step synthesis of a chemiluminescent substance can be completed in 25 min

$Na_2S_2O_4 \cdot 2\ H_2O$

Sodium hydrosulfite dihydrate
MW 210.15

The Light-Producing Reaction

Dissolve the first crop of moist luminol (dry weight about 40–60 mg) in 2 mL of 10% sodium hydroxide solution and 18 mL of water; this is stock solution A. Prepare a second stock solution, B, by mixing 4 mL of 3% aqueous potassium ferricyanide, 4 mL of 3% hydrogen peroxide, and 32 mL of water. Now dilute 5 mL of solution A with 35 mL of water and, in a dark place, pour

2. Dilute 3.12 g of the commercial 64% hydrazine solution to a volume of 25 mL.

3. The reason for adding hot water and then cooling rather than adding cold water is that the solid is then obtained in more easily filterable form.

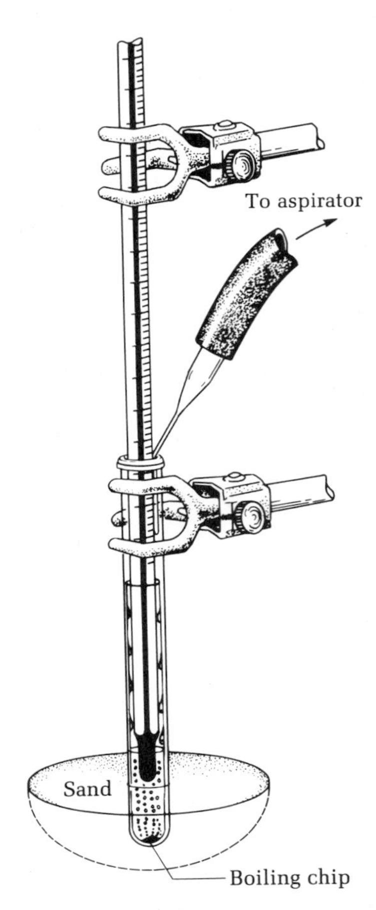

FIG. 45.1 Apparatus for synthesis of luminol.

this solution and solution B simultaneously into an Erlenmeyer flask. Swirl the flask and, to increase the brilliance, gradually add further small quantities of alkali and ferricyanide crystals.

Ultrasonic sound can also be used to promote this reaction. Prepare stock solutions A and B again but omit the hydrogen peroxide. Place the combined solutions in an ultrasonic cleaning bath or immerse an ultrasonic probe into the reaction mixture. Spots of light are seen where the ultrasonic vibrations produce hydroxyl radicals.

And the sanguinary can mix solutions A and B, omitting the ferricyanide from solution B. Light can be generated by adding blood dropwise to the reaction mixture.

Tetraphenylcyclopentadienone 46

PRELAB EXERCISE: *Write a detailed mechanism for the formation of tetraphenylcyclopentadienone from benzil and 1,3-diphenylacetone. To which general class of reactions does this condensation belong?*

$$C_6H_5C{=}O \text{ / } C_6H_5C{=}O \;+\; C_6H_5CH_2\text{—}C{=}O\text{—}CH_2C_6H_5 \xrightarrow[-2\,H_2O]{C_6H_5CH_2N^+(CH_3)_3OH^-} \text{Tetraphenylcyclopentadienone}$$

Benzil
MW 210.22
mp 96°C

1,3-Diphenylacetone
MW 210.26
mp 35°C

Tetraphenylcyclopentadienone
MW 384.45, mp 219°C

Cyclopentadienone

Cyclopentadienone is an elusive compound that has been sought for many years but with little success. Molecular orbital calculations predict that it should be highly reactive, and so it is; it exists only as the dimer. The tetraphenyl derivative of this compound is to be synthesized in this experiment. This derivative is stable, and reacts readily with dienophiles. It is used not only for the synthesis of highly aromatic, highly arylated compounds, but also for examination of the mechanism of the Diels-Alder reaction itself. Tetraphenylcyclopentadienone has been carefully studied by means of molecular orbital methods in attempts to understand its unusual reactivity, color, and dipole moment. In Chapter 48 this highly reactive molecule is used to trap the fleeting benzyne to form tetraphenylnaphthalene. Indeed this reaction constitutes evidence that benzyne does exist.

The literature procedure for condensation of benzil with 1,3-diphenylacetone in ethanol with potassium hydroxide as basic catalyst suffers from the low boiling point of the alcohol and the limited solubility of both potassium hydroxide and the reaction product in this solvent. Triethylene glycol is a better solvent and permits operation at a higher temperature. In the procedure that follows, the glycol is used with benzyltrimethylammonium hydroxide, a strong base readily soluble in organic solvents, which serves as catalyst.

Experiment

Short reaction period

Caution! Triton B is toxic

Into a 10 × 100 mm reaction tube place 42 mg of pure benzil (free of benzoin), 42 mg of 1,3-diphenylacetone, and 0.4 mL of triethylene glycol, using the solvent to wash the walls of the tube. Clamp the tube over a hot sand bath, insert a thermometer, and heat the solution until the benzil is dissolved. Remove the tube from the heat and then, using a 1-mL syringe, add to the solution 0.20 mL of a 40% solution of benzyltrimethylammonium hydroxide in methanol (Triton B) when the temperature of the solution reaches exactly 100°C. Stir once to mix. Crystallization usually starts in 10–20 s. Let the mixture cool to near room temperature and then cool it in cold water. Add 0.5 mL of methanol, cool the tube in ice, and collect the product.

If the crystals are large enough collection can be done by inserting a Pasteur pipette into the tube and removing the solvent between the tip of the pipette and the bottom of the tube (Fig. 46.1). If the crystals are small then collect them on the Hirsch funnel or the micro Büchner funnel. In either case, wash the crystals with cold methanol until the washings are purple-pink, not brown. The yield of deep purple crystals is about 60 mg. If either the crystals are not well formed or if the melting point is low, place the material in a reaction tube, add 0.6 mL of triethylene glycol, stir with a thermometer, and raise the temperature to 220°C to bring the solid into solution. Let it stand for crystallization (if initially pure material is recrystallized, the recovery is 92%).

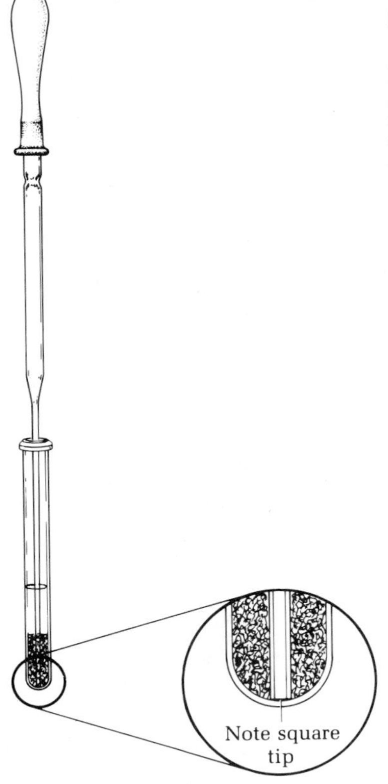

FIG. 46.1 Pasteur pipette filtration technique. Solvent is removed between pipette tip and bottom of tube, leaving crystals in reaction tube.

Question

Draw the structure of the dimer of cyclopentadienone. Why doesn't tetraphenylcyclopentadienone undergo dimerization?

Hexaphenylbenzene and Dimethyl Tetraphenylphthalate 47

PRELAB EXERCISE: *Explain the driving force behind the loss of carbon monoxide from the intermediates formed in these two reactions.*

This experiment illustrates two examples of the Diels-Alder reaction, which synthesizes molecules that would be extremely difficult to synthesize in any other way. Both reactions employ as the diene the tetraphenylcyclopentadienone prepared in the previous chapter. Although the Diels-Alder reaction is reversible, the intermediate in each of these reactions spontaneously loses carbon monoxide (why?) to form the products.

Hexaphenylbenzene melts at 465°C without decomposition. Few completely covalent organic molecules have higher melting points. Lead melts at 327.5°C. As is often the case, high melting point also means limited solubility. The solvent used to recrystallize hexaphenylbenzene, diphenylether, is very high boiling (bp 259°C) and has superior solvent power.

In the first experiment the dienone is condensed with dimethyl acetylenedicarboxylate using as the solvent 1,2-dichlorobenzene. This solvent is chosen for its solvent properties as well as its high boiling point, which guarantees the reaction is completed in a minute or two.

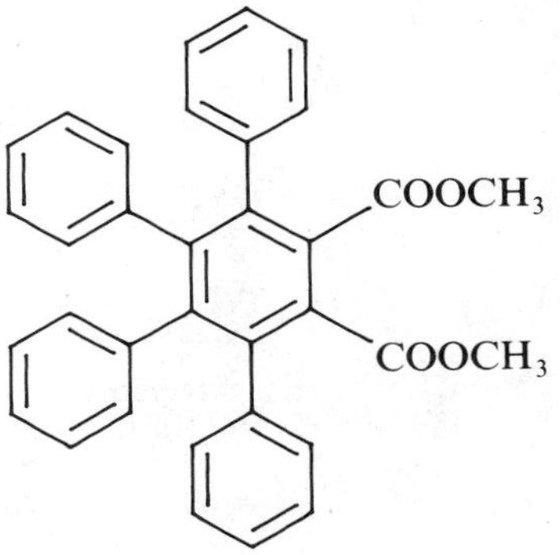

1. Dimethyl Tetraphenylphthalate

$COOCH_3$ $COOCH_3$

Dimethyl tetraphenylphthalate
MW 474.53, mp 258°C

^{-}CO

C_6H_5 C_6H_5 O C_6H_5 C_6H_5

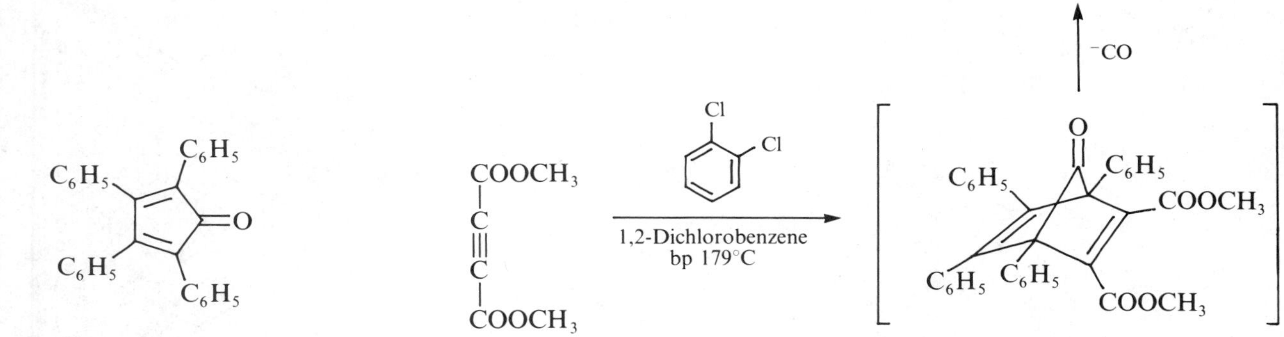

Tetraphenylcyclopentadienone
MW 384.45, mp 219°C

$COOCH_3$–C≡C–$COOCH_3$

Dimethyl acetylenedicarboxylate
MW 142.11, bp ~300°C

Cl Cl

1,2-Dichlorobenzene
bp 179°C

O C_6H_5 C_6H_5 $COOCH_3$ C_6H_5 C_6H_5 $COOCH_3$

Measure into a reaction tube 50 mg of tetraphenylcyclopentadienone (prepared in Chapter 46), 0.4 mL of 1,2-dichlorobenzene, and 27.5 mg of dimethyl acetylenedicarboxylate. Clamp the tube over a hot sand bath, insert a thermometer, and raise the temperature to the boiling point (180–185°C). Boil gently until there is no further color change and let the rim of condensate rise just high enough to wash the walls of the tube. The pure adduct is colorless, and if the starting ketone is adequately pure the color changes from purple to pale tan. A 5-min boiling period should be sufficient. Cool the tube to 100°C, slowly stir in 0.6 mL of 95% ethanol, and let crystallization proceed. After the mixture has cooled to near room temperature, cool it in ice. Then collect the product, either by removal of the solvent with a Pasteur pipette, or, if the crystals are too small for this technique, on the Hirsch funnel. Wash the crystals with cold methanol. The yield of dimethyl tetraphenylphthalate should be about 50 mg.

2. Hexaphenylbenzene

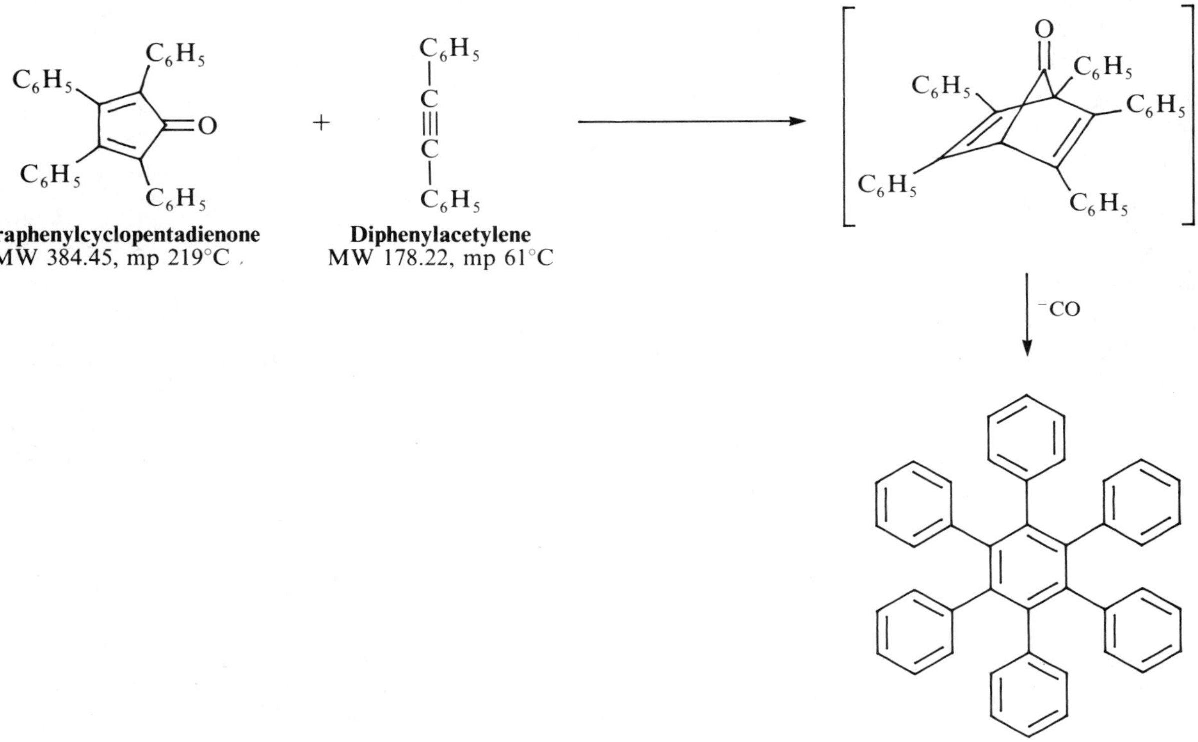

C_6H_5 C_6H_5 C_6H_5 C_6H_5 =O + C_6H_5–C≡C–C_6H_5 → [bridged ketone adduct, C_6H_5 ×6] $\xrightarrow{-CO}$

Tetraphenylcyclopentadienone
MW 384.45, mp 219°C

Diphenylacetylene
MW 178.22, mp 61°C

Hexaphenylbenzene
MW 534.66°C
mp 465°C

Diphenylacetylene is a less reactive dienophile than dimethylacetylenedicarboxylate; but when heated with tetraphenylcyclopentadienone without solvent a temperature (*ca.* 380–400°C) suitable for reaction can be attained. In the following procedure the dienophile is taken in large excess to serve as solvent. Since refluxing diphenylacetylene (bp about 300°C) keeps the temperature below the melting point of the product, removal of the diphenylacetylene lets the reaction mixture melt, which ensures completion of the reaction.

Procedure

Place 50 mg each of tetraphenylcyclopentadienone (prepared in Chapter 46) and diphenylacetylene (prepared in Chapter 58) in a reaction tube, clamp it upright, and heat the mixture with the flame of a microburner. This is the only experiment in this text that requires a flame to conduct the reaction. Use care to ensure that no flammable solvents are nearby. Soon after the reactants have melted with strong bubbling, the white product becomes visible. Let the diphenylacetylene reflux briefly on the walls of the tube and then remove some of this excess diphenylacetylene by inserting a Pasteur pipette into the vapors above the solid and drawing the hot vapors into the pipette. Repeat this operation until it is possible, by strong heating with the flame, to melt the mixture completely. Then let the melt cool and solidify. Add 0.25 mL of diphenyl ether and heat the mixture *carefully* to dissolve the solid and then let the product crystallize slowly. Extinguish the flame and, when the tube is cold, add 0.5 mL of toluene to dilute the mixture. Collect the product using the Pasteur pipette method or on the Hirsch funnel by vacuum filtration, and wash it with toluene. The yield of colorless plates is about 60 mg. Using a Mel-Temp apparatus *equipped with a 500°C thermometer* determine the melting point. The product should melt at 465°C.

Question

What two factors probably contribute to the very high melting points of these two hexasubstituted benzenes?

48 1,2,3,4-Tetraphenylnaphthalene via Benzyne

PRELAB EXERCISE: *Speculate on the reasons for the stability of the iodonium ion. What type of strain exists in the benzyne molecule?*

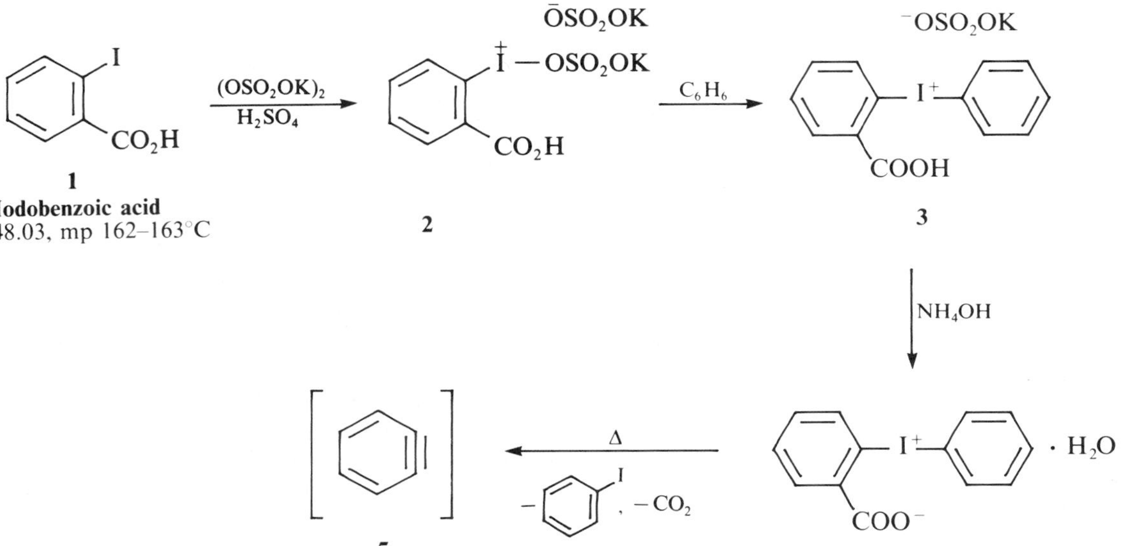

I, CO_2H —$(OSO_2OK)_2$ / H_2SO_4→ $\bar{O}SO_2OK$, $\overset{+}{I}-OSO_2OK$, CO_2H —C_6H_6→ $^{-}OSO_2OK$, I^+, COOH

1
2-Iodobenzoic acid
MW 248.03, mp 162–163°C

2

3

NH_4OH

I^+, COO^- · H_2O —Δ, $-C_6H_5I$, $-CO_2$→

5
Benzyne

Diphenyliodonium-2-carboxylate monohydrate
MW 341.13

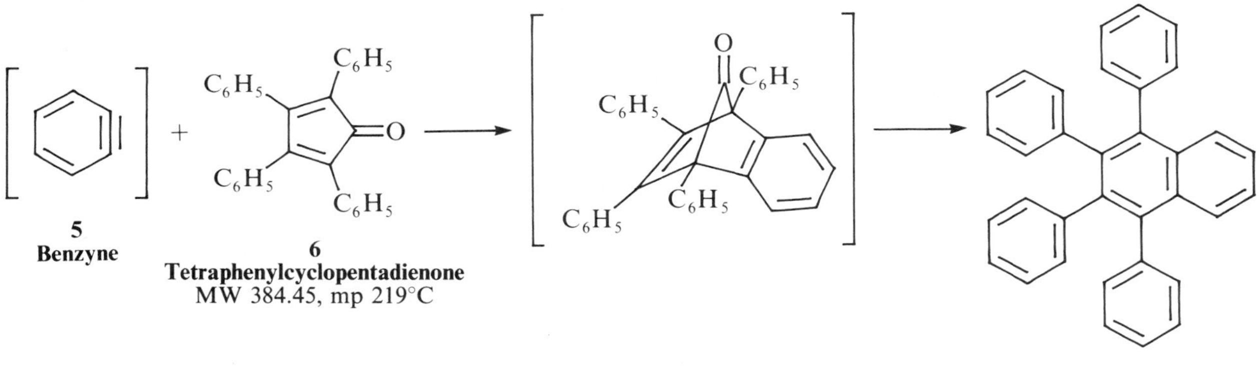

C_6H_5, C_6H_5, C_6H_5, C_6H_5, =O + → O, C_6H_5, C_6H_5, C_6H_5, C_6H_5 →

5
Benzyne

6
Tetraphenylcyclopentadienone
MW 384.45, mp 219°C

7
1,2,3,4-Tetraphenylnaphthalene
MW 432.53, mp 219–220°C

This synthesis of 1,2,3,4-tetraphenylnaphthalene (**7**) demonstrates the transient existence of benzyne (**5**), a hydrocarbon which has not been isolated as such. The precursor, diphenyliodonium-2-carboxylate (**4**), is heated in an inert solvent to a temperature at which it decomposes to benzyne, iodobenzene, and carbon dioxide in the presence of tetraphenylcyclopentadienone (**6**) as trapping agent. The preparation of the precursor **4** illustrates oxidation of a derivative of iodobenzene to an iodonium salt (**2**) and the Friedel-Crafts-like reaction of the substance with benzene to form the diphenyliodonium salt (**3**). Neutralization with ammonium hydroxide then liberates the precursor, inner salt, **4**, which when obtained by crystallization from water is the monohydrate.

Synthetic use of an intermediate not known as such

The procedure affords about twice the amount of precursor **4** required for the synthesis of **7**. Hence samples of both **4** and **7** should be submitted for evaluation.

Experiments

1. Diphenyliodonium-2-carboxylate monohydrate (4)

Measure 0.85 mL of concentrated sulfuric acid into a reaction tube and place it in an ice bath to cool. In a 10-mL Erlenmeyer flask that is clamped in an ice bath, place 200 mg of 2-iodobenzoic acid, prepared in Chapter 42, and 260 mg of potassium persulfate.[1] Remove the tube containing the cold sulfuric acid from the ice bath, wipe it dry, and add the acid to the two solids in the flask. Stir the mixture in the ice bath for 4–5 min to produce an even suspension and then remove it and note the time. The reaction mixture foams somewhat and acquires a succession of colors. After it has stood at room temperature for 20 min, swirl the flask vigorously in an ice bath for 2–3 min, add 0.2 mL of benzene (*caution!*) and swirl and cool until the benzene freezes. Then remove and wipe the flask and note the time at which the benzene melts. Warm the flask in the palm of the hand and swirl frequently at room temperature for 20 min to promote completion of reaction in the two-phase mixture. If available, magnetic stirring of the reaction mixture is convenient, but not necessary.

$K_2S_2O_8$

Potassium persulfate
(Potassium peroxydisulfate)
MW 270.33

1st step: about 20 min

Caution! *Benzene is a mild carcinogen. Carry out this experiment in the hood.*

While the reaction is going to completion, place three reaction tubes in an ice bath to chill: one containing 1.9 mL of distilled water, another 2.3 mL of concentrated ammonium hydroxide solution, and another 4 mL of dichloromethane. At the end of the 20-min reaction period, chill the reaction mixture thoroughly in an ice bath and add to it dropwise, with vigorous swirling and mixing, the 1.9 mL of ice-cold water that has been chilling. The solid that separates is the potassium bisulfate salt of diphenyliodonium-2-carboxylic acid, **3**.

2nd step: about 30 min

1. The fine granular material supplied by Fisher Scientific Co. is satisfactory. Persulfate in the form of large prisms should be finely ground prior to use.

Note for the instructor

Pour the chilled dichloromethane into the flask for extracting the product liberated when the ammonia is added. While swirling the flask vigorously, add the ammonia dropwise over a 10-min period. The mixture must be alkaline (pH 9). Test with indicator paper and add more ammonia if necessary. Using a Pasteur pipette, remove the dichloromethane layer, transfer the aqueous layer to a 4-in. test tube, and extract it with two 1.5-mL portions of dichloromethane, which are combined with the original dichloromethane extract. Dry the dichloromethane extracts over anhydrous sodium sulfate and decant into a tared 10-mL flask. Rinse the drying agent with more solvent and then evaporate the dichloromethane under a stream of air in the hood. Connect the flask to an aspirator and heat it on the steam bath until the weight is constant. The yield is about 240 mg.

For crystallization, dislodge the solid with a spatula and dissolve it in 2.8 mL of boiling water. Use a wood boiling stick to prevent bumping of the solution. If desired the tan solution may be decolorized by the addition of about 15 pieces of Norit pellets. On cooling the solution after removal from the Norit diphenyliodonium-2-carboxylate monohydrate (**4**), the benzyne precursor, separates in colorless, rectangular prisms, mp 219–220°C. The yield should be about 200 mg.

2. Preparation of 1,2,3,4-Tetraphenylnaphthalene (7)

Diels-Alder reaction

Use hood; CO is evolved

Place 100 mg of the diphenyliodonium-2-carboxylate monohydrate just prepared and 100 mg of tetraphenylcyclopentadienone, prepared in Chapter 46, in a reaction tube. Add 0.6 mL of triethylene glycol dimethyl ether (bp 222°C) so that the solvent rinses the walls of the tube. Support the tube vertically, insert a thermometer, and heat it over a hot sand bath. When the temperature reaches 200°C, remove the tube from the heat and note the time. Then keep the mixture at 200–205°C by intermittent heating until the purple color is discharged, the evolution of gas (CO_2 and CO) subsides, and a pale yellow solution is obtained. In case a purple or red color persists after 3 min at 200–205°C, add additional small amounts of the benzyne precursor and continue to heat until all the solid is dissolved.

Cool the hot solution to below 100°C and add to it 0.6 mL of ethanol. Bring the solution to boiling, using a wood stick to prevent bumping. If shiny prisms do not separate at once, add water dropwise at the boiling point of the ethanol until crystals begin to separate. Let crystallization proceed at room temperature and then at 0°C. Collect the product using the Pasteur pipette method or on the Hirsch funnel. Wash the crystals with cold methanol. The yield of dry product should be about 80 mg. The pure hydrocarbon **7** exists in two crystalline forms (allotropes) and has a double melting point, the first of which is in the range 196–199°C. Let the melting point bath cool to about 195°C, remove the sample, and let the sample solidify. Then determine the second melting point, which, for the pure hydrocarbon, is 203–204°C.

Questions

1. To what general class of compounds does potassium persulfate belong?
2. Calculate the volume of carbon monoxide, at standard temperature and pressure, released during this reaction.

49 Triptycene via Benzyne

PRELAB EXERCISE: *Outline the preparation of triptycene, writing balanced equations for each reaction, including the reactions that are used to remove excess anthracene.*

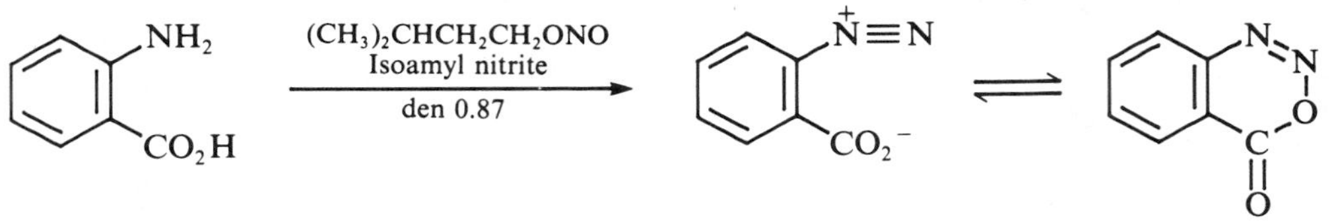

NH_2 / CO_2H — $(CH_3)_2CHCH_2CH_2ONO$ Isoamyl nitrite, den 0.87 → $\overset{+}{N}\equiv N$ / CO_2^- ⇌ N=N–O–C=O

1
Anthranilic acid
MW 137.14

2a

2b

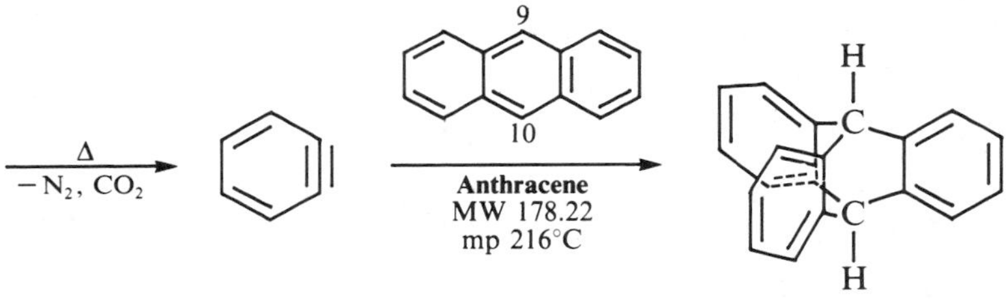

$\xrightarrow[-N_2,\ CO_2]{\Delta}$ **3** **Benzyne** → (9, 10) **Anthracene** MW 178.22 mp 216°C → **4** **Triptycene** MW 254.31, mp 225°C

This interesting cage-ring hydrocarbon results from 9,10-addition of benzyne to anthracene. In one procedure presented in the literature benzyne is generated under nitrogen from *o*-fluorobromobenzene and magnesium in the presence of anthracene, but the work-up is tedious and the yield only 24%. Diazotization of anthranilic acid to benzenediazonium-2-carboxylate (**2a**) or to the covalent form (**2b**) gives another benzyne precursor, but the isolated substance can be kept only at a low temperature and is sometimes explosive. However, isolation of the precursor is not necessary. On slow addition of anthranilic acid to a solution of anthracene and isoamyl nitrite in an aprotic solvent, the precursor **2** reacts with anthracene as fast as the precursor is formed. If the anthranilic acid is all present at the start, a side reaction of this substance with benzyne drastically reduces the yield. A low-boiling solvent

Aprotic means "no protons"; HOH *and* ROH *are protic*

(CH_2Cl_2, bp 41°C) is used, in which the desired reaction goes slowly, and a solution of anthranilic acid is added dropwise over a period of 4 h. To bring the reaction time into the limits of a laboratory period, the higher-boiling solvent 1,2-dimethoxyethane (bp 83°C, water soluble) is specified in this procedure and a large excess of anthranilic acid and isoamyl nitrite is used. Treatment of the dark reaction mixture with alkali removes acidic by-products and most of the color, but the crude product inevitably contains anthracene. However, brief heating with maleic anhydride at a suitable temperature leaves the triptycene untouched and converts the anthracene into its maleic anhydride adduct. Treatment of the reaction mixture with alkali converts the adduct into a water-soluble salt and affords colorless, pure triptycene.

Removal of anthracene

Experiment

Place 100 mg of anthracene,[1] 0.1 mL of isoamyl nitrite, 1 mL of dimethoxyethane and a boiling chip in a 10 × 100 mm reaction tube. In a small filter paper place 130 mg of anthranilic acid and push this about halfway down the reaction tube. Bring the mixture in the tube to a gentle boil and note that the anthracene does not all dissolve. Add to the filter paper 0.5 mL of dimethoxyethane dropwise to leach the acid into the solution below over a period of not less than 20 min. Remove the filter paper, add to the solution 0.1 mL more isoamyl nitrite, add 130 mg more anthranilic acid to the filter paper, replace the paper in the reaction tube, and again leach out the anthranilic acid with 0.5 mL of dimethoxyethane added dropwise over a 20-min period. Reflux for 10 min more and then add 0.5 mL of ethanol and a solution of 0.15 g of sodium hydroxide in 2 mL of water to produce a suspension of solid in a brown alkaline solvent. Cool the mixture thoroughly in ice, and also cool a 4 : 1 methanol–water mixture for rinsing.

Carry out reaction in hood

Technique for slow addition of a solid

Reaction time: 1 h

Collect the solid on the Hirsch funnel by vacuum filtration and wash it with the chilled solvent to remove brown mother liquor. Transfer the moist, nearly colorless solid to a tared, clean reaction tube and evacuate on the steam bath until the weight is constant; the anthracene-triptycene mixture (mp about 190–230°C) weighs about 100 mg.

Add 50 mg of maleic anhydride and 1 mL of triethylene glycol dimethyl ether ("triglyme," bp 222°C), reflux the mixture for 5 min. Cool to about 100°C, add 0.5 mL of ethanol and a solution of 150 mg of sodium hydroxide in 2 mL of water; then cool in ice, along with 1.25 mL of 4 : 1 methanol–water for rinsing. Triptycene separates as nearly white crystals from the slightly brown alkaline solution. The washed and dried product weighs about 75 mg and melts at 255°C. Recrystallize it from methylcyclohexane (23 mL/g). If an insoluble black impurity is seen in the hot solution, transfer the clear solution to a clean tube using a Pasteur pipette, leaving the impurity behind. Slow cooling will produce flat, rectangular, laminated prisms that can be isolated by removing the solvent with a Pasteur pipette.

1. Practical grade anthracene and anthranilic acid are satisfactory.

FIG. 49.1 ^{1}H nmr spectrum of anthracene.

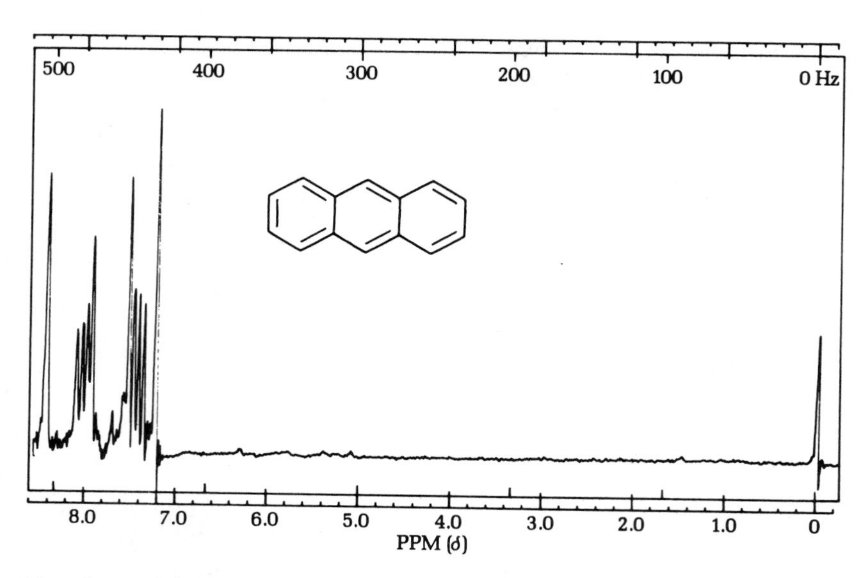

Optional Projects

In Chapter 48 diphenyliodonium-*o*-carboxylate is utilized as a benzyne precursor in the synthesis of 1,2,3,4-tetraphenylnaphthalene. For reasons unknown, the reaction of anthracene with benzyne generated in this way proceeds very poorly and gives only a trace of triptycene. What about the converse proposition? Would benzyne generated by diazotization of anthranilic acid in the presence of tetraphenylcyclopentadienone afford 1,2,3,4-tetraphenylnaphthalene in satisfactory yield? In case you are interested in exploring this possibility, plan and execute a procedure and see what you can discover.

If benzyne is generated from anthranilic acid in the absence of anthracene, as in this experiment, then dibenzocyclobutadiene is formed.[2]

Devise a procedure for the isolation of this interesting hydrocarbon.

Questions

1. Write equations showing how benzyne might be generated from *o*-fluorobromobenzene and magnesium.

2. Study the mechanism of diazonium salt formation in Chapter 42 and then propose a mechanism for diazotization using isoamyl nitrite.

2. (a) F. M. Logullo, A. H. Seitz, and L. Friedman, *Org. Syn.* **48**, 12 (1968). (b) L. Friedman and F. M. Logullo, *J. Org. Chem.*, **34**, 3089 (1969).

Oxidative Coupling of Alkynes: 2,7-Dimethyl-3,5-octadiyn-2,7-diol

50

PRELAB EXERCISE: *Write a balanced equation for the coupling of 2-methyl-3-butyn-2-ol. What role does cuprous chloride play in this reaction?*

$$2\ CH_3\underset{\underset{OH}{|}}{\overset{\overset{CH_3}{|}}{C}}-C\equiv CH \xrightarrow[CuCl-Pyridine]{O_2} CH_3\underset{\underset{OH}{|}}{\overset{\overset{CH_3}{|}}{C}}-C\equiv C-C\equiv C-\underset{\underset{OH}{|}}{\overset{\overset{CH_3}{|}}{C}}CH_3$$

2-Methyl-3-butyn-2-ol
MW 84.11, den 0.868, bp 103°C

2,7-Dimethyl-3,5-octadiyn-2,7-diol
MW 166.21, mp 130°C

The starting material, 2-methyl-3-butyn-2-ol, is made commercially from acetone and acetylene and is convertible into isoprene. This experiment illustrates the oxidative coupling of a terminal acetylene to produce a diacetylene, the Glaser reaction.

Although Glaser reported the coupling of acetylenes using cuprous ion in 1869, the details of the mechanism of this reaction are still obscure. It appears to involve both Cu^+ and Cu^{2+} ions; as noted by Glaser, the cuprous acetylide is oxidized by oxygen, but

$$C_6H_5C\equiv CH \xrightarrow[NH_4OH]{CuCl} C_6H_5C\equiv CCu \xrightarrow{air} C_6H_5C\equiv C-C\equiv CC_6H_5$$

Cu^{2+} the actual oxidizing agent. The amine seems to be needed to keep the acetylide in solution.

The reaction is very useful synthetically in the synthesis of polyenes, vitamins, fatty acids, and the annulenes. Baeyer used the reaction in his historic synthesis of indigo as long ago as 1882. The reaction allowed the unequivocal establishment of the carbon skeleton of this dye:

$$\text{o-}NO_2C_6H_4\text{-}C{\equiv}C\text{-}COOH \xrightarrow[-CO_2]{CuCl,\ NH_4OH} \text{o-}NO_2C_6H_4\text{-}C{\equiv}C\text{-}Cu \xrightarrow[\text{(oxidizing agent)}]{K_3Fe(CN)_6} \text{o-}NO_2C_6H_4\text{-}C{\equiv}C\text{-}C{\equiv}C\text{-}C_6H_4NO_2\text{-o} \xrightarrow[(NH_4)_2S]{H_2SO_4} \text{Indigo}$$

Indigo

In the present experiment the reaction time is shortened by use of excess catalyst and by supplying oxygen under the pressure of a balloon; the state of the balloon provides an index of the course of the reaction.

Experiment

Measure pyridine in hood

The reaction vessel is a 5-mL short-necked round-bottomed flask that is fitted with an addition port, rubber pipette bulb, and septum, all of which are wired on with two turns of copper wire (Fig. 50.1). Add 0.5 mL of 2-methyl-3-butyn-2-ol, 0.5 mL of methanol, 0.15 mL of pyridine, and 75 mg of copper(I) chloride. The flask is flushed out with oxygen; the reaction is about twice as fast in an atmosphere of oxygen as in air. Insert the oxygen delivery syringe into the flask with the needle under the surface of the liquid, open the valve, and let oxygen bubble through the solution in a brisk stream for 2 min.[1] Close the valve and quickly attach the addition port, which bears a rubber bulb and septum. Wire the rubber connections and insert the oxygen delivery needle into the septum and run in oxygen at 10 lb/sq in. until you have produced a balloon about 7.5 cm in dia.

Close the valve and withdraw the needle, note the exact diameter of the balloon and the time, and swirl the reaction mixture. The rate of oxygen uptake depends on the efficiency of mixing of the liquid and gas phases. By vigorous and continuous swirling you can complete the reaction and deflate the balloon to a constant size in 50–60 min, although the amount of oxygen consumed in the reaction may not make a dramatic difference in the size of the balloon. The reaction mixture becomes slightly warm and turns deep green in color. A pair of calipers is helpful in recognizing the constant size of the balloon and thus the end of the reaction.[2]

1. The valves of a cylinder of oxygen should be set to deliver gas at a pressure of 10 lb/sq in. when the terminal valve is opened. The barrel of a 2.5-mL plastic syringe is cut off and thrust into the end of a $\frac{1}{4} \times \frac{3}{16}$-in. rubber delivery tube. Read about the handling of compressed gas cylinders in Chapter 2. Be sure the cylinder is secured to a bench or wall.

2. A magnetic stirrer does not materially shorten the reaction time.

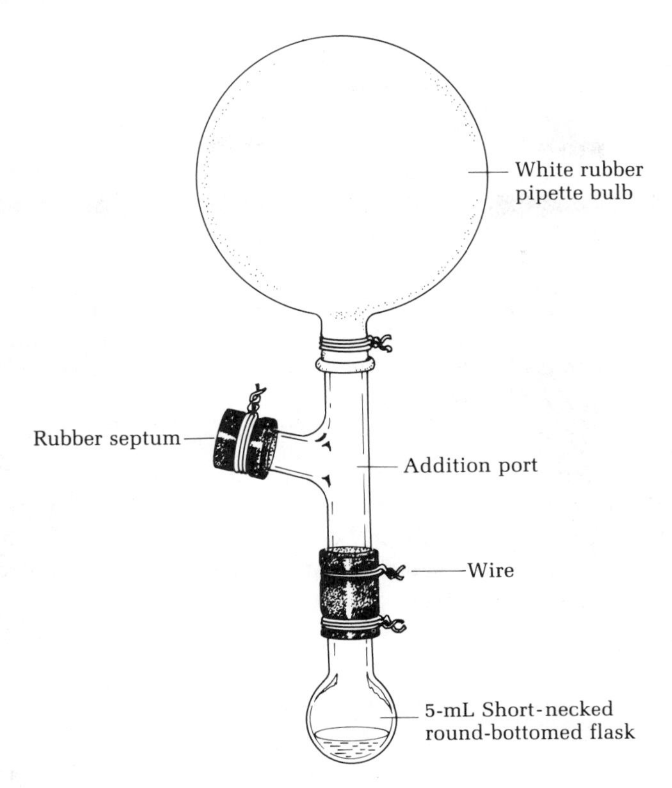

FIG. 50.1 Balloon technique of oxygenation. About 10 lb/in^2 pressure of oxygen needed to inflate pipette bulb. Identical apparatus is used for hydrogenation.

Release excess oxygen by piercing the septum with an empty syringe needle, open the flask, add 0.25 mL of concentrated hydrochloric acid to neutralize the pyridine and keep copper compounds in solution, and cool the flask; the color changes from green to yellow. Transfer the solution to a reaction tube using 1.5 mL of ether to complete the transfer. Shake the mixture thoroughly, draw off the ether, and repeat the extraction with two further 1-mL portions of ether. Dry the ether by shaking with saturated sodium chloride solution followed by the addition of sufficient anhydrous sodium sulfate to prevent clumping. Transfer the ether to a tared reaction tube, evaporate the ether, and recrystallize the product from toluene. 2,7-Dimethyl-3,5-octadiyn-2,7-diol crystallizes as colorless needles, mp 129–130°C. The yield should be about 300 mg. Analyze the product by thin-layer chromatography and infrared spectroscopy. Compare the diyn with the starting material to determine whether the product is pure.

Questions

1. Write a balanced equation for this reaction.
2. What volume of oxygen at standard temperature and pressure is consumed in this reaction?

51 Sugars

PRELAB EXERCISE: *Write a balanced equation for the hydrolysis of the disaccharide sucrose, taking care to draw the stereochemistry of the products. Will the anomeric forms of glucose give different phenylosazones? Will they give different methyl glucosides?*

The term sugar applies to mono-, di-, and oligosaccharides, which are all soluble in water and thereby distinguished from polysaccharides. Many natural sugars are sweet, but data of Table 51.1 show that sweetness varies greatly with stereochemical configuration and is exhibited by compounds of widely differing structural type.

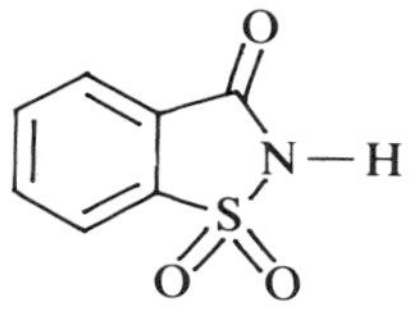

Saccharin

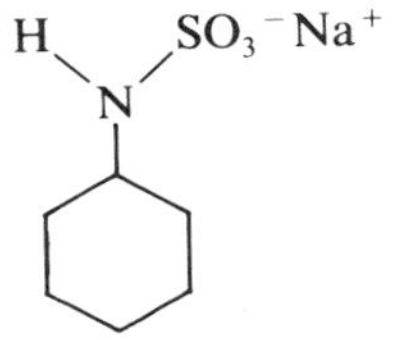

Sodium cyclamate

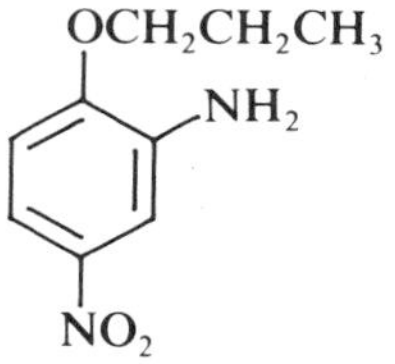

2-Amino-4-nitro-1-*n*-propoxybenzene

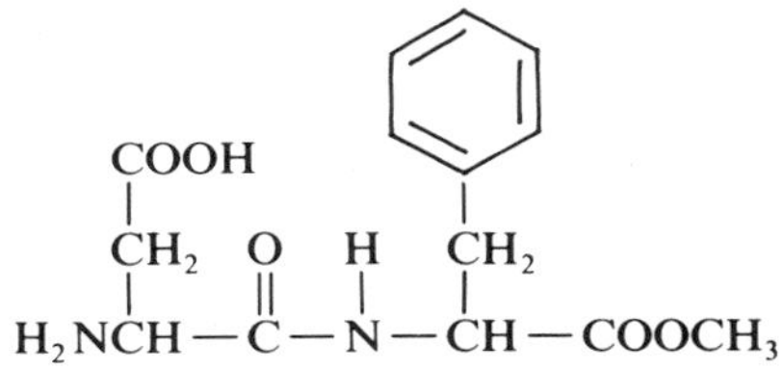

Aspartame
L-Aspartyl-L-phenylalanine methyl ester

Sugars are neutral and combustible and these properties distinguish them from other water-soluble compounds. Some polycarboxylic acids and some lower amines are soluble in water, but the solutions are acidic or basic. Water-soluble amine salts react with alkali with liberation of the amine, and sodium salts of acids are noncombustible.

One gram of sucrose dissolves in 0.5 mL of water at 25°C and in 0.2 mL at the boiling point, but the substance has marked, atypical crystallizing properties. In spite of the high solubility it can be obtained in beautiful, large crystals (rock candy). More typical sugars are obtainable in crystalline form only with difficulty, particularly in the presence of a trace of impurity, and even then give small and not well-formed crystals. Alcohol is often added to a water solution to decrease solubility and thus to induce crystallization. The amounts of 95% ethanol required to dissolve 1-g samples at 25° are sucrose, 170 mL; glucose, 60 mL; fructose, 15 mL. Some sugars have never been obtained in crystalline condition and are known only as viscous syrups. With

TABLE 51.1 Relative Sweetness of Sugars and Sugar Substitutes

Compound	Sweetness To man	Sweetness To bees
Monosaccharides		
D-Fructose	1.5	+
D-Glucose	0.55	+
D-Mannose	Sweet, then bitter	−
D-Galactose	0.55	−
D-Arabinose	0.70	−
Disaccharides		
Sucrose (glucose, fructose)	1	+
Maltose (2 glucose)	0.3	+
α-Lactose (glucose, galactose)	0.2	−
Cellobiose (2 glucose)	Indifferent	−
Gentiobiose (2 glucose)	Bitter	−
Synthetic sugar substitutes		
Aspartame (NutraSweet®)	180	
Saccharin	550	
2-Amino-4-nitro-1-*n*-propoxybenzene	4000	

phenylhydrazine many sugars form beautiful crystalline derivatives called osazones. Osazones are much less soluble in water than the parent sugars, because the molecular weight is increased by 178 units and the number of hydroxyl groups reduced by one. It is easier to isolate an osazone than to isolate the sugar, and sugars that are syrups often give crystalline osazones. Osazones of the more highly hydroxylic disaccharides are notably more soluble than those of monosaccharides.

Osazones

Some disaccharides do not form osazones, but a test for formation or nonformation of the osazone is ambiguous, because the glycosidic linkage may suffer hydrolysis in a boiling solution of phenylhydrazine and acetic acid, with formation of an osazone derived from a component sugar and not from the disaccharide. If a sugar has reducing properties it is also capable of osazone formation; hence an unknown sugar is tested for reducing properties before preparation of an osazone is attempted. Three tests for differentiation between reducing and nonreducing sugars are described below; two are classical and the third modern.

Reducing sugars have an aldehyde form

Experiments

1. Fehling's Solution[1]

$Cu^{2+} \rightarrow Cu_2O$

Blue **Red**

The reagent is made just prior to use by mixing equal volumes of Fehling's solution I, containing copper(II) sulfate, with solution II, containing tartaric acid and alkali. The copper, present as a deep blue complex anion, if reduced by a sugar from the copper(II) to the copper(I) state, precipitates as red copper(I) oxide. If the initial step in the reaction involved oxidation of the aldehydic group of the aldose to a carboxyl group, a ketose should not reduce Fehling's solution, or at least should react less rapidly than an aldose, but the comparative experiment that follows will show that this supposition is not the case. Hence, attack by an alkaline oxidizing agent must attack the α-ketol grouping common to aldoses and hexoses, and perhaps proceeds through an enediol, the formation of which is favored by alkali. A new α-ketol grouping is produced, and thus oxidation proceeds down the carbon chain.

$$\begin{array}{c} | \\ C{=}O \\ | \\ CHOH \\ | \\ CHOH \\ | \\ \textbf{α-Ketol} \end{array} \rightleftharpoons \begin{array}{c} | \\ C{-}OH \\ \| \\ C{-}OH \\ | \\ CHOH \\ | \\ \textbf{Enediol} \end{array} \xrightarrow{[O]} \begin{array}{c} | \\ C{=}O \\ | \\ C{=}O \\ | \\ CHOH \\ | \\ \textbf{α-Ketol} \end{array}$$

One milliliter of mixed solution will react with 5 mg of glucose; the empirically determined ratio is the basis for quantitative determination of the sugar. The Fehling's test is not specific to reducing sugars, because ordinary aldehydes reduce the reagent although by a different mechanism and at a different rate.

The following sugars are to be tested: 0.1 M solutions of glucose, fructose, lactose, maltose, sucrose (cane sugar).[2]

Introduce 5 drops of the 0.1 M solutions to be tested into each of five reaction test tubes carrying some form of serial numbers resistant to heat and water (rubber bands), and prepare a beaker of hot water in which all the tubes can be heated at once. Measure 2.5 mL of Fehling's solution I into a 10 mL flask and wash the pipette before using it to measure 2.5 mL of solution II into the same flask. Mix until all precipitate dissolves, measure 1 mL of mixed solution into each of the five test tubes, shake, put the tubes in the heating bath, note the time, and record any reaction as a function of time. Empty,

Note for the instructor

1. *Solution I:* 34.64 g of $CuSO_4 \cdot 5H_2O$ dissolved in water and diluted to 500 mL. *Solution II:* 173 g of sodium potassium tartrate (Rochelle salt) and 65 g of sodium hydroxide dissolved in water and diluted to 500 mL.

2. To prepare 0.1 M test solutions dissolve the following amounts of substance in 100 mL of water each: D-glucose monohydrate, 1.98 g; fructose, 1.80 g; α-lactose monohydrate, 3.60 g; maltose monohydrate, 3.60 g; sucrose, 3.42 g; *n*-butanol, 0.72 g.

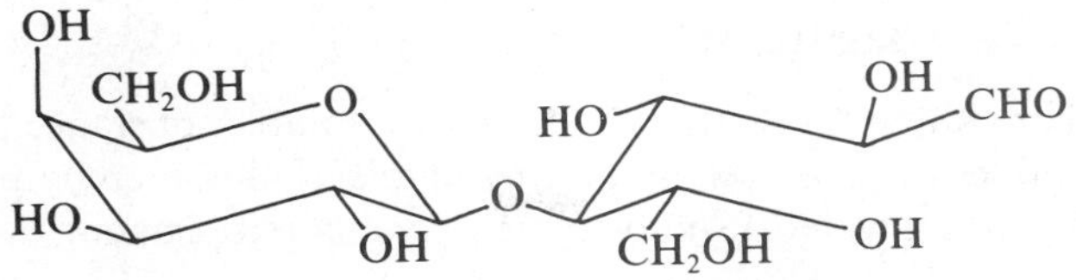

OH CH$_2$OH O HO OH CHO HO O OH CH$_2$OH OH

Lactose (aldehyde form)

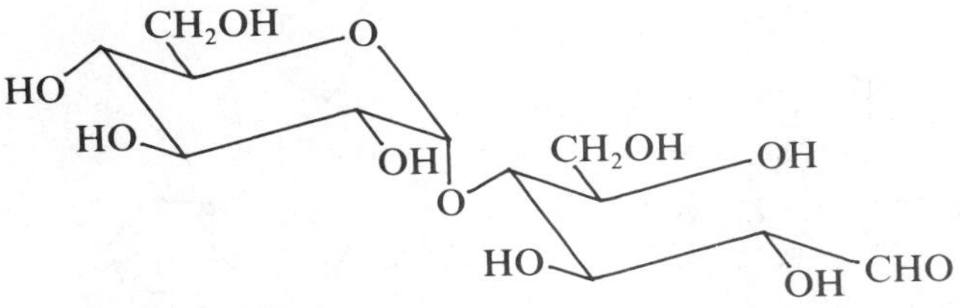

CH$_2$OH O HO HO OH CH$_2$OH OH O HO OH CHO

Maltose (aldehyde form)

CH$_2$OH O HOCH$_2$ O HO HO HO OH O CH$_2$OH OH

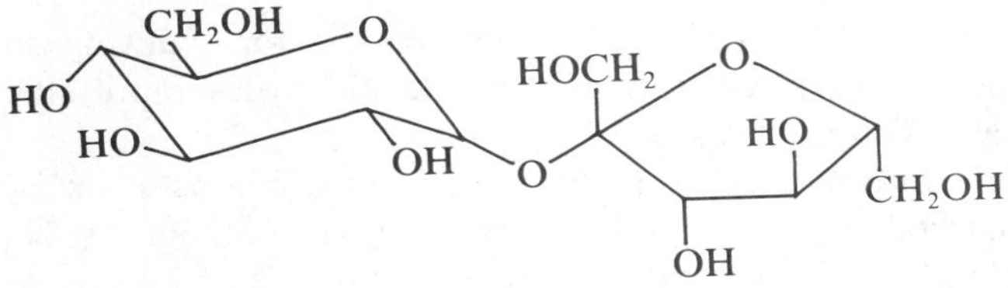

Sucrose

and wash the tubes with water and then with dilute acid (leave the markers in place and continue heating the beaker of water for the next experiment).

2. Tollens' Reagent

Tollens' reagent is a solution of a silver ammonium hydroxide complex that is reduced by aldoses and ketoses as well as by simple aldehydes. See Chapter 30 for the preparation of this reagent and the procedure for carrying out the test. The test is more sensitive than the Fehling's test and better able to reveal small differences in reactivity, but it is less reliable in distinguishing between reducing and nonreducing sugars.

Prepare the five reaction tubes according to Chapter 30. Into each tube put one drop of 0.1 M solution of glucose, fructose, lactose, maltose, and *n*-butanol. (See Experiment 1, footnote 2.) Add 1 mL of Tollens' reagent to each tube and let the reaction proceed at room temperature. Watch closely and try to define the order of reactivity as measured not by the color of solution but by the time of the first appearance of metal. At the completion of the reaction destroy all Tollens' reagent by adding nitric acid until the solution is clear.

3. Red Tetrazolium[3]

A sensitive test for reducing sugars

The reagent (RT) is a nearly colorless, water-soluble substance that oxidizes aldoses and ketoses, as well as other α-ketols, and is thereby reduced. The reduced form is a water-insoluble, intensely colored pigment, a diformazan.

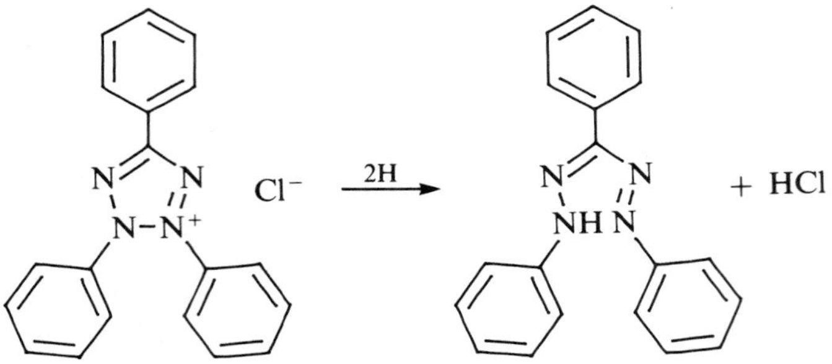

Cl^- $\xrightarrow{2H}$ + HCl

Red tetrazolium **RT-Diformazan**

Red tetrazolium affords a highly sensitive test for reducing sugars and distinguishes between α-ketols and simple aldehydes more sharply than Fehling's and Tollens' tests.

Put one drop of each of the five 0.1 M test solutions of section 1 in the cleaned, marked test tubes, and to each tube add 1 mL of a 0.5% aqueous solution of red tetrazolium and one drop of 10% sodium hydroxide solution. Put the tubes in the beaker of hot water, note the time, and note the time of development of color for each tube.

For estimation of the sensitivity of the test, use the substance that you regard as the most reactive of the five studied. Dilute 1 mL of the 0.1 M solution with water to a volume of 100 mL, and run a test with RT on 0.2 mL of the diluted solution.

4. Phenylosazones

Conduct procedure in hood. Avoid skin contact with phenylhydrazine.

Prepare 5 mL of stock phenylhydrazine reagent as in Chapter 30 and put 0.33-mL portions of it into four of the cleaned, numbered reaction tubes. Add 1-mL portions of 0.1 M solutions of glucose, fructose, lactose, and maltose and heat the tubes in the beaker of hot water for 20 min. Shake the tubes occasionally to relieve supersaturation and note the times at which osazones separate. If after 20 min no product has separated, cool and scratch the test tube to induce crystallization. (Save unused phenylhydrazine reagent.)

Note for the instructor

3. 2,3,5-Triphenyl-2H-tetrazolium chloride. Available from Aldrich and Eastman. Freshly prepared aqueous solutions should be used in tests. Any unused solution should be acidified and discarded.

Collect and save the products for possible later use in identification of unknowns. Since osazones melt with decomposition, the bath in a mp determination should be heated at a standard rate (0.5°C per second).

Glucose + Phenylhydrazine ($H_2NNH-C_6H_5$) → Glucose phenylhydrazone; + 2 $H_2NNH-C_6H_5$ → $NH_3 + H_2O$ + aniline ($C_6H_5NH_2$) + Glucose phenylosazone

Glucose **Phenylhydrazine** **Glucose phenylhydrazone**

Glucose phenylosazone

5. Hydrolysis of a Disaccharide

Sucrose is heated with dil HCl

The object of this experiment is to determine conditions suitable for hydrolysis of a typical disaccharide. Put 1-mL portions of a 0.1 M solution of sucrose in each of five numbered test tubes and add 5 drops of concentrated hydrochloric acid to tubes 2–5. Let tube 2 stand at room temperature and heat the other four tubes in the hot water bath for the following periods of time: tube 3, 2.5 min; tube 4, 5 min; tubes 1 and 5, 15 min. As each tube is removed from the bath, it is cooled to room temperature, and if it contains acid, adjusted to approximate neutrality by addition of 15 drops of 10% sodium hydroxide. Measure one drop of each neutral solution into a new numbered test tube, add 1 mL of red tetrazolium solution and a drop of 10% sodium hydroxide, and heat the five tubes together for 2 min and watch them closely.

In which of the tubes was hydrolysis negligible, incomplete, and extensive? Does the comparison indicate the minimum heating period required for complete hydrolysis? If not, return the stored solutions to the numbered test tubes, treat each with 1 mL of stock phenylhydrazine reagent, and heat the tubes together for 5 min. On the basis of your results, decide upon a hydrolysis procedure to use in studying unknowns; the same method is applicable to the hydrolysis of methyl glycosides.

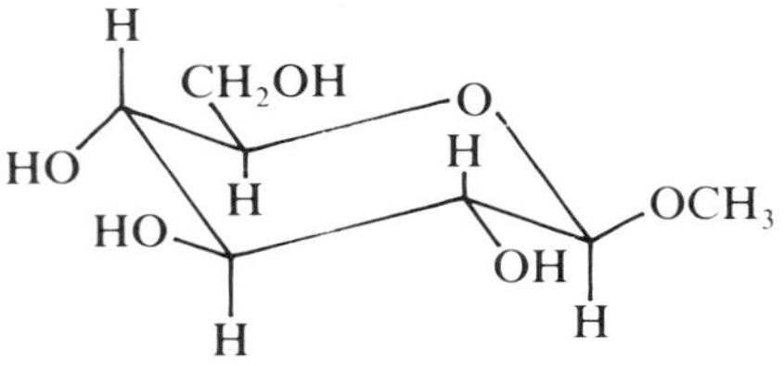

Methyl β-D-glucoside

6. Evaporation Test

Few solid derivatives suitable for identification of sugars are available. Osazones are not suitable since the same osazone can form from more than one sugar. Acetylation, in the case of a reducing sugar, is complicated by the possibility of formation of the α- or the β-anomeric form, or a mixture of both. (See Chapter 52.)

Test for lactose, maltose

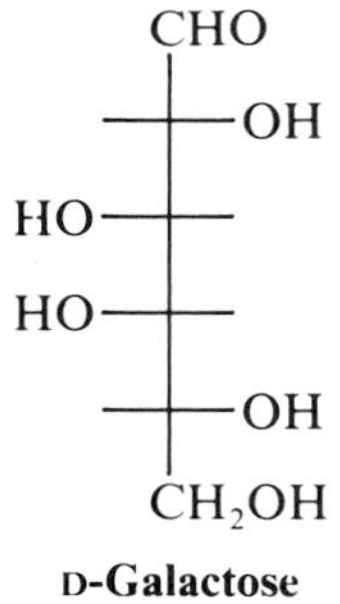

On thorough evaporation of all the water from a solution of glucose, fructose, mannose, or galactose, the sugar is left as a syrup that appears as a glassy film on the walls of the container. Evaporation of solutions of lactose or maltose gives white solid products, which are distinguishable because the temperature ranges at which they decompose differ by about 100°C.

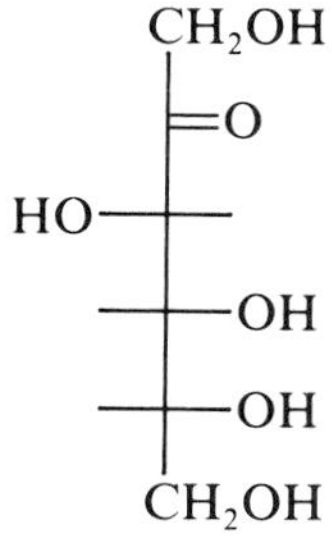

D-Galactose

Measure 0.5 mL of a 0.1 M solution of either lactose or maltose into a reaction tube, add a boiling chip and an equal volume of cyclohexane (to hasten the evaporation). Connect the test tube through a filter trap to the aspirator. Then turn the water running through the aspirator on at full force and rest the tube horizontally in the steam bath with all but the largest ring removed, so that the whole tube will be heated strongly. If evaporation does not occur rapidly, check the connections and trap to see what is wrong. If a water layer persists for a long time, disconnect and add 1–2 mL of cyclohexane to hasten evaporation. When evaporation appears to be complete, disconnect, rinse the walls of the tube with 0.5 mL of methanol, and evaporate again, when a solid should separate on the walls. Rinse this down with methanol and evaporate again to produce a thoroughly anhydrous product. Then scrape out the solid and determine the melting point, or actually the temperature range of decomposition.

D-Fructose

Anhydrous α-maltose decomposes at about 100–120°C, and anhydrous α-lactose at 200–220°C. Note that in the case of an unknown a temperature of decomposition in one range or the other is valid as an index of identity only if the substance has been characterized as a reducing sugar. Before applying the test to an unknown, perform a comparable evaporation of a 0.1 M solution of glucose, fructose, galactose, or mannose.

7. Unknowns

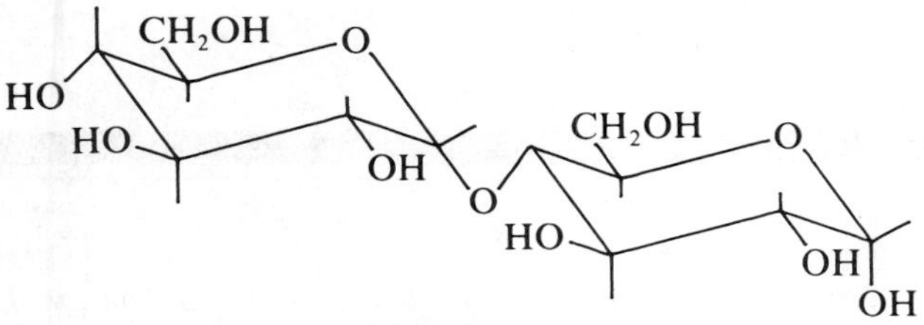

Maltose

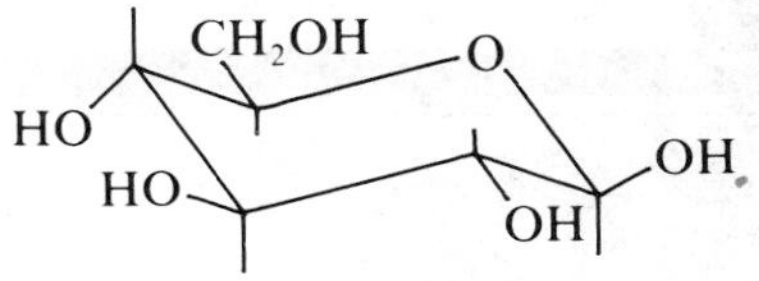

β-D-Glucose

The unknown, supplied as a 0.1 M solution, may be any one of the following substances:

D-Glucose	Maltose
D-Fructose	Sucrose
D-Galactose	Methyl β-D-glucoside
Lactose	

You are to devise your own procedure of identification.

Questions

1. What, do you conclude, is the order of relative reactivity in the RT test of the compounds studied?
2. Which test do you regard as the most reliable for distinguishing reducing from nonreducing sugars, and which for differentiating an α-ketol from a simple aldehyde?
3. Write a mechanism for the acid-catalyzed hydrolysis of a disaccharide.

52 Synthesis of Vitamin K_1: Quinones

PRELAB EXERCISE: *Write balanced equations for the reduction of Orange II to aminonaphthol with sodium hydrosulfite and for the oxidation of aminonaphthol to 1,2-naphthoquinone using ferric chloride.*

In the first experiment of this chapter the dye Orange II is reduced in aqueous solution with sodium hydrosulfite to water-soluble sodium sulfanilate and 1-amino-2-naphthol (**2**), which precipitates. This intermediate is purified as the hydrochloride and oxidized to 1,2-naphthoquinone (**3**).

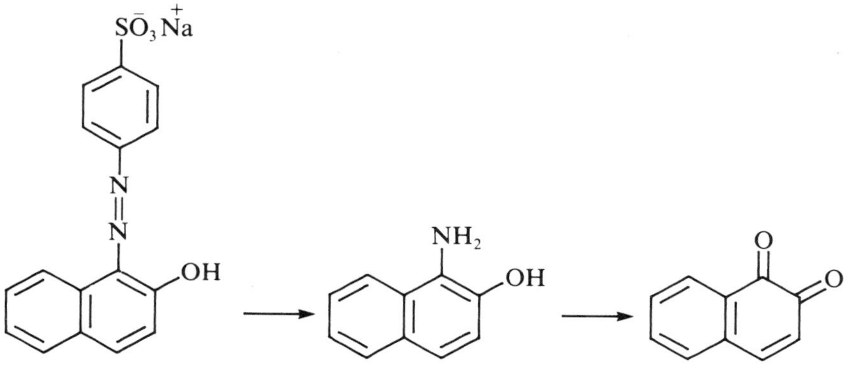

1
Orange II
MW 350.34

2
1-Amino-2-naphthol
MW 159.18

3
1,2-Naphthoquinone
MW 158.15, mp 145–147°C dec

The next three experiments of this chapter are a sequence of steps for the synthesis of the antihemorrhagic vitamin K_1 (or an analog) starting with a coal-tar hydrocarbon.

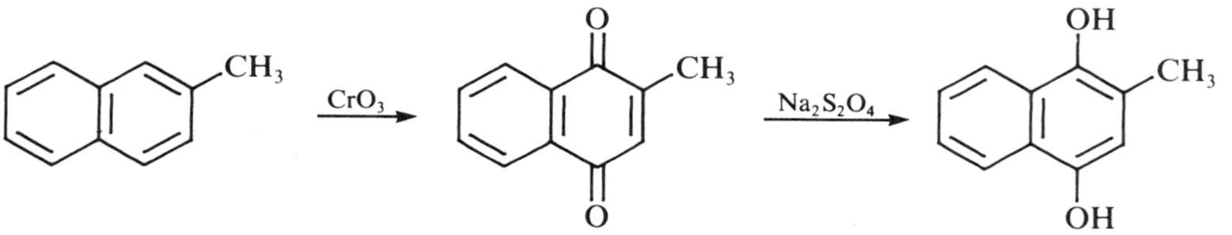

4
2-Methylnaphthalene
MW 142.19, mp 35°C

5
2-Methyl-1,4-naphthoquinone
MW 172.17, mp 105–106°C

6
2-Methyl-1,4-naphthohydroquinone
MW 174.19

2-Methylnaphthalene (**1**) is oxidized with chromic acid to 2-methyl-1,4-naphthoquinone (**5**), the yellow quinone is purified and reduced to its hydroquinone, **6**, by shaking an ethereal solution of the substance with aqueous hydrosulfite solution, the colorless hydroquinone is condensed with phytol, and the substituted hydroquinone, **7**, oxidized to vitamin K_1, **9**.

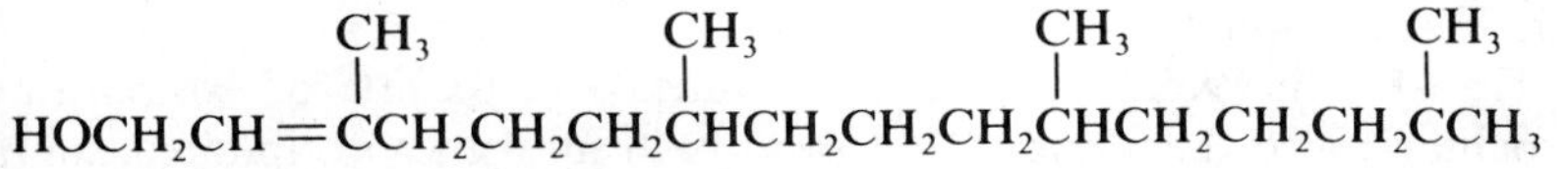

$$HOCH_2CH{=}C(CH_3)CH_2CH_2CH_2CH(CH_3)CH_2CH_2CH_2CH(CH_3)CH_2CH_2CH_2C(CH_3)H_3$$

Phytol
MW 296.52, bp 145°C at 0.03–0.04 torr

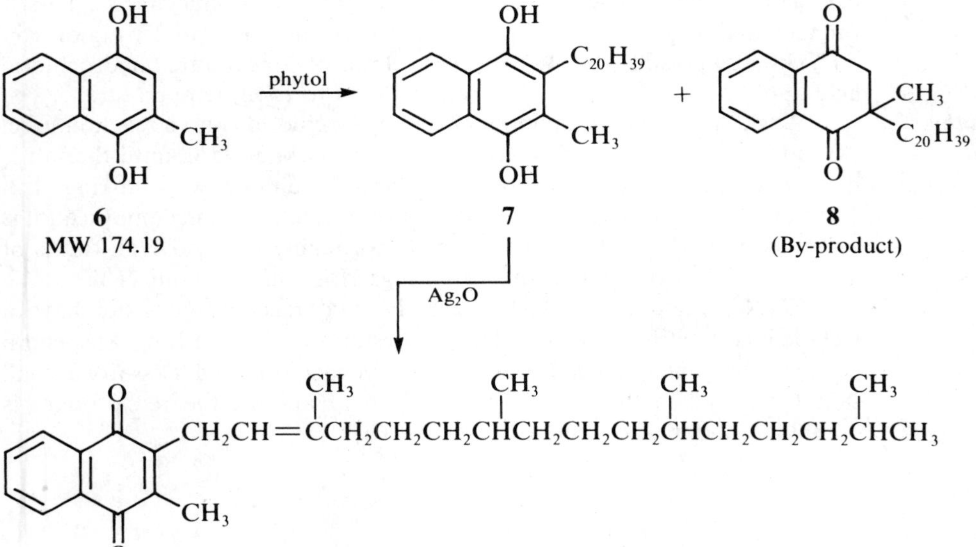

6 (OH, OH, CH_3)
MW 174.19

$\xrightarrow{\text{phytol}}$ **7** (OH, OH, $C_{20}H_{39}$, CH_3) + **8** (O, O, CH_3, $C_{20}H_{39}$)
(By-product)

7 $\xrightarrow{Ag_2O}$ **9**

$$CH_2CH{=}C(CH_3)CH_2CH_2CH_2CH(CH_3)CH_2CH_2CH_2CH(CH_3)CH_2CH_2CH_2CH(CH_3)CH_3$$

(O, O, CH_3)

9
Vitamin K_1
MW 450.68

An additional or alternative experiment is conversion of 2-methyl-1,4-naphthoquinone (**5**) through the oxide into the 3-hydroxy compound, phthiocol, which has been isolated from human tubercle bacilli after saponification, probably as a product of cleavage of vitamin K_1 (see experiment 5).

Experiments

1. 1,2-Naphthoquinone (3)

In a reaction tube dissolve 195 mg of the dye Orange II in 2.5 mL of water and warm the solution to 40–50°C. Add 225 mg of sodium hydrosulfite dihydrate and swirl until the red color is discharged and a cream-colored or pink precipitate of 1-amino-2-naphthol separates. To coagulate the product, heat the mixture nearly to boiling until it begins to froth, then cool in an ice bath, collect the product on a Hirsch funnel, and wash the residue with water. Prepare a solution of 0.05 mL of concentrated hydrochloric acid, 1 mL of water, and an estimated 2.5 mg of tin(II) chloride (antioxidant); transfer the precipitate of aminonaphthol to this solution and wash in material adhering to the Hirsch funnel.

Two short-time reactions

Transfer the solid to a reaction tube, add 0.2 mL of concentrated hydrochloric acid, heat over a hot sand bath until the precipitated aminonaphthol hydrochloride has been brought into solution, and then cool thoroughly in an ice bath. Collect the crystalline, colorless hydrochloride and wash it with a mixture of 0.05 mL of concentrated hydrochloric acid and 0.2 mL of water. Leave the air-sensitive crystalline product in the Hirsch funnel while preparing a solution for its oxidation. Dissolve 275 mg of iron (III) chloride crystals ($FeCl_3$–$6H_2O$) in 0.1 mL of concentrated hydrochloric acid and 0.5 mL of water by heating and cool to room temperature.

Discard iron and tin solutions in the container provided

Wash the crystalline aminonaphthol hydrochloride into a reaction tube, stir, add a few drops more water, and warm to about 35°C until the salt is dissolved and then add the iron(III) chloride solution with stirring. 1,2-Naphthoquinone separates at once as a voluminous precipitate and is collected on a Hirsch filter and washed thoroughly to remove all traces of acid. The yield from pure, salt-free Orange II is usually about 75%.

1,2-Naphthoquinone, highly sensitive and reactive, does not have a well-defined melting point but decomposes at about 145–147°C. Suspend a sample in hot water and add concentrated hydrochloric acid. Dissolve a small sample in cold methanol and add a drop of aniline; the red product is 4-anilino-1,2-naphthoquinone.

The Synthesis of Vitamin K_1

2. 2-Methyl-1,4-naphthoquinone (5)

CrO_3 is a carcinogen. Work in a hood with this material and do not breathe the dust.

Reaction time: 1.25 h

In a 10-mL Erlenmeyer flask dissolve 2.5 g of chromium(VI) oxide (CrO_3, chromic anhydride), in 1.75 mL of water and dilute the dark-red solution with 1.75 mL of acetic acid.[1] In a 25-mL Erlenmeyer flask prepare a mixture of 0.71 g of 2-methylnaphthalene and 7.5 mL of acetic acid, and without cooling add portions of the oxidizing solution. Stir with a thermometer until the temperature rises to 60°C. At this point ice cooling may be required to prevent a further rise in temperature. By alternate addition of reagent and

1. The chromium(VI) oxide is hygroscopic; weigh it quickly in the hood and do not leave the bottle unstoppered. The substance dissolves very slowly in acetic acid–water mixtures, and solutions are prepared by adding the acetic acid only after the substance has been completely dissolved in water.

cooling, maintain the temperature close to 60°C throughout the addition, which can be completed in about 5 min. When the temperature begins to drop spontaneously heat the solution gently on the steam bath (85–90°C) for 1 h to complete the oxidation.

Dilute the dark-green solution with water nearly to the top of the flask, stir well for a few minutes to coagulate the yellow quinone, collect the product on a Hirsch funnel, and wash it thoroughly with water to remove chromium(III) acetate. The crude material can be crystallized from methanol (2 mL) while still moist, and gives 0.3–0.35 g of satisfactory 2-methyl-1,4-naphthoquinone, mp 105–106°C. Save the product for the preparation of **6**, **9**, and **11**. This quinone must be kept away from light, which converts it into a pale-yellow, sparingly soluble polymer.

Discard chromium ion-containing solutions in the container provided

3. 2-Methyl-1,4-naphthohydroquinone (6)

Short-time reaction

In a reaction tube dissolve 100 mg of 2-methyl-1,4-naphthoquinone (**5**) in 1.75 mL of ether by warming on a steam bath. Add a fresh solution of 200 mg of sodium hydrosulfite in 1.5 mL of water. After passing through a brown phase (quinhydrone), the solution should become colorless or pale yellow in a few minutes; if not, add more hydrosulfite solution. After removing the aqueous layer, shake the ethereal solution with a mixture of 1.25 mL of saturated sodium chloride solution and 0.1 mL of saturated hydrosulfite solution to remove the bulk of the water. Dry the ether solution with anhydrous sodium sulfate. Transfer the dry ether solution to a tared reaction tube and evaporate the filtrate until nearly all the solvent has been removed, cool, and add petroleum ether. The hydroquinone separating as a white or grayish powder is collected, washed with petroleum ether, and dried; the yield is about 90 mg (the substance has no sharp mp).

4. Vitamin K_1 (2-Methyl-3-phytol-1,4-naphthoquinone, 9)

Explanation

Phytol, being an allylic alcohol, is reactive enough to condense with 2-methyl-1,4-naphthohydroquinone (**6**) under mild conditions of acid catalysis as specified in the following procedure. Overheating must be avoided or the alcohol is dehydrated to phytadiene. The reaction mixture is diluted with water and extracted with ether and unchanged starting material removed by extraction with aqueous alkali containing hydrosulfite to protect the hydroquinones from oxidation by air. The hydroquinone of vitamin K_1 (**7**) has a methyl group adjacent to one hydroxyl group and the long phytol side chain adjacent to the other hydroxyl group; it is a cryptophenol (hidden phenolic properties), insoluble in aqueous alkali. It is separated from the nonhydroxylic by-product, **8**, by crystallization from petroleum ether and oxidized to vitamin K_1 (**9**).

Procedure

Dioxane is a mild carcinogen. Carry out this experiment in a hood.

Place 150 mg of phytol[2] and 1 mL of dioxane (*caution!*) in a reaction tube and warm to 50°C on a sand or steam bath. Prepare a solution of 150 mg of 2-methyl-1,4-naphthohydroquinone (**6**) and 0.15 mL of boron trifluoride etherate in 1 mL of dioxane. Add this in portions with a capillary dropper in the course of 15 min with constant swirling and while maintaining a temperature of 50°C (do not overheat). Continue in the same way for 20 min longer. Cool to 25°C, transfer the material to a 6-in. test tube or a 25-mL separatory funnel with 4 mL of ether, and wash the orange-colored ethereal solution with three 3-mL portions of water to remove boron trifluoride and dioxane.

Extract the unchanged hydroquinone from the ether solution with a freshly prepared solution of 0.2 g of sodium hydrosulfite in 4 mL of 2% aqueous sodium hydroxide and 1 mL of a saturated sodium chloride solution (which helps break the resulting emulsion). Shake vigorously for a few minutes, during which time any red color should disappear and the alkaline layer should acquire a bright yellow color. After releasing the pressure through the stopcock, allow the layers to separate, keeping the funnel stoppered as a precaution against oxidation. Draw off the yellow liquor and repeat the extraction a second and a third time, or until the alkaline layer remains practically colorless. Separate the faintly colored ethereal solution, dry it over anhydrous sodium sulfate, filter into a tared flask, and evaporate the filtrate on the stream bath—eventually with evacuation at the aspirator. The total oil, which becomes waxy on cooling, amounts to 170–190 mg.

Add 1 mL of petroleum ether (bp 20–40°C) and boil and manipulate with a spatula until the brown mass has changed to a white paste. Wash the paste into a small centrifuge tube with 1–2 mL of fresh petroleum ether, cool well in ice, and centrifuge. Decant the brown supernatant liquor into the original tared flask, fill the tube with fresh solvent, and stir the white sludge to an even suspension. Then cool, centrifuge, and decant as before. Evaporation of the decanted liquor and washings gives 110–130 mg of residual oil, which can be discarded. Dissolve the portions of washed white sludge of vitamin K_1 hydroquinone in a total of 1–1.5 mL of absolute ether, and add decolorizing charcoal for clarification, if the solution is pink or dark. Add 0.1 g of silver oxide and 0.1 g of anhydrous sodium sulfate. Shake for 20 min, filter into a tared flask, and evaporate the clear yellow solution on the steam bath, removing traces of solvent at the aspirator. Undue exposure to light should be avoided when the material is in the quinone form. The residue is a light yellow, rather mobile oil consisting of pure vitamin K_1; the yield is about 60–90 mg. A sample for preservation is transferred with a capillary dropper to a small specimen vial wrapped in metal foil or black paper to exclude light.

Note for the instructor

2. Supplier: Aldrich Chemical Co. Geraniol can be used instead of expensive phytol; it reacts with 2-methyl-1,4-naphthohydroquinone to give a product similar in chemical and physical properties to the natural vitamin and with pronounced antihemorrhagic activity.

To observe a characteristic color reaction, transfer a small bit of vitamin on the end of a stirring rod to a test tube, stir with 0.1 mL of alcohol, and add 0.1 mL of 10% alcoholic potassium hydroxide solution; the end pigment responsible for the red color is phthiocol.

5. Phthiocol (2-Methyl-3-hydroxy-1,4-naphthoquinone, 11)

CH_3 HOONa CH_3 O H H_2SO_4 CH_3 OH

5 **10** **Oxide** MW 188.17, mp 93.5–94.5°C **11** **Phthiocol** MW 188.17, mp 172–173°C

Two short-time reactions

In a reaction tube dissolve 100 mg of 2-methyl-1,4-naphthoquinone (**5**) in 1 mL of alcohol by heating, and let the solution stand while the second reagent is prepared by dissolving 20 mg of anhydrous sodium carbonate in 0.5 mL of water and adding (cold) 0.10 mL of 30% hydrogen peroxide solution. Cool the quinone solution under the tap until crystallization begins, add the peroxide solution all at once, and cool the mixture. The yellow color of the quinone should be discharged immediately. Fill the tube with water, transfer the mixture to a 25-mL flask, cool in ice, and collect the colorless, crystalline epoxide, **10**; yield about 90 mg, mp 93.5–94.5°C (pure 95.5–96.5°C).

Avoid contamination of hydrogen peroxide with metal salts, organic compounds, or metal spatulas: explosive

To 100 mg of **10** in a reaction tube add 0.5 mL of concentrated sulfuric acid; stir if necessary to produce a homogeneous, deep red solution, and after 10 min cool this in ice and slowly add 2 mL of water. The precipitated phthiocol can be collected, washed, and crystallized by dissolving in methanol (2.5 mL), adding a few drops of hydrochloric acid to give a pure yellow color, treating with decolorizing charcoal, concentrating the filtered solution, and diluting to the saturation point. Alternatively, the yellow suspension is washed into a separatory funnel and the product extracted with a mixture of 2.5 mL each of toluene and ether. The organic layer is dried over anhydrous sodium sulfate and evaporated to a volume of about 1 mL for crystallization. The total yield of pure phthiocol (**11**), mp 172–173°C, is about 80 mg.

Questions

1. To what general class of compounds does phytol belong?
2. Write a mechanism for the reaction of 2-methyl-1,4-naphthoquinone-epoxide (**10**) with sulfuric acid to form phthiocol.

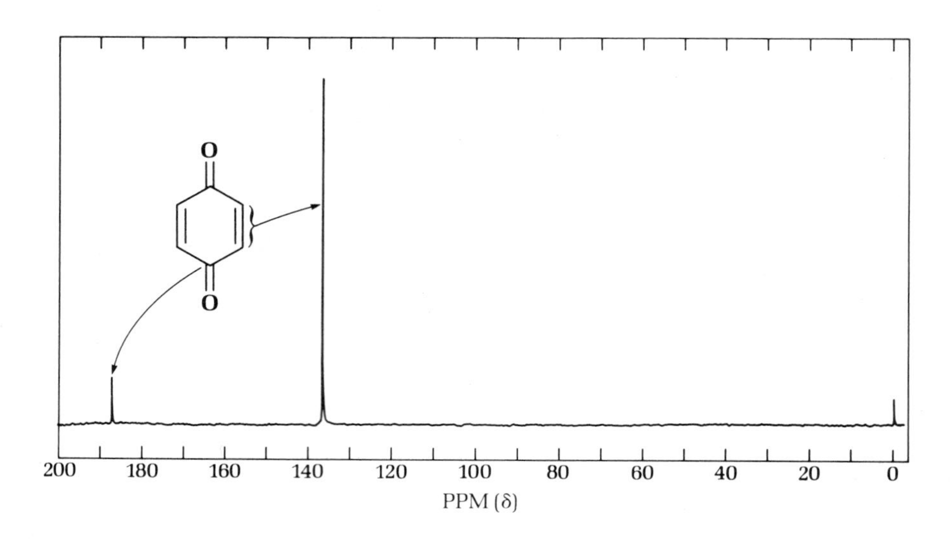

FIG. 52.1 ^{13}C nmr spectrum of quinone.

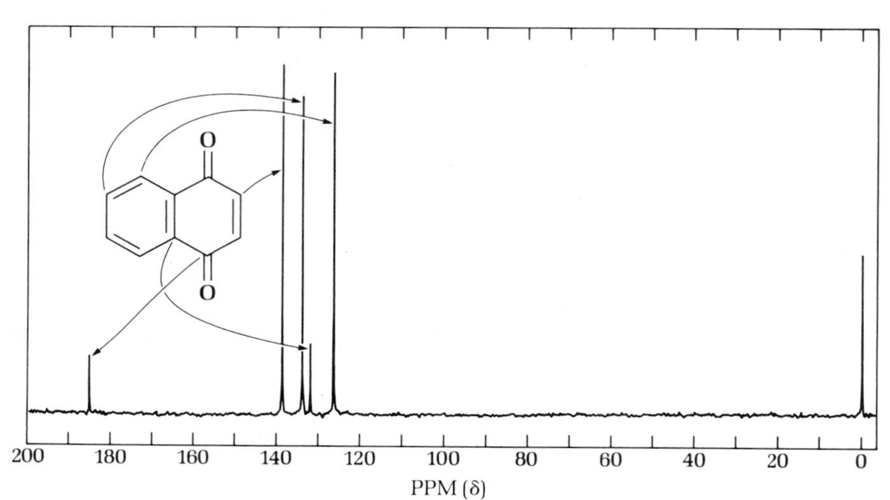

FIG. 52.2 ^{13}C nmr spectrum of 1,4-naphthoquinone.

The Friedel-Crafts Reaction: 2-Methylanthraquinone

53

PRELAB EXERCISE: *Give the mechanism for the cyclization of the benzoylbenzoic acid to 2-methylanthraquinone using concentrated sulfuric acid.*

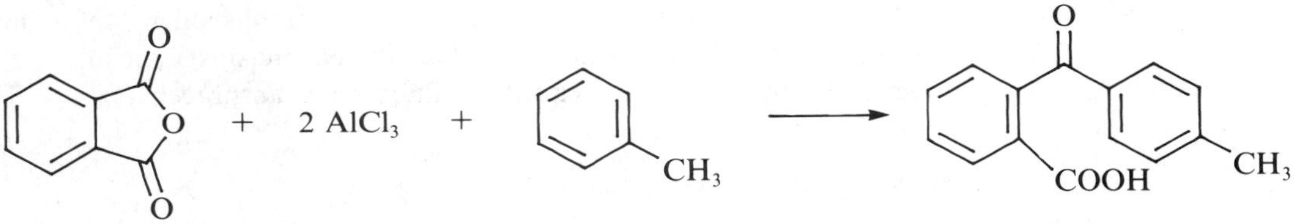

O
O
O
+ 2 $AlCl_3$ + CH_3 → O COOH CH_3

Phthalic anhydride
MW 148.11, mp 132°C

Aluminum chloride
MW 133.34

Toluene
MW 92.14
bp 111°C

2-(4-Methylbenzoyl)benzoic acid
MW 240.25, mp 138–139°C

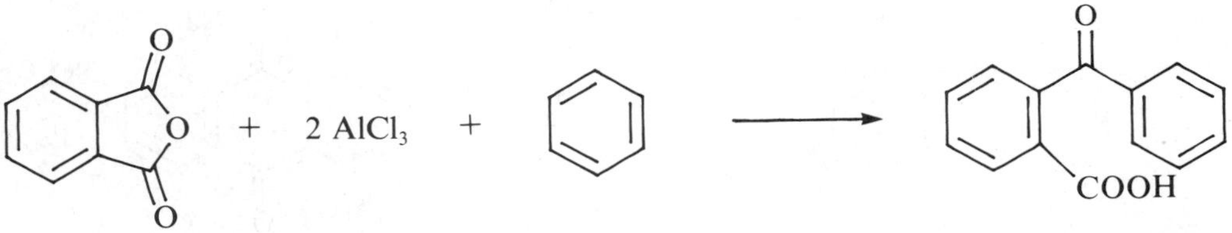

O
O
O
+ 2 $AlCl_3$ + → O COOH

Benzene
MW 78.11
bp 80°C

2-Benzoylbenzoic acid
MW 226.22, mp 127°C

The Friedel-Crafts reaction of phthalic anhydride with excess toluene as solvent and two equivalents of aluminum chloride proceeds rapidly and gives a complex salt of 2-(4-methylbenzoyl)benzoic acid in which one mole of aluminum chloride has reacted with the acid function to form the salt $RCOO^-AlCl_2^+$ and a second mole is bound to the carbonyl group. On addition of ice and hydrochloric acid the complex is decomposed and basic aluminum salts are brought into solution.

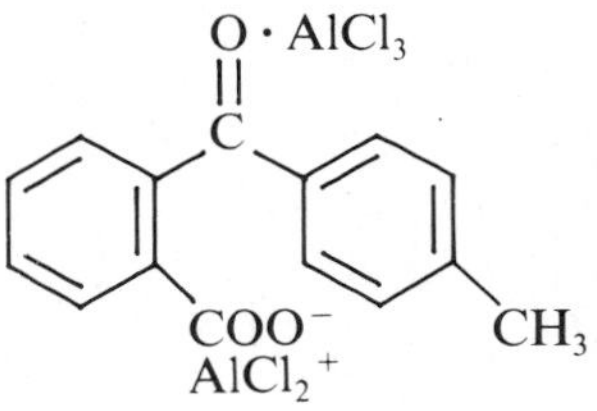

O · $AlCl_3$
C
COO^-
$AlCl_2^+$
CH_3

Complex salt

Treatment of the 2-(4-methylbenzoyl)benzoic acid with polyphosphoric acid effects cyclodehydration to 2-methylanthraquinone in a 2-h reaction. Because unsubstituted anthraquinone can be sulfonated only under forcing conditions, a high temperature can be used to shorten cyclodehydration of the parent molecule, 2-benzoylbenzoic acid, in an alternate experiment. The reaction conditions are so adjusted that anthraquinone separates from the hot solution in crystalline form favoring rapid drying.

$Na_2S_2O_4$
Sodium hydrosulfite

Reduction of anthraquinone to anthrone can be accomplished rapidly on a microscale with tin(II) chloride in acetic acid solution. A second method, which involves refluxing anthraquinone with an aqueous solution of sodium hydroxide and sodium hydrosulfite, is interesting to observe because of the sequence of color changes: anthraquinone is reduced first to a deep-red liquid containing anthrahydroquinone diradical dianion; the red color then gives place to a yellow color characteristic of anthranol radical anion; as the alkali is neutralized by the conversion of $Na_2S_2O_4$ to $2\,NaHSO_3$, anthranol ketonizes to the more stable anthrone. The second method is preferred in industry, because sodium hydrosulfite costs less than half as much as tin(II) chloride and because water is cheaper than acetic acid and no solvent recovery problem is involved.

Reduction of anthrone to anthracene is accomplished by refluxing in aqueous sodium hydroxide solution with activated zinc dust. The method has the merit of affording pure, beautifully fluorescent anthracene.

Fieser's Solution

Oxygen scavenger

A solution of 2 g of sodium anthraquinone-2-sulfonate and 15 g of sodium hydrosulfite in 100 mL of a 20% aqueous solution of potassium hydroxide affords a blood-red solution of the diradical dianion:

O O $SO_3^-Na^+$ $\underset{\text{Oxygen}}{\xrightleftharpoons{Na_2S_2O_4,\ NaOH}}$ O^-Na^+ O^-Na^+ $SO_3^-Na^+$

This solution has a remarkable affinity for oxygen. It is used to remove traces of oxygen from gases such as nitrogen or argon when it is desirable to render them absolutely oxygen-free. This solution has a capacity of about 800 mL of oxygen. The color fades and the solution turns brown when it is exhausted.

Experiments

1. 2-(4-Methylbenzoyl)benzoic Acid

Into a dry 10 × 100 mm reaction tube place 150 mg of phthalic anhydride and 0.75 mL of dry toluene. Cool the mixture in an ice bath and then add 300 mg of anhydrous aluminum chloride.[1] Cap the tube with a septum and

1. Aluminum chloride should be weighed in a small, dry, stoppered vial or reaction tube. The aluminum chloride should be from a freshly opened bottle and should be weighed and transferred in the hood. It is very hygroscopic and releases hydrogen chloride upon reaction with water. Weigh the reagent rapidly to avoid exposure of the compound to air. The quality of the aluminum chloride determines the success of this experiment.

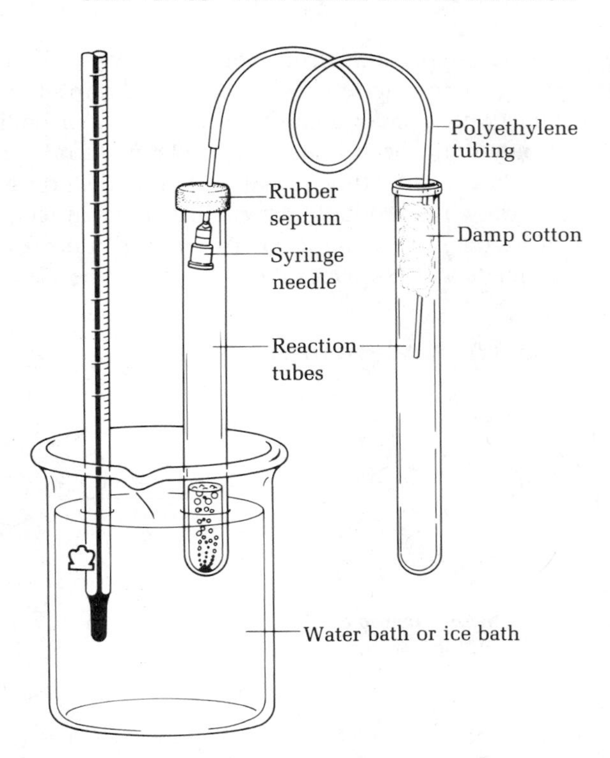

FIG. 53.1 HCl trap for Friedel-Crafts reaction.

a syringe needle connected to polyethylene tubing leading to another reaction tube, which contains a piece of damp cotton to act as a trap for the hydrogen chloride liberated in the reaction (Fig. 53.1). Mix the contents of the tube thoroughly by flicking the tube. Warm the tube by the heat of the hand. If the reaction does not start, warm the tube *very gently* in a beaker of hot water or hold it over the sand bath or steam bath for a few seconds. At the first sign of vigorous boiling, or evolution of hydrogen chloride, hold the tube over the ice bath in readiness to cool it if the reaction becomes too vigorous. Continue this gentle, cautious heating until the reaction proceeds smoothly enough to reflux it on the hot sand bath. This will take about 5 min.

Heat the reaction mixture on the sand bath until evolution of hydrogen chloride almost ceases, then cool it in ice, and add 1 g of ice in small pieces. Allow each little piece to react before adding the next. Mix the reaction mixture well during this hydrolysis, using a glass rod. After the reaction subsides, add 0.1 mL of concentrated hydrochloric acid and 1 mL of water. Mix the contents of the tube thoroughly and make sure the mixture is at room temperature. Add 0.5 mL more water, mix the solution well, ascertain that it is at room temperature, and then add 1.5 mL of ether and break up any lumps in the tube with a stirring rod. Stopper the tube and shake it vigorously to complete the hydrolysis and extraction. Allow the layers to separate and then remove the aqueous layer with a Pasteur pipette. Add 0.1 mL of concentrated hydrochloric acid and 0.15 mL of water. Shake the mixture vigorously again, and remove the aqueous layers. Transfer the organic layers to a small test tube containing anhydrous sodium sulfate. Allow the solution to dry for 5 min, then transfer it back to the clean dry reaction tube, rinsing the sodium sulfate with a small quantity of ether. Add a boiling chip and boil off the solvents until the volume in the tube is 0.5 mL; then add ligroin until the solution is slightly turbid, indicating that the product is beginning to crystallize. Allow the product to crystallize at room temperature and then at 0°C. Collect the crystals on the Hirsch funnel. Pure 2-(4-methylbenzoyl)benzoic acid melts at 138–139°C; the yield should be about 0.2 g.

2. 2-Methylanthraquinone

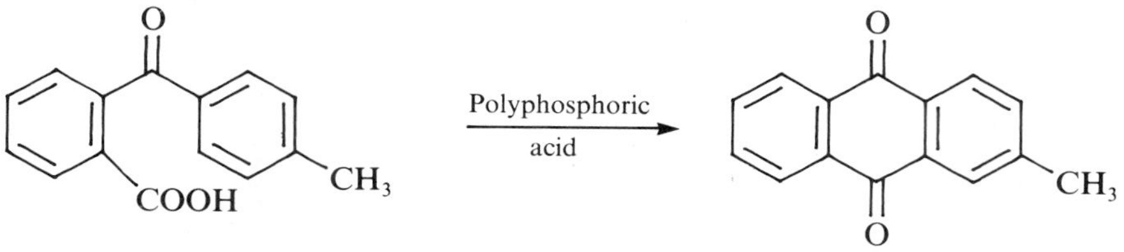

O COOH CH_3 —Polyphosphoric acid→ O O CH_3

2-(4-Methylbenzoyl)benzoic acid
MW 240.24, mp 138–139°C

2-Methylanthraquinone
MW 222.23, mp 175°C

Place 100 mg of 2-(4-methylbenzoyl)benzoic acid in a reaction tube and add to it 1.25 mL of polyphosphoric acid. Cap the tube with a rubber septum pierced with a syringe needle to help prevent moisture from diffusing into the reaction mixture. Heat the tube in a boiling water bath for 2 h, cool the tube in ice, and add 2 mL of water with stirring. Remove the solvent using the Pasteur pipette and wash the product with water. Add 0.5 mL of concentrated ammonium hydroxide and boil the mixture for a few min to remove unchanged starting material as the water-soluble ammonium salt. Remove the ammonia, wash the product with water, and recrystallize the residue from ethanol. The yield of plates of 2-methylanthraquinone should be about 70 mg.

3. Anthraquinone

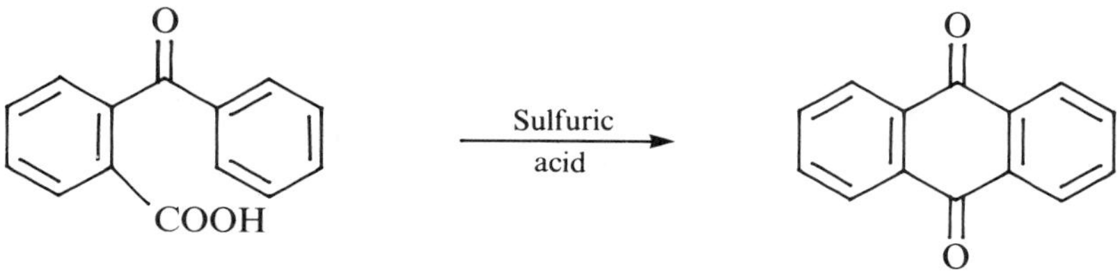

O COOH —Sulfuric acid→ O O

2-Benzoylbenzoic acid
MW 226.22, mp 127°C

Anthraquinone
MW 208.20
mp 286°C

In a reaction tube dissolve 100 mg of 2-benzoylbenzoic acid[2] in 0.5 mL of concentrated sulfuric acid by gently heating and stirring. Immerse a thermometer in the reaction mixture and heat it at 150–155°C for 5 min. Allow the tube to cool to below 100°C and then, using extreme caution, add a very small drop of water to the mixture. Mix the contents of the tube and continue adding water in minute drops to the mixture. This will cause the product to crystallize, and if done slowly enough the crystals will be large enough to collect easily by filtration. Fill the tube with water and collect the product by filtration on the Hirsch funnel. Wash it well with water, dry, determine the weight, and then calculate the percent yield. Determine the melting point.

$SnCl_2$ or $Na_2S_2O_3$

Anthraquinone
MW 208.20
mp 286°C

Anthrone
MW 194.22
mp 156°C

4. Anthrone

(a) Tin(II) Chloride Reduction

In a reaction tube put 50 mg of crude anthraquinone from the previous experiment, 0.40 mL of acetic acid, and a solution made by warming 0.13 g of tin(II) chloride dihydrate with 0.13 mL of concentrated hydrochloric acid. Add a boiling stone, note the time, and reflux gently until crystals of anthraquinone have completely disappeared (8–10 min); then reflux 15 min longer and record the total time. Then add water (about 0.12 mL) dropwise until the solution is saturated. Let the solution stand for crystallization. Collect and dry the product and take the melting point (156°C given). The yield of pale-yellow crystals is about 40 mg.

Procedure 1

$SnCl_2 \cdot 2H_2O$
Tin(II) chloride dihydrate

(b) Hydrosulfite Reduction

In a 5-mL long-necked round-bottomed flaskput 50 mg of anthraquinone, 60 mg of sodium hydroxide, 150 mg of sodium hydrosulfite, and 1.3 mL of water. Heat over a hot sand bath and swirl for a few minutes to convert the anthraquinone into the deep-red anthrahydroquinone anion. Note that particles of different appearance begin to separate even before the anthraquinone has all dissolved. Add an empty distilling column to the flask and reflux for 45 min; cool, filter the product, and wash it well and let dry. Determine the weight and melting point of the crude material and crystallize it from 95%

Procedure 2

$Na_2S_2O_4$
Sodium hydrosulfite

2. Available from the Aldrich Chemical Co.

ethanol. Record the approximate volume of solvent used and, if the first crop of crystals recovered is not satisfactory, concentrate the mother liquor and secure a second crop.

5. Comparison of Results

Compare your results with those obtained by neighbors using the alternative procedure with respect to yield, quality of product, and working time. Which is the better laboratory procedure? Then consider the cost of the three solvents concerned, the cost of the two reducing agents (current prices can be looked up in a catalog), the relative ease of recovery of the organic solvents, and the prudent disposal of by-products, and decide which method would be preferred as a manufacturing process.

6. Fluorescent Anthracene

O H H Zn dust (Cu)

Anthrone
MW 194.22, mp 156°C

Anthracene
MW 178.22, mp 216°C

Put 100 mg of zinc dust into a 5-mL round-bottomed flask and activate the dust by adding 0.6 mL of water and 0.1 mL of copper(II) sulfate solution (Fehling's solution I) and swirling for a minute or two. Add 40 mg of anthrone, 100 mg of sodium hydroxide, and 1 mL of water, attach the empty distilling column as a reflux condenser, heat to boiling on the sand bath, note the time, and start refluxing the mixture. Anthrone at first dissolves as the yellow anion of anthranol, but anthracene soon begins to separate as a white precipitate.

In about 15 min the yellow color initially observed on the walls disappears, but refluxing should be continued for a full 30 min. Then remove the heat, rinse down any anthracene in the condenser with a few drops of water, and cool the flask in ice. Remove the solvent with a Pasteur pipette and add very carefully and dropwise 0.2 mL of concentrated hydrochloric acid to dissolve the remaining zinc. Again cool the mixture and add water to wash down any unreacted zinc; after all has reacted remove the acid and wash the product liberally with water. Remove all water possible, then wash once with ice-cold methanol to remove as much of the water as possible from the tube and the product.

Hydrogen liberated

Methanol removes water without dissolving much anthracene

The product need not be dried before crystallization from about 0.5 mL of boiling toluene. Slow cooling of the toluene solution will give thin, colorless, beautifully fluorescent plates; yield about 20 mg. In washing the equipment with acetone, you should be able to observe the striking fluorescence of very dilute solutions. The fluorescence is quenched by a bare trace of impurity.

Questions

1. Calculate the number of moles of hydrogen chloride liberated in the synthesis of 2-(4-benzoyl)benzoic acid. If this gaseous acid were dissolved in water, hydrochloric acid would be formed. How many milliliters of concentrated hydrochloric acid would be formed in this reaction? The acid is 12 molar in HCl.
2. Write a mechanism for the formation of anthraquinone from 2-benzoylbenzoic acid, indicating clearly the role of sulfuric acid. What is the name commonly given to this type of reaction?

Derivatives of 1,2-Diphenylethane— A Multistep Synthesis[1]

Procedures are given in the next seven chapters for rapid preparation of small samples of twelve related compounds starting with benzaldehyde and phenylacetic acid. The quantities of reagents specified in the procedures are often such as to provide somewhat more of each intermediate than is required for completion of subsequent steps in the sequence of reactions. If the experiments are dovetailed, the entire series of preparations can be completed in very short working time. For example, one can start the preparation of benzoin (record the time of starting and do not rely on memory), and during the reaction period start the preparation of α-phenylcinnamic acid; this requires refluxing for 35 min, and while it is proceeding the benzoin preparation can be stopped when the time is up and the product let crystallize. The α-phenylcinnamic acid mixture can be let stand (and cool) until one is ready to work it up. Also, while a crystallization is proceeding one may want to observe the crystals occasionally but should utilize most of the time for other operations.

Points of interest concerning stereochemistry and reaction mechanisms are discussed in the introductions to the individual chapters. Since several of the compounds have characteristic ultraviolet or infrared absorption spectra, pertinent spectroscopic constants are recorded and brief interpretations of the data are presented.

Question. Starting with 150 mg of benzaldehyde and assuming an 80% yield on each step, what yield of diphenylacetylene, in grams, might you expect? From the information given and assuming an 80% yield on the last two reactions, what yield of stilbene dibromide would you expect employing 150 mg of benzaldehyde?

Note for the instructor

1. If the work is well organized and proceeds without setbacks, the experiments can be completed in about four laboratory periods. The instructor may elect to name a certain number of periods in which the student is to make as many of the compounds as possible; the instructor may also decide to require submission only of the end products in each series.

Benzaldehyde → (Chap. 54, KCN or thiamine) → **Benzoin** → (Chap. 55, HNO_3) → **Benzil** → (Chap. 57, Ac_2O) → ***E*- and *Z*-Diacetates**

Benzoin → (Chap. 58, Zn/Hg, HCl + EtOH) → ***E*-Stilbene** → (Chap. 58, Br_2) → ***meso*-Stilbene dibromide** → (Chap. 58, KOH) → **Diphenylacetylene**

Benzil → (Chap. 56, $NaBH_4$) → ***meso*-Stilbenediol**

Benzaldehyde + $C_6H_5CH_2COOH$ (Chap. 59) → ***E*- and *Z*-Phenylcinnamic acid** → (Chap. 60, copper chromite catalyst) → ***Z*-Stilbene** → (Chap. 60, Br_2) → ***dl*-Stilbene dibromide**

54 The Benzoin Condensation: Cyanide Ion and Thiamine Catalyzed

PRELAB EXERCISE: *What purpose does the sodium hydroxide serve in the thiamine-catalyzed benzoin condensation?*

The reaction of two moles of benzaldehyde to form a new carbon-carbon bond is known as the benzoin condensation. It is catalyzed by two rather different catalysts, cyanide ion and the vitamin thiamine, which on close examination are seen to function in exactly the same way.

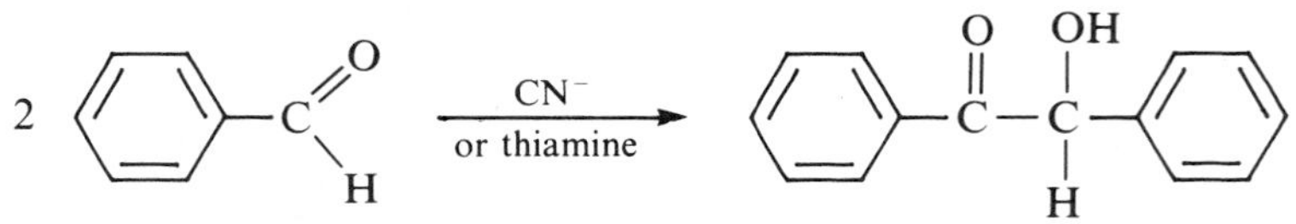

$$2\, C_6H_5-CHO \xrightarrow[\text{or thiamine}]{CN^-} C_6H_5-CO-CH(OH)-C_6H_5$$

Benzaldehyde
MW 106.12
bp 178°C

Benzoin
MW 212.24
mp 135°C

Consider first the cyanide ion–catalyzed reaction. The cyanide ion attacks the carbonyl oxygen to form a stable cyanohydrin, mandelonitrile, a liquid of bp 170°C that under the basic conditions of the reaction loses a proton to give a resonance-stabilized carbanion, **A**. The carbanion attacks another molecule of benzaldehyde to give **B**, which undergoes a proton transfer and loses cyanide to give benzoin. Evidence for this mechanism lies in the failure

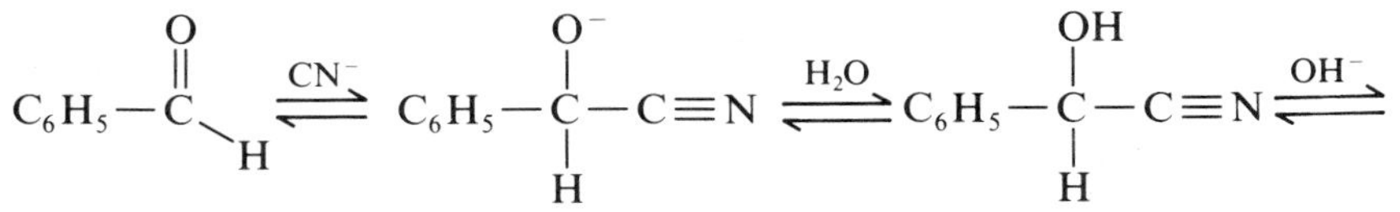

$$C_6H_5-C(=O)-H \overset{CN^-}{\rightleftharpoons} C_6H_5-C(O^-)(H)-C\equiv N \overset{H_2O}{\rightleftharpoons} C_6H_5-C(OH)(H)-C\equiv N \overset{OH^-}{\rightleftharpoons}$$

Benzaldehyde

Mandelonitrile
bp 170°C

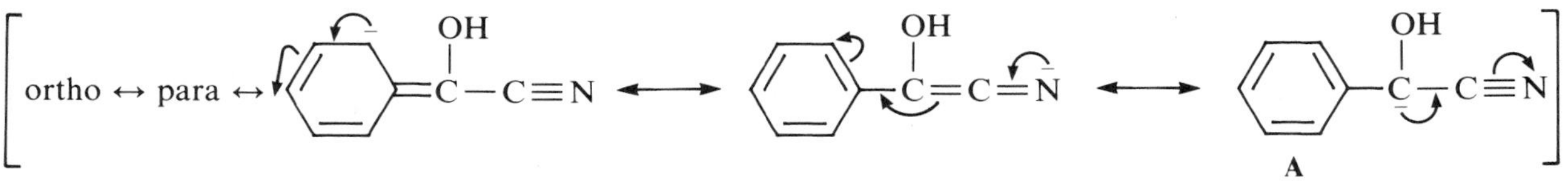

$$\left[\text{ortho} \leftrightarrow \text{para} \leftrightarrow C_6H_5{=}C(OH)-C\equiv N \longleftrightarrow C_6H_5-C(OH){=}C{=}N^- \longleftrightarrow C_6H_5-\bar{C}(OH)-C\equiv N\right]$$

A

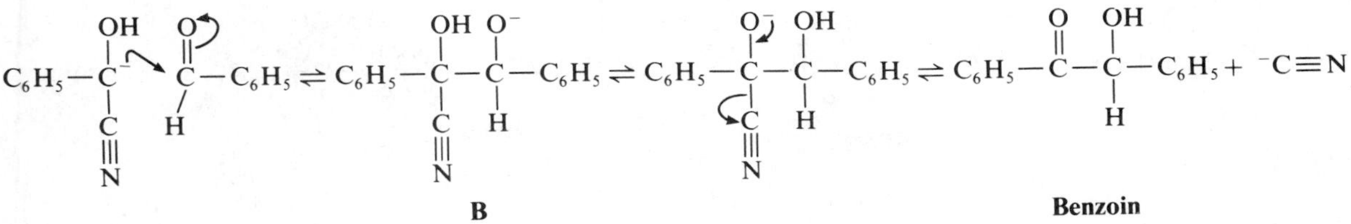

$$C_6H_5-\underset{\underset{N}{\overset{|}{\underset{|||}{C}}}}{\overset{OH}{\overset{|}{C^-}}}\;\overset{O}{\overset{\|}{C}}-C_6H_5 \rightleftharpoons C_6H_5-\underset{\underset{N}{\overset{|}{\underset{|||}{C}}}}{\overset{OH}{\overset{|}{C}}}-\underset{H}{\overset{O^-}{\overset{|}{\underset{|}{C}}}}-C_6H_5 \rightleftharpoons C_6H_5-\underset{\underset{N}{\overset{|}{\underset{|||}{C}}}}{\overset{O^-}{\overset{|}{C}}}-\underset{H}{\overset{OH}{\overset{|}{\underset{|}{C}}}}-C_6H_5 \rightleftharpoons C_6H_5-\overset{O}{\overset{\|}{C}}-\underset{H}{\overset{OH}{\overset{|}{\underset{|}{C}}}}-C_6H_5 + {}^-C\equiv N$$

B **Benzoin**

of 4-nitrobenzaldehyde to undergo the reaction, because the nitro group reduces the nucleophilicity of the anion in **A**. On the other hand a strong electron-donating group in the 4-position of the phenyl ring makes the loss of the proton from the cyanohydrin very difficult, and thus 4-dimethylamino-benzaldehyde also does not undergo the benzoin condensation with itself.

A number of biochemical reactions bear a close resemblance to the benzoin condensation but are not, obviously, catalyzed by the highly toxic cyanide ion. Some 30 years ago Breslow proposed that vitamin B_1, thiamine hydrochloride, in the form of the coenzyme thiamine pyrophosphate, can function in a manner completely analogous to cyanide ion in promoting reactions like the benzoin condensation. The resonance-stabilized conjugate base of the thiazolium ion, thiamine, and the resonance-stabilized carbanion, **C**, which it forms, are again the keys to the reaction. Like the cyanide ion, the thiazolium ion has just the right balance of nucleophilicity, ability to stabilize the intermediate anion, and good leaving group qualities.

Cyanide ion binds irreversibly to hemoglobin, rendering it useless as a carrier of oxygen

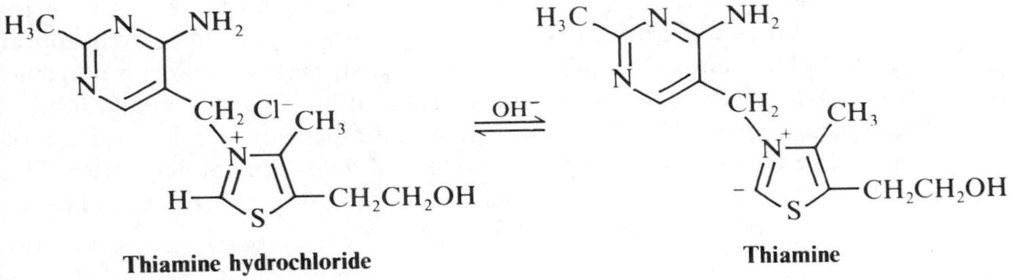

H_3C, N, NH_2, N, CH_2 Cl^- CH_3, $\overset{+}{N}$, H, S, CH_2CH_2OH $\xrightleftharpoons{OH^-}$ H_3C, N, NH_2, N, CH_2, CH_3, $\overset{+}{N}$, $-$, S, CH_2CH_2OH

Thiamine hydrochloride **Thiamine**

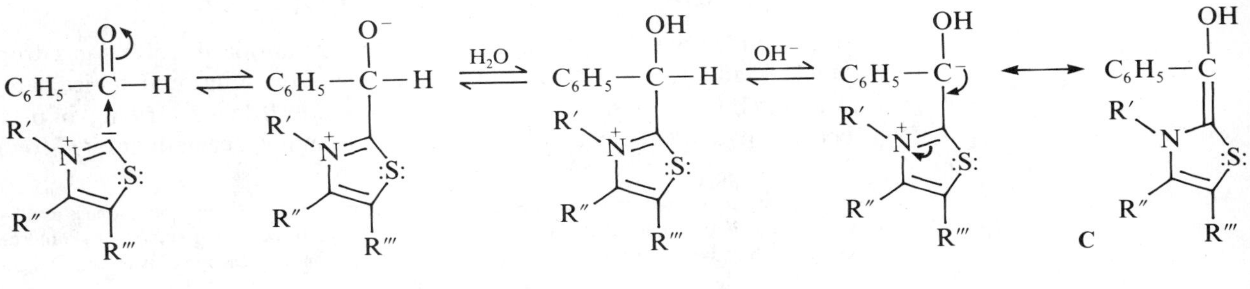

$C_6H_5-\overset{O}{\overset{\|}{C}}-H$ (R′, $\overset{+}{N}$, :S:, R″, R‴) $\rightleftharpoons$ $C_6H_5-\overset{O^-}{\overset{|}{C}}-H$ (R′, $\overset{+}{N}$, :S:, R″, R‴) $\xrightleftharpoons{H_2O}$ $C_6H_5-\overset{OH}{\overset{|}{C}}-H$ (R′, $\overset{+}{N}$, :S:, R″, R‴) $\xrightleftharpoons{OH^-}$ $C_6H_5-\overset{OH}{\overset{|}{C^-}}$ (R′, $\overset{+}{N}$, :S:, R″, R‴) $\longleftrightarrow$ $C_6H_5-\overset{OH}{\overset{|}{C}}$ (R′, N, :S:, R″, R‴)

C

$$C_6H_5-C(OH)(\text{thiazolium})^{-} + O{=}C(H)-C_6H_5 \rightleftharpoons \mathbf{B'} \rightleftharpoons C_6H_5-C(O^-)(\text{thiazolium})-C(OH)(H)-C_6H_5 \rightleftharpoons C_6H_5-C(=O)-C(OH)(H)-C_6H_5 + \textbf{Thiamine}$$

(Thiazolium ring substituents: R′, R″, R‴; ring atoms $\overset{+}{N}$ and :S:)

The importance of thiamine is evident in that it is a vitamin, an essential substance that must be provided in the diet to prevent beriberi, a nervous system disease.

In the reactions that follow, cyanide ion functions as a fast and efficient catalyst, although in large quantities it is highly toxic. The amount of potassium cyanide used in the present experiment, 15 mg, is about eight times less than the average fatal dose, a difference that underlines the advantage of carrying out organic experiments on a microscale.

The thiamine-catalyzed reaction is much slower but the catalyst is edible.

Experiments

1. Cyanide Ion–Catalyzed Benzoin Condensation

Potassium cyanide ***Poison!*** *Do not handle if you have open cuts on your hands. Never acidify a cyanide solution (HCN gas is evolved). Wash hands after handling cyanide. Dispose of cyanide ion–containing solutions in the container provided.*

In a reaction tube place 15 mg of potassium cyanide (*poison!*), dissolve it in 0.15 mL of water, add 0.30 mL of 95% ethanol and from a small, accurate syringe 0.15 mL (or 157 mg) of pure benzaldehyde,[1] introduce a boiling chip, and reflux the solution gently on a warm sand bath or a steam bath for 30 min. Remove the tube, cool it in an ice bath, and, if no crystals appear within a few minutes, withdraw a drop on a stirring rod and rub it against the inside of the tube to induce crystallization. When crystallization is complete, remove the solvent using a Pasteur pipette and while keeping it on ice, wash the crystals with a 1 : 1 mixture of 95% ethanol and water. This material is usually colorless and has a melting point of 134–135°C, indicating it is pure. The usual yield is 100–120 mg.

2. Thiamine-Catalyzed Benzoin Condensation

In a reaction tube place 26 mg of thiamine hydrochloride, dissolve it in a drop of water, add 0.30 mL of 95% ethanol, and cool the solution in an ice bath. Add 0.05 mL of 3 M sodium hydroxide followed by 0.15 mL (157 mg) of pure benzaldehyde[1], stir well, and heat the mixture in a water bath at 60°C for 1

Note for the instructor

1. Commercial benzaldehyde inhibited against autoxidation with 0.1% hydroquinone is usually satisfactory. If the material available is yellow or contains benzoic acid crystals it should be shaken with equal volumes of 5% sodium carbonate solution until carbon dioxide is no longer evolved and the upper layer dried over calcium chloride and distilled (bp 178–180°C), with avoidance of exposure of the hot liquid to air. Instant microdistillation (Chapter 5) can be used.

to 1.5 h. The progress of the reaction can be followed by thin-layer chromatography. Alternatively the reaction mixture can be stored at room temperature for at least 24 h, although a week will do no harm.

Cool the reaction mixture in an ice bath. If crystallization does not occur, withdraw a drop of solution on a stirring rod and rub it against the inside surface of the tube to induce crystallization. Remove the solvent using a Pasteur pipette and, while keeping the mixture on ice, wash the crystals with a 1 : 1 ice-cold mixture of 95% ethanol and water. The product should be colorless and of sufficient purity (mp 134–135°C) to use in subsequent reactions; the usual yield is 100–120 mg. If desired, the moist product can be recrystallized from 95% ethanol (7 mL per g) or methanol (11 mL per g) with 90% recovery.

Questions

1. Speculate on the structure of the compound formed when 4-dimethylaminobenzaldehyde is condensed with 4-chlorobenzaldehyde.
2. Why might the presence of benzoic acid be deleterious to the benzoin condensation?
3. How many π-electrons are in the thiazoline ring of thiamine hydrochloride? of thiamine?

55 Nitric Acid Oxidation. Preparation of Benzil from Benzoin. Synthesis of a Heterocycle: Diphenylquinoxaline

PRELAB EXERCISE: *Write a detailed mechanism for the formation of 2,3-dimethylquinoxaline.*

Benzoin can be oxidized to the α-diketone, benzil, very efficiently by nitric acid or by copper(II) sulfate in pyridine. On oxidation with sodium dichromate in acetic acid the yield is lower because some material is converted into benzaldehyde by cleavage of the bond between two oxidized carbon atoms and activated by both phenyl groups (a). Similarly, hydrobenzoin on oxidation with dichromate or permanganate yields chiefly benzaldehyde and only a trace of benzil (b).

(a) $C_6H_5\text{–}C(=O)\text{–}CH(OH)\text{–}C_6H_5$ **Benzoin** $\xrightarrow{2\ OH^-}$ 2 $C_6H_5\text{–}CHO$ **Benzaldehyde**

(b) $C_6H_5\text{–}CH(OH)\text{–}CH(OH)\text{–}C_6H_5$ **Hydrobenzoin** $\xrightarrow{\text{Oxid.}}$ 2 $C_6H_5\text{–}CHO$ **Benzaldehyde**

Test for the Presence of Unoxidized Benzoin. Dissolve about 0.5 mg of crude or purified benzil in 0.5 mL of 95% ethanol or methanol and add one drop of 10% sodium hydroxide. If benzoin is present the solution soon acquires a purplish color owing to a complex of benzil with a product of autoxidation of benzoin. If no color develops in 2–3 min, an indication that the sample is free from benzoin, add a small amount of benzoin, observe the color that develops, and note that if the test tube is stoppered and shaken vigorously the color momentarily disappears; when the solution is then let stand, the color reappears.

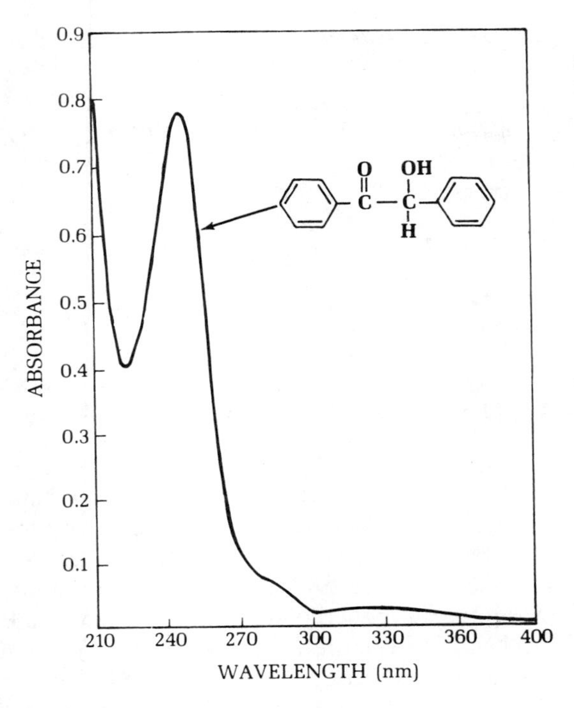

FIG. 55.1 The ultraviolet spectrum of benzoin. λ_{max}^{EtOH} 247 nm (ϵ = 13,200). Concentration: 12.56 mg/L = 5.92×10^{-5} mole/L. See Chapter 21 (Ultraviolet Spectroscopy) for the relationship between the extinction coefficient, ϵ, absorbance, A, and concentration, C. The absorption band at 247 nm is attributable to the presence of the phenyl ketone group,

$$C_6H_5—\overset{\displaystyle O}{\overset{\|}{C}}—,$$

in which the carbonyl group is conjugated with the benzene ring. Aliphatic α,β-unsaturated ketones, R—CH=CH—C=O, show selective absorption of ultraviolet light of comparable wavelength.

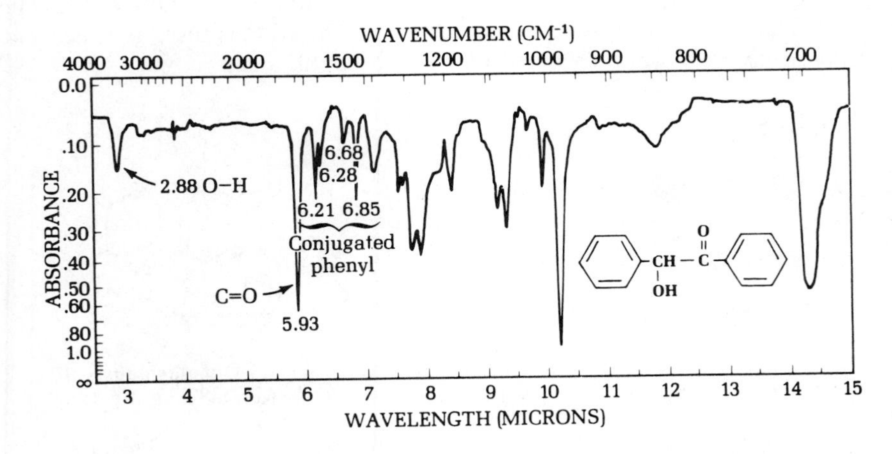

FIG. 55.2 Infrared spectrum of benzoin in $CHCl_3$.

Experiments

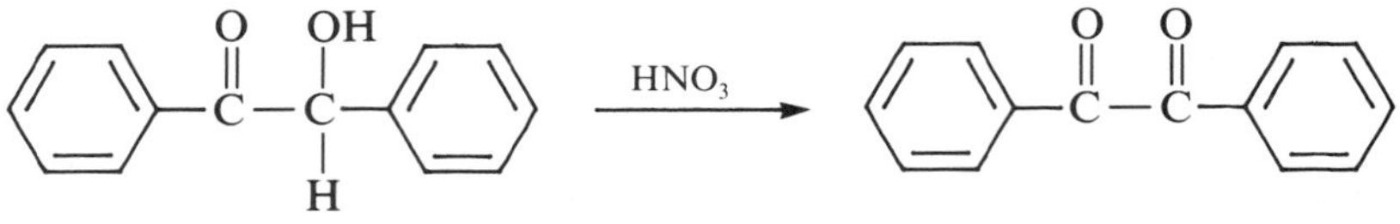

HNO$_3$

Benzoin
MW 212.24, mp 135°C

Benzil
MW 210.23, mp 94–95°C

1. Nitric Acid Oxidation of Benzoin

Heat a mixture of 100 mg of benzoin and 0.35 mL of concentrated nitric acid on the steam bath or in a small beaker of boiling water for 11 min. Carry out the reaction in the hood or use an aspirator tube near the top of the flask to remove nitrogen oxides. Add 2 mL of water to the reaction mixture, cool to room temperature, and stir the mixture for a minute or two to coagulate the precipitated product; remove the solvent with a Pasteur pipette and wash the solid with 2 mL more water. Dissolve the solid in 0.5 mL of hot ethanol and add water dropwise to the hot solution until the solution appears to be cloudy, indicating it is saturated. Heat to bring the product completely into solution and allow it to cool slowly to room temperature. Cool the tube in ice and remove the solvent with a Pasteur pipette. Scrape the benzil onto a piece of filter paper, squeeze out excess solvent, and allow the solid to dry. Record the percent yield, the crystalline form, color, and mp of the product.

2. Benzil Quinoxaline Preparation

A reaction that characterizes benzil as an α-diketone is a condensation reaction with 1,2-phenylenediamine to the quinoxaline derivative. The aromatic heterocyclic ring formed in the condensation is fused to a benzene ring to give a bicyclic system analogous to naphthalene.

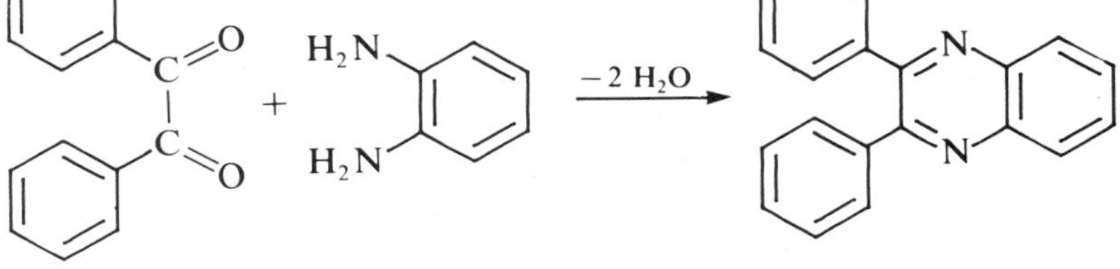

$-2\ H_2O$

Benzil
MW 210.22, mp 96°C

1,2-Phenylenediamine
MW 108.14
mp 103°C

2,3-Dimethylquinoxaline
MW 282.33, mp 126°C

Procedure

Commercial 1,2-phenylenediamine is usually badly discolored (air oxidation) and gives a poor result unless purified as follows. Place 100 mg of material in a reaction tube, evacuate the tube at full aspirator suction, clamp it in a horizontal position, and heat the bottom of the tube with a hot sand bath to distill or sublime colorless 1,2-phenylenediamine from the dark residue into the upper half of the tube. Let the tube cool until the melt has solidified, and scrape out the white solid.

Handle 1,2-phenylenediamine with care. Similar compounds (hair dyes) are mild carcinogens. Carry out reaction in hood.

Weigh 105 mg of benzil and 54 mg of your purified 1,2-phenylenediamine into a reaction tube and heat in a steam bath for 10 min, which changes the initially molten mixture to a light tan solid. Dissolve the solid in hot methanol (about 2.5 mL) and let the solution stand undisturbed. If crystallization does not occur within 10 min, reheat the solution and dilute it with a little water to the point of saturation. The crystals should be filtered as soon as formed, for brown oxidation products accumulate on standing. The quinoxaline forms colorless needles, mp 125–126°C; yield is about 90 mg.

Reaction time: 10 min

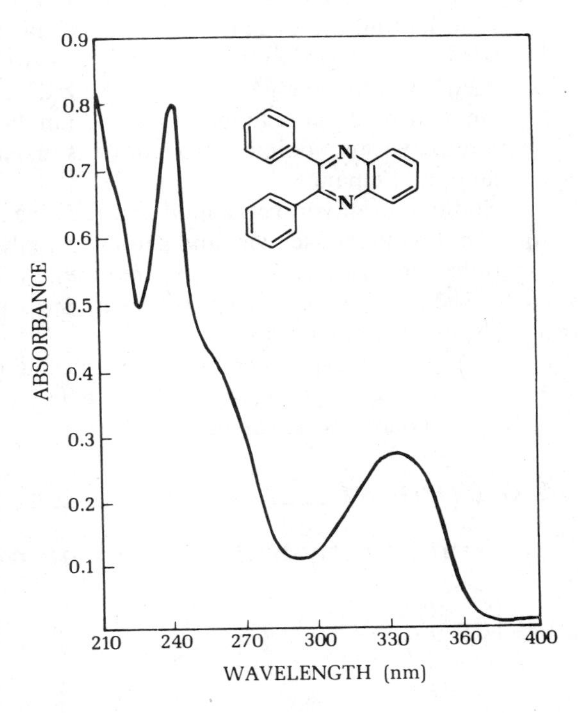

FIG. 55.3 Ultraviolet spectrum of quinoxaline derivative. λ_{max}^{EtOH} 244 nm (ϵ = 37,400), 345 nm (ϵ = 12,700). Spectrum recorded on Cary Model 17 spectrometer.

56 Borohydride Reduction of a Ketone: Hydrobenzoin from Benzil

PRELAB EXERCISE: *Compare the reductive abilities of lithium aluminum hydride with those of sodium borohydride.*

Sodium borohydride was discovered in 1943 by H. I. Schlesinger and H. C. Brown. Brown devoted his entire scientific career to this reagent, making it and other hydrides the most useful and versatile of reducing reagents. He received a Nobel prize for his work.

Considering the extreme reactivity of most hydrides (such as sodium hydride and lithium aluminum hydride) toward water, sodium borohydride is somewhat surprisingly sold as a stabilized aqueous solution 14 molar in sodium hydroxide containing 12% sodium borohydride. Unlike lithium aluminum hydride, sodium borohydride is insoluble in ether and soluble in methanol and ethanol.

Sodium borohydride is a mild and selective reducing reagent. In ethanol solution it reduces aldehydes and ketones rapidly at 25°C but is inert toward functional groups that are readily reduced by lithium aluminum hydride: carboxylic acids and esters, epoxides, lactones, nitro groups, nitriles, azides, amides, and acid chlorides.

The present experiment is a typical sodium borohydride reduction. These same conditions and isolation procedures could be applied to hundreds of other ketones and aldehydes.

Experiment

Sodium Borohydride Reduction of Benzil

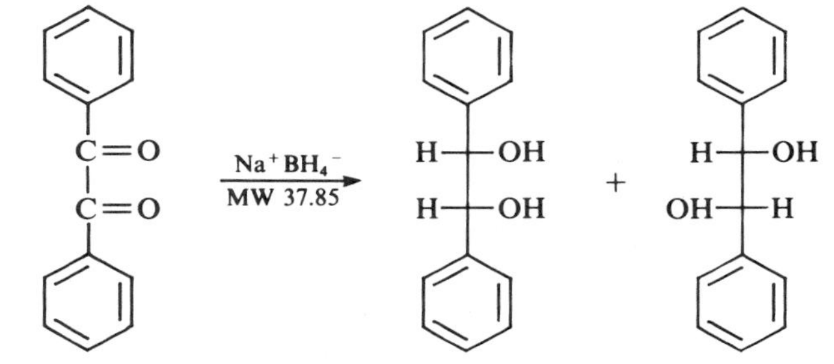

$Na^+BH_4^-$ (MW 37.85)

Benzil
MW 210.22

(1*R*,2*S*)-(*meso*)-Hydrobenzoin
mp 137°C, MW 214.25

(1*R*,2*R*) and (1*S*, 2*S*)-Hydrobenzoin
mp 120°C

Addition of two atoms of hydrogen to benzoin or of four atoms of hydrogen to benzil gives a mixture of stereoisomeric diols, of which the predominant isomer is the nonresolvable (2*R*,3*S*)-hydrobenzoin, the *meso* isomer, accompanied by the enantiomeric (2*R*,3*R*) and (2*S*,3*S*) compounds. The reaction proceeds rapidly at room temperature; the intermediate borate ester is hydrolyzed with water to give the product alcohol.

$$4\,R_2C{=}O + Na^+BH_4^- \longrightarrow (R_2CHO)_4B^-Na^+$$

$$(R_2CHO)_4B^-Na^+ + 2\,H_2O \longrightarrow 4\,R_2CHOH + Na^+BO_2^-$$

The procedure that follows specifies use of benzil rather than benzoin because you can then follow the progress of the reduction by the discharge of the yellow color of the benzil.

Procedure

In a 10 × 100-mm reaction tube dissolve 50 mg of benzil in 0.5 mL of 95% ethanol and cool the solution in ice to produce a fine suspension. Add to this suspension 10 mg of sodium borohydride (a large excess). The benzil dissolves, the mixture warms up, and the yellow color disappears in 2–3 min. After a total of 10 min add 0.5 mL of water, heat the solution to the boiling point, filter the solution in case it is not clear (usually not necessary), and dilute the hot solution with hot water to the point of saturation (cloudiness), a process requiring about 1 mL of water. (2*R*,3*S*)-Hydrobenzoin separates in lustrous thin plates, mp 136–137°C, and can be isolated by withdrawing the solvent using a Pasteur pipette while cooling the reaction tube in ice. The yield is about 35 mg.

Questions

1. Draw and name, using the *R*,*S* system of nomenclature, all of the isomers of hydrobenzoin.
2. Calculate the theoretical weight of sodium borohydride needed to reduce 50 mg of benzil.

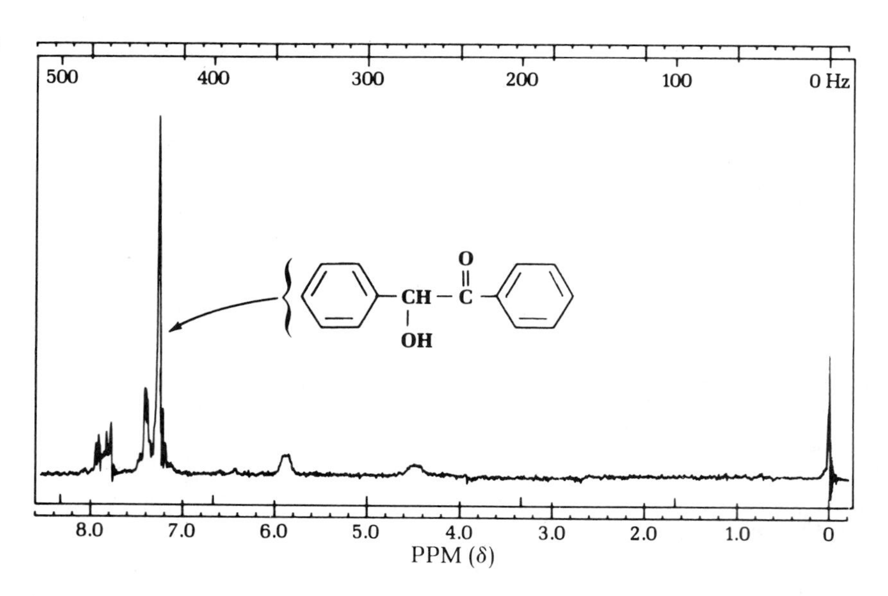

FIG. 56.1 ^{1}H nmr spectrum of benzoin.

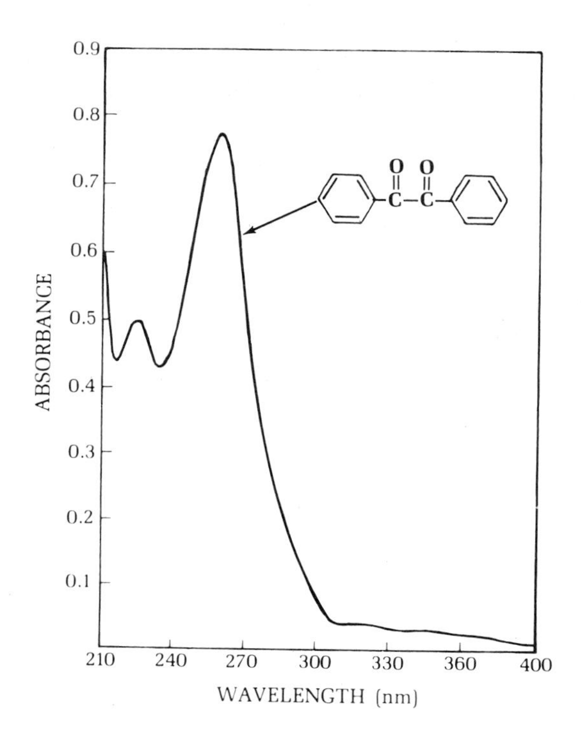

FIG. 56.2 Ultraviolet spectrum of benzil. λ_{max}^{EtOH} 260 nm (ϵ = 19,800). One-cm cells and 95% ethanol have been employed for all the uv spectra in this chapter.

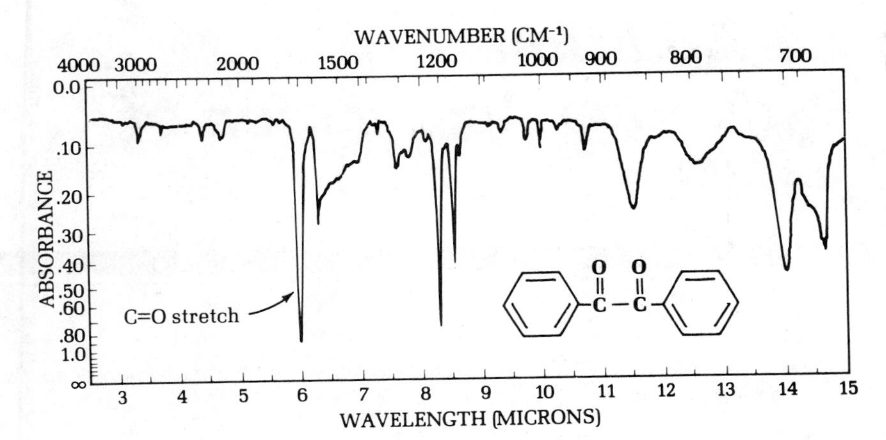

FIG. 56.3 Infrared spectrum of benzil in $CHCl_3$.

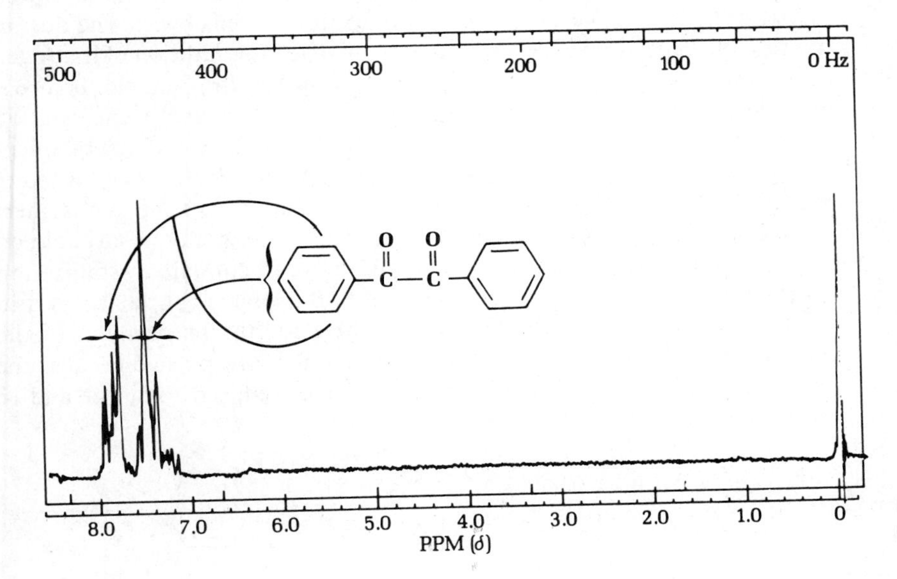

FIG. 56.4 Nmr spectrum of benzil.

57 1,4-Addition: Reductive Acetylation of Benzil

PRELAB EXERCISE: *Write the complete mechanism for the reaction of acetic anhydride with an alcohol.*

In one of the first demonstrations of the phenomenon of 1,4-addition, Johannes Thiele (1899) established that reduction of benzil with zinc dust in a mixture of acetic anhydride–sulfuric acid involves 1,4-addition of hydrogen to the α-diketone grouping and acetylation of the resulting enediol before it can undergo ketonization to benzoin. The process of reductive acetylation results in a mixture of the *E*- and *Z*-isomers **1** and **2**. Thiele and subsequent investigators isolated the more soluble, lower-melting *Z*-stilbenediol diacetate (**2**) in only impure form, mp 110°C. Separation of the two isomers by chromatography is not feasible because they are equally adsorbable on alumina. However, separation is possible by fractional crystallization (described in the following procedure) and both isomers can be isolated in pure condition. In the method prescribed here for the preparation of the isomer mixture, hydrochloric acid is substituted for sulfuric acid because the latter acid gives rise to colored impurities and is reduced to sulfur and to hydrogen sulfide.[1]

Benzil $\xrightarrow[\text{Zn, HCl}]{\text{2 H (1,4-addition)}}$ $C_6H_5C(OH)=C(OH)C_6H_5$ $\xrightarrow{Ac_2O,\ H^+}$

E-Stilbenediol diacetate (1)
mp 155°C, λ_{max}^{EtOH} 271 nm (ϵ = 23,400)
MW 296.31
1

\+

Z-Stilbenediol diacetate (2)
mp 119°C, λ_{max}^{EtOH} 265 nm (ϵ = 12,800)
MW 296.31
2

AC_2O is acetic anhydride

$CH\overset{O}{\overset{\|}{C}}O\overset{O}{\overset{\|}{C}}CH_3$

1. If acetyl chloride (2 mL) is substituted for the hydrochloric acid–acetic anhydride mixture in the procedure, the *Z*-isomer is the sole product.

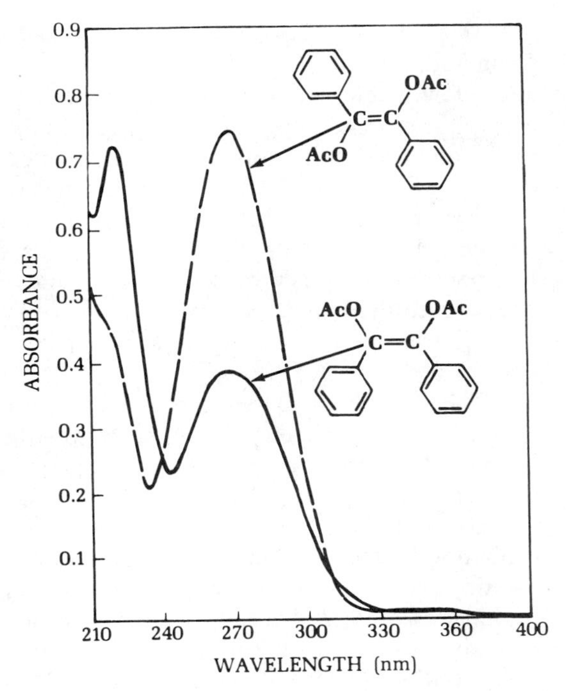

FIG. 57.1 Ultraviolet spectra of Z- and E-stilbene diacetate. Z: λ_{max}^{EtOH} 223 nm (ϵ = 20,500), 269 nm (ϵ = 10,800); E: λ_{max}^{EtOH} 272 nm (ϵ = 20,800). In the Z-diacetate the two phenyl rings cannot be coplanar, which prevents overlap of the p-orbitals of the phenyl groups with those of the central double bond. This steric inhibition of resonance accounts for the diminished intensity of the Z-isomer relative to the E. Both spectra were run at the same concentration.

Assignment of configuration

The configurations of this pair of geometrical isomers remained unestablished for over 50 years, but the tentative inference that the higher-melting isomer has the more symmetrical E configuration eventually was found to be correct. Evidence of infrared spectroscopy is of no avail; the spectra are nearly identical in the interpretable region (2–8 μ) characterizing the acetoxyl groups but differ in the fingerprint region (8–12 μ). However, the isomers differ markedly in ultraviolet absorption (Fig. 57.1); and, in analogy to E- and Z-stilbene (see Chapter 58, Experiment 1), the conclusion is justified that the higher-melting isomer, because it has an absorption band at longer wavelength and higher intensity than its isomer, does indeed have the configuration **1**.

Experiment

Procedure

Place one reaction tube containing 0.7 mL of acetic anhydride and another containing 0.1 mL of concentrated hydrochloric acid in an ice bath and, when both are thoroughly chilled, transfer the acid to the anhydride dropwise by means of a Pasteur pipette. Transfer the acidic anhydride solution into a reaction tube containing 100 mg of pure benzil (prepared in Chapter 55) and 100 mg of zinc dust, and stir the mixture for 2–3 min in an ice bath. Remove the tube and allow it to warm up slowly, cooling in ice if necessary.

When there is no further exothermic effect, let the mixture stand for 5 min and then add 2.5 mL of water. Swirl, break up any lumps of product, and allow a few minutes for hydrolysis of excess acetic anhydride. Then collect the mixture of product and zinc dust on the Hirsch funnel, wash with water, and press and apply suction to the filter cake until there is no further drip. Digest the solid (drying is not necessary) with 7.0 mL of ether to dissolve the organic material, add about 1 g of anhydrous sodium sulfate, and swirl briefly; using a Pasteur pipette, transfer the ether solution into a 10-mL Erlenmeyer flask, wash the drying agent, and concentrate the ether solution (steam bath, boiling stone, water aspirator) to a volume of approximately 1.5 mL, transfer to a reaction tube, and let the tube stand, corked and undisturbed.

The *E*-diacetate (**1**) soon begins to separate in prismatic needles, and after 20–25 min crystallization appears to stop. Remove the mother liquor from the reaction tube with a Pasteur pipette and evaporate it to dryness in another reaction tube.

Dissolve the white solid left in the second reaction tube in 1.0 mL of methanol, let the solution stand undisturbed for about 10 min, and drop in one tiny crystal of the *E*-diacetate. This should give rise, in 20–30 min, to a second crop of the *E*-diacetate. Remove the mother liquor from these crystals and concentrate it to 0.7 mL, let cool to room temperature as before, and again seed with a crystal of *E*-diacetate; this usually affords a third crop of the *E*-diacetate. The three crops might be 30 mg, mp 154–156°C, 5 mg, mp 153–154°C, and 5 mg, mp 153–155°C.

At this point the mother liquor should be rich enough in the more soluble *Z*-diacetate (**2**) for its isolation. Concentrate the methanol mother liquor and washings from the third crop of **1** to a volume of 0.4 mL, stopper the tube, and let the solution stand undisturbed overnight. The *Z*-diacetate (**2**) sometimes separates spontaneously in large rectangular prisms of great beauty. If the solution remains supersaturated, addition of a seed crystal of **2** causes prompt separation of the *Z*-diacetate in a paste of small crystals (e.g., 20 mg, mp 118–119°C, then a second crop, 7 mg, mp 116–117°C). This experiment points up the great convenience of column chromatography in isomer separations when it can be employed. Separating compounds by fractional crystallization is rarely necessary.

Synthesis of an Alkyne from an Alkene. Bromination and Dehydrobromination: Stilbene and Diphenylacetylene

58

PRELAB EXERCISE: *Calculate the theoretical quantities of thionyl chloride and of sodium borohydride needed to convert benzoin to E-stilbene.*

In this experiment benzoin, prepared in Chapter 54, is converted to the alkene *trans*-stilbene (*E*-stilbene), which is in turn brominated and dehydrobrominated to form the alkyne, diphenylacetylene.

Experiments

1. Stilbene

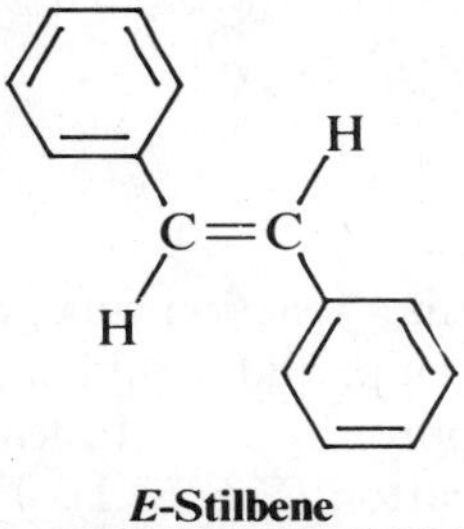

E-Stilbene
mp 125°C, MW 180.24
λ_{max}^{EtOH} 301 nm (ϵ = 28,500)
226 nm (ϵ = 17,700)

Heat of hydrogenation, −20.1 kcal/mole

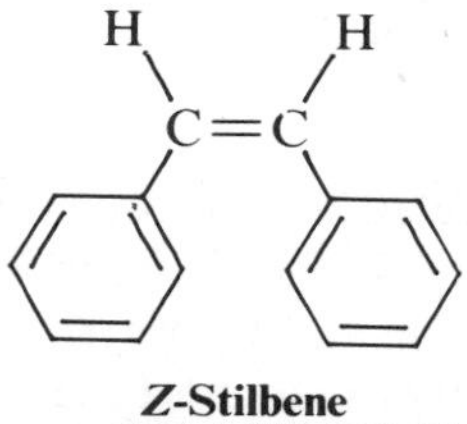

Z-Stilbene
mp 6°C, MW 180.24
λ_{max}^{EtOH} 280 nm (ϵ = 13,500)
223 nm (ϵ = 23,500)

Heat of hydrogenation, −25.8 kcal/mole

One method of preparing *E*-stilbene is reduction of benzoin with zinc amalgam in ethanol–hydrochloric acid, presumably through an intermediate:

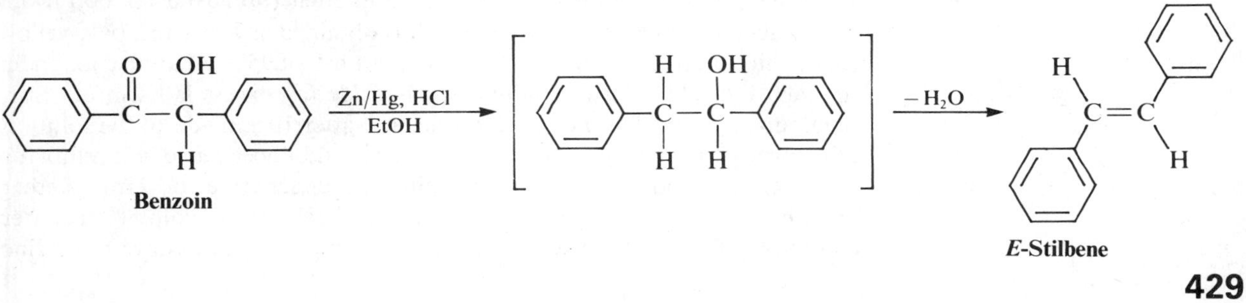

Benzoin **E-Stilbene**

The procedure that follows is quick and affords very pure hydrocarbon. It involves three steps: (1) replacement of the hydroxyl group of benzoin by

Benzoin
MW 212.24

$SOCl_2$
Thionyl chloride
MW 113.97

Desyl chloride
mp 68°C

$NaBH_4$
Sodium borohydride
MW 37.85

Zn/HOAc

Mixture of *R* and *S*-isomers

E-Stilbene

chlorine to form desyl chloride, (2) reduction of the keto group with sodium borohydride to give what appears to be a mixture of the two diastereoisomeric chlorohydrins, and (3) elimination of the elements of hypochlorous acid with zinc and acetic acid. The last step is analogous to the debromination of an olefin dibromide.

Procedure

Place 100 mg of benzoin (crushed to a powder) in a reaction tube, cover it with 0.15 mL of thionyl chloride, add a boiling chip, and a rubber septum bearing an empty syringe needle connected to a polyethylene tube leading to another reaction tube bearing a plug of damp cotton (Fig. 58.1). This will serve to trap the hydrogen chloride and sulfur dioxide evolved in this reaction. Warm the reaction mixture gently on the steam bath or on the warm sand bath until the solid has dissolved, and then more strongly for 5 min.

Caution: If the mixture of benzoin and thionyl chloride is let stand at room temperature for an appreciable time before being heated, an undesired reaction intervenes[1] and the synthesis of *E*-stilbene is spoiled.

To remove excess thionyl chloride (bp 77°C), evacuate at the aspirator for a few minutes, add 0.5 mL of petroleum ether (bp 30–60°C), boil it off, and evacuate again. Desyl chloride is thus obtained as a viscous, pale-yellow oil (it will solidify if let stand). Dissolve it in 1 mL of 95% ethanol, cool under the tap, and add 9 mg of sodium borohydride (an excess is harmful). Stir, break up any lumps of the borohydride, and after 10 min add to the solution of chlorohydrins 60 mg of zinc dust and 0.15 mL of acetic acid and reflux for 1 h. Then cool under the tap. When white crystals separate add 2 mL of ether and then wash the ether solution once with an equal volume of water containing a drop of concentrated hydrochloric acid (to dissolve basic zinc

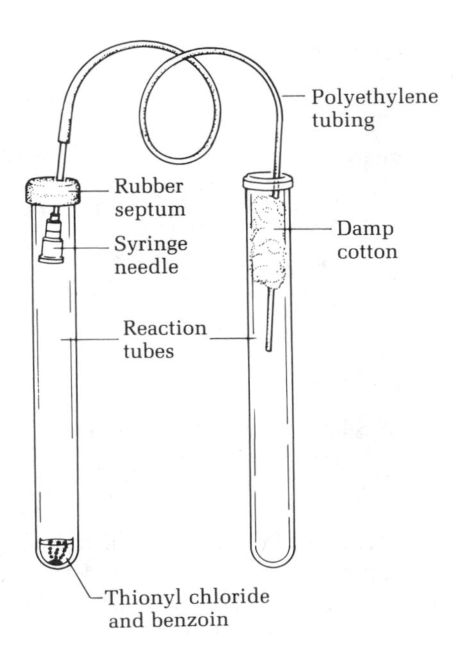

FIG. 58.1 Trap for hydrogen chloride and sulfur dioxide.

salts) and then, in turn, water and saturated sodium chloride solution. Dry the ether over anhydrous sodium sulfate, which you add until it no longer clumps together. Remove the ether from the drying agent with a Pasteur pipette (Fig. 58.2), wash the drying agent with ether, evaporate the ether to dryness, dissolve the residue in the minimum amount of hot 95% ethanol (about 1 mL), and let the product crystallize. *E*-Stilbene separates in diamond-shaped iridescent plates, mp 124–125°C; the yield is about 50 mg.

2. *meso*-Stilbene Dibromide

E-Stilbene reacts with bromine predominantly by the usual process of *trans*-addition and affords the optically inactive, nonresolvable *meso*-dibromide; the much lower melting enantiomeric mixture of dibromides is a very minor product of the reaction.

In this experiment the brominating agent will be pyridinium hydrobromide perbromide,[2] a crystalline, nonvolatile, odorless complex of high molecular weight (319.84), which, in the presence of a bromine acceptor such as an alkene, dissociates to liberate one mole of bromine. For microscale experiments it is much more convenient and agreeable to measure and use than free bromine.

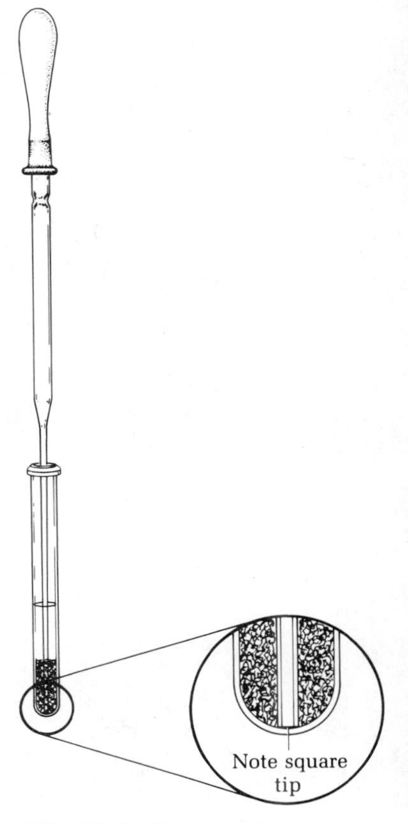

FIG. 58.2 Pasteur pipette filtration technique.

1. $$\begin{array}{l} C_6H_5C{=}O \\ \quad | \\ C_6H_5CHOH \end{array} \xrightarrow{SOCl_2} \begin{array}{l} C_6H_5C{-}O \\ \quad \| \qquad \; SO \\ C_6H_5C{-}O \end{array} \xrightarrow{NaBH_4} \begin{array}{l} C_6H_5CO \\ \quad | \\ C_6H_5CH_2 \end{array}$$

Desoxybenzoin

2. Can be purchased from Aldrich or see Chapter 13 for synthesis.

Procedure

Pyridine + HBr + Br_2
↓
Pyridinium hydrobromide perbromide

E-Stilbene + Pyridinium hydrobromide perbromide $\xrightarrow[Br_2]{\textit{trans}\text{-addition of}}$ (1*R*,2*S*)-Stilbene dibromide (*meso* isomer)

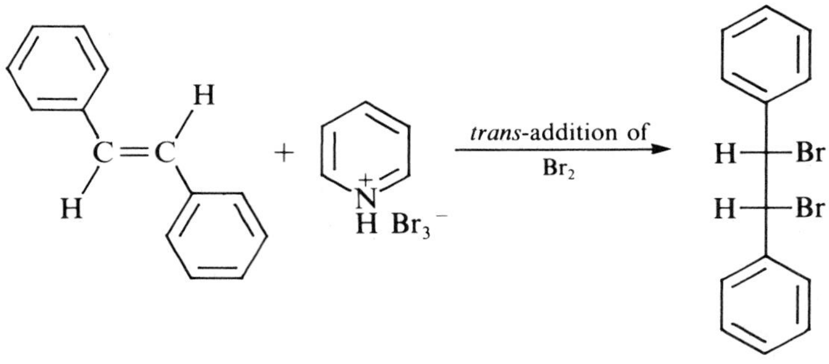

E-Stilbene MW 180.24

Pyridinium hydrobromide perbromide MW 319.86

(1*R*,2*S*)-Stilbene dibromide (*meso* isomer) mp 238°C, MW 340.07

Total time required: 10 min

Carry out procedure in hood

In a reaction tube dissolve 50 mg of *E*-stilbene in 1 mL of acetic acid, by heating on the steam bath, and then add 100 mg of pyridinium hydrobromide perbromide. Mix by swirling, if necessary rinse crystals of reagent down the walls of the flask with a little acetic acid, and continue the heating for 1–2 min longer. The dibromide separates almost at once in small plates. Cool the mixture under the tap, collect the product, and wash it with methanol; yield of colorless crystals, mp 236–237°C, 80 mg. Use this material for the preparation of diphenylacetylene after determining the percent yield and melting point.

3. Diphenylacetylene

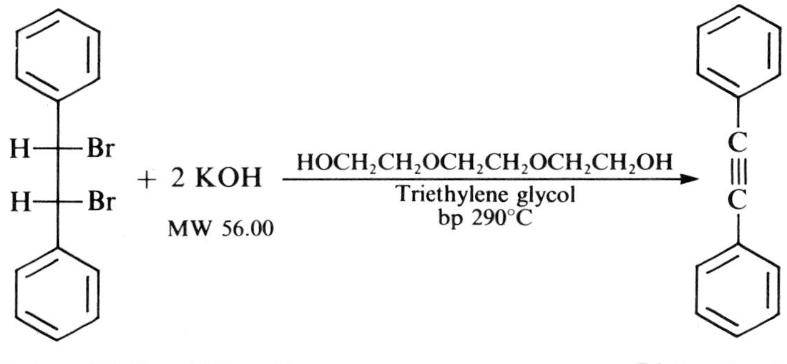

(1*R*,2*S*)-(*meso*)Stilbene dibromide + 2 KOH $\xrightarrow[\text{Triethylene glycol, bp 290°C}]{HOCH_2CH_2OCH_2CH_2OCH_2CH_2OH}$ Diphenylacetylene

(1*R*,2*S*)-(*meso*)Stilbene dibromide MW 340.07

KOH MW 56.00

Diphenylacetylene mp 61°C, MW 178.22

One method for the preparation of diphenylacetylene involves oxidation of benzil dihydrazone with mercuric oxide; the intermediate diazo compound loses nitrogen as formed to give the hydrocarbon:

Benzil + 2 H_2NNH_2 (Hydrazine) → Benzil hydrazone $\xrightarrow[C_6H_6]{HgO}$ [$C(=\overset{+}{N}=\overset{-}{N})$–$C(=\overset{+}{N}=\overset{-}{N})$] $\xrightarrow{-2\,N_2}$ Diphenylacetylene

Benzil **Hydrazine** **Benzil hydrazone** **Diphenylacetylene**

The method used in this procedure involves dehydrohalogenation of *meso*-stilbene dibromide. An earlier procedure called for refluxing the dibromide with 43% ethanolic potassium hydroxide in an oil bath at 140°C for 24 h. In the following procedure the reaction time is reduced to a few minutes by use of the high-boiling triethylene glycol as solvent to permit operation at a higher reaction temperature.

Procedure

Reaction time: 5 min

Caution! *Potassium hydroxide is corrosive to skin. Handle ether away from flames.*

In a reaction tube place 80 mg of *meso*-stilbene dibromide, 40 mg of potassium hydroxide[3] and 0.5 mL of triethylene glycol. Heat the mixture on a hot sand bath to a temperature of 160°C, when potassium bromide begins to separate. By intermittent heating, keep the mixture at 160–170°C for 5 min more, then cool to room temperature, remove the thermometer and add 2 mL of water. The diphenylacetylene that separates as a nearly colorless, granular solid is collected using the Pasteur pipette. The crude product need not be dried but can be crystallized directly from 95% ethanol. Let the solution stand undisturbed in order to observe the formation of beautiful, very large spars of colorless crystals. After a first crop has been collected, the mother liquor on concentration affords a second crop of pure product; total yield, 35 mg; mp 60–61°C.

3. Potassium hydroxide pellets are 85% KOH and 15% water.

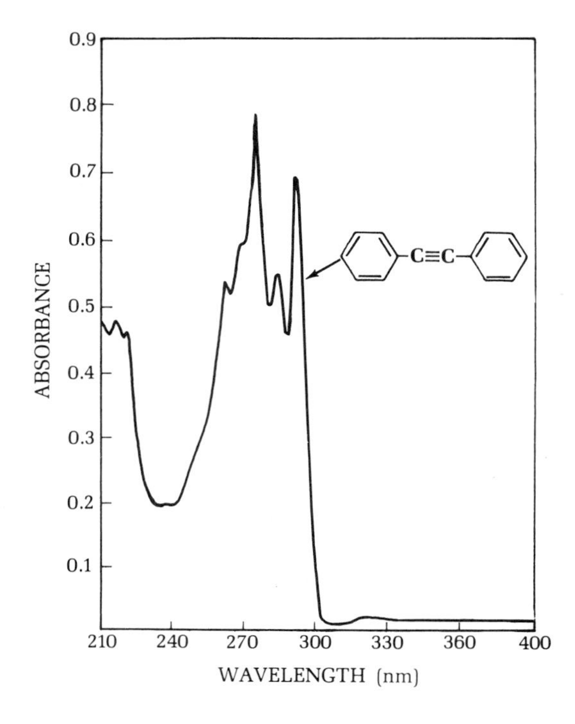

FIG. 58.3 Ultraviolet spectrum of diphenylacetylene. λ_{max}^{EtOH} 279 nm (ϵ = 31,400). This spectrum is characterized by considerable fine structure (multiplicity of bands) and a high extinction coefficient.

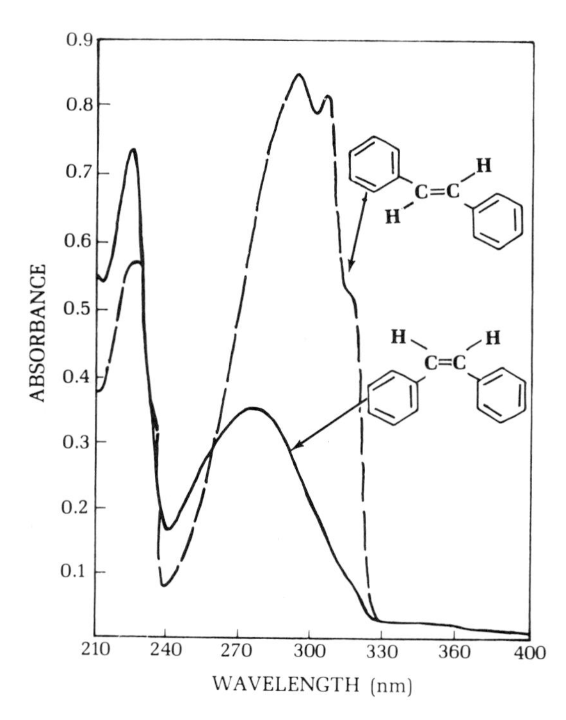

FIG. 58.4 Ultraviolet spectra of *Z*- and *E*-stilbene. *Z*: λ_{max}^{EtOH} 224 nm (ϵ = 23,300), 279 nm (ϵ = 11,100); *E*: λ_{max}^{EtOH} 226 nm (ϵ = 18,300), 295 nm (ϵ = 27,500). Like the diacetates, steric hindrance and lack of coplanarity in these hydrocarbons cause the long wavelength absorption of the *Z*-isomer to be of diminished intensity relative to the *E*-isomer.

The Perkin Reaction: Synthesis of α-Phenylcinnamic Acid

59

PRELAB EXERCISE: At the end of this reaction the products are present as mixed anhydrides. What are these, how are they formed, and how are they converted to product?

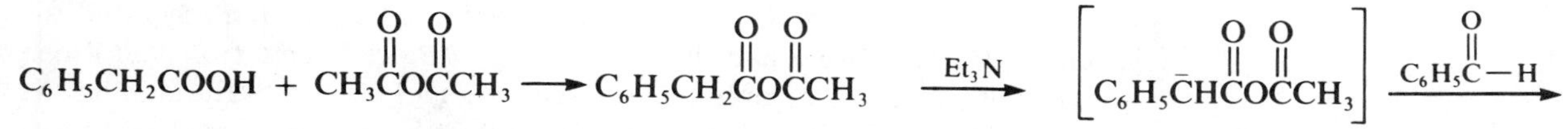

$$C_6H_5CH_2COOH + CH_3\overset{O}{\overset{\|}{C}}O\overset{O}{\overset{\|}{C}}CH_3 \longrightarrow C_6H_5CH_2\overset{O}{\overset{\|}{C}}O\overset{O}{\overset{\|}{C}}CH_3 \xrightarrow{Et_3N} \left[C_6H_5\bar{C}H\overset{O}{\overset{\|}{C}}O\overset{O}{\overset{\|}{C}}CH_3\right] \xrightarrow{C_6H_5\overset{O}{\overset{\|}{C}}-H}$$

Phenylacetic acid
mp 77°C, bp 265°C
MW 136.14

Acetic anhydride
bp 138–140°C
MW 102.09

Triethylamine
bp 89.5°C, den 0.729
MW 101.19

Benzaldehyde
bp 179°C, den 1.046
MW 106.12

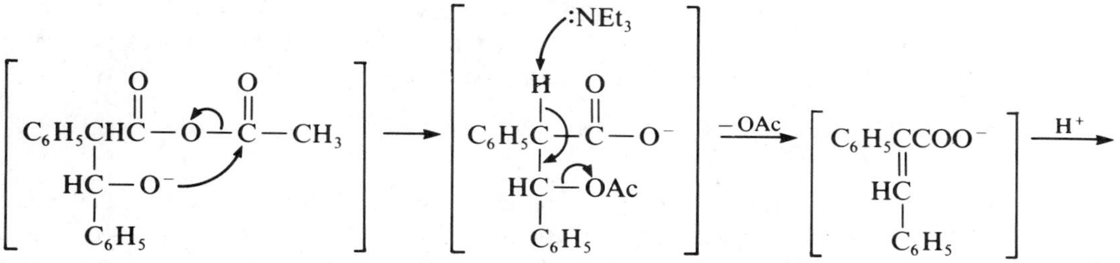

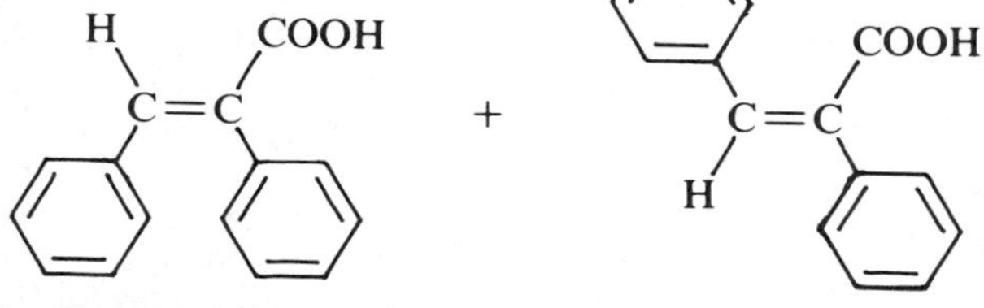

$$\left[\begin{array}{c} C_6H_5CH\overset{O}{\overset{\|}{C}}-O-\overset{O}{\overset{\|}{C}}-CH_3 \\ HC-O^- \\ C_6H_5 \end{array}\right] \longrightarrow \left[\begin{array}{c} :NEt_3 \\ H\quad O \\ C_6H_5C-\overset{\|}{C}-O^- \\ HC-OAc \\ C_6H_5 \end{array}\right] \xrightarrow{-OAc} \left[\begin{array}{c} C_6H_5CCOO^- \\ \| \\ HC \\ C_6H_5 \end{array}\right] \xrightarrow{H^+}$$

E-α-Phenylcinnamic acid[1] + *Z*-α-Phenylcinnamic acid

E-α-Phenylcinnamic acid[1]
mp 174°C, pK_a 6.1
MW 224.25

Z-α-Phenylcinnamic acid
mp 138°C, pK_a 4.8
MW 224.25

1. pK_a measured in 60% ethanol.

The reaction of benzaldehyde with phenylacetic acid to produce a mixture of the α-carboxylic acid derivatives of *Z*- and *E*-stilbene,[2] a form of aldol condensation known as the Perkin reaction, is effected by heating a mixture of the components with acetic anhydride and triethylamine. In the course of the reaction the phenylacetic acid is probably present both as anion and as the mixed anhydride resulting from equilibration with acetic anhydride. A reflux period of 5 h specified in an early procedure has been shortened by a factor of 10 by restriction of the amount of the volatile acetic anhydride, use of an excess of the less expensive, high-boiling aldehyde component, and use of a condenser that permits some evaporation and consequent elevation of the reflux temperature.

E-Stilbene is a by-product of the condensation, but experiment has shown that neither the *E*- nor *Z*-acid undergoes decarboxylation under the conditions of the experiment.

At the end of the reaction the α-phenylcinnamic acids are present in part as the neutral mixed anhydrides, but these can be hydrolyzed by addition of excess hydrochloric acid. The organic material is taken up in ether and the acids extracted with alkali. Neutralization with acetic acid (pK_a 4.76) then causes precipitation of only the less acidic *E*-acid (see pK_a values under the formulas); the *Z*-acid separates on addition of hydrochloric acid.

Whereas *Z*-stilbene is less stable and lower melting than *E*-stilbene, the reverse is true of the α-carboxylic acids, and in this preparation the more stable, higher-melting *E*-acid is the predominant product. Evidently the steric interference between the carboxyl and phenyl groups in the *Z*-acid is greater than that between the two phenyl groups in the *E*-acid. Steric hindrance is also evident from the fact that the *Z*-acid is not subject to Fischer esterification (ethanol and an acid catalyst) whereas the *E*-acid is.

Procedure

Reflux time: 35 min

Measure into a reaction tube 250 mg of phenylacetic acid, 0.30 mL of benzaldehyde, 0.2 mL of triethylamine, and 0.2 mL of acetic anhydride. Insert a boiling stone, and reflux the mixture for 35 min. Cool the yellow melt, add 0.4 mL of concentrated hydrochloric acid, mix, whereupon the mixture sets to a stiff paste. Add 2.5 mL of ether, warm to dissolve the bulk of the solid, and wash the ethereal solution twice with water and then extract it with

2. Stereochemistry of alkenes can be designated by the *E,Z*-system of nomenclature [see J. L. Blackwood, C. L. Gladys, K. L. Loening, A. E. Petrarca, and J. E. Rush, *J. Am. Chem. Soc.*, **90**, 509 (1968)] in which the groups attached to the double bond are given an order of priority according to the Cahn, Ingold, and Prelog system [see R. S. Cahn, C. K. Ingold, and V. Prelog, *Experentia*, **12**, 81 (1956)]. The atom of highest atomic number attached directly to the alkene is given highest priority. If two atoms attached to the alkene are the same, one goes to the second or third atom, etc., away from the alkene carbons. When the two groups of highest priority are on adjacent sides of the double bond, the stereochemistry is designated as *Z* (German *zusammen*, together). When the two groups are on opposite sides of the double bond, the stereochemistry is designated *E* (German *entgegen*, opposed).

four portions of a mixture of 7.5 mL of water and 1.5 mL of 10% sodium hydroxide solution.[3] Discard the dark-colored ethereal solution.[4] Acidify the combined, colorless alkaline extract to pH 6 by adding 0.5 mL of acetic acid, collect the *E*-acid that precipitates, and save the filtrate and washings. The yield of *E*-acid, mp 163–166°C, is usually about 290 mg. Crystallize 30 mg of material by dissolving it in 0.8 mL of ether, adding 0.8 mL of petroleum ether (bp 30–60°C), heating briefly to the boiling point, and letting the solution stand. Silken needles form, mp 173–174°C.

Addition of 0.5 mL of concentrated hydrochloric acid to the aqueous filtrate from precipitation of the *E*-acid produces a cloudy emulsion, which on standing for about one-half hour coagulates to crystals of *Z*-acid: 30 mg, mp 136–137°C.[5]

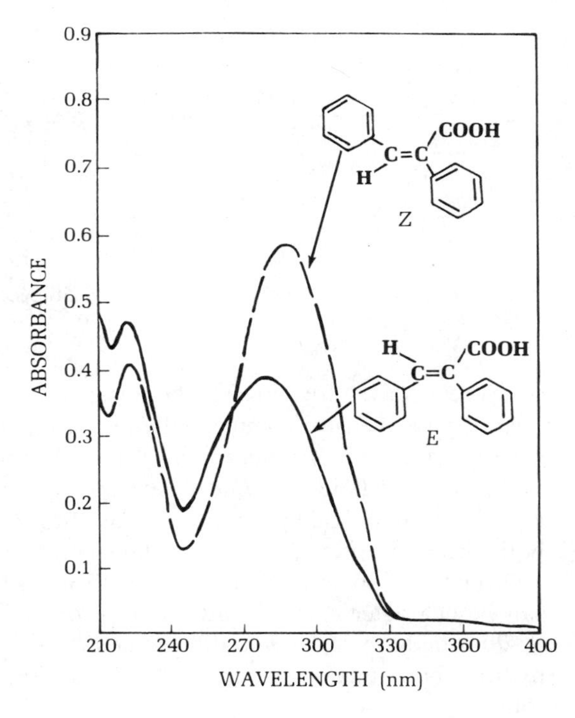

FIG. 59.1 Ultraviolet spectra of *Z*- and *E*-α-phenylcinnamic acid, run at identical concentrations. *E*: λ_{max}^{EtOH} 222 nm (ϵ = 18,000), 282 nm (ϵ = 14,000); *Z*: λ_{max}^{EtOH} 222 nm (ϵ = 15,500), 292 nm (ϵ = 22,300).

3. If stronger alkali is used the sodium salt may separate.

4. For isolation of stilbene, wash this ethereal solution with saturated sodium bisulfite solution for removal of benzaldehyde, dry, evaporate, and crystallize the residue from a little methanol. Large, slightly yellow spars, mp 122–124°C, separate (9 mg).

5. The *E*-acid can be recrystallized by dissolving 0.3 g in 5 mL of ether, filtering if necessary from a trace of sodium chloride, adding 10 mL of petroleum ether (bp 30–60°C), and evaporating to a volume of 5 mL; the acid separates as a hard crust of prisms, mp 138–139°C.

60 Decarboxylation: Synthesis of cis-Stilbene

PRELAB EXERCISE: *Calculate the volume of carbon dioxide evolved in this reaction.*

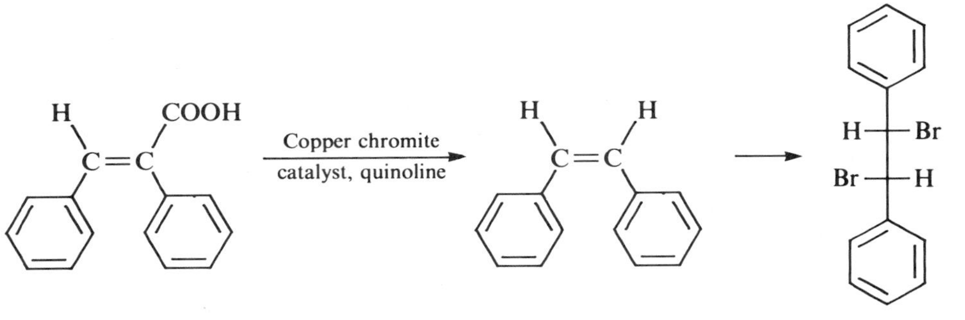

H, COOH, C=C, Copper chromite catalyst, quinoline, H, H, C=C, H–Br, Br–H

E-α-Phenylcinnamic acid
MW 224.25

Z-Stilbene
mp 4°C, bp 82–84°C/0.4 mm
MW 180.24

± Stilbene dibromide
mp 114°C
MW 340.07

The catalyst for this reaction is copper chromite, $2\,CuO \cdot Cr_2O_3$, a relatively inexpensive commercially available catalyst used for both hydrogenation and dehydrogenation as well as decarboxylation.

Decarboxylation of E-α-phenylcinnamic acid is effected by refluxing the acid in quinoline in the presence of a trace of copper chromite catalyst; both the basic properties and boiling point (237°C) of quinoline make it a particularly favorable solvent. Z-Stilbene, a liquid at room temperature, can be characterized by *trans* addition of bromine to give the crystalline ±-dibromide. A little *meso*-dibromide derived from E-stilbene in the crude hydrocarbon starting material is easily separated by virtue of its sparing solubility.

Although free rotation is possible around the single bond connecting the chiral carbon atoms of the stilbene dibromides and hydrobenzoins, evidence from dipole-moment measurements indicates that the molecules tend to exist predominantly in the specific shape or conformation in which the two phenyl groups repel each other and occupy positions as far apart as possible. The optimal conformations of the + or − dibromide and the *meso*-dibromide are represented in Fig. 60.1 by Newman projection formulas, in which the molecules are viewed along the axis of the bond connecting the two chiral carbon atoms. In the *meso*-dibromide the two repelling phenyl groups are on opposite sides of the molecule, and so are the two large bromine atoms.

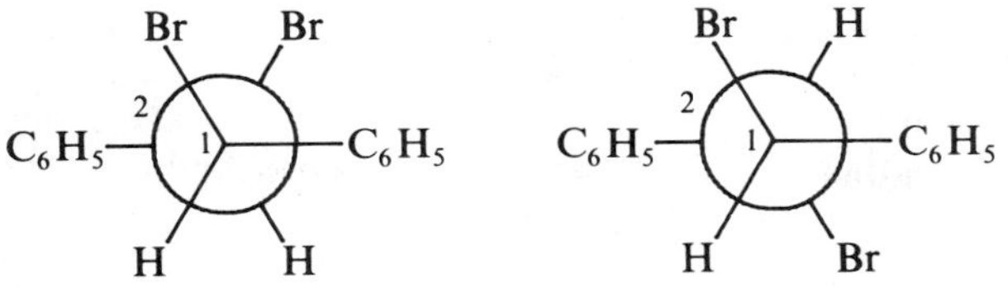

FIG. 60.1 Favored conformations of stilbene dibromide.

+-(or −)-Dibromide

(1*R*,2*S*)*meso*-Dibromide

Hence, the structure is much more symmetrical than that of the + (or −) dibromide. X-ray diffraction measurements of the dibromides in the solid state confirm the conformations indicated in Fig. 60.1 The Br-Br distances found are *meso*-dibromide, 4.50 Å; ±-dibromide, 3.85 Å. The difference in symmetry of the two optically inactive isomers accounts for the marked contrast in properties:

	Mp (°C)	Solubility in ether (18°C)
±-Dibromide	114	1 part in 3.7 parts
meso-Dibromide	237	1 part in 1025 parts

Procedure

Because a trace of moisture causes troublesome spattering, the reactants and catalyst are dried prior to decarboxylation. Stuff 250 mg of crude, "dry" *E*-α-phenylcinnamic acid and 40 mg of copper chromite catalyst[1] into a reaction tube, add 0.3 mL of quinoline[2] (bp 237°C), and let it wash down the solids. Make connection with a rubber stopper to the aspirator and turn it on full force. Make sure that you have a good vacuum (pressure gauge) and heat the tube strongly on the steam bath with most of the rings removed. Heat and evacuate for 5–10 min to remove all traces of moisture. Then wipe the outside walls of the test tube dry, insert a thermometer, clamp the tube over a hot sand bath, raise the temperature to 230°C and note the time. Then maintain a temperature close to 230°C for 10 min. Cool the yellow solution containing suspended catalyst to 25°C, add 3 mL of ether, and filter the solution by gravity (use more ether for rinsing). Transfer the solution to a reaction tube and remove the quinoline by extraction twice with about 1.5 mL of water containing 0.35 mL of concentrated hydrochloric acid. Then shake the ethereal solution well with water containing a little sodium hydroxide solution, draw off the alkaline liquor, and acidify it. A substantial precipitate will show that decarboxylation was incomplete, in which case the starting material can be recovered and the reaction repeated. If there is only

Copper chromite is toxic; weigh it in hood

Reaction time in first step: 10 min

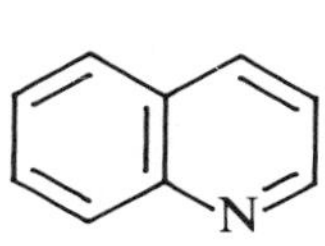

Quinoline, bp 237°C

1. The preparations described in *J. Am. Chem. Soc.*, **54**, 138 (1932) and *ibid.*, **72**, 2626 (1950) are both satisfactory.

2. Material that has darkened in storage should be redistilled over a little zinc dust.

a trace of precipitate, shake the ethereal solution with saturated sodium chloride solution for preliminary drying, dry the ethereal solution over sodium sulfate, remove the drying agent with a Pasteur pipette and evaporate the ether. The residual brownish oil (about 150 mg) is crude Z-stilbene containing a little *E*-isomer formed by rearrangement during heating.

Second step requires about one-half hour

Carry out procedure in hood

Dissolve the crude *Z*-stilbene (e.g., 150 mg) in 1 mL of acetic acid and, in subdued light, add double the weight of pyridinium hydrobromide perbromide (e.g., 300 mg). Warm on the steam bath until the reagent is dissolved, and then cool under the tap and scratch to effect separation of a small crop of plates of the *meso*-dibromide. Filter the solution by suction if necessary, dilute extensively with water, and extract with ether. Wash the solution twice with water and then with 5% sodium bicarbonate solution until neutral; shake with saturated sodium chloride solution, dry over sodium sulfate, and evaporate to a volume of about 1 mL. If a little more of the sparingly soluble *meso*-dibromide separates, remove it and then evaporate the remainder of the solvent. The residual $\pm$-dibromide is obtained as a dark oil that readily solidifies when rubbed with a rod. Dissolve it in a small amount of methanol and let the solution stand to crystallize. The $\pm$-dibromide separates as colorless prismatic plates, mp 113–114°C; yield about 60 mg.

Questions

1. Draw the mechanism that shows how the bromination of *E*-stilbene produces *meso*-stilbene dibromide.
2. Can ^{1}H nmr spectroscopy be used to distinguish between *meso*- and $\pm$-stilbene dibromide?

Martius Yellow 61

PRELAB EXERCISE: *Prepare a time-line for this experiment, indicating clearly which experiments can be carried out simultaneously.*

This experiment was introduced by Louis Fieser of Harvard University over half a century ago, long before serious thought was given to teaching organic laboratory on a microscale; yet anyone who has carried out this experiment knows it to be a test of microscale manipulation and laboratory skill. It can also be the basis for an interesting laboratory competition. The present author was a winner a third of a century ago.

The quantities of materials specified in this experiment have been reduced by half from those specified by Fieser. The use of the reaction tube for crystallization and the Pasteur pipette method for the isolation of crystals makes this possible. A skilled student could carry out the experiments on one-half the quantities specified.

Starting with 2.5 g of 1-naphthol, a skilled operator familiar with the procedures can prepare pure samples of the seven compounds in 3–4 h. In a first trial of the experiment, a particularly competent student, who plans his or her work in advance, can complete the program in two laboratory periods (6 h).

The first compound of the series, Martius Yellow, a mothproofing dye for wool (1 g of Martius Yellow dyes 200 g of wool) discovered in 1868 by Karl Alexander von Martius, is the ammonium salt of 2,4-dinitro-1-naphthol (**1**), shown on page 442. Compound **1** in the series of reactions is obtained by sulfonation of 1-naphthol with sulfuric acid and treatment of the resulting disulfonic acid with nitric acid in aqueous medium. The exchange of groups occurs with remarkable ease, and it is not necessary to isolate the disulfonic acid. The advantage of introducing the nitro groups in this indirect way is that 1-naphthol is very sensivitive to oxidation and would be partially destroyed on direct nitration. Martius Yellow is prepared by reaction of the acidic phenolic group of **1** with ammonia to form the ammonium salt. A small portion of this salt (Martius Yellow) is converted by acidification and crystallization into pure 2,4-dinitro-1-naphthol (**1**), a sample of which is saved. The rest is suspended in water and reduced to diaminonaphthol with sodium hydrosulfite according to the equation:

$$\text{Martius Yellow} + 6\,Na_2S_2O_4 + 8\,H_2O \longrightarrow \text{2,4-Diamino-1-naphthol} + 11\,NaHSO_3 + Na(NH_4)SO_4$$

Martius Yellow **2,4-Diamino-1-naphthol**

The diaminonaphthol separates in the free condition, rather than as an ammonium salt, because the diamine, unlike the dinitro compound, is a very weakly acidic substance.

1-Naphthol, MW 144.16 — H_2SO_4, HNO_3 → **1** MW 234.16 — 1. $Na_2S_2O_4$ 2. HCl → [2,4-diamino-1-naphthol dihydrochloride, $NH_3^+Cl^-$]

$FeCl_3$ → **2** MW 208.65; Ac_2O, H_2O, NaOAc → **4** MW 258.27

2 — Ac_2O, NaOAc → **3** MW 256.25

2 — NH_3, H_2O → **7** MW 173.16

4 — $FeCl_3$ → **5** MW 215.25 — H_2SO_4 → **6** MW 173.16

Since 2,4-diamino-1-naphthol is exceedingly sensitive to air oxidation as the free base, it is at once dissolved in dilute hydrochloric acid. The solution of diaminonaphthol dihydrochloride is clarified with decolorizing charcoal and divided into equal parts. One part on oxidation with iron(III) chloride affords the fiery red 2-amino-1,4-naphthoquinonimine hydrochloride (**2**). Since this substance, like many other salts, has no melting point, it is converted for identification to the yellow diacetate, **3**. Compound **2** is remarkable in that it is stable enough to be isolated. On hydrolysis it affords the orange 4-amino-1,2-naphthoquinone (**7**).

The other part of the diaminonaphthol dihydrochloride solution is treated with acetic anhydride and then sodium acetate; the reaction in aqueous solution effects selective acetylation of the amino groups and affords 2,4-diacetylamino-1-naphthol (**4**). Oxidation of **4** is attended with cleavage of the acetylamino group at the 4-position and the product is 2-acetylamino-1,4-naphthoquinone (**5**). This yellow substance is hydrolyzed by sulfuric acid to the red 2-amino-1,4-naphthoquinone (**6**), the last member of the series. The reaction periods are brief and the yields high; however, remember to scale down quantities of reagents and solvents if the quantity of starting material is less than that called for.[1]

Experiments

1. Preparation of 2,4-Dinitro-1-naphthol (1)

Place 2.5 g of pure 1-naphthol[2] in a 50-mL Erlenmeyer flask, add 5 mL of concentrated sulfuric acid, and heat the mixture with swirling on the steam bath for 5 min, when the solid should have dissolved and an initial red color should be discharged. Cool in an ice bath, add 13 mL of water, and cool the solution rapidly to 15°C. Measure 3 mL of concentrated nitric acid into a test tube and transfer it with a Pasteur pipette in small portions (0.25 mL) to the chilled aqueous solution while keeping the temperature in the range 15–20°C by swirling the flask vigorously in the ice bath. When the addition is complete

Use care when working with hot concentrated sulfuric and nitric acids

OH
NO_2
NO_2
1

1. This series of reactions lends itself to a laboratory competition, the rules for which might be as follows: (1) No practice or advance preparation is allowable except collection of reagents not available at the contestant's bench (ammonium chloride, sodium hydrosulfite, iron(III) chloride solution, acetic anhydride). (2) The time scored is the actual working time, including that required for bottling the samples and cleaning the apparatus and bench; labels can be prepared out of the working period. (3) Time is not charged during an interim period (overnight) when solutions are let stand to crystallize or solids are let dry, on condition that during this period no adjustments are made and no cleaning or other work is done. (4) Melting point and color test characterizations are omitted. (5) Successful completion of the contest requires preparation of authentic and macroscopically crystalline samples of all seven compounds. (6) Judgment of the winners among the successful contestants is based upon quality and quantity of samples, technique and neatness, and working time. (Superior performance 3–4 h.)

2. If the 1-naphthol is dark it can be purified by distillation at atmospheric pressure. The colorless distillate is most easily pulverized before it has completely cooled and hardened.

and the exothermic reaction has subsided (1–2 min), warm the mixture gently to 50°C (1 min), when the nitration product should separate as a stiff yellow paste. Apply the full heat of the steam bath for 1 min more, fill the flask with water, break up the lumps and stir to an even paste, collect the product **1** on a small Büchner funnel, wash it well with water, and then wash it into a 250-mL beaker with water (50 mL). Add 75 mL of hot water and 2.5 mL of concentrated ammonia solution (den 0.90), heat to the boiling point, and stir to dissolve the solid. Filter the hot solution by suction if it is dirty, add 5 g of ammonium chloride to the filtrate to salt out the ammonium salt (Martius Yellow), cool in an ice bath, collect the orange salt, and wash it with water containing 1–2% of ammonium chloride. The salt does not have to be dried (dry weight 3.8 g, 88.5%).

Avoid contact of the yellow product and its orange NH_4^+ *salt with the skin*

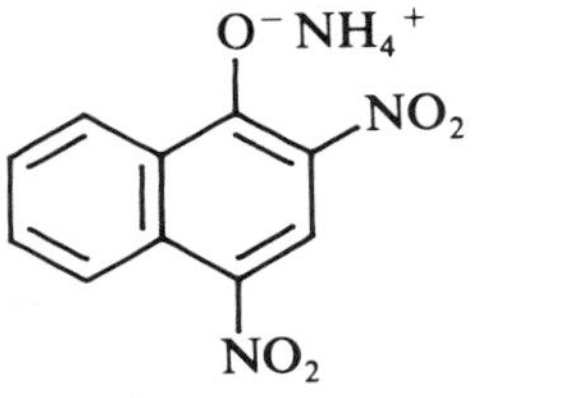

$O^-NH_4^+$, NO_2, NO_2

Martius Yellow

Set aside an estimated 150 mg of the moist ammonium salt. This sample is to be dissolved in hot water, the solution acidified (HCl), and the free 2,4-dinitro-1-naphthol (**1**) crystallized from methanol or ethanol (use decolorizing charcoal if necessary); it forms yellow needles, mp 138°C.

Preparation of Unstable 2,4-Diamino-1-naphthol

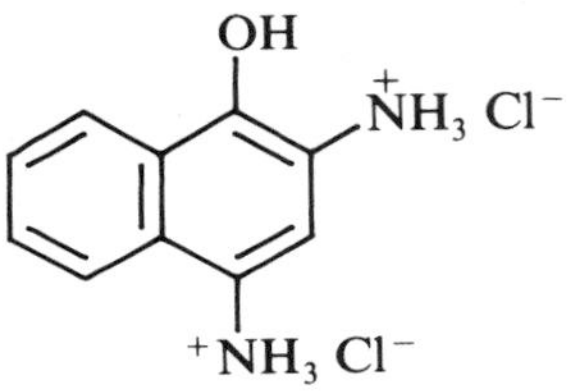

OH, $\overset{+}{N}H_3\ Cl^-$, $^+NH_3\ Cl^-$

2,4-Diamino-1-naphthol dihydrochloride

Wash the rest of the ammonium salt into a beaker with a total of about 100 mL of water, add 20 g of sodium hydrosulfite, stir until the original orange color has disappeared and a crystalline tan precipitate has formed (5–10 min), and cool in ice. Make ready a solution of 1 g of sodium hydrosulfite in 50 mL of water for use in washing and a 250-mL beaker containing 3 mL of concentrated hydrochloric acid and 12 mL of water. In collecting the precipitate by suction filtration on a small Büchner funnel, use the hydrosulfite solution for rinsing and washing, avoid even briefly sucking air through the cake after the reducing agent has been drained away, and wash the solid at once into the beaker containing the dilute hydrochloric acid and stir to convert all the diamine to the dihydrochloride.

The acid solution, often containing suspended sulfur and filter paper, is clarified by filtration by suction through a moist charcoal bed made by shaking 1 g of powdered decolorizing carbon with 13 mL of water in a stoppered flask to produce a slurry and pouring this on the paper of a 50-mm Büchner funnel. Pour the water out of the filter flask and then filter the solution of dihydrochloride. Divide the pink or colorless filtrate into approximately two equal parts and at once add the reagents for conversion of one part to **2** and the other part to **4**.

2. Preparation of 2-Amino-1,4-naphthoquinonimine Hydrochloride (2)

To one-half of the diamine dihydrochloride solution add 12.5 mL of 1.3 M iron(III) chloride solution,[3] cool in ice, and, if necessary, initiate crystalli-

Note for the instructor

3. Dissolve 45 g of $FeCl_3 \cdot 6H_2O$ (MW 270.32) in 50 mL of water and 50 mL of concentrated hydrochloric acid by warming, cool and filter (124 mL of solution).

zation by scratching. Rub the liquid film with a glass stirring rod at a single spot slightly above the surface of the liquid. If efforts to induce crystallization are unsuccessful, add more hydrochloric acid. Collect the red product and wash with dilute HCl. Dry weight is 1.2–1.35 g.

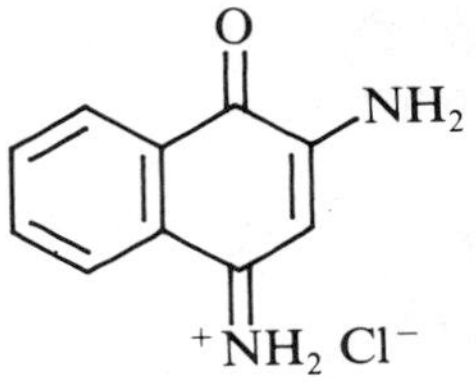

Divide the moist product into three equal parts and spread out one part to dry for conversion to **3**. The other two parts can be used while still moist for conversion to **7** and for recrystallization. Dissolve the other two-thirds by gentle warming in a little water containing 2–3 drops of hydrochloric acid, shake for a minute or two with decolorizing charcoal, filter by suction, and add concentrated hydrochloric acid to decrease the solubility.

O, NH_2, $^+NH_2$ Cl^-

2

3. Preparation of 2-Amino-1,4-naphthoquinonimine Diacetate (3)

A mixture of 0.25 g of the dry quinonimine hydrochloride (**2**), 0.25 g of sodium acetate (anhydrous), and 1.5 mL of acetic anhydride is stirred in a reaction tube and warmed gently on a sand or steam bath. With thorough stirring the red salt should soon change into yellow crystals of the diacetate. The solution may appear red, but as soon as particles of red solid have disappeared the mixture can be poured into about 5 mL of water. Stir until the excess acetic anhydride has either dissolved or become hydrolyzed, collect and wash the product (dry weight 250 mg), and (drying is unnecessary) crystallize it from ethanol or methanol; yellow needles, mp 189°C.

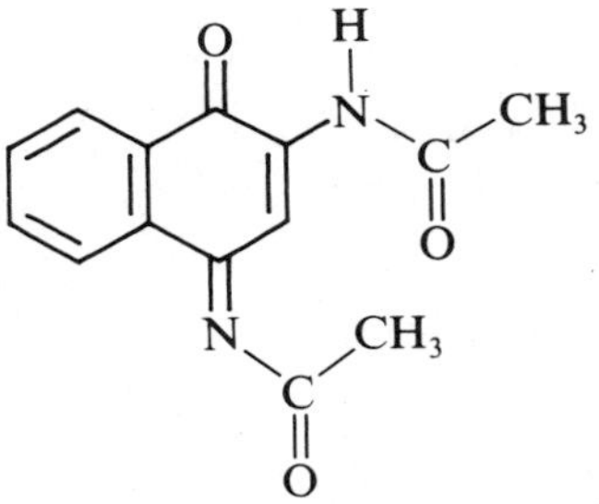

O, H, N, C, CH_3, O, N, CH_3, C, O

3

4. Preparation of 2,4-Diacetylamino-1-naphthol (4)

To one-half of the diaminonaphthol dihydrochloride solution saved from Section 1 add 1.5 mL of acetic anhydride, stir vigorously, and add a solution of 1.5 g of sodium acetate (anhydrous) and about 50 mg of sodium hydrosulfite in 10–15 mL of water. The diacetate may precipitate as a white powder or it may separate as an oil that solidifies when chilled in ice and rubbed with a rod. Collect the product and, to hydrolyze any triacetate present, dissolve it in 2.5 mL of 10% sodium hydroxide and 25 mL of water by stirring at room temperature. If the solution is colored, a few milligrams of sodium hydrosulfite may bleach it. Filter by suction and acidify by gradual addition of well-diluted hydrochloric acid (1 mL of concentrated acid). The diacetate tends to remain in supersaturated solution; and hence, either to initiate crystallization or to ensure maximum separation, it is advisable to stir well, rub the walls with a rod, and cool in ice. Collect the product, wash it with water, and divide it into thirds (dry weight 1–1.3 g).

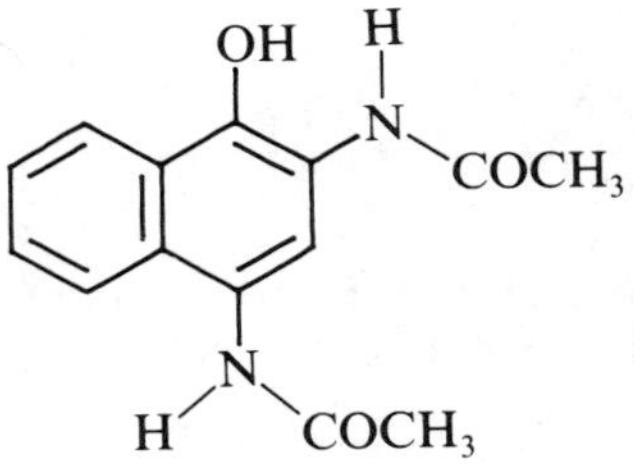

OH, H, N, $COCH_3$, N, H, $COCH_3$

4

Two-thirds of the material can be converted without drying into **5** and the other third used for preparation of a crystalline sample. Dissolve the third reserved for crystallization (moist or dry) in enough hot acetic acid to bring about solution, add a solution of a small crystal of tin(II) chloride in a few drops of dilute hydrochloric acid to inhibit oxidation, and dilute gradually with 5–6 volumes of water at the boiling point. Crystallization may be slow,

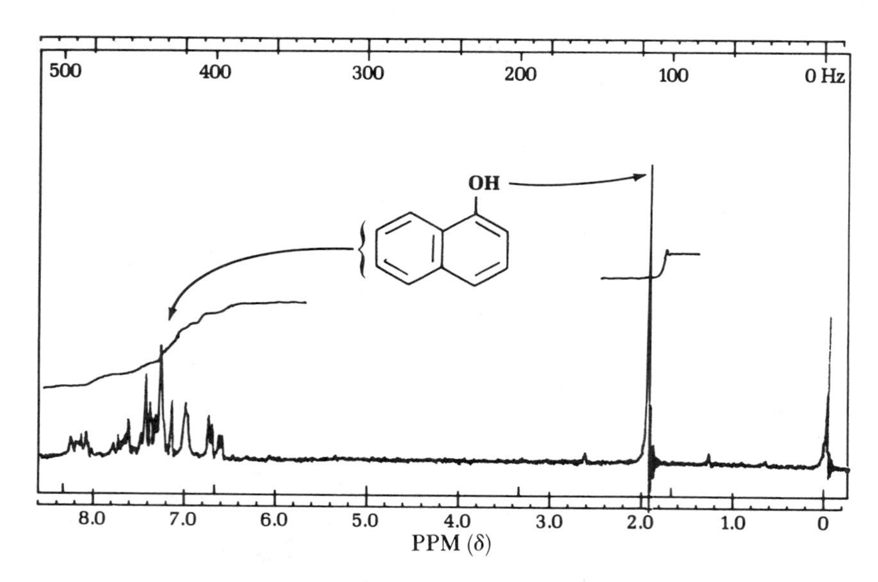

FIG. 61.1 ^{1}H nmr spectrum of α-naphthol.

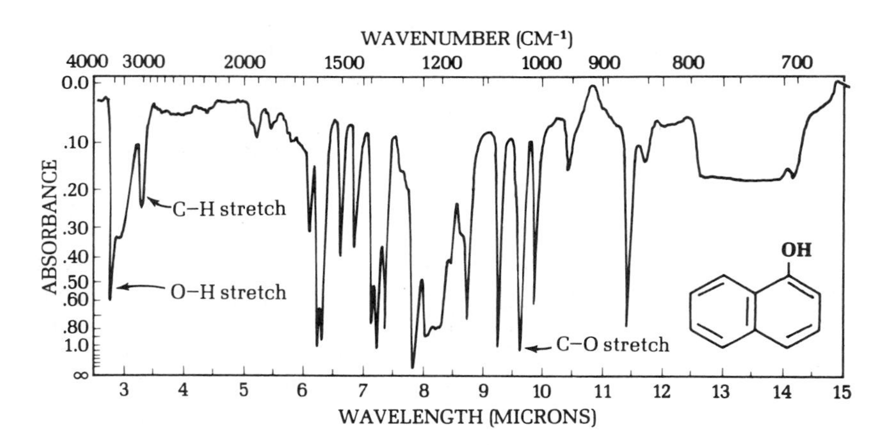

FIG. 61.2 Infrared spectrum of α-naphthol.

and cooling and scratching may be necessary. The pure diacetate forms colorless prisms, mp 224°C, dec.

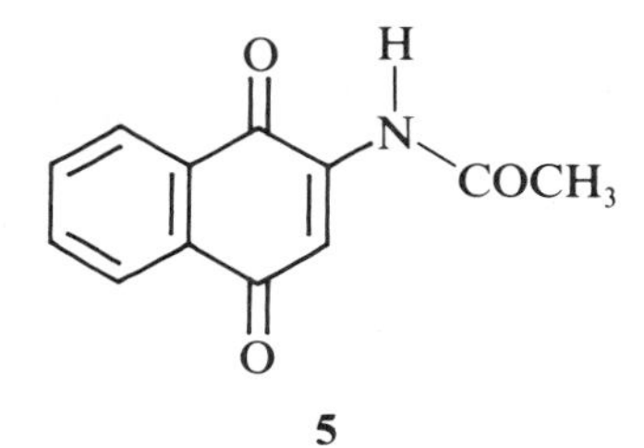

5. Preparation of 2-Acetylamino-1,4-naphthoquinone (5)

Dissolve 0.75 g of the moist diacetylaminonaphthol (**4**) in 5 mL of acetic acid (hot), dilute with 10 mL of hot water, and add 5 mL of 0.13 M iron(III) chloride solution. The product separates promptly in flat, yellow needles, which are collected (after cooling) and washed with a little alcohol; the yield is usually 0.6 g. Dry one-half of the product for conversion to **6** and crystallize the rest from 95% ethanol; mp 204°C.

6. Preparation of 2-Amino-1,4-naphthoquinone (6)

To 0.25 g of dry 2-acetylamino-1,4-naphthoquinone (**5**) contained in a reaction tube add 1 mL of concentrated sulfuric acid and heat the mixture on the steam bath with swirling to promote rapid solution (1–2 min). After 5 min cool the deep-red solution, dilute extensively with water, and collect the precipitated product; wash it with water and crystallize the moist sample (dry weight about 200 mg) from alcohol or alcohol-water; red needles, mp 206°C.

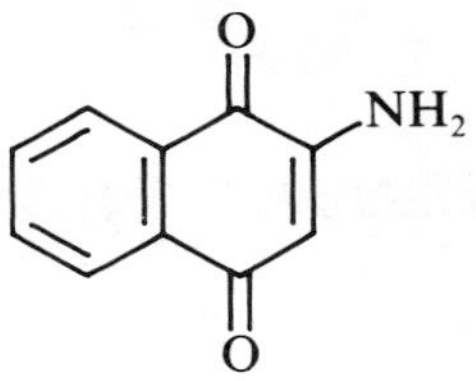

O
NH₂
O
6

7. Preparation of 4-Amino-1,2-naphthoquinone (7)

*Moist **2** is satisfactory*

Dissolve 0.5 g of the aminonaphthoquinonimine hydrochloride (**2**) reserved from Experiment 2 in 12 mL of water, add 1 mL of concentrated ammonia solution (den 0.90), and boil the mixture for 5 min. The free quinonimine initially precipitated is hydrolyzed to a mixture of the aminoquinone **7** and the isomer **6**. Cool, collect the precipitate, and suspend it in about 25 mL of water, and add 12.5 mL of 10% sodium hydroxide solution. Stir well, remove the small amount of residual 2-amino-1,4-naphthoquinone (**6**) by filtration and acidify the filtrate with acetic acid. The orange precipitate of **7** is collected, washed, and crystallized while still wet from 250–300 mL of hot water (the separation is slow). The yield of orange needles, dec about 270°C, is about 200 mg.

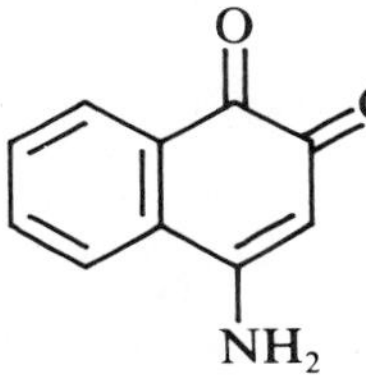

O
O
NH_2
7

Question

Write a balanced equation for the preparation of 2-amino-1,4-naphthoquinonimine hydrochloride (**2**).

62 Catalytic Hydrogenation

PRELAB EXERCISE: *Calculate the volume of hydrogen gas generated when 3 mL of 1 M sodium borohydride reacts with concentrated hydrochloric acid. Write a balanced equation for the reaction of sodium borohydride with platinum chloride. Calculate the volume of hydrogen that can be liberated by reacting 1 g of zinc with acid.*

COOH
COOH
$NaBH_4-HCl$
$PtCl_4$
COOH
COOH

***cis*-Norbornene-5,6-*endo*-dicarboxylic acid**
mp 180–190°C, dec
MW 182.17

***cis*-Norbornane-5,6-*endo*-dicarboxylic acid**
mp 170–175°C, dec
MW 184.19

Hydrogen generated in situ

This experiment presents two methods for carrying out catalytic hydrogenation on a micro scale. Conventional procedures for hydrogenation in the presence of a platinum catalyst employ hydrogen drawn from a cylinder of compressed gas and require elaborate equipment. H. C. Brown and C. A. Brown introduced the simple procedure of generating hydrogen *in situ* from sodium borohydride and hydrochloric acid and prepared a highly active supported catalyst by reduction of platinum chloride and sodium borohydride in the presence of decolorizing carbon. The special apparatus described by these authors is here dispensed with in favor of a balloon technique also employed for catalytic oxygenation (Chapter 50).

Experiments

1. Brown[1] Hydrogenation

Handle platinum chloride carefully to prevent any waste

The reaction vessel is a 5-mL short-necked round-bottomed flask fitted with an addition port with a white rubber pipette bulb or an ordinary balloon and a rubber septum (wired on) (Fig 62.1). Introduce 1.5 mL of water, 0.15 mL of platinum(IV) chloride solution,[2] and 75 mg of powdered decolorizing

1. H. C. Brown and C. A. Brown, *J. Am. Chem. Soc.*, **84**, 1495 (1962).
2. A solution of 200 mg of $PtCl_4$ in 4 mL of water.

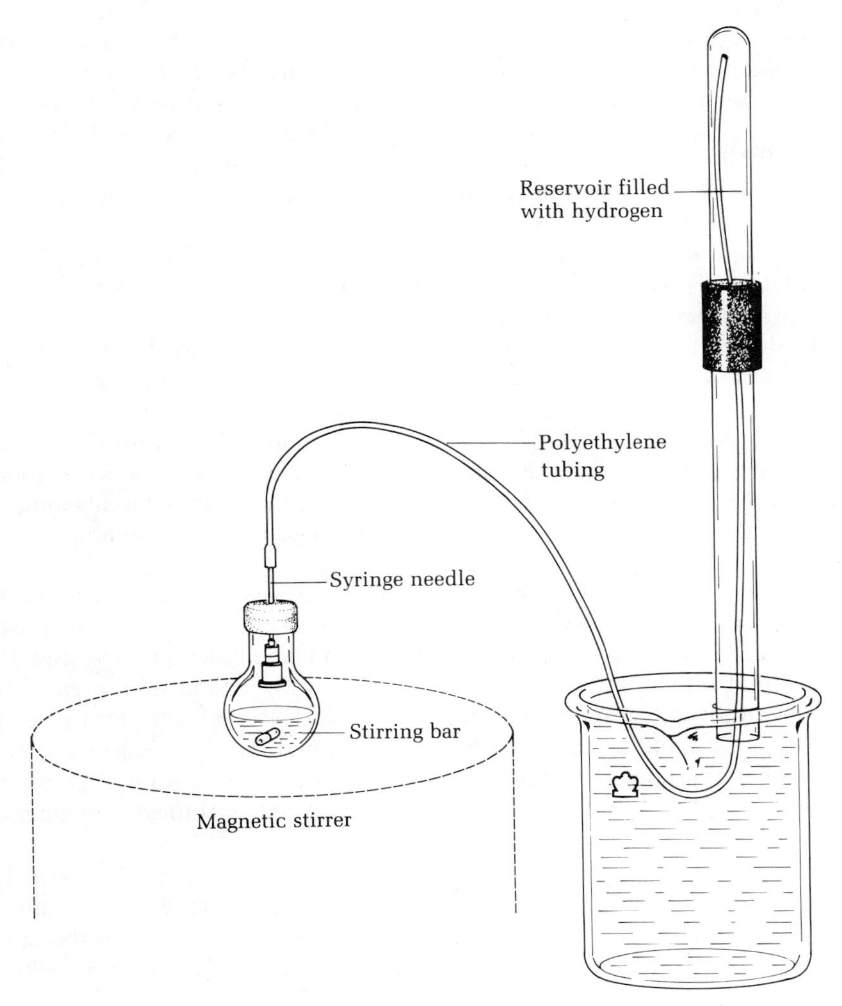

FIG. 62.1 Apparatus for microscale hydrogenation.

charcoal and swirl during addition of 0.45 mL of stabilized 1 M sodium borohydride solution.[3] While allowing 5 min for formation of the catalyst, dissolve 150 mg of *cis*-norbornene-5,6-*endo*-dicarboxylic acid in 1.5 mL of hot water. Pour 0.6 mL of concentrated hydrochloric acid into the reaction flask, followed by the hot solution of the unsaturated acid. Cap the flask with a connector and wire it on. Draw 0.23 mL of the stabilized sodium borohydride solution into the barrel of a plastic syringe, thrust the needle through the center of the stopper, and add the solution dropwise with swirling. The initial uptake of hydrogen is so rapid that the balloon may not inflate

Reaction time: about 15 min

3. Dissolve 0.8 g of sodium borohydride and 0.15 g of sodium hydroxide (stabilizer) in 20 mL of water. When not in use, the solution should be stored in a refrigerator. If left for some time at room temperature in a tightly stoppered container, gas pressure may develop sufficient to break the vessel.

until you start injecting a second 0.23 mL of borohydride solution through the stopper. When the addition is complete and the reaction appears to be reaching an end point (about 5 min), heat the flask on the steam bath with swirling and try to estimate the time at which the balloon is deflated to a constant size (about 5 min). When balloon size is constant, heat and swirl for 5 min more and then release the pressure by injecting the needle of an open syringe through the stopper.

Filter the hot solution by suction on the Hirsch funnel and place the catalyst in a jar marked "Catalyst Recovery."[4] Cool the filtrate and extract it with three 2.5-mL portions of ether. The combined extracts are to be washed with saturated sodium chloride solution and dried over anhydrous sodium sulfate. Evaporation of the ether in a 10-mL Erlenmeyer flask gives about 120 mg of white solid.

Common ion effect

The only solvent of promise for crystallization of the saturated *cis*-diacid is water, and the diacid is very soluble in water and crystallizes extremely slowly and with poor recovery. However, the situation is materially improved by addition of a little hydrochloric acid to decrease the solubility of the diacid.

Scrape out the bulk of the solid product and transfer it to a reaction tube. Add 0.2 mL of water to the 10-mL flask, heat to boiling to dissolve residual solid, and transfer the solution into the reaction tube. Bring the material into solution at the boiling point with a total of not more than 0.45 mL of water. With a Pasteur pipette add 1 drop of concentrated hydrochloric acid and let the solution stand for crystallization. Clusters of heavy prismatic needles soon separate; the recovery is about 90%. The product should give a negative test for unsaturation with acidified permanganate solution.

Observe what happens when a sample of the product is heated in a melting point capillary to about 170°C. Account for the result. You may be able to confirm your inference by letting the oil bath cool until the sample solidifies and then noting the mp temperature and behavior on remelting.

2. Catalytic Hydrogenation Using Palladium on Charcoal

In this experiment one of the most widely used techniques for catalytic hydrogenation is employed: the use of commercially available 10% palladium on carbon. This catalyst currently sells for between $2 and $3 per gram; we will use a few cents worth. The apparatus can be modified in a number of ways. If you merely want to hydrogenate a double bond and are not concerned about the quantitative aspects of the reaction, fit the apparatus with a balloon that holds the hydrogen. As soon as the balloon has reached a constant size, hydrogen uptake is over and the product can be isolated.

In the modification of the apparatus employed in the present experiment, the hydrogen uptake can be followed as water rises in the hydrogen

4. Used catalyst can be sent for recovery of the platinum to Engelhard Industries, 865 Ramsey St., Hillside, NJ 07205.

reservoir. When the reaction is complete the amount of hydrogen used can be measured volumetrically. In a slightly more elaborate version of this apparatus the reaction tube and distilling column are replaced with an inverted 10-mL pipette. With this apparatus the volume of hydrogen absorbed as a function of time can be measured.

In research grade apparatus the hydrogen is stored in a gas burette over mercury. The volume of hydrogen can be determined very accurately so that quantitative estimates of the numbers of double bonds in unknowns can be made, or so that selective hydrogenation of the more reactive of several double bonds can be carried out. The reaction is stopped when the theoretical amount of hydrogen has been absorbed.

A

COOH, COOH $\xrightarrow[10\%\ Pd/C]{H_2}$ COOH, COOH

***cis*-Norbornene-5,6-*endo*-dicarboxylic acid**
mp 180–190°C, dec
MW 182.17

***cis*-Norbornane-5,6-*endo*-dicarboxylic acid**
mp 170–175°C, dec
MW 184.19

B

$$CH_3O\overset{\overset{O}{\|}}{C}(CH_2)_7\overset{\overset{H}{|}}{C}{=}\overset{\overset{H}{|}}{C}(CH_2)_7CH_3 \xrightarrow[10\%\ Pd/C]{H_2} CH_3O\overset{\overset{O}{\|}}{C}(CH_2)_{16}CH_3$$

Methyl oleate
Methyl *cis*-9-octadecanoate
MW 296.50, bp 170°C/2 mm,
mp −20°C

Methyl stearate
Methyl octadecanoate
MW 298.51
mp 39°C

The apparatus in Fig. 62.1 is assembled. The reservoir is filled with hydrogen and the reaction flask is flushed with hydrogen. The stirrer is started and as soon as the water level in the reservoir reaches a constant level the compound to be hydrogenated is injected through the septum of the reaction flask. It can either be a pure liquid or a solid dissolved in methanol. The reservoir holds about 10 mL. Remember that a millimole of hydrogen at standard temperature and pressure has a volume of 22.4 mL and that this will hydrogenate a millimole of a compound that has one double bond. This apparatus holds enough hydrogen to hydrogenate 0.4 mmole of a compound that takes up about 9 mL of gas. For highly accurate work we would take into account the temperature, the atmospheric pressure, and the vapor pressure of the methanol in the reaction flask. We would be careful to equalize the internal and external pressures, so that precise volume measurements could be made.

The hydrogen to fill the apparatus can come from a variety of sources. A tank of hydrogen is most convenient for an individual, but perhaps not for a large group of users. The hydrogen could be generated by reacting acid with borohydride as in the previous experiment, but it is cheaper to use zinc and sulfuric acid.

The apparatus for hydrogen generation is seen in Fig. 62.2. The 5-mL long-necked flask contains 1 g of zinc. As hydrogen is needed sulfuric acid is injected through the septum.

Procedure

Assemble the apparatus shown in Fig. 62.1. In the 5-mL short-necked round-bottomed flask place 20 mg of 10% palladium on carbon, a small magnetic stirring bar and 1.5 mL of methanol. Cap the flask with a rubber septum bearing an inverted 18 gauge needle.

Fill a 250-mL beaker to near the rim with water. Attach a reaction tube with a connector to the empty distilling column, fill this reservoir with water, and clamp it in an inverted position over the beaker of water. Alternatively a 10- or 25-mL graduated cylinder can be used.

In the 5-mL round-necked flask place 1 g of granulated or mossy zinc and close it with a septum containing an inverted 18 gauge needle connected to a 12-in. piece of polyethylene tubing (Fig. 62.2). Inject a few drops of 6 M (33%) sulfuric acid to generate hydrogen and displace the air from the flask, then thread the tubing into the bottom of the reservoir clamped over a beaker of water (Fig. 62.1). Inject more acid onto the zinc and fill the reservoir with hydrogen. Pull out the tube and connect it to the needle protruding from the top of the reaction flask after adding another needle to the top of the reaction flask to act as an outlet.

Generate more hydrogen and flush out the reaction flask. Alternatively some of the hydrogen in the reservoir could be used for this purpose. During this time turn on the stirrer briefly in order to saturate the methanol with hydrogen. Disconnect the hydrogen inlet tube from the top of the reaction flask, pull out the pressure relief needle, and connect the reaction flask to an 18-in. length of polyethylene tubing. Thread this tube to near the top of the reservoir (Fig. 62.1 shows the apparatus at this stage). Turn on the stirrer and note any changes in the water level in the reservoir. If it comes to the same level as the water on the outside, there is a leak in the system.

Once the pressure in the system has stabilized with the stirrer off, inject 0.4 mmole of a compound that has one olefinic double bond to be hydrogenated. For example, inject (**A**) 73 mg of *cis*-norbornene-5,6-*endo*-dicarboxylic acid dissolved in 0.6 mL of methanol or (**B**) 119 mg of methyl oleate dissolved in 0.6 mL of methanol. If a larger reservoir has been used, e.g., a 25-mL graduated cylinder, a proportionately larger quantity of olefin can be hydrogenated. Turn on the stirrer and note the height of the hydrogen in the reservoir as a function of time. Be sure the end of the polyethylene tube is always above the water level in the reservoir. When the water level no longer changes the reaction is complete. Remove the polyethylene tube from

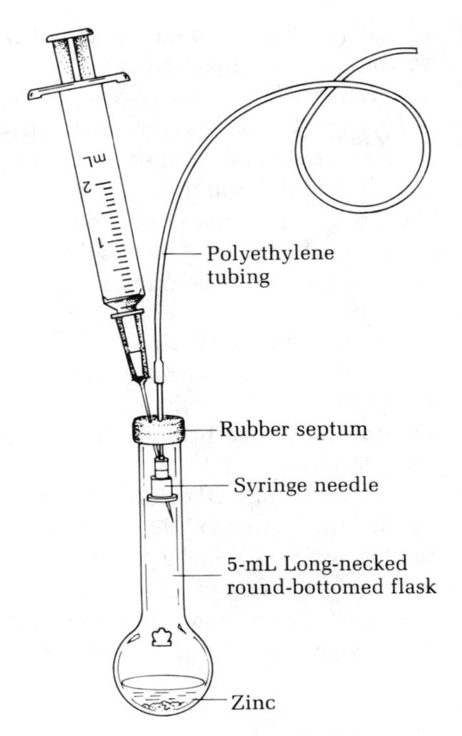

FIG. 62.2 Hydrogen generator.

the reservoir, place a finger under the open end of the reservoir and invert it. Measure carefully the volume of water in the reservoir. This is the volume of hydrogen consumed in the reaction if the reservoir was full of hydrogen at the beginning of the experiment.

Filter the reaction mixture in a Pasteur pipette: push a piece of cotton firmly into a Pasteur pipette that is placed above a reaction tube. With a second pipette transfer the reaction mixture to the filter pipette and allow the solution to filter through the cotton to remove the charcoal. If necessary force the solution through the cotton with a rubber bulb. Rinse the reaction flask, stirring bar, and transfer pipette with a few drops of methanol, and filter this solution also. If the filtrate is gray refilter the solution in the same pipette without changing the cotton filter. Alternatively, filter the solution in the pressure filtration apparatus (Fig. 17.2).

A

If the bicyclic compound has been hydrogenated evaporate the methanol to dryness and dissolve the residue in no more than 0.3 mL of hot water. Add a drop of hydrochloric acid to the solution to decrease the solubility of the product and allow the product to crystallize slowly. Isolate the crystals by removal of the solvent while the reaction tube is in an ice bath. Then scrape the product onto a piece of filter paper, squeeze out excess solvent, and allow the product to dry in air.

Observe the behavior of the sample when heated in a melting point capillary to about 170°C. Account for the result. It may be possible to let the oil bath cool until the sample solidifies and determine the melting point again to confirm your hypothesis.

B If methyl oleate was hydrogenated concentrate the methanol solution to 1.7 mL in a beaker and then cool it in ice. After crystallization of the product is complete, remove the solvent, isolate and dry the crystals, and determine the melting point and percent yield of the product. Although it will appear there is a large quantity of product, you will find it difficult to obtain a high yield because the crystals dissolve readily in the solvent as it warms up. Add water dropwise to the filtrate and try to isolate a second crop of crystalline product.

Evaporate a few drops of the filtrate to dryness, dissolve the solid in a drop or two of dichloromethane, and do the same with a few milligrams of the product. Test both the product and the residue from the filtrate for unsaturation using bromine or permanganate solutions (see Chapter 13 for these tests).

A freshly filtered catalyst is pyrophoric (spontaneously flammable in air) so do not dispose of the filter pipette immediately. Since the amount of catalyst used is so small, the chances of any fire inside the pipette are remote, but care is warranted. Clean the syringe used to inject the sulfuric acid with water and then acetone, disassemble it, and allow it to dry. Do the same with the syringe used to inject the sample.

Questions

1. How would you convert the reduced bicyclic diacid product to the corresponding bicyclic anhydride?
2. Why does the addition of hydrochloric acid cause the solubility of the bicyclic diacid in water to decrease?
3. Why is decolorizing charcoal added to the hydrogenation reaction mixture in the first experiment?
4. Why is methyl stearate a solid whereas methyl oleate is a liquid?

Dichlorocarbene 63

***PRELAB EXERCISE:** Propose a synthesis of* C_6H_5–O–$CH_2CH_2CH_2CH_3$ *using the principles of phase transfer catalysis. Why is phase transfer catalysis particularly suited to this reaction?*

Dichlorocarbene is a highly reactive intermediate of bivalent carbon with only six valence electrons around the carbon. It is electrically neutral and a powerful electrophile. As such it reacts with alkenes, forming cyclopropane derivatives by *cis*-addition to the double bond.

There are a dozen or so ways by which dichlorocarbene may be generated. In this experiment thermal decomposition of anhydrous sodium trichloroacetate in an aprotic solvent in the presence of *cis,cis*-1,5-cyclooctadiene generates dichlorocarbene to give 5,5,10,10-tetrachlorotricyclo[7.1.0.$0^{4,6}$]decane, **1**.

Experiments

1. Thermal Decomposition of Sodium Trichloroacetate; Reaction of Dichlorocarbene with 1,5-Cyclooctadiene

The thermal decomposition of sodium trichloroacetate initially gives the trichloromethyl anion (Eq. 1). In the presence of a proton-donating solvent (or moisture) this anion gives chloroform; in the absence of these reagents the anion decomposes by loss of chloride ion to give dichlorocarbene (Eq. 2).

The conventional method for carrying out the reaction is to add the salt portionwise, during 1–2 h, to a magnetically stirred solution of the olefin in diethylene glycol dimethyl ether ("diglyme") at a temperature maintained in a bath at 120°C. Under these conditions the reaction mixture becomes almost black, isolation of a pure product is tedious, and the yield is low.

$CH_3OCH_2CH_2OCH_2CH_2OCH_3$

Diglyme MW 134.17, bp 161°C
miscible with water

Tetrachloroethylene, a nonflammable solvent widely used in the dry cleaning industry, boils at 121°C and is relatively inert toward electrophilic dichlorocarbene. On generation of dichlorocarbene from either chloroform or sodium trichloroacetate in the presence of tetrachloroethylene, the yield of hexachlorocyclopropane (mp 104°C) is only 0.2–10% (W. R. Moore, 1963; E. K. Fields, 1963). The first idea for simplifying the procedure[1] was to use

---

1. L. F. Fieser and David H. Sachs, *J. Org. Chem.*, **29**, 1113 (1964).

$$Cl_3C{-}CO_2^-\,Na^+ \xrightarrow[-CO_2]{\Delta} Cl_3C^- \xleftarrow{OH^-} + \; H{-}CCl_3 \quad (1)$$

Sodium trichloroacetate MW 185.39

Trichloromethyl anion

Chloroform MW 119.39, bp 61°C

$$RO^- + CHCl_3 \xleftarrow{ROH} Cl_3C^- \longrightarrow Cl_2C{:} + Cl^- \quad (2)$$

Dichlorocarbene

$$\text{cis,cis-1,5-Cyclooctadiene} + 2\; Cl_2C{:} \longrightarrow \mathbf{1} + \mathbf{2} \quad (3)$$

***cis,cis*-1,5-Cyclo-octadiene** MW 108.14 bp 149–150°C

1 *cis*, mp 176°C

2 ***trans***, mp 230°C dec

5,5,10,10-Tetrachlorotricyclo[7.1.0.$0^{4,6}$]decane MW 274.03

tetrachloroethylene to control the temperature to the desired range, but sodium trichloroacetate is insoluble in this solvent and no reaction occurs. Diglyme, or an equivalent, is required to provide some solubility. The reaction proceeds better in a 7 : 10 mixture of diglyme to tetrachloroethylene than in diglyme alone, but the salt dissolves rapidly in this mixture and has to be added in several small portions and the reaction mixture becomes very dark. The situation is vastly improved by the simple expedient of decreasing the amount of diglyme to a 2.5 : 10 ratio. The salt is so sparingly soluble in this mixture that it can be added at the start of the experiment and it dissolves slowly as the reaction proceeds. The boiling and evolution of carbon dioxide provide adequate stirring, the mixture can be left unattended, and what little color develops is eliminated by washing the crude product with methanol.

Rationale of the procedure

The main reaction product crystallizes from ethyl acetate in beautiful prismatic needles, mp 174–175°C, and this was the sole product encountered in runs made in the solvent mixture recommended. In earlier runs made in diglyme with manual control of temperature, the ethyl acetate mother liquor material on repeated crystallization from toluene afforded small amounts of a second isomer, mp 230°C, dec. Analyses checked for a pair of *cis-trans* isomers and both gave negative permanganate tests. To distinguish between them, the junior author of the paper undertook an X-ray analysis, which showed that the lower melting isomer is *cis* and the higher melting isomer is *trans*.

X-ray analysis

A striking reaction of a bis dihalocarbene adduct

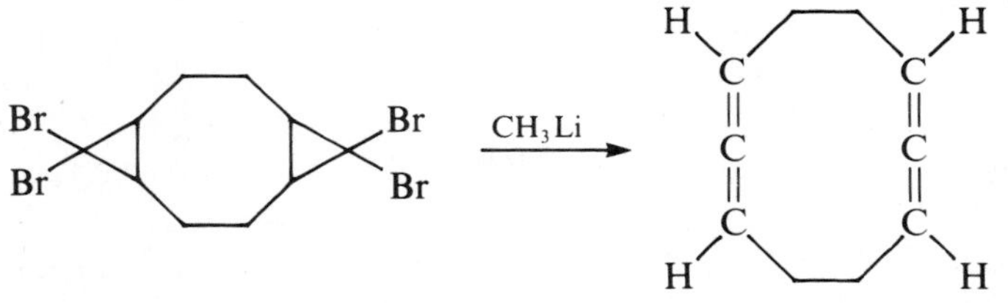

CH_3Li →

4
5,5,10,10-Tetrabromotricyclo-[7.10.0^{4,6}]decane

5
Cyclodeca-1,2,6,7-tetraene

The bis adduct (**4**) of *cis,cis*-1,5-cyclooctadiene with dibromocarbene is described as melting at 174–180°C and may be a *cis-trans* mixture. Treatment of the substance with methyl lithium at −40°C gave a small amount of the ring-expanded bisallene (**5**).

Procedure

Sodium Trichloroacetate. Use commercial sodium trichloroacetate (dry) or prepare it as follows. Place 0.8 g of trichloroacetic acid in a 25-mL Erlenmeyer flask, dissolve 0.20 g of sodium hydroxide pellets in 0.75 mL of water in a reaction tube, cool the solution thoroughly in an ice bath, and swirl the flask containing the acid in the ice bath while slowly dropping in about nine-tenths of the alkali solution. Then add a drop of 0.04% Bromocresol Green solution to produce a faint yellow color, visible when the flask is dried and placed on white paper. With a Pasteur pipette, titrate the solution to an end point where a single drop produces a change from yellow to blue. If the end point is overshot, add a few crystals of acid and titrate more carefully. Close the flask with a rubber stopper, connect to an aspirator, and place the flask within the rings of the steam bath and wrap a towel around both for maximum heat. Turn on the water at full force for maximum suction. The evaporation requires no further attention and should be complete in 15–20 min. When you have an apparently dry white solid, scrape it out with a spatula and break up the large lumps. If you see any evidence of moisture, or in case the weight exceeds the theory, place the solid in a reaction tube and rest this on its side on a drying tray mounted 5 cm above the base of a 70-watt hot plate and let the drier operate overnight.

Use caution in handling the corrosive trichloroacetic acid. Carry out procedure in hood.

Make sure that the salt is completely dry

2. Dichlorocarbene Reaction

Place 0.91 g of *dry* sodium trichloroacetate in a 5-mL round-bottomed short-necked flask mounted over a hot sand bath and add 1 mL of tetrachloroethylene, 0.25 mL of diglyme, and 0.25 mL (220 mg) of *cis,cis*-1,5-cyclo-

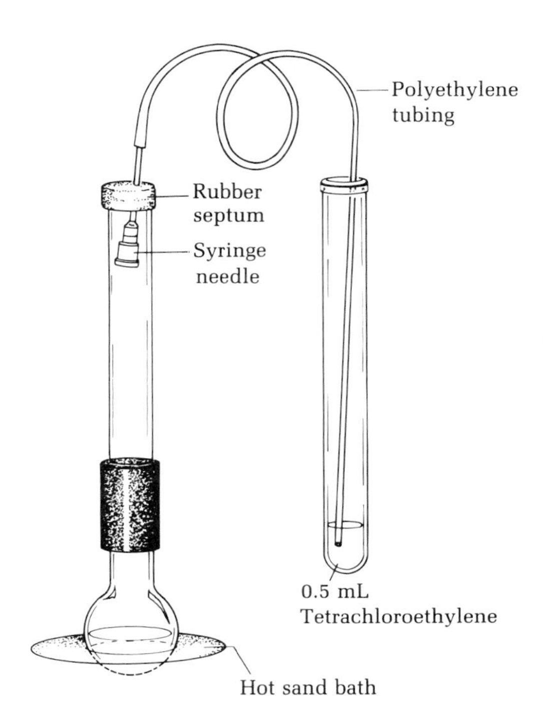

FIG. 63.1 Carbon dioxide bubbler, reflux air condenser and reaction flask for dichlorocarbene synthesis.

octadiene.[2] Attach the empty distilling column as an air condenser and cap it with a septum bearing an inverted 22-gauge syringe needle that is connected to polyethylene tubing, which in turn dips below the surface of 0.5 mL of trichloroethylene contained in a reaction tube (Fig. 63.1). This bubbler will show when the evolution of carbon dioxide ceases. The solvent should be the same as that of the reaction mixture, should there be a suckback. Heat the reaction mixture to boiling, note the time, and reflux gently until the reaction is complete. You will notice foaming, due to liberated carbon dioxide, and separation of finely divided sodium chloride. Inspection of the bottom of the flask will show lumps of undissolved sodium trichloroacetate, which gradually disappear. You need a large flask because it will later serve for removal of tetrachloroethylene by steam distillation.

When the reaction is complete, add 3.5 mL of water to the hot mixture, connect an adapter and a distilling head to the flask (Fig. 63.2), add a boiling chip, heat the mixture to boiling, and steam distill until the tetrachloroethylene is eliminated and the product separates as an oil or semisolid. If necessary, inject more water through the septum to complete the steam distillation. Cool the flask to room temperature, transfer the contents to a reaction tube using dichloromethane, and remove the aqueous layer. Dry the

Note for the instructor

2. Store diglyme, tetrachloroethylene, and 1,5-cyclooctadiene over Linde 5A molecular sieves or anhydrous magnesium sulfate to guarantee that the reagents are dry. The reaction will not work if any reagent is wet.

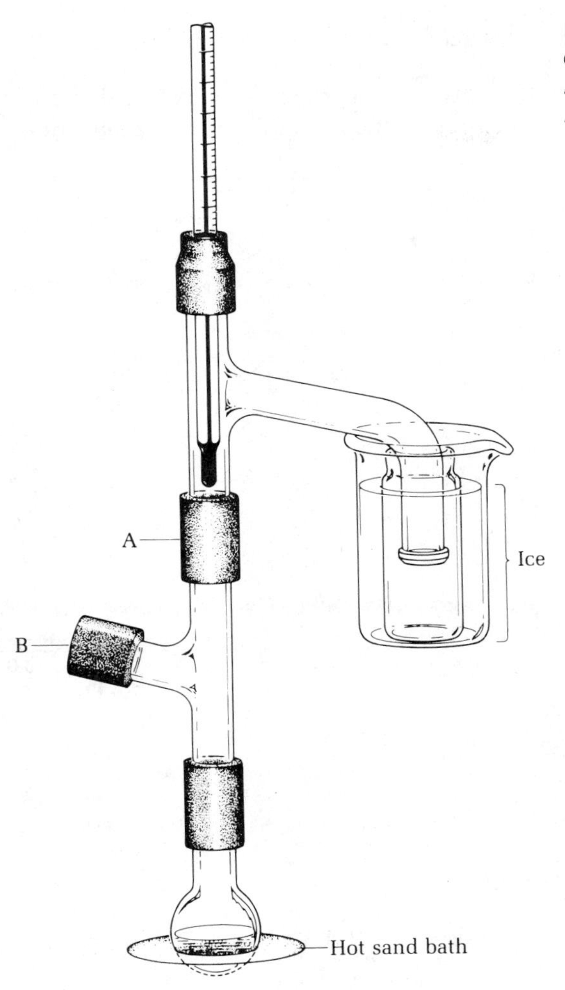

FIG. 63.2 Microscale steam distillation apparatus. Clamp at A with three-prong microclamp. Add water via syringe at B.

solution over anhydrous sodium sulfate, remove the solvent, wash the drying agent, and evaporate the solvent in a reaction tube. The weight of the crude product should be about 400 mg.

Cover the crude product with methanol, break up the cake with a flattened stirring rod, and crush the lumps. Cool in ice, collect, and wash the product with methanol. The yield of colorless, or nearly colorless, 5,5,10,10-tetrachlorotricyclo[7.1.0.0^{4,6}]-decane is about 300 mg. This material, mp 174–175°C, consists almost entirely of the *cis* isomer. Dissolve it in ethyl acetate (1.5–2 mL) and let the solution stand undisturbed for crystallization at room temperature. The pure *cis* isomer separates in large, prismatic needles, mp 175–176°C.

Question

If water is not rigorously excluded from the sodium trichloroacetate dichlorocarbene synthesis, what side reaction occurs?

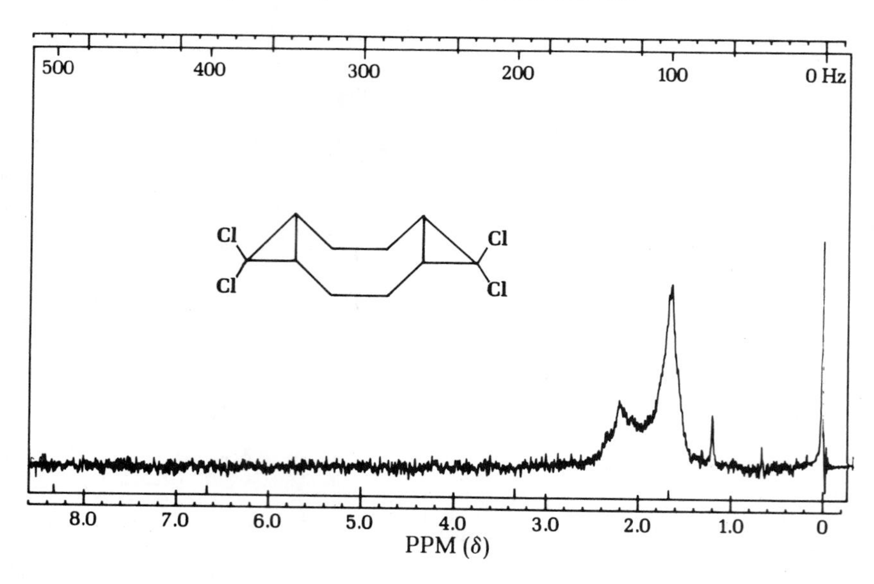

FIG. 63.3 ^{1}H nmr spectrum of 5,5,10,10-tetrachlorotricyclo-[7,1,0,0^{4,6}]decane.

Enzymatic Reduction: A Chiral Alcohol from a Ketone

64

PRELAB EXERCISE: *Study the biochemistry of the conversion of glucose to ethanol (Chapter 18). Which enzyme might be responsible for the reduction of ethyl acetoacetate?*

Reduction of an achiral ketone with the usual laboratory reducing agents such as sodium borohydride or lithium aluminum hydride will not give a chiral alcohol because the chances for attack on each side of the planar carbonyl group are equal. However, if the reducing agent is chiral, there is the possibility of obtaining a chiral alcohol. Organic chemists in recent years have devised a number of such chiral-reducing agents, but few of them are as efficient as those found in nature.

In this experiment we will use the enzymes found in baker's yeast to reduce ethyl acetoacetate to *S*(+)-ethyl-3-hydroxybutanoate. This compound is a very useful synthetic building block. At least eight chiral natural product syntheses are based on this hydroxyester.[1]

There are a large number of different enzymes present in yeast. The primary ones responsible for the conversion of glucose to ethanol are discussed in Chapter 18. While this fermentation reaction is taking place, certain ketones can be reduced to chiral alcohols.

Whenever a chiral product is produced from an achiral starting material, the chemical yield as well as the optical yield is important—in other words, how stereoselective the reaction has been. The usual method for recording this is to calculate the enantiomeric excess (ee). A sodium borohydride reduction will produce 50% *R* and 50% *S* alcohol with no enantiomer in excess. If 93% *S*(+) and 7% *R*(−) isomer are produced, then the enantiomeric excess is 86%. In the present yeast reduction of ethyl acetoacetate various authors have reported enantiomeric excesses ranging from 70% to 97%. This optical yield is distinct from the chemical yield, which depends on how much material is isolated from the reaction mixture.

The use of enzymes to carry out stereospecific chemical reactions is not new, but it is not always possible to predict if an enzymatic reaction (unlike a purely chemical reaction) will occur, or how stereospecific it will be if it

---

1. R. Amstutz, E. Hungerbühler, and D. Seebach, *Helv. Chim. Acta*, **64**, 1796 (1981) and D. Seebach, *Tetrahedron Lett.* 159 (1982).

does. Because this experiment is easily carried out, it might be an interesting research project to explore the range of possible ketones that yeast will reduce to chiral alcohols. For example, butyrophenone can be reduced to the corresponding chiral alcohol. For a review see Sih and Rosazza.[2] The present experiment is based on the work of Seebach,[3] Mori,[4] and Ridley.[5]

Experiments

1. Enzymatic Resolution

Procedure

Keep the mixture warm; the reaction is slow at low temperatures

Extinguish all flames when working with ether

In a 25-mL flask (see Chapter 18, Biosynthesis of Ethanol) dissolve 2.3 g of sucrose and 15 mg of disodium hydrogen phosphate in 8.5 mL of warm (35°C) tap water. Add 0.5 g of dry yeast and swirl to suspend the yeast throughout the solution. After 15 min, while fermentation is progressing vigorously, add 150 mg of ethyl acetoacetate. Store the flask in a warm place, ideally at 30–35°C for at least 48 h (a longer time will do no harm). At the end of this time add 0.5 g of Celite filtration aid and remove the yeast cells by filtration on a Hirsch funnel (see Chapter 18). Wash the cells with 1.5 mL of water; then saturate the filtrate with sodium chloride in order to decrease the solubility of the product. See a handbook for the solubility of sodium chloride in water to decide approximately how much to use. Extract the resulting solution five times with 1.5-mL portions of ether in a test tube, taking care to shake hard enough to mix the layers, but not so hard as to form an emulsion between the ether and water. Addition of a small amount of methanol may help to break up emulsions. Dry the ether layer by adding anhydrous sodium sulfate until it no longer clumps together. After approximately 15 min remove the ether solution and evaporate it in portions in a tared reaction tube. The residue should weigh about 100 mg. It should, unlike the starting material, give a negative iron(III) chloride test (see p. 515). It can be analyzed by TLC (use dichloromethane as the solvent) to determine whether unreacted ethyl acetoacetate is present. Infrared spectroscopy should show the presence of the hydroxyl group and may show unreduced methyl ketone. The nmr spectrum of the product is easily distinguished from the starting material. The optical purity of the product can be determined by measuring the optical rotation in a polarimeter. The specific rotation, $[\alpha]_D^{25}$, of *S*(+)-ethyl-3-hydroxybutanoate has been reported to vary from +31.3° to +41.7° in chloroform. The specific rotation, $[\alpha]_D^{25}$, of 37.2° (chloroform, C 1.3) corresponds to an enantiomeric excess of 85%.

2. C. J. Sih and J. P. Rosazza, *Application of Biochemical Systems in Organic Chemistry*, Part 1 (J. B. Jones, C. J. Sih, and D. Perlman, Eds.), pp. 71–78, Wiley, New York, 1976.

3. D. Seebach, *Tetrahedron Lett.* 159 (1982).

4. K. Mori, *Tetrahedron*, **37**, 1341 (1981).

5. D. D. Ridley and M. Stralow, *Chem. Commun.*, **400** (1975).

2. Preparation of 3,5-Dinitrobenzoate

Pyridine
MW 79.10
bp 115°C
den 0.978

+

S(+)-Ethyl-3-hydroxybutanoate
MW 132.16
bp 71–73°C/12 mm

+

3,5-Dinitrobenzoyl chloride
MW 230.56
m.p. 71–74°C

→

3,5-Dinitrobenzoate
mp 154°C

Since the quantity of product is so small, one of the best ways of purifying it is to convert it to a crystalline derivative, the 3,5-dinitrobenzoate. 3,5-Dinitrobenzoates are common alcohol derivatives (see Chapter 71); they are easily prepared, the two nitro groups add considerably to the molecular weight, and they are easily crystallized.

The 3,5-dinitrobenzoates of a racemic mixture of *R*- and *S*-ethyl-3-hydroxybutanoate, like almost all racemates, crystallize together to give crystals that in this case melt at 146°C. No amount of recrystallization causes one enantiomer to crystallize out while the other remains in solution since they are, after all, mirror images of each other. However if one enantiomer is in large excess it is possible to effect a separation by crystallization. In the present case the *S* enantiomer predominates and it crystallizes out, leaving most of the *R* + *S* racemate in solution. Repeated crystallization increases the melting point and the purity to the point at which the optical purity of the crystalline product is almost 100%. At this point the 3,5-dinitrobenzoate has a melting point of 154°C. Treatment of the derivative with excess acidified ethanol regenerates 100% ee *S*(+)-ethyl-3-hydroxybutanoate.

Procedure

To 100 mg of the *S*(+)-ethyl-3-hydroxybutanoate in a reaction tube add 175 mg of pure[6] 3,5-dinitrobenzoyl chloride and 1 mL of pyridine. Reflux the mixture for 15 min and then transfer it with a Pasteur pipette to 3.5 mL of water in another reaction tube. Remove the solvent from the crystals and shake the crystals with 2 mL of 5% sodium carbonate solution to remove dinitrobenzoic acid. Remove the bicarbonate solution and recrystallize the derivative from ethanol. Dry a portion and determine the melting point. If it is not near 154°C repeat the crystallization. Dry the pure derivative and calculate the percent yield.

Question

What can you say about the purity of the reduction product whose ^{1}H nmr spectrum is illustrated in Fig. 64.1? Assign the peaks of ethyl-3-hydroxybutanoate to specific protons in the molecule.

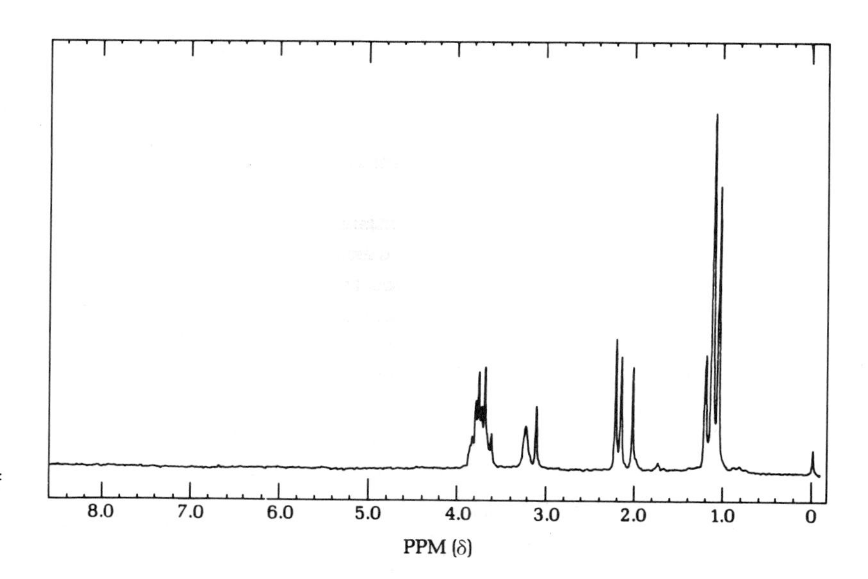

FIG. 64.1 ^{1}H nmr spectrum of yeast reduction reaction mixture.

6. Check the melting point of the 3,5-dinitrobenzoyl chloride. If it is below 70°C it should be recrystallized from dichloromethane.

Enzymatic Resolution of DL-Alanine

65

PRELAB EXERCISE: *Discuss at least three other methods for resolving DL-alanine into its enantiomers. Judging from its name, would you expect acylase to be equally effective on the acetyl derivatives of other amino acids? Explain why the enzyme acylase will hydrolyze only one of a pair of enantiomers and how a mixture of L-alanine and N-acetyl-D-alanine can be separated at the end of the reaction.*

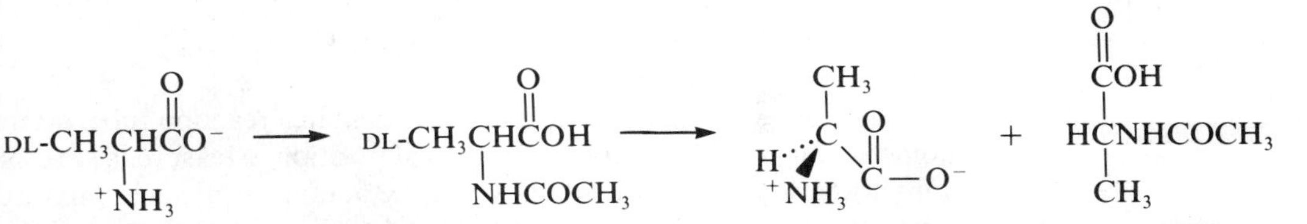

DL-$CH_3CH(^+NH_3)CO^-$ ⟶ DL-$CH_3CH(NHCOCH_3)COOH$ ⟶ L-alanine + $HC(CH_3)(NHCOCH_3)COOH$

DL-Alanine (1)
mp 295°C
MW 89.10

N-Acetyl-DL-alanine (2)
mp 137°C
MW 131.13

L-Alanine (3)
mp 297°C
α_D + 14.4°

N-Acetyl-L-alanine (4)
mp 125°C
α_D + 66.5°

Resolution of DL-alanine (**1**) is accomplished by heating the N-acetyl derivative (**2**) in weakly alkaline solution with acylase, a proteinoid preparation from porcine kidney containing an enzyme that promotes rapid hydrolysis of N-acyl derivatives of natural L-amino acids but acts only immeasurably slowly on the unnatural D-isomers. N-Acetyl-DL-alanine (**2**) can thus be converted into a mixture of L-(+)-alanine (**3**) and N-acetyl-D-alanine (**4**). The mixture is easily separable into the components, because the free amino acid (**3**) is insoluble in ethanol and the N-acetyl derivative (**4**) is readily soluble in this solvent. Note that, in contrast to the weakly levorotatory D-(−)-alanine (α_D − 14.4°), its acetyl derivative is strongly dextrorotatory.

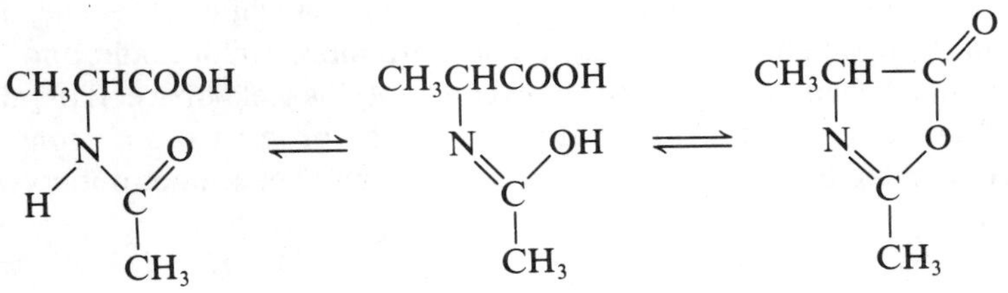

$CH_3CH(COOH)-NH-C(=O)CH_3$ (**4**) ⇌ $CH_3CH(COOH)-N=C(OH)CH_3$ (**5**) ⇌ $CH_3CH-C(=O)-O-C(CH_3)=N$ (ring) (**6**)

The acetylation of an α-amino acid presents the difficulty that, if the conditions are too drastic, the N-acetyl derivative (**4**) is converted in part through the enol (**5**) to the azlactone (**6**).[1] However, under critically controlled conditions of concentration, temperature, and reaction time, N-acetyl-DL-alanine can be prepared easily in high yield.

Experiments

1. Acetylation of DL-Alanine.

$$CH_3CH(^{+}NH_3)\overset{O}{\overset{\|}{C}}O^{-} + CH_3\overset{O}{\overset{\|}{C}}O\overset{O}{\overset{\|}{C}}CH_3 \xrightarrow{CH_3COOH} CH_3CH(NHCOCH_3)\overset{O}{\overset{\|}{C}}OH + CH_3COOH$$

DL-Alanine (1)
MW 89.10
mp 295°C

Acetic anhydride
MW 102.09
bp 140°C

N-Acetyl DL-alanine (2)
MW 131.13
mp 137°C

Check the pressure gauge

Place 200 mg of DL-alanine and 0.5 mL of acetic acid in a reaction tube, insert a thermometer, and clamp the tube in a vertical position. Measure 0.3 mL of acetic anhydride, which is to be added when the alanine/acetic acid mixture is at exactly 100°C. Heat the tube on a hot sand bath with stirring, until the temperature of the suspension has risen a little above 100°C. Stir the suspension, let the temperature gradually fall, and when it reaches 100°C add the 0.3-mL portion of acetic anhydride and note the time. In the course of 1 min the temperature falls (91–95°C, cooled by added reagent), rises (100–103°C, the acetylation is exothermic), and begins to fall with the solid largely dissolving. Stir to facilitate reaction of a few remaining particles of solid, let the temperature drop to 80°C, pour the solution into a tared 10-mL Erlenmeyer flask, and rinse the thermometer and test tube with a little acetone. Add 1 mL of water to react with excess anhydride, connect the flask to the aspirator operating at full force, put the flask *inside* the rings of the steam bath and wrap the flask with a towel. Evacuation and heating for about 5–10 min should remove most of the acetic acid and water and leave an oil or thick sirup. Add 2 mL of cyclohexane and evacuate and heat as before for 5–10 min. Traces of water in the sirup are removed as a cyclohexane/water azeotrope. If the product has not yet separated as a white solid or semisolid, determine the weight of the product, add 2 mL more cyclohexane, and repeat the process. When the weight becomes constant, the yield of acetyl DL-alanine should be close to the theoretical amount. The product has a pronounced tendency to remain in supersaturated solution and hence does not crystallize readily.

1. The azlactone of DL-alanine is known only as a partially purified liquid.

2. Enzymatic Resolution of N-Acetyl-DL-Alanine

Add 1.0 mL of distilled water[2] to the reaction flask, grasp this with a clamp, swirl the mixture over a hot sand bath to dissolve all the product, and cool under the tap. Remove a drop of the solution on a stirring rod, add it to 0.5 mL of a 0.3% solution of ninhydrin in water, and heat to boiling. If any unacetylated DL-alanine is present a purple color will develop and should be noted. Pour the solution into a reaction tube and rinse the flask with a little water. Add 0.15 mL of concentrated ammonia solution, stir to mix, check the pH with Hydrion paper, and if necessary adjust to pH 8 by addition of more ammonia with a capillary dropping tube. Add 5 mg of commercial acylase,[3] mix with a stirring rod, rinse the rod with distilled water and make up the volume until the tube is about half full. Then stopper the tube, mark it for identification and let the mixture stand at room temperature overnight, or at 37°C[4] for 4 h.

At the end of the incubation period add 0.3 mL of acetic acid to denature the enzyme and if the solution is not as acidic as pH 5 add more acid. Rinse the cloudy solution into a 10-mL Erlenmeyer flask, add 10 mg of granulated decolorizing carbon, heat and swirl over a hot sand bath for a few moments to coagulate the protein, and filter the solution. Transfer the solution to a 25-mL round-bottomed flask and evaporate on a rotary evaporator under vacuum, or add 2 mL of cyclohexane (to prevent frothing) and a boiling stone and evaporate on the steam bath under vacuum to remove water and acetic acid as completely as possible. Remove the last traces of water and acid by adding 2 mL of cyclohexane and evaporating again to remove water and acetic acid as azeotropes. The mixture of L-alanine and acetyl-D-alanine separates as a white scum on the walls. Add 1.5 mL of 95% ethanol, digest on the steam bath, and dislodge some of the solid with a spatula. Cool well in ice for a few minutes, and then scrape as much of the crude L-alanine as possible onto a suction funnel, and wash it with ethanol. Save the ethanol mother liquor.[5] To recover the L-alanine retained by the flask, add 0.2 mL of water and warm on the steam bath until the solid is all dissolved, then transfer the solution to a reaction tube by means of a Pasteur pipette, rinse the flask with 0.2 mL more water, and transfer in the same way. Add the

Work-up time $\frac{1}{2}-\frac{3}{4}$ h

2. Tap water may contain sufficient heavy metal ion to deactivate the enzyme.

3. Commerical porcine kidney acylase is available from Schwarz/Mann, Division of Becton, Dickinson and Co., Orangeburg, NY 10962, or from Sigma Chemical Co., P.O. Box 14503, St. Louis, MO 63172.

Note for the instructor

4. A reasonably constant heating device that will hold 30 or more tubes is made by partially filling a 1-L beaker with water, adjusting to 37°C, and maintaining this temperature by the heat of a 250-watt infrared drying lamp shining horizontally on the beaker from a distance of about 40 cm.

5. In case the yield of L-alanine is low, evaporation of this mother liquor may reveal the reason. If the residue solidifies readily and crystallizes from acetone to give acetyl-DL-alanine, mp 130°C or higher, the acylase preparation is recognized as inadequate in activity or amount. Acetyl-D-alanine is much more soluble and slow to crystallize.

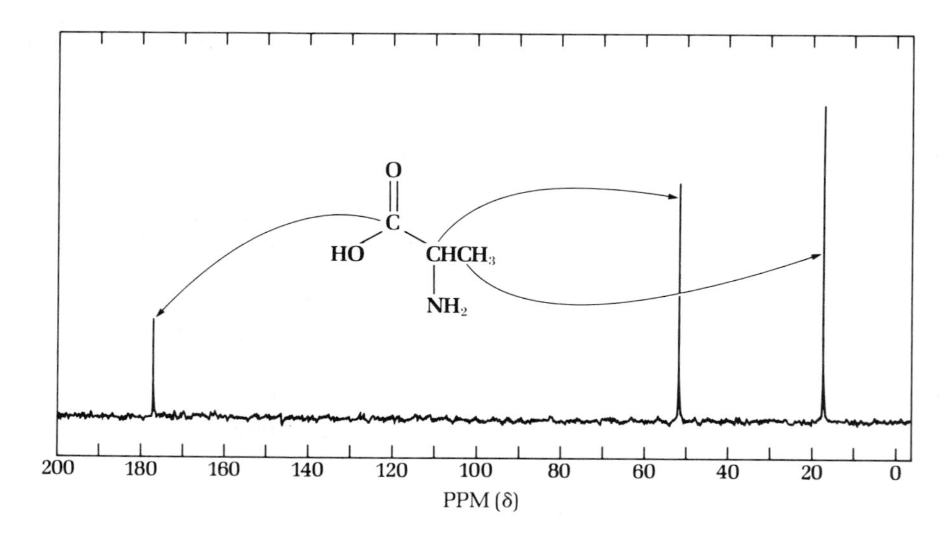

FIG. 65.1 ^{13}C nmr spectrum of L-alanine.

filtered L-alanine, dissolve by warming, and filter the solution into a reaction tube using the pressure apparatus. Rinse the flask and funnel with 0.1 mL of water and then with 0.5 mL of warm 95% ethanol. Then heat the filtrate on the steam bath and add more 95% ethanol (1–1.5 mL) in portions until crystals of L-alanine begin to separate from the hot solution. Let crystallization proceed. Collect the crystals and wash with ethanol. The yield of colorless needles of L-alanine, α_D + 13.7 to + 14.4°[6] (in 1 N hydrochloric acid) varies from 40 to 50 mg, depending on the activity of the enzyme.

Questions

1. The melting point of N-acetyl-DL-alanine is 137°C and that of N-acetyl-D-alanine is 125°C. What would you expect the melting point of N-acetyl-L-alanine to be, or is this impossible to predict?

2. Would the ^{13}C nmr spectrum of D-alanine differ from that of L-alanine (Fig. 65.1)?

6. Determination of optical activity can be made in the student laboratory with a Zeiss Pocket Polarimeter, which requires no monochromatic light source and no light shield. For construction of a very inexpensive polarimeter, see W. H. R. Shaw, *J. Chem. Ed.*, **32**, 10 (1955).

Dyes and Dyeing[1] 66

PRELAB EXERCISE: *Operating on the simple hypothesis that the intensity of a dye on a fiber will depend on the number of strongly polar or ionic groups in the fiber molecule, predict the relative intensities produced by methyl orange when it is used to dye a variety of different fibers, such as are found in the Multifiber Fabric.*

Since prehistoric times man has been dyeing cloth. The "wearing of the purple" has long been synonymous with royalty, attesting to the cost and rarity of Tyrian purple, a dye derived from the sea snail *Murex brandaris*. The organic chemical industry originated with William Henry Perkin's discovery of the first synthetic dye, Perkin's Mauve, in 1856.

In this experiment several dyes will be synthesized and these and other dyes will be used to dye a representative group of natural and man-made fibers. You will receive several $2\frac{1}{2}''$ squares of Multifiber Fabric 10,[2] which has six strips of different fibers woven into it: spun acetate rayon, cotton, a spun polyamide (Nylon 6,6), Dacron, Orlon, and wool (see Fig. 66.1).

Acetate rayon is cellulose (from any source) in which about two of the hydroxyl groups in each unit have been acetylated. This renders the polymer soluble in acetone from which it can be spun into fiber. The smaller number of hydroxyl groups in acetate rayon compared to cotton makes direct dyeing of rayon more difficult than of cotton.

Cotton is pure cellulose. Nylon is a polyamide and made by polymerizing adipic acid and hexamethylenediamine. The nylon polymer chain can be prepared with one acid and one amine group at the termini, or with both acids or both amines. Except for these terminal groups, there are no polar centers in nylon and consequently it is difficult to dye. Similarly Dacron, a polyester made by polymerizing ethylene glycol and terephthalic acid, has few polar centers within the polymer and consequently is difficult to dye. Even more difficult to dye is Orlon, a polymer of acrylonitrile. Wool and silk are polypeptides crosslinked with disulfide bridges. The acidic and basic amino acids (e.g., glutamic acid and lysine) provide many polar groups in wool and silk to which a dye can bind, making these fabrics easy to dye. In this experiment note the marked differences in shade produced by the same dye on different fibers.

---

1. For a detailed discussion of the chemistry of dyes and dyeing see *Topics in Organic Chemistry*, by Louis F. Fieser and Mary Fieser, Reinhold Publishing Corp., New York, 1963, pp. 357–417.

2. Obtained from Testfabrics, Inc., P.O. Box 118, 200 Blackford Ave., Middlesex, NJ 08846. Cut into $2\frac{1}{2}''$ squares that include all six fibers.

FIG. 66.1 Multifiber Fabric 10.

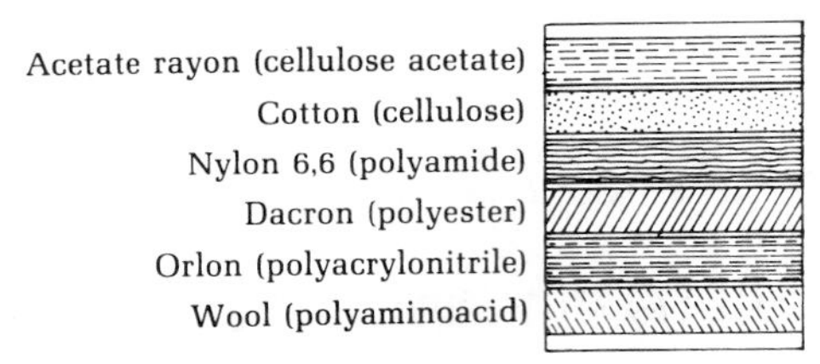

Cellulose (Cotton, R=H)
Acetylated Cellulose (Acetate rayon, R=OAc)

Wool (R=amino acid residue)

Polyethyleneglycol terephthalate (Dacron)

Nylon

Polyacrylonitrile (Orlon)

Part 1. Dyes

Azo group

The most common dyes are the azo dyes, formed by coupling diazotized amines to phenols. The dye can be made in bulk, or, as we shall see, the dye molecule can be developed on and in the fiber by combining the reactants in the presence of the fiber.

One dye, Orange II, is made by coupling diazotized sulfanilic acid with 2-naphthol in alkaline solution; another, Methyl Orange, is prepared by coupling the same diazonium salt with N,N-dimethylaniline in a weakly acidic solution. Methyl Orange is used as an indicator as it changes color at pH 3.2–4.4. The change in color is due to transition from one chromophore (azo group) to another (quinonoid system).

You are to prepare one of these two dyes and then exchange samples with a neighbor and do the tests with both dyes. Both substances dye wool, silk, and skin, and you must work carefully to avoid getting them on your hands or clothes. The dye will eventually wear off your hands or they can be cleaned by soaking them in warm, slightly acidic (H_2SO_4) permanganate solution until heavily stained with manganese dioxide and then removing the stain in a bath of warm, dilute bisulfite solution.

Experiments

1. Diazotization of Sulfanilic Acid

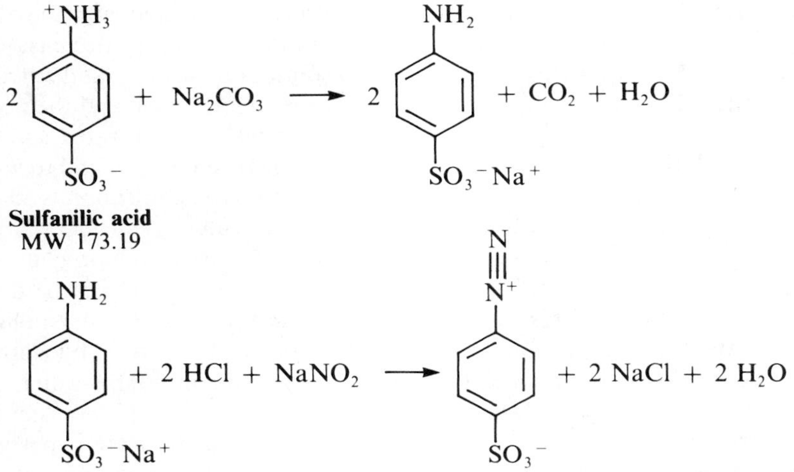

$$2\ {}^{+}H_3N{-}C_6H_4{-}SO_3^- + Na_2CO_3 \longrightarrow 2\ H_2N{-}C_6H_4{-}SO_3^-Na^+ + CO_2 + H_2O$$

Sulfanilic acid
MW 173.19

$$H_2N{-}C_6H_4{-}SO_3^-Na^+ + 2\ HCl + NaNO_2 \longrightarrow N{\equiv}N^+{-}C_6H_4{-}SO_3^- + 2\ NaCl + 2\ H_2O$$

In a 10 × 100 mm reaction tube dissolve, by boiling, 120 mg of sulfanilic acid monohydrate in 1.25 mL of 2.5% sodium carbonate solution (or use 35 mg of anhydrous sodium carbonate and 1.25 mL of water). Cool the solution to room temperature, add 50 mg of sodium nitrite, and stir until it is dissolved. Cool the tube in ice and add to it with thorough stirring a mixture of 0.75 g of ice and 0.125 mL of concentrated hydrochloric acid. In a minute or two a powdery white precipitate of the diazonium salt should separate and the material is then ready for use. The product is not collected but is used in the preparation of the dyes Orange II and/or Methyl Orange while in suspension. It is more stable than most diazonium salts and will keep for a few hours.

2. Orange II (1-*p*-Sulfobenzeneazo-2-naphthol Sodium Salt)

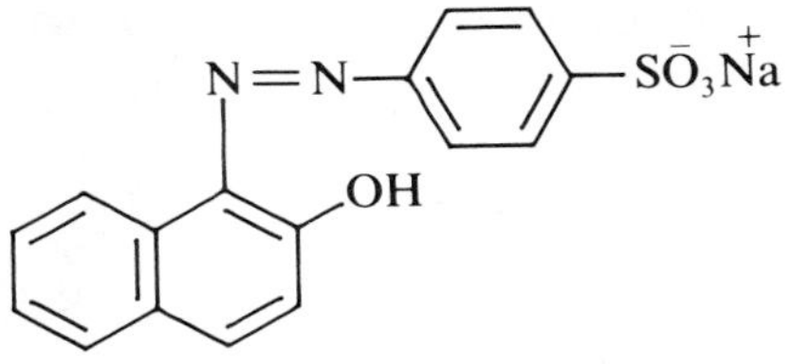

$$\text{2-HO-C}_{10}\text{H}_6\text{-1-N{=}N{-}C}_6\text{H}_4\text{{-}}\overset{-}{\text{S}}\text{O}_3\overset{+}{\text{Na}}$$

Orange II

In a 10 × 100 mm reaction tube dissolve 90 mg of 2-naphthol in 0.5 mL of 10% sodium hydroxide solution and transfer to this solution, with *thorough* stirring, the suspension of diazotized sulfanilic acid prepared in the previous experiment. Rinse all of the diazonium salt into the naphthol

Handle 2-naphthol with care, in the hood. Do not breathe the dust or allow skin contact. Carcinogen.

solution with a few drops of cold water. Coupling occurs very rapidly and the dye, being a sodium salt, separates easily from the solution because a considerable excess of sodium ion from the carbonate, the nitrite, and the alkali is present. Stir the crystalline paste thoroughly to effect good mixing and, after 5–10 min, heat the mixture in a beaker of boiling water until the solid dissolves. Add 0.25 g of sodium chloride to further decrease the solubility of the product, bring this all into solution by heating and stirring. Allow the tube to cool to near room temperature undisturbed and then cool it in ice. Collect the product on the Hirsch funnel. Use saturated sodium chloride solution rather than water to rinse out the reaction tube and to wash the filter cake free of the dark-colored mother liquor. The filtration is somewhat slow and transfer to the Hirsch funnel is difficult.

The product dries slowly and it contains about 20% sodium chloride. The crude yield is thus not significant, and the material need not be dried before being purified. This particular azo dye is too soluble to be crystallized from water; it can be obtained in a fairly satisfactory form by adding saturated sodium chloride solution to a hot, filtered solution in water and cooling, but the best crystals are obtained from aqueous ethanol. Transfer the filter cake to a reaction tube, wash the material from the filter paper and funnel with 1 mL of water, and bring all of the solid into solution at the boiling point. It may be necessary to add another 0.25 mL (no more) of water in the course of filtering this hot solution through the micro Büchner funnel equipped with a polyethylene frit (Fig. 17.2). Collect the filtrate in a reaction tube, add 2.5–3 mL of ethanol, and allow crystallization to proceed as the tube cools slowly to room temperature. Cool well in ice before collecting the product. Rinse the tube with some of the filtrate and finally wash the product with a small quantity of ethanol. The yield of pure, crystalline material should be about 150 mg. Orange II separates from aqueous alcohol as the dihydrate, containing two molecules of water of crystallization, and allowance for this should be made in calculating the yield. If the water of hydration is eliminated by drying at 120°C the material becomes fiery red.

3. Methyl Orange (*p*-Sulfobenzeneazo-4-dimethylaniline Sodium Salt)

$^{-}O_3S$–C_6H_4–$\overset{+}{N}\equiv N$ + C_6H_5–$N(CH_3)_2$ ⟶ Na^+ $^{-}O_3S$–C_6H_4–$N{=}N$–C_6H_4–$N(CH_3)_2$

Diazotized sulfanilic acid
4-Diazobenzenesulfonic acid

N,N-Dimethylaniline
MW 121.18
bp 194°C

Methyl orange
4-[4-(Dimethylamino)phenylazo]benzenesulfonic acid, sodium salt
MW 327.34
λ_{max} 507 nm

In a reaction tube mix 75 mg of dimethylaniline and 65 mg of acetic acid. To the suspension of diazotized sulfanilic acid add, with stirring, the solution of dimethylaniline acetate, rinsing out the last portions with a few drops of water. Stir the mixture thoroughly and within a few minutes the red, acid-stable form of the dye should separate. A stiff paste should result in 5–10 min, at which time 1 mL of 10% sodium hydroxide is added to produce the orange sodium salt. Heat the mixture to the boiling point with constant stirring (to avoid bumping), when a large part of the dye should dissolve. This might better be done by placing the tube in a small beaker of boiling water. Allow the tube to cool slowly to room temperature and then cool it thoroughly in ice before collecting the product by vacuum filtration on the Hirsch funnel. Use saturated sodium chloride solution rather than water to rinse the tube and to wash the dark mother liquor from the filter cake.

The crude product need not be dried but can be crystallized from water after making preliminary solubility tests to determine the proper conditions. The yield should be about 125 to 150 mg. Recrystallized material need not necessarily be used for the dyeing process.

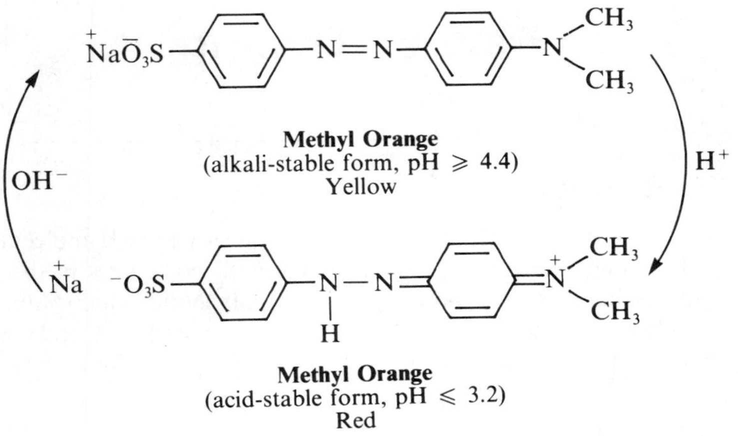

$\overset{+}{Na}\bar{O}_3S-C_6H_4-N{=}N-C_6H_4-N(CH_3)_2$

Methyl Orange
(alkali-stable form, pH ⩾ 4.4)
Yellow

OH^- ⇄ H^+

$\overset{+}{Na}\ {}^-O_3S-C_6H_4-N(H)-N{=}C_6H_4{=}\overset{+}{N}(CH_3)_2$

Methyl Orange
(acid-stable form, pH ⩽ 3.2)
Red

Methyl Orange is an acid-base indicator

Tests

Solubility and Color. Compare the solubility in water of Orange II and Methyl Orange and account for the difference in terms of structure. Treat the first solution with alkali and note the change in shade due to salt formation; to the other solution alternately add acid and alkali.

Reduction. Characteristic of an azo compound is the ease with which the molecule is cleaved at the double bond by reducing agents to give two amines. Since amines are colorless, the reaction is easily followed by the color change. The reaction is of use in preparation of hydroxyamino and similar compounds, in analysis of azo dyes by titration with a reducing agent, and in identification of azo compounds from an examination of the cleavage products.

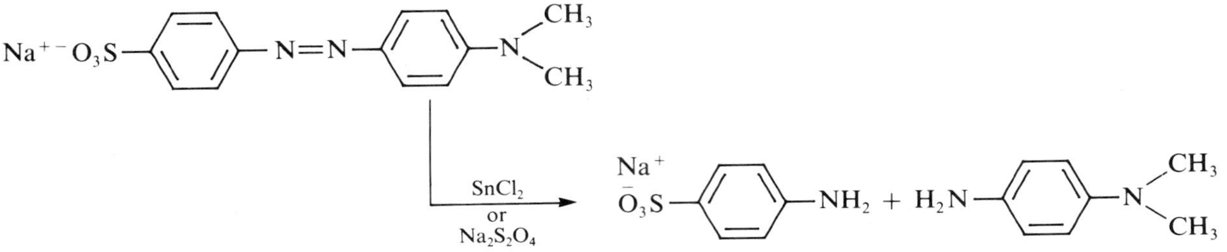

$SnCl_2$ or $Na_2S_2O_4$

Dissolve about 0.1 g of tin(II) chloride in 0.2 mL of concentrated hydrochloric acid, add a small quantity of the azo compound (20 mg) and heat. A colorless solution should result and no precipitate should form on adding water. The aminophenol or the diamine products are present as the soluble hydrochlorides; the other product of cleavage, sulfanilic acid, is sufficiently soluble to remain in solution.

Part 2. Dyeing

Experiments

1. Direct Dyes

With good laboratory technique your hands will not be dyed; use care

The sulfonate groups on the Methyl Orange and Orange II molecules are polar and thus enable these dyes to combine with polar sites in the fibers. Wool and silk have many polar sites on their polypeptide chains and hence bind strongly to a dye of this type. Martius Yellow, picric acid, and eosin are also highly polar dyes and thus dye directly to wool and silk.

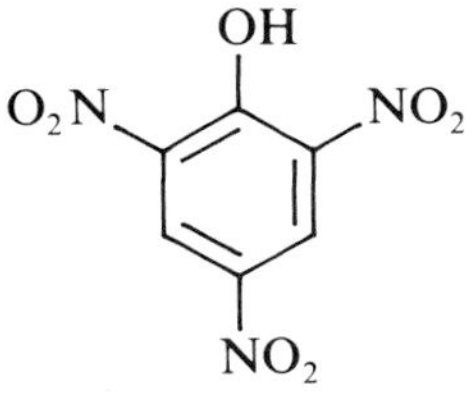

Picric Acid

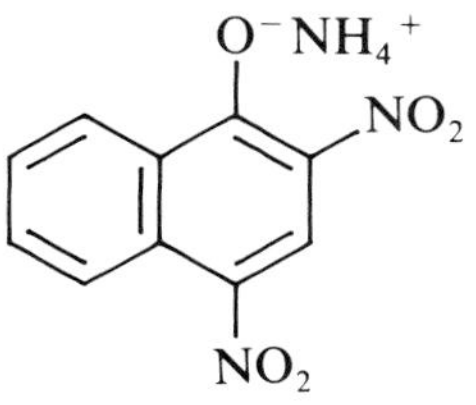

Martius Yellow

Orange II or Methyl Orange

The dye bath is prepared from 50 mg of Orange II or Methyl Orange, 0.5 mL of 10% sodium sulfate solution, 15 mL of water, and 5 drops of 10% sulfuric acid in a 25-mL beaker. Place a piece of test fabric, a strip 3/4-in. wide, in the bath for 5 min at a temperature near the boiling point. Remove the fabric from the dye bath, allow it to cool and then wash it thoroughly under running water before drying it.

Dye untreated test fabric and one or more of the pieces of test fabric that have been treated with a mordant following this same procedure. See part 3 for application of mordants.

Picric Acid or Martius Yellow

Dissolve 50 mg of one of these acidic dyes in 15 mL of hot water to which a few drops of dilute sulfuric acid have been added. Heat a piece of test fabric in this bath for one minute, then remove it with a stirring rod, rinse well, wring, and dry. Describe the results.

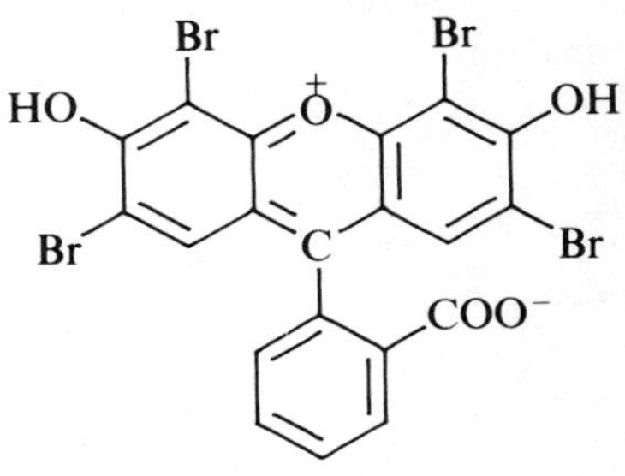

Eosin

Dissolve 10 mg of sodium eosin in 20 mL of water and dye a piece of test fabric by heating it with the solution for about 10 min. Eosin is the dye used in red ink. Also dye pieces of mordanted cloth in eosin (see part 3).

Eosin A
(λ_{max} 516,483 nm)

2. Substantive Dyes

Cotton and the rayons do not have the anionic and cationic carboxyl and amine groups of wool and silk and hence do not dye well with direct dyes, but they can be dyed with substances of rather high molecular weight showing colloidal properties. Such dyes probably become fixed to the fiber by hydrogen bonding. Such a dye is Congo Red, a substantive dye.

Congo Red, a Benzidine Dye

Dissolve 10 mg of Congo Red in 40 mL of water, add about 0.1 mL each of 10% solutions of sodium carbonate and sodium sulfate, heat to a temperature just below the boiling point, and introduce a piece of test fabric. At the end of 10 min remove the fabric and wash in warm water as long as the dye is removed. Place pieces of the dyed material in very dilute hydrochloric acid solution and observe the result. Rinse and wash the material with soap.

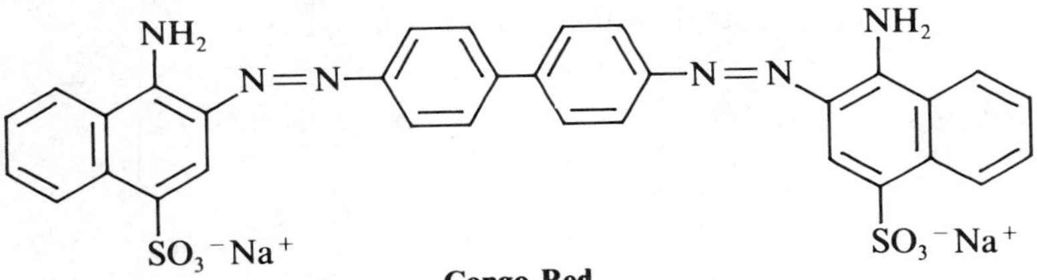

Congo Red
(λ_{max} 497 nm)

3. Mordant Dyes

One of the oldest known methods of producing wash-fast colors involves the use of metallic hydroxides, which form a link, or mordant (L. *mordere*, to bite), between the fabric and the dye. Other substances, such as tannic acid, also function as mordants. The color of the final product depends on both the dye used and the mordant. For instance, the dye Turkey Red is red with an aluminum mordant, violet with an iron mordant, and brownish-red with a chromium mordant. Some important mordant dyes possess a structure based on triphenylmethane, as do Crystal Violet and Malachite Green.

Chromium functioning as a mordant

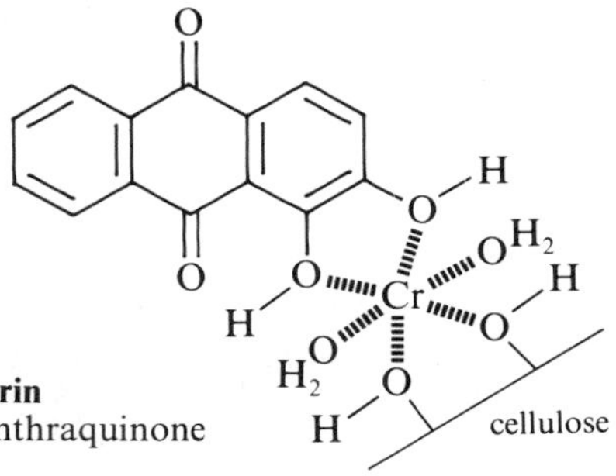

Alizarin
2,3-Dihydroxyanthraquinone

cellulose

Applying Mordants—Tannic acid, Fe, Sn, Cr, Cu, Al

Mordant pieces of test fabric by allowing them to stand in a hot (nearly boiling) solution of 0.1 g of tannic acid in 50 mL of water for 30 min. The tannic acid mordant must now be fixed to the cloth; otherwise it would wash out. For this purpose, transfer the cloth to a hot bath made from 20 mg of potassium antimonyl tartrate (tartar emetic) in 20 mL of water. After 5 min, wring the cloth and dry it as much as possible over a warm hot plate.

Mordant 1/2-in. strips of test cloth in the following mordants, which are 0.1 M solutions of the indicated salts. Immerse pieces of cloth in the solutions, which are kept near the boiling point, for about 15 to 20 min or longer. The mordants are ferrous sulfate, stannous chloride, potassium dichromate, copper sulfate, and potassium aluminum sulfate (alum). The alum and dichromate solutions should also contain 0.05 M oxalic acid. These mordanted pieces of cloth will be dyed with alizarin (1,2-dihydroxyanthraquinone) and either Methyl Orange or Orange II.

Synthesis of Crystal Violet, a Triphenylmethane Dye

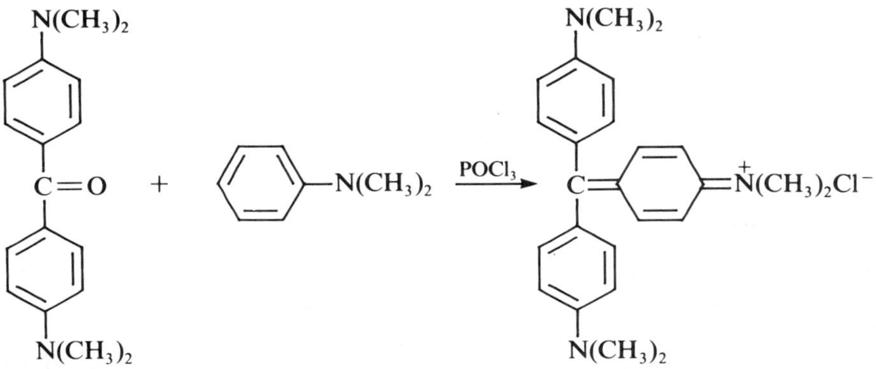

$POCl_3$

Michler's ketone

4,4′-Bis(dimethylamino) benzophenone

N,N-Dimethylaniline

Crystal Violet
(λ_{max} 591,540 nm)

Place 50 mg of 4,4′-bis(dimethylamino)benzophenone (Michler's ketone), 3 drops of freshly distilled dimethylaniline, and 1 drop of phosphorus oxychloride in a reaction tube, and heat the tube in boiling water for 0.5 h. Add 4 mL of water and stir. Add several drops of this solution to 20 mL of water and treat with a few drops of ammonium hydroxide solution. Let stand until the color has disappeared and then add dilute hydrochloric acid. Account for the color changes noted.

If the original solution is allowed to stand overnight, crystals of Crystal Violet should separate.

Dyeing with a Triphenylmethane Dye—Crystal Violet or Malachite Green

A dye bath is prepared by dissolving 10 mg of either Crystal Violet or Malachite Green in 20 mL of boiling water. Dye the mordanted cloth in this bath for 5–10 min at a temperature just below the boiling point. Dye another piece of cloth that has not been mordanted and compare the two. In each case allow as much of the dye to drain back into the beaker as possible and then, using glass rods, wash the dyed cloth under running water, blot and dry.

Note: The stains on glass produced by triphenylmethane dyes can be removed with concentrated hydrochloric acid and washing with water, as HCl forms a di- or trihydrochloride more soluble in water than the original monosalt.

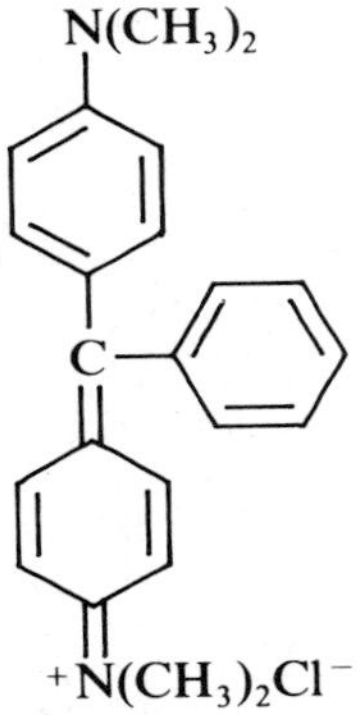

$N(CH_3)_2$

C

$^+N(CH_3)_2Cl^-$

Malachite Green
(λ_{max} 617 nm)

4. Developed Dyes

A superior method of applying azo dyes to cotton, patented in England in 1880, is that in which cotton is soaked in an alkaline solution of a phenol and then in an ice-cold solution of a diazonium salt; the azo dye is developed directly on the fiber. The reverse process (ingrain dyeing) of impregnating cotton with an amine, which is then diazotized and developed by immersion in a solution of the phenol, was introduced in 1887. The first ingrain dye was Primuline Red, obtained by coupling the sulfur dye Primuline, after application to the cloth and diazotization, with 2-naphthol. Primuline (substantive to cotton) is a complex thiazole, prepared by heating *p*-toluidine with sulfur and then introducing a solubilizing sulfonic acid group.

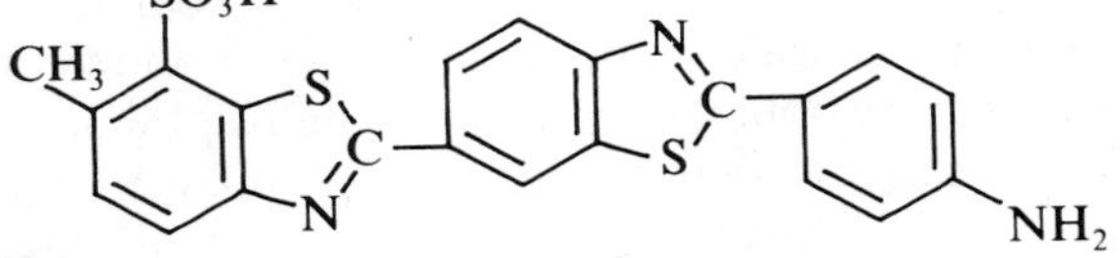

SO_3H CH_3 S C N N C S NH_2

Primuline

Primuline Red

Dye three pieces of cotton cloth in a solution of 20 mg of Primuline and 0.5 mL of sodium carbonate solution in 50 mL of water, at a temperature just below the boiling point for 15 min. Wash the cloth twice in about 50 mL of water. Prepare a diazotizing bath by dissolving 20 mg of sodium nitrite in 50 mL of water containing a little ice and, just before using the bath, add 0.5 mL of concentrated hydrochloric acid. Allow the cloth dyed with Primuline to stay in this diazotizing bath for about 5 min. Now prepare three baths for the coupling reaction. Dissolve 10 mg of 2-naphthol in 0.2 mL of 5% sodium hydroxide solution and dilute with 10 mL of water; prepare similar baths from phenol, resorcinol, Naphthol AS, or other phenolic substances.

Transfer the cloth from the diazotizing bath to a beaker containing about 50 mL of water and stir. Put one piece of cloth in each of the developing baths and allow them to stay for 5 min.

Para Red, an Ingrain Color

2-Naphthol + **4-Nitrobenzene diazonium chloride** → **Para Red**

A solution is prepared by suspending 300 mg of 2-naphthol in 10 mL of water, stirring well and adding 10% sodium hydroxide solution, a drop at a time, until the naphthol just dissolves. Do not add excess alkali. The material to be dyed is first soaked in or painted with this solution and then dried, preferably in an oven.

Avoid contact with all the amines used here. 2-Naphthol is a carcinogen. Handle it with gloves in the hood.

Prepare a solution of 4-nitrobenzenediazonium chloride as follows: dissolve 140 mg of 4-nitroaniline in a mixture of 3 mL of water and 0.6 mL of 10% hydrochloric acid by heating. Cool the solution in ice (the hydrochloride of the amine may crystallize), add all at once a solution of 80 mg of sodium nitrite in about 0.5 mL of water, and stir. In about 10 min a clear solution of the diazonium salt will be obtained. Just before developing the dye on the cloth add a solution of 80 mg of sodium acetate in 0.5 mL of cold water. Stir in the acetate well, add 30 mL of water, and immediately add the cloth. The diazonium chloride solution may also be painted onto the cloth.

Good results can be obtained by substituting Naphthol-AS for 2-naphthol; in this case it is necessary to warm the Naphthol-AS with alkali and to break the lumps with a flattened stirring rod in order to bring the naphthol into solution.

5. Vat Dyes

Vat dyeing depends upon the reduction of some dyes (e.g., indigo) to a colorless, or leuco, derivative, which is soluble in dilute alkali. If fabric is immersed in this akaline solution, the leuco compound is adsorbed by hydrogen bonding. On exposure to air the leuco compound is oxidized to the dye, which remains fixed to the cloth. Vat dyes are all quinonoid substances that are readily reduced to hydroquinonoid compounds reoxidizable by oxygen in the air.

The indigo so formed is very insoluble in all solvents. However, it is not covalently bound to the cotton, only adhering to the surface of the fiber. Hence, it is subject to removal by abrasion. This explains why the knees and other parts of blue jeans (dyed exclusively with indigo) subject to wear will gradually turn white.

Dyeing with a Vat Dye

Use 100 mg of one of the following dyes: Indigo, Indanthrene Brilliant Violet, or Indanthrene Yellow. Boil the dye with 50 mL of water, 2.5 mL of 10% sodium hydroxide solution, and about 0.5 g of sodium hydrosulfite until the dye is reduced. Introduce a piece of cloth and boil the solution gently for 10 min. Rinse the cloth well in water and allow it to dry. To increase the intensity of the dye, repeat the process several times with no drying.

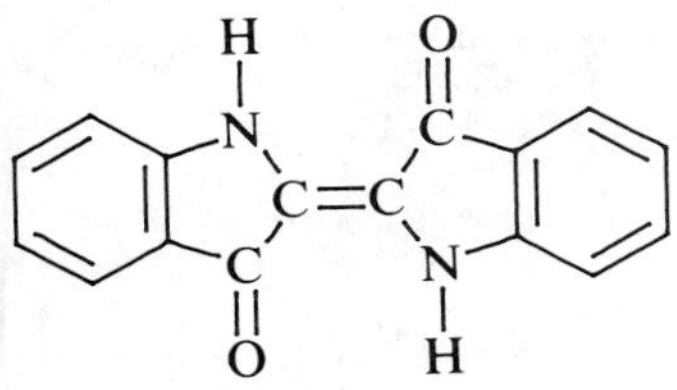

Indigo

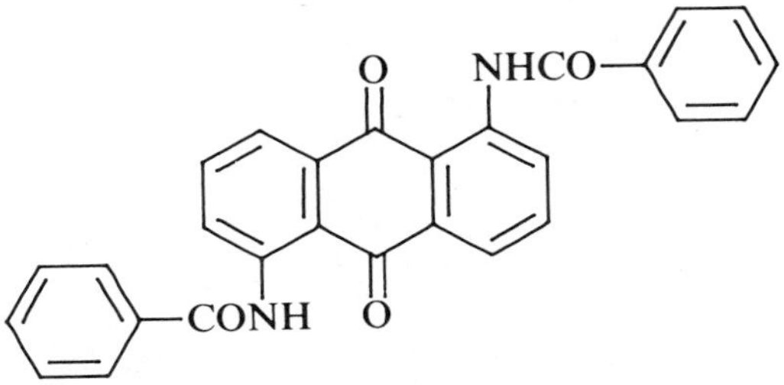

Indanthrene Yellow

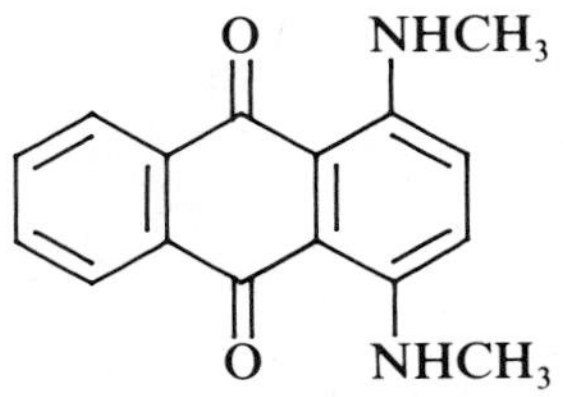

6. Disperse Dyes

Fibers such as Dacron, acetate rayon, Nylon, and polypropylene are difficult to dye with conventional dyes because they contain so few polar groups. These fibers are dyed with substances that are insoluble in water but which at elevated temperatures (pressure vessels) are soluble in the fiber as true solutions. They are applied to the fiber in the form of a dispersion of finely divided dye (hence the name). The Cellitons are typical disperse dyes.

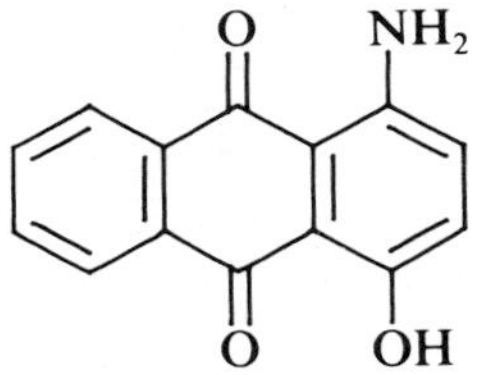

In this experiment Disperse Red, a brillant red dye used commercially, is synthesized.

Celliton Fast Blue B

Celliton Fast Pink B

Diazotization of 2-Amino-6-methoxybenzothiazole

$$\text{2-Amino-6-methoxybenzothiazole} \xrightarrow[0^\circ C]{NaNO_2,\ HCl} \text{6-methoxybenzothiazole-2-diazonium } (-\overset{+}{N}{\equiv}N)\ Cl^-$$

2-Amino-6-methoxybenzothiazole
MW 180.23
mp 165–167°C

To 135 mg (0.75 mmole) of 2-amino-6-methoxybenzothiazole in 1.5 mL of water in a reaction tube add 0.175 mL of concentrated hydrochloric acid, then cool the solution to 0–5°C. To this mixture add, dropwise, an *ice-cold* solution of 55 mg of sodium nitrite that has been dissolved in 0.75 mL of water. The reaction mixture changes color and some of the diazonium salt crystallizes out, but it should not foam. Foaming, caused by the evolution of nitrogen, is an indication the mixture is too warm. Keep the mixture ice-cold until used in the coupling reaction.

Disperse Red

$$CH_3O\text{-benzothiazole-}C{-}\overset{+}{N}{\equiv}N\ Cl^- + C_6H_5{-}N(CH_2CH_2OH)_2 \longrightarrow CH_3O\text{-benzothiazole-}C{-}N{=}N{-}C_6H_4{-}N(CH_2CH_2OH)_2$$

N-Phenyldiethanolamine
MW 181.24
mp 56–58°C

To 135 mg (0.75 mmole) of N-phenyldiethanolamine in 0.75 mL of hot water add just enough 10% hydrochloric acid to bring the amine into solution. This amount is less than 0.5 mL. Cool the resulting solution to 0°C in ice and add to it, dropwise and with *very thorough mixing*, the diazonium chloride solution. Mix the solution well by drawing into the Pasteur pipette and expelling it into the cold reaction tube. Allow the mixture to come to room temperature over a period of 10 min, then add 225 mg of sodium chloride, and heat the mixture to boiling. The sodium chloride decreases the solubility of the product in water. Allowing the hot solution to cool slowly to room temperature should afford easily filterable crystals. Collect the dye on the Hirsch funnel, wash it with a few drops of saturated sodium chloride solution, and press it dry on the funnel.

Often the reaction gives a noncrystalline product that looks like purple tar. This is the dye and it can be used to dye the multifiber test cloth, so don't discard the reaction mixture.

7. Fiber Reactive Dyes

Among the newest of the dyes are the fiber reactive compounds which form a covalent link to the hydroxyl groups of cellulose. The reaction involves an amazing and little understood nucleophilic displacement of a chloride ion from the triazine part of the molecule by the hydroxyl groups of cellulose; yet the reaction occurs in aqueous solution.

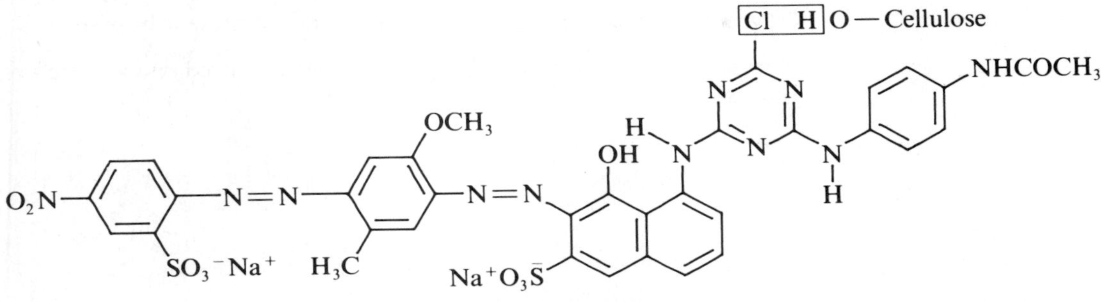

Chlorantin Light Blue 8G

8. Optical Brighteners—Fluorescent White Dyes

Most modern detergents contain a blue-white fluorescent dye that is adsorbed on the cloth during the washing process. These dyes fluoresce, that is, absorb ultraviolet light and reemit light in the visible blue region of the spectrum. This blue color counteracts the pale yellow color of white goods, which develops because of a buildup of lipid soil. The modern-day use of optical brighteners has replaced a past custom of using bluing (ferriferrocyanide).

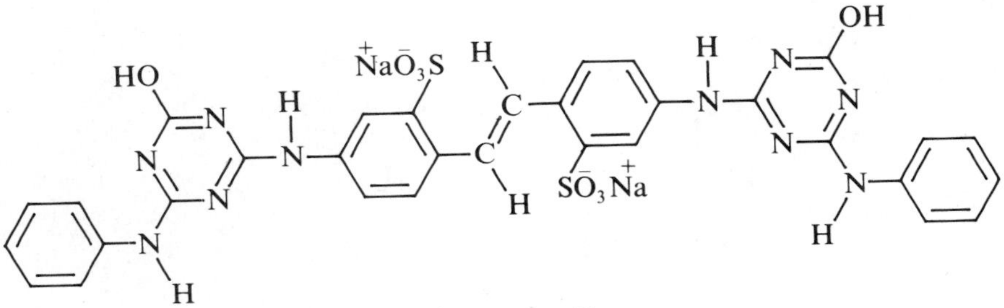

Blankophor B
an optical brightener

Dyeing with Detergents

Immerse a piece of test fabric in a hot solution (0.5 g of detergent, 200 mL of water) of a commercial laundry detergent which you suspect may contain an optical brightener (e.g., Tide and New Blue Cheer) for 15 min. Rinse the fabric thoroughly, dry, and compare with an untreated fabric sample under an ultraviolet lamp.

Questions

1. Write reactions showing how nylon can be synthesized such that it will react with (a) basic dyes and (b) acidic dyes.
2. Draw the resonance form of dimethylaniline that is most prone to react with diazotized sulfanilic acid.
3. Draw a resonance form of indigo that would be present in base.
4. Draw a resonance form of indigo that has been reduced and is therefore colorless.

Polymers 67

PREALAB EXERCISE: *In the preparation of nylon by interfacial polymerization sebacoyl chloride is synthesized from decanedioic acid and thionyl chloride. What volume of hydrogen chloride is produced in this reaction? What volume of sulfur dioxide is produced?*

Polymers are ubiquitous. Natural polymers such as proteins (polyamino acids), DNA (polynucleotides), and cellulose (polyglucose) are the basic building blocks of plant and animal life. Synthetic organic polymers, or plastics, are now among our most common structural materials. In the United States we make and use more synthetic polymers than we do steel, aluminum, and copper combined—in 1984, 46 billion pounds worth $18 billion dollars.

We use more polymers than steel, aluminum, and copper combined

Polymers, from the Greek meaning "many parts," are high-molecular-weight molecules made up of repeating units of smaller molecules. Most polymers consist of long chains held together by hydrogen bonds, van der Waals forces, and the tangling of the long chains. When heated, the covalent bonds of some polymers, which are thermoplastic, do not break but the chains slide over one another to adopt new shapes. These shapes can be films, sheets, extrusions, or molded parts in a myriad of forms.

Nitrocellulose

The first man-made plastic was nitrocellulose, made in 1862 by nitrating the natural polymer, cellulose. Nitrocellulose, when mixed with a plasticizer such as camphor to make it more workable, was originally used as a replacement for ivory in billiard balls and piano keys and to make Celluloid collars. This material from which the first movie film was made is notoriously flammable.

Cellulose acetate

Cellulose acetate, made by treating cellulose with acetic acid and acetic anhydride, was originally used as a waterproof varnish to coat the fabric of airplanes during World War I. It later became important as a photographic film base and as acetate rayon.

Bakelite

The first completely synthetic organic polymer was Bakelite, named for its discoverer Leo Baekeland. He was a Belgian chemistry professor who invented the first successful photographic paper, Velox. He came to America at the age of 35 and sold his invention to George Eastman for $1,000,000 in 1899. He then turned his attention to finding a replacement for shellac, which comes from the Asian lac beetle. At the time shellac was coming into great demand in the fledgling electrical industry as an insulator. The polymer Baekeland produced is still used for power plugs, distributor caps in automobiles, switches, and the black handles and knobs on pots and pans. It has superior electrical insulating properties and very high heat resistance. It is made by the base-catalyzed reaction of excess formaldehyde with phenol. In a low-molecular-weight form it is used to glue together the plies of plywood or mixed with a filler such as sawdust. When it is heated to a high temperature, crosslinking occurs as the polymer "cures."

Thermoplastic polymers

Most polymers are amorphous, linear macromolecules that are thermoplastic and soften at high temperature. In Bakelite the polymer crosslinks to form a three-dimensional network and the polymer becomes a dark, insoluble, infusible substance. Such polymers are said to be thermosetting. Natural rubber is thermoplastic. It becomes a thermosetting polymer when heated with sulfur, as Charles Goodyear discovered. With 2% sulfur the rubber becomes crosslinked but is still elastic; at 30% sulfur it can be made into bowling balls. Some other important thermosetting polymers are urea-formaldehyde resins and melamine-formaldehyde resins. The latter are among the hardest of polymers and take on a high-gloss finish. Melamine is used extensively to manufacture plastic dinnerware.

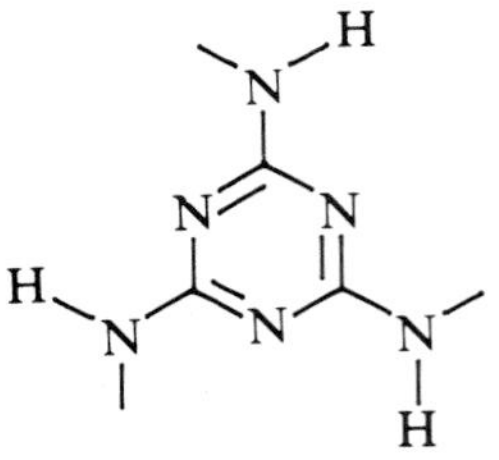

Melamine, a thermosetting polymer

Even though vinyl chloride was discovered in 1835, polyvinyl chloride was not produced until 1912. It is now one of our commonest polymers; production in 1984 was over 6 billion pounds. The monomer is made by the pyrolysis of dichloroethylene, formed by chlorination of ethylene. Free radical polymerization follows Markovnikov's rule to give the head-to-tail polymer with high specificity:

Vinyl chloride

$$n\mathrm{CH_2{=}CHCl} \longrightarrow -(\mathrm{CH_2CHCl})_n-$$

Vinyl chloride **Polyvinyl chloride**

Pure polyvinyl chloride (PVC) is an extremely hard polymer. It along with some other polymers can be modified by the addition of plasticizers; the greater the ratio of plasticizer to polymer, the greater the flexibility of the polymer. PVC pipe is rigid and contains little plasticizer, while shower curtains contain a large percentage. The most common plasticizer is di-(2-ethylhexyl)phthalate, which can be added in concentrations of up to 50%.

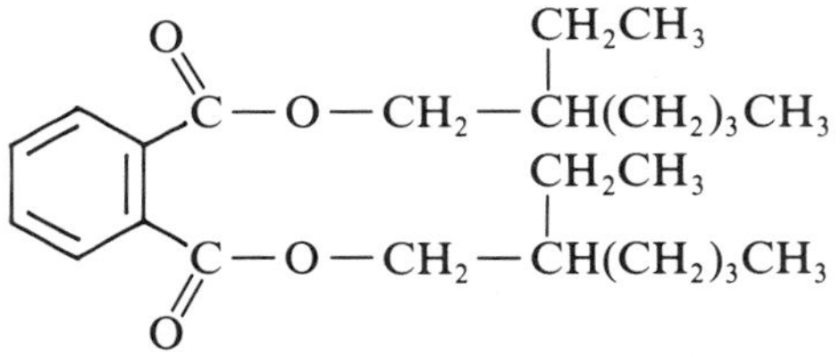

Plasticizers

Di(2-ethylhexyl)phthalate (Dioctylphthalate)

PVC is used for raincoats, house siding, and artificial leather for handbags, briefcases, and inexpensive shoes. It is found in garden hose, floor covering, swimming pool liners, and automobile upholstery. When vinyl upholstery is exposed to high temperatures, as in the interior of an automobile, the plasticizer distills out. The result is an opaque, difficult-to-remove film on the insides of the windows and upholstery that is hard and brittle.

Polymerization methods

Monomers can be polymerized in the gas phase, in bulk, as suspensions, and as emulsions. The most common method of making PVC is by emulsifying the monomer, vinyl chloride, in water with surfactants (soaps), water-soluble catalysts, and heat. The monomer is polymerized to solid particles, which are suspended in the aqueous phase. This product can be centrifuged and dried or used as such. Chemists can control the average molecular weight, which can become very high in emulsion polymerization. A high molecular weight means a more rigid and stronger polymer, but also one that is more difficult to work with. An emulsion of polyvinyl acetate is sold as latex paint. When the vehicle, water, evaporates, the polymer is left as a hard film. A thicker emulsion of polyvinyl acetate is an excellent adhesive, the familiar white glue. When vinyl acetate and vinyl chloride are polymerized together, a copolymer results that has properties all of its own. This copolymer is particularly good for detailed moldings and is used to make phonograph records.

Polyvinylidene chloride is primarily extruded as a film that has low permeability to water vapor and air and is therefore used as the familiar clinging plastic food wrap, Saran Wrap.

Teflon

Polytetrafluoroethylene, Teflon, another of the halogenated polymers, has a number of unique properties. It has a very high melting point, 327°C, it does not dissolve in any solvent, and nothing sticks to it. It is also an extremely good electrical insulator. A product of the DuPont company, it is one of the densest of the polymers and also one of the most expensive. The surface of the polymer must be etched with metallic sodium to form free radicals, to which glue can adhere. The polymer has an extremely low coefficient of friction, which makes it useful for bearings. Its chemical inertness makes it an ideal liner for chemical reagent bottle caps, and its no-stick property is ideal for the coating on the inside of frying pans. At 380°C Teflon is still so viscous that it cannot be injection-molded. Instead it is molded by pressing the powdered polymer at high temperature and pressure, a process called sintering.

$-(CH_2CHCl)_n-$	$-(CH_2CCl_2)_n-$	$-(CF_2CF_2)_n-$	$-(CH_2CH_2)_n-$
Polyvinyl chloride	**Polyvinylidene chloride**	**Polytetrafluoroethylene**	**Polyethylene**

The polymer produced in highest volume is polyethylene. Invented by the British, who call it polythene, and put into production in 1939, it could for a long time only be produced by the oxygen-catalyzed polymerization of ethylene at pressures near 40,000 lb/in.^{2}. Such pressures are expensive and dangerous to maintain on an industrial scale. The polyethylene produced has a low density and is used primarily to make film for bags of all types—from sandwich bags to trash can liners. The opaque appearance of polyethylene is due to crystallites, regions of order in the polymer that resemble crystals. In the 1950s Karl Ziegler and Guilio Natta developed catalysts composed of $TiCl_4$, alkyl aluminum, and transition metal halides with which ethylene can be polymerized at pressures of just 450 lb/in.^{2}. The resulting product has a higher density and a 20° C higher softening temperature than the low-density material. The catalysts which Ziegler and Natta developed and for which they received the 1963 Nobel prize, cause stereoregular polymerization and thus a crystalline product. High-density polyethylene is as rigid as polystyrene and yet has high impact resistance. It is used to mold very large articles such as luggage, the cases for domestic appliances, trash cans, and soft drink crates.

High-density polyethylene

Polystyrene

Polystyrene is a brilliantly clear, high-refractive-index polymer familiar in the form of disposable drinking glasses. It is brittle and produces sharp, jagged edges when fractured. It softens in boiling water and it burns readily with a very smoky flame. But it foams readily and makes a very good insulator; witness the disposable white hot drink cup. It is used extensively for insulation when properly protected from ignition. The addition of a small quantity of butadiene to the styrene makes a polymer that is no longer transparent but that has high impact resistance. Blends of acrylonitrile, butadiene, and styrene (ABS) have excellent molding properties and are used to make car bodies. One formulation can be chrome-plated for automobile grills and bumpers.

$CH_2{=}CHCN$ — **Acrylonitrile**

$C_6H_5CH{=}CH_2$ — **Styrene**

$CH_2{=}CH{-}CH{=}CH_2$ — **1,3-Butadiene**

Rubber

Joseph Priestley, the discoverer of oxygen, named rubber for its ability to remove lead pencil marks. Rubber is an elastomer, defined as a substance that can be stretched to at least twice its length and return to its original size. The Germans, cut off from a supply of natural rubber, began manufacturing synthetic rubber during World War I. Called "buna" for *bu*tadiene and sodium, *Na*, the polymerization catalyst, it was not the ideal substitute. Cars with buna tires had to be jacked up when not in use because their tires would develop flat spots. The addition of about 25% styrene greatly improved the qualities of the product; styrene-butadiene synthetic rubber now dominates the market, a principal outlet being automobile tires. Addition of 30% acrylonitrile to butadiene produces nitrile rubber, which is used to make conveyor belts, tank liners, rubber hose, and gaskets.

Synthetic rubber, a styrene-butadiene copolymer

The chemist classifies polymers in several ways. There are thermosetting plastics such as Bakelite and melamine and the much larger category of thermoplastic materials, which can be molded, blown, and formed after polymerization. There are the arbitrary distinctions made among plastics, elastomers, and fibers. And there are the two broad categories formed by the polymerization reaction itself: (1) addition polymers (e.g., vinyl polymerizations), in which a double bond of a monomer is transformed into a single bond between monomers, (2) condensation polymers (e.g., Bakelite), in which a small molecule, such as water or alcohol, is split out as the polymerization reaction occurs.

Addition polymers
Condensation polymers

Nylon

One of the most important condensation polymers is nylon, a name so ingrained into our language it has lost trademark status. It was developed by Wallace Carothers, director of organic chemicals research at DuPont, and was the outgrowth of his fundamental research into polymer chemistry. Introduced in 1938, it was the first totally synthetic fiber. The most common form of nylon is the polyamide formed by the condensation of hexamethylene diamine and adipic acid:

$$nH_2N(CH_2)_6NH_2 + nHOOC(CH_2)_4COOH$$

Hexamethylene diamine **Adipic acid**

$$\downarrow$$

$$nH_3\overset{+}{N}(CH_2)NH_3^+ \quad ^-O\overset{O}{\overset{\|}{C}}(CH_2)_4\overset{O}{\overset{\|}{C}}O^-$$

$$\downarrow \text{Pressure, 280 C}$$

$$H{+}NH(CH_2)_6NH\overset{O}{\overset{\|}{C}}(CH_2)_4\overset{O}{\overset{\|}{C}}{+}_nOH + (2n - 1)H_2O$$

Nylon 6.6

The reactants are mixed together to form a salt that melts at 180°C. This is converted into the polyamide by heating to 280°C under pressure, which eliminates water. Nylon 6.6 is used to make textiles; while nylon 6.10, from the 10-carbon diacid, is used for bristles and high-impact sports equipment. Nylon can also be made by interfacial and by ring-opening polymerization, both of which are used in the following experiments.

The condensation polymer made by reacting ethylene glycol with 1,4-benzene dicarboxylic acid (terephthalic acid) produces a polymer that is almost exclusively converted into the fiber Dacron. The polymerization is run as an ester interchange reaction using the methyl ester of terephthalic acid:

$$H_3CO\overset{O}{\overset{\|}{C}}-C_6H_4-\overset{O}{\overset{\|}{C}}OCH_3 + HOCH_2CH_2OH$$

$$\downarrow$$

$$H_3CO(\overset{O}{\overset{\|}{C}}-C_6H_4-\overset{O}{\overset{\|}{C}}OCH_2CH_2O)_nH + 2n\ CH_3OH$$

Polyethylene glycol terephthalate (Dacron)

The structure of a polymer is not simple. For example, polymerization of styrene produces a chiral carbon at each benzyl position. We can ask whether the phenyl rings are all on the same side of the long carbon chain, whether they alternate positions, or whether they adopt some random configuration. In the case of copolymers we can ask whether the two components alternate: ABABABABAB . . . or whether they adopt a random configuration: AABBBABAAB . . . or whether they polymerize as short chains of one and then the other: AAAABBBBBAAABBBB . . . . These questions are important because the physical properties of the resulting polymer depend on them. The polymer chemist is concerned with finding the answers and discovering catalysts and reaction conditions that can control these parameters.

In the experiments that follow, the nitration of cellulose, while easy to carry out, gives variable results. Nylon by interfacial polymerization is a spectacular and reliable experiment easily carried out in one afternoon. The synthesis of Bakelite works well; it requires overnight heating in an oven to complete the polymerization. Nylon by ring-opening polymerization requires skill and care because of the high temperatures involved. The polymerization of styrene also requires care, but is somewhat easier to carry out.

Experiments

1. The Nitration of Cellulose

Cellulose is composed of several thousand anhydroglucose units connected at the 1 and 4 positions and having the beta configuration at C-1:

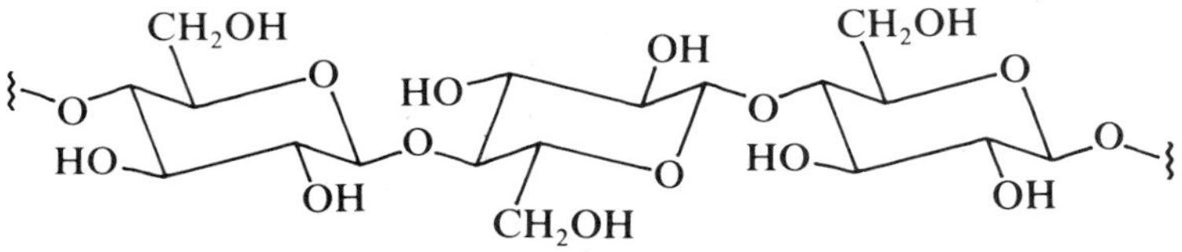

Cellulose

With one primary and two secondary hydroxyls in each glucose unit, cellulose is held together by many hydrogen bonds and hence is crystalline. In cotton the cellulose chains are parallel and closely packed but the nitronium ion can penetrate the crystal and form a nitro ester with great rapidity:

$$HNO_3 + H_2SO_4 \longrightarrow NO_2^+ + HSO_4 + H_2O$$

$$ROH + NO_2^+ \longrightarrow RONO_2$$

The number of nitro groups introduced is controlled by the time and temperature of the reaction and the concentration of reagents. If two hydroxyl groups in each glucose molecule are, on the average, nitrated, the final product will contain 12.5% nitrogen. Cellulose with a nitrogen content of about 11% is used to make plastics, of about 11.5% to make lacquers, and of more than 12% to make smokeless gunpowder. Cellulose in which all three hydroxyls on each glucose are nitrated contains 14.1% nitrogen.

Procedure

Cellulose nitrate

In a 10-mL beaker mix 2 mL of concentrated nitric acid and 0.3 mL of water and then add 2 mL of concentrated sulfuric acid. Cool the mixture to 20°C and then introduce 50 mg of cotton that has been "fluffed up" to give maximum exposure to the acid. Leave the cotton in the acid for 10 min with occasional stirring and then, with a glass rod, transfer it to a beaker of water. The reaction temperature is critical; be sure to maintain it at 20°C. Thoroughly rinse the cotton under running water and then once in a few milliliters of ethanol. Dissolve the cellulose nitrate in a mixture of 1.2 mL of diethyl ether and 0.6 mL of 95% ethanol. Try to avoid stirring in air bubbles. Dissolve 15 mg of camphor in the mixture and place the thick gel on a watchglass to dry until the next laboratory period. Check the solubility of unnitrated cotton in a mixture of ether and ethanol.

At the end of the experiment pour the mixture of acids into the container provided. In the report on the experiment describe the properties of the dry cellulose nitrate. Attach a small sample of the product to your report.

2. Nylon by Interfacial Polymerization[1]

$$\underset{\textbf{Thionyl chloride}}{SOCl_2} \quad \underset{\textbf{Sebacic acid}}{HOOC(CH_2)_8COOH} \longrightarrow \underset{\textbf{Sebacoyl chloride}}{ClC(=O)(CH_2)_8C(=O)Cl} + HCl + SO_2$$

$$ClC(=O)(CH_2)_8C(=O)Cl + \underset{\textbf{Hexane-1,6-diamine}}{H_2N(CH_2)_6NH_2} \longrightarrow \underset{\textbf{Nylon 6.10}}{\left[C(=O)(CH_2)_8C(=O)NH(CH_2)_6NH\right]_n} + HCl$$

In this experiment a diamine dissolved in water is carefully floated on top of a solution of a diacid chloride dissolved in an organic solvent. Where the two solutions come in contact (the interface), an S_N2 reaction occurs to form a film of a polyamide. The reaction stops there unless the polyamide is removed. In the case of nylon 6.10, the product of this reaction, the film is so strong that it can be picked up with a wire hook and continuously removed in the form of a rope.

This reaction works because the diamine is soluble in both water and carbon tetrachloride, the organic solvent used. As the diamine diffuses into the organic layer, reaction occurs immediately to give the insoluble polymer. The HCl produced reacts with the sodium hydroxide in the aqueous layer. The acid chloride does not hydrolyze before reacting with the amine because it is not very soluble in water. The acid chloride is conveniently prepared using thionyl chloride.

1. P. W. Morgan and S. L. Kwolek, *J. Chem. Ed.*, **36**, 182 (1959).

Procedure[2]

In a reaction tube fitted with a gas trap (Fig. 68.1) place 0.25 g of sebacic acid (1,8-octane dicarboxylic acid, decanedioic acid), 0.25 mL of thionyl chloride and 12 mg of N,N-dimethylformamide. Heat the tube to 60–70°C in a water bath in the hood. This is best accomplished by putting very hot water in a beaker and placing the beaker on a steam bath to keep it at 60–70°C. As the reaction proceeds the product forms a liquid layer on the bottom of the tube. Use this liquid to wash down unreacted acid (if necessary) as the reaction proceeds. When the acid has all reacted and gas evolution has ceased (about 10–15 min) transfer the product, which should be a clear liquid at this point, to a 30-mL beaker that has a thin coat of silicone oil or grease on the inside using 12 mL of dichloromethane. Carefully pour on to the top of this dichloromethane solution 0.25 g of hexane-1,6-diamine (hexamethylene-diamine) that has been dissolved in 6 mL of water containing 125 mg of sodium hydroxide. Pick up the polymer film at the center with a copper wire and lead it over glass rods to allow it to drop into a 5% hydrochloric acid solution. The polymer can also be wrapped on the outer surface of a bottle, beaker, or graduated cylinder as it is removed. Remove as much of the polymer as possible, wash it thoroughly in water, and press it as dry as possible. After the polymer has dried determine its weight and calculate the yield. Try dissolving the polymer in two or three solvents. Mix the contents of the reaction beaker thoroughly before disposing of its contents. Attach a piece of the polymer to your laboratory report.

3. The Condensation Polymerization of Phenol and Formaldehyde: Bakelite

Condensation of phenol with formaldehyde is a base-catalyzed process in which one resonance form of the phenoxide ion attacks formaldehyde. The resulting trimethylol phenol is then crosslinked by heat, presumably by dehydration with the intermediate formation of benzylcarbocations. The resulting polymer is Bakelite. Since the cost of phenol is relatively high and the polymer is somewhat brittle, it is common practice to add an extender such as sawdust to the material before crosslinking. The mixture is placed in molds and heated to form the polymer. The resulting polymer, like other thermosetting polymers, is not soluble in any solvent and does not soften when heated.

:Ö—H + $NH_4^+OH^-$ → [:Ö:⁻ ↔ :Ö] ; H–C(H)=Ö: → :Ö, H, H–C(H)–Ö:⁻ →

2. G. C. East and S. Hassell, *J. Chem. Ed.*, **60**, 69 (1933).

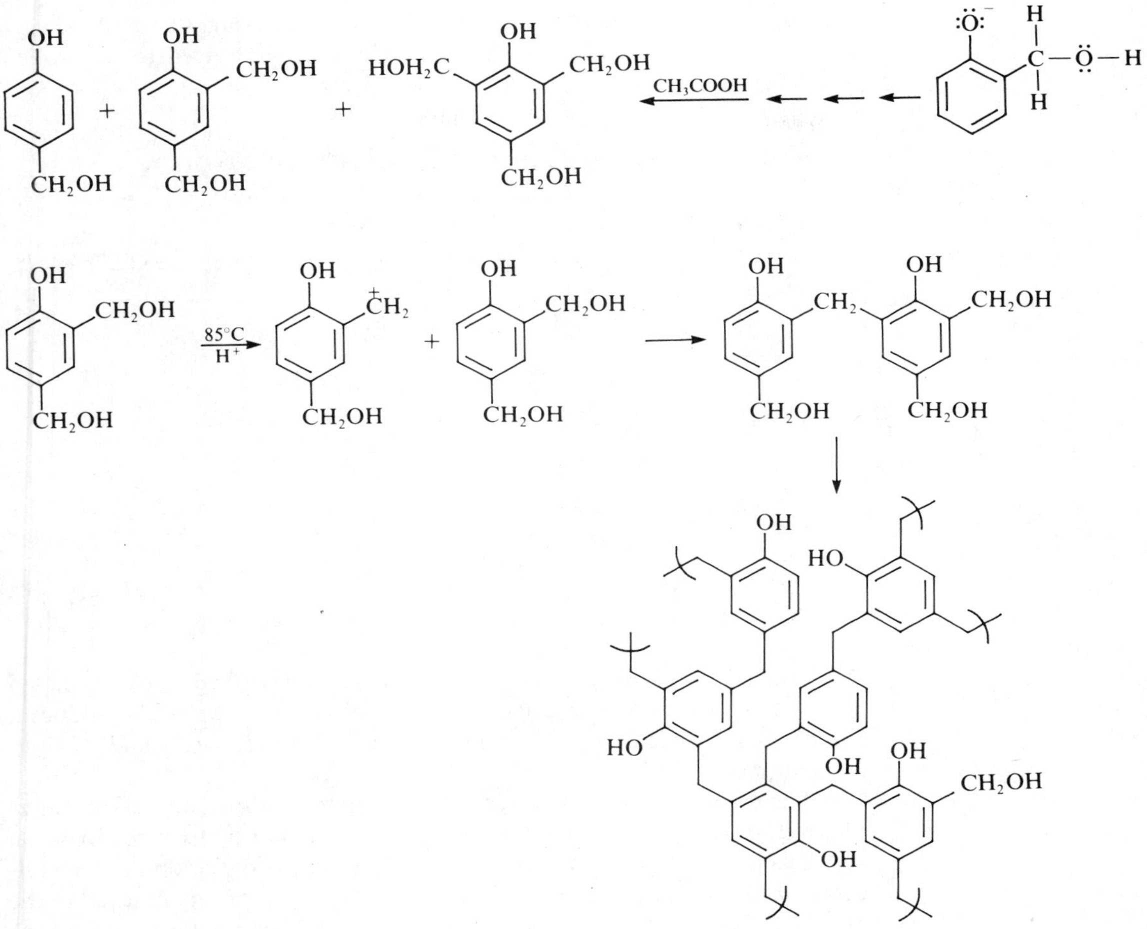

Bakelite

Procedure

Bakelite

In a 5-mL short-necked round-bottomed flask place 0.75 g of phenol and 2.5 mL of 37% by weight aqueous formaldehyde solution. The formaldehyde solution contains 10–15% methanol, which has been added as a stabilizer to prevent the formaldehyde from polymerizing. Add 0.37 mL of concentrated ammonium hydroxide to the solution, attach an empty distilling column as an air condenser, and reflux it for 5 min beyond the point at which the solution turns cloudy, a total reflux time of about 10 min. In the hood pour the warm solution into a small, disposable tube and draw off the upper layer. Immediately clean the flask with a small amount of acetone. Warm the viscous milky lower layer on the steam bath and add acetic acid dropwise with thorough mixing until the layer is clear, even when the polymer is cooled to room temperature. Heat the tube on a water bath at 60–65°C for 30 min and then, after placing a wood stick in the polymer to use as a handle, leave the tube, with your name attached, in an 85°C oven overnight or until the next laboratory period. To free the polymer the tube may need to be broken (*Caution*). Attach a piece of the polymer to your lab report.

4. Nylon by Ring-Opening Polymerization[3]

$$N^{-}\ (\mathbf{1}) + N{-}COCH_3\ (\mathbf{2}) \longrightarrow N{-}CO{-}(CH_2)_5{-}\bar{N}{-}COCH_3$$

$$N{-}CO{-}(CH_2)_5{-}\bar{N}{-}COCH_3 + NH \longrightarrow$$

$$N{-}CO{-}(CH_2)_5{-}NH{-}COCH_3\ (\mathbf{3}) + N^{-}\ (\mathbf{4})$$

While interfacial polymerization in the manner described above is not a commercial process, the ring-opening of caprolactam is. The nylon produced, nylon 6, is used extensively in automobile tire cord and for gears and bearings in small mechanical devices.

The catalyst used in this reaction is sodium hydride; therefore this is referred to as an anionic polymerization. The sodium hydride removes the acidic lactam proton to form an anion, **1**, that attacks the coinitiator, acetylcaprolactam, **2**, which has an electron-attracting acetyl group attached to the nitrogen. The ring of the acetylcaprolactam is attacked by the anion and the acetylcaprolactam ring opens, forming a substituted caprolactam, **3**, that still has an electron-attracting group attached to nitrogen. A proton transfer reaction occurs, generating a new caprolactam anion, **4**, and the reaction is repeated.

Ordinarily anionic polymerizations must be run in the absence of oxygen, but the addition of polyethylene glycol serves to complex with the sodium ion just as 18-crown-6 does and enhances the catalytic activity of the sodium hydride.

Procedure

Set the heater to its maximum setting on your sand bath. The sand must be quite hot before beginning this experiment. Heating can also be done over a very small Bunsen burner flame.

3. L. J. Mathias, R. A. Vaidya, and J. B. Canterbury, *J. Chem. Ed.*, **61**, 805 (1984).

Into a disposable 10 × 75 mm test tube place 1 g of caprolactam, 60 mg of polyethylene glycol, and 1 small drop of N-acetylcaprolactam. Heat the mixture and as soon as it has melted remove it from the heat and add 20 mg of gray (not white) sodium hydride (50% dispersion in mineral oil). Mix the catalyst with the reactants by stirring with a Pasteur pipette and heat the mixture *rapidly* to boiling (200–230°C). This should take place over a 2-min period. Polymerization takes place rapidly as indicated by an increase in viscosity. If polymerization has not occurred within 3 min remove the tube from the heat, cool it somewhat, and add another 15 mg of sodium hydride. When the solution is so viscous that it will barely flow, insert a wood stick and with help from a neighbor draw fibers from the melt. After it cools the nylon-6 can usually be removed from the tube as one cylindrical piece. Try dissolving a piece of the nylon or the fibers in various solvents. Test the physical properties of the fibers by stretching them to the breaking point. Describe your observations and attach a piece of fiber to your report.

5. Polystyrene by Free-Radical Polymerization

Polystyrene, the familiar crystal-clear brittle plastic used to make disposable drinking glasses and when foamed the lightweight, white cups for hot drinks, is usually made by free-radical polymerization. Commercially an initiator is not used because polymerization begins spontaneously at elevated temperatures. At lower temperatures a variety of initiators could be used (e.g., 2,2′-azobis-(2-methylpropionitrile)), which was used in the free-radical chlorination of 1-chlorobutane. In this experiment we use benzoyl peroxide as the initiator. On mild heating it splits into two benzoyloxy radicals:

$$C_6H_5-\overset{O}{\overset{\|}{C}}-O-O-\overset{O}{\overset{\|}{C}}-C_6H_5 \longrightarrow 2\ C_6H_5-C-O\cdot \equiv R\cdot$$

Benzoyl peroxide
MW 242.23
dec 106°C

Benzoyloxy radical

which react with styrene through initiation, propagation, and termination steps to form polystyrene:

Initiation:

$$R\cdot + CH_2{=}\underset{C_6H_5}{\underset{|}{CH}} \longrightarrow RCH_2\underset{C_6H_5}{\underset{|}{CH}}\cdot$$

Propagation:

$$RCH_2\underset{C_6H_5}{\underset{|}{CH}}\cdot + CH_2{=}\underset{C_6H_5}{\underset{|}{CH}} \longrightarrow RCH_2\underset{C_6H_5}{\underset{|}{CH}}-CH_2\underset{C_6H_5}{\underset{|}{CH}}\cdot + CH_2{=}\underset{C_6H_5}{\underset{|}{CH}} \longrightarrow RCH_2\underset{C_6H_5}{\underset{|}{CH}}-CH_2\underset{C_6H_5}{\underset{|}{CH}}-CH_2\underset{C_6H_5}{\underset{|}{CH}}\cdot, \text{ etc.}$$

Termination:

$$2R\cdot \longrightarrow R-R, \quad RCH_2CH(C_6H_5)\cdot + R\cdot \longrightarrow RCH_2CH(C_6H_5)-R, \quad 2RCH_2CH(C_6H_5)\cdot \longrightarrow RCH_2CH(C_6H_5)-CH(C_6H_5)CH_2R$$

$$H_2C=CH(C_6H_5) \longrightarrow R-[CH_2-CH(C_6H_5)]_n-CH_2-CH(C_6H_5)-H$$

Styrene
MW 104.15
bp 145–146°C

Polystyrene
MW 300,000–25,000,000

The final polymer has about 3000 monomer units in a single chain, but it can be made with up to 240,000 monomers per chain.

To prevent styrene from polymerizing in the bottle in which it is sold, the manufacturer adds 10 to 15 parts per million of 4-*tert*-butylcatechol, a radical inhibitor (a particularly good chain terminator). This must be removed by passing the styrene through a column of alumina before the styrene can be polymerized.

Procedure

Styrene is flammable, an irritant, and has a bad odor. Work with it in the hood.

Benzoyl peroxide is flammable and may explode on heating or on impact. There is no need for more than a gram or two in the laboratory at any one time. Clean up spills immediately with a damp sponge and do not discard the solid in a waste paper container since it can start fires.

In a Pasteur pipette loosely place a very small piece of cotton followed by 2.5 g of alumina. Add to the top of the pipette 1.5 mL of styrene and collect 1 mL in a reaction tube. Add to the reaction tube 50 mg of benzoyl peroxide and a thermometer and heat the tube over a hot sand bath. When the temperature reaches about 135°C polymerization begins and, since it is an exothermic process, the temperature rises. Keep the reaction under control by cautious heating. The temperature rises, perhaps to 180°C, well above the boiling point of styrene (145°C); the viscosity also increases. Pull the thermometer from the melt from time to time to form fibers; when a cool fiber is found to be brittle remove the thermometer and immediately pour the viscous liquid onto a watch glass or into a mold to cool. Objects can be cast into the hot polymer. Should the cast polymer be sticky the polymerization can be completed in an oven overnight at a temperature of about 85°C.

Questions

1. What might the products be from an explosion of smokeless gunpowder? How many moles of CO_2 and H_2O would come from one mole of trinitroglucose? Does the molecule contain enough oxygen for the production of these two substances?
2. Write a balanced equation for the reaction of sebacoyl chloride with water.
3. In the final step in the synthesis of Bakelite the partially polymerized material is heated at 85°C for several hours. What other product is produced in this reaction?
4. Give the detailed mechanism of the first step of the ring-opening polymerization of caprolactam.

68 Epoxidation of Cholesterol

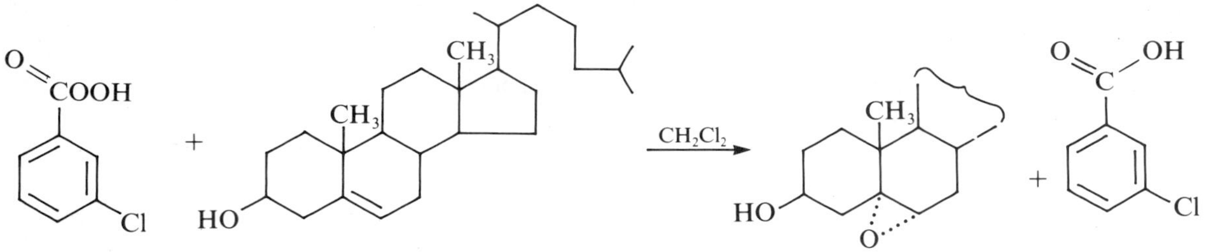

3-Chloroperoxybenzoic acid
MW 172.57
mp 92–94°C (dec)

Cholesterol
MW 386.66
mp 149°C

$\xrightarrow{CH_2Cl_2}$

5α,6α-Epoxycholestan-3β-ol
MW 402.66

3-Chlorobenzoic acid
MW 156.57
mp 157°C

This experiment carries out an epoxidation reaction on cholesterol, which is a representative of a very important group of molecules, the steroids. The rigid cholesterol molecule gives products of well-defined stereochemistry. The epoxidation reaction is stereospecific and the product can be used to carry out further stereospecific reactions.

Cholesterol itself is the principal constituent of gallstones and can be readily isolated from them (see Chapter 22). The average person contains about 200 g of cholesterol primarily in brain and nerve tissue. The closing of arteries by cholesterol leads to the disease arteriosclerosis (hardening of the arteries).

Certain naturally occurring and synthetic steroids have powerful physiological effects. Progesterone and estrone are the female sex hormones and testosterone is the male sex hormone; they are responsible for the development of secondary sex characteristics. The closely related synthetic steroid, norethisterone, is an oral contraceptive, and addition of four hydrogen atoms (reduction of the ethynyl group to the ethyl group) and a methyl group gives an anabolic steroid, ethyltestosterone. This muscle-building steroid is now outlawed for use by Olympic athletes. Fluorocortisone is used to treat inflammations such as arthritis, and ergosterol on irradiation with ultraviolet light is converted to vitamin D_2.

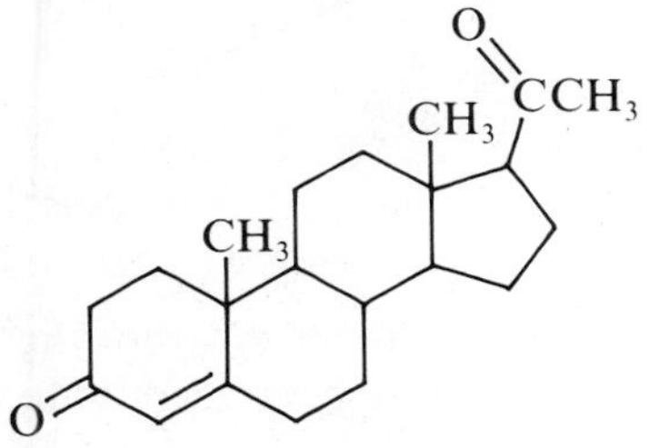

Progesterone

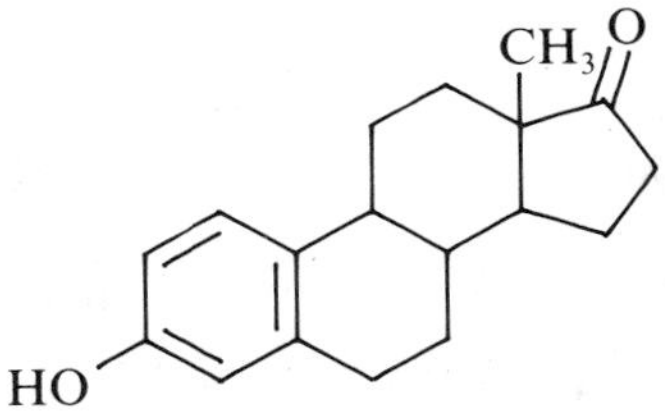

Estrone

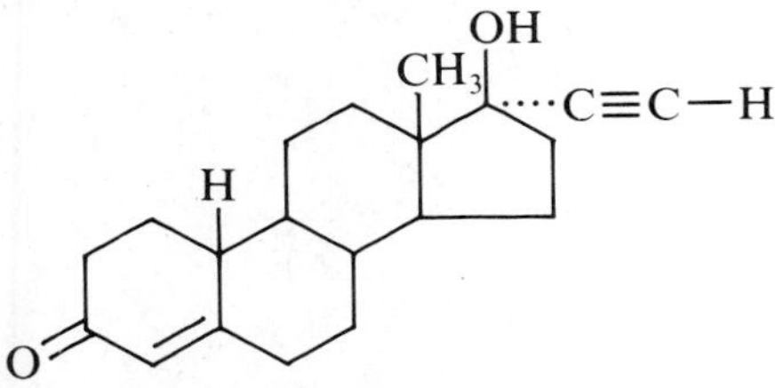

Norethisterone

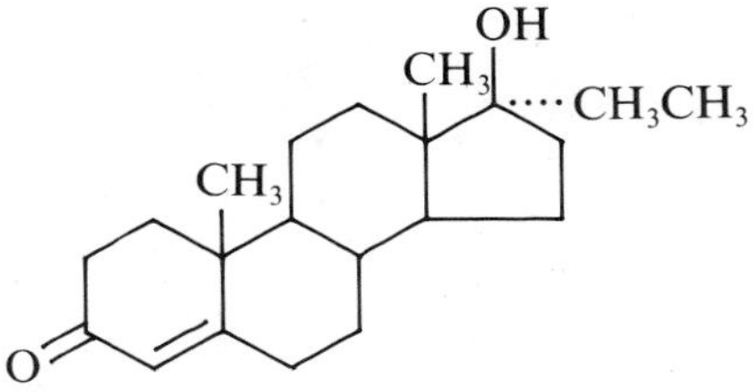

Ethyltestosterone

Much of our present knowledge about the stereochemistry of reactions was developed from steroid chemistry. In this experiment the double bond of cholesterol is stereospecifically converted to the $5\alpha,6\alpha$ epoxide. The α designation indicates the epoxide is on the backside of the molecule. A substituent on the topside is designated β. Study of molecular models reveals that the angular methyl group prevents topside attack on the double bond by the perbenzoic acid; hence the epoxide forms exclusively on the back, or α side of the molecule.

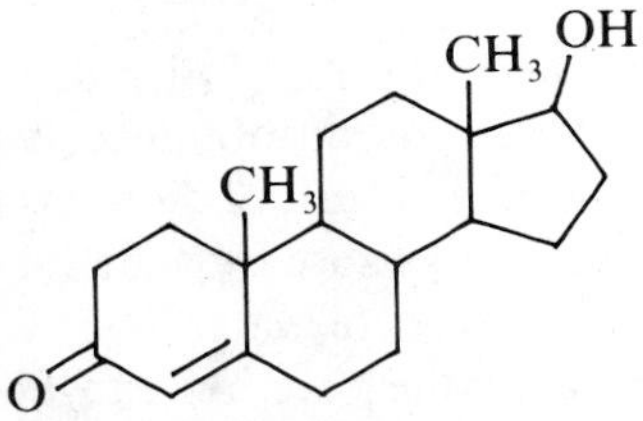

Testosterone

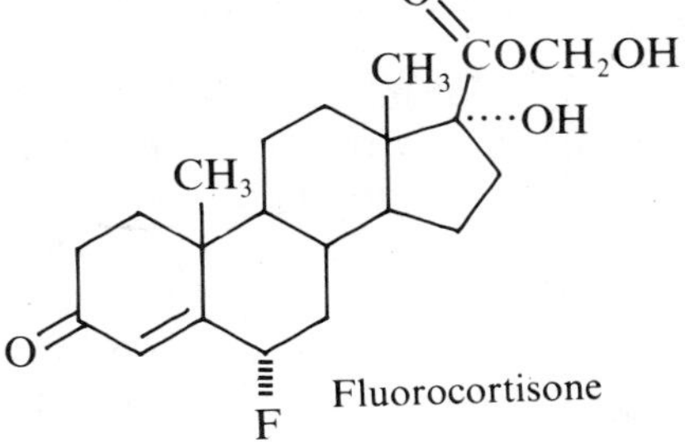

Fluorocortisone

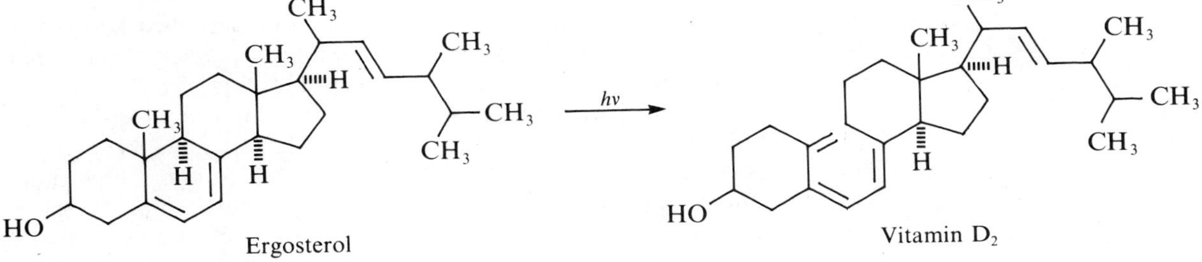

Ergosterol $\xrightarrow{h\nu}$ Vitamin D_2

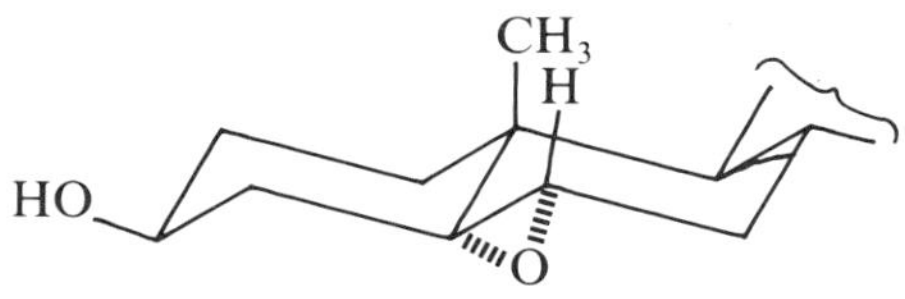

Epoxides are most commonly formed by reaction of a peroxycarboxylic acid with an olefin at room temperature. It is a one-step cycloaddition reaction:

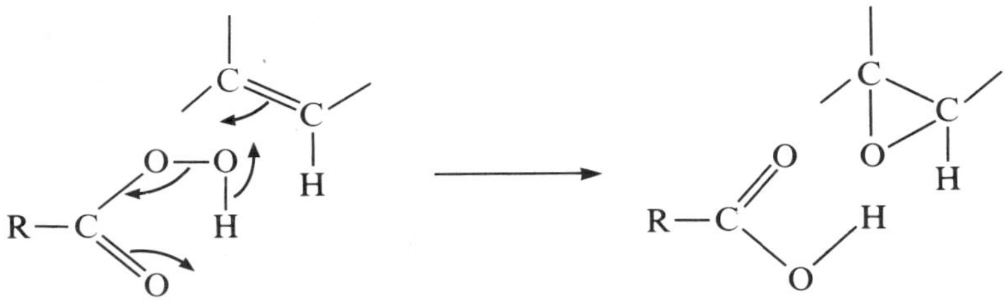

Some peroxyacids are explosive; the reagent used in the present experiment is a particularly stable and convenient peroxycarboxylic acid.

The reaction is carried out in an inert solvent, dichloromethane, and the product is isolated by chromatography. No great care is required in the chromatography to collect fractions because the 3-chlorobenzoic acid, being polar, is adsorbed strongly onto the alumina while the relatively nonpolar product is eluted easily by ether. After removal of ether the product is easily recrystallized from a mixture of acetone and water.

Experiment

To 194 mg of cholesterol dissolved by gentle warming in 0.8 mL of dichloromethane in a 10 × 100 mm reaction tube is added a solution obtained by gently warming 117 mg of 3-chloroperoxybenzoic acid in 0.8 mL of dichloromethane. The two solutions must be cool before mixing because the reaction is exothermic. Place the stoppered reaction tube in a beaker of water at 40°C for 30 min to complete the reaction. The progress of the reaction can be followed by thin-layer chromatography on silica gel plates using ether as the eluent.

The reaction mixture is pipetted onto a chromatography column prepared from 3 g of alumina following the procedure described in Chapter 10, except that ether is used to fill the column and to prepare the alumina slurry. The 3-chlorobenzoic acid will be adsorbed strongly by the alumina. The product is eluted with 30 mL of ether collected in a tared (previously weighed) 50-mL round-bottomed flask. The ether can flow through the column by gravity or can be forced out using pressure from a rubber bulb.

Most of the ether is removed on the rotary evaporator and the last traces are removed using the apparatus depicted in Fig. 10.4 for drying a solid under reduced pressure. The residue should weigh more than 150 mg. If it does not, pass more ether through the column and collect the product as before. Dissolve the product in 1.5 mL of warm acetone and, using a Pasteur pipette, transfer it to a reaction tube. Add 0.2 mL of water to the solution, warm the mixture to bring the solid into solution, and then let the tube and contents cool slowly to room temperature. Cool the mixture in ice and collect the product on a Hirsch funnel. Press the solid down on the filter to squeeze solvent from the crystals and then wash the product with 0.25 mL of ice-cold 90% acetone. Spread the product out on a watch glass to dry. Determine the weight and mp of the product and calculate the percent yield.

Questions

1. What are the numbers of moles of the two reactants? Assume the 3-chloroperoxybenzene acid is 80% pure.
2. What simple test could you perform to show that 3-chlorobenzoic acid is not eluted from the chromatography column?

69 Glass Blowing

PRELAB EXERCISE: *Grasp a pencil in the left hand, palm down, and grasp another in the right hand, palm up. Touch the erasers together and rotate both pencils at the same rate. This technique is needed to heat a glass tube, without twisting it, prior to bending or blowing a bulb in the tube.*

A knowledge of elementary glass blowing techniques is useful to every chemist and scientist, and glass blowing can be an enjoyable pastime as well.

In the laboratory you will encounter primarily Kimax or Pyrex glass, a borosilicate glass with a low coefficient of thermal expansion, which gives it remarkable resistance to thermal shock. It cannot easily be worked in a Bunsen burner flame because it has a working temperature of 820°C. Consequently many laboratories stock soft or lime glass tubing and rod, which, with a softening point of about 650°C, can be worked in a Bunsen burner. However, except for very simple bends and joints, soft glass is not used for laboratory glass blowing because of the ease with which it breaks on being subjected to a thermal gradient. Since the two types of glass are not compatible it is important to be able to distinguish between them. This is easily done by immersing the glass object in a solution having exactly the same refractive index as the glass. In such a solution the glass will seem to disappear, whereas a glass of different refractive index will be plainly visible. A solution of 14 parts (by volume) of methanol and 86 parts of toluene has a refractive index of 1.474, which is the same as that of Pyrex 7740 glass, the most common of the various Pyrexes. In this solution Pyrex will not be visible. Store the solution in a wide-mouth jar with a close-fitting cap. It will keep indefinitely.

Glass Tubing

To cut a glass tube, first make a fine straight scratch, extending about a quarter of the way around the tube, with a glass scorer. This is done by applying firm pressure on the scorer and rotating the glass tube slightly (Fig. 69.1). Only one scratch should be made; in no case should you try to saw a groove in the tube. The tube is then grasped with the scratch away from the body and the thumbs pressed together at the near side of the tube just opposite the scratch, with the arms pressed tightly against the body (Fig. 69.2). A straight, clean break will result when *slight* pressure is exerted with the thumbs and a strong force applied to pull the tube apart. It is a matter of 90% pull and 10% bend.

Break by pulling

Fire Polishing

The sharp edges that result from breaking a glass rod or tube will cut both you and the corks and rubber stoppers and tubing being fitted over them. Remove these sharp edges by holding the end of the rod (or tube) in a Bunsen burner flame and rotating the rod until the sharp edges melt and disappear. This fire-polishing process can be done even for Pyrex glass if the flame is hot enough. Open the air inlet at the bottom of the burner barrel to its maximum; the hottest part of the flame is about 7 mm above the inner blue cone. A stirring rod with a flattened head, useful for crushing lumps of solid against the bottom of a flask, is made by heating a glass rod until a short section at the end is soft and quickly pressing the end onto a smooth metal surface.

Bends

The secret to successful glass working is to have the glass thoroughly and uniformly heated before an operation. Since Pyrex glass softens at 820°C and soft glass at 650°C, the best way to work Pyrex is with a gas–oxygen torch; but with patience it can be satisfactorily heated over an ordinary Bunsen burner with a wing top attached (Fig. 69.3). Stopper the tube at the left-hand end, grasp in the left hand with palm down and in the right hand with palm up so you can swing the open end of the tube into position for blowing without interruption of the synchronous rotation of the two ends. Adjust the air intake of the burner for the maximum amount of air possible (too much will blow out the flame) and rotate the tube constantly, holding it about 7 mm above the inner blue cone. A bit of coordination is needed to rotate both ends at the same speed once the glass begins to soften; when the flame is thoroughly tinged with yellow (from sodium ions escaping from the hot glass) and the tube begins to sag, remove the tube from the flame and bend it in the vertical plane with the ends upward and the bend at the bottom. Should the tube become constricted at the bend, blow into the open end immediately upon completion of the bend to expand the glass to its full size.

Rotate constantly

The Gas–Oxygen Torch

The following operations are best carried out using Pyrex tubing and a gas–oxygen torch. To light the torch turn on the gas first to give a large luminous flame. *Gradually* turn on the oxygen until a long thin blue flame with a clearly defined inner blue cone is formed. The hottest part of the flame is at the tip of the inner blue cone. To turn off the flame always turn off the oxygen first. Wear glass blower's didymium goggles to protect the eyes from the blinding glare of hot Pyrex.

Flaring

It will often be necessary to flare the end of a tube in order to make a joint. Heat the end of the tube until the glass begins to sag, then remove the glass

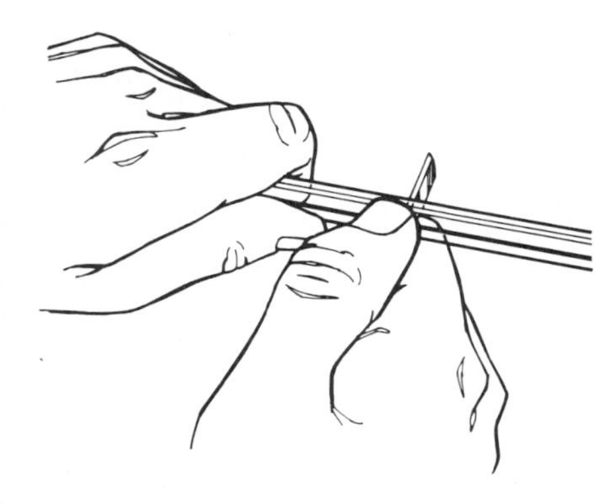

FIG. 69.1 Scratching glass tubing with glass scorer prior to breaking. The scratch is about one-fourth the tube's circumference in length.

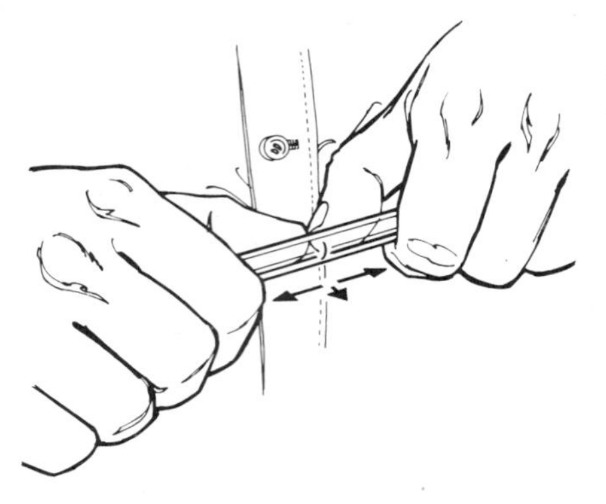

FIG. 69.2 Breaking glass tubing. Thumbs are opposite the scratch. Pull, about 90%; bend, about 10%.

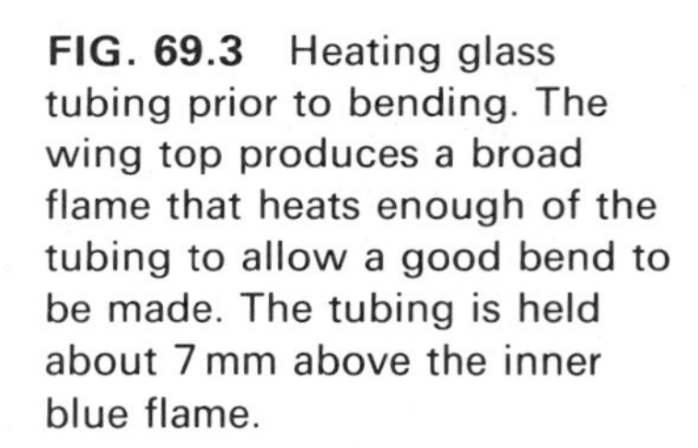

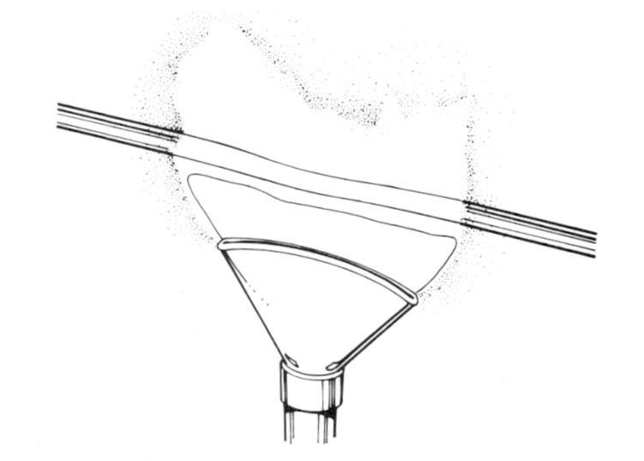

FIG. 69.3 Heating glass tubing prior to bending. The wing top produces a broad flame that heats enough of the tubing to allow a good bend to be made. The tubing is held about 7 mm above the inner blue flame.

from the fire. While rotating the tube, insert a tool such as the tine of a file or a carbon rod and press it sufficiently to form the flare (Fig. 69.4).

Test Tube Ends

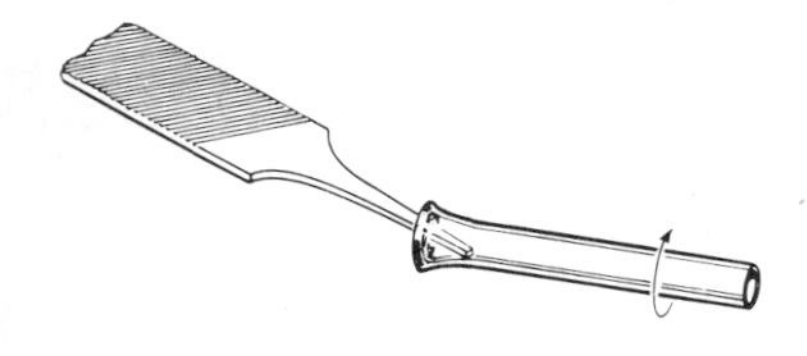

FIG. 69.4 Flaring a glass tube.

To close a tube heat it strongly at some convenient point while rotating both ends simultaneously. When the glass is soft remove the tube from the flame and pull the ends rapidly for a few inches while maintaining the rotation. Allow the glass to cool slightly, then heat the tube at A (Fig. 69.5) and pull the tube into two pieces. Heat the point B with a sharp flame to collapse it and blow it out slightly as in C. Heat the whole end of the tube until it shrinks as in D and finally blow it to a uniform hemisphere as in E. Maintain uniform rotation while carrying out all of these operations.

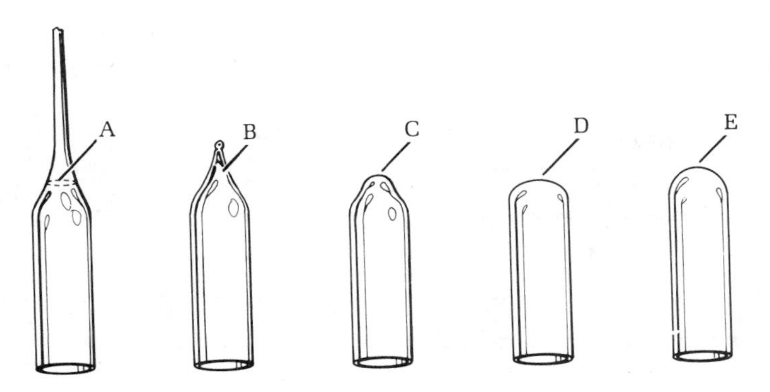

FIG. 69.5 Blowing a test tube end.

Straight Seal

To join two tubes end-on, cork one and hold it with the palm down in the left hand. Hold the open tube in the right hand, palm up. Rotate both tubes simultaneously in the flame so that the glass becomes soft just at the end and forms a rounded edge, but it does not constrict. Move the tubes to a cool part of the flame, press the ends lightly and evenly together on the same axis, and then with no hesitation pull the tubes apart slightly to reduce the thickness of the glass (Fig. 69.6). While rotating the tubing with both hands, heat the joint and cause it to shrink to about half its original diameter. Remove the tubing from the flame and blow gently to expand the joint slightly larger than the tubing diameter. Heat the joint once more and pull the tubing if necessary to restore the glass thickness to that of the original tubing. The joint when finished should resemble ordinary tubing closely.

T-Seal

Cork the end of a tube and heat a small spot on the side of the tube with a sharp flame (Fig. 69.7). Blow a small bulge on the tube. Reheat this bulge carefully at its tip and blow sharply to form a very thin walled bulb, which is broken off. Cork the other end of the tube, heat uniformly the edges of the opening as well as the end of the side tube, which has previously been slightly flared. After the edges of the openings are fairly soft remove the tubes from the flame, press them together, and then pull slightly as soon as complete contact has been made. Blow slightly to remove any irregularities. If necessary reheat, shrink, and blow until all irregularities are removed. Finally, heat the whole joint to obtain the correct angles between the tubes.

Ring Seals

Ring seals can be made in two different ways. In one method (Fig. 69.8) a flared tube of the appropriate length is dropped inside a test tube and centered with a smaller diameter tube through a cork (Fig. 69.8A). Heat the

FIG. 69.6 Making a straight seal.

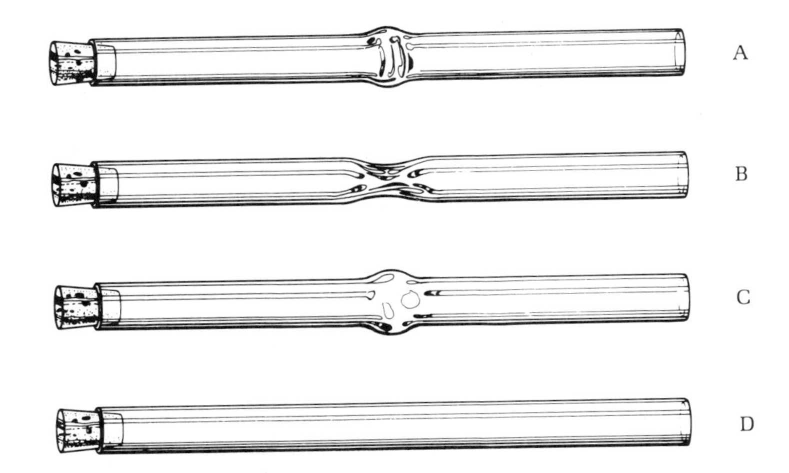

FIG. 69.7 Making a T-seal.

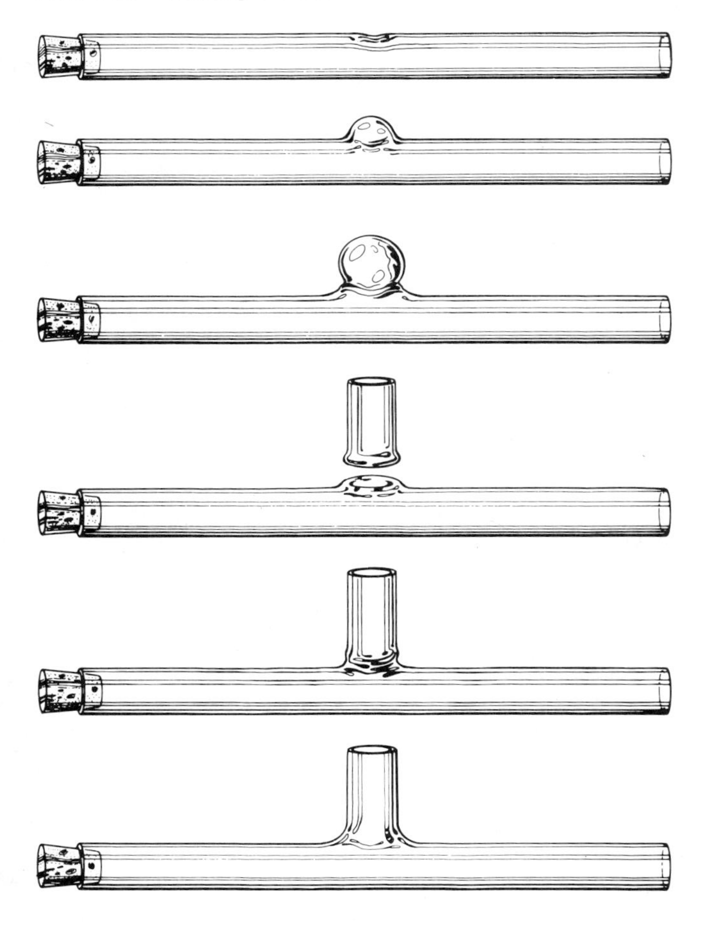

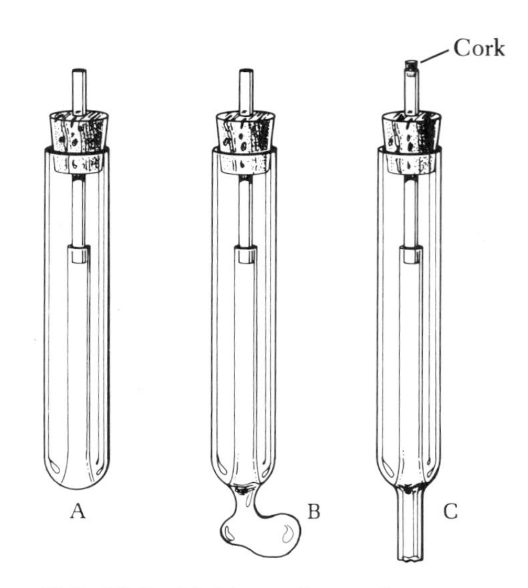

FIG. 69.8 Making a ring seal, first method.

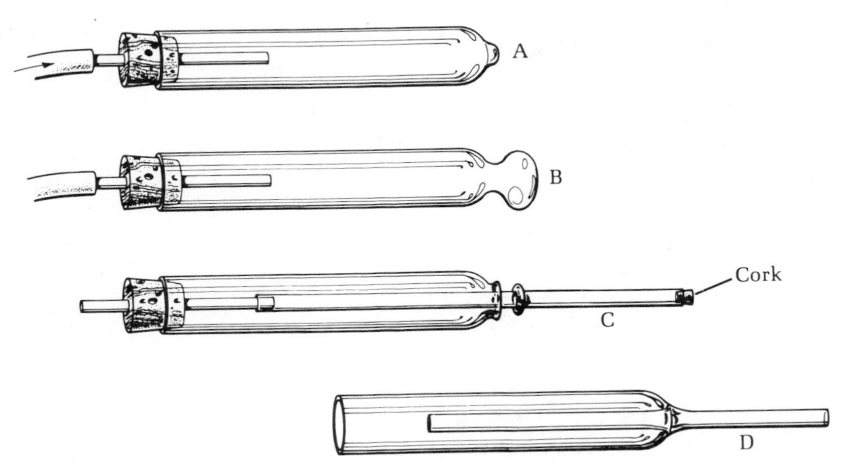

FIG. 69.9 Making a ring seal, second method.

bottom of the test tube until the inner tube forms a seal, then blow out a bulb as in Fig. 69.8B. Cork the guide tube and attach another flared tube to give the finished seal as seen in Fig. 69.8C.

In the second method for making ring seals a test tube is heated at the bottom and a small bulge is blown (Fig. 69.9A). This is heated again and a thin bulb blown out (Fig. 69.9B) and broken off. The resulting hole should be the same size as the tube to be sealed in. At an appropriate place in the tube to be sealed in, blow a small bulge and assemble the two pieces (Fig. 69.9C). Heat the joint with a small flame until it is completely sealed then blow and shrink the glass alternately until irregularities are removed (Fig. 69.9D).

Blowing a Bulb in a Tube

Cork one end of a 6-mm tube, grasp it left hand palm down, right hand palm up, and heat it uniformly over a length of about 25 mm while maintaining constant rotation. Surface tension will cause the glass to thicken as the ends of the tube are brought closer. When the tube has the appearance of Fig. 69.10A, carefully blow a bulb (Fig. 69.10B). The success of this procedure is governed by the first operation. It is necessary to have enough hot glass gathered in the hot section such that the bulb, when blown, will have a wall thickness equal to that of the tubing.

Annealing

The rapid cooling of hot glass will put strains in the glass. In the case of soft glass these strains will often cause the glass to crack. With Pyrex the strains will make the glass mechanically weak where they occur, even if the glass does not crack. These strains are relieved by cooling the glass slowly from the molten state. On a small scale this can be done by turning off the oxygen of

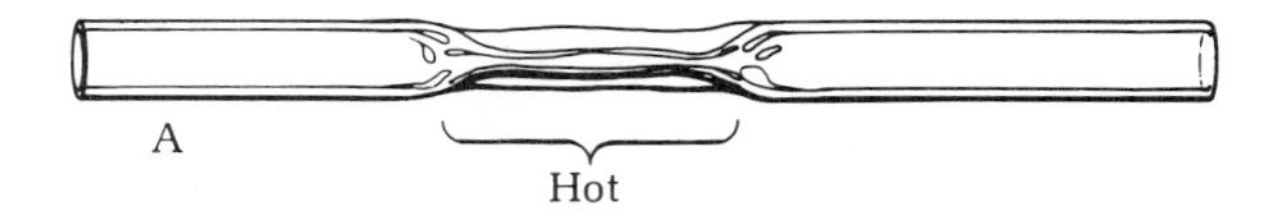

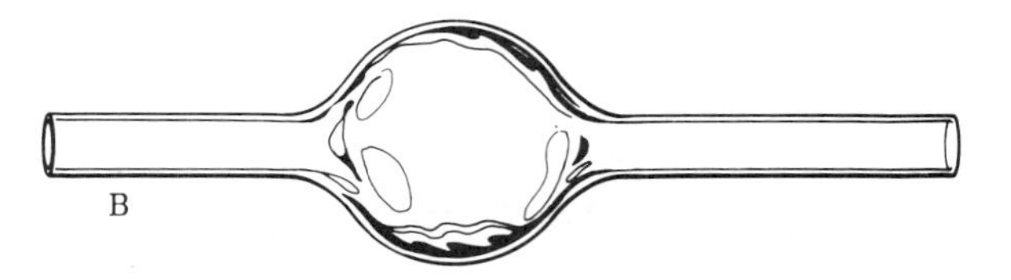

FIG. 69.10 Blowing a bulb.

FIG. 69.11 Squaring the end of a jagged tube.

the burner and holding the hot glass joint in the relatively cool luminous flame until the joint is coated with an even layer of soot. During the minute or two this requires many of the strains will be relieved. Large and complex pieces are annealed in an oven with a controlled temperature drop.

Squaring a Jagged Break

It is often desired to square the end of a tube that has a jagged end prior to carrying out the fire-polishing operation. This can be accomplished by stroking the glass with a 13-cm square of wire screen over a waste container (Fig. 69.11). The glass is removed as very small chips and dust. Wear safety glasses when doing this. Much potentially hazardous apparatus can be repaired in this way, followed by fire polishing of the resulting opening.

Qualitative Organic Analysis 70

PRELAB EXERCISE: *In the identification of an unknown organic compound certain procedures are more valuable than others. For example, much more information is obtained from an infrared spectrum than from a refractive index measurement. Outline, in order of priority, the steps you will employ in identifying your unknown.*

Identification and characterization of the structures of unknown substances are an important part of organic chemistry. It is often, of necessity, a micro process, e.g. in drug analyses. It is sometimes possible to establish the structure of a compound on the basis of spectra alone (ir, uv, and nmr), but these spectra must usually be supplemented with other information about the unknown: physical state, elementary analysis, solubility, and confirmatory tests for functional groups. Conversion of the unknown to a solid derivative of known melting point will often provide final confirmation of structure.

Procedures

1. Physical State

Check for Sample Purity. Distill or recrystallize as necessary. Constant bp and sharp mp are indicators, but beware of azeotropes and eutectics. Check homogeneity by TLC, gas, HPLC, or paper chromatography.

Note the Color. Common colored compounds include nitro and nitroso compounds (yellow), α-diketones (yellow), quinones (yellow to red), azo compounds (yellow to red), and polyconjugated olefins and ketones (yellow to red). Phenols and amines are often brown to dark-purple because of traces of air oxidation products.

Note the Odor. Some liquid and solid amines are recognizable by their fishy odors; esters are often pleasantly fragrant. Alcohols, ketones, aromatic hydrocarbons, and aliphatic olefins have characteristic odors. On the unpleasant side are thiols, isonitriles, and low-molecular-weight carboxylic acids.

Caution! *Do not taste an unknown compound. Smell it only once.*

Make an Ignition Test. Heat a small sample on a spatula; first hold the sample near the side of a microburner to see if it melts normally and then burns. Heat it in the flame. If a large ashy residue is left after ignition, the unknown is probably a metal salt. Aromatic compounds often burn with a smoky flame.

2. Spectra

Obtain infrared and nuclear magnetic resonance spectra following the procedures of Chapters 19 and 20. If these spectra indicate the presence of conjugated double bonds, aromatic rings, or conjugated carbonyl compounds obtain the ultraviolet spectrum following the procedures of Chapter 21. Interpret the spectra as fully as possible by reference to the sources cited at the end of the various spectroscopy chapters.

Explanation

3. Elementary Analysis, Sodium Fusion

This method for detection of nitrogen, sulfur, and halogen in organic compounds depends on the fact that fusion of substances containing these elements with sodium yields NaCN, Na_2S, and NaX (X = Cl, Br, I). These products can, in turn, be readily identified. The method has the advantage that the most usual elements other than C, H, and O present in organic compounds can all be detected following a single fusion, although the presence of sulfur sometimes interferes with the test for nitrogen. Unfortunately, even in the absence of sulfur the test for nitrogen is sometimes unsatisfactory (nitro compounds in particular). Practicing organic chemists rarely perform this test. Either they know what elements their unknowns contain, or they have access to a mass spectrometer or atomic absorption instrument.

Rarely performed by professional chemists

***Caution!** Manipulate sodium with a knife and forceps; never touch it with the fingers. Wipe it free of kerosene with a dry towel or filter paper; return scraps to the bottle or destroy scraps with methyl or ethyl alcohol, never with water. Safety glasses! Hood!*

Do not use $CHCl_3$ *or* CCl_4 *as samples in sodium fusion. They react extremely violently.*

Place a 3-mm cube of sodium[1] (30 mg, no more)[2] in a 10 × 75 mm Pyrex test tube and support the tube in a vertical position (Fig. 70.1). Have a microburner with small flame ready to move under the tube, place an estimated 20 mg of solid on a spatula or knife blade, put the burner in place, and heat until the sodium first melts and then vapor rises 1.5–2.0 cm in the tube. Remove the burner and at once drop the sample onto the hot sodium. If the substance is a liquid add 2 drops of it. If there is a flash or small explosion the fusion is complete; if not, heat briefly to produce a flash or a charring. Then let the tube cool to room temperature, be sure it is cold, add a drop of methanol, and let it react (heat effect). Repeat until 10 drops have been added. With a stirring rod break up the char to uncover sodium. When you are sure that all sodium has reacted, empty the tube into a 13 × 100 mm test tube, hold the small tube pointing away from you or a neighbor, and pipette into it 1 mL of water. Boil and stir the mixture and pour the water into the larger tube; repeat with 1 mL more water. Then transfer the solution with a Pasteur pipette to a 2.5-cm funnel (fitted with a fluted filter paper) resting

Notes for the instructor

1. Sodium spheres $\frac{1}{16}''$ to $\frac{1}{4}''$ (CB1035) from Matheson, Coleman, and Bell are convenient.
2. A dummy 3-mm cube of rubber can be attached to the sodium bottle to indicate the correct amount.

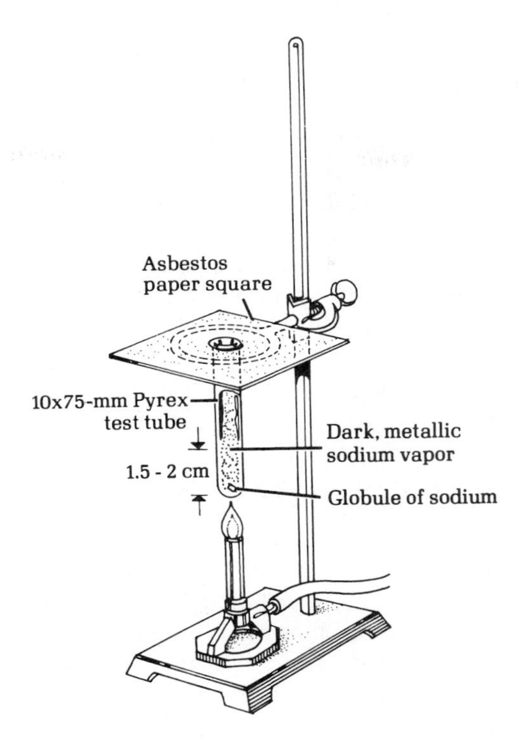

FIG. 70.1 Sodium fusion, just prior to addition of sample.

in a second 13 × 100 mm test tube. Portions of the alkaline filtrate are used for the tests that follow:

Run each test on a known and an unknown

(a) Nitrogen. The test is done by boiling the alkaline solution with iron(II) sulfate and then acidifying. Sodium cyanide reacts with iron(II) sulfate to produce ferrocyanide, which combines with iron(III) salts, inevitably formed by air oxidation in the alkaline solution, to give Prussian Blue, $NaFe^{3+}[Fe^{2+}(CN)_6]$. Iron(II) and iron(III) hydroxide precipitate along with the blue pigment but dissolve on acidification.

Place 50 mg of powdered iron(II) sulfate (this is a large excess) in a reaction tube, add 0.5 mL of the alkaline solution from the fusion, heat the mixture gently with shaking to the boiling point, and then—without cooling—acidify with dilute sulfuric acid (hydrochloric acid is unsatisfactory). A deep blue precipitate indicates the presence of nitrogen. If the coloration is dubious, filter through a 2.5-cm funnel and see if the paper shows blue pigment.

(b) Sulfur. (1) Dilute one drop of the alkaline solution with 1 mL of water and add a drop of sodium nitroprusside; a purple coloration indicates the presence of sulfur. (2) Prepare a fresh solution of sodium plumbite by adding 10% sodium hydroxide solution to 0.2 mL of 0.1 M lead acetate solution until the precipitate just dissolves, and add 0.5 mL of the alkaline test solution. A

$Na_2(NO)Fe(CN)_6 \cdot 2H_2O$

Sodium nitroprusside

black precipitate or a colloidal brown suspension indicates the presence of sulfur.

Differentiation of the halogens

Do not waste silver nitrate

(c) Halogen. Acidify 0.5 mL of the alkaline solution from the fusion with dilute nitric acid (indicator paper) and, if nitrogen or sulfur has been found present, boil the solution (hood) to expel HCN or H_2S. On addition of a few drops of silver nitrate solution, halide ion is precipitated as silver halide. Filter with minimum exposure to light on a 2.5-cm funnel, wash with water, and then with 1 mL of concentrated ammonia solution. If the precipitate is white and readily soluble in ammonium hydroxide solution it is AgCl; if it is pale yellow and not readily soluble it is AgBr; if it is yellow and insoluble it is AgI. Fluorine is not detected in this test since silver fluoride is soluble in water.

4. Beilstein Test for Halogens

A fast, easy and reliable test

Heat the tip of a copper wire in a burner flame until no further coloration of the flame is noticed. Allow the wire to cool slightly then dip it into the unknown (solid or liquid) and again heat it in the flame. A green flash is indicative of chlorine, bromine, and iodine; fluorine is not detected since copper fluoride is not volatile. The Beilstein test is very sensitive; halogen-containing impurities may give misleading results. Run the test on a compound known to contain halogen for comparison with your unknown.

It is good practice to run tests on knowns in parallel with unknowns for all qualitative organic reactions. In this way, interpretations of positive reactions are clarified and defective test reagents can be identified and replaced.

5. Solubility Tests

Like dissolves like; a substance is most soluble in that solvent to which it is most closely related in structure. This statement serves as a useful classification scheme for all organic molecules. The solubility measurements are done at room temperature with 1 drop of a liquid, or 5 mg of a solid (finely crushed), and 0.2 mL of solvent. The mixture should be rubbed with a rounded stirring rod and shaken vigorously. Lower members of a homologous series are easily classified; higher members become more like the hydrocarbons from which they are derived.

Weigh and measure carefully

If a very small amount of the sample fails to dissolve when added to some of the solvent, it can be considered insoluble; and, conversely, if several portions dissolve readily in a small amount of the solvent, the substance is obviously soluble.

If an unknown seems to be more soluble in dilute acid or base than in water, the observation can be confirmed by neutralization of the solution; the original material will precipitate if it is less soluble in a neutral medium.

If both acidic and basic groups are present, the substance may be amphoteric and therefore soluble in both acid and base. Aromatic amino-carboxylic acids are amphoteric, like aliphatic ones, but they do not exist as zwitterions. They are soluble in both dilute hydrochloric acid and sodium hydroxide, but not in bicarbonate solution. Aminosulfonic acids exist as zwitterions; they are soluble in alkali but not in acid.

The solubility tests are not infallible and many borderline cases are known.

Carry out the tests according to the scheme of Fig. 70.2 and the Notes to Solubility Tests (below) and tentatively assign the unknown to one of the groups I–X.

6. Classification Tests

After the unknown is assigned to one of the solubility groups (Fig. 70.2) on the basis of solubility tests, the possible type should be further narrowed by application of classification tests, e.g., for alcohols, or methyl ketones, or esters.

7. Complete Identification—Preparation of Derivatives

Once the unknown has been classified by functional group, the physical properties should be compared with those of representative members of the group (see tables at the end of this chapter). Usually, several possibilities present themselves, and the choice can be narrowed by preparation of derivatives. Select derivatives that distinguish most clearly among the possibilities.

Notes to Solubility Tests

See chart on next page

1. Groups I, II, III (soluble in water). Test the solution with pH paper. If the compound is not easily soluble in cold water, treat it as water-insoluble but test with indicator paper.
2. If the substance is insoluble in water but dissolves partially in 5% sodium hydroxide, add more water; the sodium salts of some phenols are less soluble in alkali than in water. If the unknown is colored, be careful to distinguish between the *dissolving* and the *reacting* of the sample. Some quinones (colored) *react* with alkali and give highly colored solutions. Some phenols (colorless) *dissolve and then* become oxidized to give colored solutions. Some compounds (e.g., benzamide) are hydrolyzed with such ease that careful observation is required to distinguish them from acidic substances.
3. Nitrophenols (yellow), aldehydophenols, and polyhalophenols are sufficiently strongly acidic to react with sodium bicarbonate.
4. Oxygen- and nitrogen-containing compounds form oxonium and ammonium ions in concentrated sulfuric acid and dissolve.
5. On reduction in the presence of hydrochloric acid these compounds form water-soluble amine hydrochlorides. Dissolve 250 mg of tin(II)

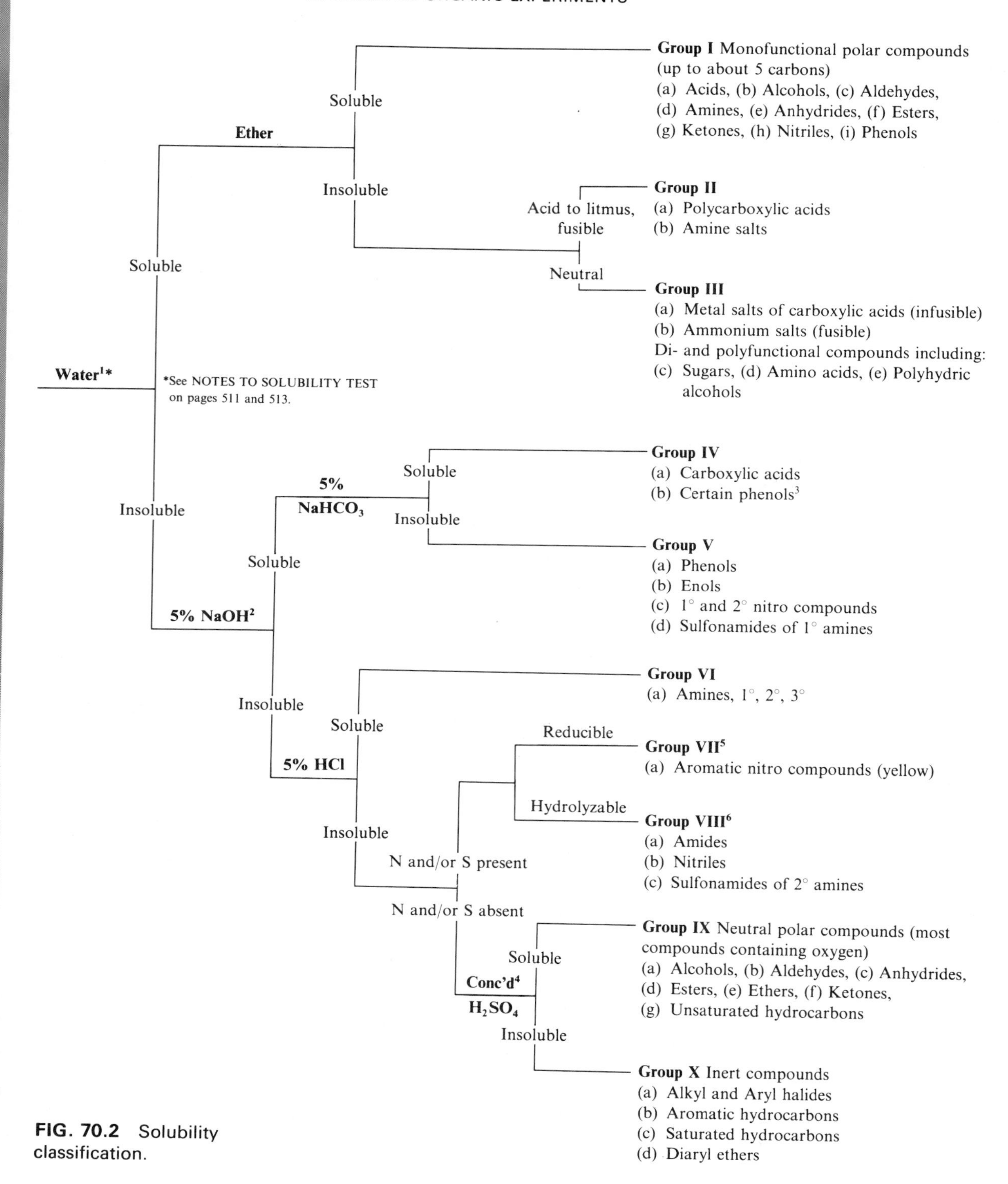

FIG. 70.2 Solubility classification.

chloride in 0.5 mL of concentrated hydrochloric acid, add 50 mg of the unknown, and warm. The material should dissolve with the disappearance of the color and give a clear solution when diluted with water.

6. Most amides can be hydrolyzed by short boiling with 10% sodium hydroxide solution; the acid dissolves with evolution of ammonia. Reflux 100 mg of the sample and 10% sodium hydroxide solution for 15–20 min under a cold finger condenser. Test for the evolution of ammonia, which confirms the elementary analysis for nitrogen and establishes the presence of a nitrile or amide.

Classification Tests

Group I. Monofunctional Polar Compounds (up to ca. 5 carbons)

(a) Acids. (Table 70.1; Derivatives, pages 520–521)
No classification test is necessary. Carboxylic and sulfonic acids are detected by testing aqueous solutions with litmus. Acyl halides may hydrolyze during the solubility test.

(b) Alcohols. (Table 70.2; Derivatives, page 521)

Jones Oxidation. Dissolve 5 mg of the unknown in 0.5 mL of pure acetone in a reaction tube and add to this solution 1 small drop of Jones reagent (chromic acid in sulfuric acid). A positive test is formation of a green color within 5 s upon addition of the orange-yellow reagent to a primary or secondary alcohol. Aldehydes also give a positive test, but tertiary alcohols do not.

Reagent: Dissolve/suspend 13.4 g of chromium trioxide in 11.5 mL of concentrated sulfuric acid and add this carefully with stirring to enough water to bring the volume to 50 mL.

Cr^{+6} dust is carcinogenic
Caution!

Cerium(IV) Nitrate Test (Ammonium Hexanitratocerium(IV) Test). Dissolve 15 mg of the unknown in a few drops of water or dioxane in a reaction tube. Add to this solution 0.25 mL of the reagent and mix thoroughly. Alcohols cause the reagent to change from yellow to red.

Reagent: Dissolve 22.5 g of ammonium hexanitratocerium(IV), $Ce(NH_4)_2(NO_3)_6$, in 56 mL of 2 N nitric acid.

(c) Aldehydes. (Table 70.3; Derivatives, page 527)

2,4-Dinitrophenylhydrazones. All aldehydes and ketones readily form bright-yellow to dark-red 2,4-dinitrophenylhydrazones. Yellow derivatives are formed from isolated carbonyl groups and orange-red to red derivatives from aldehydes or ketones conjugated with double bonds or aromatic rings.

Dissolve 10 mg of the unknown in 0.5 mL of ethanol and then add 0.75 mL of 2,4-dinitrophenylhydrazine reagent. Mix thoroughly and let sit for a few minutes. A yellow to red precipitate is a positive test.

Reagent: Dissolve 1.5 g of 2,4-dinitrophenylhydrazine in 7.5 mL of concentrated sulfuric acid. Add this solution, with stirring, to a mixture of 10 mL of water and 35 mL of ethanol.

Schiff Test. Add 1 drop (30 mg) of the unknown to 1 mL of Schiff's reagent. A magenta color will appear within 10 min with aldehydes. Compare the color of your unknown with that of a known aldehyde.

Reagent: Prepare 50 mL of a 0.1 percent aqueous solution of *p*-rosaniline hydrochloride (fuchsia). Add 2 mL of a saturated aqueous solution of sodium bisulfite. After 1 h add 1 mL of concentrated hydrochloric acid.

Bisulfite Test. Follow the procedure on page 280, Chapter 30. Nearly all aldehydes and most methyl ketones form solid, water-soluble bisulfite addition products.

Destroy used Tollen's reagent with nitric acid.

Tollen's Test. Follow the procedure on page 279, Chapter 30. A positive test, deposition of a silver mirror, is given by most aldehydes, but not by ketones.

(d) Amines. (Tables 70.5 and 70.6; Derivatives, page 521)

Hinsberg Test. Follow the procedure on page 336, Chapter 40, using benzenesulfonyl chloride to distinguish between primary, secondary, and tertiary amines.

(e) Anhydrides and Acid Halides. (Table 70.7; Derivatives, page 522)

Anhydrides and acid halides will react with water to give acidic solutions, detectable with litmus paper. They easily form benzamides and acetamides.

Acidic Iron(III) Hydroxamate Test. With iron(III) chloride alone a number of substances give a color which can interfere with this test. Dissolve 2 drops (or about 30 mg) of the unknown in 1 mL of ethanol and add 1 mL of 1 N hydrochloric acid followed by 1 drop of 10% aqueous iron(III) chloride solution. If any color except yellow appears you will find it difficult to interpret the results from the following test.

Add 2 drops (or about 30 mg) of the unknown to 0.5 mL of a 1 N solution of hydroxylamine hydrochloride in alcohol. Add 2 drops of 6 M hydrochloric acid to the mixture, warm it slightly for 2 min, and boil it for a few seconds. Cool the solution and add 1 drop of 10 percent ferric chloride solution. A red-blue color is a positive test.

(f) Esters. (Table 70.8, page 531. Derivatives prepared from component acid and alcohol obtained on hydrolysis)

Esters, unlike anhydrides and acid halides, do not react with water to give acidic solutions and do not react with acidic hydroxylamine hydrochloride. They do, however, react with alkaline hydroxylamine.

Alkaline Iron(III) Hydroxamate Test. First test the unknown with iron(III) chloride alone. (See under Group I(e), Acidic Iron(III) Hydroxamate Test.)

To a solution of one drop (30 mg) of the unknown in 0.5 mL of 0.5 N hydroxylamine hydrochloride in ethanol add two drops of 20% sodium hydroxide solution. Heat the solution to boiling, cool slightly, and add 1 mL of 1 N hydrochloric acid. If cloudiness develops add up to 1 mL of ethanol. Add 10% iron(III) chloride solution dropwise with thorough mixing. A red-blue color is a positive test. Compare your unknown with a known ester.

(g) Ketones. (Table 70.14; Derivatives; page 535)

2,4-Dinitrophenylhydrazone. See under Group I(c), Aldehydes. All ketones react with 2,4-dinitrophenylhydrazine reagent.

Iodoform Test for Methyl Ketones. Follow the procedure on page 279, Chapter 30. A positive iodoform test is given by substances containing the $CH_3\overset{O}{\overset{\|}{C}}-$ group or by compounds easily oxidized to this group, e.g., CH_3COR, CH_3CHOHR, CH_3CH_2OH, CH_3CHO, $RCOCH_2COR$. The test is negative for compounds of the structure CH_3COOR, CH_3CONHR, and other compounds of similar structure that give acetic acid on hydrolysis. It is also negative for $CH_3COCH_2CO_2R$, CH_3COCH_2CN, $CH_3COCH_2NO_2$.

Bisulfite Test. Follow the procedure on page 279, Chapter 30. Aliphatic methyl ketones and unhindered cyclic ketones form bisulfite addition products. Methyl aryl ketones, such as acetophenone, $C_6H_5COCH_3$, fail to react.

(h) Nitriles. (Table 70.15, page 536. Derivatives prepared from the carboxylic acid obtained by hydrolysis)

At high temperature nitriles (and amides) are converted to hydroxamic acids by hydroxylamine:

$$RCN + 2\ H_2NOH \longrightarrow RCONHOH + NH_3$$

The hydroxamic acid forms a red-blue complex with iron(III) ion. The unknown must first give a negative test with hydroxylamine at lower temperature (Group I(f), Alkaline Hydroxamic Acid Test) before trying this test.

Hydroxamic Acid Test for Nitriles (and Amides). To 1 mL of a 1 M hydroxylamine hydrochloride solution in propylene glycol add 15 mg of the unknown dissolved in the minimum amount of propylene glycol. Then add 0.5 mL of 1 N potassium hydroxide in propylene glycol and boil the mixture for 2 min. Cool the mixture and add 0.1 to 0.25 mL of 10% iron(III) chloride solution. A red-blue color is a positive test for almost all nitrile and amide groups, although benzanilide fails to give a positive test.

(i) Phenols. (Table 70.17, page 537)

Iron(III) Chloride Test. Dissolve 15 mg of the unknown compound in 0.5 mL of water or water-alcohol mixture and add one or two drops of 1% iron(III) chloride solution. A red, blue, green or purple color is a positive test.

A more sensitive test for phenols consists of dissolving or suspending 15 mg of the unknown in 0.5 mL of chloroform and adding 1 drop of a solution made by dissolving 0.1 g of iron(III) chloride in 10 mL of chloroform. (*Caution!* $CHCl_3$ is a carcinogen.) Addition of a drop of pyridine, with stirring, will produce a color if phenols or enols are present.

Group II. Water-soluble Acidic Salts, Insoluble in Ether

(a) Amine Salts. [Table 70.5 (1° and 2° amines); Table 70.6 (3° amines)]

The free amine can be liberated by addition of base and extraction into ether. Following evaporation of the ether the Hinsberg test, Group I(d), can be applied to determine if the compound is a primary, secondary, or tertiary amine.

The acid iron(III) hydroxamate test, Group I(d), can be applied directly to the amine salt.

Group III. Water-soluble Neutral Compounds, Insoluble in Ether

(a) Metal Salts of Carboxylic Acids. (Table 70.1, carboxylic acids; Derivatives, page 524)

The free acid can be liberated by addition of acid and extraction into an appropriate solvent, after which the carboxylic acid can be characterized by mp or bp before proceeding to prepare a derivative.

(b) Ammonium Salts. (Table 70.1, carboxylic acids; Derivatives, page 524)

Ammonium salts on treatment with alkali liberate ammonia, which can be detected by its odor and the fact it will turn red litmus, blue. A more sensitive test utilizes the copper(II) ion, which is blue in the presence of ammonia [see Group VIII a(i)]. Ammonium salts will not give a positive hydroxamic acid test (Ih) as given by amides.

(c) Sugars. See page 279, Chapter 30, for Tollen's test and page 394, Chapter 51, for phenylosazone formation.

(d) Amino Acids. Add 2 mg of the suspected amino acid to 1 mL of ninhydrin reagent, boil for 20 s, and note the color. A blue color is a positive test.

Reagent: Dissolve 0.2 g of ninhydrin in 50 mL of water.

(e) Polyhydric Alcohols. (Table 70.2; Derivatives, page 526)

Periodic Acid Test for vic-Glycols.[3] Vicinal glycols (hydroxyl groups on adjacent carbon atoms) can be detected by reaction with periodic acid. In

3. R. L. Shriner, R. C. Fuson, D. Y. Curtin, and T. C. Morill, *The Systematic Identification of Organic Compounds*, 6th ed., John Wiley & Sons, Inc., New York, 1980.

addition to 1,2-glycols a positive test is given by α-hydroxy aldehydes, α-hydroxy ketones, α-hydroxy acids, and α-amino alcohols, as well as 1,2-diketones.

To 2 mL of periodic acid reagent add one drop (no more) of concentrated nitric acid and shake. Then add one drop or a small crystal of the unknown. Shake for 15 s and add 1 or 2 drops of 5% aqueous silver nitrate solution. Instantaneous formation of a white precipitate is a positive test.

Reagent: Dissolve 0.25 g of paraperiodic acid (H_5IO_6) in 50 mL of water.

Group IV. Certain Carboxylic Acids, Certain Phenols, and Sulfonamides of 1° Amines

(a) Carboxylic Acids. Solubility in both 5% sodium hydroxide and sodium bicarbonate is usually sufficient to characterize this class of compounds. Addition of mineral acid should regenerate the carboxylic acid. The neutralization equivalent can be obtained by titrating a known quantity of the acid (ca. 50 mg) dissolved in water-ethanol with 0.1 N sodium hydroxide to a phenolphthalein end point.

(b) Phenols. Negatively substituted phenols such as nitrophenols, aldehydophenols, and polyhalophenols are sufficiently acidic to dissolve in 5% sodium bicarbonate. See Group I(i), page 515, for the iron(III) chloride test for phenols; however, this test is not completely reliable for these acidic phenols.

Group V. Acidic Compounds, Insoluble in Bicarbonate

(a) Phenols. See Group I(i).

(b) Enols. See Group I(i).

(c) 1° and 2° Nitro Compounds. (Table 70.16, Derivatives, page 536)

Iron (II) Hydroxide Test. To a small vial (capacity 1–2 mL) add 5 mg of the unknown to 0.5 mL of freshly prepared ferrous sulfate solution. Add 0.4 mL of a 2 N solution of potassium hydroxide in methanol, cap the vial, and shake it. The appearance of a red-brown precipitate of iron(III) hydroxide within 1 min is a positive test. Almost all nitro compounds give a positive test within 30 s.

Reagents: Dissolve 2.5 g of ferrous ammonium sulfate in 50 mL of deoxygenated (by boiling) water. Add 0.2 mL of concentrated sulfuric acid and a piece of iron to prevent oxidation of the ferrous ion. Keep the bottle tightly stoppered. The potassium hydroxide solution is prepared by dissolving 5.6 g of potassium hydroxide in 50 mL of methanol.

(d) Sulfonamides of 1° Amines. An extremely sensitive test for sulfonamides (Feigl, *Spot Tests in Organic Analysis*) consists of placing a drop of a suspension or solution of the unknown on sulfonamide test paper followed by a drop of 0.5% hydrochloric acid. A red color is a positive test for sulfonamides.

The test paper is prepared by dipping filter paper into a mixture of equal volumes of a 1% aqueous solution of sodium nitrite and a 1% methanolic solution of N,N-dimethyl-1-naphthylamine. Allow the filter paper to dry in the dark.

Caution! *Handle the naphthylamine with great care. It is carcinogenic. Carry out this test in a hood.*

Group VI. Basic Compounds, Insoluble in Water, Soluble in Acid

(a) Amines. See Group I(d).

Group VII. Reducible, Neutral N- and S-Containing Compounds

(a) Aromatic Nitro Compounds. See Group V(c).

Group VIII. Hydrolyzable, Neutral N- and S-Containing Compounds (Identified through the acid and amine obtained on hydrolysis)

(a) Amides. Unsubstituted amides are detected by the hydroxamic acid test, Group I(h).

(1) Unsubstituted Amides. Upon hydrolysis, unsubstituted amides liberate ammonia, which can be detected by reaction with cupric ion (Group III(b)).

To 1 mL of 20% sodium hydroxide solution add 25 mg of the unknown. Cover the mouth of the reaction tube with a piece of filter paper moistened with a few drops of 10% copper(II) sulfate solution. Boil for 1 min. A blue color on the filter paper is a positive test for ammonia.

(2) Substituted Amides. The identification of substituted amides is not easy. There are no completely general tests for the substituted amide groups and hydrolysis is often difficult.

Hydrolyze the amide by refluxing 250 mg with 2.5 mL of 20% sodium hydroxide for 20 min. Isolate the primary or secondary amine produced, by extraction into ether, and identify as described under Group I(d). Liberate the acid by acidification of the residue and isolate by filtration or extraction and characterize by bp or mp and the mp of an appropriate derivative.

Hot sodium hydroxide solution is corrosive; use care

(3) Anilides. Add 50 mg of the unknown to 1.5 mL of concentrated sulfuric acid. Carefully stopper the reaction tube with a rubber stopper and shake vigorously. **Caution.** Add 25 mg of finely powdered potassium dichromate. A blue-pink color is a positive test for an anilide that does not have substituents on the ring (e.g., acetanilide).

Use care in shaking concentrated sulfuric acid

Dichromate dust is carcinogenic, when inhaled. Cr^{+6} is not a carcinogen when applied to the skin or ingested.

(b) Nitriles. See Group I(h).

(c) Sulfonamides. See Group V(d), page 517.

Group IX. Neutral Polar Compounds, Insoluble in Dilute Hydrochloric Acid, Soluble in Concentrated Sulfuric Acid (most compounds containing oxygen)

(a) Alcohols. See Group I(b).

(b) Aldehydes. See Group I(c).

(c) Anhydrides. See Group I(e).

(d) Esters. See Group I(f).

(e) Ethers. (Table 70.9, page 532)
Ethers are very unreactive. Care must be used to distinguish ethers from those hydrocarbons that are soluble in concentrated sulfuric acid.

Ferrox Test. In a dry test tube grind together, with a stirring rod, a crystal of iron(III) ammonium sulfate (or iron(III) chloride) and a crystal of potassium thiocyanate. Iron(III) hexathiocyanatoferrate(III) will adhere to the stirring rod. In a clean tube place 3 drops of a liquid unknown or a saturated toluene solution of a solid unknown and stir with the rod. The salt will dissolve if the unknown contains oxygen to give a red to red-purple color, but it will not dissolve in hydrocarbons or halocarbons. Diphenyl ether does not give a positive test.

$Fe[Fe(SCN)_6]$

Iron(III) hexathiocyanatoferrate(III)

Alkyl ethers are generally soluble in concentrated hydrochloric acid; alkyl aryl and diaryl ethers are not soluble.

(f) Ketones. See Group I(g).

(g) Unsaturated Hydrocarbons. (Table 70.12, page 533)

Bromine in Carbon Tetrachloride. Dissolve 1 drop (20 mg) of the unknown in 0.5 mL of carbon tetrachloride. Add a 2% solution of bromine in carbon tetrachloride dropwise with shaking. If more than 2 drops of bromine solution are required to give a permanent red color, unsaturation is indicated. The bromine solution must be fresh.

Use care in working with the bromine solution

Potassium Permanganate Solution. Dissolve 1 drop (20 mg) of the unknown in reagent grade acetone and add a 1% aqueous solution of potassium permanganate dropwise with shaking. If more than one drop of reagent is required to give a purple color to the solution, unsaturation or an easily oxidized functional group is present. Run parallel tests on pure acetone and, as usual, a compound known to be an alkene.

Group X. Inert Compounds. Insoluble in Concentrated Sulfuric Acid

(a) Alkyl and Aryl Halides. (Table 70.10, pages 532–533)

Do not waste silver nitrate.

Alcoholic Silver Nitrate. Add one drop of the unknown (or saturated solution of 10 mg of unknown in ethanol) to 0.2 mL of a saturated solution of silver nitrate. A precipitate which forms within 2 min is a positive test for an alkyl bromide, or iodide, or a tertiary alkyl chloride, as well as allyl halides.

If no precipitate forms within 2 min, heat the solution to boiling. A precipitate of silver chloride will form from primary and secondary alkyl chlorides. Aryl halides and vinyl halides will not react.

(b) Aromatic Hydrocarbons. (Table 70.13; Derivatives, page 532)

Aromatic hydrocarbons are best identified and characterized by UV and nmr spectroscopy, but the Friedel-Crafts reaction produces a characteristic color with certain aromatic hydrocarbons.

Keep moisture away from aluminum chloride

Chloroform is carcinogenic. Carry out this test in a hood.

Friedel-Crafts Test. Heat a reaction tube containing about 50 mg of anhydrous aluminum chloride in a hot flame to sublime the salt up onto the sides of the tube. Add a solution of about 10 mg of the unknown dissolved in a drop of chloroform to the cool tube in such a way that it comes into contact with the sublimed aluminum chloride. Note the color that appears.

Nonaromatic compounds fail to give a color with aluminum chloride, benzene and its derivatives give orange or red colors, naphthalenes a blue or purple color, biphenyls a purple color, phenanthrene a purple color, and anthracene a green color.

(c) Saturated Hydrocarbons.

Saturated hydrocarbons are best characterized by nmr and ir spectroscopy, but they can be distinguished from aromatic hydrocarbons by the Friedel-Crafts test (Group X(b)).

(d) Diaryl Ethers.

Because they are so inert, diaryl ethers are difficult to detect and may be mistaken for aromatic hydrocarbons. They do not give a positive Ferrox Test, Group IX(e), for ethers, and do not dissolve in concentrated sulfuric acid. Their infrared spectra, however, are characterized by an intense C—O single-bond, stretching vibration in the region 1270–1230 cm^{-1}.

Derivatives

1. Acids. (Table 70.1)

Caution! Benzene is a mild carcinogen. Carry out test in a hood.

***p*-Toluidides and Anilides.** Reflux a mixture of the acid (100 mg) and thionyl chloride (0.5 mL) in a reaction tube for 0.5 h. Cool the reaction mixture and add 0.25 g of aniline or *p*-toluidine in 3 mL of benzene (*caution!*). Warm the mixture on the steam bath for 2 min, and then wash with 1-mL portions of water, 5% hydrochloric acid, 5% sodium hydroxide and water. The benzene is filtered through a cone of anhydrous sodium sulfate and evaporated in the hood; the derivative is recrystallized from water or ethanol-water.

Amides. Reflux a mixture of the acid (100 mg) and thionyl chloride (0.5 mL) for 0.5 h. Transfer the cool reaction mixture into 1.4 mL of ice-cold concentrated ammonia. Stir until reaction is complete, collect the product by filtration, and recrystallize it from water or water-ethanol.

Thionyl chloride is irritating. Use it in a hood.

2. Alcohols. (Table 70.2)

3,5-Dinitrobenzoates. Gently boil 100 mg of 3,5-dinitrobenzoyl chloride and 100 mg of the alcohol for 5 min. Cool the mixture, pulverize any solid that forms, and add 2 mL of 2% sodium carbonate solution. Continue to grind and stir the solid with the sodium carbonate solution (to remove 3,5-dinitrobenzoic acid) for about a minute, filter, and wash the crystals with water. Dissolve the product in about 2.5–3 mL of hot ethanol, add water to the cloud point, and allow crystallization to proceed. Wash the 3,5-dinitrobenzoate with water-alcohol and dry.

Note to instructor: Check to ascertain that the 3,5-dinitrobenzoyl chloride has not hydrolyzed. The mp should be > 70°C.

Phenylurethanes. Mix 100 mg of anhydrous alcohol (or phenol) and 100 mg of phenyl isocyanate (or α-naphthylurethane) and heat on the steam bath for 5 min. (If the unknown is a phenol add a drop of pyridine to the reaction mixture.) Cool, add about 1 mL of ligroin, heat to dissolve the product, filter hot to remove a small amount of diphenylurea which usually forms, and cool the filtrate in ice, with scratching, to induce crystallization.

3. Aldehydes. (Table 70.3)

Semicarbazones. See page 278, Chapter 30. Use 0.5 mL of the stock solution and an estimated 1 millimole of the unknown aldehyde (or ketone).

2,4-Dinitrophenylhydrazones. See page 277, Chapter 30. Use 1 mL of the stock solution of 0.1 M 2,4-dinitrophenylhydrazine and an estimated 0.1 millimole of the unknown aldehyde (or ketone).

4. Primary and Secondary Amines. (Table 70.5)

Benzamides. Add about 0.25 g of benzoyl chloride in small portions with vigorous shaking and cooling to a suspension of 0.5 millimole of the unknown amine in 0.5 mL of 10% aqueous sodium hydroxide solution. After about 10 min of shaking the mixture is made pH 8 (pH paper) with dilute hydrochloric acid. The lumpy product is removed by filtration, washed thoroughly with water, and recrystallized from ethanol-water.

Picrates. Add a solution of 30 mg of the unknown in 1 mL of ethanol (or 1 mL of a saturated solution of the unknown) to 1 mL of a saturated solution of picric acid (2,4,6-trinitrophenol, a strong acid) in ethanol, and heat the solution to boiling. Cool slowly, remove the picrate by filtration, and wash with a small amount of ethanol. Recrystallization is not usually necessary; in the case of hydrocarbon picrates the product is often too unstable to be recrystallized.

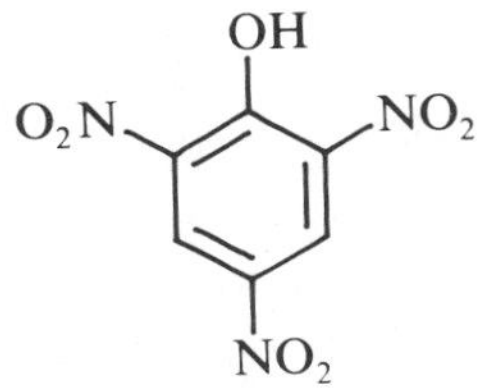

OH, O_2N, NO_2, NO_2

Picric acid (2,4,6-Trinitrophenol)

Handle pure acid with care (explosive). It is sold as a moist solid. Do not allow to dry out.
Acetic anhydride is corrosive. Work with this in a hood.

Acetamides. Reflux about 0.5 millimole of the unknown with 0.2 mL of acetic anhydride for 5 min, cool, and dilute the reaction mixture with 2.5 mL of water. Initiate crystallization by scratching, if necessary. Remove the crystals by filtration and wash thoroughly with dilute hydrochloric acid to remove unreacted amine. Recrystallize the derivative from alcohol-water. Amines of low basicity, e.g., *p*-nitroaniline, should be refluxed for 30 to 60 min with 1 mL of pyridine as a solvent. The pyridine is removed by shaking the reaction mixture with 5 mL of 2% sulfuric acid solution; the product is isolated by filtration and recrystallized.

$$R_3N + CH_3I \longrightarrow R_3\overset{+}{N}CH_3I^-$$

Methyl iodide is a cancer suspect agent.

5. Tertiary Amines. (Table 70.6)

Picrates. See under Primary and Secondary Amines.

Methiodides. Reflux 100 mg of the amine and 100 mg of methyl iodide for 5 min on the steam bath. Cool, scratch to induce crystallization, and recrystallize the product from ethyl alcohol or ethyl acetate.

6. Anhydrides and Acid Chlorides. (Table 70.7)

Acids. Reflux 40 mg of the acid chloride or anhydride with 1 mL of 5% sodium carbonate solution for 20 min or less. Extract unreacted starting material with 1 mL of ether, if necessary, and acidify the reaction mixture with dilute sulfuric acid to liberate the carboxylic acid.

Amides. Since the acid chloride (or anhydride) is already present, simply mix the unknown (50 g) and 0.7 mL of ice-cold concentrated ammonia until reaction is complete, collect the product by filtration, and recrystallize it from water or ethanol-water.

***Caution!** Benzene is a mild carcinogen. Carry out preparation in a hood.*

Anilides. Reflux 40 mg of the acid halide or anhydride with 100 mg of aniline in 2 mL of benzene (*caution!*) for 10 min. Wash the benzene solution with 5-mL portions each of water, 5% hydrochloric acid, 5% sodium hydroxide and water. The benzene solution is filtered through a cone of anhydrous sodium sulfate and evaporated; the anilide is recrystallized from water or ethanol-water.

7. Aryl Halides. (Table 70.11)

Nitration. Add 0.4 mL of concentrated sulfuric acid to 100 mg of the aryl halide (or aromatic compound) and stir. Add 0.4 mL of concentrated nitric acid dropwise with stirring and shaking while cooling the reaction mixture in water. Then heat and shake the reaction mixture in a water bath at about 50°C for 15 min, pour into 2 mL of cold water, and collect the product by filtration. Recrystallize from methanol to constant melting point.

Use great care when working with fuming nitric acid

To nitrate unreactive compounds use fuming nitric acid in place of concentrated nitric acid.

Sidechain Oxidation Products. Dissolve 0.2 g of sodium dichromate in 0.6 mL of water and add 0.4 mL of concentrated sulfuric acid. Add 50 mg of the unknown and boil for 30 min. Cool, add 0.4 to 0.6 mL of water and then remove the carboxylic acid by filtration. Wash the crystals with water and recrystallize from methanol-water.

8. Hydrocarbons: Aromatic. (Table 70.13)

Nitration. See preceding, under Aryl Halides.

Picrates. See preceding, under Primary and Secondary Amines (page 521)

9. Ketones. (Table 70.14)

Semicarbazones and 2,4-dinitrophenylhydrazones. See preceding directions under Aldehydes.

10. Nitro Compounds. (Table 70.16)

Reduction to Amines. Place 100 mg of the unknown in a reaction tube, add 0.2 g of tin, and then—in portions—2 mL of 10% hydrochloric acid. Reflux for 30 min, add 1 mL of water, then add slowly, with good cooling, sufficient 40% sodium hydroxide solution to dissolve the tin hydroxide. Extract the reaction mixture with three 1-mL portions of ether, dry the ether extract over anhydrous sodium sulfate, and evaporate the ether to leave the amine. Determine the boiling point or melting point of the amine and then convert it into a benzamide or acetamide as described under Primary and Secondary Amines.

11. Phenols. (Table 70.17)

α-Naphthylurethane. Follow the procedure for preparation of a phenylurethane under Alcohols, page 521.

Bromo Derivative. In a reaction tube dissolve 160 mg of potassium bromide in 1 ml of water. *Carefully* add 100 mg of bromine. In a separate flask dissolve 20 mg of the phenol in 0.2 mL of methanol and add 0.2 mL of water. Add about 0.3 mL of the bromine solution with swirling (hood); continue the addition of bromine until the yellow color of unreacted bromine persists. Add 0.6 to 0.8 mL of water to the reaction mixture and shake vigorously. Remove the product by filtration and wash well with water. Recrystallize from methanol-water.

Use great care when working with bromine. Work in a hood and wear disposable gloves. Should any touch the skin wash it off with copious quantities of water.

TABLE 70.1 Acids

Bp	Mp	Compound	Derivatives		
			p-Toluidide[a]	*Anilide*[b]	*Amide*[c]
			Mp	*Mp*	*Mp*
101		Formic acid	53	47	43
118		Acetic acid	126	106	79
139		Acrylic acid	141	104	85
141		Propionic acid	124	103	81
162		*n*-Butyric acid	72	95	115
163		Methacrylic acid		87	102
165		Pyruvic acid	109	104	124
185		Valeric acid	70	63	106
186		2-Methylvaleric acid	80	95	79
194		Dichloroacetic acid	153	118	98
202–203		Hexanoic acid	75	95	101
237		Octanoic acid	70	57	107
254		Nonanoic acid	84	57	99
	31–32	Decanoic acid	78	70	108
	43–45	Lauric acid	87	78	100
	47–49	Bromoacetic acid		131	91
	47–49	Hydrocinnamic acid	135	92	105
	54–55	Myristic acid	93	84	103
	54–58	Trichloroacetic acid	113	97	141
	61–62	Chloroacetic acid	162	137	121
	61–62.5	Palmitic acid	98	90	106
	67–69	Stearic acid	102	95	109
	68–69	3,3-Dimethylacrylic acid		126	107
	71–73	Crotonic acid	132	118	158
	77–78.5	Phenylacetic acid	136	118	156
	101–102	Oxalic acid dihydrate		257	400 (dec)
	98–102	Azelaic acid (nonanedioic)	164 (di)	107 (mono) 186 (di)	93 (mono) 175 (di)
	103–105	*o*-Toluic acid	144	125	142
	108–110	*m*-Toluic acid	118	126	94
	119–121	DL-Mandelic acid	172	151	133
	122–123	Benzoic acid	158	163	130
	127–128	2-Benzoylbenzoic acid		195	165
	129–130	2-Furoic acid	107	123	143
	131–133	DL-Malic acid	178 (mono) 207 (di)	155 (mono) 198 (di)	163 (di)
	131–134	Sebacic acid	201	122 (mono) 200 (di)	170 (mono) 210 (di)

a. For preparation, see page 520.
b. For preparation, see page 520.
c. For preparation, see page 521.

TABLE 70.1 continued

Bp	Mp	Compound	Derivatives		
			p-Toluidide[a]	*Anilide*[b]	*Amide*[c]
			Mp	*Mp*	*Mp*
	134–135	*E*-Cinnamic acid	168	153	147
	134–136	Maleic acid	142 (di)	198 (mono) 187 (di)	260 (di)
	135–137	Malonic acid	86 (mono) 253 (di)	132 (mono) 230 (di)	
	138–140	2-Chlorobenzoic acid	131	118	139
	140–142	3-Nitrobenzoic acid	162	155	143
	144–148	Anthranilic acid	151	131	109
	147–149	Diphenylacetic acid	172	180	167
	152–153	Adipic acid	239	151 (mono) 241 (di)	125 (mono) 220 (di)
	153–154	Citric acid	189 (tri)	199 (tri)	210 (tri)
	157–159	4-Chlorophenoxyacetic acid		125	133
	158–160	Salicylic acid	156	136	142
	163–164	Trimethylacetic acid		127	178
	164–166	5-Bromosalicylic acid		222	232
	166–167	Itaconic acid		190	191 (di)
	171–174	D-Tartaric acid		180 (mono) 264 (di)	171 (mono) 196 (di)
	179–182	3,4-Dimethoxybenzoic acid		154	164
	180–182	4-Toluic acid	160	145	160
	182–185	4-Anisic acid	186	169	167
	187–190	Succinic acid	180 (mono) 255 (di)	143 (mono) 230 (di)	157 (mono) 260 (di)
	201–203	3-Hydroxybenzoic acid	163	157	170
	203–206	3,5-Dinitrobenzoic acid		234	183
	210–211	Phthalic acid	150 (mono) 201 (di)	169 (mono) 253 (di)	149 (mono) 220 (di)
	214–215	4-Hydroxybenzoic acid	204	197	162
	225–227	2,4-Dihydroxybenzoic acid		126	228
	236–239	Nicotinic acid	150	132	128
	239–241	4-Nitrobenzoic acid	204	211	201
	299–300	Fumaric acid		233 (mono) 314 (di)	270 (mono) 266 (di)
	> 300	Terephthalic acid		334	

TABLE 70.2 Alcohols

Bp	Mp	Compound	Derivatives	
			3,5-Dinitrobenzoate[a]	*Phenylurethane*[b]
			Mp	*Mp*
65		Methanol	108	47
78		Ethanol	93	52
82		2-Propanol	123	88
83		*t*-Butyl alcohol	142	136
96–98		Allyl alcohol	49	70
97		1-Propanol	74	57
98		2-Butanol	76	65
102		2-Methyl-2-butanol	116	42
104		2-Methyl-3-butyn-2-ol	112	
108		2-Methyl-1-propanol	87	86
114–115		Propargyl alcohol		63
114–115		3-Pentanol	101	48
118		1-Butanol	64	61
118–119		2-Pentanol	62	
123		3-Methyl-3-pentanol	96(62)	43
129		2-Chloroethanol	95	51
130		2-Methyl-1-butanol	70	31
132		4-Methyl-2-pentanol	65	143
136–138		1-Pentanol	46	46
139–140		Cyclopentanol	115	132
140		2,4-Dimethyl-3-pentanol		95
146		2-Ethyl-1-butanol	51	
151		2,2,2-Trichloroethanol	142	87
157		1-Hexanol	58	42
160–161		Cyclohexanol	113	82
170		Furfuryl alcohol	80	45
176		1-Heptanol	47	60(68)
178		2-Octanol	32	oil
178		Tetrahydrofurfuryl alcohol	83	61
183–184		2,3-Butanediol		201 (di)
183–186		2-Ethyl-1-hexanol		34
187		1,2-Propanediol		153 (di)
194–197		Linaloöl		66
195		1-Octanol	61	74
196–198		Ethylene glycol	169	157 (di)
204		1,3-Butanediol		122
203–205		Benzyl alcohol	113	77
204		1-Phenylethanol	95	92
219–221		2-Phenylethanol	108	78
230		1,4-Butanediol		183 (di)
231		1-Decanol	57	59

a. For preparation, see page 521.
b. For preparation, see page 521.

TABLE 70.2 continued

Bp	Mp	Compound	Derivatives	
			3,5-Dinitrobenzoate	*Phenylurethane*
			Mp	*Mp*
259		4-Methoxybenzyl alcohol		92
	33–35	Cinnamyl alcohol	121	90
	38–40	1-Tetradecanol	67	74
	48–50	1-Hexadecanol	66	73
	58–60	1-Octadecanol	77(66)	79
	66–67	Benzhydrol	141	139
	147	Cholesterol	195	168

TABLE 70.3 Aldehydes

Bp	Mp	Compound	Derivatives	
			Semicarbazone[a]	*2,4-Dinitrophenylhydrazone*[b]
			Mp	*Mp*
21		Acetaldehyde	162	168
46–50		Propionaldehyde	89(154)	148
63		Isobutyraldehyde	125(119)	187(183)
75		Butyraldehyde	95(106)	123
90–92		3-Methylbutanal	107	123
98		Chloral	90	131
104		Crotonaldehyde	199	190
117		2-Ethylbutanal	99	95(130)
153		Heptaldehyde	109	108
162		2-Furaldehyde	202	212(230)
163		2-Ethylhexanal	254	114(120)
179		Benzaldehyde	222	237
195		Phenylacetaldehyde	153	121(110)
197		Salicylaldehyde	231	248
204–205		4-Tolualdehyde	234(215)	232
209–215		2-Chlorobenzaldehyde	146(229)	213
247		2-Ethoxybenzaldehyde	219	
248		4-Anisaldehyde	210	253
250–252		*E*-Cinnamaldehyde	215	255
	33–34	1-Naphthaldehyde	221	
	37–39	2-Anisaldehyde	215	254
	42–45	3,4-Dimethoxybenzaldehyde	177	261
	44–47	4-Chlorobenzaldehyde	230	254
	57–59	3-Nitrobenzaldehyde	246	293
	81–83	Vanillin	230	271

a. For preparation, see page 278.
b. For preparation, see page 277.

TABLE 70.4 Amides

Bp	Mp	Name of Compound	Mp	Name of Compound
153		N,N-Dimethylformamide	127–129	Isobutyramide
164–166		N,N-Dimethylacetamide	128–129	Benzamide
210		Formamide	130–133	Nicotinamide
243–244		N-Methylformanilide	177–179	4-Chloroacetanilide
	26–28	N-Methylacetamide		
	79–81	Acetamide		
	109–111	Methacrylamide		
	113–115	Acetanilide		
	116–118	2-Chloroacetamide		

TABLE 70.5 Primary and Secondary Amines

Bp	Mp	Compound	Derivatives		
			Benzamide[a]	*Picrate*[b]	*Acetamide*[c]
			Mp	*Mp*	*Mp*
33–34		Isopropylamine	71	165	
46		*t*-Butylamine	134	198	
48		*n*-Propylamine	84	135	
53		Allylamine		140	
55		Diethylamine	42	155	
63		*s*-Butylamine	76	139	
64–71		Isobutylamine	57	150	
78		*n*-Butylamine	42	151	
84		Di-isopropylamine		140	
87–88		Pyrrolidine	oil	112	
106		Piperidine	48	152	
111		Di-*n*-propylamine		75	
118		Ethylenediamine	244 (di)	233	172 (di)
129		Morpholine	75	146	
137–139		Di-isobutylamine		121	86
145–146		Furfurylamine		150	
149		N-Methylcyclohexylamine	85	170	
159		Di-*n*-butylamine		59	
182–185		Benzylamine	105	199	60
184		Aniline	163	198	114
196		N-Methylaniline	63	145	102
199–200		2-Toluidine	144	213	110
203–204		3-Toluidine	125	200	65
205		N-Ethylaniline	60	138(132)	54

a. For preparation, see page 521.
b. For preparation, see page 521.
c. For preparation, see page 522.

TABLE 70.5 continued

Bp	Mp	Compound	Derivatives		
			Benzamide[a]	*Picrate*[b]	*Acetamide*[c]
			Mp	*Mp*	*Mp*
208–210		2-Chloroaniline	99	134	87
210		2-Ethylaniline	147	194	111
216		2,6-Dimethylaniline	168	180	177
218		2,4-Dimethylaniline	192	209	133
218		2,5-Dimethylaniline	140	171	139
221		N-Ethyl-*m*-toluidine	72		
225		2-Anisidine	60(84)	200	85
230		3-Chloroaniline	120	177	72(78)
231–233		2-Phenetidine	104		79
241		4-Chloro-2-methylaniline	142		140
242		3-Chloro-4-methylaniline	122		105
250		4-Phenetidine	173	69	137
256		Dicyclohexylamine	153(57)	173	103
	35–38	N-Phenylbenzylamine	107	48	58
	41–44	4-Toluidine	158	182	147
	49–51	2,5-Dichloroaniline	120	86	132
	52–54	Diphenylamine	180	182	101
	57–60	4-Anisidine	154	170	130
	57–60	2-Aminopyridine	165 (di)	216(223)	
	60–62	N-Phenyl-1-naphthylamine	152		115
	62–65	2,4,5-Trimethylaniline	167		162
	64–66	3-Phenylenediamine	125 (mono) 240 (di)	184	87 (mono) 191 (di)
	66	4-Bromoaniline	204	180	168
	68–71	4-Chloroaniline	192	178	179(172)
	71–73	2-Nitroaniline	110(98)	73	92
	97–99	2,4-Diaminotoluene	224 (di)		224 (di)
	100–102	2-Phenylenediamine	301	208	185
	104–107	2-Methyl-5-nitroaniline	186		151
	107–109	2-Chloro-4-nitroaniline	161		139
	112–114	3-Nitroaniline	157(150)	143	155(76)
	115–116	4-Methyl-2-nitroaniline	148		99
	117–119	4-Chloro-2-nitroaniline			104
	120–122	2,4,6-Tribromoaniline	198(204)		232
	131–133	2-Methyl-4-nitroaniline			202
	138–140	2-Methoxy-4-nitroaniline	149		
	138–142	4-Phenylenediamine	128 (mono) 300 (di)		162 (mono) 304 (di)
	148–149	4-Nitroaniline	199	100	215
	162–164	4-Aminoacetanilide			304
	176–178	2,4-Dinitroaniline	202(220)		120

TABLE 70.6 Tertiary Amines

Bp	Compound	Derivatives	
		Picrate[a]	*Methiodide*[b]
		Mp	*Mp*
85–91	Triethylamine	173	280
115	Pyridine	167	117
128–129	2-Picoline	169	230
143–145	2,6-Lutidine	168(161)	233
144	3-Picoline	150	92(36)
145	4-Picoline	167	149
155–158	Tri-*n*-propylamine	116	207
159	2,4-Lutidine	180	113
183–184	N,N-Dimethylbenzylamine	93	179
216	Tri-*n*-butylamine	105	186
217	N,N-Diethylaniline	142	102
237	Quinoline	203	133(72)

a. For preparation, see page 521.
b. For preparation, see page 522.

TABLE 70.7 Anhydrides and Acid Chlorides

Bp	Mp	Compound	Derivatives			
			Acid[a]		*Amide*[b]	*Anilide*[c]
			Bp	*Mp*	*Mp*	*Mp*
52		Acetyl chloride	118		82	114
77–79		Propionyl chloride	141		81	106
102		Butyryl chloride	162		115	96
138–140		Acetic anhydride	118		82	114
167		Propionic anhydride	141		81	106
198–199		Butyric anhydride	162		115	96
198		Benzoyl chloride		122	130	163
225		3-Chlorobenzoyl chloride		158	134	122
238		2-Chlorobenzoyl chloride		142	142	118
	32–34	*cis*-1,2-Cyclohexanedicarboxylic anhydride		192		
	35–37	Cinnamoyl chloride		133	147	151
	39–40	Benzoic anhydride		122	130	163
	54–56	Maleic anhydride		130	181 (mono) 266 (di)	173 (mono) 187 (di)
	72–74	4-Nitrobenzoyl chloride		241	201	211

a. For preparation, see page 522.
b. For preparation, see page 522.
c. For preparation, see page 522.

TABLE 70.7 continued

Bp	Mp	Compound	Derivatives: Acid[a] Bp	Acid[a] Mp	Amide[b] Mp	Anilide[c] Mp
	119–120	Succinic anhydride		186	157 (mono) 260 (di)	148 (mono) 230 (di)
	131–133	Phthalic anhydride		206	149 (mono) 220 (di)	170 (mono) 253 (di)
	254–258	Tetrachlorophthalic anhydride		250		
	267–269	1,8-Naphthalic anhydride		274		250–282 (di)

TABLE 70.8 Esters

Bp	Mp	Compound	Bp	Mp	Compound
34		Methyl formate	169–170		Methyl acetoacetate
52–54		Ethyl formate	180–181		Dimethyl malonate
72–73		Vinyl acetate	181		Ethyl acetoacetate
77		Ethyl acetate	185		Diethyl oxalate
79		Methyl propionate	198–199		Methyl benzoate
80		Methyl acrylate	206–208		Ethyl caprylate
85		Isopropyl acetate	208–210		Ethyl cyanoacetate
93		Ethyl chloroformate	212		Ethyl benzoate
94		Isopropenyl acetate	217		Diethyl succinate
98		Isobutyl formate	218		Methyl phenylacetate
98		*t*-Butyl acetate	218–219		Diethyl fumarate
99		Ethyl propionate	222		Methyl salicylate
99		Ethyl acrylate	225		Diethyl maleate
100		Methyl methacrylate	229		Ethyl phenylacetate
101		Methyl trimethylacetate	234		Ethyl salicylate
102		*n*-Propyl acetate	268		Dimethyl suberate
106–113		*s*-Butyl acetate	271		Ethyl cinnamate
120		Ethyl butyrate	282		Dimethyl phthalate
127		*n*-Butyl acetate	298–299		Diethyl phthalate
128		Methyl valerate	298–299		Phenyl benzoate
130		Methyl chloroacetate	340		Dibutyl phthalate
131–133		Ethyl isovalerate		56–58	Ethyl *p*-nitrobenzoate
142		*n*-Amyl acetate		88–90	Ethyl *p*-aminobenzoate
142		Isoamyl acetate		94–96	Methyl *p*-nitrobenzoate
143		Ethyl chloroacetate		95–98	*n*-Propyl *p*-hydroxybenzoate
154		Ethyl lactate		116–118	Ethyl *p*-hydroxybenzoate
168		Ethyl caproate (ethyl hexanoate)		126–128	Methyl *p*-hydroxybenzoate

TABLE 70.9 Ethers

Bp	Mp	Compound	Bp	Mp	Compound
32		Furan	215		4-Bromoanisole
33		Ethyl vinyl ether	234–237		Anethole
65–67		Tetrahydrofuran	259		Diphenyl ether
94		*n*-Butyl vinyl ether	273		2-Nitroanisole
154		Anisole	298		Dibenzyl ether
174		4-Methylanisole		50–52	4-Nitroanisole
175–176		3-Methylanisole		56–60	1,4-Dimethyoxybenzene
198–203		4-Chloroanisole		73–75	2-Methoxynaphthalene
206–207		1,2-Dimethoxybenzene			

TABLE 70.10 Halides

Bp	Compound	Bp	Compound
34–36	2-Chloropropane	100–105	1-Bromobutane
40–41	Dichloromethane	105	Bromotrichloromethane
44–46	Allyl chloride	110–115	1,1,2-Trichloroethane
57	1,1-Dichloroethane	120–121	1-Bromo-3-methylbutane
59	2-Bromopropane	121	Tetrachloroethylene
68	Bromochloromethane	123	3,4-Dichloro-1-butene
68–70	2-Chlorobutane	125	1,3-Dichloro-2-butene
69–73	Iodoethane	131–132	1,2-Dibromoethane
70–71	Allyl bromide	140–142	1,2-Dibromopropane
71	1-Bromopropane	142–145	1-Bromo-3-chloropropane
72–74	2-Bromo-2-methylpropane	146–150	Bromoform
74–76	1,1,1-Trichloroethane	147	1,1,2,2-Tetrachloroethane
81–85	1,2-Dichloroethane	156	1,2,3-Trichloropropane
87	Trichloroethylene	161–163	1,4-Dichlorobutane
88–90	2-Iodopropane	167	1,3-Dibromopropane
90–92	1-Bromo-2-methylpropane	177–181	Benzyl chloride
91	2-Bromobutane	197	(2-Chloroethyl)benzene
94	2,3-Dichloro-1-propene	219–223	Benzotrichloride
95–96	1,2-Dichloropropane	238	1-Bromodecane
96–98	Dibromomethane		

TABLE 70.11 Aryl Halides

Bp	Mp	Compound	Derivatives			
			Nitration Product[a]		*Oxidation Product*[b]	
			Position	*Mp*	*Name*	*Mp*
132		Chlorobenzene	2, 4	52		
156		Bromobenzene	2, 4	70		
157–159		2-Chlorotoluene	3, 5	63	2-Chlorobenzoic acid	141
162		4-Chlorotoluene	2	38	4-Chlorobenzoic acid	240
172–173		1,3-Dichlorobenzene	4, 6	103		
178		1,2-Dichlorobenzene	4, 5	110		
196–203		2,4-Dichlorotoluene	3, 5	104	2,4-Dichlorobenzoic acid	164
201		3,4-Dichlorotoluene	6	63	3,4-Dichlorobenzoic acid	206
214		1,2,4-Trichlorobenzene	5	56		
279–281		1-Bromonaphthalene	4	85		
	51–53	1,2,3-Trichlorobenzene	4	56		
	54–56	1,4-Dichlorobenzene	2	54		
	66–68	1,4-Bromochlorobenzene	2	72		
	87–89	1,4-Dibromobenzene	2, 5	84		
	138–140	1,2,4,5-Tetrachlorobenzene	3	99		
			3, 6	227		

a. For preparation, see page 522.
b. For preparation, see page 522.

TABLE 70.12 Hydrocarbons: Alkenes

Bp	Compound	Bp	Compound
34	Isoprene	149–150	1,5-Cyclooctadiene
83	Cyclohexene	152	DL-α-Pinene
116	5-Methyl-2-norbornene	160	Bicyclo[4.3.0]nona-3,7-diene
122–123	1-Octene	165–167	(−)-β-Pinene
126–127	4-Vinyl-1-cyclohexene	165–169	α-Methylstyrene
132–134	2,5-Dimethyl-2,4-hexadiene	181	1-Decene
141	5-Vinyl-2-norbornene	181	Indene
143	1,3-Cyclooctadiene	251	1-Tetradecene
145	4-Butylstyrene	274	1-Hexadecene
145–146	Cycloctene	349	1-Octadecene
145–146	Styrene		

TABLE 70.13 Hydrocarbons: Aromatic

Bp	Mp	Compound	Melting Point of Derivatives		
			Nitro[a]		*Picrate*[b]
			Position	*Mp*	*Mp*
80		Benzene	1, 3	89	84
111		Toluene	2, 4	70	88
136		Ethylbenzene	2, 4, 6	37	96
138		*p*-Xylene	2, 3, 5	139	90
138–139		*m*-Xylene	2, 4, 6	183	91
143–145		*o*-Xylene	4, 5	118	88
145		4-*t*-Butylstyrene	2, 4	62	
145–146		Styrene			
152–154		Cumene	2, 4, 6	109	
163–166		Mesitylene	2, 4	86	97
			2, 4, 6	235	
165–169		α-Methylstyrene			
168		1,2,4-Trimethylbenzene	3, 5, 6	185	97
176–178		*p*-Cymene	2, 6	54	
189–192		4-*t*-Butyltoluene			
197–199		1,2,3,5-Tetramethylbenzene	4, 6	181(157)	
203		*p*-Diisopropylbenzene			
204–205		1,2,3,4-Tetramethylbenzene	5, 6	176	92
207		1,2,3,4-Tetrahydronaphthalene	5, 7	95	
240–243		1-Methylnaphthalene	4	71	142
	34–36	2-Methylnaphthalene	1	81	116
	50–51	Pentamethylbenzene	6	154	131
	69–72	Biphenyl	4, 4′	237(229)	
	80–82	1,2,4,5-Tetramethylbenzene	3, 6	205	
	80–82	Naphthalene	1	61(57)	149
	90–95	Acenaphthene	5	101	161
	99–101	Phenanthrene			144(133)
	112–115	Fluorene	2	156	87(77)
			2, 7	199	
	214–217	Anthracene			138

a. For preparation, see page 522.
b. For preparation, see page 521.

TABLE 70.14 Ketones

Bp	Mp	Compound	Derivatives: *Semicarbazone*[a] *Mp*	Derivatives: *2,4-Dinitrophenylhydrazone*[b] *Mp*
56		Acetone	187	126
80		2-Butanone	136, 186	117
88		2,3-Butanedione	278	315
100–101		2-Pentanone	112	143
102		3-Pentanone	138	156
106		Pinacolone	157	125
114–116		4-Methyl-2-pentanone	132	95
124		2,4-Dimethyl-3-pentanone	160	88, 95
128–129		5-Hexen-2-one	102	108
129		4-Methyl-3-penten-2-one	164	205
130–131		Cyclopentanone	210	146
133–135		2,3-Pentanedione	122 (mono) 209 (di)	209
145		4-Heptanone	132	75
145		5-Methyl-2-hexanone	147	95
145–147		2-Heptanone	123	89
146–149		3-Heptanone	101	
156		Cyclohexanone	166	162
162–163		2-Methylcyclohexanone	195	137
169		2,6-Dimethyl-4-heptanone	122	66, 92
169–170		3-Methylcyclohexanone	180	155
173		2-Octanone	122	58
191		Acetonylacetone	185 (mono) 224 (di)	257 (di)
202		Acetophenone	198	238
216		Phenylacetone	198	156
217		Isobutyrophenone	181	163
218		Propiophenone	182	191
226		4-Methylacetophenone	205	258
231–232		2-Undecanone	122	63
232		*n*-Butyrophenone	188	
232		4-Chloroacetophenone	204	236
235		Benzylacetone	142	127
	35–37	4-Chloropropiophenone	176	223
	35–39	4-Phenyl-3-buten-2-one	187	227
	36–38	4-Methoxyacetophenone	198	228
	48–49	Benzophenone	167	238
	53–55	2-Acetonaphthone	235	262
	60	Desoxybenzoin	148	204
	76–78	3-Nitroacetophenone	257	228
	78–80	4-Nitroacetophenone		257
	82–85	9-Fluorenone	234	283
	134–136	Benzoin	206	245
	147–148	4-Hydroxypropiophenone		240

a. For preparation, see page 273.
b. For preparation, see page 277.

TABLE 70.15 Nitriles

Bp	Mp	Compound	Bp	Mp	Compound
77		Acrylonitrile	212		3-Tolunitrile
83–84		Trichloroacetonitrile	217		4-Tolunitrile
97		Propionitrile	233–234		Benzyl cyanide
107–108		Isobutyronitrile	295		Adiponitrile
115–117		*n*-Butyronitrile		30.5	4-Chlorobenzyl cyanide
174–176		3-Chloropropionitrile		32–34	Malononitrile
191		Benzonitrile		38–40	Stearonitrile
205		2-Tolunitrile		46–48	Succinonitrile
				71–73	Diphenylacetonitrile

TABLE 70.16 Nitro Compounds

Bp	Mp	Compound	Amine Obtained by Reduction of Nitro Groups			
					Acetamide[a]	*Benzamide*[b]
			Bp	*Mp*	*Mp*	*Mp*
210–211		Nitrobenzene	184		114	160
225		2-Nitrotoluene	200		110	146
225		2-Nitro-*m*-xylene	215		177	168
230–231		3-Nitrotoluene	203		65	125
245		3-Nitro-*o*-xylene	221		135	189
245–246		4-Ethylnitrobenzene	216		94	151
	34–36	2-Chloro-6-nitrotoluene	245		157(136)	173
	36–38	4-Chloro-2-nitrotoluene		21	139(131)	
	40–42	3,4-Dichloronitrobenzene		72	121	
	43–50	1-Chloro-2,4-dinitrobenzene		91	242 (di)	178 (di)
	52–54	4-Nitrotoluene		45	147	158
	55–56	1-Nitronaphthalene		50	159	160
	83–84	1-Chloro-4-nitrobenzene		72	179	192
	88–90	*m*-Dinitrobenzene		63	87 (mono)	125 (mono)
					191 (di)	240 (di)

a. For preparation, see page 522.
b. For preparation, see page 521.

TABLE 70.17 Phenols

Bp	Mp	Compound	Derivatives	
			α-Naphthylurethane[a]	*Bromo*[b]
			Mp	*Mp*
175–176		2-Chlorophenol	120	48 (mono) 76 (di)
181	42	Phenol	133	95 (tri)
202	32–34	*p*-Cresol	146	49 (di) 108 (tetra)
203		*m*-Cresol	128	84 (tri)
228–229		3,4-Dimethylphenol	141	171 (tri)
	32-33	*o*-Cresol	142	56 (di)
	42–43	2,4-Dichlorophenol		68
	42–45	4-Ethylphenol	128	
	43–45	4-Chlorophenol	166	90 (di)
	44–46	2,6-Dimethylphenol	176	79
	44–46	2-Nitrophenol	113	117 (di)
	49–51	Thymol	160	55
	62–64	3,5-Dimethylphenol		166 (tri)
	64–68	4-Bromophenol	169	95 (tri)
	74	2,5-Dimethylphenol	173	178 (tri)
	92–95	2,3,5-Trimethylphenol	174	
	95–96	1-Naphthol	152	105 (di)
	98–101	4-*t*-Butylphenol	110	50 (mono) 67 (di)
	104–105	Catechol	175	192 (tetra)
	109–110	Resorcinol	275	112 (tri)
	112–114	4-Nitrophenol	150	142 (di)
	121–124	2-Naphthol	157	84
	133–134	Pyrogallol	173	158 (di)

a. For preparation, see page 521.
b. For preparation, see page 523.

71 Searching the Chemical Literature

In planning a synthesis or investigating the properties of compounds, the organic chemist is faced with the problem of locating information on the chemical and physical properties of substances that have been previously prepared as well as the best methods and reagents for carrying out a synthesis. With over 7,000,000 organic compounds known and new ones being reported at the rate of more than a third of a million per year, the task might seem formidable. However, the field of chemistry has developed one of the best information retrieval systems of all the sciences. In a university chemistry library it it possible to locate information on almost any known compound in a few minutes. Even a modest library will have information on several million different substances. The ultimate source of information is the primary literature: articles written by individual chemists and published in journals. The secondary literature consists of compilations of information taken from these primary sources. For example, to find the melting point of benzoic acid, one would naturally turn to a secondary reference containing a table of physical constants of organic compounds. If, on the other hand, one wished to have information about the phase changes of benzoic acid under high pressure, one would need to consult the primary literature. If the piece of information, such as a melting point, is crucial to an investigation, the primary literature should be consulted. Transcription errors occur in the preparation of secondary references.

Handbooks

For rapid access to information such as mp, bp, density, solubility, optical rotation, λ max and crystal form, one turns first to the *Handbook of Chemistry and Physics* where information is found on some 15,000 organic compounds, including the Beilstein reference (see below) to each compound. These compounds are well known and completely characterized. The majority are commercially available. The *Merck Index* contains information on nearly 10,000 compounds, especially those of pharmaceutical interest. In addition to the usual physical properties, information and literature references to synthesis, isolation, and medicinal properties, such as toxicity data, are found. The last third of the book is devoted to such items as a long cross index of names (which is very useful for looking up drugs), a table of organic name reactions, an excellent section on first aid for poisons, a list of chemical poisons, and a listing of the locations of many poison control centers.

A useful reference is the *Aldrich Catalog Handbook of Fine Chemicals*, available without charge from the Aldrich Company. Not only does this catalog list the prices of the more than 12,000 chemicals (primarily organic) as well as the molecular weights, melting points, boiling points, and optical rotations, but it also gives reference to ir and nmr spectral data and to Beilstein, *Merck Index*, and Fiesers' *Reagents*. In addition, for each chemical the catalog provides the reference to the *Registry of Toxic Effects of Chemical Substances* (RTECS No.) and, if appropriate, the reference to Sax, *Dangerous Properties of Industrial Materials*. Special hazards are noted ("severe poison," "lachrymator," "corrosive"), and a reference is made to one of thirty different disposal methods for each chemical. The catalog also gives pertinent information (uses, physiological effects, etc.) with literature references for many compounds. A complete list of all chemicals sold commercially and the addresses of the companies making them is found in *Chem Sources*, published yearly.

After consulting these three single-volume references, one would turn to more comprehensive multivolume sources such as the *Dictionary of Organic Compounds*, Fifth Edition. This dictionary, still known as "Heilbron," the name of its former editor, now comprises seven volumes of specific information, with primary literature references, on the synthesis, reactions, and derivatives of more than 50,000 compounds. Rodd's *Chemistry of Carbon Compounds*, another valuable multivolume work with primary literature references, is organized by functional group rather than in dictionary form. Elsevier's *Encyclopedia of Organic Compounds* in about twenty volumes is an incomplete reference work on the chemical and physical properties of compounds. It is useful for those areas it covers. References to Elsevier are found in the *Handbook of Chemistry and Physics*.

Beilstein's *Handbüch der organischen Chemie* is certainly not a "handbook" in the American sense—it can occupy an entire alcove of a chemistry library! And the currently produced volumes must be among the most expensive contemporary works purchased by a library, for each book now costs more than $700. However, with characteristic German thoroughness, this reference covers every well-characterized organic compound that has been reported in the literature up to 1949. Along with a main reference set come three supplements covering the periods 1910–1919, 1920–1928, and 1930–1949. The fourth supplement covering 1950–1959 is not yet complete. Although written in German, this reference can provide much information even to those with no knowledge of the language. Physical constants and primary literature references are easy to pick out. Beilstein is organized around a complex classification scheme that is explained in *The Beilstein Guide* by O. Weissback, but the casual user can gain access through the *Handbook of Chemistry and Physics* and the *Aldrich Catalog*. In the *Handbook* the reference for 2-iodobenzoic acid is listed as $\mathbf{B9}^2$, 239, which means the reference will be found on p. 239 of Vol. (Band) 9 of the second supplement (Zweites Ergänzungwerk, EII). In the *Aldrich Catalog* the reference is given as Beil **9**, 363, which indicates information can be found in the main series on p. 363, Vol. 9. The system number assigned to each compound can be traced through the supplements. The easiest access to Beilstein itself is through the formula index (General Formelregister) of the second supplement. A rudimentary knowledge of German will enable one to pick out iodo (Jod) benzoic acid (Saure), for example.

Chemical Abstracts

Secondary references are incomplete, some suffer from transcription errors, and the best—Beilstein—is twenty years behind in its survey of the organic literature, so one must often turn to the primary chemical literature. However, there are more than 14,000 periodicals where chemical information might appear. Chemists are fortunate in having an index, *Chemical Abstracts*, that covers this huge volume of literature very promptly and publishes biweekly abstracts of each article. The information in each abstract is then compiled into author, chemical substance, general subject, formula, and patent indexes. Since publication began in 1907, nomenclature has changed. Now no trivial names are used, so acetone appears as 2-propanone and *o*-cresol as benzene, 1-methyl, 2-hydroxy. It takes some experience to adapt to these changing names and even now *Chemical Abstracts* does not follow the IUPAC rules exactly. The formula index is useful because it provides not only reference to specific abstracts but also a correct name that can be found in the chemical substances index. Currently indexes are published twice each year. These are then grouped into five-year collective indexes. The one for the period 1977–1981 occupies 74 volumes and costs $5000. The yearly subscription price is now $7800.

A complete search of the chemical literature would entail use of *Chemischer Zentralblatt*, the oldest abstract journal, which first appeared in 1830. Collective and formula indexes started in 1930 go up to 1969. These collective indexes are more reliable than those of *Chemical Abstracts* for the period 1930–1939.

Title Indexes

As long as a year can elapse between the time a paper appears and the time an abstract is published. Three publications help fill the gap by providing much the same information: a list of the titles, authors, and keywords that appear in titles of papers recently published—or, in some cases, papers about to be published. These publications are *Chemical Titles, Current Chemical Papers*, and *Current Contents*.

Science Citation Index

Science Citation Index is unique in that it allows for a type of search not possible with any other index—a search forward in time. For example, if one wanted to learn what recent applications have been made of the cuprous chloride catalyzed coupling of terminal acetylenes (Chapter 50) first reported by Stansbury and Proops, one would look up their paper [*J. Org.* **27**, 320 (1962)] in *Science Citation Index* and find there a list of those articles subsequent to 1962 in which an author cited the work of Stansbury and Proops in a footnote.

Planning a Synthesis

In planning a synthesis the organic chemist is faced with at least three overlapping considerations: the chemical reactions, the reagents, and the experimental procedure to be employed. For students an advanced textbook such as Carey and Sundberg, *Advanced Organic Chemistry*, Vols. 1 and 2, or March's *Advanced Organic Chemistry*, 3rd Ed., 1985, which has literature references, might be a place to start. It is also instructive for the beginning synthetic chemist to read about some elegant and classical syntheses that have been carried out in the past. For this purpose see Anand, Bindra, and Ranganathan, *Art in Organic Synthesis*, for some of the most elegant syntheses carried out through 1970. Similarly, Bindra and Bindra, *Creativity in Organic Synthesis*, Vol. 1, and Fleming, *Selected Organic Syntheses*, are compendia of elegant syntheses. For natural product synthesis see the excellent series by Ap Simon, *The Total Synthesis of Natural Products*, Vols. 1–6. House's *Modern Synthetic Reactions*, 2nd ed., is a more comprehensive source of information with several thousand references to the original literature.

If it appears that a particular reaction will be employed in the synthesis, then *Organic Reactions* should be consulted. Over 100 preparative reactions, with examples of experimental details, are covered in great detail in some thirty volumes. Theilheimer's 40 volumes of *Synthetic Methods of Organic Chemistry* is organized according to the types of bonds being made or broken and, because it is published annually, serves as a means for continually updating all synthetic procedures. Buehler and Pearson, *Survey of Organic Synthesis*, in two volumes, is a good summary of synthetic methods, classified by functional group. Also the annual *Newer Methods of Preparative Organic Chemistry* and *Annual Reports in Organic Synthesis* should be consulted. The latter is an excellent review of new reactions in a given year, organized by reaction type. The Fiesers' unique *Reagents for Organic Synthesis*, a biannual series since 1967, critically surveys the reagents employed to carry out organic synthesis. Included are references to the original literature, *Organic Syntheses* (see below), the critical reviews that have been written about various reagents, and suppliers of reagents. Smith's comprehensive indices (1987) make the series even more valuable. The index of reagents according to type of reaction is very useful when planning a synthesis. Other good references are *Compendium of Organic Synthetic Methods*, Vols. 1–6, and Nakanishi et al., *Natural Products Chemistry*, Vols. 1 and 2.

In order to operate on one functional group without affecting another functional group it is often necessary to put on a protective group. Two books deal with this rather specialized subject: McOmie's *Protective Groups in Organic Chemistry* and Greene's *Protective Groups in Organic Synthesis.*

Before carrying out the synthesis itself one should consult *Organic Syntheses*, an annual series since 1920, which is grouped in four collective volumes with reaction and reagent indexes. *Organic Syntheses* gives detailed procedures for carrying out more than 1000 different reactions. Each synthesis is submitted for review and then the procedure is sent to an independent laboratory to be checked. The reactions, unlike many that are reported in the primary literature, are carried out a number of times and therefore can be relied upon to work. Laboratory techniques are covered in Bates and Schaefer's *Research Techniques in Organic Chemistry* and in the fourteen volumes of Weissberger et al., *Technique of Organic Chemistry*, an uneven, multiauthor compendium. An excellent advanced text on the detailed mechanisms of many representative reactions is Lowry and Richardson's *Mechanism and Theory in Organic Chemistry*, 2nd ed.

Modern Chemistry Library

Although the chemistry library as we now know it probably will not change markedly in the next decade, in one aspect it has already changed. It is now possible to search a large part of the chemical literature using a personal computer connected via telephone lines to a commercial data base held in a large computer. By rapidly and efficiently calling into a data base containing *Chemical Abstracts, Science Citation Index*, and several other indexes, a librarian trained to use the system can aquire information for the chemist much more rapidly than he or she could do by a manual search. In some cases, the information obtained would be impossible to get by conventional means. For example, to locate all references to the nmr spectra of insulin during the period 1976–1981 using the Lockheed DIALOG® system, one first asks for the number of references to nmr found in *Chemical Abstracts* during that period. The answer is typed out within a few seconds: 12,000. Next one asks how many references to insulin. The answer: 20,000. The next query asks the computer to cross the two lists for references common to both. The answer: 5. One can then ask that the references be typed out. At some expense, the entire abstract for each reference can be typed out. It is less expensive to have the abstracts typed off-line at the computer center and sent by mail.

Wiswesser Line Notation

Using a computer it is also possible to search for partial structures, an option not available with a manual search. For example, one could ask for a list of all compounds having a four-membered ring with a carbonyl group in the ring. These compounds, when fused to another ring system, will not appear in an index under the name cyclobutanone. Wiswesser line notation, a linear computer-intelligible way of encoding organic structures, can be searched by computer for partial structures. The notation for cyclobutanone is L4V BHJ.

The V is the symbol for a carbonyl group; the 4 indicates a four-membered ring. These elements will be found in the line notation of bigger and more complex molecules that have four-membered rings containing a carbonyl group.

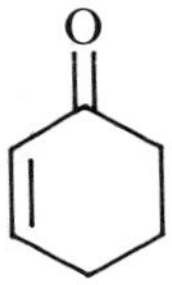

$CH_3CH_2CH_2CH_3$

L4V BHJ LGV BUTJ 4H

Wiswesser line notation

Index

Note: Principle references are italicized.